B — Standard Normal Distribution

Numerical entries represent the probability that a standard normal random variable is between $-\infty$ and z where $z = \dfrac{x - \mu}{\sigma}$.

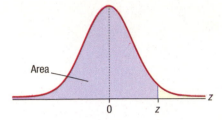

Area

z	0.00	0.01	0.02	0.03	0.04	0.05	0.06	0.07	0.08	0.09
0.0	0.5000	0.5040	0.5080	0.5120	0.5160	0.5199	0.5239	0.5279	0.5319	0.5359
0.1	0.5398	0.5438	0.5478	0.5517	0.5557	0.5596	0.5636	0.5675	0.5714	0.5753
0.2	0.5793	0.5832	0.5871	0.5910	0.5948	0.5987	0.6026	0.6064	0.6103	0.6141
0.3	0.6179	0.6217	0.6255	0.6293	0.6331	0.6368	0.6406	0.6443	0.6480	0.6517
0.4	0.6554	0.6591	0.6628	0.6664	0.6700	0.6736	0.6772	0.6808	0.6844	0.6879
0.5	0.6915	0.6950	0.6985	0.7019	0.7054	0.7088	0.7123	0.7157	0.7190	0.7224
0.6	0.7257	0.7291	0.7324	0.7357	0.7389	0.7422	0.7454	0.7486	0.7517	0.7549
0.7	0.7580	0.7611	0.7642	0.7673	0.7704	0.7734	0.7764	0.7794	0.7823	0.7852
0.8	0.7881	0.7910	0.7939	0.7967	0.7995	0.8023	0.8051	0.8078	0.8106	0.8133
0.9	0.8159	0.8186	0.8212	0.8238	0.8264	0.8289	0.8315	0.8340	0.8365	0.8389
1.0	0.8413	0.8438	0.8461	0.8485	0.8508	0.8531	0.8554	0.8577	0.8599	0.8621
1.1	0.8643	0.8665	0.8686	0.8708	0.8729	0.8749	0.8770	0.8790	0.8810	0.8830
1.2	0.8849	0.8869	0.8888	0.8907	0.8925	0.8944	0.8962	0.8980	0.8997	0.9015
1.3	0.9032	0.9049	0.9066	0.9082	0.9099	0.9115	0.9131	0.9147	0.9162	0.9177
1.4	0.9192	0.9207	0.9222	0.9236	0.9251	0.9265	0.9279	0.9292	0.9306	0.9319
1.5	0.9332	0.9345	0.9357	0.9370	0.9382	0.9394	0.9406	0.9418	0.9429	0.9441
1.6	0.9452	0.9463	0.9474	0.9484	0.9495	0.9505	0.9515	0.9525	0.9535	0.9545
1.7	0.9554	0.9564	0.9573	0.9582	0.9591	0.9599	0.9608	0.9616	0.9625	0.9633
1.8	0.9641	0.9649	0.9656	0.9664	0.9671	0.9678	0.9686	0.9693	0.9699	0.9706
1.9	0.9713	0.9719	0.9726	0.9732	0.9738	0.9744	0.9750	0.9756	0.9761	0.9767
2.0	0.9772	0.9778	0.9783	0.9788	0.9793	0.9798	0.9803	0.9808	0.9812	0.9817
2.1	0.9821	0.9826	0.9830	0.9834	0.9838	0.9842	0.9846	0.9850	0.9854	0.9857
2.2	0.9861	0.9864	0.9868	0.9871	0.9875	0.9878	0.9881	0.9884	0.9887	0.9890
2.3	0.9893	0.9896	0.9898	0.9901	0.9904	0.9906	0.9909	0.9911	0.9913	0.9916
2.4	0.9918	0.9920	0.9922	0.9925	0.9927	0.9929	0.9931	0.9932	0.9934	0.9936
2.5	0.9938	0.9940	0.9941	0.9943	0.9945	0.9946	0.9948	0.9949	0.9951	0.9952
2.6	0.9953	0.9955	0.9956	0.9957	0.9959	0.9960	0.9961	0.9962	0.9963	0.9964
2.7	0.9965	0.9966	0.9967	0.9968	0.9969	0.9970	0.9971	0.9972	0.9973	0.9974
2.8	0.9974	0.9975	0.9976	0.9977	0.9977	0.9978	0.9979	0.9979	0.9980	0.9981
2.9	0.9981	0.9982	0.9982	0.9983	0.9984	0.9984	0.9985	0.9985	0.9986	0.9986
3.0	0.9987	0.9987	0.9987	0.9988	0.9988	0.9989	0.9989	0.9989	0.9990	0.9990
3.1	0.9990	0.9991	0.9991	0.9991	0.9992	0.9992	0.9992	0.9992	0.9993	0.9993
3.2	0.9993	0.9993	0.9994	0.9994	0.9994	0.9994	0.9994	0.9995	0.9995	0.9995
3.3	0.9995	0.9995	0.9995	0.9996	0.9996	0.9996	0.9996	0.9996	0.9996	0.9997
3.4	0.9997	0.9997	0.9997	0.9997	0.9997	0.9997	0.9997	0.9997	0.9997	0.9998

Discovering
Business Statistics

Quinton J. Nottingham
James S. Hawkes

Executive Editor: Margaret Gibbs

Executive Project Manager: Kimberly Cumbie

Vice President, Research and Development: Marcel Prevuznak

Editorial Assistants: Danielle C. Bess, Susan Niese, Claudia Vance, Nina Waldron

Copy Editors: Taylor Jones, Jason Ling, Joseph Randich

Review Coordinator: Lisa Young

Senior Design Specialists: Jennifer Moran, Tee Jay Zajac

Design Specialists: Kaylen Saxon

Art: Jennifer Moran, Kaylen Saxon, Tee Jay Zajac

Cover Design: Tee Jay Zajac

Photograph Credits:

BigStockPhoto.com, iStockPhoto.com, and Digital Vision

A division of Quant Systems, Inc.

546 Long Point Road, Mt. Pleasant, SC 29464

10 9 8 7 6 5 4 3

Library of Congress Control Number: 2012941873

Printed in the United States of America

ISBN: 978-1-935782-87-2

To Jacqueline, Sir, and Jacob

Table of Contents

1 Decision Making Using Statistics

2 Data, Reality, and Problem Solving

3 Organizing, Displaying, and Interpreting Data

4 Numerical Descriptive Statistics

5 Probability, Randomness, and Uncertainty

6 Discrete Probability Distributions: Information about the Future

7   Continuous Random Variables

8   Samples and Sampling Distributions

9 Estimating Means and Proportions: Single Samples

10 Hypothesis Testing: Single Samples

11   Inferences about Means and Proportions: Two Samples

12   Analysis of Variance (ANOVA)

13 Regression, Inference, and Model Building

14 Multiple Regression

15 Looking for Relationships in Qualitative Data

16 Nonparametrics

17 Statistical Process Control

A   Appendix

Appendix A

Appendix B

Appendix C

Answer Key

Index

Preface

We have taught Business Statistics for roughly thirty years. We have always wanted to write a textbook that portrays our teaching style – one that introduces the topic being discussed as it relates to the business world and then presents the methodologies in a manner that helps the students know when, where, and how these methodologies are used in the business world. This textbook does just that. As you work your way through the book, you'll see that in each chapter, the topics are introduced with examples of real-world applications. Then, as subsequent sections are covered in each chapter, the examples are presented and worked through step-by-step in a manner that will keep the students interested. When we set out to write this text more than two years ago, our goal was to discuss topics at a level that would engage the students, present examples that would interest them, and write the book in a relaxed, conversational style – we believe we have accomplished those goals with this text.

We have also integrated TI-84 Plus, Microsoft Excel 2010, and Minitab 16 support in the form of Discovering Technology sections at the end of each chapter, along with getting started guides for both Excel and Minitab. Although the text includes quite a few manual computations with many of its examples, learning to use technology is helpful in analyzing the book's many business-related data sets and cases (presented as Discovery Projects at the ends of the chapters). In addition, we have included solutions to odd-numbered exercises and a tear-out formula and table reference that we hope students will find helpful. Further, the material in this text is integrated with the *Hawkes Learning Systems: Discovering Business Statistics* software. Through the use of this text along with the accompanying software, it is our hope that students will discover the many advantages of statistical thinking and understanding, both in business environments and everyday situations.

Quinton J. Nottingham

James S. Hawkes

Acknowledgements

We would like to thank our reviewers for giving us their time and feedback throughout the editorial process.

Scott Bailey, Troy University

Sangit Chatterjee, Northeastern University

Dr. Gregory Colman, Pace University

Dr. Jerry Dake, University of North Texas

Kim Gilbert, University of Georgia

Dr. R. Martin Jones, Virginia Tech

Dr. Lara Khansa, Virginia Tech

Dr. Onur Seref, Virginia Tech

Dr. Rick Simmons, Texas A&M University – Central Texas

Dr. Mark Skean, Mayville State University

Dr. Judith Woerner Mills, Southern Connecticut State University

Dr. Zhiwei Zhu, University of Louisiana – Lafayette

There are quite a few people at Hawkes Learning Systems that have made significant contributions to the creation of this book. First and foremost, we would like to thank our editor, Margaret Gibbs, who has worked tirelessly on the development of this book with her editorial expertise. Without her thorough proofreading, we are sure it would have taken significantly longer to complete this project. We would also like to thank Kimberly Cumbie, Danielle C. Bess, Susan Niese, Claudia Vance, Nina Waldron, Jennifer Moran, Kaylen Saxon, and Tee Jay Zajac for their editorial assistance.

About the Authors

Dr. Quinton J. Nottingham

Quinton J. Nottingham is Associate Professor of Business Information Technology (BIT) in the Pamplin College of Business at Virginia Tech. He earned his B.S. (1989), M.S. (1991), and Ph.D. (1995) degrees in Statistics from Virginia Tech. Professor Nottingham has served on the faculty at Virginia Tech since 1995. He primarily teaches the Quantitative Methods courses and the MBA-level Managerial Statistics. As a Master Online Instructor, he has also delivered the Quantitative Methods courses online with the use of the *Hawkes Learning Systems: Discovering Business Statistics* software.

Professor Nottingham has published numerous articles in the areas of applied statistics, regression, nonparametric regression, logistic regression, time series analysis, artificial intelligence, and security. He has served on the board of the Southeastern Decision Sciences Institute (SEDSI) as the VP of Student Liaison, VP Finance, Program Chair, as well as the President (2011-2012). He is a member of or has held membership in the Institute for Operations Research and Management Science (INFORMS), Decision Sciences Institute (DSI), American Statistical Association (ASA), Institute of Mathematical Sciences, and the Virginia Academy of Sciences.

Professor Nottingham enjoys spending time with his family and friends, vacationing at various beaches in the southeast, playing golf and basketball, and lifting weights. As a former Virginia Tech basketball player, he is an avid supporter of Virginia Tech Athletics and can be found in attendance at many Virginia Tech sporting events.

Dr. James S. Hawkes

James S. Hawkes earned his Bachelor's degree at the University of Richmond (1969), a Master's of Business Administration at New York University (1972), and his Ph.D. in Management Science at Clemson University (1978). As a faculty member at the College of Charleston from 1977 to 1998, he primarily taught Business Statistics. In 1978, he created Hawkes Learning Systems (HLS) where he is currently CEO. In 1986, Dr. Hawkes designed a breakthrough teaching technology called *Adventures in Statistics*. This software was the first to use expert system methodology to teach problem solving in statistics. Successive generations of the software expanded the content and improved the expert system. The software was renamed *Hawkes Learning Systems: Statistics* and is the basis for the software that accompanies this textbook, *Discovering Business Statistics*.

Other accomplishments include the founding of Quant Systems India (QSI), a software development company located in Vishakaputnam, India in 1994. QSI focuses on software development and quality assurance of HLS software. Dr. Hawkes also co-founded Automated Trading Desk in 1988, a proprietary stock trading company that traded approximately six percent of the NYSE and NASDAQ market places before it was sold to Citi Group in 2008. Most recently, he created The Writer's Muse which was an outgrowth of his hobby of collecting interesting phrases. This company has produced a phrase thesaurus which is available for smart phones, tablets and personal computers. Dr. Hawkes enjoys composing music, playing bridge, and collecting hats.

Features and Pedagogy

This text includes a number of special features developed to engage and challenge students while ensuring that their knowledge of statistics is growing. The following list outlines a number of these features.

Discovering the Real World

Each Discovering the Real World section introduces the main topic of the chapter in the context of a real-world application of the topic. These applications are designed to pique students' interest and get them excited about learning the material in the chapter.

An Example-Driven Text

Examples are relevant and engaging for students, and are presented in a step-by-step manner that is easy to follow. Real-world contexts are used within examples to illustrate the advantages of the statistical methods being taught. The text is very example-driven – and most of the concepts are taught by example. Examples are clearly presented within the layout of the text, making them easy for students to recognize and learn from.

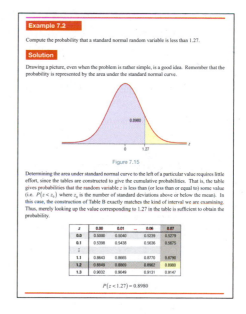

Definition

Statistical measures which are not affected by outliers are said to be **resistant**.

The arithmetic mean is not a **resistant measure**.

Weighted Mean

The weighted mean is similar to the arithmetic mean except it allows you to give different weights (or importance) to each data value. The weighted mean gives you the flexibility to assign weights when you find it inappropriate to treat each observation the same. The weights are usually positive numbers that sum to one, with the largest weight being applied to the observation with the greatest importance. The weights can be determined in a variety of ways, such as the number of employees, market value of a company, or some other objective or subjective method. There are occasions in which it is easier to assign the weights without worrying that they will sum to one. If you are concerned about your weights summing to one, you can make your weights sum to one by dividing each weight by the sum of all the weights.

Formula

Weighted Mean

The **weighted mean** of a data set with values $x_1, x_2, x_3, ..., x_n$ is given by

$$\bar{x} = \frac{w_1 x_1 + w_2 x_2 + ... + w_n x_n}{w_1 + w_2 + ... + w_n} = \frac{\sum w_i x_i}{\sum w_i}$$

where w_i is the weight of observation x_i.

Important Definitions, Formulas, Properties, Procedures, and Theorems Highlighted

Definitions, formulas, properties, procedures, and theorems are clearly presented and easy to reference for students. The color coding will help students zone in on the most important components of the chapters. Definition boxes are green, formula boxes are purple, and properties, procedures, and theorem boxes are blue.

Informational Sidebars

Interesting and engaging sidebars are presented in the margins to stimulate student interest in statistical topics. The sidebars highlight the origins of the statistical topics being covered, as well as real-world applications of statistical concepts.

Gauss
1777 – 1855
In statistics Karl Friedrich Gauss is best known for invoking the normal error distribution in conjunction with solving the least squares regression problem. Although modern statistical historians find much to criticize in Gauss' work, it was his publication of the normal error curve in relation to the regression problem that led Pierre Simon LaPlace to see the connection between the central limit theorem and linear estimation. For several generations of statistics, linear estimation based on the normal error distribution was a fundamental tool of data analysis.

Basic Concepts Questions

Each section has a set of exercises that begins with a list of questions geared towards testing students' knowledge of the basic concepts being taught. Students benefit from expressing the ideas they are learning through the written word. These questions both review the key definitions from the section and help students examine the parameters and limitations of the statistical methods they are applying.

Exercise Sets

Exercise sets are given at the end of each section, which range in difficulty and cover a wide variety of subject matter. Utilizing the exercises in this text, students can easily get the practice they need to master the topics being taught. Care has been taken to present students with a variety of exercise types to encourage a broad understanding of the material presented. Additionally, exercises have been created using data that is relevant to students' lives and their entry into the business world.

Discovering Technology

Discovering Technology sections at the end of each chapter highlight easy-to-follow directions for using a TI-84 Plus calculator, Microsoft Excel 2010, or Minitab 16 to apply the concepts learned in the chapter. Though the directions were written to accompany Excel 2010 and Minitab 16, many of the steps can also be followed when using earlier versions of these software products. Additionally, many of the calculator instructions can be applied to graphing calculators other than the TI-84 Plus. Screen shots are presented alongside the step-by-step instructions to aid in student learning. Although the text includes quite a few manual computations with many of its examples, learning to use technology is helpful in analyzing the book's many business-related data sets and cases (presented as Discovery Projects at the ends of chapters). In addition, Getting Started guides are included in Appendices B and C for Excel 2010 and Minitab 16, respectively, to get students started with spreadsheet technology and data analysis.

Chapter Review

This end-of-chapter feature includes a list of the key terms and ideas from the chapter as well as a list of the chapter's key formulas. By gathering the key ideas and formulas in one place, students have a great resource to use while studying, as they can more easily identify the concepts they may need to review.

Additional Exercises

At the end of each chapter, additional exercises have been included to provide students with additional practice on the skills covered. Also, by providing students with a compilation of problems from the whole chapter, the guidance given in the section exercises is removed, and students are forced to evaluate which method to use to solve each of the different question types.

Discovery Projects

Discovery Projects at the ends of chapters engage students with more in-depth questions regarding the chapter content. Large, real-world data sets that accompany the Discovery Projects are available for download in Excel or Minitab format. The projects also give instructors the opportunity to enhance student learning through individual or group projects and class discussions. Through these projects, instructors have the freedom to supplement the material as they see fit with a longer-term, more difficult assignment.

Answer Key and Formula Sheets

The answer key contains solutions for odd-numbered exercises for student reference. In addition, this text contains a tear-out formula sheet and table reference that students may use throughout the course.

Answer Key

 Decision Making Using Statistics

Additional Exercises

19. a. Internet users **b.** The amount of time users view photos of products on websites **c.** Inferential
citizens **b.** Self-esteem after reading a news story **c.** 276 elderly citizens **d.** Answers will vary. **e.**
23. a. Mobile phone users **b.** The percentage of people who use their mobile phones to access the Intern
reasons **c.** 500 American adults 18 years of age and older **d.** Answers will vary. **e.** Inferential **25.**
couples **b.** The percentage of married couples that met online **c.** 7000 adults married in the past 5 yea
vary. **27. a.** U.S. states with coastlines **b.** Coastline length

2 **Data, Reality, and Problem Solving**

2.1 Quality of Measurements

EXERCISES

5. a. Well-defined **b.** Well-defined **c.** Not well-defined **d.** Well-defined **e.** Not well-defined **7.**
well-defined scale to measure cleanliness or aesthetics. Answers will vary.

2.2 Science and Data

EXERCISES

5. Answers will vary. **7.** Inferential. Data were collected from a sample and the information was used t
claim about the population of the company's bulbs.

2.4 Collecting Data

EXERCISES

15. a. By randomly assigning women to two groups and using one of the groups as a "control" group, the
produce data that will reveal the impact of the different diets. **b.** Difference in diet (The first group recei
per day diet for the entire period whereas the second group received a 420 calorie per day diet for 16 week
shifted to a 1200 calorie per day diet for the rest of the experimental period.) **c.** Weight loss. **d.** Yes, t
the 1200 calorie per day diet represent a control group. **e.** Observational studies are subject to self-selec
not necessarily know the cause of the weight reduction. **17. a.** Phase 1: Gather information about the ph
studied. **b.** Controlled experiment. **c.** Number of major attacks of Multiple Sclerosis. **d.** Bovine my
individuals in the early stages of MS fed bovine myelin. **f.** Fifteen individuals in the early stages of MS
placebo. **19.** Jacob's knee could feel better simply because he took a week break from playing basketbal
vary. **21.** There are many factors that affect whether or not someone is happy. Additionally, since both o
or no reply there is no way to quantify happiness or going to church. Answers will vary. **23.** Generally p
money will seek the help of a financial advisor since advisors are paid for their services. Answers will var

viii

HAWKES LEARNING SYSTEMS:
Discovering Business Statistics Software

Overview

Hawkes Learning Systems specializes in interactive mathematics courseware with a unique mastery-based approach to student learning. ***Hawkes Learning Systems: Discovering Business Statistics*** is designed to help students develop a solid foundation and understanding of statistics. It has been proven to promote student success. The courseware is comprised of three learning modes: Instruct, Practice, and Certify. Supplemental Instructional videos are also available via HawkesTV.

Getting Started

Step 1: Install the software	**Step 2:** Find your Course ID
Download at: **hawkeslearning.com/download** OR Insert the enclosed disc (PC only), and follow the on-screen directions.	Visit: **hawkeslearning.com/CourseID**

Step 3: Get your Access Code	**Step 4:** Enroll in your course
Register your license number or purchase an access code at: **hawkeslearning.com/AccessCode**	Log in to the software to enroll (must be online). Select your instructor's name and section.

Minimum Requirements

In order to run ***Hawkes Learning Systems: Discovering Business Statistics***, you will need:

PC: Windows® XP (with Service Pack 3) or later and Internet Explorer 7.0   **Mac:** OS X v10.5.8 or later, Intel x86 systems

- 1 GHz or faster processor

- 256 MB memory (RAM)

- 500 MB hard drive space (compact install), or up to 1.5 GB (complete install)

Exploring the Courseware

Instruct Mode

Instruct is a multimedia presentation of each lesson and correlates with the textbook. Instruct provides students with:

- Definitions, rules, properties
- Example problems
- Video tutorials and audio narration
- Clickable HotWords provide additional information on the topic

Practice Mode

Practice offers students an unlimited number of algorithmically generated problems and feedback on incorrect answers. Within practice the Interactive Tutor encourages active learning.

Interactive Tutor Features

- Instruct: Multimedia presentation of the topic.
- Explain Error: Uses artificial intelligence to diagnose errors made and provide specific feedback.
- Step-by-step: Breaks each problem down into smaller steps offering feedback and solutions at each point.
- Solution: Unlimited guided solutions to every problem.

Certify Mode

Certify hold students accountable for mastering the material at a defined proficiency level while removing learning aids. This motivational, non-penalty approach allows students to demonstrate mastery of the topic. A Certification Code is provided upon completion and uploaded directly to your Progress Report.

HawkesTV

View instructional videos anytime, anywhere at **hawkesTV.com**

Support

Feel free to contact support for questions or technical help with *Hawkes Learning Systems: Discovering Business Statistics* courseware.

Support Center: Support.hawkeslearning.com

Chat: hawkeslearning.com/chat

Email: support@hawkeslearning.com

Phone: 843.571.2825

Chapter One

Decision Making Using Statistics

1

D   Discovering the Real World

Statistics and Your Health

The Centers for Disease Control and Prevention estimates that more than 400,000 people die in the United States each year from smoking and related diseases. That is considerably more deaths than those attributed to alcohol, illegal drugs, murders, suicides, car accidents, and HIV combined. Surprisingly, the evidence relating smoking to cancer is largely statistical, most of which could be digested after an introductory statistics course.

Cigarettes did not always carry warnings such as "Smoking causes lung cancer, heart disease, emphysema, and may complicate pregnancy" or "Quitting smoking now greatly reduces serious risks to your health." In the early 1960's, congress debated the surgeon general's plan to label all packages of cigarettes with a warning that smoking could be hazardous to your health. Hearings were held and testimonies were gathered from interested parties. The tobacco companies did not want a health notice placed on their product, so they sponsored a significant amount of investigation into the health effects of smoking. On the other side of the issue was a large group of medical researchers who had analyzed the effects of smoking on animals and had collected considerable data on the health of both smoking and nonsmoking populations. The evidence presented seemed unquestionable: heavy smokers had a much higher incidence of lung cancer, about thirty times higher than nonsmokers. Furthermore, researchers found smoking is associated with cancers of the mouth, pharynx, larynx, esophagus, pancreas, uterus, cervix, kidney, and bladder. Even suicide rates were at higher levels for smokers.

The tobacco companies countered with their own experts who pointed out that the underlying cause of both disease and smoking could be attributed to heredity. That is, whatever caused people to have higher rates of lung cancer, heart disease, and so forth, could also cause them to smoke. If this were the case, then the proposed warning on packages of cigarettes would be untruthful. So, the pivotal problem was, how could someone know whether smoking or heredity caused cancer?

One method of attacking this problem is to perform a **controlled experiment**. In a controlled experiment the researcher collects data that will hopefully clarify the relationship between variables. A simple method of accomplishing this task would be to create two groups, smokers and nonsmokers, and compare them. The nonsmokers would be the **control group** and the smokers the **treatment group**. Each subject is randomly assigned to one of two groups, in this case smokers and nonsmokers, to avoid bias. The smoking group would be required to smoke a pack of cigarettes per day (for say 15 years) and at the end of the time period their health would be compared to the health of the nonsmokers. By randomly choosing members of the population, the hereditary pool of both groups should be similarly mixed and thus any subsequent differences in health between the two groups could be attributed to the **treatment** (smoking) that was applied to one group. This method of experimentation has one drawback; it is impossible to get human beings to volunteer for the experiment. As a result, the experiment has never been done.

However, researchers have performed controlled statistical experiments on many different kinds of animals, and have discovered that smoking does cause problems similar to those observed in human beings. In 1964, the surgeon general produced a historic report linking tobacco smoking to cancer. Since January 1, 1966, all cigarette packages sold in the United States have been required to carry the health warning. Since January 1, 1971, all cigarette television advertising has been banned. The percentage of adults who smoke in the United States has decreased significantly. The prevalence of smoking has diminished in the United States since the exposure of the statistical evidence, from about 40% in 1965 to about 29% in 1987, 23% in 2002, and 21% in 2008. However, this is not the case in many parts of the world.

Statistics made a difference. They helped people make informed decisions about their choice to smoke or not. Fortunately, it was not speculation, opinion, or special interest that decided the issue. Research on the effects of smoking continues and of the 4000 substances found in tobacco smoke, about 60 are now believed to be associated with cancer or tumor formation.

So does smoking cause cancer? There is no absolute proof, but the evidence overwhelmingly suggests that it does. In 1998, tobacco companies reached a $206 billion agreement with 46 states to settle a lawsuit on the health damage done by cigarette smoking.

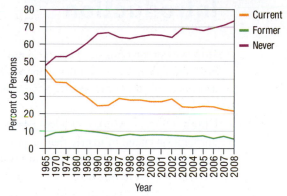

Percent of Young Adults (Ages 18-24) Who Where Current, Former, or Never Smokers

Source: National Center for Health Statistics

Figure 1.1

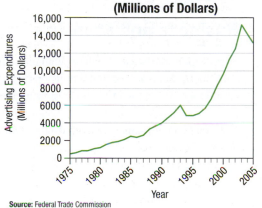

U.S. Advertising and Promotional Expenditures for Cigarettes (Millions of Dollars)

Source: Federal Trade Commission

Figure 1.2

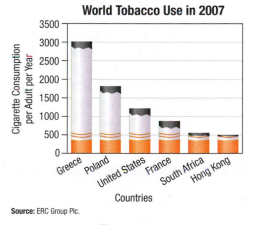

World Tobacco Use in 2007

Source: ERC Group Plc.

Figure 1.3

Introduction

How we conduct business and make business decisions in the real world is intensely data driven. In business today, almost all decisions are made based on the analysis of data. In fact, decisions in all aspects of our lives are based on the diverse and ubiquitous data that are available to us.

CEOs, CFOs, Presidents, Vice Presidents, and many leaders and executives of multinational corporations need to justify their decisions. Almost all of the time, these decisions are supported by statistical models, summary statistics (graphs and tables), and other statistical tools and concepts that we will discuss in this text.

Decision makers use statistics to collect data, analyze the data, and make interpretations – all relevant to their profession. They build models with data and then make decisions using those models. They describe data using graphs and summary measurements. They develop methods of designing experiments and gathering data that are cost effective and diminish bias.

Business and Statistics

Simply put, statistics will help managers make better business decisions. Managers use data to help them oversee multiple divisions of their company, countless products, and thousands of employees. Even for small businesses, understanding data via the use of statistics is critical in today's competitive environment. Even though many business decisions are made based on expert intuition and experience, ignoring <u>all</u> relevant information available (such as the use of statistical techniques) would be senseless.

Thus, the use of statistical techniques must be viewed as an important part of the decision-making process. Combining intuition, experience, and statistical analysis arms the manager with a wealth of information giving him or her an advantage over the competition. If not an advantage, then this combination of tools certainly allows the manager, company, or corporation to maintain its competitive status in its respective industry.

For example, the development of Six Sigma can be traced back to the eighteenth century when Carl Frederick Gauss introduced it as a metric for the normal curve. Later Walter Shewhart began showing how three sigma deviations from the mean required a process correction. In the 1980s, a Motorola engineer coined the term Six Sigma (and copyrighted it) for improving quality management. Today Six Sigma is viewed as a concept that incorporates many statistical techniques to improve processes and make business decisions.

1.1   Who Will Be Our Next President?

Once every four years the media extravaganza known as the American presidential election is intensely carried over the airways and newspapers for a period of at least six months. We speculate on who the candidates will be, watch the primary campaigns, see the conventions, look at the debates, put up with slick image-making commercials, and endure negative advertisements. Finally we cast our votes and watch the results on the evening news.

Often times, the results of the election are known well in advance. How can this be the case? With statistics! By sampling registered voters prior to the election, one can get an estimate of the outcome and thus, who will win the election. Of course, even though there is a chance that the candidate receiving the highest percentage of votes may not win, it is expected that if the sample is large enough, then the estimate of votes that each candidate receives will be close to the outcome of the actual election.

Recent evidence indicates the statistical forecasts based on sample estimates are becoming increasingly accurate. Table 1.1 provides a list of the predicted percentages of popular vote for the winning candidate the day prior to the election and the actual vote received by the candidate.

Table 1.1 – Presidential Polls			
Year	Candidates	Winner's Predicted %	Actual Winning %
1976	Carter Ford	48%	50.1%
1980	Reagan Carter	47%	50.8%
1984	Reagan Mondale	59%	59.2%
1988	Bush Dukakis	56%	53.0%
1992	Clinton Bush, Perot	49%	43.3%
1996	Clinton Dole, Perot	52%	50.1%
2000	Bush Gore, Nader	47%	47.9%
2004	Bush Kerry	49%	50.7%
2008	Obama McCain	55%	52.8%

Source: Gallup.com

What methods do statisticians use to make predictions? The most difficult part of the process is finding a sample that accurately reflects those people who plan to vote in the election. Those who will vote on Election Day define a group of people that we are interested in studying. In statistics the group we wish to study is called the **population**.

> ### Definition
>
> A **population** is the set of all subjects or elements about which we are interested in making inferences.

Populations are defined by what a researcher is studying and can come in all shapes and sizes. If someone is studying houses foreclosed upon in 2008, then all the houses foreclosed upon in 2008 would constitute the population. If someone is studying banks in the United States, then all banks in the United States would represent the population. Citizens who are registered to vote in a presidential election constitute a population of considerable interest to presidential candidates. The parties spend hundreds of millions of dollars finding out what people think (sampling) and adopt or reject political positions consistent with the preponderance of the data. In the world of political science this "data-driven" approach to politics is called populism.

Populations in the NYSE

As a student studying business (whether it be management, marketing, finance, or management science), your ultimate goal is to make money by making good business decisions. What makes up the New York Stock Exchange? The NYSE is made up of many companies and organizations, large and small. As someone attempting to make money on the NYSE, there are many populations to study. One can study the daily closing price of a particular stock; the stocks valued at less than a certain dollar amount, say $10; or the number of profitable companies on the NYSE. As you can see, since there are more than 2700 companies valued at more than 28.5 trillion dollars, one needs to make informed decisions in order to be profitable.

> ### Definition
>
> A list containing all members of the population is referred to as a **frame**.

According to the Census Bureau there are about 309 million people in the United States and about 6.8 billion people around the world. The **frame** for the population of the U.S. would be a rather long list containing about 309 million names. Although a previous census would be a good start in developing a frame for the U.S. population, it is doubtful that an exact frame could ever be developed at a given point in time since there is one new birth every 8 seconds, one death every 12 seconds, and one new immigrant every 36 seconds. There are just too many people being born, dying, and immigrating over a 10-year period to get an exact frame for the U.S. population. But for problems that deal with smaller populations, frames are easily developed. For example, if your business statistics class were the population under consideration, the class roll would be the frame for the population.

Note: A strict definition of a **census** is a survey that includes all the elements or units in the **frame**.

> ### Definition
>
> **Population parameters** are facts about the population. Since parameters are descriptions of the population, a population can have many parameters.

For a presidential election, some population parameters in which candidates and pollsters will be interested are:

- The percentage of eligible voters who will vote on Election Day.
- The percentage of voters who will vote for a specific candidate.
- The percentage of men who favor a candidate.
- The percentage of women who favor a candidate.
- The percentage of people in the 18-25 age group who favor a candidate.
- The average income of voters who favor a candidate.

The parameters mentioned in this instance are either averages or percentages. However, other measures such as the maximum or minimum value of the population measurement as well as other characteristics would also be considered population parameters. For a specific population at a specific point in time, population parameters do not change; they are fixed numbers. But seldom will the value of a population parameter be known since the value involves all the population measurements which are usually too expensive or time consuming to collect. It is the statistician's job to discover these values. This is done by taking a sample and using the sample measurements to estimate the desired population measurement.

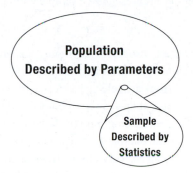

The percentage of votes a presidential candidate receives on Election Day is a parameter presidential candidates spend enormous amounts of money trying to estimate. Sample data are used to estimate this population parameter. Sample estimates and population parameters for the percentage of votes the winning candidate received in an election appear in Table 1.1.

Definition

A **sample** is a subset of the population which is used to gain insight about the population. Samples are used to represent a larger group, the population.

Definition

A **statistic** is a fact or characteristic about a sample.

In Table 1.1 the winner's predicted percentage is a statistic since it is computed from a sample. The elected president's actual percentage is a parameter, since it is computed using the votes of all who participated in the election (the population in this instance).

For any given sample a statistic is a fixed number. Because there are lots of different samples that could be drawn from a population, statistics will vary depending on the sample selected. Statistics are used as estimates of the population parameters. See Figure 1.4.

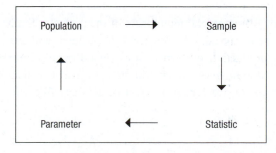

Figure 1.4

Studying an entire population can be an expensive proposition. In 2000 the census cost the government $39.98 per household. Total expenditure during the 2000 census cycle was approximately 4.5 billion dollars. The estimated cost of the 2010 census was around 14.5 billion dollars, or $117.33 per household. Because of the enormous expense, even the United States government with all its resources does not undertake a census of its citizens but once every ten years. Yet, amazingly accurate information about the population, even large populations like the United States, can be found by using a small sample. If the sample is a good representation of the population, then the conclusions reached using sample data will likely be reasonable for the population as a whole. A statistician faces the interesting problem of developing a representative sample without spending an inordinate amount of time or money.

Where Did the Term *Statistics* Come From?

LATIN
The evolution of the word statistics comes from the Latin phrase *statisticum collegium* (lecture about state affairs).

GERMAN
The German term *Statistik* was first used by Gottfried Achenwall in 1749.

ENGLISH
The anglicized form of the German term was introduced by John Sinclair (a Presbyterian pastor) who published the *Statistical Account of Scotland*, a 21 volume compilation. This work was published at various periods during the 1790s.

Political pollsters spend much of their resources developing a representative sample of Americans who will cast their votes on Election Day. After the sample is identified, the problem becomes one of obtaining candidate preferences from the selected voters. Tallying the votes and computing the actual percentages will take a computer at most a few seconds. After the estimates of the population parameters are calculated, one of the interesting statistical questions is, *how good are the estimates?* As Table 1.1 demonstrates, none of the estimates of the percentage of votes each candidate would receive were exactly correct, but most were very close. If we cannot determine how much faith we should place in our estimates, it will be difficult to use the estimates to make decisions. The process of selecting samples and determining the reliability of our estimates is a large part of what statistics is about.

1.2   Statistics and Quality

"Statistical thinking is critical to improvement of a system."

-Mary Walton, *The Deming Management Method.*

> **Definition**
>
> A **process** is a series of actions that changes inputs to outputs.

The idea of a process is closely tied to quality control. For a physician a process is a treatment performed in a series of steps designed to improve a patient's condition. In a business context a process is a series of steps that produces a product or service. Closely monitoring and continuously improving processes produce high quality products. Monitoring the process means taking measurements of key variables over time. Improving processes means reducing process variation by finding the causes of variation and eliminating them.

In order to improve a process there must be an understanding of how the process is currently performing. This requires definition and monitoring of the process. Statistics helps with decisions about how the data will be collected, what data will be needed, and the analysis of the data. In addition to ferreting out production problems, **Statistical Process Control** (SPC) is a group of statistical methods designed to monitor and control processes. SPC is helpful in detecting problems in a process before they create a defective product or service. We will study this subject more extensively in Chapter 17.

1.3   Descriptive versus Inferential Statistics

The science of statistics is divided into two categories, **descriptive** and **inferential**. Descriptive methods describe and summarize data, while inferential methods aid in drawing conclusions and making decisions and predictions about populations and processes for which it is impractical to obtain measurements on each member.

Descriptive Statistics

The emphasis in **descriptive statistics** is analyzing observed measurements, usually from a sample. With descriptive statistics we try to answer questions such as:

- What is a typical value for the measurements?

- How much variation do the measurements possess?

- What is the shape or distribution of the measurements?

- Are there any extreme values in the measurements and, if so, what does that tell us?

- What is the relative position of a particular measurement in the group of data?

- What kind of relationship exists, if any, when there are two variables and how strong is the relationship?

> **Definition**
>
> **Descriptive statistics** is the collection, organization, analysis, and presentation of data.

Examples of Descriptive Statistics

- Frequency Distribution
- Measures of Central Tendency:
  - Mean
  - Median
  - Mode
- Measures of Dispersion:
  - Range
  - Variance
  - Standard Deviation

The application of descriptive statistical tools is usually *ad hoc*, that is, the exact method of analysis changes from one problem to the next. Sometimes the application of descriptive statistics can raise as many questions as it answers. And when that happens, statistics is working at its best as a problem-solving or process-improvement tool.

The importance of descriptive statistics as an information-producing tool relates to the amount of data to be comprehended. If there are only two observations, say 6 and 4, then comprehending the data in its entirety is not difficult and descriptive statistical aids are of little value. However, the 100 observations of the revenues of the top 100 U.S. companies in Table 1.2 suggest a different story. Individually inspecting 100 revenues would produce very little useful knowledge. To comprehend a large set of data, it must be summarized. That is the function of descriptive statistical techniques. Descriptive techniques are the most common statistical applications.

Table 1.2 – Profits of the Top 100 Companies by Revenues in the Fortune 500 (Millions of Dollars)			
14,335	1707	−1098	1054
19,280	1312	3021	−329
10,483	2661	−21,553	740
11,025	2488	1056	202
6276	4745	5704	−71,969
4858	2006	3307	2468
12,535	12,266	6134	−338
2717	767	3638	−1237
11,728	1280	98	12,901
7660	14,569	1686	1755
8055	3829	4369	−537
−1606	1433	1277	2130
3651	13,385	683	−2915
823	8635	3124	−459
−10,949	1463	895	381
1151	1783	−2436	204
3696	2152	854	−115
12,275	3024	2394	6342
13,425	1003	1346	1935
3822	648	1023	828
13,436	−2855	6824	−887
70	235	1040	3622
503	28	2153	1161
1086	5946	5746	902
−1982	−2246	−3378	−30,860

Source: CNNMoney.com

Inferential Statistics

It would be preferable to have measurements of the entire population, but in most cases these data are either not obtainable or would be much too costly to obtain. For example, to be absolutely certain that all car air bags will inflate in head-on collisions would require each new car to be crash tested in a head-on crash. If 100 percent inspection were a requirement, cars would be a scarce commodity. Fortunately for automobile manufacturers, statistical sampling techniques can reliably estimate, with a relatively small sample, the fraction of air bags that will inflate.

Use Sample Data to Make Inferences about Unknown Population Characteristics

Definition
The objective of **inferential statistics** is to make reasonable estimates about population characteristics using sample data.

1.4   The Consequences of Statistical Illiteracy

Part of being an intelligent human being is the desire to learn the truth about the world we live in. But as Oscar Wilde said:

"The truth is rarely pure and never simple."

-Oscar Wilde, *The Importance of Being Earnest.*

Being statistically illiterate puts one (or one's organization) at a competitive disadvantage compared to companies that possess and use statistical knowledge and analytical tools. Statistics and its uses cannot be avoided. Therefore, learning and using statistical tools will give you and your organization more flexibility when making decisions.

To intelligently appreciate or produce statistical information, you must be statistically literate to defend yourself from a persuasive but fallacious statistical argument, to decrease your vulnerability to pseudo-sciences, and to diminish the chances of making poor and sometimes injurious business decisions.

A statistically literate person understands the language of statistics and understands statistical concepts and reasoning. To become statistically literate one should be able to think "statistically". This will involve asking questions like:

- Where did the data come from?
- How was the sample taken and is the sample large enough?
- How reliable or accurate were the measures used to generate the reported data?
- Are the reported statistics appropriate for this kind of data?
- Is a graph drawn appropriately?
- How was this probabilistic statement calculated?
- Do the claims make sense?
- Should there be additional information?
- Are there alternative interpretations?

Liar or Statistician?

In his book *How to Tell the Liars from the Statisticians*, Robert Hooke sheds light on our exposure to misleading statistics in everyday life. In the preface he writes, "The science of statistics has made great progress in this century, but progress has been accompanied by a corresponding increase in the misuse of statistics. The public, whether it gets its information from television, newspapers, or news magazines, is not well prepared to defend itself against those who would manipulate it with statistical arguments. Many people either believe everything they hear or come to believe in nothing statistical, which is even worse." Throughout the remaining chapters, Hooke uses examples from politics, economics, entertainment, and the medical community to illustrate the dangers of being statistically illiterate. You might be surprised to learn the ways in which the misuse of statistics affects you every day. In order to digest the plethora of statistical information you encounter, you must first become statistically literate.

Source: Hooke, Robert. *How to Tell the Liars from the Statisticians.* New York, New York: Marcel Dekker Inc., 1983. Print.

R   Chapter 1 Review

Key Terms and Ideas

- Controlled Experiment
- Control Group
- Treatment Group
- Treatment
- Population
- Frame
- Census
- Parameter

- Sample
- Statistic
- Process
- Statistical Process Control
- Descriptive Statistics
- Inferential Statistics
- Statistical Literacy

AE   Additional Exercises

Basic Concepts

1. What are three objectives of statistical methods?

2. Briefly describe the role of statistics in managerial decision making.

3. What is a controlled experiment?

4. Explain the purpose of a control group.

5. Explain the purpose of a treatment.

6. How is it possible to know the results of a presidential election before Election Day?

7. What is a population?

8. What is a frame?

9. What is a population parameter?

10. Why is it often difficult to determine the exact value of a population parameter?

11. What is a sample?

12. What is a statistic?

13. Describe the relationships between populations, samples, parameters, and statistics.

14. Define statistical inference.

15. Describe the role of statistics in the quality movement.

16. What is a process. How are processes improved?

17. What is the difference between descriptive and inferential statistics?

18. What are the consequences of being statistically illiterate? How could this put you at a disadvantage in business?

Exercises

19. Jakob Nielsen, a website consultant, recently conducted an eye-tracking survey to determine whether people ignore purely decorative images when viewing web pages. An aspect of the study compared a set of products on Pottery Barn's furniture website and a page of televisions on Amazon.com. The study found that consumers tended to ignore the televisions on Amazon.com because they were generic, making the product image less inviting. When consumers viewed Pottery Barn's website, they were more inclined to view the photos of the bookcases for longer periods of time because they were images of the actual products for sale.

Source: New York Times

a. Identify the population of interest.

b. What characteristic of the population is being measured?

c. Is the purpose of the data collection to perform descriptive or inferential statistics?

20. A recent study at Britain's Oxford and Exeter Universities explored whether a woman's diet before conception affects the gender of her child. Researchers studied the eating habits of 740 women during their first-time pregnancies and found that higher caloric intake prior to conception can significantly increase the chances of having a son while more restricted diets are more likely to produce daughters. They found that high potassium diets (eating bananas) and calcium rich diets (cereal and milk) were associated with having a baby boy. Researchers concluded that eating a bowl of cereal for breakfast can increase the chances of a male birth. "Of women eating cereals daily, 59 percent had boys, compared with only 43 percent who bore boys in the group eating less than a bowlful per week."

Source: CNNHealth.com

a. Identify the population of interest.

b. What characteristic of the population is being measured?

c. Identify the sample.

d. Is the purpose of the data collection to perform descriptive or inferential statistics?

e. What are some problems that could be associated with collecting data in this study?

21. Researchers at The Ohio State University and Zeppelin University Friedrichshafen, in Germany, recently conducted a study regarding the elderly and negative news coverage. The researchers presented 276 subjects with several stories (with photos) about either old or young people. Participants were presented with one of two versions of each story. In one version the main character was painted in a positive light and in the other the same character was described negatively. After the participants finished reading their self-esteem was measured. The study found that older readers were more inclined to read the negative stories about youth. In addition, they found that the more negative stories older people read about younger individuals, the higher their self-esteem tended to be. This could explain the prominence of negative media coverage on networks with an older audience such as Fox News and MSNBC.

Source: Huffpost Media

a. Identify the population of interest.

b. What characteristics of the population are being measured?

c. Identify the sample.

d. What are some problems that could be associated with this study?

e. Is the purpose of this study to perform descriptive or inferential statistics?

22. Researchers at Pepperdine University's Graziadio School of Business and Management recently conducted a study to determine whether the capital crunch is affecting small businesses' abilities to expand. The companies studied included alternative lenders, venture capitalists, and private equity firms, among others. The project surveyed 559 privately held businesses and 1430 lenders and investors nationwide. The study found that 78 percent of businesses had solid growth strategies but only 40 percent had access to the resources needed to grow. Lenders and investors rejected 90 percent of loan applications or investment proposals that would be secured by a business's real estate holdings and 73 percent of loan applications or investment proposals that are based on a business's cash flow. According to the survey author John Paglia, "The study shows private business owners feel they are being constrained by access to financial capital. Owners currently expect a 10 percent revenue growth over the next 12 months. If they were to receive additional capital, they estimate their revenue growth rate to jump to 25 percent."

    Source: smallbiztrends.com

    a. Identify the population of interest.

    b. What characteristics of the population are being measured?

    c. Identify the sample.

    d. Is the purpose of this study to perform descriptive or inferential statistics?

23. Ruder Finn, one of the world's largest independent public relations agencies, recently announced the *Mobile Intent Index* which studies mobile phone user habits and explores the underlying reasons that people have for accessing the internet on mobile devices. The *Mobile Intent Index* asked 500 American adults 18 years of age and older how often they use their mobile phones to access the internet for 295 reasons. The study found that 91% of mobile phone users go online to socialize compared to 79% of desktop users. In addition, 60% of mobile internet users go online to manage finances compared to only 45% of desktop users. They also found that mobile users were less likely to use the internet for educational purposes, only 42% compared to 92% of desktop users. Finally, unsurprisingly, mobile phones are not used for creative purposes; only 42% of mobile users personally express themselves online compared to 54% of desktop users.

    Source: PR Newswire

    a. Identify the population of interest.

    b. What characteristic of the population is being measured?

    c. Identify the sample.

    d. What are some problems that could be associated with this study?

    e. Is the purpose of this study to perform descriptive or inferential statistics?

24. A personnel director is interested in determining how effective a new reading course will be in improving the reading comprehension of her company's employees. The director randomly selects twenty employees and determines the average reading comprehension both before and after instruction in the reading course.

    a. Identify the population.

    b. What characteristic of the population is being measured?

    c. Identify the sample.

    d. Is the purpose of the data collection to perform descriptive or inferential statistics?

25. In 2009 and 2010 the market research firm Chadwick Martin Bailey conducted several studies to provide insights into recent dating behavior in the United States. The data were collected through research via an online Consumer Research Panel. In the "Marriage Survey" (a survey of recently married people), 7000 adults 18 years of age and older who were married in the past 5 years were polled. The study found that 17% of couples married within the last 3 years met each other on an online dating site. This is compared to 26% that met their significant other through a friend or family member, 36% who met through work or school, 4% who met through a church or place of worship, 11% who met through bars, clubs, or social events, and 7% who met in some other manner.

Source: Chadwick Martin Bailey/Match.com

    a. Identify the population of interest.

    b. What characteristics of the population are being measured?

    c. Identify the sample.

    d. What are some problems that could be associated with this study?

26. In a recent study, seat belt users were found to have 20% fewer fatalities than those who do not wear seat belts. Do these results prove that seat belts reduce the chances of a fatality?

27. States having an abundance of coastline have an obvious advantage over landlocked states, or states with little coastline, in that their economies may profit from an extensive fishing industry, tourism, shipping, or other water related activities. Alaska, the leader by far in miles of coastline, has a total of 6640 miles, of which 5580 miles border the Pacific Ocean and 1060 miles border the Arctic. Florida, the leader in the continental United States, has a total of 1350 miles with 580 miles on the Atlantic Ocean and 770 miles on the Gulf of Mexico. Of all states with some coastline, New Hampshire, with 13 miles of coastline, is in last place.

    a. Identify the population.

    b. What characteristic is being measured?

28. A young actuary (statistician usually working in the insurance industry) has been asked to summarize the number of automobile accident claims by region for his company. He randomly selects 50 automobile accident claims which his company has settled in the last year and counts the number of accidents in each region: North, South, East, and West. He summarizes the counts by region in a chart and gives the results to his supervisor.

    a. Identify the population.

    b. What characteristic of the population is being measured?

    c. Identify the sample.

    d. Is the purpose of the data collection to perform descriptive or inferential statistics?

P   Discovery Project

The American Recovery and Reinvestment Act

On February 13, 2009, Congress passed the American Recovery and Reinvestment Act of 2009 at the urging of President Obama, who signed it into law four days later. A direct response to the economic crisis, the Recovery Act has three immediate goals:

1. Create new jobs and save existing ones.

2. Spur economic activity and invest in long-term growth.

3. Foster unprecedented levels of accountability and transparency in government spending.

The Recovery Act intends to achieve these goals by:

1. Providing $288 billion in tax cuts and benefits for millions of working families and businesses.

2. Increasing federal funds for education and healthcare as well as entitlement programs (such as extending unemployment benefits) by $224 billion.

3. Making $275 billion available for federal contracts, grants, and loans.

4. Requiring recipients of recovery funds to report quarterly on how they are using the money.

Data are posted online (www.recovery.gov) so the public can track the recovery funds. In addition to offering financial aid directly to local school districts, expanding the Child Tax Credit, and underwriting a process to computerize health records to reduce medical errors and save on healthcare costs, the Recovery Act is targeted at infrastructure development and enhancement. For instance, the act plans investment in the domestic renewable energy industry and the weatherizing of 75 percent of federal buildings as well as more than one million private homes around the country. Construction and repair of roads and bridges as well as scientific research and expansion of broadband and wireless service are also included among the many projects that the Recovery Act will fund. While many projects are focused more immediately on jumpstarting the economy, others, especially those involving infrastructure improvements, are expected to contribute to economic growth for many years.

There are (and were) many skeptics of the act. However, before the act was passed, specific reporting requirements were set for recipients of recovery funds, as well as agency reporting requirements. With that said, opponents to the act want to check on the progress and statistics of recovery. How will the opponents, as well as its supporters, carry out such checks?

Go to the website www.recovery.gov and discuss the following:

1. What would you define as the population of recipients for the recovery funds?

2. What variable(s) is/are of interest for those seeking information about the Recovery Act? Support your response with data from the website.

3. Do you think the Recovery Act served its purpose? Support your response with data from the website or other sources (news articles, tables, etc.).

Chapter Two
Data, Reality, and Problem Solving

2

D   # Discovering the Real World

Management and Measurement

The main objective of any business is to earn a profit and consequently companies depend heavily on measurement, data, and statistical thinking. In fact, there is an old management adage, *You can't manage what you can't measure.* Perhaps this adage would be better stated as *You can't manage unless you know what "reality" is.* That is why there are so many important measurement systems in a business. The accounting system is designed to measure profitability and to inform management of potential problems. The inventory system is a measurement tool designed to measure the status of inventory, indicate when orders should be placed, and spot potential inventory theft. The cash flow system is a forecasting system that measures the company's need for cash.

Your career will essentially be a choice of the kinds of problems you desire to solve. The more difficult the problems you decide to solve, the more you will depend upon data and measurement to solve them.

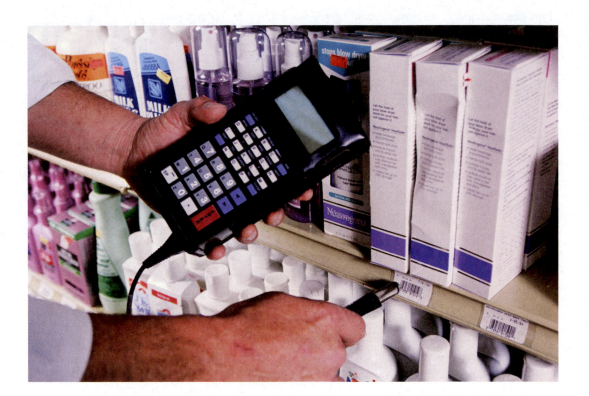

Introduction

For most people, the words "data" and "measurement" are words that generate about as much enthusiasm as watching dust settle. So why should you be interested in measurement and data?

The poem *Under Ben Bulben*, written in 1939, is one of the last poems of the Irish poet W. B. Yeats. The poem contains the insightful line "*Measurement began our might.*" Long ago, our species learned that if you do not know what "reality" is, it is difficult to predictably change it to a more desirable state. Measurement is the first step in understanding the "reality" of any circumstance you wish to change in a predictable way. Thus, measurement is a fundamentally important link in controlling our environment. That is why measurement is so pervasive in our culture. We seem to want to measure just about everything in the physical world: temperature, weight, distance, pressure, hardness (Mohs scale of hardness), wind speed (Beaufort wind scale), earthquakes (Richter scale), and so on. We even try to measure feelings, like love.

Type "measurement" into an Internet search engine and you will be surprised by the number of organizations that are devoted to measurement. One of those groups, The International Society for Measurement and Control (ISA), plays a prominent role in setting worldwide standards. Without measurement, standards would not be possible. Without complex measurements and standards, just about all of the conveniences that we take for granted (telephones, automobiles, refrigerators, televisions, and computers) would not exist.

Measurements can be quite costly. A company looking for oil invests substantial sums of money to obtain seismic measurements, the principal discovery tool in the hunt for oil. For companies that do oil exploration, seismic data are the company's crown jewels. Data can be just as important on the personal level.

2.1   Quality of Measurements

Measurement and selling the resulting data is big business. Did you know...

- ...that most prescriptions are recorded in a database at the local pharmacy and then sold (without names attached) to firms that collect and summarize this information? The data are then resold to pharmaceutical companies, enabling them to measure how often specific doctors prescribe their drugs. As a consequence, they are able to measure the effectiveness of their salespersons.

- ...that when you are late on a car payment, house payment, or credit card payment, the information may wind up in a national credit database? This database is used by companies in the credit business to assess credit worthiness. It is also used by many businesses in employment screening.

- ...that banks and other institutions that issue credit cards keep data on your spending habits? The data is used to build models to help prevent credit card fraud. If the bank's model determines that you have used your credit card to make "unusual" purchases, it's possible that someone from the credit card company will call and confirm that you did make the purchases. This proactive use of data has prevented an enormous amount of credit card fraud.

- ...that when you order something from a mail order company, your name and what you ordered are recorded and frequently sold to other businesses? Even your grocery store gives you a card so it can collect and sell data on what items you buy and how often you buy them.

Measurement

A large part of using statistics to make good business decisions is developing an ability to appraise the quality of measurements. For many problems, what you measure and how you measure it is more important than how you analyze the data. Thus, it is not surprising that the science of statistics is just as concerned with producing good data as it is with interpreting it.

When you encounter data, ask yourself the following questions.

Procedure

Do we have good measurements?

1. Is the concept under study adequately reflected by the proposed measurements?

2. Are the data measured accurately?

3. Is there a sufficient quantity of the data to draw a reasonable conclusion?

If each of these questions can be answered affirmatively, the data possess good properties.

The development of a suitable measurement involves two essential questions: *What should be measured?* and *How should the concept be measured?* Measures (sometimes called metrics) are developed from the field of study, not statistics. Suppose you decide to measure the speed of an object. The concept of speed is well-defined. It is measured in distance per unit of time (for example, 60 miles per hour). Therefore, determining the speed of an object will require two variables to be measured, one distance and the other time. There are well-defined standard measures (such as feet, yards, miles) to use for distance traveled, as well as time, so the measurement is relatively easy to obtain. However, measuring someone's intelligence is another story. How do you define intelligence? Are there standard measures that can be applied? While IQ tests exist, what do they really measure? The more well-defined a concept, the easier it is to develop measurements for it. Our ability to comprehend the world we live in is rooted in good measurement.

2.1   Exercises

Basic Concepts

1.   What are the two fundamental problems of measurement?

2.   When measurements are used to help solve a problem, what desirable characteristics should the measurements possess?

3.   Name and briefly describe three measurement systems commonly used in business.

4.   When you encounter any type of data, what three questions should you ask to determine the quality of the measurements?

Exercises

5.   Specify whether the following variables are well-defined or not. Justify your answer.

   a.   Height

   b.   Weight

   c.   Hot

   d.   Temperature

   e.   Beauty

6.   A researcher has developed a test that reportedly measures intelligence. The test includes questions such as:

   What is the lowest common denominator of the fractions $\dfrac{5}{32}$ and $\dfrac{6}{9}$?

   Who invented the digital computer?

   Is it reasonable to measure intelligence with these questions? Discuss.

7.   A hotel manager is interested in getting feedback from guests. Two variables of interest to the manager are cleanliness and aesthetics of the rooms. Discuss what problems you would encounter when measuring those variables.

2.2   Science and Data

Measurement and data are an integral part of science. Methods for exploring research problems have been developed over a long period of time and have become standards in the scientific community. These methods are collectively known as the **scientific method**.

Procedure

The Scientific Method

1.   Gather information about the phenomenon being studied.

2.   On the basis of the data, formulate a preliminary generalization or hypothesis.

3.   Collect further data to test the hypothesis.

4.   If the data and other subsequent experiments support the hypothesis, it becomes a law. If the experiment does not support the hypothesis, construct a new hypothesis and start the method over again. Sometimes, even if the hypothesis is supported, you may want to test it again but in a different way.

Statistics and data are fundamental to the scientific method. Data from carefully designed experiments are the ultimate evidence that support or discredit new theories.

The data collection process in steps one and three of the scientific method can be quite different. The first step of the scientific method is exploratory, finding out what "reality" is about the subject under consideration. Since the data in this phase need not produce convincing evidence, whatever data are available are used to generate ideas. However, the third step begins the validation of a hypothesis. Scientists are trained to be critical thinkers. If a new idea is to be accepted by the scientific community, convincing evidence must be developed at the third stage. The manner in which the data are collected is an important part of that evidence. If the evidence is to be persuasive, a data gathering strategy (an experimental design) that will produce data without the unwelcome influences of **confounding variables** (i.e., an uncontrollable or unaccounted for variable that damages the integrity of the experiment or the other data) is required. For example, a manager wants to know if a new marketing plan has increased sales. In the analysis of the data, the manager must ensure that any increase in sales is due to the marketing plan and not other factors such as an increase in the sales force or a reduction in the cost of the product, for example. These "other factors" would be considered confounding variables.

Definition

A **confounding variable** is a variable that was not controlled or accounted for by the researcher and thus damaged the integrity of the experiment.

Because of science's emphasis on data, statistics has become inextricably linked with the scientific method. Experiments are designed to yield data with maximum information. There are two branches of statistics: descriptive and inferential. The branch of **descriptive statistics** focuses on exploratory methods for examining data that yield the hypothesis mentioned in step two. The branch of **inferential statistics** develops theories to test the hypothesis using data collected from an experiment to make formal conclusions about a population or parameter.

2.2   Exercises

Basic Concepts

1. What is the scientific method?

2. What is a confounding variable?

3. How does statistics interact with the steps in the scientific method?

4. Name and briefly describe the two main branches of statistics.

Exercises

5. Suppose you want to determine the proportion of college students in the state of Virginia that pays more than $500 per year on textbooks. Using the scientific method, how would you conduct the experiment?

6. The manager of an electronics company was interested in determining the reason for the increase in sales volume over the last three years. The manager randomly selected data on the advertising budget, number of salespeople, and average product costs. When examining the data, the manager found that her average product costs were fairly stable but the advertising budget steadily increased over the last two years along with the number of salespeople. Are there any confounding variables in this study? If so, what are they and why do you consider them confounding?

7. A company that produces bulbs for projectors wanted to conduct an experiment to determine the length of life of its bulbs. The company's leading competitor's bulbs have an average life of 1000 hours. The company sampled its bulbs and found that the average life of the bulbs was 1200 hours. Thus, the company has concluded and advertises that its bulbs last longer than the competition by at least 100 hours. Were the results of this experiment an example of descriptive or inferential statistics? Explain your answer.

2.3   Decision Making and Data

Collecting data is a natural part of our lives. For example, consider the everyday question, *What will I have for dinner?* Although virtually no one formally applies the scientific method to such a problem, most people perform experiments and collect data (by eating). This leads to generalizations such as *I hate asparagus* or *I like ice cream*. After sufficient experimentation, these generalizations become personal preference laws.

Selecting the evening meal does not have incredible consequences, yet important problems, by definition, do.

Procedure

The Decision-Making Method

1. Clearly define the problem and any influential variables.

2. Decide upon objectives and decision criteria for choosing a solution.

3. Create alternative solutions.

4. Compare alternatives using the criteria established in the second step.

5. Implement the chosen alternative.

6. Check the results to make sure the desired results are achieved.

Notice the first step in the decision-making method is to define the problem. This is important because almost *any solution to the right problem is better than the best solution to the wrong problem*.

Measurement, Data, and Problem Definition

When you go to a physician what is the first thing the physician does? Most physicians measure your weight, temperature, and blood pressure before he or she meets with you. During the examination the doctor will elicit more data. If there are insufficient data to make a diagnosis, more data (perhaps the dreaded blood test) will be ordered. In the medical world it is extremely important to know what reality is. That is why medical professionals gather so much data. Data are just as important in solving business problems.

Consider a trucking company manager who is beginning to hear a few complaints about freight being delivered late. If the operations manager of the trucking company keeps data on the number of shipments delivered late each week, then statistics can be used to help assess the magnitude of the problem. For example, if the number of late deliveries is plotted (see Figure 2.1), the data in the graph confirms a disturbing trend. Late deliveries are on the rise. Left unsolved, this problem can jeopardize jobs and the existence of the business.

Figure 2.1

Suppose a nurse working the late shift takes a patient's temperature once each hour. After recording the measurements in the patient's chart, the data are plotted (see Figure 2.2). The data reveal a potential problem. Undoubtedly, more data will be needed.

Figure 2.2

Graphical and numerical summaries are frequently useful in discovering the existence of a problem as well as in shedding light on what some of the potential causes may be. In many instances, problems are caused by systems that do not operate as they are designed. Collecting data (finding out system "reality") and using simple statistical tools to monitor a system are the most common ways of insuring a system performs properly.

Problems are not always the result of a diagnosis of some complaint or system malfunction. A "problem" may well be the result of an inquiry into unexpectedly good system performance. That is, often times, improving a process does not imply that something was wrong with the process. In this context, a problem presents itself as an opportunity for improvement. Suppose, for example, a finance instructor develops a new method for teaching introductory finance. Measurements are kept on his students as well as students taking subsequent finance courses. Using statistical methods to compare students using the new method to those using the old method can be valuable in pointing out potential educational improvements that may be used at other institutions. As another example, consider the development of the Intel chip. Improving the speed of its processors/chipsets is a constant goal for research and development, regardless of the current speed of the chip. It is not a matter of the chips being too slow, but as technology and hardware advance, the chipsets need to "keep up" so that computers using the chips maintain (or improve) their processing speeds.

Statistics as Criteria

The second step in the decision-making method suggests defining objectives and developing criteria in order to evaluate various alternative decisions. Not all statistics are simple means, proportions, or standard deviations. Managers and researchers often develop their own statistical measures for summarizing some aspect of a phenomenon. In baseball, for example, teams regularly compute the earned run average (ERA) of each pitcher. This statistic summarizes the average number of earned runs per nine innings pitched. ERA is used as a criterion for comparing and evaluating pitchers as well as for salary negotiations. The mean time between failure (MTBF) is a statistic that is used to compare the reliability of various equipment or components. The Consumer Price Index (CPI) is a summary statistic that describes the overall price level in the United States. This statistic is an economic measure of inflation and is used in labor contracts to escalate wages as well as to calculate cost of living increases in social security payments.

What Should Be Measured and How Should It Be Measured?

What should be measured depends on the problem to be solved. Sometimes what should be measured is obvious and relatively easy. If the problem is to maintain or improve a system, key variables are monitored and decisions are made based on the level of these variables. For example, if you are responsible for a machine manufacturing pistons, then some of the variables that should be measured and controlled are: the diameter of the piston, the length of the piston, the width of the piston wall, and the number of defects on the piston's surface. However, if you are trying to design a system to perform automated stock trading, deciding what variables are important could take years to discover.

A precisely defined concept is usually easy to measure. The less precise (the fuzzier) the concept, the more difficult the measurement becomes. There is a vast difference in measuring the height of a person and measuring his or her intelligence. Defining height is relatively simple. It is nothing more than the vertical length of an object. There are standard scales, such as inches and feet, that everyone agrees upon, that can be used to measure height. The National Institute of Standards and Technology maintains rods which define a government standard for distance measures (feet and inches). Because these standards are widely accepted, if ten different people measure a person's height, there should not be large differences in their measurements. This is not true when measuring intelligence. The National Institute of Standards and Technology does not have a measuring rod for intelligence. It is unlikely that ten randomly selected people could agree on a definition of intelligence, much less on how it should be measured. Intelligence is a fuzzy concept because there is no universally accepted definition and hence there can be no universally accepted standard of measure. If a concept cannot be precisely defined, it cannot be precisely measured. How do you measure fuzzy concepts?

Measuring Fuzzy Concepts

Fuzzy concept definitions produce fuzzy measurements. One could devote an enormous amount of time developing measuring instruments for fuzzy concepts such as love, rivalry, and prejudice, and still have a poor measurement. These concepts are fuzzy because they are perceptions. No person can be sure that their perception of love, rivalry, or prejudice is the same as someone else's. Science that relies on fuzzy measurements usually makes the assumption that everyone's interpretation of the concept is more or less alike.

Often when measuring fuzzy concepts, the instrument used to measure the concept ends up defining the concept. The Wechsler Adult Intelligence Scale (WAIS) is a test that is often used to measure intelligence – an IQ test. There could be a long debate over what constitutes intelligence. For most researchers studying intelligence, developing a new instrument to measure intelligence, however it is defined, is simply not a practical alternative. If a measuring device is needed for a fuzzy concept, using an established instrument like the WAIS is usually the method of choice. However, the necessity for using a measuring instrument does not validate the instrument.

There has been a great deal of controversy in recent years over whether the SAT Reasoning Test accurately measures scholastic aptitude. Still the SAT is an important part of the college admission process. Why? Because college admission committees apparently believe there is no better alternative. When measurements purport to represent what you believe is a fuzzy concept, it is important to think critically about the measurements. Are they valid measures of the intended concept? This issue is important and will again be addressed later in the chapter in the levels of measurement section.

2.3   Exercises

Basic Concepts

1. What is the decision-making method?

2. What is different between the scientific method and the decision-making method?

3. Are problems that can be solved by collecting data always the result of a system malfunction? Explain.

4. Give an example of how statistics can be used to improve a process.

5. What are fuzzy concepts? What are the measurement problems associated with fuzzy concepts?

6. Give an example of a tool that has been widely accepted as an instrument used to measure a fuzzy concept.

2.4   Collecting Data

Essentially there are two ways to obtain data: **observation** and **controlled experiments**. The data collection method is related to the nature of the problem to be solved and the ethical and practical constraints of collecting data in some particular environment. There are many instances in which controlled experiments that would produce particularly appropriate data are not practical or would be unethical. For example, scientists cannot exclude data in an analysis for the purpose of validating their theory or viewpoint.

There are many instances in which data that have been collected provide either no information or misleading information about the effect under study.

Suppose a high school physics teacher wants to determine if there is any beneficial effect of studying calculus before taking physics. He obtains the records of his physics students and compares the group of students that had calculus or are taking it concurrently with those who have not. The average physics grade for those having had calculus is a great deal higher than those that had not. The conclusion reached from these data is: The study of calculus improves one's understanding of physics. Do the data and the manner in which they were collected support such a conclusion?

Students who elect to take calculus generally have above average skills in mathematics. Are the higher physics grades of the calculus students due to their average mathematical skills or to the content of the calculus course? Because of the data collection method, the data are not of sufficient quality to reach a conclusion as to the benefits of taking calculus prior to taking the physics course. A confounding variable (the exceptional mathematical ability of students electing to take calculus) makes it impossible to distinguish the effects we wish to study.

Fuzzy Data May Produce Misleading Conclusions

In a statistical analysis, it is usually not possible to recover from poorly measured concepts or badly collected measurements. Unfortunately, during your lifetime you will be bombarded with statistics derived from poorly measured concepts, confounded measurements, and simply fictitious data. When confronted with statistical evidence of any kind, regardless of whether or not the statistical analysis is done in good faith, it is ultimately up to you to ask reasonable questions about the data. The conclusions suggested by statistics can be no stronger than the quality of the measurements which produced the statistical evidence. Fuzzy or confounded measurements must produce fragile conclusions.

Controlled Experiments: Data by Design

Suppose you wanted to know whether an agricultural crop would have higher yields if a different amount of fertilizer was used. This kind of question would be ideal for a **controlled experiment**. The purpose of controlled experiments is to reveal the response of one variable (crop yield, the response variable) to changes in another variable (level of fertilization, the explanatory variable). In a controlled experiment the researcher attempts to control the environment of the experiment so that the effect of one variable on another can be isolated and measured. In these studies there is a **control group** and an **experimental group**. Ideally there is no initial difference between the two groups. During the experiment a **treatment** is applied to the experimental group. The exact form of the treatment will depend on the particular experiment. If the experimental group were an agricultural crop, a treatment might be applying a given level of fertilizer. The treatment changes the level of the **explanatory variable** in the experiment. The effect of the treatment can be measured by comparing the **response variable** in the control and experimental groups.

> ### Definition
>
> The **response variable** is the variable of interest in an experiment.

> ### Definition
>
> An **explanatory variable** is a variable that affects the variable of interest (response variable) in an experiment.

Comparative Experiments

Isolating the effects of one variable on another means anticipating potentially confounding variables and designing a controlled experiment to produce data in which the values of the confounding variables are regulated. In the physics example, the confounding variable is the mathematical ability of those students taking calculus before physics. To control for this *bias*, students could be randomly assigned to take the calculus course. By randomly assigning students to take the calculus course before the physics course, the mathematical ability of the two groups should be equalized. Since mathematical ability is controlled, any difference in physics scores could be attributed to a beneficial effect of the calculus course. An experimental design in which experimental units (students in this case) are randomly assigned to two different treatments is called a **completely randomized design**.

For this example of a completely randomized design:

Response Variable: Students' grades in the physics course.

Explanatory Variable: Whether students take calculus before physics.

Randomization is often used as a method of controlling bias and is an important principle in the design of experiments.

Flowchart for the Calculus before Physics Experiment

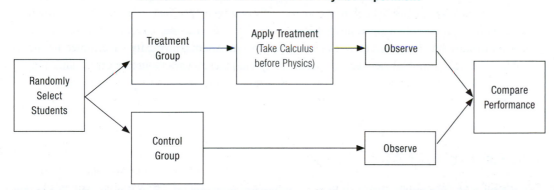

Example 2.1

Suppose a new species of tomato has been genetically engineered to increase yields. The question: Is there a relationship between the use of the new species and yield?

Solution

In this experimental design, the plots of tomatoes will be divided into two groups, one that contains the new species and another that contains the unaltered variety. The plots containing the genetically engineered variety will be called the **treatment group**, and the plots containing the standard variety are called the **control group**. If the experiment is properly performed, any change in the response variable (yield) can be attributed to the explanatory variable (genetic engineering) and not to other variables that are controlled. The untangling of variables at the data-gathering stage makes the analysis of the data much easier. There is no better way to establish a causal relationship.

Variables like rainfall, amount of sunshine, or type of soil can affect the yield of tomatoes. These variables must be controlled in the experiment so that differences in yields between plots can be attributed just to the genetically engineered species. One possible method of controlling for these variables is to create small plots in the same field. Several plots of each type of tomato are planted. Each plot contains only one type of tomato. Rainfall is measured and supplemented with irrigation to assure that each plot has the same amount of water. By controlling for these variables in this manner, we can say that any differences noted in the yield (response variable) between the genetically engineered variety (treatment group) and the unaltered variety (control group) are due to genetic improvements (explanatory variable).

The Before and After Study

The before and after study also contains a comparative experiment. The control group and the experimental group are initially identical. The response variable is measured in the control group at the beginning of the study, and then a treatment is applied to the control group. After the treatment is applied, the control group becomes the experimental group. The response variable is again measured after the treatment has been applied. If the treatment affects the response variable then there should be a difference between the value of the response variable for the control and experimental groups, presumably caused by the treatment. An example of a before and after study is given in Example 2.2.

Example 2.2

Does an SAT preparation course improve performance on the SAT Reasoning Test?

Solution

A group of high school students take the SAT. Then they are given an SAT preparation course. They retake the SAT. If the group's second SAT performance improves, then it may be related to the SAT preparation course.

For this example of a before and after study:

Explanatory variable: SAT preparation course.

Response variable: SAT scores.

In this experiment there is only one group of students. Suppose there were two groups of students, one that had taken the course (treatment group) and one that had not (the control group). Could you assign any difference in results solely to the SAT preparation course? Because of potential differences in the cognitive abilities of the two groups of students, this experiment would be much more vulnerable to the justifiable criticism that the differences in student cognitive ability caused the difference in the group's SAT performance rather than the SAT preparation course. However, by using a before and after study we have only one group of students and cognitive ability is controlled.

The Placebo Effect

A difficult problem arises in experiments involving people. In clinical trials of some drug or medical treatment, patients often respond favorably to any treatment, including a "dummy" or "fake" treatment. These fake treatments are called **placebos**. In medical research, a placebo is a pill that contains none of the drug that is being tested. It has been shown in several pain studies that placebos relieve pain in 30 to 40 percent of patients, even though the placebo has no active ingredients. The placebo effect is not confined to medicine. Similar effects have been noticed in psychological research in which the subjects seem to try to help the researcher prove some conclusion.

One of the more interesting experiments contaminated by this effect was the Hawthorne Study conducted at the Western Electric Company's Hawthorne Works in Chicago between 1927 and 1932. The studies were initiated to determine the effect of lighting on worker productivity. Lighting was increased in stages, and the investigators found that each time lighting increased worker output increased. The investigators were suspicious that another effect might be causing worker productivity to improve. So, workers were told lighting was to be increased when, in fact, it was decreased. Despite the decrease, worker output increased again. Clearly there was some other variable affecting worker output.

The workers wanted the study to be successful, and their desire was confounding the experiment. Instead of discovering the expected relationship between worker output and lighting, the investigators found that the social system and the employees' roles within that system had a great deal more to do with worker productivity than lighting. The Hawthorne Study has been credited with introducing psychology to the workplace. It also points out the hazards of measurement, even in a controlled experiment.

Placebos

One might think that the effectiveness of placebos should be close to zero; this, however, is not the case. Studies have shown that up to 62% of headache sufferers, 58% of those suffering from sea sickness, and even 39% of those suffering from postoperative wound pain showed symptom relief given a placebo. This is a stunning revelation of the effect of psychology on our mind and body.

An intriguing but ethical question for you to ponder is this:
If placebos can be 62% effective for curing headaches, would or should a physician treat a headache with a placebo only?

Double Blind Studies

The placebo effect is prevalent in medical studies. **Double blind studies** are used to counteract this effect. In a double blind study the subjects are not told whether they are members of the treatment group or the control group. The evaluators (the persons that measure the response variable) are also not told whether their subjects are members of the control or experimental group. An example of a double blind study is given in Example 2.3.

Example 2.3

Until recently, ulcers in the upper intestine were a rather common illness. A new treatment for ulcers was proposed. This treatment involved gastric freezing and required the patient to swallow a deflated balloon with tubes attached. A refrigerated solution was then pumped through the balloon for an hour. The idea behind this therapy was to cool the stomach wall and reduce the amount of acid produced. Initially, the results looked promising and the treatment was used for several years. However, none of the initial studies were double blind.

In the double blind study half of the patients were given the procedure and the other half were given a placebo treatment, which included swallowing the balloon, but no cooling solution was injected into the balloon. Patients that received the placebo treatment actually did better than the ones receiving the refrigerated solution. Gastric freezing was eventually abandoned as a treatment for upper intestinal ulcers.

Observational Studies

Observational data comes about by measuring "what is." If you are trading stocks, the market data you receive are observational. The data show simply what is happening in the marketplace at the time. Census data are observational; they are a measure of how things are in a specific geographic area at a given point in time. There is no experimentation to see how manipulating one variable will affect another variable or variables. Virtually all of the data we routinely encounter is observational. Examples regularly appearing in the newspaper include:

- Stock, commodity, bond, option, and currency market data

- Almost all federal government data, including census, economic, and educational data

- Virtually all local and state government data

- Sports data (scores, outcomes, etc.)

The data described are often collected to satisfy state and government regulation as well as for business purposes, such as to examine trends in a particular stock price or interest rates. These data values are not the result of a designed experiment.

Observational studies can be extremely valuable. For example, the effect of vitamin C has been the subject of controversy for some time. Numerous claims of beneficial health effects from regularly consuming vitamin C have been made. However, Enstrom, Kemin, and Klein ("Vitamin C Intake and Mortality," *Epidemiology*, May 1992) in an observational study with 10,000 participants noted that males in their study who supplemented their diets with 500 milligrams of vitamin C each day (on average) lived approximately six years longer than those who did not take the vitamin. Females taking this dosage of vitamin C (on average) lived approximately one year longer. Although this study was quite large, it does not prove that vitamin C causes increased longevity. If the data had come from a controlled experiment the strength of the study's conclusion would be much more powerful. The vitamin C study is an observational study because the subjects were not chosen randomly for the control and treatment groups. If the experiment were conducted as a controlled experiment, the control group would be given a placebo and the treatment group would be given vitamin C.

Untangling Variables in an Observational Study

In 1973, the Graduate Division at the University of California, Berkeley, carried out an observational study on gender bias in admissions to the Graduate School. There were 8442 men and 4321 women who applied for admission. Of the men that applied, 3714 were subsequently accepted, and 1512 of the women were accepted. Given that almost twice as many men applied to the graduate program, it is reasonable to expect that more men would be accepted. To make a reasonable comparison, the first thing that is needed is to adjust for the difference in application rates between men and women. This is achieved by comparing the percentage of each group that was accepted. Nearly 44% of the men and 35% of the women were admitted. These admission statistics suggested there was rather substantial support for the idea of discrimination against women.

The admission process at Berkeley was done by major. If there was discrimination against women, those departments that were discriminating would stand out when the admissions data were examined separately. But when the data were examined, the investigator did not find what was expected.

Table 2.1 – College Acceptance Rates				
	Men		Women	
Major	Number of Applicants	Percent Admitted	Number of Applicants	Percent Admitted
I	825	62	108	82
II	560	63	25	68
III	325	37	593	34
IV	417	33	375	35
V	191	28	393	24
VI	373	6	341	7

Source: P.J. Bickel, E.A. Hammel, and J.W. O'Connell. Sex Bias in Graduate Admissions: Data from Berkeley. Science, 187:398-404, 1975.

Suppose the data in Table 2.1 represent the six largest majors on the Berkeley campus.

In four of the six majors, women were admitted more frequently than men. Not only were women not being discriminated against, but there appears to be potential discrimination against men in Major I. How could this completely opposite conclusion be true? A close examination of the data shows that in the majors with the largest percentage admitted (the easiest to get into) there were a large number of male applicants and very few female applicants. The majors that had very low acceptance rates (difficult to be admitted) had relatively very few men and a large number of women applying. Thus, the variable major field of study was **confounding** the variable gender in the original analysis and **biasing** the original conclusion.

The original conclusion was the result of the fact that women were applying to the most difficult departments for admission, not because of sex discrimination. By separating the data by major, the analyst was able to **control** for the confounding variable, choice of major, and remove the bias. When it is possible to remove the effect of one variable, we are said to be **controlling for** that variable. The Berkeley data illustrates a subtle problem in the comparison of two or more proportions.

In this instance the analyst was able to untangle the two variables, but this is not always the case. Unless the data are gathered with a controlled experiment, it may not be possible to untangle the effects of the causal factors. This is an example of **Simpson's Paradox** where the results are counterintuitive. This paradox is often seen with reports based on frequency data when making inferences about relationships between two or more variables. For the Berkeley example, the relationship between the admissions data for men and women is reversed when the majors are combined and when they are considered separately.

Surveys

A great deal of the statistical information presented to us is the result of surveys. Often we will see in the news that one of the major polling organizations, Gallup, Harris, ABC-Washington Post, or NBC-New York Times, is reporting findings on various topics, from the approval rating of the President to the popularity of gun control.

In some instances, the purpose of a survey is purely descriptive, as those described above. However, in many cases the researcher is interested in discovering a relationship. Because virtually all surveys produce observational data, survey research belongs to steps one and two of the scientific method. Sometimes a plausible relationship is discovered and a designed experiment is undertaken to more convincingly demonstrate the relationship.

A famous observational study, known as the Framingham Study, recorded various data on 4500 middle-aged men. The men were followed for many years with the hope of uncovering what factors relate to the development of heart disease. It was discovered that the development of heart disease seemed to be associated with obesity, heavy smoking, and high blood pressure. Because of the large number of participants and the researcher's ability to control for potentially confounding variables after the data were collected, this research influenced many physicians to work with their patients to control the three **causal factors** found in the study.

Steps to Consider When Conducting a Survey

1. Have specific goals.
2. Consider alternatives for collecting data.
3. Select samples to represent the population.
4. Match question wording to the concepts being measured.
5. Pretest questionnaires.
6. Construct quality checks.
7. Use statistical analysis and reporting techniques.
8. Disclose all methods used to conduct the survey.

Source: American Association for Public Opinion Research

Even Hollywood studios do survey research. The survey shown in Figure 2.3 is used by a major studio. Looking at the questions that are posed, you can see they want data on *why you came to the movie and what you liked about it.* Collecting data on a wide variety of movies gives the studio insight into what makes a movie successful, the importance of stars, and what other factors have universal appeal.

SAMPLE MOVIE SURVEY

Please complete the first 9 questions of this survey before the movie begins and questions 10 through 14 after the movie. Please give the completed questionnaire to the persons collecting them at the exits.

1. Indicate the <u>one item</u> in the list below that was most influential in your decision to see the movie. (Check one)

Newspaper:	Ad	( )1	Radio:	Ad	( )1
	Review	( )2		Review	( )2
	Article	( )3	Around Town:		
Magazine:	Ad	( )4		Received a Complimentary Pass	( )3
	Review	( )5		Coming Attractions	( )4
	Article	( )6		Poster or Billboard	( )5
Television:	Commercial	( )7		The soundtrack album	( )6
	Review	( )8		Recommended from a friend or relative	( )7
	MTV or other Music Videos	( )9		Computer on-line service	( )8
	Other cable	( )0		Internet	( )9
	Talk Show, other	( )x		Other:	( )0

2. Which of the following were important to you in deciding to come to this movie? (Mark as many as apply)

Tony Curtis	( )1	Billy Wilder, the producer	( )4	The visual effects	( )7
Marilyn Monroe	( )2	The story	( )5	The reviews	( )8
Jack Lemmon	( )3	The drama	( )6	Other_____	( )9

3. What is your age? (Check one)

Under 12	( )1	30 to 34	( )7
12 to 14	( )2	35 to 39	( )8
15 to 17	( )3	40 to 44	( )9
18 to 20	( )4	45 to 49	( )0
21 to 24	( )5	50 to 59	( )x
25 to 29	( )6	60 & over	( )y

4. What is the last grade of school you completed?

Some high school or less	( )1
Completed high school	( )2
Some college/currently in college	( )3
Completed 2 year college	( )4
Completed 4 year college	( )5
Currently in/completed post-grad	( )6

5. Are you…(Check One)

| Male | ( )1 |
| Female | ( )2 |

6. What is your ethnic background? (Check One)

African-American	( )1	Caucasian	( )4
Asian	( )2	Native American	( )5
Latino	( )3		

7. What is your marital status? (Check one)

Single	( )1
Married	( )2
Divorced/Separated	( )3
Widowed	( )4

8. Before today, how familiar were you with the storyline? Very familiar…( )1   Somewhat familiar…( )2   Not at all familiar…( )3

9. How many times before today have you seen the movie?

None…( )1   Once…( )2   Twice…( )3   Three or more times…( )4

COMPLETE QUESTIONS 10 THROUGH 14 AFTER THE MOVIE

10. How would you rate the movie?

Excellent	( )1
Very Good	( )2
Good	( )3
Fair	( )4
Poor	( )5

11. Would you recommend the movie?

Definitely	( )1
Probably	( )2
Probably Not	( )3
Definitely Not	( )4

12. How did the movie measure up to your expectations?

Better than expected	( )1
About what I expected	( )2
Not as good as expected	( )3

13. Which of the following words or phrases best describe the movie you just saw? (Mark as many as apply)

Entertaining	( )1	Worn-out theme	( )1	Well acted	( )1
Boring/dull	( )2	Surprising	( )2	Not enough drama	( )2
Dramatic	( )3	Too long	( )3	Has a good story	( )3
Interesting settings	( )4	Different/original	( )4	Unrealistic	( )4
Too slow in parts	( )5	Not my kind of movie	( )5	Educational	( )5
Offensive	( )6	Controversial	( )6	Nothing new/done before	( )6
Confusing	( )7	Moved just right	( )7	Thought provoking	( )7
Interesting characters	( )8	Humorous	( )8	Too predictable	( )8
Too silly/stupid	( )9	Good cast	( )9	Believable	( )9
Action packed	( )0	In bad taste	( )0	Depressing	( )0

14. Would you pay to see this movie again?   Yes…( )1   No…( )2

Figure 2.3

2.4 Exercises

Basic Concepts

1. What are the two ways of obtaining data?

2. What are the dangers of making conclusions based on poorly collected data?

3. How do you treat the problem of a confounding variable?

4. Explain the difference between the control group and the experimental group in a controlled experiment.

5. What is an explanatory variable?

6. What is a response variable?

7. What is bias? How can it be controlled?

8. What is a completely randomized design? What are the advantages of using a completely randomized design?

9. What is a before and after study?

10. What is the placebo effect? Give an example.

11. What is a double blind study?

12. How do observational studies differ from controlled experiments?

13. What kinds of problems can be associated with an observational study?

14. Researchers use surveys for two main purposes. Name and give an example of each.

Exercises

15. The health and social problems associated with obesity can be a severe hindrance in attaining many of life's goals. Methods for treating obesity were compared in "One Year Behavioral Treatment of Obesity: Comparison of Moderate and Severe Caloric Restriction and the Effect of Weight Maintenance Therapy," in the *Journal of Consulting and Clinical Psychology*. In the study, a group of 25 women, each of whom was at least 25 kilograms (kg) overweight, were randomly split into two groups. The first group received behavior therapy and was placed on a 1200 calorie per day diet for a period of one year. The second group received behavior therapy and was placed on a 420 calorie per day diet for the first 16 weeks of the year. Then they returned to a 1200 calorie per day diet for the remainder of the year. At the end of a 26-week period, the average weight lost was 11.86 kg for the first group and 21.45 kg for the second group. But after 52 weeks, the average weight lost was 10.94 kg for the first group and 12.18 kg for the second group.

    a. Why is this study an example of a controlled experiment?

    b. What is the explanatory variable?

    c. What is the response variable?

    d. Is there a control group in the study? Explain.

    e. Suppose that the data were gathered from an observational study instead of from a controlled experiment. How would this affect the conclusions that might be made from the study?

16. An article appearing in the *New England Journal of Medicine* investigated whether the academic performance of asthmatic children being treated with the drug Theophylline was inferior to a non-asthmatic group. In one part of the study, 72 children were identified as being treated for asthma. For each child with asthma, a non-asthmatic sibling was also identified. (The use of sibling controls allows for control of family environment and certain genetic factors on academic achievement.) All 144 children were then given a test to measure academic achievement. There were no significant differences on the test between the two groups.

    a. Why is this study an example of a controlled experiment?

    b. What is the explanatory variable?

    c. What is the response variable?

    d. Is there a control group in the study? Explain.

    e. Suppose that the data were gathered from an observational study instead of from a controlled experiment. How would this affect the conclusions that might be made from the study?

17. A small clinical pilot study was conducted by a research team from Harvard Medical School and the School of Public Health. Fifteen individuals in the early stages of Multiple Sclerosis were fed bovine myelin, a substance containing two antigens thought to be the target of the immune system's attack in Multiple Sclerosis. Another fifteen were given a placebo. In the study, fewer members of the group fed bovine myelin had major attacks of the disease.

Source: *Science*, Vol. 259, No. 5099

    a. Which phase of the Scientific Method best describes this study?

    b. Is this an observational study or a controlled experiment?

    c. What is the response variable?

    d. What is the explanatory variable?

    e. Which group is the treatment group?

    f. Which group is the control group?

18. London scientists conducted a study to determine if chocolate can trigger migraines. Twelve migraine-prone subjects were given a peppermint-laced chocolate candy and eight migraine-prone subjects were given a peppermint-laced placebo made of carob, peppermint, and vegetable fat. Five subjects from the group given chocolate developed a migraine headache within one day. No one from the group given the placebo developed a migraine in the same time period.

Source: *Self* magazine

    a. Which phase of the Scientific Method best describes this study?

    b. Is this an observational study or a controlled experiment?

    c. What is the response variable?

    d. What is the explanatory variable?

    e. Which group is the treatment group?

    f. Which group is the control group?

19. Jacob normally plays basketball three days a week and has begun to develop patellar tendinitis, which is inflammation in the patellar tendon and results in nagging knee pain. In an effort to relieve his knee pain, Jacob decides to take a week away from playing basketball and rest his knee. However, after about four days, his friend offers him an analgesic rub and insists that his knee will feel better in two to three days. After using the analgesic rub for a couple of days, Jacob's knee begins to feel better. Did the analgesic rub work? Explain how confounding variables might have played a role on Jacob's knee getting better.

20. The Nurse's Health Study conducted on 87,245 women at Boston's Brigham and Women's Hospital revealed that women who eat a cup of beta carotene-rich food a day have 40 percent fewer strokes and 22 percent fewer heart attacks than those who consume a quarter of a cupful per day.

    Source: *Self* magazine

    a. Which phase of the Scientific Method best describes this study?

    b. Is this an observational study or a controlled experiment?

    c. What is the response variable?

    d. What is the explanatory variable?

    e. Which group is the treatment group?

    f. Which group is the control group?

21. A religious group conducted a survey with two of the questions asking "Do you go to church?" and "Are you happy?" After conducting the survey, the group concluded that those who go to church are generally happier than those that do not go to church. Do you think going to church makes one happier? Describe how confounding variables could play a role with the conclusion drawn by the religious group.

22. In May 2011, Internet Explorer reversed its trend in the United States and gained usage share (the percentage of users using a particular Internet browser). In June of 2011, the trend reversal became global. Internet Explorer gained 0.57% in June across all operating systems with Internet Explorer 8.0 gaining 0.86% globally. The gains for Internet Explorer came primarily at the expense of Mozilla Firefox (−0.51%). Google Chrome's pace of usage share gains slowed to +0.2% for June. The gains for IE were the largest in Europe and Asia:

    Internet Explorer in Europe: +0.88%

    Internet Explorer in Asia: +0.81%

    This increase may be the result of a marketing campaign. In early June, Microsoft launched their "Confidence" campaign aimed at showing the security features of Internet Explorer 8.

    Source: netmarketshare.com

    a. Are the results stated above likely to have come from an observational study?

    b. How can Microsoft (and other companies) benefit from this information?

23. A survey was conducted by an investment firm asking participants the following questions: "Are you financially secure?" and "Do you independently make decisions about your investments?" After analyzing the data from the survey, the firm concluded that people who make investment decisions independently tend to be not as financially secure as those who make decisions with the help of an investment advisor. What confounding variables could have played a role in this conclusion?

2.5   Levels of Measurement

Like most things in life, data come in different qualities. Some measurements are purely numerical and are based on well-defined standards, such as pounds, inches, dollars, and percentages. On the other hand, some measurements are exceedingly "fuzzy" and the standard of measure is ill-defined, if defined at all. For example, consider the response to the following question.

What is your opinion of the President's performance?

1. Extremely disappointing

2. Disappointing

3. Satisfactory

4. Good

5. Extraordinary

Someone's *good* response may be equal to someone else's *extraordinary*. There can be no guarantee of a common scale and thus the level of consistency between measurements is unreliable. This type of data is much lower quality than measurements on some standard scale, such as pounds or inches.

> **Definition**
>
> The quality of data is referred to as the **level of measurement**.

The terms used to describe the quality of data are **nominal**, **ordinal**, **interval**, and **ratio**. When analyzing data you must be exceedingly conscious of the data's level of measurement because many statistical analyses can only be applied to data that possess a certain level of measurement.

Because of different levels of measurement, not all data are created equally. Unfortunately, once data are in numerical form, many believe that a number is a number. Everyone knows that you can add, subtract, multiply, and divide numbers. But for some measurements, adding, subtracting, multiplying, and dividing are simply meaningless.

Standard mathematical operations (such as addition and subtraction) are not defined for nominal and ordinal data, and, consequently, many forms of statistical analysis (descriptive and inferential) are not appropriate.

Nominal Data

Sex (male or female) and hair color (blond, brunette, or redhead) are examples of nominal variables. If the gender variable is coded numerically, say Female = 1, Male = 0, then what do the numbers mean?

> **Definition**
>
> Data that represent whether a variable possesses some characteristic are called **nominal** data.

Is it meaningful to add, subtract, multiply, or divide these numbers? Let's try addition:

1	+	0	=	1	
Female	+	Male	=	Female	?

Is it meaningful to add a female and a male and get a female? If it is, it is certainly a bizarre interpretation.

Let's try subtraction.

1	−	0	=	1	
Female	−	Male	=	Female	?

Is it meaningful to subtract a male from a female and get a female? If this kind of mathematics works, it doesn't say a lot for males!

Similarly absurd conclusions can be reached for multiplication and division. For strictly nominal data, none of the arithmetic operators can be applied. Statistically speaking, only some graphical and very few numerical statistical procedures can be applied to nominal data. Keep in mind that the nominal level of measurement identifies the variable in name only.

Ordinal Data

Consider a response to the fill in the blank question.

Frosty Pops taste _____.
1. Very bad   **2.** Bad   **3.** Fair   **4.** Good   **5.** Very good

The response to this question is often referred to as being measured on the Likert Scale, in which one rates the taste of Frosty Pops on a scale from 1 to 5. These responses are ordered on the basis of an individual's impression of the goodness of Frosty Pops. The response *bad* is perceived to be better than *very bad*, *good* better than *fair*, and so forth. If numerical codes are used to represent the responses, should the properties of addition, subtraction, multiplication, and division be applied to them? Let's try addition.

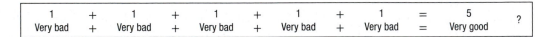

1	+	1	+	1	+	1	+	1	=	5	
Very bad	+	Very bad	+	Very bad	+	Very bad	+	Very bad	=	Very good	?

This means that if you thought that a Frosty Pop tasted *very bad* (1) you should eat five of them and doing so would be equivalent to eating something that you thought was *very good* (5). Addition of ordinal values is not reasonable. And for that matter, neither are subtraction, division, and multiplication. So the only difference between ordinal and nominal data is that ordinal data possess order. Here again, like nominal data, only a very limited number of statistical analyses can be performed.

Note that ordinal data are also nominal, but they possess the additional property of ordinality (or ranking).

Interval Data

Definition

If the data can be ordered and the arithmetic difference is meaningful, the data are **interval**.

One example of interval data is temperature (measured on the Fahrenheit scale).

48 degrees	–	45 degrees	=	3 degrees
72 degrees	–	69 degrees	=	3 degrees

In the case above, the difference between the temperatures is three degrees, and it is true that the difference in kinetic energy between 48 and 45 degrees is the same as the difference between 72 and 69 degrees.

Interval data also have another interesting property, an arbitrary zero value. You don't have to be a mathematician to appreciate the usual meaning of the zero concept, having zero dollars means that you do not have any money. However, a temperature of zero degrees on the Celsius or Fahrenheit scale does not mean there is no kinetic energy, and thus in the case of temperature, zero has been arbitrarily selected. In fact, the Celsius temperature scale places the value of zero at a temperature equivalent to 32 degrees Fahrenheit.

One implication of an arbitrary zero point is that the ratio of two variables has no meaning. For example, the kinetic energy associated with a temperature of four degrees on the Fahrenheit scale is not twice as great as the kinetic energy associated with a temperature of two degrees. The property that distinguishes interval data is the notion that equal intervals represent equal amounts. For example, the interval between four degrees and one degree represents the same difference in kinetic energy as the difference between 74 degrees and 71 degrees.

Interval data are numerical data that possess both the property of ordinality (ranking) and the interval property. However, interval data do not possess a meaningful origin (zero value).

Ratio Data

Definition

Ratio data are similar to interval data, except that they have a meaningful zero point and the ratio of two data points is meaningful.

The operations of addition, subtraction, multiplication, and division are reasonable on ratio data. Many of the variables we commonly encounter are ratio variables: volumes, heights, weights, pressure. Being aware of the data's level of measurement is an extremely important part of any statistical analysis, since many statistical measures that are meaningful for interval and ratio data are not meaningful for ordinal and nominal measurements. Ratio data are the most desired and meaningful type of data for decision makers. We will explore analytical techniques for each type of data in subsequent chapters of this text.

Get Out of Here Aristotle

"Aristotle maintained that women have fewer teeth than men; although he was twice married it never occurred to him to verify this statement by examining his wives' mouths."
    –Bertrand Russell

The Greek philosopher Aristotle held that one could develop all the laws that govern the universe by pure thought and that it was unnecessary to obtain measurements that would confirm the validity of these laws. Perhaps this explains why he never bothered to count his wife's teeth. The goal of Aristotelian science was to explain why things happen. Galileo's approach originated the method of controlled experiments which form the basis of scientific investigation. Controlled experiments yield data. Statistics allows people to think with data. If Aristotle had been right, there wouldn't be a very large demand for statistics.

Example 2.4

Is money a ratio variable?

Solution

Say a friend has $40 and you have $20.

$$\frac{\$40}{\$20} = 2$$

According to the ratio we just computed your friend has twice as much money as you. Is this really true? Money is a ratio variable because ratios (quotients) are meaningful. If someone does have $40 and you have $20, they do have twice as much money as you.

2.5   Exercises

Basic Concepts

1. What is a level of measurement?

2. What are the four levels of measurement? Give an example of each.

3. For which level(s) of measurement is arithmetic appropriate?

4. What is the primary difference between nominal and ordinal data?

5. What is an arbitrary zero value? Which level of measurement has this property?

6. What is the fundamental difference between interval and ratio data?

7. Decision makers usually prefer to consider data that possess which level of measurement?

Exercises

8. Determine the level of measurement (nominal, ordinal, interval, or ratio) for each of the following variables.

   a. The temperature (in degrees Fahrenheit) of patients with pneumonia.

   b. The age at which the average male marries.

   c. Client satisfaction survey responses: Poor, Average, Good, and Excellent.

   d. The region of the U.S. in which an individual lives: North, South, East, or West.

   e. The number of people with a Type A personality.

9. Determine the level of measurement (nominal, ordinal, interval, or ratio) for each of the following variables.

   a. The time it takes for a student to complete an exam.

   b. Majors of randomly selected students at a university.

   c. The category which best describes how frequently a person eats chocolate: Frequently, Occasionally, Seldom, Never.

   d. The number of pounds of snack food eaten by an individual in his or her lifetime.

10. Given the table below on browser usage (see Exercise 22 in Section 2.4), what is the highest level of measurement that these data could have? Justify your answer.

Table for Exercise 10 – Browser Usage Share (%)				
Month	Microsoft Internet Explorer	Mozilla Firefox	Google Chrome	Apple Safari
July 2010	60.74	22.91	7.16	5.09
August 2010	60.48	22.90	7.50	5.15
September 2010	59.62	22.97	7.99	5.27
October 2010	59.18	22.83	8.50	5.36
November 2010	58.44	22.76	9.26	5.55
December 2010	57.08	22.81	9.98	5.89
January 2011	56.00	22.75	10.70	6.30
February 2011	56.77	21.74	10.93	6.36
March 2011	55.92	21.80	11.57	6.61
April 2011	55.11	21.63	11.94	7.15
May 2011	54.27	21.71	12.52	7.28
June 2011	54.84	21.20	12.72	7.41

2.6   Data Classifications

Data or variables can be classified as **qualitative** or **quantitative**. If data or variables are quantitative, they can be further identified as **discrete** or **continuous**. Since the kind of data available affects the kind of analyses that can be performed, it is important to recognize data attributes.

Qualitative and Quantitative Data

> **Definition**
>
> **Qualitative data** are measurements that can change in kind, but not in degree. Qualitative measurements often consist of labels or descriptions and do not have naturally occurring numerical values.

Qualitative data are measured on the nominal and ordinal scales. Some examples of qualitative data are gender (M/F), occupation (doctor, lawyer, professor, etc.) and eye color (brown, blue, and green). As discussed earlier, regardless of how the measurements are coded, arithmetic operations cannot be performed with qualitative data.

> **Definition**
>
> **Quantitative data** are measurements that change in magnitude from trial to trial such that some order or ranking can be applied. Quantitative variables can be measured using a naturally occurring numerical scale.

Quantitative data are measured on the interval or ratio scale and we can perform arithmetic operations with quantitative variables. Some examples of quantitative data are the number of students in class today, the number of dependents claimed on a tax return, the number of fans at a sporting event, and the time it takes to complete a particular task.

Discrete and Continuous Data

Quantitative data or variables can be categorized as either discrete or continuous. To illustrate, suppose that observations are taken on two variables: the number of stocks and bonds in a mutual fund and the annual yield for the fund. In looking at the data, one difference is that the number of stocks and bonds is a whole number, whereas the annual yield is decimal-valued. Since it is impossible for a mutual fund to have 10.167 stocks and bonds, or any fractional number for that matter, there are gaps in the values that the variable "Number of Stocks" can assume.

Table 2.2—Mutual Funds					
Number of Stocks	10	20	30	40	50
Annual Yield (%)	4.25	3.86	13.52	15.00	9.62

The stock data given in Table 2.2 are **discrete**. While the data representing stocks and bonds can only assume integer values, it would be misleading to create the impression that a discrete variable can only assume integer values. Discrete data may assume decimal values. For example, a variable that takes only the values 1, 1.5, 2, 2.5 is discrete.

> **Definition**
>
> Data in which the observations are restricted to a set of values (such as 1, 2, 3, 4) that possesses gaps are **discrete**.

The annual yield data given in Table 2.2 are **continuous**. It may appear that the annual yields are discrete, since values appear to only take on values up to the second decimal place. However, it is only the inadequacy of the measuring instrument that creates this illusion. The first annual yield is listed as 4.25%, but the actual annual yield of the first mutual fund may be 4.2531642%. If the measuring device can only detect differences in the hundredths place, then all measurements will be given to the hundredths place. Any digits beyond the hundredths place will be ignored. However, just because these digits are ignored doesn't mean they don't exist.

> **Definition**
>
> Data that can take on any value within some interval are **continuous**.

Example 2.5

State whether each of the following variables is qualitative or quantitative. If the variable is quantitative, indicate whether it is discrete or continuous.

a. Political party affiliation

b. Amount of time spent on the computer

c. Marital status

d. Number of online orders made per month

Solution

a. Examples of political party affiliation are Republican, Democrat, Independent, etc. These are qualitative measurements because they identify the party by name only and we are unable to perform any arithmetic operations or establish distances between parties. Thus, political party affiliation is a qualitative variable.

b. Examples of time spent on the computer could be any value from a few seconds to a large number of hours. Also, depending on the quality of the measurement, time could be recorded (or reported) in any number of significant digits. With that said, time spent on the computer is a quantitative, continuous random variable.

c. For marital status, examples could be single, married, or divorced. These are qualitative measurements since one classification cannot be considered to be more than another and we also cannot perform any arithmetic operations on this variable.

d. The number of online orders is countable, starting at 0. Thus, this is a quantitative, discrete random variable.

2.6   Exercises

Basic Concepts

1. What are qualitative data? Give an example.

2. What are quantitative data? Give an example.

3. Which levels of measurement are associated with qualitative data? Which levels are associated with quantitative data?

4. If data are quantitative, they can be further classified into two categories. Name and briefly describe these categories.

5. What is the difference between discrete and continuous data?

Exercises

6. Identify the following variables as discrete or continuous.

    a. The number of doctors who wash their hands between patient visits.

    b. The amount of liquid consumed by the average American each day.

    c. The weight of a newborn baby at a local hospital.

    d. The time it takes a person to react to a stimulus.

    e. The number of voters who favor a particular candidate.

7. Identify the following variables as discrete or continuous.

    a. The number of on-time flights at the Hartsfield International Airport in Atlanta.

    b. The height of skyscrapers in New York City.

    c. The price of General Electric's common stock.

    d. The temperature of U.S. cities.

    e. The number of alcoholics who are men.

8. The results of a study investigating the nutritional status of mid-nineteenth century Americans were reported in "The Height and Weight of West Point Cadets: Dietary Changes in Antebellum America," in the *Journal of Economic History*. The data are based upon physical examination lists for West Point applicants from 1843 to 1894. Some of the information obtained from each cadet were his height, weight, the state from which the cadet was appointed, the occupation of the father, the income of the parents, and the type of home residence (city, town, or rural) of the cadet.

    a. List the different variables measured on the cadets.

    b. Which variables are quantitative and which are qualitative?

    c. Give the levels of measurement for these variables.

    d. Why is some method of data summary necessary here?

9. The major television networks regularly conduct polls in order to ascertain the feelings of Americans on current political issues. In May of 1993, such a poll was conducted by ABC concerning United States involvement in Bosnia. The respondent's gender, political affiliation, and opinion (approve, disapprove, or no opinion) on how President Clinton was handling the situation in Bosnia represented some of the information supplied by the respondent on the survey. Each respondent was also asked to rate the job that the news media had done (excellent, good, not so good, poor) in covering the situation in Bosnia.

    a. List the different variables measured on the respondents.

    b. Which variables are quantitative and which are qualitative?

    c. Give the levels of measurement for these variables.

    d. What are some problems associated with collecting data in polls such as the one described in this exercise?

10. Under most states' auto lemon laws, dealers or car makers must replace defective autos that aren't successfully repaired after three attempts or that remain in the shop for 30 days. The table below shows data for Hawaii for the year 2010, weighing car makers' lemons against statewide market share. Assume the "lemon index" is the share of the complaints divided by the total market share for each manufacturer.

Table for Exercise 10 – Lemon Index: Hawaii, 2010			
Best	Lemon Index	Worst	Lemon Index
Toyota (includes Lexus)	0.212	Chrysler (includes Dodge and Jeep)	6.512
Honda	0.462	Kia	2.750
Ford (includes Lincoln)	0.868	GM (includes Chevrolet, GMC, Buick)	2.375
Nissan (includes Infinity)	1.056	BMW	2.129
Mazda	1.833	Hyundai	2.000

Source: Hawaii.gov

Answer the following questions for the variable "Lemon Index".

a. Are the data quantitative or qualitative? Why?

b. What is the highest level of measurement these data could have?

| 2.7 | **Time Series Data** |

Recall from Chapter 1, the science of statistics is divided into two categories: descriptive statistics and inferential statistics. Fundamental to the concept of statistical inference is the notion of population – the total collection of measurements. *Time series data originate as measurements usually taken from some process over equally spaced intervals of time.* Because measurements are taken over time, the concept of a population gets a little blurry. Suppose we want to examine the divorce rate in the United States since 1900. What is the population we are studying?

Presumably the members of the population under study would be the residents of the United States. But in 1900 there were only about 76 million people in this country. Now there are over 317 million. Moreover, none of the 76 million that were around in 1900 are alive today, so the population from which the divorce data are drawn is certainly not fixed. In addition, cultural and sociological conditions prevailing in 1900 are substantially different from those in the 2011s. Why is this important? If a population doesn't contain a fixed set of members or subjects, how can inferences be made about it? The concept of population is not sufficiently broad to cope with time series measurements.

Table 2.3 – Divorce Rate in the United States 1900 – 2008 (Per 1000 People)			
Year	**Number of Divorces**	**Year**	**Number of Divorces**
1900	0.7	1985	5.0
1905	0.8	1990	4.7
1910	0.9	1995	4.4
1915	1.0	1996	4.3
1920	1.6	1997	4.3
1925	1.5	1998	4.2
1930	1.6	1999	4.1
1935	1.7	2000	4.1
1940	2.0	2001	4.0
1945	3.5	2002	3.9
1950	2.6	2003	3.8
1955	2.3	2004	3.7
1960	2.2	2005	3.6
1965	2.5	2006	3.7
1970	3.5	2007	3.6
1975	4.8	2008	3.5
1980	5.2		

Source: Centers for Disease Control and Prevention

Time series data originate from processes. Processes can be divided into two categories: stationary and nonstationary. All time series that are interesting vary, and the nature of the variability determines how the process is characterized. In a **stationary process** the time series varies around some central value and has approximately the same variation over the series. In a **nonstationary process** the time series possesses a trend – the tendency for the series to either increase or decrease over time.

Example 2.6

A company's cash operations are defined as its revenues minus all operating expenses, but calculated through a series of adjustments to net income. Table 2.4 contains cash operations data for Applied Materials, a technology manufacturing company. The company's cash operations range from approximately −$185 million to more than $850 million. Determine if this time series is stationary or nonstationary.

Table 2.4 – Applied Materials (AMAT) Cash Operations October 1999 – October 2009 (Millions of Dollars)			
Period	**Cash Operations**	**Period**	**Cash Operations**
Oct. 1999	413.85	Jan. 2005	236.98
Jan. 2000	184.05	Apr. 2005	475.97
Apr. 2000	369.18	Jul. 2005	336.41
Jul. 2000	405.36	Oct. 2005	197.78
Oct. 2000	693.22	Jan. 2006	411.73
Jan. 2001	−4.32	Apr. 2006	502.96
Apr. 2001	838.62	Jul. 2006	361.94
Jul. 2001	294.02	Oct. 2006	659.09
Oct. 2001	451.98	Jan. 2007	419.18
Jan. 2002	101.10	Apr. 2007	475.15
Apr. 2002	20.40	Jul. 2007	637.56
Jul. 2002	182.30	Oct. 2007	677.40
Oct. 2002	188.33	Jan. 2008	389.73
Jan. 2003	87.45	Apr. 2008	874.00
Apr. 2003	313.41	Jul. 2008	320.42
Jul. 2003	−97.06	Oct. 2008	126.31
Oct. 2003	497.95	Jan. 2009	−185.21
Jan. 2004	217.29	Apr. 2009	84.30
Apr. 2004	220.27	Jul. 2009	193.70
Jul. 2004	682.63	Oct. 2009	239.87
Oct. 2004	507.08		

Solution

The cash operations data are plotted in Figure 2.4. Notice that the time series seems to fluctuate around some central value and the dispersion around the central value is reasonably constant throughout this time frame. Thus, the time series appears to be stationary.

Figure 2.4

Example 2.7

Determine if the time series corresponding to the divorce data given in Table 2.3 is stationary or nonstationary.

Solution

Figure 2.5 shows the divorce data from Table 2.3. There is strong evidence of an upward trend. Thus, the time series is nonstationary. (Note these data are not affected by population increases over the years since it is given in divorces per 1000 people.)

Figure 2.5

Example 2.8

Consider the following data regarding hourly compensation for manufacturing durable goods. Determine if the time series associated with these data is stationary or nonstationary.

Table 2.5 – Hourly Compensation for Manufacturing Durable Goods 1990 – 2010 (U.S. Dollars)			
Year	**Hourly Compensation**	**Year**	**Hourly Compensation**
1990	89.12	2001	133.57
1991	94.83	2002	140.10
1992	100.00	2003	153.01
1993	101.96	2004	152.52
1994	104.52	2005	157.24
1995	105.27	2006	160.72
1996	106.42	2007	168.19
1997	108.27	2008	173.10
1998	115.27	2009	182.34
1999	120.33	2010	182.35
2000	129.81		

Source: Bureau of Labor Statistics

Solution

As can be seen in Figure 2.6, hourly compensation for manufacturing durable goods has been on a steady increase since 1990. Given that the time series increases from year to year, this is obviously not a stationary series.

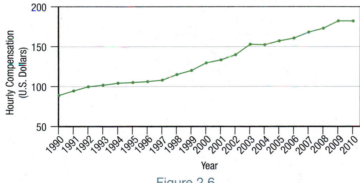

Hourly Compensation for Manufacturing Durable Goods

Figure 2.6

2.7   Exercises

Basic Concepts

1. What are time series measurements?

2. What problems are associated with the concept of population when studying time series data?

3. What is a stationary process?

4. What is a nonstationary process?

5. What is a trend? If a time series has an 'upward trend' what does this mean?

Exercises

6. Consider the following graph of long-term interest rates (10-year treasury notes) and inflation rates.

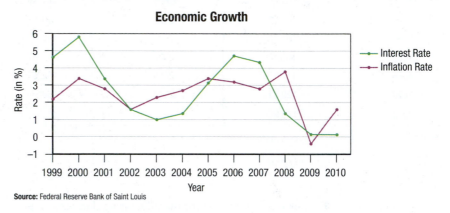

Source: Federal Reserve Bank of Saint Louis

   a. Are the interest rate data presented above time series data?

   b. Are the inflation rate data presented above time series data?

   c. For each of parts **a.** and **b.** if the data is time series data, does the series appear to be stationary or nonstationary?

7. Consider the following graph of total exports.

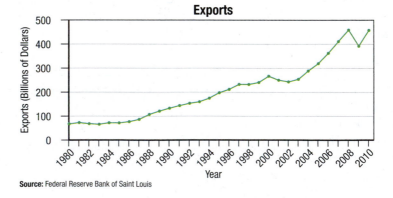

Source: Federal Reserve Bank of Saint Louis

   a. Are these data time series data?

   b. If the data are time series data, does the series appear to be stationary or nonstationary?

8. Using a newspaper, journal, or website as your source, give an example of time series data. Be sure to reference your source and give a brief description of the data.

2.8   Cross-Sectional Data

Cross-sectional data are measurements created at approximately the same period of time. For example, consider the life expectancy at birth (in 2010) for selected countries given in Table 2.6. These data represent cross-sectional measurements since the measurements were made in the same time period (2010). People in Andorra are expected to live on average until age 82, 4 years longer than the average for Americans. Developed countries such as Australia and the United States generally have higher life expectancies than developing countries such as Botswana and Kenya. But according to the World Health Organization, life expectancies in developing countries are on the rise due to medical interventions based on advanced technology and drugs. In fact, developing countries are expected to experience a 200 to 300 percent increase in their elderly populations in the next 35 years.

Table 2.6 – Life Expectancy at Birth: 2010	
Country	**Life Expectancy (Years)**
Afghanistan	45
Andorra	82
Australia	82
Botswana	61
Egypt	72
Guatemala	71
Kenya	59
Sri Lanka	75
Sweden	81
United Kingdom	80
United States	78

Source: CIA World Fact Book

Another example of cross-sectional data is greenhouse gas emissions of each state on a per capita basis. The air that surrounds the earth consists of a mixture of oxygen and nitrogen, interspersed with small amounts of carbon dioxide, methane, and other trace gases. The trace gases capture heat as the sun warms the earth and holds part of that heat in the atmosphere in what is known as the "greenhouse effect."

These gases prevent the sun's heat from simply hitting the ground and being rechanneled back into space. For most of the last several thousand years, the earth's greenhouse gases have been stable. The most abundant trace gas, carbon dioxide, was processed by plants and maritime organisms at approximately the same rate as it was given off by other organisms, until about the time of the Industrial Revolution. The natural world's ability to absorb carbon dioxide has been unable to keep up, and consequently the level of carbon dioxide in the atmosphere has been rising. A time series of the atmospheric carbon dioxide concentrations at Mauna Loa Observatory reveals a disturbing trend, as seen in Figure 2.7.

Source: Carbon Dioxide Research Group, Scripps Institution of Oceanography, University of California

Figure 2.7

The U.S. produces 19.9% of the world's carbon dioxide. It would be interesting to examine the greenhouse emissions of each state on a per capita basis.

Table 2.7 – Per Capita Carbon Dioxide Emissions by State in 2007 (Metric Tons)					
State	**Emissions**	**State**	**Emissions**	**State**	**Emissions**
Wyoming	126.67	Ohio	23.39	Nevada	16.20
North Dakota	82.26	Mississippi	23.07	Arizona	16.03
West Virginia	64.76	Arkansas	22.62	Maine	15.64
Alaska	64.63	Pennsylvania	22.21	New Jersey	15.37
Louisiana	42.46	Tennessee	20.53	New Hampshire	14.50
Montana	39.38	Colorado	20.42	Florida	14.12
Indiana	37.09	Delaware	20.00	Maryland	12.32
Kentucky	37.07	South Carolina	19.85	Washington	12.64
Alabama	31.48	Georgia	19.47	Massachusetts	12.32
Oklahoma	30.54	Minnesota	19.38	Oregon	11.65
New Mexico	30.07	Illinois	19.05	Connecticut	11.55
Kansas	29.04	Hawaii	19.03	California	11.10
Iowa	28.84	Wisconsin	18.74	Idaho	10.81
Texas	26.83	Michigan	18.24	Vermont	10.48
Utah	26.43	South Dakota	17.31	New York	10.31
Nebraska	25.09	North Carolina	17.10	Rhode Island	10.24
Missouri	23.79	Virginia	16.71	District of Columbia	4.26

Source: Environmental Protection Agency

2.8   Exercises

Basic Concepts

1. What are cross-sectional data?

2. What is the difference between cross-sectional data and time series data?

Exercises

3. The following table shows the annual average crude oil price from 1946 through 2011. Prices are adjusted for inflation to April 2011 prices using the Consumer Price Index (CPI-U) as presented by the Bureau of Labor Statistics. Inflation adjusted prices were at an all-time high in 1980, reaching $102.26 dollars per barrel. Crude oil prices reached an all-time low in 1998 (lower than the price in 1946!) when the price per barrel dipped to $16.44. Using the data in the table, discuss if the data set contains time series or cross-sectional data. Also, discuss the data and make some inferences. That is, can you explain some of the fluctuations in the oil prices?

Table for Exercise 3 – Annual Average Domestic Crude Oil Prices ($ per Barrel)					
Year	Nominal	Inflation Adjusted (April 2011)	Year	Nominal	Inflation Adjusted (April 2011)
1946	1.63	18.49	1979	25.10	77.05
1947	2.16	21.73	1980	37.42	102.26
1948	2.77	25.92	1981	35.75	88.55
1949	2.77	26.17	1982	31.83	74.24
1950	2.77	25.90	1983	29.08	65.69
1951	2.77	24.00	1984	28.75	62.26
1952	2.77	23.47	1985	26.92	56.28
1953	2.92	24.50	1986	14.44	29.62
1954	2.99	25.04	1987	17.75	35.13
1955	2.93	24.57	1988	14.87	28.32
1956	2.94	24.35	1989	18.33	33.24
1957	3.14	25.12	1990	23.19	39.80
1958	3.00	23.38	1991	20.20	33.36
1959	3.00	23.15	1992	19.25	30.85
1960	2.91	22.15	1993	16.75	26.09
1961	2.85	21.44	1994	15.66	23.76
1962	2.85	21.19	1995	16.75	24.73
1963	2.91	21.39	1996	20.46	29.32
1964	3.00	21.75	1997	18.64	26.12
1965	3.01	21.47	1998	11.91	16.44
1966	3.10	21.48	1999	16.56	22.30
1967	3.12	21.04	2000	27.39	35.76
1968	3.18	20.53	2001	23.00	29.23
1969	3.32	20.36	2002	22.81	28.50
1970	3.39	19.65	2003	27.69	33.86

Table for Exercise 3 – Annual Average Domestic Crude Oil Prices ($ per Barrel) Cont.

Year	Nominal	Inflation Adjusted (April 2011)	Year	Nominal	Inflation Adjusted (April 2011)
1971	3.60	20.00	2004	37.66	44.81
1972	3.60	21.44	2005	50.04	57.57
1973	4.75	23.87	2006	58.30	65.03
1974	9.35	42.58	2007	64.20	69.51
1975	12.21	51.00	2008	91.48	95.25
1976	13.10	51.78	2009	53.48	55.96
1977	14.40	53.41	2010	71.21	73.44
1978	14.95	51.58	2011 (Partial)	86.84	–

Source: www.inflationdata.com

4. Do you think the pay of executives working for digital companies increases/decreases as the company's stock price increases/decreases? Examine the following table.

Table for Exercise 4 – CEO Compensation and Stock Performance

Exec	Salary/Bonus ($)	Stock/Options ($)	Other Non-Equity Compensation ($)	Total 2007 Compensation ($)	Change from 2006 Compensation (%)	2007 Stock Performance (%)
Tom Rogers (Tivo)	800,000	6,200,000	495,075	7,495,075	+102	+32
Mel Karmazin (Sirius)	5,250,000	–	18,743	5,268,743	+23	−23
Paul Sagan (Akamai)	403,651	3,554,264	497,362	4,455,277	−40	−48
Reed Hastings (Netflix)	850,000	1,568,307	270	2,418,577	+5	−6
Rob Glaser (RealNetworks)	1,169,384	643,400	354,200	2,166,984	−26	−45
Bobby Kotick (Activision Blizzard)	899,560	1,188,467	–	2,088,027	+6	+49
Magid M. Abraham (comScore)	421,952	1,125,000	–	1,546,952	+185	−16
Barry Diller (IAC)	500,000	–	927,429	1,427,429	+270	+21
John S. Riccitiello (Electronic Arts)	750,000	–	625,350	1,375,350	−37	−38
Steve Ballmer (Microsoft)	1,340,833	–	10,001	1,350,834	N/A	0
Wayne T. Gattinella (WebMD)	830,000	–	9214	839,214	+6	+10

Source: paidContent.org

What type of data is in the Salary/Bonus column? What do you think about executive salaries as a function of the company's stock performance? Justify your responses.

2.9   Data Resources

We live in a data rich society. Anyone with access to a personal computer can access thousands of different databases throughout the Internet. These databases are full of observational data.

Here are some interesting sites that you might wish to explore:

- Data Interchange Standards Association – www.disa.org
- The World Bank – data.worldbank.org
- The Population Reference Bureau – www.prb.org/DataFinder.aspx
- The World Health Organization – www.who.int

The United States government is one of the more prolific producers of data in the world.

Check out a few of the many U.S. government's websites which publish a wide array of national statistics:

- Federal Statistics – www.fedstats.gov
- The National Center for Education Statistics – www.nces.ed.gov
- Centers for Disease Control and Prevention – www.cdc.gov/nchs
- The Whitehouse – www.whitehouse.gov
- The United States Census Bureau – www.census.gov/compendia/statab
- The Bureau of Labor Statistics – www.bls.gov

T Discovering Technology

Note: If you are not familiar with Excel, Appendix B is a tutorial section that will familiarize you with the controls in Excel.

Using Excel

Line Chart

For this exercise, use the information from Example 2.6.

1. Enter the time period data into Column A – label as "Period".

2. Enter the data for the cash operations into Column B – label as "Cash Operations".

3. Highlight the data in Column A and Column B and under the "Insert" tab click on the **Line** button. Select **Line with Markers** (the first graph in the second row under "2-D line") so that you can see each data point that is plotted.

4. You now see a time series plot of the cash operations data. To change the interval of the x-axis, select the values on the x-axis, right click and select **Format Axis**. In the "Axis Options" dialog box, next to "Major unit" select the radio button for **Fixed**. Enter **6** in the dialog box and make sure that **Months** is selected. Click **Close**.

5. To format the y-axis, select the values on the y-axis, right click and select **Format Axis**. Since there are negative values in this data set, the chart may be easier to read if the horizontal axis crosses at a negative value, say -400.0 rather than 0. Under the heading "Horizontal axis crosses," select the radio button next to **Axis value** and enter **−400.0** in the dialog box. Click **Close**. Now the entire line chart is above the x-axis.

6. Under the "Layout" tab you can change the chart title, axis titles, and legend. Add the horizontal axis label "Period" by clicking **Axis Titles**, **Primary Horizontal Axis Title**, and **Title Below Axis**. Add the vertical axis label "Cash Operations (Millions of Dollars)" by clicking **Axis Titles**, **Primary Vertical Axis Title**, and **Rotated Title**.

7. Click on the **Legend** button and select **none**. The legend is now removed and the chart is easier to read. There are numerous ways to format your chart in Microsoft Excel. Try creating some other charts using Excel for the various data sets presented in this chapter.

Figure 2.8

Using Minitab

Time Series Plot

For this exercise, use the information from Example 2.8.

1. Enter the data for the years in Column C1 – label as "Year".

2. Enter the corresponding data for the hourly compensation in Column C2 – label as "Hourly Compensation".

3. Click **Graph** and select **Time Series Plot**.

4. Select the "Simple" time series plot. Click **OK**.

5. In the "Time Series Plot – Simple" dialog box, highlight "Hourly Compensation" in the box on the left-hand side and click **Select**. This moves "Hourly Compensation" into the "Series" box on the right-hand side.

6. Now click **Time/Scale**. This will bring up the "Time Series Plot – Time/Scale" dialog box.

7. On the right-hand side under the heading "Time Scale", select **Calendar**. In the drop down menu next to "Calendar" select **Year**.

8. Under the heading "Start Values", select **One set for all variables**. In the white box below the heading "Year" enter **1990**. This tells Minitab that time on the *x*-axis should begin at the year 1990.

9. In the box next to "Data Increment" enter **1**. This tells Minitab that the labels on the *x*-axis should be in increments of 1 year. Click **OK**.

10. Now in the "Time Series Plot – Simple" dialog box, click **OK** to display the graph.

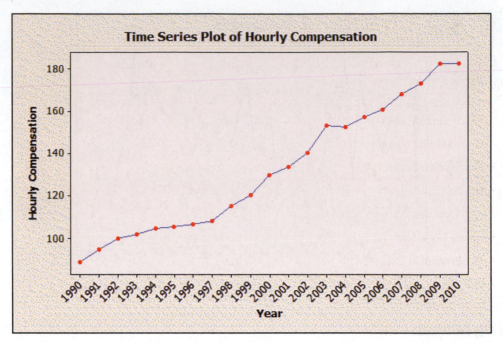

Figure 2.9

R   Chapter 2 Review

Key Terms and Ideas

- The Scientific Method
- Confounding Variable
- Descriptive Statistics
- Inferential Statistics
- The Decision-Making Method
- Fuzzy Concepts
- Observational Experiment
- Controlled Experiment
- Control Group
- Experimental Group
- Treatment
- Explanatory Variable
- Response Variable
- Completely Randomized Design
- Before and After Study
- The Placebo Effect
- Double Blind Study
- Observational Study

- Bias
- Simpson's Paradox
- Surveys
- Level of Measurement
- Nominal Data
- Ordinal Data
- Interval Data
- Ratio Data
- Qualitative Data
- Quantitative Data
- Discrete Data
- Continuous Data
- Time Series Data
- Stationary Process
- Nonstationary Process
- Trend
- Cross-Sectional Data

AE   Additional Exercises

1. Suppose you were the administrator of a public school system. What kinds of variables would you measure and how would you collect the measurements on the following subjects:

   a. Student learning

   b. School discipline

   c. Teacher preparation

   d. Absenteeism (pupil and teacher)

   e. Cafeteria food quality

2. The head of the Veterans Administration has been receiving complaints from a Vietnam Veterans organization concerning disability checks. The organization claims that checks are continually late. The checks are to arrive no later than the tenth of each month.

   a. What variables would you measure to explore this problem?

   b. How would you collect measurements on these variables?

3. A family member has unexpectedly bequeathed you a sizable sum of money.

   a. What criteria might you wish to evaluate in deciding how to invest the money?

   b. What data might be useful in your considerations?

4. Flying Eagle Airlines advertises that it surpasses all other airlines in flights that arrive on time. A competitor states that it has a better on-time record than any other airline. Can they both be correct? Explain.

5. Two local grocery stores both claim to have the lowest prices in town. Develop a measurement that you believe could be used as a criterion to determine which store actually has the lowest prices.

6. At the end of 2001, the United States had 32.9 million people living in poverty according to the Census Bureau (www.census.gov). This was an increase of 1.3 million from the previous year. Poverty was defined by the Census Bureau as having a cash income less than $14,255 a year. The Census Bureau does not include in their income measurement any part of $167 billion spent on Medicaid, a federal program by which medical care is provided to the poor. The Census Bureau only includes $34.9 billion out of the $205 billion spent annually on public welfare. Forty percent of those classified as impoverished own their own homes. How do you think poverty should be defined?

7. The quality movement has compelled American businesses to address the problem of measuring customer satisfaction. How would you measure customer satisfaction if you owned a car dealership?

8.   Identify the following variables as discrete or continuous.

   a.   Average test score on a test ranging from 0 to 100

   b.   Number of boot errors on a computer

   c.   Investment ratios for earnings per share

   d.   Energy usage in a production process

9.   Determine the level of measurement for each of the following variables.

   a.   Golf score in relation to par

   b.   SAT score

   c.   Rating from 1 to 5 of quality of service in a restaurant

   d.   Make and model of a vehicle

   e.   The number of students with a business major

10.   According to a Danish researcher, if you drop your average daily activity level by taking elevators instead of stairs, by parking your car in the closest space, or by never walking to run errands, you increase your risk of diabetes, heart disease, and premature death. The researcher studied two groups of healthy men (eight in the first group with an average age of 27 and an average body mass index (BMI) of 22.9, which is well within the normal range; and ten in the second group with an average age of 23.8 years and a BMI of 22.1). In addition to age and BMI, researchers also collected information such as number of steps per day (each group of men was fitted with pedometers), height, weight, and race. With the first group of men, the researchers asked that they reduce their daily activity (steps) by taking cars on short trips and elevators instead of stairs. The insulin levels were also measured for each group and the researchers found that with the reduced activity, insulin levels rose by nearly 60 percent after two weeks of inactivity, thus increasing the risk of diabetes and heart disease. However, the good news is that by increasing activity over a two-week period of time, one can begin to reduce his or her risk of diabetes and heart disease.

   Source: U.S. News and World Report

   a.   List the different variables measured in this study.

   b.   Which variables are quantitative and which are qualitative?

   c.   Of the variables that are quantitative, are they discrete or continuous?

   d.   Give the levels of measurement for these variables.

   e.   Why is some method of data summary necessary here?

11.   Consider the world production of crude oil given in millions of barrels per day.

Table for Exercise 11 – World Production of Crude Oil			
Year	Total World Production (Millions of Barrels per Day)	Year	Total World Production (Millions of Barrels per Day)
1980	63.987	1995	70.274
1981	60.602	1996	71.919
1982	58.098	1997	74.160
1983	57.934	1998	75.656
1984	59.568	1999	74.853
1985	59.172	2000	77.768
1986	61.407	2001	77.686
1987	62.086	2002	76.994
1988	64.380	2003	79.598
1989	65.508	2004	83.105
1990	66.426	2005	84.595
1991	66.399	2006	84.661
1992	66.564	2007	84.543
1993	67.091	2008	85.507
1994	68.590	2009	84.389

Source: Energy Information Administration

a.   What is the level of measurement of the data?

b.   Are the data time series or cross-sectional? If the data are time series, plot the data. Does the series appear to be stationary or nonstationary? Explain your answer.

12.   Consider the graph of the number of respondents (in percentages) who think things in the U.S. are now on the wrong track versus those that think the economy is going in the right direction. The data were collected using a survey asking the question, *In general, are you satisfied or dissatisfied with the way things are going in the United States at this time?*
Source: Gallup Poll

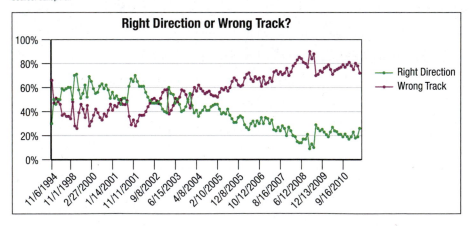

a.   Are the opinions on the outlook of the economy presented in time series or cross-sectional data? Justify your answer.

b.   If the data are time series data, does the series appear to be stationary or nonstationary? Explain your answer.

13.   Can you think of a process that would yield measurements that did not have any variability? Would studying such a process be very interesting?

14. One of the measurements that population experts use in predicting trends in population growth is the fertility rate. The total fertility rate is sometimes defined as the number of likely births one woman will have in her lifetime. The accompanying table gives the fertility rate from 1934 to 2005. If the data are time series data, plot the data in a line chart. Make observations based on the graph as to whether the series is stationary or nonstationary. If the time series is nonstationary, identify any noticeable trends.

\multicolumn{6}{c}{**Table for Exercise 14** – Fertility Rates}
Year	**Fertility Rate**	**Year**	**Fertility Rate**	**Year**	**Fertility Rate**
1934	2.294	1958	3.693	1982	1.829
1935	2.250	1959	3.705	1983	1.803
1936	2.207	1960	3.654	1984	1.806
1937	2.236	1961	3.629	1985	1.843
1938	2.288	1962	3.474	1986	1.836
1939	2.238	1963	3.333	1987	1.871
1940	2.301	1964	3.208	1988	1.933
1941	2.399	1965	2.928	1989	1.977
1942	2.628	1966	2.736	1990	2.081
1943	2.718	1967	2.573	1991	2.073
1944	2.568	1968	2.477	1992	2.065
1945	2.491	1969	2.465	1993	2.046
1946	2.943	1970	2.480	1994	2.036
1947	3.274	1971	2.267	1995	2.019
1948	3.109	1972	2.010	1996	2.040
1949	3.110	1973	1.879	1997	2.000
1950	3.091	1974	1.835	1998	2.030
1951	3.267	1975	1.774	1999	2.070
1952	3.355	1976	1.738	2000	2.056
1953	3.418	1977	1.790	2001	2.034
1954	3.537	1978	1.760	2002	2.013
1955	3.574	1979	1.808	2003	2.043
1956	3.682	1980	1.840	2004	2.046
1957	3.760	1981	1.815	2005	2.054

Source: U.S. National Center for Health Statistics

15. One of the problems associated with the management of solid waste is the NIMBY (not in my backyard) syndrome. In separate surveys taken in 1988, 1989, and 1990 the National Solid Waste Management Association asked, *Would you object to a new landfill in your community?* The percentage response is given in the table below.

Survey Date	Don't Object	Object	Not Sure
March 1990	36	59	5
February 1989	23	65	12
February 1988	30	62	8

Table for Exercise 15 – Survey Results

   a. What is the level of measurement of the survey data?

   b. Are the data time series or cross-sectional?

   c. What other information would be useful in evaluating the results of the study?

16. In a recent study of four leading anesthetics, three hundred patients were randomly selected and assigned to be given one of the four products during a surgery. One of the products performed significantly better than the rest. Is this an observational study or a controlled experiment?

P   Discovery Project

Bachelor's Degrees by Gender

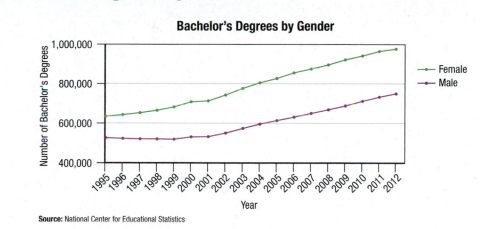

Source: National Center for Educational Statistics

1. What is the level of measurement of the data displayed in the graph above?

2. Are the data above time series or cross-sectional data?

3. Are the data in the graph above part of a stationary or nonstationary process? Why?

4. According to the graph, is the difference between the number of women and men earning bachelor's degrees increasing, decreasing, or remaining constant? Try to think of some reasons why this is true.

5. Do you believe the data? Elaborate.

Chapter Three

Organizing, Displaying, and Interpreting Data

Introduction

Statistics is about understanding data. Graphical images are universally regarded as a powerful form of communicating data because the eyes and brain process visual data with amazing speed. The processing of a visual image begins in the retina of the eye where a few hundred million neurons pass their output to about a million ganglion cells, which in turn pass the visual data into the brain. This system enables the equivalent of approximately 90 million bytes per second of data to be transmitted to the brain from the eyes. The brain creates three-dimensional color models (our visual reality) from the data. The brain's ability to process an enormous quantity of data and to create and interpret images in their totality is an astonishing processing feat.

Although the eyes, in conjunction with our visual processing system, are a fabulous sensory system, they digest text or numbers slowly in comparison to the comprehension of a visual image. Reading is a tremendously complex transformation. When we read words or numbers, the brain must recognize the letters, decide what word they form, and think about the definition and how it relates to previous words. Recognizing letters and words is such an integral part of our lives that it is performed without conscious thought. Nevertheless, this process is slow (perhaps 50 bytes a second) compared to absorbing a picture or graphical display. Therefore, graphical displays are used daily in business meetings and presentations to communicate data quickly and effectively. For example, an advertising agency may use a chart showing an upward trend in sales to convince a potential client that its services are beneficial.

In financial markets, technical analysts (sometimes called technicians) use graphs, tables, and other tools to identify price patterns, trends, and relationships in an effort to forecast price movements over certain periods of time.

3.1   Frequency Distributions

Statistics exists because of variation. The statistician's job is to comprehend variation by looking for structure. Frequency distributions are one method of examining a data set's structure. To examine structural characteristics, ask questions such as, *Where are most of the observations located? Do the data cluster around one central point or are there several points that data seem to cluster around? Do the data seem to be uniformly spread out over some interval or bunched in some range?* These questions all relate to the concept of "distribution."

> **Definition**
>
> A **frequency distribution** summarizes data into classes and provides in tabular form a list of the classes along with the number of observations in each class.

The process of refining information is interesting. The analyst begins with raw data, then organizes that data by counting the number of observations in each classification. In Table 3.1, the raw data consist of population counts in each state for the years 2000 and 2008. By comparing the populations in 2008 with the populations in 2000, a percentage population growth can be computed for each state over the 8-year period.

When the growth rate data is classified in the frequency distribution (see Table 3.2), the actual magnitudes of the data values disappear. Losing information may not seem to be a desirable result, but without some lumping together, it is difficult to comprehend large amounts of data.

State	2000 Population	2008 Population	Growth	Percent Growth	State	2000 Population	2008 Population	Growth	Percent Growth
Table 3.1 – Population Counts of Individual States in 2000 and 2008 (Rounded to the nearest 100,000)									
Alabama	4447	4662	215	4.83	Montana	902	967	65	7.21
Alaska	627	686	59	9.41	Nebraska	1711	1783	72	4.21
Arizona	5131	6500	1369	26.68	Nevada	1998	2600	602	30.13
Arkansas	2673	2855	182	6.81	New Hampshire	1236	1316	80	6.47
California	33872	36757	2885	8.52	New Jersey	8414	8683	269	3.20
Colorado	4302	4939	637	14.81	New Mexico	1819	1984	165	9.07
Connecticut	3406	3501	95	2.79	New York	18977	19490	513	2.70
Delaware	784	873	89	11.35	North Carolina	8047	9222	1175	14.60
D.C.	572	592	20	3.50	North Dakota	642	641	−1	−0.16
Florida	15983	18328	2345	14.67	Ohio	11353	11486	133	1.17
Georgia	8187	9686	1499	18.31	Oklahoma	3451	3642	191	5.53
Hawaii	1212	1228	16	1.32	Oregon	3421	3790	369	10.79
Idaho	1294	1524	230	17.77	Pennsylvania	12281	12448	167	1.36
Illinois	12420	12902	482	3.88	Rhode Island	1048	1051	3	0.29
Indiana	6081	6377	296	4.87	South Carolina	4012	4480	468	11.67
Iowa	2926	3003	77	2.63	South Dakota	755	804	49	6.49
Kansas	2689	2802	113	4.20	Tennessee	5689	6215	526	9.25
Kentucky	4042	4269	227	5.62	Texas	20852	24327	3475	16.67
Louisiana	4469	4411	−58	−1.30	Utah	2233	2736	503	22.53
Maine	1275	1316	41	3.22	Vermont	609	621	12	1.97
Maryland	5297	5634	337	6.36	Virginia	7079	7769	690	9.75
Massachusetts	6349	6498	149	2.35	Washington	5894	6549	655	11.11
Michigan	9938	10003	65	0.65	West Virginia	1808	1814	6	0.33
Minnesota	4919	5220	301	6.12	Wisconsin	5364	5628	264	4.92
Mississippi	2845	2939	94	3.30	Wyoming	494	533	39	7.89
Missouri	5597	5912	315	5.63					

Source: U.S. Census Bureau

Table 3.2 – Frequency Distribution of State 8-Year Population Growth Rates	
Percentage Population Growth 2000–2008	Frequency
Less than or equal to 0	2
0.01 to 5	21
5.01 to 10	15
10.01 to 15	7
15.01 to 20	3
20.01 to 25	1
25.01 to 30	1
Greater than 30	1

With the frequency distribution, we are able to see the broader structure of the data. It is now easy to see that most state population 8-year growth rates are between 0 and 15 percent. Further, growth rates above 20% are uncommon. Without the organization that a frequency distribution provides, these conclusions would be more difficult to establish. If there were 10,000 data values instead of 51, it would be much more difficult to make similar conclusions.

There are only two steps in the construction of a frequency distribution.

Step 1: Choose the classifications.

Step 2: Count the number in each class.

For simple data, such as the results from tossing a coin, the choice of classifications is easy. Heads is one category and tails the other. However, for continuous data, such as weights, heights, and volumes, the choice of classification scheme becomes less obvious, since there are an enormous number of possibilities. There are two requirements that should be met when setting up the categories for classification: the categories must be both mutually exclusive and exhaustive. Essentially, this means categories should not overlap and should cover all possible values.

Since choosing the classification depends on whether the data are qualitative (nominal or ordinal) or quantitative (interval or ratio), the discussion of frequency distributions will be presented on the basis of these data types. This section will present how to construct a frequency distribution for qualitative data, and Section 3.4 will introduce constructing frequency distributions for quantitative data.

Constructing Frequency Distributions for Qualitative Data

To construct a frequency distribution for qualitative data, choose the categories to classify the data. In many instances, the problem at hand will suggest the classification scheme. For instance, in the coin-tossing example, there are only two classes: heads and tails. If we are classifying students by gender, again there are only two classes: male and female. If we are classifying a company's size based on its market capitalization, we could have three classes: small cap, mid cap, and large cap. For qualitative data, it would be unusual if a reasonable set of categories is *not* relatively obvious.

After the categories have been chosen, count the items belonging to each class in order to construct the frequency distribution.

Example 3.1

Apple Inc. introduced its third generation iPad in March of 2012. After months of anticipation, owners were thrilled with their new iPads. Two of the main features that separated the 3rd generation iPad from its predecessors were an improved camera and a higher resolution display. In spite of the excitement of the launch, there were still things that owners did not like about the new device. The following table contains the responses from a survey of 30 new iPad owners when asked what they dislike about the iPad.

Table 3.3 – Survey Responses		
Cost of Device	Size/Weight	Excessive Heat from Device
Battery Life Too Short	Size/Weight	Cost of Device
Cost of Device	Battery Life Too Short	Excessive Heat from Device
Cost of Wireless Data Plan	Amount of Flash Memory Storage	Size/Weight
Cost of Device	Battery Life Too Short	Cost of Device
Amount of Flash Memory Storage	Cost of Device	Integration with Other Devices
Amount of Flash Memory Storage	Battery Life Too Short	Integration with Other Devices
Cost of Wireless Data Plan	Cost of Device	Excessive Heat from Device
Amount of Flash Memory Storage	Cost of Wireless Data Plan	Cost of Device
Cost of Device	Battery Life Too Short	Cost of Wireless Data Plan

Create a frequency distribution using the data in Table 3.3 to summarize what iPad owners dislike about their devices.

Solution

Examining the table of responses, one can see that there are seven responses (or categories) that were given by the 30 new iPad owners. Summarizing the responses in a table (by counting the number of times that each reason was given), we can create a frequency distribution.

Table 3.4 – Frequency Distribution of Survey Responses	
Response	**Frequency**
Cost of Device	9
Battery Life Too Short	5
Cost of Wireless Data Plan	4
Amount of Flash Memory Storage	4
Size/Weight	3
Excessive Heat from Device	3
Integration with Other Devices	2

The frequency distribution makes it easy to see the greatest concern for iPad owners. From the table, it is evident that in spite of the anticipation and the rush for many consumers to purchase the new iPad, the cost of the device was the number one reason that new owners disliked it. We can also see that iPad integration with other devices is not much of a concern for new customers. These conclusions would have been much more difficult to make if we considered the raw data alone without consolidating the responses into a frequency distribution.

Example 3.2

A Model to Determine Virtual Security Practices

Our increased reliance on digital information and our expansive use of the Internet for a steadily rising number of tasks requires that more emphasis be placed on digital information security. The importance of securing digital information is apparent but the success in persuading individual users to adopt and utilize tools to improve security has been arguably more difficult. A recent study conducted by Information Security Associates wanted to determine the extent to which students are aware of security measures available to protect personal information on their computers and the importance of such measures. One of the survey items was, "I am aware that there are measures that I can take to help protect my personal information on my personal computer." The frequency of each response category is shown in Table 3.5. The summary tables are much more informative than looking at more than a thousand observations for each question (note that the number of observations for each question is different because every student did not answer every question on the survey). The frequency distribution for the second item, "I am aware that I can reduce exposure to system compromise by restricting who uses my personal computer," is given in Table 3.6.

Table 3.5 – Frequency Distribution of Response	
"I am aware that there are measures that I can take to help protect my personal information on my personal computer."	
Strongly Agree	419
Slightly Agree	327
Neutral	250
Slightly Disagree	124
Strongly Disagree	85

Table 3.6 – Frequency Distribution of Response	
"I am aware that I can reduce exposure to system compromise by restricting who uses my personal computer."	
Strongly Agree	520
Slightly Agree	435
Neutral	310
Slightly Disagree	115
Strongly Disagree	90

In Table 3.7 and Table 3.8, relative frequency distributions are calculated. In these tables, the frequencies are converted into percentages. These are defined as **relative frequencies**, i.e. the proportion relative to the total. They are valuable in assessing the data quickly in terms we use frequently. (See Section 3.4 for an additional discussion of relative frequency distributions.)

Table 3.7 – Relative Frequency Distribution of Response	
"I am aware that there are measures that I can take to help protect my personal information on my personal computer."	
Strongly Agree	35%
Slightly Agree	27%
Neutral	21%
Slightly Disagree	10%
Strongly Disagree	7%

Table 3.8 – Relative Frequency Distribution of Response	
"I am aware that I can reduce exposure to system compromise by restricting who uses my personal computer."	
Strongly Agree	35%
Slightly Agree	30%
Neutral	21%
Slightly Disagree	8%
Strongly Disagree	6%

As one can see in Tables 3.5 through 3.8, the majority of students are aware that there are measures that they can take to protect the information on their personal computers. Summarizing the qualitative data (via a frequency distribution table or relative frequency distribution table) allows the researcher to make conclusions about the data without having to view each observation.

3.1   Exercises

Basic Concepts

1. From a comprehension standpoint, what are the advantages of visual images over the written word?

2. Describe two situations in which graphical displays are used in business.

3. Describe the purpose of a frequency distribution.

4. What are the basic questions to ask when examining the structure of a data set?

5. What are the two steps to constructing a frequency distribution?

6. In the construction of a frequency distribution, what are the two requirements that the classification categories must meet?

Exercises

7. In order to help him decide when and where to advertise, a local repairman decided to pull his invoices for the month of June and tally what types of machines he had worked on. There were forty-eight items repaired that month.

Office copier	Washing machine	Air conditioner
Air conditioner	Fan	Lawn mower
Lawn mower	Air conditioner	Fan
DVD Player	Fan	Air conditioner
Air conditioner	Lawn mower	Washing machine
Lawn mower	Air conditioner	Stereo
Exercise bike	DVD Player	Air conditioner
Air conditioner	Lawn mower	Lawn mower
Lawn mower	Air conditioner	Fan
Radio	Washing machine	Air conditioner
Air conditioner	Radio	Stereo
Fan	Air conditioner	Lawn mower
Washing machine	Lawn mower	Air conditioner
Air conditioner	Fan	Fan
Lawn mower	Air conditioner	DVD player
Washing machine	Washing machine	Air conditioner

a. What level of measurement does the data possess?

b. Are the data qualitative or qualitative?

c. Construct a frequency distribution for the data. Any machine types worked on three or less times are classified as miscellaneous.

8. Parkinsonism is an affliction of the aged and is frequently caused by Parkinson's disease, Alzheimer's disease, or other illnesses. The results from a recent study on Parkinsonism were reported in "Prevalence of Parkinsonian Signs and Associated Mortality in a Community Population of Older People," *New England Journal of Medicine*. A sample of 467 people, all 65 years of age or older, was selected from East Boston, Massachusetts. Each person was clinically evaluated and various signs of Parkinsonism, if any, were noted. The following table is a frequency distribution for some of the signs of Parkinsonism.

Table for Exercise 8 – Signs of Parkinsonism	
Sign	**Frequency**
Reduced arm swing	210
Prolonged turning	153
Right leg rigidity	141
Left leg rigidity	154
Slow finger taps	197
Shuffling gait	83

a. What level of measurement does the data possess?

b. What percent of the sample suffered from left leg rigidity? Round your answer to two decimal places.

c. Add up the frequencies. Why does the sum of the frequencies exceed the total sample size of 467?

d. Suppose 30 people suffer from both left leg rigidity and right leg rigidity. How many people in the sample suffer from rigidity in at least one of their legs?

9. A small commuter airline in the West keeps records of complaints received from its customers. Complaints for March and July are listed in the following table.

Table for Exercise 9 – Customer Complaints		
Type of Complaint	**March**	**July**
Tickets cost too much	11	15
Stewardess did not provide blankets	8	3
Schedules not convenient	12	17
Plane often late	17	16
Seats too stiff	3	3
Airplane too hot	6	20
Airplane too cold	8	5
Poor reservation system	5	5
Plane interior looks shabby	5	6

a. Classify the items by the following categories: comfort, price, service, and schedule, and develop a qualitative frequency distribution.

b. Classify the items by the following categories: plane, personnel, building/equipment, and other, and develop a qualitative frequency distribution.

c. Would another person necessarily assign the same items to the same categories as you have? Discuss the implications of this when reviewing data collected and distributed by someone else for open answer questions.

d. Do the categories chosen in parts **a.** and **b.** meet the requirement that categories be mutually exclusive and exhaustive? Discuss.

3.2   The Value of Graphs

Graphical analysis is a trade-off. We lose sight of the individual observations (the raw data). In return, we are able to see a representation of the totality of observations. The trade is almost always beneficial since a well-designed graph gives our visual processing system the kind of image it processes best, a picture.

Because a set of data can be graphically represented in many different ways, selecting and creating graphical displays requires a certain amount of artistic judgment. Fortunately, the development of graphics software has made the creation of sophisticated graphs quite easy.

Several types of graphs and tabular displays will be discussed in this chapter: bar charts (2-D and 3-D), pie charts, line charts, stem-and-leaf diagrams, several types of frequency distributions, and histograms. An examination of publications such as *Time, USA Today, The Wall Street Journal, Scientific American,* or *Forbes* provides convincing evidence of the frequent and beneficial usage of these graphical display techniques.

3.2   Exercises

Basic Concepts

1. What are some benefits of graphing?

2. What is the major disadvantage of graphing?

3.3   Displaying Qualitative Data Graphically

Bar charts, stacked bar charts, 3-D bar charts, and pie charts are effective, visually appealing methods of graphically displaying qualitative data. These tools will help us visually describe the distribution of the data.

Bar Charts

Bar charts are often used to illustrate a frequency distribution for qualitative data.

> **Definition**
>
> The **bar chart** is a simple graphical display in which the length of each bar corresponds to the number of observations in a category.

Bar charts are valuable as presentation tools and are especially effective at reinforcing differences in magnitudes, since they permit the visual comparison of data by displaying the magnitude of each category by a vertical or horizontal bar. Figure 3.1 is a bar chart constructed from majors of the students in a business statistics course.

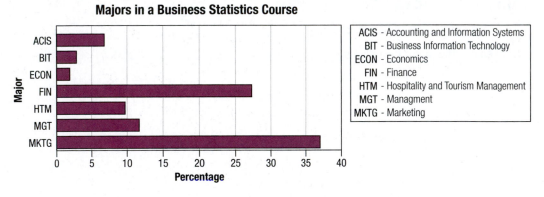

Figure 3.1

The Aesthetics of Bar Chart Construction

Bar chart construction requires numerous layout decisions such as size, use of color, and labeling locations. These decisions are frequently made on a trial-and-error basis. However, certain conventions have been developed that improve the quality and effectiveness of charts. They are presented below as suggestions, not rules. Actually, several of the points are general in nature and would serve as useful guidelines in the construction of any graph.

- Bar charts can be constructed horizontally or vertically. Customarily, horizontal orientation is used for categories that are descriptively labeled, and vertical (or columnar) orientation is used for categories that are numerical. It is important to remember that this idea is only a suggestion. If you believe that a vertical bar chart is more appealing, use it.

- If the categories have some associated order, they maintain that order in the bar chart. Otherwise, the categories may be listed alphabetically, in either ascending or descending order, or in some other pattern related to the nature of the data.

- Miscellaneous or "other" categories should be listed at the bottom of the chart (if oriented horizontally) or at the far right (if oriented vertically).

- The difference in bar length is the principal visual feature in comparing differences in category amounts. Consequently, scales for the axes should be chosen that will most effectively allow for the desired comparisons. **Unless there is a good reason, the axis used to measure the bars should start at zero. Otherwise the axis can be stretched to exaggerate differences in the bar lengths.** For example, suppose the data in the following table were plotted.

Table 3.9 – Sales Performance	
Salesperson	Total Sales (Thousands of Dollars)
Susan	187
William	201
Beth	207
Rob	193

Figure 3.2

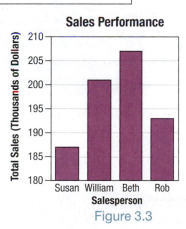

Figure 3.3

Figures 3.2 and 3.3 are plots of the same data (Table 3.9). What a difference axis selection can make on perception! Figure 3.3 begins the y-axis at 180 instead of zero. If you want to emphasize similarity, use Figure 3.2 to do the job. If you want to emphasize differences, use Figure 3.3. However, it is difficult to imagine any legitimate reason for using Figure 3.3 to represent the data. Axis stretching is often employed to mislead. **When you see an axis that does not start at zero, you should be a bit skeptical as to the conclusions the author intends for you to make.**

- Bar widths should be chosen that are visually pleasing and should not be allowed to vary within a particular chart.

- Appropriate shading, crosshatching, and coloring of the bars can help in presenting data. Many spreadsheet programs incorporate sophisticated graphing programs which make changes in shading, color, and crosshatching patterns extremely easy.

- The spacing between bars can dramatically affect the perception of the graph. Spacing should be set at approximately one-half the width of a bar. This, however, is not a rigid rule. Artistic judgment is needed.

- Gridlines extended into the body of the chart are often useful and may be included if deemed helpful. Study Figure 3.4 and Figure 3.5. Both bar charts illustrate the same data. Notice that the readability of the graph in Figure 3.5 is improved by adding gridlines.

Figure 3.4

Figure 3.5

- Labels should be provided for each bar (category) and for each axis.

- Notes on sources of data or other footnotes should be given below the chart.

Bar charts used to be flat and simple. Now, computer graphics packages offer three-dimensional, stacked, side-by-side, colored, and other variations of bars. These capabilities offer the user the chance to create spectacular, multi-dimensional, eye-catching graphics.

Rotating the axes on a bar chart can create a very different graphical perspective. Figures 3.6 and 3.7 are identical graphs with a slight difference in perspective. (Note also the change in perspective by making the graphs 3-D. These are the same data that are graphed in Figures 3.4 and 3.5.) If you are using a large number of graphics in a document, even 3-D graphs can be burdensome. It would be more visually interesting to change the perspective of the graphic from time to time.   Making the changes from 2-D to 3-D and performing the rotation are simple operations in most graphics programs.

Figure 3.6

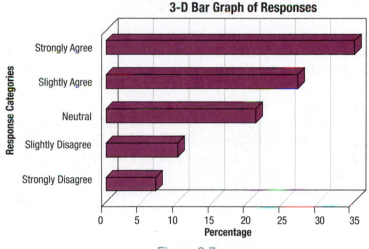

Figure 3.7

One of the themes of this chapter is that a graph can be an analytical device and a presentation tool. Simple graphics are fine for analytical purposes, but if you are trying to make a point, then a higher standard is required to create visual impact. Designing effective graphics is not just about making things look better; graphics can help the reader comprehend information. Using graphics well can emphasize meaning and organize content. It pays to know your audience's likes and dislikes in viewing graphics so that the maximum impact can be achieved.

Stacked Bar Charts

Stacked bar charts are an interesting variation on the standard bar chart. The number of medals (gold, silver, bronze) won during the 2010 Winter Olympics for selected countries is given in Figure 3.8.

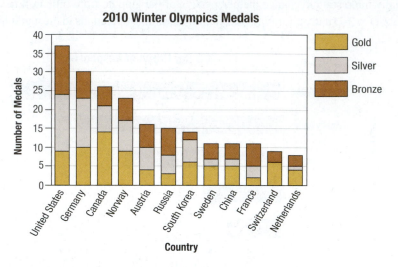

Figure 3.8

Using the stacked bar chart enables the reader to compare the total number of medals for each country as well as observe the number of each type of medal won.

Without the stacked bar chart, the reader would have to view either three different charts tallying medal counts for the 2010 Winter Olympics or a much "busier" chart plotting each medal on the horizontal axis above each country.

Figure 3.9 is another example of stacked bar chart. The chart displays the number of grandchildren by age group living with their grandparents based on which parent is in the household. From this chart we can see that the majority of grandchildren living with their grandparents live in the same household with their mother only. We can also see that it is fairly rare that grandchildren live in the same household with their grandparents and their father only. Stacked bar charts are useful when there are three components to the data (in this case the percentage of children, the age group, and the parents in the household).

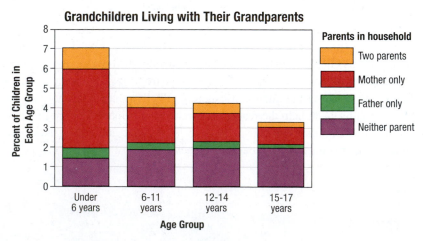

Figure 3.9

3-D Bar Charts

Another interesting way of looking at the Olympic medal data is to plot a three-dimensional bar chart (see Figure 3.10). Using this graph, the totals for each type of medal are graphed for each country.

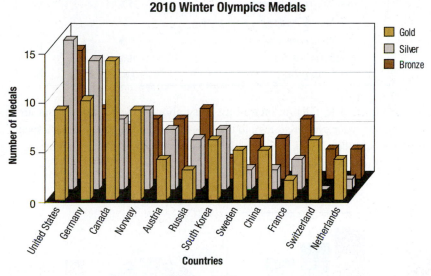

Figure 3.10

When constructing a 3-D graph there are a number of perspective issues that affect the visual quality of the graph. For example, in Figure 3.10, we are barely able to see the number of bronze medals won by Germany, Canada, or South Korea. All the major graphics programs permit you to spin and tilt the graph until you find a perspective you like, and that allows you to see all of the categories.

Pie Charts

We have just seen how bar charts are used as a means of expressing frequency distributions. Pie charts can perform the same function.

The circle represents the total "pie" available, and the slices are proportional to the amount in each category. Each slice of the pie represents the proportion of total observations belonging to the category. One of the advantages of the pie chart is the ability to easily compare the total in each of the classifications to the total number of observations.

One common use of pie charts is to display how some set of assets is spent. The biggest asset pie in the world is the budget of the United States government. In 2009, government outlays were in the neighborhood of 3.52 trillion dollars. If the budget is split into six categories, the percentages spent in 2009 in each category are provided in Table 3.10. While the table provides the information, it is not visually interesting. A pie chart will enable you to improve the presentation of the information in Table 3.10.

| Table 3.10 – Percentage Spent by the Federal Government in 2009 ||
Category	Percentage Spent
Social Security	20%
Non-Defense Discretionary	12%
National Defense	23%
Other Entitlements	21%
Medicare and Medicaid	19%
Net Interest	5%

The pie chart in Figure 3.11 tells an interesting story about how our tax dollars are spent. In a glance at the pie chart, your eyes are drawn to the biggest slice of the pie, the 23% spent on National Defense. If you would like to look at current information on how government monies are spent go to www.whitehouse.gov/omb/budget.

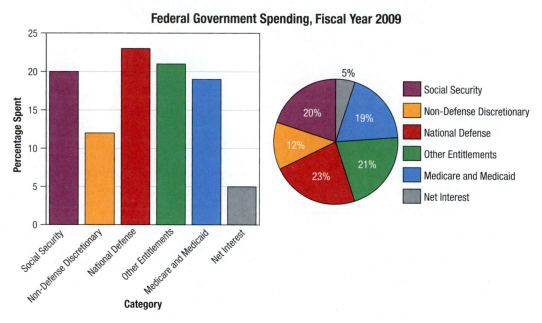

Figure 3.11

A three-dimensional version of the pie chart (see Figure 3.12) adds some spice to any presentation. If you are working with a computer application that can perform graphics, changing the image from a two-dimensional pie chart to a three-dimensional chart takes only a few mouse clicks. A 3-D pie chart showing where government revenues originate is given in Figure 3.12.

In order to determine from the pie chart how much was spent in a particular category, multiply the total amount by the proportion given for that category in the pie chart. For example, to find the dollar amount of government receipts contributed by individual income taxes, multiply the total amount of government receipts ($2105 billion) by the proportion contributed by individual income taxes (0.43). This means that $2105 \cdot 0.43 = \$905.15$ billion of all government receipts come from individual income taxes.

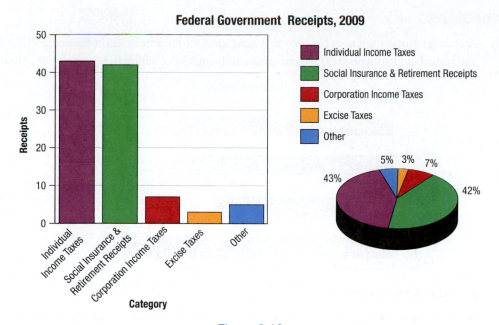

To learn more how to make these charts in Microsoft Excel, see Appendix B.

Figure 3.12

3.3   Exercises

Basic Concepts

1. Describe the types of data that a bar chart would be useful in displaying.

2. Where should miscellaneous categories be displayed in a bar chart?

3. Explain how axis scales on bar charts can be misleading.

4. What is a stacked bar chart?

5. Why would a stacked bar chart be preferred over a normal bar chart?

6. What is one disadvantage of using a 3-D chart?

7. What is a pie chart?

8. What is the main advantage of using a pie chart?

Exercises

9. A consumer magazine uses bar charts to compare four popular brands of automobiles. This particular bar chart represents a comparison of the miles per gallon (mpg) for the four brands.

Vehicle Mileage

a. What is wrong with this picture?

b. Evaluate the bar chart using the guidelines suggested in the section on the aesthetics of bar chart construction.

10. The following bar chart presents the median income of U.S. employees by education level and gender. Evaluate the bar chart using the guidelines suggested in the section on the aesthetics of bar chart construction.

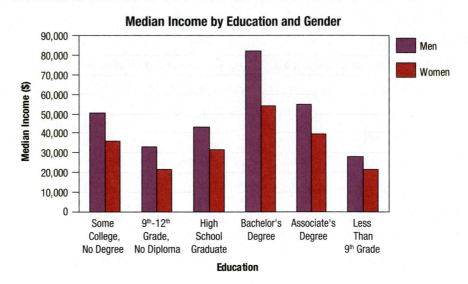

Median Income by Education and Gender

11. Consider the following data regarding the number of wildfires in the U.S. categorized by the size class in acres and the cause of the fire.

Table for Exercise 11 – Number of Wildfires in the U.S.		
Size Class (Acres)	Lightning-Caused	Person-Caused
0.25 or less	4637	2367
0.26 – 9	1940	1904
10 – 99	219	571
100 – 299	44	103
300 – 999	26	52
1000 – 4999	43	17
5000 +	21	9
Total	6930	5023

   a. Construct a bar chart for the number of wildfires caused by lightning.

   b. Construct a bar chart for the number of wildfires caused by people.

   c. Construct a stacked bar chart for the number of wildfires caused by lightning and the number of wildfires caused by people.

   d. Construct a pie chart for the number of wildfires caused by lightning.

   e. Construct a pie chart for the number of wildfires caused by people.

   f. What did you learn from the charts created in parts **a.** through **e.**?

12. Consider the following data regarding the average spending on healthcare per person for various countries.

Table for Exercise 12 – Healthcare Costs Around the World per Capita, 2009	
Country	Average Cost ($)
Australia	2886
Canada	2998
Denmark	2743
Finland	2104
France	3048
Germany	2983
Iceland	3159
Ireland	2455
Japan	2249
Sweden	2745
Switzerland	3847
United Kingdom	2317
United States	5711

Source: www.creditloan.com

   a. Construct a bar chart for the average healthcare cost per person for the various countries.

   b. What did you learn from the chart?

13. Consider the following data regarding professions with high projected growth rates for the years 2008 through 2018.

Table for Exercise 13 – Occupation Growth Rates	
Occupation	**Projected Increase 2008 – 2018**
Biomedical engineers	72%
Network systems and data communications analysts	53%
Home health aides	50%
Personal and home care aides	46%
Financial examiners	41%
Medical scientists, except epidemiologists	40%
Physician assistants	39%
Skin care specialists	38%

Source: Bureau of Labor Statistics

a. Construct a bar chart for the projected growth rates of the various occupations.

b. What did you learn from the chart?

14. Consider the following data regarding the methods which consumers use to pay for items purchased in a particular store.

Table for Exercise 14 – Payment Methods	
Method of Payment	**Relative Frequency**
Cash	81.1%
Checks	7.6%
General-Purpose Credit Cards	5.5%
Proprietary Credit Cards	5.3%
Debit Cards	0.5%

a. Construct a bar chart for the relative frequencies of the various methods of payment.

b. Construct a pie chart for the relative frequencies of the various methods of payment.

c. Comment on any information about the relative frequencies of the various methods which you were able to ascertain by examining the charts.

| 3.4 | **Constructing Frequency Distributions for Quantitative Data** |

When data are qualitative (nominal or ordinal), selecting the categories for display is relatively easy. However, selecting categories for quantitative data is not particularly obvious, since using different schemes can dramatically affect the perception of the data. The fundamental decision in constructing the frequency distribution is selecting the number of classes.

Example 3.3

A data set containing the revenues (in millions of dollars) of the top 100 companies on the Fortune 500 is listed in Table 3.11.

Table 3.11 – Revenues of the Top 100 Companies on the Fortune 500 (Millions of Dollars)				
116	408	31	26	25
29	77	99	118	35
72	66	45	36	71
37	32	49	53	100
25	105	45	31	30
36	25	38	44	96
45	115	27	50	27
41	43	87	80	99
140	107	61	65	31
47	28	108	45	44
25	164	62	157	41
26	70	1	35	109
112	26	63	35	58
103	32	285	30	31
69	30	33	27	65
36	30	32	35	31
37	32	25	27	25
29	32	25	27	25
45	60	28	68	26
25	150	40	53	34

Of course, looking at the raw numbers is not very revealing. Given the number of observations, it is difficult to get a good idea of the measurements. Instead of trying to examine or analyze all 100 values, a frequency distribution is given in Table 3.12. Examining the frequency distribution, an analyst can look at a table divided into ten categories and ten frequencies, thus reducing its complexity.

3

Table 3.12 – Frequency Distribution	
Revenue (Millions of Dollars)	Number of Companies
0 to 40	50
41 to 81	30
82 to 122	14
123 to 163	3
164 to 204	1
205 to 245	0
246 to 286	1
287 to 327	0
328 to 368	0
369 to 409	1

With the frequency distribution, we are able to see the broader structure of the data. It is now easy to see that the overwhelming majority of revenues are between $0 and $122 million. Further, revenues of $164 million or more are uncommon. These conclusions would be considerably more difficult to establish without the organization that the frequency distribution provides.

The purpose of the frequency distribution is to condense the set of data into a meaningful summary form. There are only two steps in the construction of a frequency distribution.

Step 1: Choose the number of classifications.

Step 2: Count the number in each class.

Selecting the Number of Classes

Choosing the number of classes is arbitrary and should depend on the amount of data available. In general, the more observations one has in a data set, the more classes or intervals that can be used in the frequency table or histogram. Consequently, only very general guidelines exist. Generally, fewer than four classes would be too much compression of the data and greater than 20 classes provides too little summary information. To start, a good rule of thumb for the number of classes to create is to round $\sqrt{n}$ or $\sqrt[3]{2n}$ to the nearest whole number.

Determining Class Width

After deciding on the desired number of classes, the next step is to specify the width of each class. Using classes that are of equal widths makes the histogram easier to interpret. (We will discuss constructing histograms in the next section.) There is really no perfect formula for class width that will work for every data set. However, a good starting point for determining class width is to divide the difference between the largest and the smallest observations by the number of classes.

$$\text{Class Width} = \frac{\text{Largest Value} - \text{Smallest Value}}{\text{Number of Classes}}$$

Suppose we wanted to create a frequency distribution from the revenue data in Example 3.3. If there are to be 10 classes, determine a class width.

$$\text{Class Width} = \frac{\text{Largest Value} - \text{Smallest Value}}{\text{Number of Classes}} = \frac{408 - 1}{10} = \frac{407}{10} = 40.7$$

Class endpoints with fractional values will make the graph slightly harder to digest. If possible, try to keep the width to an integer value (i.e. round up to the next largest integer). If the calculated class width is 40.7, you might try a class width of 41. An interval width of 41 is used for the revenue data in Example 3.3.

Generally, class widths should be equal. We want the lower limit of the first interval to be 0 or a multiple of the class width. In the example, we have selected the starting point as 0 because our smallest value is 1, and obviously we want the first interval to contain the smallest observation. Adding the class width to the lower class limit creates the next interval. That is, we want the lower limit of each subsequent interval to be a multiple of the class width. If this pattern is followed, the result will be intervals that will not overlap and will capture all the data (i.e. the classes will be mutually exclusive and exhaustive).

Class boundaries are used to determine in which interval an observation falls. To find the class boundaries, subtract 0.5 from the lower class limit and add 0.5 to the upper class limit. An observation falls in an interval if it is greater than the lower class boundary and less than or equal to the upper class boundary.

The frequency distribution given in Table 3.12 is just one way of organizing the data. There are three other distributions that can be calculated from the frequency distribution: the relative frequency, cumulative frequency, and cumulative relative frequency distributions. Each gives a slightly different perspective on the data.

Relative Frequency Distribution

Relative frequency represents the proportion of the total observations in a given class. The **relative frequency distribution** enables the reader to view the number in each category in relation to the total number of observations. Relative frequency is a standardizing technique. Converting the frequency in each class to a proportion in each class enables us to compare data sets with different numbers of observations.

Formula

Relative Frequency

The **relative frequency** of any class is the number of observations in the class divided by the total number of observations.

$$\text{Relative Frequency} = \frac{\text{Number in Class}}{\text{Total Number of Observations}}$$

A relative frequency distribution of the revenue data from Example 3.3 is given in Table 3.13. Notice that the relative frequencies are obtained by dividing the frequencies in Table 3.12 by the total number of observations, which is 100.

Table 3.13 – Relative Frequency Distribution of Revenue Data	
Revenue (Millions of Dollars)	Relative Frequency
0 to 40	$\frac{50}{100} = 0.50$
41 to 81	$\frac{30}{100} = 0.30$
82 to 122	$\frac{14}{100} = 0.14$
123 to 163	$\frac{3}{100} = 0.03$
164 to 204	$\frac{1}{100} = 0.01$
205 to 245	$\frac{0}{100} = 0.00$
246 to 286	$\frac{1}{100} = 0.01$
287 to 327	$\frac{0}{100} = 0.00$
328 to 368	$\frac{0}{100} = 0.00$
369 to 409	$\frac{1}{100} = 0.01$

Cumulative Frequency Distribution

The cumulative frequency distribution gives the reader an opportunity to look at any category and determine immediately the number of observations that belong to a particular category and all categories below it.

Definition

The **cumulative frequency** is the sum of the frequency of a particular class and all preceding classes.

Table 3.14 – Cumulative Frequency Distribution of Revenue Data		
Revenue (Millions of Dollars)	Frequency	Cumulative Frequency
0 to 40	50	50
41 to 81	30	80
82 to 122	14	94
123 to 163	3	97
164 to 204	1	98
205 to 245	0	98
246 to 286	1	99
287 to 327	0	99
328 to 368	0	99
369 to 409	1	100

In this example, the reader can easily see in Table 3.14 that 97 out of 100 revenues are less than or equal to $163 million.

Cumulative Relative Frequency

To obtain the cumulative relative frequency, add the relative frequencies of all preceding classes to the relative frequency of the current class.

Definition

The **cumulative relative frequency** is the proportion of observations in a particular class and all preceding classes.

Table 3.15 – Cumulative Relative Frequency Distribution of Revenue Data

Revenue (Millions of Dollars)	Frequency	Relative Frequency	Cumulative Relative Frequency
0 to 40	50	0.50	0.50
41 to 81	30	0.30	0.80
82 to 122	14	0.14	0.94
123 to 163	3	0.03	0.97
164 to 204	1	0.01	0.98
205 to 245	0	0.00	0.98
246 to 286	1	0.01	0.99
287 to 327	0	0.00	0.99
328 to 368	0	0.00	0.99
369 to 409	1	0.01	1.00

From the cumulative relative frequency in Table 3.15, it is easy to see that 97% of the revenues are less than or equal to $163 million.

3.4   Exercises

Basic Concepts

1. What are the fundamental decisions in constructing frequency distributions for quantitative data?

2. Describe the general guidelines for selecting the number of classes for a quantitative frequency distribution.

3. What is a good starting point for determining the class width?

4. What is a relative frequency distribution? How do you calculate relative frequencies from raw frequencies?

5. What is a cumulative frequency distribution?

6. What is a cumulative relative frequency distribution?

Exercises

7. A business magazine was conducting a study into the amount of travel required for mid-level managers across the U.S. Seventy-five managers were surveyed for the number of days they spent traveling each year.

Table for Exercise 7 – Mid-Level Manager Travel	
Days Traveling	**Frequency**
0 – 6	15
7 – 13	21
14 – 20	27
21 – 27	9
28 – 34	2
35 and above	1

a. Construct a relative frequency distribution.

b. Construct a cumulative frequency distribution.

8. The closing prices (in pence) for selected stocks trading on the London Stock Exchange were as follows. Construct a frequency distribution for the stock prices.

Table for Exercise 8 – Closing Prices	
Stock	**Closing Price (Pence)**
Allied Lyons	439
Babcock	208
Barclays Bank	543
Bass Ltd	992
British GE	238
Cadbury Sch	257
Guinness	379
Hanson Trust	169
Lucas Indus	655
Reed Int'l	467
STC	318
Tate & Lyle	833
Thorm EMI	741
Utd. Biscuit	326

9. Every year, the average temperatures of 100 selected U.S. cities are published by the National Oceanic and Atmospheric Administration. The average temperature (°F) for the month of October for 15 randomly selected cities from the list of 100 are listed in the following table.

Table for Exercise 9 – Average Temperatures (°F)				
68.5	50.9	67.5	57.5	56.0
47.1	50.1	65.8	51.5	49.5
75.2	56.0	62.3	53.0	46.1

a. Construct a frequency distribution for the average temperatures for the month of October.

b. Construct a relative frequency distribution for the average temperatures for the month of October.

c. Construct a cumulative frequency distribution for the average temperatures for the month of October.

10. Consider the assets (in billions of dollars) of the 10 largest life insurance companies listed in the following table.

Table for Exercise 10 – Assets (Billions of Dollars)				
148.4	110.8	55.6	52.4	50.4
42.7	41.7	36.2	35.7	35.7

a. Construct a frequency distribution for the assets (in billions of dollars) of the 10 largest life insurance companies.

b. Construct a relative frequency distribution for the assets (in billions of dollars) of the 10 largest life insurance companies.

c. Construct a cumulative frequency distribution for the assets (in billions of dollars) of the 10 largest life insurance companies.

3.5   Histograms

A histogram is a common graphical method that reveals the distribution of the data. Histograms are often constructed based on frequency distributions of quantitative data. Histograms look similar to bar graphs, but are used to analyze quantitative data rather than qualitative data.

Definition

A **histogram** is a bar graph of a frequency or relative frequency distribution in which the height of each bar corresponds to the frequency or relative frequency of each class.

Each of the classes in the frequency distribution is represented by a vertical bar whose height is proportional to the frequency of the interval. The horizontal boundaries of each vertical bar correspond to the class boundaries. Once the frequency distribution has been calculated, all the information necessary for plotting a histogram is available. In Figure 3.13, the histogram is created from the frequency distribution of the revenue data in Table 3.12.

Figure 3.13

Histograms are one of the more frequently used statistical tools. A histogram is not only easy to interpret; it also reveals a great deal about the structure of the data. By examining the histogram, one can determine the shape of the distribution of the data. That is, the data could be symmetric or skewed (positively or negatively). A **symmetric distribution** is one in which if a line were drawn down the middle of the distribution, the two sides would mirror each other. A **skewed distribution** is represented by a group of observations that is not equal on both sides (sometimes called asymmetric). One can also get an idea about the modality of a distribution by examining the histogram. The mode of a set of observations is the value that occurs with the greatest frequency. (We will discuss the mode in further detail in Section 4.1.)

You can quickly see from Figure 3.13 that most of the revenues are in the first and second categories (0 to 40 and 41 to 81). The center of the data appears to be near 41 and the distribution of the data appears to be skewed to the right.

You will see many histograms throughout this text. When we look at a histogram, what features are important?

1.   Is the distribution symmetric or skewed (one tail is longer than the other)?

skewed right   skewed left

2.   Is it bell-shaped?

3.   Does the distribution have several peaks or modes?

4.   Where is the center of the distribution?

5.   Are there outliers (data values that are very different from the others)?

There are many statistical software programs that can be used to construct histograms. The Discovering Technology section at the end of this chapter explains how to create a histogram from raw data using Microsoft Excel and Minitab. In spite of the popularity of using computer software packages to summarize data, there may be times when you need to manually create a histogram. Although there isn't a unified method for constructing a histogram, some steps are provided in the following box that can help you create one.

Procedure

Constructing a Histogram

1. Determine the number of classes (or intervals) for the histogram. There isn't a rule of thumb for the number of classes to use, and most of the time the number of classes is determined by trial and error. However, a good starting point could be to use $\sqrt{n}$ or $\sqrt[3]{2n}$.

2. Determine the largest and the smallest observations. Depending on the size of the data set, it may be a good idea to sort the observations from smallest to largest. This will also make it easier later when trying to determine the interval in which an observation falls.

3. Calculate the difference between the smallest and largest observations by subtracting the smallest observation from the largest observation. (Note that this measure is called the range, which will be discussed further in Chapter 4.)

4. Calculate the class width, using the following formula.

$$\text{Class Width} = \frac{\text{Largest Value} - \text{Smallest Value}}{\text{Number of Classes}}$$

5. Calculate the limits of the class intervals. The rule of thumb for determining the class intervals is to choose the lower class limit of the first class interval so that it is either 0 or a multiple of the class width, so that the smallest observation falls in the first class interval. Also note that the lower class limit of all subsequent class intervals will be a multiple of the class width.

6. Calculate the class boundaries. To find the class boundaries, we subtract 0.5 from the lower class limit and add 0.5 to the upper class limit. The class boundaries will be used to determine in which interval an observation falls. Lastly, to determine if an observation falls in an interval, the rule is as follows.

$$\text{Lower Class Boundary} < x_i \leq \text{Upper Class Boundary}$$

    where x_i is observation i.

7. Tally each observation to determine in which interval (based on class boundaries) it should fall. This will give you the class frequency $\left(\text{denoted by} f_i\right)$, which is the number of observations in each interval.

8. All of the information for a histogram can be displayed in a frequency distribution table or a relative frequency distribution table, as can be seen in the following example.

9. Construct the histogram with the class boundaries on the horizontal axis and the frequency (or relative frequency) on the vertical axis. The heights of the bars will represent the class frequencies (or relative frequencies).

Example 3.4

Suppose you are given a random sample of 28 applicants who took an examination designed to measure their aptitude for a job in sales. The following scores (measured in percentages) were obtained. Note that the observations are already sorted from smallest to largest.

Table 3.16 – Aptitude Scores (%)						
38	49	53	56	58	58	60
62	66	67	69	69	71	74
75	76	77	77	77	78	78
81	82	83	84	87	88	88

Construct a relative frequency distribution table for the test scores using 6 class intervals, and construct a histogram.

Solution

To solve this problem, we will work through the 9-step procedure for constructing a histogram.

1. First, we need to determine the number of classes for the histogram. We are given that we want to construct a histogram using 6 intervals, so there will be 6 classes.

2. The second step is to identify the largest and the smallest observations in the data set. This is relatively straightforward in this case since the data are ordered from smallest to largest. From the data in Table 13.16, the smallest test score is 38 and the largest score is 88.

3. Next, we need to calculate the difference between the smallest and the largest observations. Subtracting the smallest score from the largest score, we have the following.

$$\text{Largest Score} - \text{Smallest Score} = 88 - 38 = 50$$

4. Now we need to calculate the class width. Given that the difference between the largest and smallest scores is 50, we have the following.

$$\text{Class Width} = \frac{\text{Largest Score} - \text{Smallest Score}}{\text{Number of Classes}} = \frac{50}{6} \approx 8.33$$

For simplicity, if the class width is not an integer we round up to the next largest integer. Thus, we will use a class width of 9 when constructing the histogram.

5. The next step is to construct the class intervals. Note that we want the lower limit of the first class interval to be either 0 or a multiple of the class width, and the first interval must include the smallest observation. The lower limit of the first class interval will not be 0 in this case, since the smallest score, 38, is much larger than the class width. Therefore, it must be a multiple of the class width, which is 9. Since the smallest observation is 38, the lower limit of the first class interval is 36, which is the largest multiple of 9 that is less than the smallest value in the data. So, the first interval is 36–44, the second is 45–53, and so on. Thus, the class limits are determined to be the following.

Table 3.17 – Class Limits
36–44
45–53
54–62
63–71
72–80
81–89

6. Once we have calculated the class limits, we calculate the class boundaries by subtracting 0.5 from the lower class limit of each interval and by adding 0.5 to the upper class limit of each interval. The class boundaries are given in Table 3.18.

Table 3.18 – Class Boundaries
35.5–44.5
44.5–53.5
53.5–62.5
62.5–71.5
71.5–80.5
80.5–89.5

7. Tally each observation to determine the interval in which each observation falls. Remember that the observation falls in the interval if it is greater than the lower class boundary and less than or equal to the upper class boundary. This gives you the class frequencies, as seen in Table 3.19.

Table 3.19 – Class Frequencies	
Class	**Frequency**
36–44	1
45–53	2
54–62	5
63–71	5
72–80	8
81–89	7

8. The relative frequency distribution table (Table 3.20) provides a summary of the data and allows the histogram to be constructed.

Table 3.20 – Relative Frequency Distribution			
Class Limits	**Class Boundaries**	**Frequency** (f_i)	**Relative Frequency**
36–44	35.5–44.5	1	0.0357
45–53	44.5–53.5	2	0.0714
54–62	53.5–62.5	5	0.1786
63–71	62.5–71.5	5	0.1786
72–80	71.5–80.5	8	0.2857
81–89	80.5–89.5	7	0.2500
	Total	28	1.0000

Keep in mind that the sum of the frequency column should sum to n, the total number of observations. Also, the sum of the relative frequency column should equal to 1. If the sum of the relative frequency column is different from 1, but fairly close, it is likely due to rounding.

9. Finally, construct the histogram.

Figure 3.14

3.5 Exercises

Basic Concepts

1. What is the main characteristic of data that a histogram reveals?

2. Describe the type of data that could be usefully described with a histogram.

3. True or false: A frequency distribution contains all of the information needed to construct a histogram.

4. List the important features to look for when studying a histogram.

Exercises

5. The closing prices (in dollars) for selected stocks trading on the New York Stock Exchange and NASDAQ Exchange were as follows.

Table for Exercise 5 – Closing Prices	
Stock	**Closing Price ($)**
Citigroup (C)	34.70
Pfizer (PFE)	22.34
Herbalife (HLF)	69.72
JP Morgan Chase (JPM)	44.34
Intel (INTC)	28.07
WalMart (WMT)	60.67
Microsoft (MSFT)	31.52
PepsiCo (PEP)	66.17
General Motors (GM)	24.81
Verizon Communications (VZ)	37.66
Southwest Airlines (LUV)	8.31
Sprint Nextel (S)	2.76
Yahoo! Inc (YHOO)	15.07
International Business Machines (IBM)	205.47

   a. Construct a frequency distribution for the closing prices.

   b. Construct a histogram for the closing prices.

6. A sample of 80 laborers is selected from a large city and their annual salaries are determined. The following histogram summarizes the data.

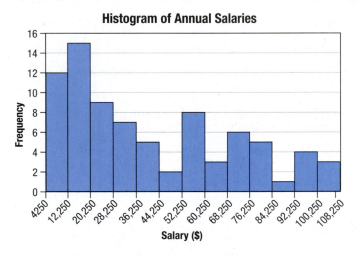

   a. What is the level of measurement of the variable?

   b. How many of the laborers earn more than $36,250?

   c. What percent of the laborers earn at most $28,250?

   d. What percent of the laborers earn more than $84,250?

7. A nutritionist is interested in knowing the percent of calories from fat which Americans consume on a daily basis. To study this, the nutritionist randomly selects 25 Americans and evaluates the percent of calories from fat consumed in a typical day. The results of the study are as follows.

Table for Exercise 7 – Calories from Fat per Day (%)				
34	18	33	25	30
42	40	33	39	40
45	35	45	25	27
23	32	33	47	23
27	32	30	28	36

a. Construct a frequency distribution for the percent of calories from fat.

b. Construct a relative frequency distribution for the percent of calories from fat.

c. Construct a histogram of the relative frequency distribution.

d. Comment on any information about the percent of calories from fat consumed by the participants in the study which you were able to ascertain by examining the distributions and the histogram.

8. Consider the assets (in billions of dollars) of the 10 largest commercial banks listed in the following table.

Table for Exercise 8 – Assets (Billions of Dollars)				
216.9	138.9	115.5	110.3	103.5
98.2	76.4	64.0	53.5	49.0

a. Construct a frequency distribution for the assets (in billions of dollars) of the 10 largest commercial banks.

b. Construct a relative frequency distribution for the assets (in billions of dollars) of the 10 largest commercial banks.

c. Construct a histogram of the relative frequency distribution.

d. Comment on any information about the assets (in billions of dollars) of the 10 largest commercial banks which you were able to ascertain by examining the distributions and the histogram.

3.6   The Stem-and-Leaf Display

The **stem-and-leaf display** is a hybrid graphical method. The display is similar to a histogram, but the data remain visible to the user. Like all graphical displays, the stem-and-leaf display is useful for both ordering and detecting patterns in the data. It is one of the few graphical methods in which the raw data are not lost in the construction of the graph. As the name implies, there is a "stem" to which "leaves" will be attached in some pattern.

Consider the following data: 97, 99, 108, 110, and 111. If we are interested in the variation of the last digit, the stems and leaves are as shown in Table 3.21 and displayed in Figure 3.15. The leaves in this case are the *ones* digits and the stems are the *tens* digits. All of the data values that have common stems are grouped together, and their leaves branch out from the common stem.

Table 3.21 – Data, Stems, and Leaves		
Data	**Stem**	**Leaf**
97	09	7
99	09	9
108	10	8
110	11	0
111	11	1

Stem-and-Leaf Display

Stem	Leaf
09	7 9
10	8
11	0 1

Figure 3.15

If we are interested in the variation of the last two digits, the stems and leaves are shown in Table 3.22 and displayed in Figure 3.16. The leaves in this case are the *last two* digits and the stems are the *hundreds* digits. Again all of the data values that have common stems are grouped together, and their leaves branch out from the common stem.

Table 3.22 – Data, Stems, and Leaves		
Data	**Stem**	**Leaf**
97	0	97
99	0	99
108	1	08
110	1	10
111	1	11

Stem-and-Leaf Display

Stem	Leaf
0	97 99
1	08 10 11

Figure 3.16

Deciding which part to make the stem and which part to make the leaf depends on the focus of the analysis.

Example 3.5

The closing stock prices for Microsoft Corporation from December 1990 through December 2011 are given in Table 3.23. Construct a stem-and-leaf display of the closing stock prices of Microsoft (rounded to the nearest dollar).

Table 3.23 – Closing Prices of Microsoft Corporation (MSFT) December 1990 through December 2011			
Period	**Closing Price (Dollars)**	**Period**	**Closing Price (Dollars)**
December 1990	1	December 2001	33
December 1991	2	December 2002	26
December 1992	3	December 2003	27
December 1993	3	December 2004	27
December 1994	4	December 2005	26
December 1995	5	December 2006	30
December 1996	10	December 2007	36
December 1997	16	December 2008	19
December 1998	35	December 2009	30
December 1999	58	December 2010	30
December 2000	22	December 2011	36

Solution

Using a stem of the *tens* unit will break the data into six classes, which seems reasonable. The stem-and-leaf display is given in Figure 3.17.

Stem-and-Leaf Display (MSFT)

Stem	Leaf
0	1 2 3 3 4 5
1	0 6 9
2	2 6 6 7 7
3	0 0 0 3 5 6 6
4	
5	8

Figure 3.17

From the stem-and-leaf display, we would guess that the average closing price for MSFT is somewhere around $25. When analyzing a stem-and-leaf display, you can guess the average of the data set by choosing the value with an equal number of leaves greater than it and less than it. To be more precise, you can find the median of the data fairly easily using a stem-and-leaf display.

3.6   Exercises

Basic Concepts

1.   Explain why the stem-and-leaf display is sometimes called a "hybrid graphical method."

2.   Identify the advantages of a stem-and-leaf display.

3.   Consider the following data value: 39. What would be the stem and leaf for this value if we identified the stem as the tens digit? What would be the steam and leaf if we identified the stem as the hundreds digit?

4.   When constructing a stem-and-leaf display, how do you determine which part to make the stem and which part to make the leaf?

Exercises

5.   Fifty hospitals in a western state were polled as to their basic daily charges for a semi-private room. The results are listed in the following table, rounded to the nearest dollar.

Table for Exercise 5 – Daily Charges for Semi-Private Rooms (Dollars)									
125	135	148	156	248	215	156	148	135	149
178	156	135	125	214	256	258	265	156	148
123	147	189	199	189	248	215	259	158	235
268	269	158	198	147	258	269	239	288	199
179	179	189	169	258	178	257	249	259	259

a.   What level of measurement does the data possess?

b.   Construct a stem-and-leaf display for the data using the tens digits as the stems.

c.   Comment on the shape of the distribution.

6.   The data in the following table are the toxic emissions (in thousands of tons) for 10 states in the United States.

Table for Exercise 6 – Toxic Emissions (Thousands of Tons)									
206	147	441	128	127	133	422	152	114	134

Source: Toxics in the Community, U.S. Environmental Protection Agency

a.   Construct a stem-and-leaf display for the data using the hundreds digits as the stems.

b.   Comment on any information about the toxic emissions (in thousands of tons) of the 10 states that you were able to ascertain by examining the stem-and-leaf display.

7. Consider the following highway miles per gallon for 19 selected models of mini-compact, sub-compact, and compact cars.

Table for Exercise 7 – Miles per Gallon									
26	46	36	31	28	28	27	38	42	36
37	33	23	29	37	34	29	40	28	

   a. Construct a stem-and-leaf display for the data.

   b. Comment on any information about the highway mpg of the selected models which you were able to ascertain by examining the stem-and-leaf display.

8. An instructor is interested in comparing exam scores for fraternity and non-fraternity males in her class. Meaningful comparisons between two sets of data can be made using a side-by-side stem-and-leaf display. To illustrate this, note the following display summarizing the scores.

Leaf (Non-Fraternity)	Stem	Leaf (Fraternity)
	0	9
2	1	4 0 8
	2	5 7 9 4 5 5 1
3 9	3	2 6 6 9 7 7 3 2 1 6 0
	4	2 7 5
5 6 4 8 9 9 0 2	5	4 7 6 7
4 4 7 8 1 0 3 2 2 6 8 9	6	6 8 9 9 5
5 4 7 8 4 3 8 8 9 1	7	3 4 2 7 8 6 7 4 3
2 9 7 4	8	4 5 3 8 9 9 6 4 2 1 1 4 5
4 2	9	4 3 5 1 6 7 7 0 3

   a. What level of measurement does the data possess?

   b. Based upon the stem-and-leaf display, compare the two groups. Think of the several ways in which this can be done.

   c. Suppose that 60% is considered a passing score on the exam. What percent of the fraternity students passed the exam? Non-fraternity students?

   d. If someone scores 90 or higher on the exam, they will be exempt from taking the next exam. What percent of the fraternity students will be exempt from taking the next exam? Non-fraternity students?

3.7   The Ordered Array

An **ordered array** is a listing of all the data in either increasing or decreasing magnitude. Data listed in increasing order are said to be listed in **rank order**. If listed in decreasing order, they are listed in **reverse rank order**. Listing the data in an ordered way can be very helpful. It allows you to scan the data quickly for the largest and smallest values, for large gaps in the data, and for concentrations or clusters of values.

Example 3.6

The personnel records for a clothing department store located in the local mall are examined, and the current ages for all employees are noted. There are 25 employees, and their ages are listed in the following table. It is desired that their ages be placed in rank order.

Table 3.24 – Ages (Raw)												
32	21	24	19	61	18	18	16	16	35	39	17	22
21	60	18	53	18	57	63	28	20	29	35	45	

Solution

Table 3.25 – Ages (Ordered)												
16	16	17	18	18	18	18	19	20	21	21	22	24
28	29	32	35	35	39	45	53	57	60	61	63	

It is always a good idea to get a look at the ordered array of the data early in your analysis. Examining the ranked data produces a good intuitive sense for the data. Looking at the ordered array, it is evident that over half of the employees are younger than 25 and only three employees are within 5 years of retirement. We can also easily see that the youngest employee is 16 years old and the oldest employee is 63 years old. Ordering the data makes it easy to analyze the data quickly and easily.

Ordered arrays are easy to create. Virtually all statistics, spreadsheet, and database programs enable the user to quickly sort the data in ascending or descending order. For example, in Microsoft Excel, the Sort tool is found under the Data tab, and allows the user sort any number of data values in ascending or descending order. If a spreadsheet or database program is not available, a stem-and-leaf display can be helpful in sorting the data.

3.7   Exercises

Basic Concepts

1. What is an ordered array?

2. What are some advantages of the ordered array?

Exercises

3. Microsoft's consumer PC sales growth for the last 16 quarters are listed in the following table. Examine the data (sales growth, in percentages) and answer the following questions.

Table for Exercise 3 – PC Sales Growth			
Quarter	Sales Growth (%)	Quarter	Sales Growth (%)
1	20	9	20
2	24	10	19
3	22	11	33
4	19	12	37
5	23	13	24
6	27	14	10
7	16	15	0
8	10	16	−4

Source: Citi Investment Research and Analysis IDC, Company Reports, May 2011

a. Construct an ordered array of the data in rank order.

b. What conclusions can you make based on the ordered array?

4. *Fortune* magazine publishes a list of the top 100 best companies to work for. For the top 10 companies on this list, the average annual employee salaries are given in the following table (in thousands of dollars).

Table for Exercise 4 – Average Salaries (Thousands of Dollars)									
121	122	136	74	118	101	114	61	95	132

a. Construct a stem-and-leaf display for the data using the tens digits as the stems.

b. Comment on any information about the average annual salaries (in thousands of dollars) of the top 10 companies which you were able to ascertain by examining the stem-and-leaf display.

c. Construct an ordered array of the average annual salaries in rank order.

d. Does the ordered array provide any additional insight into the nature of the data?

3.8   Dot Plots

A **dot plot** is a graph where each data value is plotted as a point (or a dot) above a horizontal axis. If there are multiple entries of the same data value, they are plotted one above another.

Example 3.7

Construct a dot plot of the closing stock prices on December 1st for Microsoft Corporation from the data presented in Section 3.6. For convenience, the data are repeated below.

Table 3.26 – Closing Prices of Microsoft Corporation (MSFT) December 1990 through December 2011			
Period	**Closing Price (Dollars)**	**Period**	**Closing Price (Dollars)**
December 1990	1	December 2001	33
December 1991	2	December 2002	26
December 1992	3	December 2003	27
December 1993	3	December 2004	27
December 1994	4	December 2005	26
December 1995	5	December 2006	30
December 1996	10	December 2007	36
December 1997	16	December 2008	19
December 1998	35	December 2009	30
December 1999	58	December 2010	30
December 2000	22	December 2011	36

Solution

We plot each data value on the horizontal axis. For values where there are multiple entries, such as $3, we stack the points on top of one another.

Dot Plot of Closing Prices for Microsoft Corporation

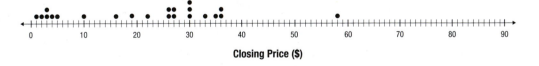

Closing Price ($)

As we can see from the dot plot, most of the prices are less than $30. The value that occurs most often is $30. $3, $26, $27, and $36 occur twice. Dot plots are useful when you are interested in where the data are clustered and which values occur most often.

3.8   Exercises

Basic Concepts

1. What is a dot plot?

2. What are some advantages of using a dot plot?

3. How can the most frequently occurring value be identified by studying a dot plot?

Exercises

4. Construct a dot plot for the following set of data.

Table for Exercise 4				
23	19	15	20	17
16	18	14	23	22
19	23	19	16	25
17	20	21	23	24

5. Listed in the following table is the number of passing attempts per game by super bowl champion Aaron Rodgers in the 2010 NFL season. Construct a dot plot for the data set.

Table for Exercise 5 – Passing Attempts by Aaron Rodgers				
31	29	45	17	46
33	34	34	34	31
35	30	11	37	28

Source: ESPN

6. The following table contains the average monthly energy consumption (in kilowatt hours) by households for nine South Atlantic states in 2007. Construct a dot plot for the data set.

Table for Exercise 6 – Monthly Energy Consumption (Kilowatt Hours)
773
960
1163
1171
1086
1143
1210
1207
1138

Source: U.S. Energy information Administration

3.9   Plotting Time Series Data

Recall our discussion of time series data from Section 2.7. When looking at time series data, we are interested in whether the process is stationary or nonstationary. To determine whether a time series is stationary or nonstationary, we use a **time series plot**. A time series plot graphs data using time as the horizontal axis. Most often, a time series plot comes in the form of a **line graph**, which connects consecutive points over time with a line. You have already seen some line graphs as examples of time series plots in Chapter 2. A good example of time series data is the census measurements taken every decade since 1790. The population of the United States has been growing steadily since the first census (as shown in Table 3.27). Interestingly, the change in population between 2000 and 2010 was 27.3 million, which is greater than the entire population of the United States in 1850.

Table 3.27 – U.S. Census Data			
Year	Population (Millions)	Year	Population (Millions)
1790	3.9	1900	76.2
1800	5.3	1910	92.2
1810	7.2	1920	106.0
1820	9.6	1930	123.2
1830	12.9	1940	132.2
1840	17.1	1950	151.3
1850	23.2	1960	179.3
1860	31.4	1970	203.3
1870	38.6	1980	226.5
1880	50.2	1990	248.7
1890	63.0	2000	281.4
		2010	308.7

Source: U.S. Census Bureau; 2010

For a quick look at the current population of the U.S. and the world, go to www.census.gov.

In a line graph, time is always labeled on the horizontal axis, with the variable being measured labeled on the vertical axis. Points are then plotted for each time period, and a line is drawn that connects each consecutive point. In Figure 3.18, each of the points is plotted with the year on the horizontal axis and the corresponding population (in millions) on the vertical axis. In Figure 3.19, a line connects consecutive points to give a line graph. From the line graph, we can easily see that the series is nonstationary, and that there is an upward trend in the data.

Plotted Points – U.S. Population (Millions)

Figure 3.18

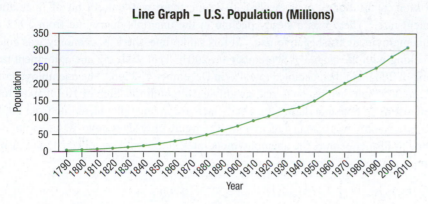

Line Graph – U.S. Population (Millions)

Figure 3.19

Line graphs can be analyzed with or without each point being visible. Another plot, called a **3-D line plot**, or a **ribbon plot**, is a three-dimensional version of a line graph (see Figure 3.20). Although it is difficult to determine the population size in any one year, the three dimensional-graph makes an interesting visual statement about population in the United States. Again, it is clear that the population has exhibited an upward trend since 1790.

Ribbon Plot – U.S. Population (Millions)

Figure 3.20

A Look at the Unemployment Rate

Another example of time series data is the unemployment rate in the United States. Civilian unemployment rate is defined as the number of unemployed people divided by the total size of the labor force. People who are jobless, looking for jobs, and available for work are considered unemployed. The labor force is made up of people that are either employed or unemployed. Figure 3.21 contains a time series plot of the civilian unemployment rate from January 1990 through November 2010.

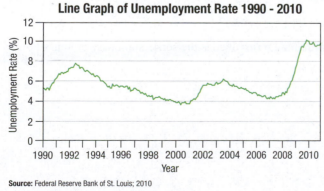

Line Graph of Unemployment Rate 1990 - 2010

Source: Federal Reserve Bank of St. Louis; 2010

Figure 3.21

As can be seen in the figure, over the last 20 years, there was quite a bit of fluctuation in the unemployment rate. There was a steady decline in the unemployment rate from 1992-2001. In 2001, the unemployment rate began to rise. At this same time, the U.S. economy was entering into a recession along with the events of September 11, 2001. Post 9/11, the unemployment rate rose to 6.3% in June of 2003 and then declined to 4.4% in December of 2006. The unemployment rate was fairly stable in 2007 but then began to rise steadily each month to a high of 10.1% in October 2009. It is also interesting to note that during the steady increase in unemployment rate from 2008 through 2009, the U.S. economy was also in a recession. The figure only shows overall unemployment rates. If one wanted to further explore the unemployment rates, data could be collected (and displayed) by worker groups, gender, and race, for example.

3.9   Exercises

Basic Concepts

1. Why is it important to plot time series data?

2. The time variable is always graphed on which axis?

3. Identify a variation on a time series plot that can make the data more visually interesting.

Exercises

4. The following line graph displays the total IRA and Keogh accounts (in billions of dollars) in the U.S., charted from June 1990 to June 2011.

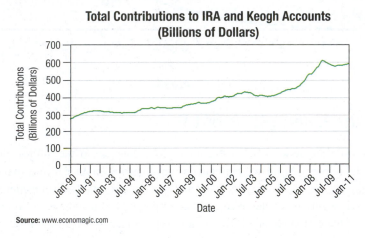

Total Contributions to IRA and Keogh Accounts (Billions of Dollars)

Source: www.economagic.com

  a. What conclusions can you make regarding the total contributed to the accounts?

  b. Are the data time series data?

  c. If the data are time series data, is the series stationary or nonstationary?

5. The following chart contains LIBOR (which stands for London Interbank Offered Rate) data for January 2011 through May 2011. LIBOR is the average interest rate that banks in London charge when lending funds to other banks. The line graphs in the figure represent 1 month, 3 month, 6 month, and 12 month interest rates.

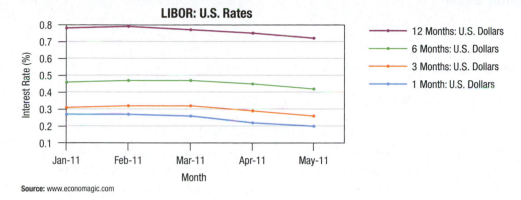

Source: www.economagic.com

    a. Examine the chart and discuss the data. What conclusions can you make?

    b. If the data are time series data, is it a stationary or nonstationary time series? Explain your reasoning.

6. The Gallup Poll frequently obtains responses to the question, *At the present time, do you think religion as a whole is increasing its influence on American life or losing its influence?* The percent of the respondents who answered "increasing" is given below for various polls.

Table for Exercise 6 – Survey Responses									
Year	2001	1995	1992	1991	1990	1988	1986	1984	1982	1980
Percent	71	38	27	27	33	36	48	42	41	35
Year	1978	1977	1975	1974	1970	1969	1968	1965	1962	1957
Percent	37	37	39	31	14	14	19	33	45	69

    a. What level of measurement do the responses to the question possess?

    b. Construct a time series plot for the data.

    c. What conclusions can you make from the plot?

7. The following table gives the number of immigrants (in thousands) and the average annual immigration rate per 1000 people in the U.S. population for the decade ending in the year given.

Table for Exercise 7 – Annual Immigration (per 1000 People)										
Year	1900	1910	1920	1930	1940	1950	1960	1970	1980	1990	2000
Number	3688	8795	5736	4107	528	1035	2515	3322	4493	7338	9095
Rate	5.3	10.4	5.7	3.5	0.4	0.7	1.5	1.7	2.1	2.9	3.2

    a. What levels of measurement do the three variables in this exercise possess?

    b. Construct a time series plot of the number of immigrants per decade.

    c. Find the percent change in the number of immigrants from the decade ending in 1900 to the decade ending in 2000.

    d. Find the percent change in the average annual immigration rate per 1000 people in the U.S. population from the decade ending in 1900 to the decade ending in 2000. Compare your answer to that which you obtained in part **c.** Can you explain why these answers are different?

T   Discovering Technology

Using Excel

Histogram

A histogram is a graphical image of a frequency distribution.

Use the data in Example 3.3 on company revenues for this exercise.

1.   Enter the data from the example into Column A of a new worksheet.

2.   In Column B, you will enter the upper class limits (Excel calls these "Bins"). For the frequency distribution in Table 3.12, these are 41, 82, 123, 164, 205, 246, 287, 328, 369, and 410. However, you can experiment with different class widths if you wish.

3.   To create a histogram, you will need the Data Analysis ToolPak add-in. If the add-in is installed, it will be found under the "Data" tab, and will be a button labeled "Data Analysis." If the add-in is already installed, proceed to step 8. If it is not installed, follow steps 4 through 7.

4.   Click on the **File** tab, and select **Options**.

5.   In the Excel Options window, select **Add-ins** from the menu on the left-hand side. In the Add-ins window, click **Go**.

6.   In the Add-ins dialog box, place a check mark next to **Analysis ToolPak** and click **OK**.

7.   A box will appear with the available Add-Ins for the Analysis ToolPak. Check the box next to the **Analysis ToolPak** option. Press **OK**. This will install the add-in.

8.   Under the **Data** tab, click on **Data Analysis**, and choose **Histogram**.

9.   The Histogram dialog box will appear. Set the Input Range to **$A$1:$A$100**. You can click on the box in the address window to drag a selection rather than typing in cell addresses if you like.

10. Set the Bin Range to **$B$1:$B$10**.

11. Click on the radio button next to **New Worksheet Ply** under Output options.

12. Check the box next to **Chart Output** to get a visual representation of the histogram.

Figure 3.22

13. Click **OK**. You should now have a new worksheet appear that looks like the one given in Figure 3.23.

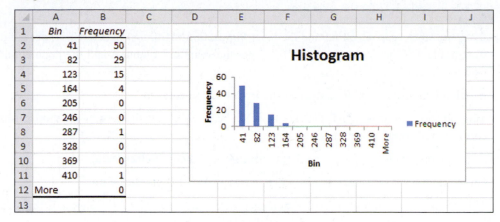

Figure 3.23

14. You will notice that Excel creates a histogram with gaps between the bars, and the upper limits are given between the tick marks below the bars. This looks different from the histograms we discussed in this chapter. First we need to reformat the *x*-axis so that the numbers appear on the tick marks rather than between them. To do this, right click on the numbers on the *x*-axis and select **Format Axis**. Under Position Axis, select the radio button next to **On tick marks**. Click **Close**.

15. Next, right click on one of the bars in the histogram and select **Format Data Series**. To remove the gap between the bars, drag the slider under Gap Width down to **No Gap**. Click **Close**.

16. Notice that when Excel creates the histogram it also creates a class called "More." This class is not very helpful here, so remove this class by highlighting cells **A12** and **B12** and pressing **Delete**. Now the histogram looks much more like the ones we saw in the chapter.

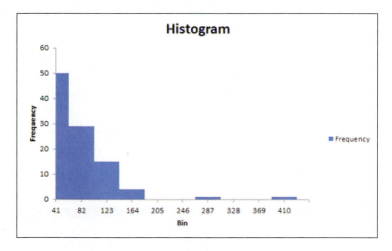

Figure 3.24

17. You can further customize your histogram by using the many tools that are available in the Chart Tools menu under the Design, Layout, and Format tabs.

Line Graph

Use the data in Table 3.27 for this exercise.

1. Enter the label **Year** in cell A1 and the label **Population (Millions)** in cell A2. Enter the data from Table 3.27 into columns A and B. Enter the years in Column A and the populations in Column B.

2. Highlight columns A and B, and under the **Insert** tab, select **Line**.

3. You can choose to make a 2-D or 3-D line graph here. Select the line graph in the top left-hand corner of the graphs given.

4. A line graph is now displayed. Under the **Design** tab, choose **Select Data**.

5. On the right-hand side under Horizontal (Category) Axis Labels click **Edit**.

6. This prompts you to enter an axis label range. Select the data values in Column A (this should appear as =Sheet1!$A$2:$A$24). Click **OK**. This makes time the horizontal axis.

7. In the left box under Legend Entries (Series), highlight **Year** and click **Remove**.

8. Click **OK**. You should now have a line graph that appears something like the one shown in Figure 3.25. You can further customize the graph by using the many tools that are available in the Chart Tools menu under the Design, Layout, and Format tabs.

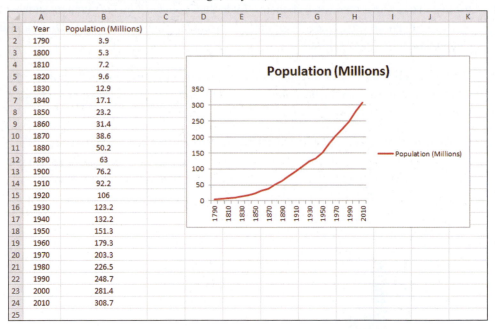

Figure 3.25

Using Minitab

Stem-and-Leaf Display

Use the information from Example 3.5 on the closing prices for Microsoft Corporation for this exercise.

1. Enter the data into Column C1.

2. Choose **Graph**, and select **Stem-and-Leaf**.

3. Select **C1** for the Graph Variables. Enter **10** for the Increment and select **OK**. This will create the Stem-and-Leaf display.

```
Stem-and-Leaf Display: Closing Price (Dollars)

Stem-and-leaf of Closing Price (Dollars)   N  = 22
Leaf Unit = 1.0


    6    0   123345
    9    1   069
   (5)   2   26677
    8    3   0003566
    1    4
    1    5   8
```

Figure 3.26

Histogram

Use the information from Example 3.4 on aptitude scores for this exercise.

1. Enter the data from Table 3.16 into Column C1.

2. Choose **Graph**, and select **Histogram**.

3. Select the **Simple Histogram**, and click **OK**.

4. Select **C1** as the Graph Variables. Click **OK**. A histogram of the data will be displayed.

Figure 3.27

R   Chapter 3 Review

Key Terms and Ideas

- Frequency Distribution
- Bar Chart
- 3-D Bar Chart
- Stacked Bar Chart
- Pie Chart
- Relative Frequency
- Cumulative Frequency
- Cumulative Relative Frequency
- Histogram
- Class Boundaries
- Symmetric Distribution

- Skewed Distribution
- Stem-and-Leaf Display
- Ordered Array
- Rank Order
- Reverse Rank Order
- Dot Plot
- Time Series Plot
- Line Graph
- 3-D Line Plot
- Ribbon Plot

Key Formulas

	Section
Number of Classes Round $\sqrt{n}$ or $\sqrt{2n}$ to the nearest whole number.	3.4
Class Width $$\frac{\text{Largest Value} - \text{Smallest Value}}{\text{Number of Classes}}$$	3.4
Relative Frequency $$\frac{\text{Number in Class}}{\text{Total Number of Observations}}$$	3.4

AE   Additional Exercises

1. Median family income has grown substantially in recent years. The table below contains median household incomes in the United States for the years 1990 through 2009.

Table for Exercise 1 – Median Family Income					
Year	Income ($)	Percent Change	Year	Income ($)	Percent Change
1990	29,943	–	2000	41,990	3.2
1991	30,126	0.6	2001	42,228	0.6
1992	30,636	1.7	2002	42,409	0.4
1993	31,241	2.0	2003	43,318	2.1
1994	32,264	3.3	2004	44,334	2.3
1995	34,076	5.6	2005	46,326	4.5
1996	35,492	4.2	2006	48,201	4.0
1997	37,005	4.3	2007	50,233	4.2
1998	38,885	5.1	2008	50,303	0.1
1999	40,696	4.7	2009	49,777	−1.0

Source: U.S. Census Bureau

 a. What graphical methods would be useful in displaying the data?

 b. Graph the data.

 c. Write a short paragraph describing the data.

2. The following data represent income of households headed by adults 25 years and older, tabulated by educational attainment.

Table for Exercise 2 – Income and Educational Attainment								
Educational Attainment	Under $5000	$5000 – $9999	$10,000 – 14,999	$15,000 – 24,999	$25,000 – 34,999	$35,000 – 49,999	$50,000 – 74,999	$75,000 & Over	Median Income
Elementary	16.3%	26.7%	17.2%	19.7%	9.9%	6.4%	2.9%	0.8%	$11,730
< 8 Years	18.4	27.9	17.3	18.1	8.9	5.9	2.7	0.7	10,884
8 Years	13.7	25.2	17.1	21.7	11.0	7.0	3.1	1.0	12,999
High School	6.8	12.4	12.0	21.8	17.8	17.0	9.5	2.8	23,382
1 – 3 Years	10.9	19.4	14.9	22.2	14.2	11.2	5.6	1.6	16,727
4 Years	5.4	10.0	11.0	21.6	19.0	18.9	10.8	3.2	25,910
College	2.6	4.3	5.8	15.1	16.3	22.7	20.1	13.2	38,337
1 – 3 Years	3.6	6.6	8.1	18.7	18.1	22.6	16.1	6.1	31,865
4 Years +	1.8	2.5	4.1	12.4	14.9	22.8	23.0	18.5	43,952

Source: U.S. Census Bureau

 a. What graphical methods would be useful in displaying the data?

 b. Use a graphics program to display the data.

 c. Discuss any conclusions you made from your graph(s).

3

3. Where do business school students most want to work? A CNNMoney list ranks companies that MBA students want to work for most after getting their degree. It can be seen that men and women have different desires when it comes to employment after business school. The table below shows seven of the top companies students want to work for and the percentage of students that ranked the particular company in their top 5.

Table for Exercise 3 – Most Desired Companies		
Company	**Percent of Men**	**Percent of Women**
Google	22.36	21.72
Goldman Sachs	17.76	5.81
Johnson & Johnson	5.68	11.41
Bain & Company	12.89	7.88
Apple Computer	12.80	11.01
Nike	8.06	9.04
J.P. Morgan	12.08	4.44

Source: CNNMoney.com

a. What graphical method do you think would be most useful in summarizing these data? Explain your answer.

b. Graph the data using the method you identified in part **a**.

c. Write a short paragraph describing the data, making conclusions from the graph you constructed in part **b**.

4. The following data give the percentage of people in the U.S. holding more than one job in a one-year period. Graph the data using a method that would contrast the difference between men and women in this situation.

Table for Exercise 4 – Percentage Holding More Than One Job		
Year	**Men**	**Women**
1970	7.0%	2.2%
1979	5.9%	3.5%
1989	6.4%	5.9%
1999	5.7%	5.6%
2003	5.2%	5.4%

Source: Bureau of Labor Statistics

5. *Billboard* magazine, in cooperation with Arbitron, produces a national radio format rating. The following data were gathered from radio listeners 12 and older.

Table for Exercise 5 – Radio Formats						
	Mon – Fri 6 AM – 10 AM	Mon – Fri 10 AM – 3 PM	Mon – Fri 3 PM – 7 PM	Mon – Fri 7 PM – 12 AM	Mon – Sun 12 AM – 6 AM	Mon – Sun 6 AM – 12 AM
Adult Contemporary	17.2%	19.7%	17.7%	15.0%	16.2%	20.0%
News/Talk	17.9	13.1	12.5	14.3	5.3	15.6
Country	13.0	13.2	13.2	10.3	11.7	14.3
Album Rock	10.0	10.4	10.9	9.8	18.7	10.2
Top 40	8.9	9.7	10.9	12.9	14.3	4.7
Urban	7.5	7.6	8.9	14.1	11.8	7.1
Oldies	6.0	6.8	6.9	6.5	4.3	10.2
Classic Rock	4.7	3.6	3.7	3.9	6.1	2.9
Spanish	4.5	4.2	3.7	2.2	4.9	4.2
Adult Standards	3.4	4.2	3.7	2.7	0.3	2.8
Religious	2.1	1.7	1.8	1.8	1.3	2.5
Classical	1.4	1.7	1.7	1.9	0.5	2.3
Easy Listening	0.9	1.1	0.9	0.8	0.2	1.2
Modern Rock	1.0	1.1	1.3	1.6	2.4	0.4
Adult Alternative	1.5	1.9	2.2	2.2	2.0	1.6

a. What kinds of graphs would be appropriate for displaying the data? Explain your choices.

b. Graph a column of the data. Briefly analyze your graph.

c. Create a graph that would be useful in visually comparing two columns of the data. Briefly analyze your graph.

6. The Caribbean has been a favorite vacation spot for affluent North Americans and Europeans, especially during the winter months. The following table lists the number of tourists during the first six months of the year for a number of Caribbean destinations.

Table for Exercise 6 – Number of Tourists			
	U.S.	Canada	Europe
Antigua & Barbuda	53,811	10,709	18,591
Aruba	94,028	1320	4681
Barbados	105,236	51,830	34,562
Bermuda	250,390	21,241	11,715
Bonaire	12,210	352	2266
Cayman Islands	81,180	3791	3025
Curacao	15,186	572	6543
Guadeloupe	15,596	10,654	25,409
Trinidad & Tobago	29,110	12,470	11,820

a. Create a stacked bar graph that shows where tourists from the U.S., Canada, and Europe travel in the Caribbean.

b. Create three separate bar charts, one for American tourists, one for Canadian tourists, and one for European tourists, that show the number of people traveling to each Caribbean destination.

7. The following table contains a list of the top 20 global corporations, ranked by the amount spent on research and development in 2009.

Table for Exercise 7 – Amount Spent on Research and Development (R&D) in 2009 (Millions of Dollars)					
Rank	Company	R&D Spending	Spending as a % of Sales	Headquarters Location	Industry
1	Roche Holding	9120	20.1	Europe	Healthcare
2	Microsoft	9010	15.4	N. America	Software and Internet
3	Nokia	8240	14.4	Europe	Computing and Electronics
4	Toyota	7822	3.8	Japan	Auto
5	Pfizer	7739	15.5	N. America	Healthcare
6	Novartis	7469	16.9	Europe	Healthcare
7	Johnson & Johnson	6986	11.3	N. America	Healthcare
8	Sanofi-Aventis	6391	15.6	Europe	Healthcare
9	GlaxoSmithKline	6187	13.9	Europe	Healthcare
10	Samsung	6002	5.5	S. Korea	Computing and Electronics
11	General Motors	6000	5.7	N. America	Auto
12	IBM	5820	6.1	N. America	Computing and Electronics
13	Intel	5653	16.1	N. America	Computing and Electronics
14	Merck	5613	20.5	N. America	Healthcare
15	Volkswagen	5359	3.7	Europe	Auto
16	Siemens	5285	5.1	Europe	Industrials
17	Cisco Systems	5208	14.4	N. America	Computing and Electronics
18	Panasonic	5143	6.4	Japan	Computing and Electronics
19	Honda	4996	5.4	Japan	Auto
20	Ford	4900	4.1	N. America	Auto

Source: Booz & Company

a. For comparative purposes, which of the two columns reporting R&D spending is more useful, and why?

b. What types of graphs would be useful in presenting these data? Explain your answers.

c. Develop a histogram for the spending as a percent of sales.

d. Use computer software to develop pie charts for the headquarters location and industry categories of the top 20 global R&D spenders.

8. In New York, a group of women challenged the state's ban on topless sunbathing. The legal issue was whether the ban was discriminatory. During the controversy, the Gallup poll conducted a survey asking the following question: *Do you think women should be permitted to sunbathe topless on public beaches, if they choose to, or do you think topless sunbathing on public beaches should be banned?*

Table for Exercise 8 – Response to Survey Question				
	Permitted	**Banned**	**No Opinion**	**Number of Interviews**
National	33%	63%	4%	1001
Gender				
Male	50	45	5	500
Female	18	79	3	501
Age				
18 – 29	47	51	2	219
30 – 49	39	58	3	411
50 – 64	18	76	6	206
65 +	18	77	5	357
Region				
East	39	59	2	247
Midwest	34	62	4	254
South	25	71	4	301
West	38	57	5	199
Community				
Urban	42	55	3	345
Suburban	35	62	3	351
Rural	23	72	5	298
Race				
White	33	64	3	871
Non-white	39	57	4	121
Education				
College Grads	46	48	6	288
Some College	35	62	3	233
No College	28	69	3	475
Sex/Education				
Male/College	56	40	4	238
Male/ No College	45	49	6	238
Female/ College	26	70	4	264
Female/ No College	13	85	2	237

a. Suggest two different types of graphs that might be useful in graphing the data.

b. Create two different graphs using the data.

c. Write a short paragraph describing the data.

9. The nation's political identification (Republican, Democrat, or Independent) changes over time. The data in the following table represent Harris poll results on political identification from 1977 to 2008.

Table for Exercise 9 – Nation's Political Identification (Percent of the Population) 1977 – 2008							
Year	Republican	Democrat	Independent	Year	Republican	Democrat	Independent
1977	21	48	25	1993	29	38	27
1978	22	43	30	1994	32	37	26
1979	22	41	31	1995	31	36	28
1980	24	41	29	1996	30	38	26
1981	28	39	28	1997	29	37	26
1982	26	40	28	1998	28	37	27
1983	26	41	27	1999	29	36	26
1984	27	40	24	2000	29	37	23
1985	30	39	26	2001	31	36	22
1986	30	39	25	2002	31	34	24
1987	29	38	28	2003	28	33	24
1988	31	39	25	2004	31	34	24
1989	33	40	23	2005	30	36	22
1990	33	38	25	2006	27	36	24
1991	32	37	26	2007	26	35	23
1992	30	36	29	2008	26	36	31

Source: Harris Interactive

a. What types of graphs would be useful in visualizing these data? Explain your answer.

b. Construct two different types of graphs from the data.

c. Examine the data and write a short paragraph on your conclusions.

10. Monaco is noted for having one of the highest population densities in the world, approximately 16,923 persons per square kilometer. Usually, dense urban areas have relatively high crime rates. This is not the case in Monaco. The following table gives crime data per 100,000 population for the year 2000 in Monaco as well as other urban areas.

Table for Exercise 10 – Crime Data per 100,000 Population					
	Monaco	**London**	**Chicago**	**New York**	**San Francisco**
Homicide	1.0	4.7	23.0	43.4	8.6
Forcible Rape	7.9	34.7	–	19.1	26.7
Robbery	–	625	635.6	352.5	409.9
Aggravated Assault	–	847.4	880.4	473.7	327.2
Burglary	–	938.9	895.4	394.6	764.8
Larceny/Theft	333.0	2675.3	3361.9	1674.2	803.0

Source: U.S. Department of Justice, CIA, BBC

a. What types of graphs would be useful in visualizing this data? Explain your answer.

b. Construct two different types of graphs from the data.

c. Examine the data and write a short paragraph on your conclusions.

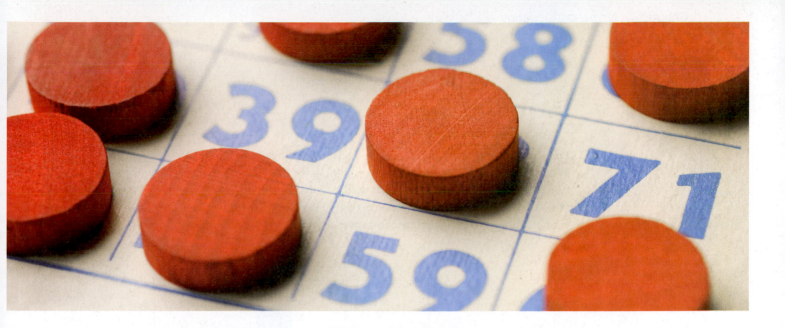

Chapter Four
Numerical Descriptive Statistics

Introduction

Frequency distributions, bar charts, pie charts, and histograms (all of which were discussed in Chapter 3) can be informative visual tools for examining the big picture when analyzing data. But there is a lack of exactness in the language that we use to describe these graphs. Suppose we say that one data set is more compact than another. This only leads to the question, *How much more compact is it?* Graphical analysis is ill-equipped to answer that question precisely.

Doctor's Office Waiting Times

Figure 4.1

Emergency Room Waiting Times

Figure 4.2

If we examine Figure 4.1 and Figure 4.2, we notice the two histograms are somewhat similar, but there is a clear difference in the underlying data. To precisely describe the differences, we need summary measures to characterize specific data attributes.

- Location: Where is the center of the data?

- Dispersion: Are the data widely scattered or tightly grouped around the central value?

- Shape: Are the data spread symmetrically about the central value? Are the data unbalanced or skewed (e.g. are the values much larger than the mean, but not much smaller)?

- Are there outliers (values that are drastically different from the mean) in the data?

- Do the data tend to cluster in several groups?

Different types of measurements will be developed for each of the attributes listed. For example, the mean (average) and the median measure the **location** or **central tendency** of a data set. Both of these statistical measures are trying to tell us something about the location of the middle of the data, but they use different ideas for defining that notion of middle.

From the morning paper to the evening news, general concepts are translated into specific statistical measures. Some examples are listed below.

- The proportion of United States residents earning below the poverty level was 14.3% in 2009.
  Source: U.S. Census Bureau

- The median price of new homes sold in 2009 was $216,700.
  Source: U.S. Census Bureau

- The average hourly manufacturing wage in the U.S. in 2010 was $18.90.
  Source: U.S. Department of Labor

Such measures are examples of **numerical descriptive statistics**.

Definition

Numerical descriptive statistics are numerical summaries of data

There is a distinction between measures that are applied to populations and measures that are applied to samples.

Definition

Measures that apply to population data are called **parameters**.

Definition

Measures that apply to sample data are called **statistics**.

In most instances a data analyst will not know what the population parameters are, since the cost and/ or feasibility of obtaining all the population data is usually prohibitive. Inferential statistics involves making conclusions regarding a population using data from a sample.

Definition

Inferential statistics is concerned with making conclusions about population parameters using sample statistics.

There are many formulas in this chapter that you will have to spend some time examining to appreciate. In most cases, the concepts which motivate these formulas are simple. Yet if these concepts are either ignored or forgotten, statistics becomes a meaningless assortment of symbols instead of a useful problem-solving and decision-making tool.

4

The data in Table 4.1 are costs (rounded to the nearest dollar) of fill-ups at a local gasoline station in December 2010 for 400 transactions. Looking at the 400 observations without using a graphical representation would be confusing. (It seems 400 measurements would contain a great deal of information, yet the sheer volume of the data obscures comprehension.) It is the old problem of not being able to see the forest for the trees.

Table 4.1 – Costs of Fill-Ups – December 2010 ($)									
22	39	25	35	43	36	52	44	37	49
32	32	35	44	33	51	44	28	29	43
37	45	44	28	51	38	47	34	52	50
33	45	28	54	38	40	37	42	38	44
30	40	36	38	32	38	28	23	51	28
43	43	25	24	47	36	22	24	42	55
39	45	36	36	38	32	43	32	34	54
45	58	49	23	55	43	29	11	34	45
48	36	22	34	18	27	37	50	39	30
57	48	36	64	62	33	33	58	31	47
36	38	33	41	26	43	28	42	43	39
50	32	47	22	42	35	21	56	46	21
50	23	52	46	31	41	35	33	30	47
54	43	43	44	51	49	31	56	27	49
34	51	38	34	28	36	43	35	29	47
31	32	37	22	44	34	19	42	47	25
29	48	32	30	25	61	30	21	33	42
41	24	42	39	25	45	36	43	49	41
31	46	55	27	57	32	56	45	66	30
30	33	37	33	41	59	33	24	29	42
44	58	53	47	50	46	30	36	47	51
17	54	59	54	41	39	60	40	36	32
41	45	43	50	53	28	49	45	43	50
51	35	67	49	38	38	41	30	31	37
11	48	37	27	32	42	34	31	39	41
36	41	30	31	44	34	39	38	52	32
32	45	55	20	48	46	35	36	42	54
48	41	31	31	54	52	35	34	46	55
28	14	41	50	43	44	47	40	44	37
43	53	28	32	47	50	38	53	44	35
34	44	40	42	37	51	39	46	26	46
19	40	72	40	37	20	43	38	33	32
45	45	33	53	46	39	42	49	37	31
43	50	26	41	36	45	27	43	37	29
47	42	55	48	54	24	30	52	43	40
35	49	25	42	38	55	47	49	43	43
37	38	34	35	48	45	43	53	28	34
45	46	21	56	35	37	46	24	19	25
41	50	45	36	31	36	60	40	36	35
23	51	43	24	51	36	25	42	25	43

Looking at a graph of the data (see Figure 4.3) is always a good first step. But, we need to learn more. To do this, we use statistical tools designed to reveal the data's fundamental characteristics. These statistical tools answer important questions such as, *Where is the center of the data?* and *How dispersed are the data?*

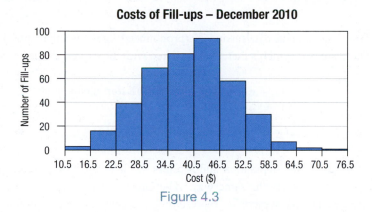

Figure 4.3

4.1 Measures of Location

Statistically speaking, the idea of location is similar to knowing the whereabouts of a person. If we think of a data set as a group of data values that cluster around some central value, then this central value provides a focal point for the data set – a location of sorts. Unfortunately, the notion of *central value* is a vague concept, which is as much defined by the way it is measured as by the notion itself. There are several statistical measures that can be used to define the notion of center: the arithmetic mean, weighted mean, trimmed mean, median, and mode.

Arithmetic Mean

The arithmetic mean is one of the more commonly used statistical measures. It appears every day in newspapers, business publications, and frequently in conversation. For example, when your instructor returns an assessment, after viewing your grade, one of the first questions asked is, *What is the average?* The word average is often associated with the mean.

Formula

Arithmetic Mean

Suppose there are n observations in a data set, consisting of the observations $x_1, x_2, \ldots, x_n$; then the **arithmetic mean** is defined to be

$$\frac{1}{n}\left(x_1 + x_2 + \cdots + x_n\right).$$

If we use some common mathematical notation (summation notation, represented by Σ), the formula can be simplified to

$$\frac{\sum x_i}{n},$$

where x_i is the ith data value in the data set and Σ (pronounced *sigma*) is a mathematical notation for adding values. There are two symbols that are associated with the expression given above:

$$\mu = \frac{1}{N}\left(x_1 + x_2 + \cdots + x_N\right) \text{ the \textbf{population mean}, and}$$

$$\bar{x} = \frac{1}{n}\left(x_1 + x_2 + \cdots + x_n\right) \text{ the \textbf{sample mean}.}$$

Here N refers to the size of the population and n refers to the size of the sample. Otherwise, the calculations are made in precisely the same way. The Greek letter μ, representing the population mean, is pronounced *mu* and the symbol $\bar{x}$, representing the sample mean, is pronounced *x-bar*.

Example 4.1

Calculate the sample mean of the following sample data values: 4, 10, 7, 15.

Solution

Note that $x_1 = 4$, $x_2 = 10$, $x_3 = 7$, $x_4 = 15$, and $n = 4$.

$$\bar{x} = \frac{1}{4}\left(4 + 10 + 7 + 15\right) = 9$$

$$= \frac{\sum x_i}{n} = \frac{4 + 10 + 7 + 15}{4} = \frac{36}{4} = 9$$

The sample mean is 9. But why does adding up a group of numbers and dividing by the number of observations measure central tendency? As unlikely as it sounds, the answer is related to balancing a scale.

Definition

Given some point A and a data point x, then $x - A$ represents how far x **deviates** from A. This difference is also called a **deviation**.

Let's calculate the deviations from the mean for the data in Example 4.1. Examining the deviations from the mean in Table 4.2, we can see the deviations on the left side (-5 and -2) and right side (1 and 6) are in balance. In fact, the mean is considered a point of centrality because the deviations from the mean on the positive side and the negative side are equal (See Figure 4.4). The sample mean can be interpreted as a center of gravity.

Table 4.2 – Deviations from the Mean	
Data (x_i)	**Deviations from the Mean $(x_i - 9)$**
4	−5
10	1
7	−2
15	6
Total	$\sum (x_i - 9) = 0$

Figure 4.4

On the other hand, if we calculate the deviations about any other value the deviations do not balance. For example, assume the central value is 8. The deviation from the alleged central value, 8, for each data value is calculated in Table 4.3 and shown in Figure 4.5.

The positive deviations (2 and 7) are not counterbalanced by the negative deviations (−4 and −1). A desirable characteristic of a central value would be to have the positive and negative deviations equal to each other in absolute value.

Table 4.3 – Deviations from Some Other Value	
Data (x_i)	**Deviations from 8 $(x_i - 8)$**
4	−4
10	2
7	−1
15	7
Total	$\sum (x_i - 8) = 4$

Figure 4.5

Although the arithmetic mean is frequently used, there are times when it should not be employed. Since the mean requires that the data values be added, it should only be used for quantitative data. Furthermore, if one of the data values is extremely large or small relative to others, this could be considered an **outlier**. An outlier is a data value that can have a dramatic impact on the value of the mean.

Definition

Statistical measures which are not affected by outliers are said to be **resistant**.

The arithmetic mean is not a **resistant measure**.

Weighted Mean

The weighted mean is similar to the arithmetic mean except it allows you to give different weights (or importance) to each data value. The weighted mean gives you the flexibility to assign weights when you find it inappropriate to treat each observation the same. The weights are usually positive numbers that sum to one, with the largest weight being applied to the observation with the greatest importance. The weights can be determined in a variety of ways, such as the number of employees, market value of a company, or some other objective or subjective method. There are occasions in which it is easier to assign the weights without worrying that they will sum to one. If you are concerned about your weights summing to one, you can make your weights sum to one by dividing each weight by the sum of all the weights.

Formula

Weighted Mean

The **weighted mean** of a data set with values x_1, x_2, x_3,..., x_n is given by

$$\bar{x} = \frac{w_1 x_1 + w_2 x_2 + ... + w_n x_n}{w_1 + w_2 + ... + w_n} = \frac{\sum (w_i x_i)}{\sum w_i}$$

where w_i is the weight of observation x_i.

Example 4.2

The following table consists of the November 2010 unemployment rates and civilian labor force sizes for the Mid-Atlantic states.

Table 4.4 – Unemployment Rates		
State	**Size of Civilian Labor Force (Thousands)**	**Unemployment Rate (%)**
Delaware	422.4	8.4
Maryland	2972.7	7.4
New Jersey	4490.6	9.2
New York	9658.4	8.3
Pennsylvania	6362.0	8.6
Virginia	4177.7	6.8
District of Columbia	331.6	9.8
West Virginia	777.6	9.3

Source: U.S. Department of Labor, Bureau of Labor Statistics

Using the weighted mean, calculate the average unemployment rate for the Mid-Atlantic states.

Solution

The average unemployment rate is calculated as follows.

$$\overline{x} = \frac{\sum (w_i x_i)}{\sum w_i}$$

$$\sum (w_i x_i) = 422.4(8.4) + 2972.7(7.4) + 4490.6(9.2) + 9658.4(8.3) + 6362.0(8.6)$$
$$+ 4177.7(6.8) + 331.6(9.8) + 777.6(9.3)$$
$$= 240627.3$$

$$\sum w_i = 422.4 + 2972.7 + 4490.6 + 9658.4 + 6362.0 + 4177.7 + 331.6 + 777.6$$
$$= 29193$$

$$\text{so,}\ \overline{x} = \frac{\sum (w_i x_i)}{\sum w_i} = \frac{240627.3}{29193} \approx 8.24\%$$

Thus, the average unemployment rate, calculated by the weighted mean, is 8.24%. It is appropriate to use the weighted mean to calculate the average unemployment rate since the size of the civilian labor force (the weight) is different for each state.

Trimmed Mean

Since outliers can have an enormous effect on the value of the mean, the mean's usefulness as a typical measure of data is diminished if the data contain outliers.

Definition

The **trimmed mean** is a modification of the arithmetic mean which ignores an equal percentage of the highest and lowest data values in calculating the mean.

Finding the 10% Trimmed Mean

Before calculating the trimmed mean, the data are arranged in ascending order of magnitude. A 10% trimmed mean uses the middle 80% of the values. It is calculated by chopping off the top 10% and the bottom 10% of the data values, and finding the arithmetic mean of the remaining values. If the data set does not contain any outliers, the mean and the trimmed mean will be similar. Note that because the trimmed mean is not affected by outliers like the arithmetic mean is, the trimmed mean is considered a resistant measure.

Measuring Figure Skating Performances

Almost every figure skating competition has some scoring controversy. The Winter Olympics of 2002 were no exception. French judge, Marie-Reine Le Gougne, said she was "pressured to vote a certain way" when she scored the Russian couple, Elena Berezhnaya and Anton Sikharulidze, over the Canadian pair, Jamie Sale and David Pelletier. In addition, very few people understood exactly how Sarah Hughes won the gold medal and how Michelle Kwan dropped to third after leading the event.

For almost a century figure skating has used a scoring method that is similar to the methodology of the trimmed mean in order to remove bias. Skaters are scored on a 0 to 6 scale. The highest and lowest scores are discarded (the data is trimmed) and the resulting score is computed. The intent of trimming the data is to avoid bias caused by judges with nationalistic or political agendas. Since the controversy, the International Skating Union has replaced this scoring method with a new system which, though it is different, still utilizes trimmed data to eliminate bias.

Example 4.3

Consider the following data.

$$15, 21, 25, 31, 35, 42, 48, 51, 54, 60$$

Find the 10% trimmed mean.

Solution

Since there are 10 observations, removing the highest 10% and the lowest 10% means removing only one observation from each end of the data.

That is,

$$10\% \text{ of } 10 = 0.1 \cdot 10 = 1.$$

Note that the data are already sorted. If the mean is calculated without including the values of 15 and 60, the resulting measure is called the 10% trimmed mean.

$$\cancel{15}, \ 21, \ 25, \ 31, \ 35, \ 42, \ 48, \ 51, \ 54, \ \cancel{60}$$

$$10\% \text{ trimmed mean} = \frac{21+25+31+35+42+48+51+54}{8} = \frac{307}{8} = 38.375.$$

If there had been 100 observations, the largest 10% and the smallest 10% (a total of 20 data values) would have been removed before the mean was calculated.

Median

The median of a set of data provides another measure of center. It is a simple idea. To find the median, place the data in ascending order and then find the observation that has an equal number of data values on either side. That is, half of the observations are less than the median and half of the observations are greater than the median. The median is the middle value.

Definition

The **median** of a set of observations is the data value in the middle of an ordered array. The same number of data values is on either side of the median value.

Finding the Median

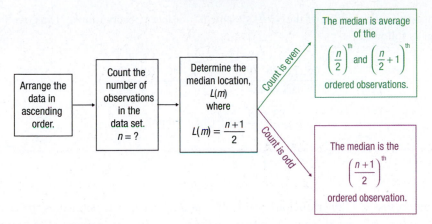

Example 4.4

Given the following eleven observations, find the median.

$$2, 4, 0, 3, 0, 1, 8, 5, 1, 5, 9$$

Solution

First, the data set must be ordered.

$$0, \ 0, \ 1, \ 1, \ 2, \ 3, \ 4, \ 5, \ 5, \ 8, \ 9$$

The number of observations, $n = 11$.

Next, calculate the median location, $L(m) = \dfrac{n+1}{2} = \dfrac{11+1}{2} = 6.$ Therefore, the median is the 6th ordered observation, which is 3.

$$\underbrace{0, \ 0, \ 1, \ 1, \ 2,}_{5 \text{ data values}} \ \boxed{3}, \ \underbrace{4, \ 5, \ 5, \ 8, \ 9}_{5 \text{ data values}}$$

Example 4.5

Consider the following ten test scores.

$$65, 98, 76, 83, 94, 79, 88, 72, 90, 85$$

Find the median.

Solution

The number of observations, $n = 10$.

The median location, $L(m) = \dfrac{10+1}{2} = \dfrac{11}{2} = 5.5.$

Kiwi Eggs Are Outliers!

Both plant and animal kingdoms offer spectacularly odd and beautiful sights. A kiwi bird (one of the many interesting life forms from New Zealand) lays eggs that are close to 25% of its body weight and sometimes lays two or three such eggs at a time. For most species of birds, eggs usually correspond to about 5% of the bird's body weight. If you draw a graph relating (log) egg weight to (log) body weight you get a so-called hummingbird-moa curve (moa is an extinct ostrich-like bird of the New Zealand area). In this curve, the kiwi bird is an outlier. Using the kiwi body weight (about 5 lb) one expects an egg weight of about 55 to 100 grams while the actual weight of kiwi eggs is about 400 to 435 grams. This egg weight matches an expected body weight of about 40 lb according to the hummingbird-moa curve. Why is this the case, and what accounts for such an anomaly?

Hummingbird – Moa Curve

Continued ...

If there is an even number of observations, average the two center values in the ordered array.

The median is the average of the $\frac{10}{2} = 5^{\text{th}}$ and 6th ordered observations. Thus, we find the median as follows.

$$65, 72, 76, 79, \boxed{83}, \boxed{85}, \underbrace{88, 90, 94, 98}$$
$$\underbrace{\qquad\qquad}_{\text{4 data values}} \qquad \underbrace{\qquad\qquad}_{\text{4 data values}}$$

$$\frac{83+85}{2} = 84 \text{ (the median).}$$

The median possesses a rather obvious notion of centrality, since it is defined as the central value in an ordered list. It is not affected by outliers and is thus a resistant measure. For example, if we replaced 98 with 200,000,000 in the data set from Example 4.5, the median would not change at all. The median does possess one limitation: it cannot be applied to nominal data. In order to calculate the median, the data must be placed in order. To accomplish this task meaningfully, the level of measurement must be at least ordinal.

Unless the data set is skewed or contains outliers, the median and the mean usually have similar values.

Example 4.6

a. Consider the following data.
$$16, 18, 20, 21, 23, 23, 24, 32, 36, 42$$
$$\text{mean} = 25.5$$
Find the median and 10% trimmed mean.

Solution

First, find the median.

The number of observations, $n = 10$.

The median location, $L(m) = \frac{10+1}{2} = \frac{11}{2} = 5.5$.

If there is an even number of observations, average the two center values in the ordered array. The median is the average of the $\frac{10}{2} = 5^{\text{th}}$ and 6th ordered observations.

$$16, 18, 20, 21, \boxed{23}, \boxed{23}, 24, 32, 36, 42$$
$$\underbrace{\qquad\qquad}_{\text{4 data values}} \qquad \underbrace{\qquad\qquad}_{\text{4 data values}}$$

$$\frac{23+23}{2} = 23 \text{ (the median)}$$

Now find the 10% trimmed mean.

Since there are 10 observations, removing the highest 10% and the lowest 10% means removing only one observation from each end of the data.

$$10\% \text{ of } 10 = 0.1 \cdot 10 = 1$$

If the mean is calculated without including the values 16 and 42 in the data, the resultant measure is the 10% trimmed mean.

$$16, 18, 20, 21, 23, 23, 24, 32, 36, \cancel{42}$$

$$\text{10\% trimmed mean} = \frac{18 + 20 + 21 + 23 + 23 + 24 + 32 + 36}{8} = 24.625$$

b. Consider the same data set, except the last data value is replaced with an outlier (490). Compare the mean, median, and 10% trimmed mean for this data set to the mean, median, and 10% trimmed mean of the original data set.

$$16, 18, 20, 21, 23, 23, 24, 32, 36, 490$$

$$\text{mean} = 70.3 \qquad \text{median} = 23 \qquad \text{10\% trimmed mean} = 24.625$$

Solution

The median and 10% trimmed mean are not affected by the addition of the outlier, while the mean increases dramatically. This illustrates why the median and 10% trimmed mean are said to be resistant measures while the arithmetic mean is not.

... Continued
The most reasonable explanation provided by biologists is that kiwis and moa birds are members of the same species except the kiwis have dwarfed through their evolutionary history. A sub-area of biology called "allometry" states that as body size decreases the internal organs decrease relatively slowly which supports the dwarfism hypothesis. The kiwis have lost body weight but not their internal womb structure which still holds large eggs. Outliers are important because they force you to think about data more seriously.

Mode

The **mode** is another measure of location. It is not used as frequently as the mean or the median, and its relation to these values is not so predictable. The mode is the only measure of location that can be used for nominal (or qualitative) data. Of the three measures of location, the mode is used the least due to the limited information it provides. Sometimes sorting the data (in ascending or descending order) makes it easier to find the mode.

Definition

The **mode** of a data set is the most frequently occurring value.

Example 4.7

Find the mode of the following data set.

$$0, 1, 4, 3, 9, 8, 10, 0, 1, 3, 0$$

Solution

Since the value of 0 occurs more than any other value, it is the mode. In this instance, as a measure of location, the modal value is not a particularly appealing choice. However, the mode does possess one very favorable property – it is the only measure of location that can be applied to nominal data. Thus, for nominal measurements like color preferences, it would be perfectly reasonable to discuss the modal color.

Suppose we added one more value to the data set in Example 4.7. If this value were a 1, then both 0 and 1 would be repeated three times and there would be two modes. When this occurs, the data is said to be **bimodal**. Any time data has more than two modes it is said to be **multimodal**. If all observations in a data set occur with the same frequency, then there is **no mode** for that data set.

The Relationship between the Mean, Median, and Mode

Often times, the shape of the data determines how the mean, median, and mode are related. For a bell-shaped distribution, the mean, median, and mode are identical.

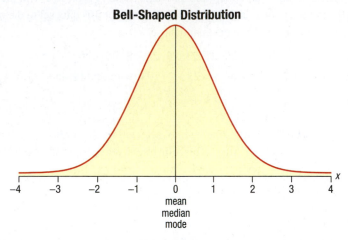

Bell-Shaped Distribution

Figure 4.6

Certainly, not all data produce distributions that follow a bell-shaped curve. If the distribution of the data has a long tail on the right, it is said to be **skewed to the right**, or **positively skewed**. Conversely, if the distribution has a long tail on the left, it is said to be **skewed to the left**, or **negatively skewed**.

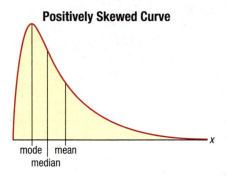

Positively Skewed Curve

Figure 4.7

If the data are positively skewed, the median will be smaller than the mean.

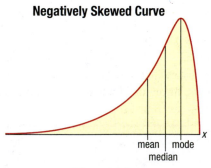

Negatively Skewed Curve

Figure 4.8

If the data are negatively skewed, the mean will be smaller than the median.

Where does the mode fall on these graphs? The area containing the greatest number of observations contains the mode. That area is represented by a large peak in the curve. In Figure 4.7 and Figure 4.8 the obvious peaks in the curves are the portions of the distributions that will contain the mode. The highest point on the curve will be the mode of the distribution. Notice that in a bell-shaped distribution (Figure 4.6), the mode is equal to the mean and median. If the distribution is positively skewed (Figure 4.7), the mode is less than the mean and median. If the distribution is negatively skewed (Figure 4.8), the mode is greater than the mean and median.

Selecting a Measure of Location

The objective of using descriptive statistics is to provide measures that convey useful summary information about the data. When selecting a statistic to represent the central value of a data set, the first thing to consider is the type of data being analyzed.

The arithmetic mean is used so frequently that its computation is almost a knee-jerk reaction to analyzing data. Unfortunately it is not always a reasonable measure of location. Table 4.5 defines the applicable levels of measurement for each measure of location. Table 4.6 defines the sensitivity to outliers for each measure of location. When the data are qualitative (nominal or ordinal), the mean should not be calculated and, if the data are quantitative and contain outliers, the mean does not convey the notion of typical value as well as some other measures. The only time in which the mean should be used without any explanation is when the distribution of the data is symmetrical or nearly so. In that event, the mean and median should be about the same value.

Table 4.5 – Applicable Levels of Measurement

| | Applicable Level of Measurement | | | |
| | Qualitative | | Quantitative | |
	Nominal	Ordinal	Interval	Ratio
Mean			✓	✓
Median		✓	✓	✓
Mode	✓	✓	✓	✓
Trimmed Mean			✓	✓

Table 4.6 – Sensitivity to Outliers

	Not Sensitive	Very Sensitive
Mean		✓
Median	✓	
Mode	✓	
Trimmed Mean	✓	

The median is also a good measure of central tendency. It is not sensitive to outliers and can be applied to data gathered from all levels of measurement except nominal.

If the level of measurement of the data is interval or ratio and there are no outliers, the mean is a reasonable choice. If the data set appears to have any unusual values, then the trimmed mean or the median would be more appropriate.

If the data's level of measurement is nominal or ordinal (the data are qualitative), appropriate measures of center are limited. If the data are ordinal, then the median is the best choice. If the data are nominal, there is only one choice, the mode. The mode is applicable to any level of data, although it is usually not very useful for quantitative data.

Time Series Data and Measures of Location

We discussed two types of time series data in Chapter 2, stationary and nonstationary. Stationary time series wobbled around some central value, so calculating a central value is perfectly reasonable, and the methods we previously discussed are applicable. A nonstationary time series is another story. Nonstationary time series possess trend. That means there is no central value for the time series. Instead, the series trends in one direction or another. Computing a central value using the methods discussed earlier would be inappropriate for such data.

Table 4.7 shows the average U.S. gas price over a 20-year period. In this nonstationary time series, the central value of the process is trending upward as shown in Figure 4.9. One way to capture this movement is with a **moving average**.

Table 4.7 – Average U.S. Gas Price 1991 – 2010 (Dollars per Gallon)			
Year	Average U.S. Gas Price	2-Period Moving Average	3-Period Moving Average
1991	1.14		
1992	1.13	1.135	
1993	1.11	1.120	1.127
1994	1.11	1.110	1.117
1995	1.15	1.130	1.123
1996	1.23	1.190	1.163
1997	1.23	1.230	1.203
1998	1.06	1.145	1.173
1999	1.17	1.115	1.153
2000	1.51	1.340	1.247
2001	1.46	1.485	1.380
2002	1.36	1.410	1.443
2003	1.59	1.475	1.470
2004	1.88	1.735	1.610
2005	2.30	2.090	1.923
2006	2.59	2.445	2.257
2007	2.80	2.695	2.563
2008	3.27	3.035	2.887
2009	2.35	2.810	2.807
2010	2.79	2.570	2.803

Source: Bureau of Labor Statistics

Figure 4.9

Definition

A **moving average** is obtained by adding consecutive observations for a number of periods and dividing the result by the number of periods included in the average.

We will assume the moving average is to be used as a method of forecasting the next level of the time series. Suppose a two-period moving average is calculated for the gas price data and is used to specify the level of the series at a given point in time. The two-period moving average for 1992 averages the values of the time series in 1991 and 1992.

$$\frac{1.14+1.13}{2}=1.135$$

Similarly, the two-period moving average for 1993 would be the average of the time series values in 1992 and 1993.

$$\frac{1.13+1.11}{2}=1.120$$

Since data are not available for 1989 or 1990, the three-period moving average for 1991 cannot be calculated. The three-period moving average associated with 1993 is the average of the time series values in 1991, 1992, and 1993.

$$\frac{1.14+1.13+1.11}{3}\approx1.127$$

The three-period moving average for 1994 would be the average of the time series values in 1992, 1993, and 1994.

$$\frac{1.13+1.11+1.11}{3}\approx1.117$$

The chart in Figure 4.10 displays the time series and the two and three-period moving averages. Both of the averages follow the time series quite closely. However, notice that the two-period moving average (the purple line) follows the actual data values (the green line) more closely than the three-period moving average (the orange line).

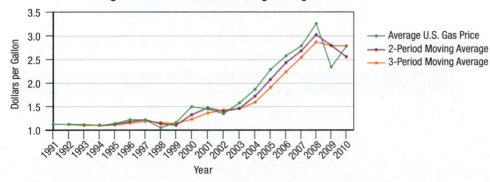

Figure 4.10

4.1   Exercises

Basic Concepts

1. Describe the difference between statistics and parameters.

2. Discuss three major attributes used in summarizing a data set.

3. What are numerical descriptive statistics and why are they important?

4. Identify and describe five measures of location. List the advantages and disadvantages of each.

5. List the types of data that are appropriate for each of the measures of location discussed in the previous question.

6. Why is the mean a measure of central tendency?

7. What is a resistant measure?

8. Describe a situation in which using the weighted mean as a measure of location would be appropriate.

9. What does it mean if we say that a data set is positively skewed? Negatively skewed?

10. Explain why the mean should not be calculated for a nonstationary time series.

11. What is a moving average? When is it useful?

Exercises

12. Calculate the mean, median, 10% trimmed mean, and mode for the following data.

Table for Exercise 12									
90.25	93.83	91.41	92.27	90.89	99.12	92.88	97.74	96.28	95.33
91.16	94.30	95.51	92.27	97.63	95.94	90.95	94.76	92.27	92.88

13. Calculate the mean, median, 20% trimmed mean, and mode for the following data.

Table for Exercise 13				
19	14	11	11	18
20	10	19	11	18
20	18	10	11	15

14. Calculate the mean, median, 20% trimmed mean, and mode for the following data.

Table for Exercise 14				
0.5380	0.8856	0.2738	0.4895	0.8422

15. Calculate the mean, median, 10% trimmed mean, and mode for the following data.

Table for Exercise 15									
2	22	6	18	10	14	12	12	16	8

16. Discuss the usefulness of each of the measures of central tendency with respect to the following situations.

   a. A company is considering a move into a regional market for specialty soft drinks. In analyzing the size containers that his competitors are currently offering, would the company be more interested in the mean, median, or mode of their containers?

   b. The creative director for an advertising agency is trying to target an ad campaign that will be shown in one city only. Would he be more interested in the mean or median family income in the city?

   c. A young economist was assigned the task of comparing the interest rates of ninety day certificates of deposit (CDs) in three major cities. Should she compare the mean, median, or modal interest for the banks in the three cities?

   d. A telephone company is interested in knowing how customers rate their service: excellent, good, average, or poor. Would the company be more interested in studying the mean, median, or mode of the customer service ratings?

17. Discuss the usefulness of each of the measures of central tendency with respect to the following situations.

   a. A doctor is interested in analyzing the increase in systolic blood pressure caused by a certain antibiotic. Would the manufacturer be more interested in the mean, median, or mode of the ratings?

   b. A car manufacturer is trying to decide in what colors it should offer its new sports coupe. In analyzing the preferred colors of other sports coupes, would the manufacturer be more interested in the mean, median, or mode of the colors?

   c. A manufacturer of chocolate bars is interested in knowing how people rate its chocolate: the best, above average, average, below average, or the worst. Would the company be more interested in the mean, median, or mode of the ratings?

   d. A realtor is interested in studying the prices of recent home sales in an area which has many diverse neighborhoods. Would the mean, median, or mode of the prices of recent home sales be the best measure of central tendency?

18. The following table contains the daily high temperatures for a southern city in July (measured in degrees Fahrenheit).

Table for Exercise 18 – High Temperatures in July (°F)									
84	85	84	88	94	100	97	102	97	89
89	90	88	95	91	95	99	93	97	99
90	94	90	88	91	88	106	99	102	85

   a. Calculate the mean of the daily high temperatures.

   b. Calculate the median of the daily high temperatures.

   c. Calculate the mode of the daily high temperatures.

   d. Calculate the 10% trimmed mean of the daily high temperatures.

   e. Which measure of central tendency do you think best describes the center of the data set? Why?

19. A tour guide informs his group that the "average" temperature at their destination is 60 degrees Fahrenheit. Once they arrive, they discover that the daytime highs are about 120 degrees Fahrenheit and the nighttime lows are about 0 degrees Fahrenheit. Do you feel the tour guide accurately described the temperatures to the group? Discuss.

20. A worker is participating in a test on a new machine. Her daily production, measured in numbers of units, for the twenty-day test is listed in the following table. On days 4 and 5, the worker was ill and went home shortly after coming to work.

Table for Exercise 20 – Daily Production									
Day	1	2	3	4	5	6	7	8	9	10
Units	100	104	117	20	20	111	105	106	115	101
Day	11	12	13	14	15	16	17	18	19	20
Units	101	102	115	116	113	103	104	119	118	108

a. What level of measurement does the data possess?

b. Compute the 10% trimmed mean and the 20% trimmed mean.

c. Considering the worker's illness, which measure computed in part **b.** best describes the production capability of the machine? Discuss.

21. Consider the following per capita greenhouse emissions (in tons of carbon dioxide equivalent per capita) for 10 randomly selected states.

Table for Exercise 21 – Greenhouse Emissions per Capita (Tons)				
11.76	15.65	22.93	24.75	21.22
18.72	22.55	27.99	12.23	114.40

a. What level of measurement does the data possess?

b. Compute the 10% trimmed mean and the 20% trimmed mean.

c. Considering the data, which measure computed in part **b.** best describes the per capita greenhouse emissions? Discuss.

22. Consider the following monthly sales for a small clothing store in a resort community.

Table for Exercise 22 – Sales	
Month	**Sales ($)**
January	100,500
February	120,000
March	133,000
April	145,000
May	160,000
June	180,000
July	200,000
August	185,000
September	175,000
October	120,000
November	180,000
December	330,000

a. Draw a line graph of the data.

b. Calculate the two-period moving averages for the data.

c. Calculate the three-period moving averages for the data.

d. Add line graphs for the two-period moving averages and three-period moving averages to the graph which you constructed in part **a.**

e. Which series of data (the original sales data, the two-period moving averages, or the three-period moving averages) do you think best represents sales for the year? Why?

23. Late in the summer of 1996, Tiger Woods became a professional golfer. This highly publicized event followed a sensational college career at Stanford University, where Tiger won three United States Amateur championships. Tiger was not a professional very long before he had his first win on the pro tour, the Las Vegas Invitational. He received a total of $297,000 for his accomplishment. The prize money (in thousands of dollars) for the top 40 finishers in the tournament are given below.

Table for Exercise 23 – Tournament Prize Money (Thousands of Dollars)							
297.0	60.2	46.2	31.3	21.4	14.5	10.7	8.5
178.2	60.2	31.3	31.3	14.5	14.5	10.7	8.5
95.7	46.2	31.3	24.7	14.5	14.5	8.5	8.5
95.7	46.2	31.3	21.4	14.5	14.5	8.5	8.5
60.2	46.2	31.3	21.4	14.5	10.7	8.5	8.5

a. Find the mean.

b. Find the median.

c. Find the mode.

d. Find the 10% trimmed mean and compare it to the mean and the median.

e. Comment on the skewness of the distribution.

4.2   Measures of Dispersion

Suppose all people looked alike, all cars looked alike, everyone wore the same kind of clothes, and there was only one kind of hamburger (plain). Without diversity it would be a boring world and a world in which statistics would be of little value. Since much of statistics is devoted to describing, analyzing, and explaining variability, understanding how variability is measured is essential to understanding statistics.

The concept of **variability** (also referred to as **dispersion** or **spread**) is as vague as the concept of central tendency. And vague concepts lead to different measurement ideas. The same issues that are important in evaluating measures of location are meaningful in evaluating measures of dispersion.

Many of the good measures of dispersion use the concept of deviation from the mean. If the mean is a focal point of base, use it as a common basis from which to measure variation. The distance that a point is from its mean is called a **deviation from the mean**. A data set and its deviations from the mean are calculated in Table 4.8.

Table 4.8 – Calculating Deviations from the Mean	
Data Set: 3, 12, 20, 15, 0 Mean = 10	
Data Values	Deviations from the Mean (Data – Mean = Deviation)
3	$3 - 10 = -7$
12	$12 - 10 = 2$
20	$20 - 10 = 10$
15	$15 - 10 = 5$
0	$0 - 10 = -10$

Because the mean is the point at which the sum of the positive deviations equals the sum of the absolute values of the negative deviations, the deviations will always sum to zero. Many of the variability measures average the deviations in some form.

Range

The range is the simplest measure of dispersion. It does not provide much depth or understanding of the measure of spread and does not use the deviation concept.

Definition

The **range** is the difference between the largest and smallest data values.

Example 4.8

Calculate the range of the following data set.

$$4, 6, 16, 9, 24, 8, 0, 12, 1$$

Solution

The largest value equals 24 and the smallest value equals 0. Thus, the range is calculated as follows.

$$\text{Range} = 24 - 0 = 24$$

The problem with the range is that it is also affected by outliers, and it does not bring all the information in the data directly to bear on the problem of measuring variation. That is, the range only uses two values (the largest and smallest) to measure spread rather than all of the observations. The other measures of dispersion discussed in this chapter are generally more appropriate measures of spread.

Mean Absolute Deviation

One of the ways of obtaining information about the spread of the data is to analyze the deviations from the mean. Instead of adding the raw deviations, suppose the absolute values of the deviations (which can be interpreted as distance from the mean) are summed and divided by the number of deviations. This new measure computes the average distance from the mean for the data set. This measure is called the **mean absolute deviation**. If data set A has a larger average deviation than B, then it is reasonable to believe that data set A has more variability than data set B.

Formula

Mean Absolute Deviation

The sample mean absolute deviation (MAD) is given by

$$\text{MAD} = \frac{\sum |x_i - \bar{x}|}{n}.$$

Example 4.9

Suppose six people participated in a 1000 meter run. Their times, measured in minutes, are given below.

$$4, 10, 9, 11, 9, 7$$

The mean time is approximately 8.3 minutes. Find the mean absolute deviation.

Solution

In Table 4.9 we do the basic calculations needed to compute the mean absolute deviation.

Note: that the percentages in Tables 4.9 and 4.11 actually add up to 99.99 due to rounding.

Table 4.9 – Calculating Mean Absolute Deviation

Time (Minutes)	Deviation $x_i - \bar{x}$	Absolute Deviation $\lvert x_i - \bar{x} \rvert$	% of Total Deviation
4	4 – 8.3	4.3	37.72
10	10 – 8.3	1.7	14.91
9	9 – 8.3	0.7	6.14
11	11 – 8.3	2.7	23.68
9	9 – 8.3	0.7	6.14
7	7 – 8.3	1.3	11.40
Total		11.4	100%

$$\text{Mean Absolute Deviation } = \frac{11.4}{6} = 1.9 \text{ minutes}$$

Thus, on average, the points are 1.9 units from the mean. Note that the contribution to the sum of the deviations is proportional to the size of the deviation. That is, if one absolute deviation is twice as large as another, it contributes twice as much to the value of the statistic. For example, compare the data point 7, which is 1.3 units from the mean, to the data point 11, which is 2.7 units from the mean. The percentage contribution to the total deviation is 11.40% for 7 and 23.68% for 11, which is in proportion to their respective distances from the mean. A variability measure in which each data value contributes proportionally to its distance from the mean seems reasonable.

Example 4.10

Suppose the value 200 is added to the data set given in Example 4.9.

The mean is drastically affected, increasing from 8.3 to 35.7. In Table 4.10 we redo the basic calculations for the mean absolute deviation. What effect, if any, does the value of 200 have on the MAD?

Solution

Table 4.10 – Calculating Mean Absolute Deviation

Data	Deviation $x_i - \bar{x}$	Absolute Deviation $\lvert x_i - \bar{x} \rvert$
4	4 – 35.7	31.7
10	10 – 35.7	25.7
9	9 – 35.7	26.7
11	11 – 35.7	24.7
9	9 – 35.7	26.7
7	7 – 35.7	28.7
200	200 – 35.7	164.3
Total		328.5

The mean absolute deviation changes dramatically, increasing to 46.9 $\left(\dfrac{328.5}{7}\right)$. Therefore, the mean absolute deviation is sensitive to outliers and is not a resistant measure. The mean absolute deviation is a very intuitive measure of variation.

Variance and Standard Deviation

The **variance** and **standard deviation** are the most common measures of variability. Since the standard deviation is computed directly from the variance, our discussion will center on the variance. Like the MAD, the variance and standard deviation provide numerical measures of how the data vary around the mean. If the data are tightly packed around the mean, the variance and standard deviation will be relatively small. On the other hand, if the data are widely dispersed about the mean, the variance and standard deviation will be relatively large.

Formula

Variance

The **variance** of a data set containing the complete set of population data is given by

$$\sigma^2 = \frac{\sum(x_i - \mu)^2}{N},$$

where μ is the population mean of the data set, N is the size of the population, and x_i is a particular value in the data set. σ^2 is pronounced *sigma squared*, and is called the **population variance**.

The **variance** of a data set containing sample data is given by

$$s^2 = \frac{\sum(x_i - \bar{x})^2}{n-1},$$

where $\bar{x}$ is the mean of the sample data, n is the size of the sample, and x_i is a particular value in the sample. s^2 is called the **sample variance**.

Both these definitions can be construed to be averages, although at first glance it may not be readily apparent. That is, for the population variance, we are adding up the sum of the squared deviations and dividing by the number of items that are added. Thus, the population variance is the average squared deviation from the mean. For the sample variance, we are dividing by $n - 1$ because it gives us an unbiased estimate of the population variance.

It is usually not necessary to compute a variance by manual methods, except to become familiar with the definition.

Example 4.11

Given the following times in minutes of six persons running a 1000 meter course, compute the sample variance.

$$4, 10, 9, 11, 9, 7$$

Solution

We previously computed the mean of this sample as 8.3. In Table 4.11 we do the basic calculations needed to compute the sample variance.

Data	Deviation $x_i - \bar{x}$	Squared Deviation $(x_i - \bar{x})^2$	% of Total Squared Deviation
4	$4 - 8.3 = -4.3$	18.49	59.00
10	$10 - 8.3 = 1.7$	2.89	9.22
9	$9 - 8.3 = 0.7$	0.49	1.56
11	$11 - 8.3 = 2.7$	7.29	23.26
9	$9 - 8.3 = 0.7$	0.49	1.56
7	$7 - 8.3 = -1.3$	1.69	5.39
Total		31.34	100%

Table 4.11 – Calculating the Sample Variance

$$s^2 = \frac{\sum (x_i - \bar{x})^2}{n-1} = \frac{31.34}{5} = 6.268 \text{ minutes squared}$$

Thus the average squared deviation of the data is 6.268 minutes squared. The phrase "minutes squared" in the last sentence may seem a bit odd. No one carries out transactions in square minutes, or square dollars, or square tons, so it is difficult to interpret the significance of the measurement in this form. That is why the standard deviation exists. It converts the measure into the original units by taking the square root of the variance.

Definition

The **standard deviation** is the square root of the variance.

But since there are two measures of variance, there will be two standard deviations, one for population data and one for sample data.

$$\sigma = \sqrt{\sigma^2} \text{ the \textbf{population standard deviation}}$$

$$s = \sqrt{s^2} \text{ the \textbf{sample standard deviation}}$$

For Example 4.11,

$$s = \sqrt{6.268 \text{ minutes}^2} \approx 2.50 \text{ minutes.}$$

It is important to remember these symbols $(\sigma$ and $s)$, since the standard deviation is a fundamental statistical concept.

Describing the standard deviation in an intuitive way is not easy. It is not the average deviation from the mean (which always equals 0), although in most cases it will be reasonably close to the mean absolute deviation. The fact is that the standard deviation is the square root of the average squared deviation. It is not an intuitive concept! Certainly a reasonable question at this juncture is, if the standard deviation is not very intuitive, why use it to measure deviation? Part of the answer lies in the fact that it is expressed in the same units as the data. The variance is an important theoretical measure of variability because it has "nice" mathematical properties (in contrast to the more intuitive MAD).

The standard deviation can be used to measure how far data values are from their mean. You will often see the majority of data values within one standard deviation of the mean. Further, relatively few data values will be more than two standard deviations from the mean.

The variance and standard deviation both suffer from the same problem as the mean: they are very sensitive to outliers. Suppose the value 200 was added to the data in Example 4.11. The sample variance would increase from 6.268 to 5253.238. The new sample variance (which includes the outlier 200) then is 838 times as large as the original variance. The standard deviation increases from 2.50 to 72.50. The presence of the outlier tarnishes the interpretation of the standard deviation as a measure of variability.

Another interesting property of the variance is that values further from the mean contribute a disproportionate amount to the value of the statistic. In Example 4.11, one data point, 4, which is 4.3 units from the mean, contributes 59% of the variation in the data (see the column labeled "% of Total Squared Deviation" in Table 4.11). Compare this to the data point 7, which is 1.3 units from the mean, yet only contributes to 5.39% of the total variation. The reason that 4 contributes so heavily to the total variation is because the deviations are squared. By squaring the deviations, values further from the mean have disproportionate affects on the sum of the squared deviations.

While there are a number of descriptive tools available for summarizing variability, the variance and standard deviation are the most frequently used statistics.

Dispersion and Time Series Data

If the time series is stationary, then the methods discussed in this section can be used to measure dispersion. However, if the series is nonstationary, the dispersion measures we have discussed are not applicable. Measuring variation in a nonstationary time series is beyond the scope of this text.

Using the Standard Deviation

Although the standard deviation is not an intuitive concept, knowing the mean and standard deviation of a data set provides a great deal of information about the data. If the histogram of the measurements is bell-shaped, the **empirical rule** describes the variability of a set of measurements. **Chebyshev's Theorem** is a more general rule describing the variability of *any* set of data regardless of the shape of its distribution.

Empirical Rule

Who is King of the Hill?

In 1961, Wilt Chamberlain was the National Basketball Association (NBA) rebounding leader with 27 rebounds per game. In 1992, the colorful Dennis Rodman won the same honor with 18.7 rebounds per game. Common sense suggests that professional basketball in the 1990's is played at a much higher level than in the 1960's. So why have the rebounding leader's number fallen? Is it another case of "less is more"?

Researchers investigating this interesting puzzle considered two other variables: the number of rebounding opportunities (this had gone down since the field goal percentage has increased historically) and the average number of minutes played per game, which has also fallen. Thus, when we adjust the actual rebounds obtained by the rebounding leaders to the number of minutes played and the total number of rebounding opportunities, we see a completely different picture. The adjusted rebound numbers for Chamberlain and Rodman are 35.42 and 51.06 respectively.

Figure 4.11 – One Sigma

One sigma rule: If the distribution of the data is bell-shaped, about 68% of the data should lie within one standard deviation of the mean.

A deviation of more than one sigma from the mean is to be expected about once in every three observations.

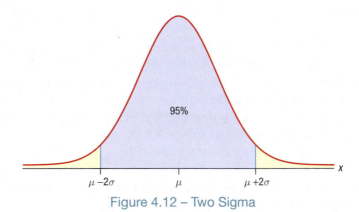

Figure 4.12 – Two Sigma

Two sigma rule: If the distribution of data is bell-shaped, about 95% of the data should lie within two standard deviations of the mean.

A deviation of more than two sigma from the mean is to be expected about once in every twenty observations.

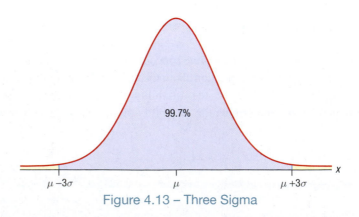

Figure 4.13 – Three Sigma

Three sigma rule: If the distribution of the data is bell-shaped, about 99.7% of the observations should lie within three standard deviations of the mean.

A deviation of more than three sigma from the mean is to be expected about once in every 333 observations, slightly less than 0.3% of the time.

Example 4.12

Suppose a group of high technology stocks has an average earnings per share of $6.26, with a standard deviation of $1.37. If the data possess a bell-shaped distribution, which interval contains 68% of the earnings? Which interval contains 95% of the earnings?

Solution

Using the one sigma rule, we will capture 68% of the observations.

$$\mu \pm 1 \cdot \sigma = \$6.26 \pm \$1.37$$

Using the one sigma rule results in an interval from $6.26 − $1.37 to $6.26 + $1.37. Doing the arithmetic produces an interval from $4.89 to $7.63.

To capture 95% of the earnings, use the two sigma rule, $6.26 \pm 2 \cdot \$1.37$. Doing the arithmetic results in an interval from $3.52 to $9.00.

Note that to increase the percentage of data captured from 68% to 95% requires an interval that is twice as large.

Chebyshev's Theorem

It is important to remember that the empirical rule applies only to bell-shaped distributions. For **any** distribution, regardless of shape, Chebyshev's Theorem may be used, although its results are much more approximate.

Theorem

Chebyshev's Theorem

The proportion of any data set lying within k standard deviations of the mean is at least

$$1 - \frac{1}{k^2}, \text{ for } k > 1.$$

For example, if $k = 2$ at least $1 - \dfrac{1}{2^2} = \dfrac{3}{4}$ (or 75%) of the data values lie within 2 standard deviations of the mean, for any data set. Similarly, if $k = 3$ at least $1 - \dfrac{1}{3^2} = \dfrac{8}{9}$ (or approximately 88.9%) of the data values lie within 3 standard deviations of the mean, for any data set.

Also note that k does not have to be an integer value. If $k = 1.5$, at least $1 - \dfrac{1}{1.5^2} = \dfrac{5}{9}$ (or approximately 55.6%) of the data values will lie within 1.5 standard deviations of the mean, for any data set.

Example 4.13

The tuition and fees of colleges and universities distribution histogram for the United States in 2009-2010 is shown in Figure 4.14. The mean of the data is \$16,422, while the standard deviation is \$10,014. What can we conclude from Chebyshev's Theorem using $k = 2$?

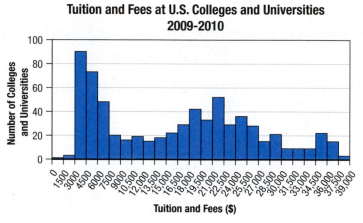

Tuition and Fees at U.S. Colleges and Universities 2009-2010

Figure 4.14

Solution

Because we are interested in $k = 2$, we will look at the values two standard deviations above and below the mean.

Two standard deviations above the mean is

$$\mu + 2\sigma = 16422 + 2(10014) = \$36,450$$

and two standard deviations below the mean is

$$\mu - 2\sigma = 16422 - 2(10014) = -\$3606.$$

Because costs (in this case) cannot be negative, we have a natural boundary, so we will consider this as zero.

Therefore, by Chebyshev's Theorem, we can say that at least 75% of the tuition and fees of colleges and universities in the United States are between \$0 and \$36,450 for 2009-2010.

4.2   Exercises

Basic Concepts

1. Describe three measures of variation. Discuss the strengths and weaknesses of each.

2. What does the standard deviation measure?

3. Why are the variance and standard deviation more commonly used as measures of variability than the MAD?

4. Explain how the variance can be construed as an average.

5. True or false: The variance and standard deviation are resistant measures.

6. When is it appropriate to calculate the variance of a time series?

7. What is the empirical rule? When is it appropriate to use the empirical rule?

8. What is Chebyshev's Theorem?

Exercises

9. Find the missing age in the following set of four student ages.

Table for Exercise 9 – Student Ages		
Student	Age	Deviation from the Mean
A	19	−4
B	20	−3
C	?	+1
D	29	+6

10. Find the missing weight in the following data set.

Table for Exercise 10 – Weights		
Person	Weight	Deviation from the Mean
A	144	−20
B	156	−8
C	?	+1
D	176	+12

11. Consider the following time until failure for 10 randomly selected car batteries (measured in years).

Table for Exercise 11 – Years until Failure for Car Batteries									
5	3	4	6	2	5	7	10	8	4

a. Calculate the sample variance of the time until failure.

b. Calculate the sample standard deviation of the time until failure.

c. Calculate the range of the time until failure.

d. What are some of the factors which might contribute to the variation in the observations?

12. Consider the following distances jumped (in feet) by 8 randomly selected long jumpers.

Table for Exercise 12 – Jump Distances (Feet)							
21	15	12	18	10	14	17	11

    a. Calculate the sample variance of the distances jumped.

    b. Calculate the sample standard deviation of the distances jumped.

    c. Calculate the range of the distances jumped.

    d. What are some of the factors which might contribute to the variation in the observations?

13. The interest rates on 30-year mortgages offered by seven randomly selected banks in a large metropolitan area are recorded in the following table.

Table for Exercise 13 – Interest Rates (%)						
7.5	8.0	7.0	7.25	8.5	8.25	7.75

    a. Calculate the sample variance of the interest rates.

    b. Calculate the sample standard deviation of the interest rates.

    c. Calculate the range of the interest rates.

    d. What are some of the factors which might contribute to the variation in the observations?

14. A researcher has hypothesized that female college students are more disciplined than male college students. The researcher believes that a reasonable measure of discipline is performance on a statistics test in terms of both absolute scores and consistency of scores. Seven male statistics students and seven female statistics students are randomly selected and their scores on a statistics test are observed.

Table for Exercise 14 – Test Scores							
Males	65	100	75	45	85	73	95
Females	75	80	95	85	82	72	49

    a. Calculate the average test score for male students and female students separately.

    b. Calculate the variance of the test scores for male students and female students separately.

    c. Calculate the standard deviation of the test scores for male students and female students separately.

    d. Do you think that the data tend to support the hypothesis that female college students are more disciplined than male college students based on the researcher's measurement?

    e. What do you think about this particular measurement of discipline?

15. Consider the following market values of two portfolios of stocks at five randomly selected times during a year.

Table for Exercise 15 – Market Values ($)					
Portfolio A	150,000	155,000	145,000	160,000	140,000
Portfolio B	130,000	175,000	100,000	150,000	195,000

   a. What statistical criteria might you use to select the better portfolio? Justify your answer.

   b. Calculate the statistics you proposed in part **a.**

   c. Which portfolio has the least amount of risk? Why?

16. Add 20 to each of the following data values.

Table for Exercise 16					
81	99	97	81	85	86
99	93	96	83	82	91

   a. Compute the mean and standard deviation for both the original data and adjusted data.

   b. Compare the mean and standard deviation of the adjusted data to the mean and standard deviation of the original data.

   c. Describe the effect on the mean and standard deviation of adding a constant to a data set.

17. Adjust the following data values by subtracting 20 from each data value.

Table for Exercise 17				
745	789	712	764	736
758	722	773	751	741

   a. Calculate the mean and variance for the original and adjusted data.

   b. Compare the mean and variance of the adjusted data to the mean and the variance of the original data.

   c. Describe the effect of subtracting a constant value from each member of a data set on the mean and variance of the data.

18. The average score on a pre-employment test is 26 with a standard deviation of 7. Using Chebyshev's Theorem, state the range in which at least 88.89% of the data will reside.

19. The daily average number of phone calls to a call center is 972 with a standard deviation of 127. Using Chebyshev's Theorem, state the range in which at least 75% of the data will reside.

20. There is an annual chowder eating contest in a small New England town. The average amount of chowder eaten at the contest was 32 ounces with a variance of 64 ounces. Given that one hundred people participated in the contest, find:

   a. the approximate number of people who ate between 24 and 40 ounces of chowder.

   b. the approximate number of people who ate between 16 and 48 ounces of chowder.

   c. What assumptions did you make about the amount of chowder eaten by each contestant in answering parts **a.** and **b.**?

21. The manager of a local diner has calculated his average daily sales to be $4500 with a standard deviation of $750.

 a. In what range can the manager expect his daily sales to be 68% of the time?

 b. In what range can the manager expect his daily sales to be 95% of the time?

 c. In what range can the manager expect his daily sales to be 99.7% of the time?

 d. What assumption did you make about daily sales when answering parts **a.**, **b.**, and **c.**?

22. A management consulting firm is evaluating the salary structure for a large insurance company. The goal of the study is to develop salary ranges for each of the possible job grades within the company. The company and the firm have agreed that a reasonable salary range for each job grade can be determined by finding the salary range in which 95% of the current salaries for that job grade fall. The average salary and the standard deviation of the salaries are listed in the following table for three of the job grades.

Table for Exercise 22 – Salaries ($)			
Job Grade	25	33	40
$\bar{x}$	22,000	35,000	45,000
s	1500	2000	5000

 a. Determine the appropriate salary ranges for the three job grades.

 b. What assumption did you make about the salaries in each of the job grades in answering part **a.**?

4.3   Measures of Relative Position

Suppose you want to know where an observation stands in relation to other values in a data set. For example, on many standardized tests such as the SAT, GMAT, and ACT, the test scores themselves are rather meaningless unless they are associated with some measure that tells you how well you did relative to others taking the same test. There are two principal methods of communicating relative position: **percentiles** and **z-scores**. Both of these methods are data transformations which change the scale of the data in some way.

Percentiles

The most commonly used measure of relative position is the percentile. In fact, we have already discussed the 50th percentile; it is the median. For example, in data sets that do not contain significant quantities of identical data, the 30th percentile is a value such that about 30 percent of the values are below it, and about 70 percent are above it.

> ### Definition
>
> Given a set of data $x_1, x_2, ..., x_n$, the **P^{th} percentile** is a value, say x, such that approximately P percent of the data is less than or equal to x and approximately $(100 - P)$ percent of the data is greater than or equal to x.

Finding the Pth Percentile

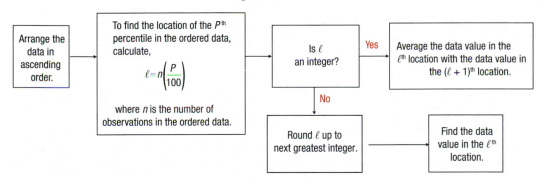

To determine the Pth percentile, perform the following steps.

> ### Procedure
>
> #### Finding the Pth Percentile
>
> 1. Form an ordered array by placing the data in order from smallest to largest.
>
> 2. To find the location of the Pth percentile in the ordered array, let
>
> $$\ell = n\left(\frac{P}{100}\right)$$
>
> where n is the number of observations in the ordered data.
>
> 3. If ℓ is not an integer, then round ℓ up to the next greatest integer. For example, if $\ell = 7.1$, then round ℓ up to 8 and find the data value in the ℓth location. If ℓ is an integer value, then average the data value in the ℓth location with the data value in the $(\ell + 1)^{th}$ location.

Interpreting Percentiles

When students take the SAT Reasoning Test, they receive a copy of their scores as well as the percentile they fall into. This percentile can sometimes be confusing. If a student receives a score of 620 on the verbal section, they might fall into the 84th percentile. This means that they received a higher score than 84 percent of the students. The same score on the math section might place the student into the 80th percentile. Receiving a score of 800 on verbal or math will put the student in the 99th percentile. This means that less than 1 percent of the students taking the SAT had the same score.

It is important to remember that when you find the value of ℓ, this result is not the percentile. It is the *location* of the percentile in the ordered array. Thus, if the result of calculating (and rounding up) ℓ is 15, then the desired percentile would be the fifteenth value in the ordered list.

Example 4.14

Find the 50th percentile for the following data set.

$$3, 5, 0, 1, 9, 2, 7$$

Solution

Number of observations, $n = 7$.

The percentile, $P = 50$.

The location of the percentile, $\ell = 7 \cdot \left(\dfrac{50}{100} \right) = 3.5$.

Since the location of the percentile is not an integer, the value is rounded up to 4.

Thus, the fourth observation in the ordered array is the 50th percentile.

$$0, \quad 1, \quad 2, \quad \boxed{3}, \quad 5, \quad 7, \quad 9$$
$$\uparrow$$
fourth observation

Therefore, the median value (which is the 50th percentile) is 3.

Example 4.15

Suppose the 40 members of your company are given a screening test for a new position. These scores are reported in Table 4.12. To inform potential employees of their screening test performance you may wish to report various percentiles for the test scores. Find the 10th and 88th percentiles for the test.

Table 4.12 – Test Scores			
67	45	18	82
45	54	61	55
63	47	21	31
58	46	43	49
35	71	69	56
54	80	73	77
27	70	41	29
66	32	44	33
21	64	52	81
48	55	57	62

Table 4.13 – Ordered Test Scores			
18	43	54	66
21	44	55	67
21	45	55	69
27	45	56	70
29	46	57	71
31	47	58	73
32	48	61	77
33	49	62	80
34	52	63	81
41	54	64	82

Solution

In order to calculate the percentiles, the data must be placed in an ordered array (Table 4.13). To compute the 10th percentile, its position in the ordered array must be determined.

The number of observations, $n = 40$.

The percentile, $P = 10$.

The location of the percentile, $\ell = 40 \cdot \left(\dfrac{10}{100} \right) = 4$. Since ℓ is an integer, the 4th and 5th observations in the array must be averaged. Since the fourth data value is 27 and the fifth data value is 29, then the 10th percentile is calculated as follows.

$$10^{th} \text{ percentile} = \frac{27 + 29}{2} = 28$$

To determine the 88th percentile, first calculate its location in the ordered array.

$$\ell = 40 \cdot \left(\frac{88}{100} \right) = 35.2$$

Since the location is not an integer, its value is rounded up to 36. The 36th observation in the ordered array will correspond to the 88th percentile. The 36th value is 73 in Table 4.13, so 73 is the 88th percentile.

A slightly different problem connected with percentiles involves taking a raw score and determining its corresponding percentile. Raw scores are usually not very meaningful. If someone scores a 56 on the screening test in the previous example, is that substantially less or about the same as someone who scored a 67 on the same test? To compare these two scores, find the percentile of each.

Formula

Percentile

The **percentile** of some data value x is given by:

$$\text{percentile of } x = \frac{\text{number of data values less than or equal to } x}{\text{total number of data values}} \cdot 100$$

Note that when finding the percentile of a specific value, if there are multiple occurrences of that value in the data, they all need to be counted in the numerator in order to calculate the percentile. To determine the percentile for a score of 56, the number of data values less than or equal to 56 must be counted. Since there are 24 data values less than or equal to 56, the resulting percentile would be

$$\text{percentile of a score of } 56 = \frac{24}{40} \cdot 100 = 60.$$

Hence a score of 56 on the screening test corresponds to the 60th percentile. Thus, approximately 60 percent of the scores are less than or equal to 56. Next, compute the percentile for a score of 67:

$$\text{percentile of a score of } 67 = \frac{32}{40} \cdot 100 = 80.$$

A score of 67 on the screening test corresponds to the 80th percentile. The score was better than or equal to approximately 80 percent of all other scores on the test. By computing percentiles, we have changed the data's scaling. We see the data from a new perspective. Using percentiles, it is clear that a score of 67 is significantly better than a score of 56. The 11 point difference in raw score is translated into a 20 percent differential on the percentile scale.

Quartiles

The 25th, 50th, and 75th percentiles are known as **quartiles** and are denoted as Q_1, Q_2, and Q_3. They serve as markers that divide the data into four equal parts. Q_1 separates the lowest 25 percent, Q_2 represents the median (50th percentile), and Q_3 marks the beginning of the top 25 percent of the data.

Since quartiles are nothing more than percentiles (25th, 50th, and 75th), the same methods used to calculate percentiles will also produce quartiles. For the screening test data in the previous example, the location of the 25th percentile would be

$$\ell = 40 \cdot \left(\frac{25}{100} \right) = 10.$$

Since the location is an integer, we average the 10th and 11th observations in the ordered data to find the 25th percentile.

$$Q_1 = 25\text{th percentile} = \frac{41 + 43}{2} = 42.$$

Therefore, we would expect 25 percent of the data to be less than or equal to 42.

The location of the 50th percentile is given by

$$\ell = 40 \cdot \left(\frac{50}{100} \right) = 20.$$

Since the location is an integer, we must average the 20th and 21st observations in order to calculate the percentile.

$$Q_2 = 50\text{th percentile} = \frac{54 + 54}{2} = 54,$$

which means approximately half the data is at or below 54.

The location of the 75th percentile is given by

$$\ell = 40 \cdot \left(\frac{75}{100} \right) = 30.$$

Since the location is an integer, average the 30th and 31st observations in the ordered array.

$$Q_3 = 75\text{th percentile} = \frac{64+66}{2} = 65$$

This means that approximately 75 percent of the data are less than or equal to 65. The quartiles are useful descriptions of data. They provide a good idea of how the data vary. The **interquartile range** is a measure of dispersion that is calculated using the first and third quartiles.

Formula
Interquartile Range
The **interquartile range** is a measure of dispersion which describes the range of the middle fifty percent of the data.
$$IQR = Q_3 - Q_1$$

For the screening test data, the interquartile range is 65 − 42 = 23, indicating that the middle 50 percent of the data spans a 23 unit range.

Box Plots – Graphing with Quartiles

A very important use of quartiles is in the construction of box plots. As the name implies, box plots are graphical summaries of the data which, when constructed, have a box-like shape. They provide an alternative method to the histogram for displaying data. A **box plot** is a graphical summary of the central tendency, the spread, the skewness, and the potential existence of outliers in the data. Figure 4.15 displays a box plot of the screening test data from Example 4.15.

Box Plot of Screening Test Scores

Figure 4.15

The box plot is constructed from five summary measures: the largest data value, the smallest data value, the 25th percentile, the 75th percentile, and the median.

The lower boundary of the box is the 25th percentile, which is 42 for the screening test data. The upper boundary of the box is the 75th percentile, which is 65 for the screening test data. The median is marked with a line through the box. The median of the test scores data is 54. Notice that the box itself represents the middle 50% of the data, and the length of the box is the interquartile range.

In Figure 4.15, a line is drawn from the 25th percentile to the smallest test score of 18, and another line is drawn from the 75th percentile to the largest score of 82. These lines are often referred to as whiskers and the box plot is often referred to as the **box and whisker plot**. The box plot for the screening test scores shows that the test score data are slightly skewed to the left. Why? Because the whisker extending from Q_1 appears to be longer than the whisker that extends from Q_3.

Although the box plot can be used to display data for a single data set, the histogram is probably more useful for this purpose. The real power of the box plot is the ease with which it allows the comparison of several data sets. Consider the number of wins per season for the New York Yankees, Los Angeles Dodgers, Atlanta Braves, and Chicago Cubs. The four data sets are displayed by box plots in Figure 4.16. It is easy to see from the box plots that the center of the Yankees' number of wins is higher than that for the Dodgers which has a higher center than the center of Braves, and finally the Cubs. Also it appears that the spread of the data, or the variation within the observed values, is not the same for all four data sets. This type of comparison will be used in later chapters to help confirm assumptions which must be made about the data in order to perform statistical inference.

Box Plots of the Number of Franchise Wins per Season 1961–2010

Figure 4.16

Detecting Outliers

The concept of an outlier is an arbitrary concept. What you consider an outlier and what someone else considers an outlier may not be the same thing. However, one definition of an outlier which has gained some acceptance is developed in the context of a box plot.

Definition

A data point is considered an **outlier** if it is 1.5 times the interquartile range above the 75th percentile or 1.5 times the interquartile range below the 25th percentile.

If there is an outlier in the data set, the whiskers are drawn to the largest or smallest data point which is within 1.5 times the interquartile range from the box, and the outliers are marked with an open circle. For example, suppose test scores of 110 and 2 were added to the screening test data.

Table 4.14 – Ordered New Test Scores			
2	43	55	67
18	44	55	69
21	45	56	70
21	45	57	71
27	46	58	73
29	47	61	77
31	48	62	80
32	49	63	81
33	52	64	82
34	54	66	110
41	54		

For the new screening test data, the 25th percentile is 41 and the 75th percentile is 66, meaning that the interquartile range is $66 - 41 = 25$. A point is considered an outlier if the data point is:

- larger than the 75th percentile + 1.5 times the interquartile range.

$$66 + 1.5 \cdot 25 = 103.5$$

- smaller than the 25th percentile − 1.5 times the interquartile range.

$$41 - 1.5 \cdot 25 = 3.5$$

Since 110 is larger than 103.5, it is considered an outlier. Since 2 is smaller than 3.5, it is also considered an outlier. Figure 4.17 shows the box plot of the screening test data with the outliers incorporated. Notice that the whiskers did not change because 82 and 18 are still the largest and smallest observations within 1.5 times the interquartile range from the box.

Box Plot with Added Test Scores

Figure 4.17

z-Scores

The **z-score** is a standardized measure of relative position, with respect to the mean and variability (as measured by the standard deviation) of the data set.

Formula

z-Score

The z-score transforms a data value into the number of standard deviations that value is from the mean.

$$z = \frac{x - \mu}{\sigma}.$$

Describing a data value by its number of standard deviations from the mean is a fundamental concept in statistics that is found throughout this book. It is used as a standardization technique to describe properties of data sets and to compare the relative values of data from different data sets.

Baseball Players, Better or Worse?

In the history of Major League Baseball (MLB), the last time someone had a batting average above .400 was Ted Williams in 1941. Before that, Ty Cobb had hit .420 in 1911. Finally, in 1980 George Brett came close but hit only .390. From this it appears that with time hitting for high averages is becoming more difficult. Is this then evidence of progress or lack of progress? The late evolutionary biologist Stephen Jay Gould, himself an avid baseball fan, had an interesting statistical angle to this riddle. He computed and observed:

(1) The yearly batting averages have remained more or less stable (around .267) over the history of the league.

(2) The annual standard deviations have declined steadily over the same period.

(3) Finally, the yearly batting averages are all normally distributed. Gould then computed the z-scores for all three players and observed they were all well above 4.0. His conclusion: Hitting has improved since the standard deviations of the averages are getting smaller.

The best hitters are still at the fence of the normal distribution and are the best of their times. Therefore, as a result of the decline in standard deviation, the apparent decrease in batting average (disappearance of .400 hitters) is actually a sign of improvement! Less can actually be more in some situations!

Example 4.16

Suppose you scored an 86 on your marketing test and a 94 on your management test. The mean and standard deviations of the two tests are given below.

Table 4.15 – Test Scores		
Course	**Mean**	**Standard Deviation**
Marketing	74	10
Management	82	11

What are the z-scores for your two tests? On which of the tests did you perform relatively better?

Solution

The z-score for the marketing test is $z = \dfrac{86 - 74}{10} = 1.20$.

The z-score for the management test is $z = \dfrac{94 - 82}{11} \approx 1.09$.

On the marketing test you scored 1.20 standard deviations above the mean, compared to only 1.09 standard deviations above the mean for the management test. Even though the raw score on the management test is larger than the raw score on the marketing test, relative to the mean of the data sets, the performance on the marketing test was slightly better. Once again, changing the scale of the data has beneficial effects. It enables the comparison of two measurements that are drawn from different populations.

If a z-score is negative, the data value is less than the mean. Conversely, if the z-score is positive, the data value is greater than the mean. The z-score is also a unit free measure. That is, regardless of the original units of measurement (whether the data are measured in centimeters, meters, or kilometers), an observation's z-score will be the same.

4.3   Exercises

Basic Concepts

1.   What are two methods for describing relative position?

2.   If a data value is calculated to be the 72nd percentile, what does this mean?

3.   Describe how to find the percentile of a particular data value.

4.   What are quartiles? Are they equivalent to percentiles? If so, how?

5.   What is the interquartile range? What does it measure?

6.   What are the advantages of using a box plot to display a data set?

7.   What are the key calculations needed in order to construct a box plot?

8. What is an outlier? How can outliers be identified?

9. What is a *z*-score? Why is it useful?

Exercises

10. The following test scores were recorded for an economics final examination.

Table for Exercise 10 – Test Scores														
60	81	100	44	90	56	71	42	64	100	69	80	90	87	94
41	78	100	50	96	77	61	38	41	68	50	69	85	47	86

   a. Calculate the 20th percentile.

   b. Calculate the 95th percentile.

   c. Interpret the meaning of each of these percentiles.

   d. Determine the percentile rank for the student who scored 56.

   e. Determine the percentile rank for the student who scored 80.

11. Copiers Etc. collects data on the number of copiers sold each day by each salesperson. The number of copiers sold for each salesperson for a small office on a randomly selected day is listed below.

Table for Exercise 11 – Numbers of Copiers Sold											
1	5	2	3	7	6	1	0	0	3	4	5

   a. Calculate the 25th percentile.

   b. Calculate the 90th percentile.

   c. Interpret the meaning of each of these percentiles.

   d. Determine the percentile rank for the salesperson who sold 5 copiers.

   e. Determine the percentile rank for the salesperson who sold 1 copier.

12. Subjects in a marketing study were shown a film and at the end of the film were given a test to measure their recall. The scores are listed in the following table.

Table for Exercise 12 – Test Scores														
97	31	61	49	61	85	35	57	31	26	27	40	86	78	28
61	87	62	92	58	38	95	81	68	64	72	45	57	84	100

   a. Calculate Q_1, the first quartile.

   b. Calculate Q_2, the second quartile.

   c. Calculate Q_3, the third quartile.

   d. Explain the meaning of these quartiles in the context of the marketing study.

   e. Calculate the interquartile range.

   f. Construct a box plot for the test scores. Are there any outliers?

   g. Compute the *z*-score for a test score of 81.

   h. Compute the *z*-score for a test score of 62.

   i. Explain what the *z*-scores in parts **g.** and **h.** are measuring.

13.   A baseball recruiter is interested in 20 perspective players. He goes to several games and determines the batting average for each player. The batting averages are displayed in the following table.

Table for Exercise 13 – Batting Averages									
.330	.260	.180	.150	.200	.400	.020	.190	.290	.200
.170	.150	.250	.270	.320	.280	.270	.220	.270	.300

a.   Calculate Q_1, the first quartile.

b.   Calculate Q_2, the second quartile.

c.   Calculate Q_3, the third quartile.

d.   Explain the meaning of these quartiles in the context of the batting averages.

e.   Calculate the interquartile range.

f.   Construct a box plot for the batting averages. Are there any outliers? (Guess which player is the pitcher.)

g.   Compute the z-score for a batting average of .020.

h.   Compute the z-score for a batting average of .330.

i.   Explain what the z-scores in parts **g.** and **h.** are measuring.

j.   Determine the percentile rank for a player who had a batting average of .270.

k.   Determine the percentile rank for a player who had a batting average of .150.

14.   Consider a set of data in which the sample mean is 64 and the sample standard deviation is 21. For the following specific values, calculate the z-score and interpret the results.

a. $x = 80$

b. $x = 64$

c. $x = 40$

15.   A statistics student scored a 75 on the first exam of the semester and an 82 on the second exam of the semester. The average score and standard deviation of scores for the two exams are given in the following table. On which exam did the student perform relatively better?

Table for Exercise 15 – Test Scores		
	First Exam	**Second Exam**
μ	74	85
σ	10	7

16. A hospital measures babies' heights when they are born in both inches and centimeters. Eight baby girls are randomly selected and the following heights are recorded in both inches and centimeters.

Table for **Exercise 16** – Newborn Heights								
Baby	**1**	**2**	**3**	**4**	**5**	**6**	**7**	**8**
Inches	17.75	18.50	19.25	19.75	20.25	20.50	20.50	20.75
Centimeters	45.09	46.99	48.90	50.17	51.44	52.07	52.07	52.71

a. Calculate the mean height in inches and centimeters for the baby girls.

b. Calculate the standard deviation of the heights of baby girls in both inches and centimeters.

c. Calculate the z-score for the height of Baby Girl 3 measured in inches.

d. For Baby Girl 3, calculate the z-score for the height measured in centimeters.

e. Consider the z-scores calculated in parts **c.** and **d.** Are the z-scores as you expected them to be? Explain.

4.4   Data Subsetting

Looking at the tuition data presented in Example 4.13 with a histogram using fewer intervals gives a slightly different picture. Figure 4.18 provides some idea about **location** and **dispersion** of the data, but nothing specific. Histograms are outstanding at defining the **shape** of the data.

Where is the Central Value of the Tuition Data?

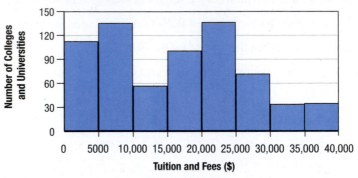

Figure 4.18

There are several measurements that can be used to describe the location or central value of the tuition data. The mode is not considered because the data are quantitative and because there are a large number of observations. Moreover, the mode is not a very informative measure of central tendency. Since the median of the data ($17,468) is substantially larger than the mean ($16,442) or trimmed mean ($16,087), the choice will make a difference. The graph suggests a somewhat multi-modal distribution given the large chunks of data between $0 and $10,000 and from $15,000 to $25,000.

Since the trimmed mean is roughly the same as the mean, the lack of outliers seems to be confirmed. Which one of the three candidate measurements should be considered as the central value of the data?

The choice is not easy. You are justified in selecting any of the three statistical measures (median, mean, or trimmed mean). However, because of the difference between the median and the other two measures, reporting two measures (the mean and median) would not be a bad idea. If only one measure is reported, identify the measure as the mean or the median rather than the average. The term "average" is a bit too ambiguous in current usage.

Describing Dispersion of the Tuition Data

Several measurements which assess variation have been discussed, including variance, standard deviation, mean absolute deviation, and percentiles. Since the variance and standard deviation are different forms of the same measurement, usually only the standard deviation is reported. For the tuition data, a reasonable description of dispersion for the 677 values would be given by the following statistics.

Table 4.16 – Tuition Data	
Measure of Location	**Value**
Mean	$16,442
Median	$17,468
10% Trimmed Mean	$16,087

Standard Deviation: $10,014
Minimum: $790
First Quartile (Q_1): $6049
Median (Q_2): $17,468
Third Quartile (Q_3): $23,800
Maximum: $38,140

Table 4.17 – Tuition Data Percentiles	
Percentile	**Value**
100th (Max)	$38,140
95th	$35,015
90th	$29,982
75th (Q_3)	$23,800
50th (Q_2)	$17,468
25th (Q_1)	$6049
10th	$4166
5th	$3558
0 (Min)	$790

The histogram in Figure 4.18 reveals an interesting characteristic of the data. There appear to be two distinct groups of values: a bundle of data between $0 and $10,000, and another bundle of data between $15,000 and $25,000. What might be responsible for the clustering? We could be looking at the difference between state supported institutions and private institutions. This supposition could be checked rather easily if the data are divided (subsetted) into private and state-supported schools.

The original data contained the annual tuition costs of private and state-funded institutions. In-state tuition was used for the state-supported schools. The structure of a batch of data is often exposed by using another variable to break the data in smaller groups. This is called **subsetting**. A natural structuring of the tuition data is a grouping of private and state-supported institutions.

State-Supported Institutions

(n = 231):

Mean Tuition: $5396

Median Tuition: $5177

Trimmed Mean Tuition: $5232

Standard Deviation: $1864

$Q_1 = \$3995$

$Q_3 = \$6162$

Minimum: $2685

Maximum: $12,844

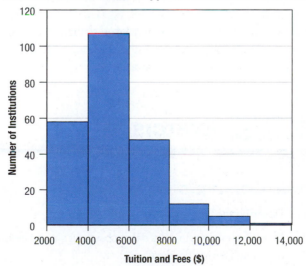

Tuition and Fees at State-Supported Institutions 2009-2010

Figure 4.19

Private Institutions
($n = 446$):

Mean Tuition: $22,163

Median Tuition: $21,630

Trimmed Mean Tuition: $22,174

Standard Deviation: $7376

$Q_1 = \$17,650$

$Q_3 = \$26,468$

Minimum: $790

Maximum: $38,140

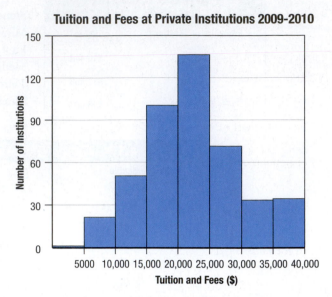

Figure 4.20

The data subsetting suggests that the data clustering revealed in the histogram in Figure 4.18 is the result of merging data from different kinds of educational institutions. Once the data are subsetted, the strong bimodal nature of the data disappears. For private institutions, the majority of the values tend to cluster in the middle. However, for state-supported institutions, it appears that the data are positively skewed with many values clustered to the left of the mean tuition and fees. The data tell quite an interesting story. From a tuition point of view, state-supported schools are substantially less expensive than private schools. There is roughly a $17,000 average annual tuition difference between private colleges and in-state tuition at state-supported schools.

Another data characteristic should stimulate our interest. One of the data points for private school tuition is $790. Is this a mistake? The z-score for this data point is

$$z = \frac{790 - 22163}{7376} \approx -2.90$$

which would indicate the point may be an outlier. Any point 2.90 standard deviations from the mean is worth investigating. In this case the school is Berea College in Berea, KY, which awards every student a 4-year tuition scholarship.

4.4   Exercises

Basic Concepts

1. Describe the purpose of data subsetting.

2. Describe a data set where data subsetting should be implemented. What are the disadvantages of not subsetting the data?

4.5   The Coefficient of Variation

> ### Formula
>
> ### Coefficient of Variation
>
> The **coefficient of variation**, another statistical measure, compares the variation in data sets.
>
> For population data, the coefficient of variation is defined as
>
> $$CV = \left(\frac{\sigma}{\mu} \cdot 100 \right) \%,$$
>
> and for sample data,
>
> $$CV = \left(\frac{s}{\bar{x}} \cdot 100 \right) \%.$$

When comparing the variation of data sets, many times the units of measure will be different. The coefficient of variation standardizes the variation measure by dividing it by the mean. The division has one interesting side effect: the unit of measure is removed from the statistic.

$$CV_{private} = \frac{\$7376}{\$22163} \cdot 100\% \approx 33.28\%$$

$$CV_{public} = \frac{\$1864}{\$5396} \cdot 100\% \approx 34.54\%$$

The coefficient of variation for the private tuition data is 33.28% of the mean value. The group with the largest variation relative to the mean is the in-state tuitions. Their variation is 34.54% of the mean. Using the criterion of variation relative to the mean, the data set with the smallest absolute variation $(s = \$1864)$ has the largest relative variation.

4.5   Exercises

Basic Concepts

1. Discuss the purpose of the coefficient of variation.

2. How is the coefficient of variation calculated?

3. Why is the coefficient of variation important?

Exercises

4. A consumer interest group is interested in comparing two brands of vitamin C. One brand of vitamin C advertises that its tablets contain 500 mg of vitamin C. The other brand advertises that its tablets contain 250 mg of vitamin C. Tablets for each brand are randomly selected and the milligrams of vitamin C for each tablet are measured with the following results.

Table for Exercise 4 – Vitamin C Content (mg)		
	Brand A (500 mg)	Brand B (250 mg)
$\bar{x}$	500	250
s	10	7

   a. Calculate the coefficient of variation for Brand A.

   b. Calculate the coefficient of variation for Brand B.

   c. Which brand more consistently produces tablets as advertised? Explain.

5. A manufacturer of bolts has two different machines. One machine is used to produce $\frac{1}{4}$-inch bolts; the other machine is used to produce $\frac{1}{2}$-inch bolts. It is very important that the machines consistently produce bolts of the correct diameters, or the bolts will not fit on the corresponding nuts. In order to compare the two machines, management randomly selects bolts produced from each machine and computes the average diameter of the bolts and the standard deviation of the bolts. The results of the study are shown in the following table.

Table for Exercise 5 – Bolt Diameter		
	Machine X $\left(\frac{1"}{4}\right)$	Machine Y $\left(\frac{1"}{2}\right)$
$\bar{x}$	0.25"	0.50"
s	0.03"	0.05"

   a. Calculate the coefficient of variation for Machine X.

   b. Calculate the coefficient of variation for Machine Y.

   c. Which machine more consistently produces bolts of the correct diameter? Explain.

4.6   Analyzing Grouped Data

All of the statistical measurements we have discussed so far presume that raw data measurements are readily available. However, there may be instances in which only a frequency distribution of the data is available. When data are presented in that form, they are called **grouped data**. It is important to be able to compute measures such as the mean and variance for this type of data. Note that because the raw data observations are not available, the measures will be approximate.

Finding the Mean and Variance of Grouped Data

The strategy for finding the mean of grouped data involves finding the midpoint of each of the classes in the frequency distribution and then weighting each of these midpoints by the number of observations in the class.

Formula

Mean of Grouped Data

The population **mean of grouped data** is given by

$$\mu = \frac{\sum (f_i M_i)}{N}$$

where

f_i = number of observations in the ith class,

N = the total number of observations in all classes, $N = \sum f_i$, and

M_i = midpoint of the ith class.

The sample mean of grouped data is given by

$$\bar{x} = \frac{\sum (f_i M_i)}{n},$$

where n is the number of observations in the sample.

We can also estimate the variance of grouped data using the following formulas.

Formula

Variance of Grouped Data

The population **variance of grouped data** is given by the following expression.

$$\sigma^2 = \frac{\sum (M_i - \mu)^2 f_i}{N}$$

The corresponding formula for the sample variance is as follows.

$$s^2 = \frac{\sum (M_i - \bar{x})^2 f_i}{n-1}$$

where n equals the total number of observations in the sample.

There are also computational formulas that can be used to calculate the variance of grouped data. These formulas may make it easier to calculate the variance by hand. We will use these formulas in Example 4.17 to calculate the variance for a grouped data set.

Formula

Computational Formulas for the Variance of Grouped Data

The computational formulas for the population and sample variances of grouped data are as follows.

$$\sigma^2 = \frac{\sum (f_i M_i^2) - \dfrac{\left(\sum (f_i M_i)\right)^2}{N}}{N} = \frac{\sum (f_i M_i^2)}{N} - \left(\frac{\sum (f_i M_i)}{N}\right)^2$$

$$s^2 = \frac{\sum (f_i M_i^2) - \dfrac{\left(\sum (f_i M_i)\right)^2}{n}}{n-1}$$

Example 4.17

Table 4.18 presents, in grouped form, the amount of cash on hand for 45 technology companies in the business software and services industry. Compute the mean and variance for these data.

Table 4.18 – Cash on Hand for Technology Companies	
Cash on Hand (Millions of Dollars)	Frequency
0 – 10	10
10 – 20	7
20 – 30	7
30 – 40	7
40 – 50	1
50 – 60	4
60 – 70	2
70 – 80	2
80 – 90	2
90 – 100	3

Solution

To compute the mean and variance, the midpoints of each interval must be calculated. The class midpoint is given by

$$\text{Midpoint} = \frac{\text{lower class boundary} + \text{upper class boundary}}{2}.$$

The midpoints as well as the other required calculations are presented in Table 4.19.

Table 4.19 – Midpoints and Other Required Calculations					
Cash on Hand (Millions of Dollars)	Midpoint M_i	Frequency f_i	$f_i \cdot M_i$	M_i^2	$f_i \cdot M_i^2$
0 – 10	5	10	50	25	250
10 – 20	15	7	105	225	1575
20 – 30	25	7	175	625	4375
30 – 40	35	7	245	1225	8575
40 – 50	45	1	45	2025	2025
50 – 60	55	4	220	3025	12,100
60 – 70	65	2	130	4225	8450
70 – 80	75	2	150	5625	11,250
80 – 90	85	2	170	7225	14,450
90 – 100	95	3	285	9025	27,075
Totals		45	1575	33,250	90,125

Assuming the data in Table 4.18 are population data, the mean cash on hand for the 45 companies is calculated as follows.

$$\mu = \frac{\sum(f_i M_i)}{N} = \frac{1575}{45} = \$35 \text{ million}$$

The variance of the grouped data is calculated as follows.

$$\sigma^2 = \frac{\sum(f_i M_i^2) - \dfrac{\left(\sum(f_i M_i)\right)^2}{N}}{N}$$

$$= \frac{90125 - \dfrac{1575^2}{45}}{45} \approx 777.7778.$$

If the data are sample data, then the variance is:

$$s^2 = \frac{\sum(f_i M_i^2) - \dfrac{\left(\sum(f_i M_i)\right)^2}{n}}{n-1}$$

$$= \frac{90125 - \dfrac{1575^2}{45}}{44} \approx 795.4545.$$

It is important to remember that the calculations of the mean and variance are approximate. That is, if the raw data are available, the actual mean and variance would differ from the measures calculated using the grouped data.

4.6   Exercises

Basic Concepts

1. When analyzing grouped data, are the measurements exact? Why or why not?

2. What calculations are required in order to analyze grouped data?

Exercises

3. A client of a commercial rose grower has been keeping records on the shelf-life of a rose. The client sent the frequency distribution to the grower. Calculate the mean and variance for the shelf-life given the following frequency distribution.

Table for Exercise 3 – Rose Shelf-Life	
Days of Shelf-Life	**Frequency**
1 – 6	2
7 – 12	3
13 – 18	9
19 – 24	6
25 – 30	3
31 – 36	1

4. An article in *Business Week* discussed the large spread between the federal funds rate and the average credit card rate. The table below is a frequency distribution of the credit card rate charged by the top 100 issuers. Note that at the time these figures were published, the average federal funds rate was well below 5%.

Table for Exercise 4 – Credit Card Rates	
Credit Card Rate	**Frequency**
19% – 24%	36
18% – 18.9%	8
17% – 17.9%	15
16% – 16.9%	12
15% – 15.9%	29

a. Calculate the average credit card rate charged by the top 100 issuers based on the frequency distribution.

b. Calculate the variance of the credit card rate charged by the top 100 issuers based on the frequency distribution.

c. Calculate the standard deviation of the credit card rate charged by the top 100 issuers based on the frequency distribution.

4.7   Proportions

The **proportion** is one of the more common summary measures.

Definition

A **proportion** measures the fraction of a group that possesses some characteristic.

To calculate a proportion, simply count the number in the group that possess the characteristic and divide the count by the total number in the group. Let

x = number of observations that possess the characteristic,

N = number of observations in the population, and

n = number of observations in the sample, then

$$p = \frac{x}{N} = \text{ the population proportion, and}$$

$$\hat{p} = \frac{x}{n} = \text{ the sample proportion.}$$

The symbol $\hat{p}$ is pronounced *p-hat*.

Example 4.18

Suppose your statistics class is composed of 48 students of which 4 are left-handed. What proportion of the class is left-handed?

Solution

There are 48 pieces of data in the class. Think of data as composed of 0s and 1s. Any left-handed person will be a 1, and any right-handed person will be a 0. In our data set, there will be four 1s and forty-four 0s.

$$\text{Assuming } x_i = \begin{cases} 1 \text{ if person is left-handed} \\ 0 \text{ if person is right-handed} \end{cases}$$

$$\text{then } \sum x_i = 1+1+1+1+0+0+\cdots+0 = 4.$$

In the notation we used earlier, x equals the number of observations that possess the characteristic. Therefore,

$$x = 4 \text{ and } p = \frac{x}{N} = \frac{4}{48} = 0.0833.$$

Thus, approximately $0.0833(100) = 8.33\%$ of people in the class are left-handed.

Note that we are using p, the population proportion in this case because we are considering the population of students in your statistics class. If we were using the data from your statistics class to estimate the proportion of all statistics students that are left-handed, we would use the symbol $\hat{p}$, the sample proportion, since we would then be computing the proportion from sample data.

Example 4.19

Suppose you have been playing softball and have kept records on each plate appearance. According to your records you have batted 216 times. Of these 216 plate appearances, you have walked 24 times, gotten on base by a fielding error 7 times, and reached base on a hit 64 times. Let's compute your batting average, which is a proportion. The batting average is the proportion of times you reached base on a hit, excluding walks and errors. In this case the number in the group of at bats we will consider is

$$N = \text{Plate appearances} - \text{Walks} - \text{Bases by fielding errors}$$
$$= 216 - 24 - 7$$
$$= 185 \text{ at bats.}$$

The proportion of times you got a hit (excluding walks and errors) is

$$p = \frac{64}{185} \approx 0.3459.$$

Hence, your batting average is .346.

This chapter has been devoted to summarizing data. Yet, with the exception of the mode, none of the summary methods discussed should be applied to nominal data. Using proportions is one of the few summary methods available for analyzing qualitative data.

4.7   Exercises

Basic Concepts

1. What is a proportion?

2. What is the difference in notation between a population proportion and a sample proportion?

3. Other than the mode, proportions are one of the few summary methods available to analyze what type of data?

Exercises

4. A survey of shoppers at a local mall was taken to study the shopping habits of consumers. The mall contains a variety of specialty stores, especially stores that specialize in electronics and gadgets. One question in particular asked, "Do you enjoy shopping for electronics?" Of the 300 men surveyed, 175 answered "Yes." Of the 200 women surveyed, 55 answered "Yes."

    a. What proportion of men enjoy shopping for electronics?

    b. What proportion of women enjoy shopping for electronics?

    c. What is the overall proportion of consumers at the Tech Mall that enjoy shopping for electronics?

5. A survey released in April 2011 conducted by the consulting firm Booz & Company regarding the automobile industry found that U.S. automotive executives were skeptical about the industry's economic recovery. Suppose that 118 original equipment manufacturers (OEM) and 82 supplier executives participated in the study. When asked whether the overall state of the industry is fairly similar to, or somewhat better than, its low point in January 2009, 59 OEMs and 28 supplier executives answered that the current state of the industry was "about the same," and 55 OEMs and 52 supplier executives answered that the current state of the industry was "somewhat better."

    Source: Booz & Company; 2011

    a. Calculate the sample proportion of original equipment manufacturers that believe the state of the automobile industry is "about the same" as in January 2009.

    b. Calculate the sample proportion of supplier executives that believe the state of the automobile industry is "about the same" as in January 2009.

    c. Calculate the sample proportion of original equipment manufacturers that believe the state of the automobile industry is "somewhat better" than in January 2009.

    d. Calculate the sample proportion of supplier executives that believe the state of the automobile industry is "somewhat better" than in January 2009.

    e. Do these responses seem to support Booz & Company's conclusion that automotive executives are skeptical about the industry's economic recovery? Discuss.

6. An experiment was conducted to study how investors selected mutual funds. Two groups of investors were selected. Group 1 consisted of investors that used online brokerages and Group 2 was made up of investors that used full-service brokerages. Of the 150 investors in Group 1, 120 indicated that they selected mutual funds on their own, while 30 stated that they selected mutual funds using recommendations of the brokerage, family, and friends. Of the 200 investors in Group 2, 25 indicated that they selected their funds on their own, while 175 indicated that they selected the mutual funds using recommendations of the brokerage, family, and friends.

    a. What proportion of Group 1 investors selected mutual funds on their own?

    b. What proportion of investors in Group 2 selected mutual funds on the recommendation of others?

    c. What does this tell you about investors that use online brokerages versus those using full-service brokerages?

7. According to a study administered by the National Bureau of Economic Research, half of Americans would struggle to come up with $2000 in the event of a financial emergency. The majority of the 1900 Americans surveyed said they would rely on more than one method to come up with emergency funds if required. In the survey, 532 people said that they "certainly" would not be able to cope with an unexpected $2000 bill if they had to come up with the money in 30 days, and 418 people said they "probably" would not be able to cope.

   Source: CNNMoney.com; 2011

   a. What percentage of Americans "certainly" would not be able to produce $2000 in the event of an emergency according to the study?

   b. What percentage of Americans would "probably" not be able to pay a $2000 bill in 30 days if required?

   c. What does this say about the savings habits of Americans?

8. What college football conference has the right to brag about putting players in the NFL? A random sample of 100 current NFL players was surveyed to determine the conference in which they played college football. The following table displays the results of the survey.

Table for Exercise 8 – NFL Players and College Football Conferences	
Conference	**Number of Players**
SEC	20
Big 12	16
Big 10	12
Pac 10	6
ACC	6
MAC	2
Big East	2
Other	36

   a. What proportion of players are from the SEC?

   b. What proportion of players are from a conference other than the first 6 listed?

   c. Is it true that a player in the SEC has a better chance of being drafted in the NFL than a player from any other conference? Explain.

9. According to a survey administered by the market research group ChangeWave in February 2011, approximately 27% of respondents reported that they plan on buying a tablet device in the future. This result was 2 percentage points higher than in a similar survey administered in November of 2010. In the February study, 3091 customers were surveyed on tablet demand and future buying trends. Suppose that in the November study, 721 people said they planned on buying a tablet device in the future.

   Source: InvestorPlace; 2011

   a. In the February study, how many people said that they planned on buying a tablet device in the future?

   b. In the November study, how many total customers were surveyed?

10. It is no secret that Wall Street firms compete aggressively to lure their clients. Having more high-end clients translates into fees and revenues that turn into profits. A survey of 150 high-end clients asked what lured them to their respective Wall Street firm. The following table shows the results.

Table for Exercise 10 – High-End Client Response	
Perk Received	**Client Response**
Pay a Kick-Back	25
Lucrative Golf Outings	12
Lavish Dinners	8
Free Private Jet Use	33
Prime Seats at Sports Events	20
Other	22
No Perk Received	30

a. Which type of perk appears to be most successful in luring clients?

b. What proportion of clients were lured to a Wall Street firm by the perk identified in part **a.**?

c. What proportion of clients did not receive a perk at all?

d. Given that these perks aren't inexpensive, what conclusion can you make about providing perks to clients? Explain.

4.8   Measures of Association between Two Variables

Often times a manager or decision maker is interested in the relationship between two variables. Such relationships could be the amount of sales and advertising expenditures, education level and income, number of contacts made and sales, sales and earnings, number of real estate foreclosures and home prices, and so on. In earlier sections, we discussed methods used to study each of these measurements individually. Now we want to study the measurements of two different variables at the same time to see if there is a relationship between them. The purpose of studying the relationship between two variables is three-fold: to describe and understand the relationship, to forecast and predict a new observation, and if one is working with a process, not understanding the relationship could be detrimental to making necessary adjustments. There are statistical tools that can aid in the discovery of relationships.

Thinking about relationships is something that everyone does. Why, for example, does an admissions counselor want to know the relationship between SAT scores and college performance? The admissions counselor's task is to select students who will be successful at that college. If college performance is related to SAT scores and the relationship can be specified explicitly, then SAT scores can be used to *predict* performance. Consequently, the relationship would be helpful in selecting students for admission. On the other hand, if SAT scores are not useful in predicting college performance, then they should not be considered very important in making admissions decisions. Discovering whether SAT scores are related to college performance could improve the admissions process. Accurate predictions of a variable often suggest methods for process improvement.

Most students view grades as important and work to make good grades for self-satisfaction and to enhance career opportunities. Because grades are perceived as important, students often wonder about the relationship between the amount of time spent studying and the resulting exam grade. If study time is related to the grade received, a student could use the relationship to predict a grade based on study time. In addition, if the student is able to predict the grade this way, the student could adjust study time to obtain whatever grade is desired. Being able to predict a low grade could lead to corrective action and an improvement in grade point average. The ability to predict often leads to the ability to control.

Bivariate Data

In earlier sections, all of the statistical summary measurements, like the mean, variance, and proportions, were concerned with describing **univariate** data (measurements of one variable). To understand the relationship between two variables, data on both variables need to be collected. This type of data is called **bivariate** data. With bivariate data, two observations are recorded from some entity.

Table 4.20 contains 60 bivariate data points of monthly rent and vacancy rates for various locations in the southeastern part of the United States.

Table 4.20 – Monthly Rents and Vacancy Rates for Various Locations

Location	Monthly Rent ($)	Vacancy Rate (%)	Location	Monthly Rent ($)	Vacancy Rate (%)
Accomack	639.00	9.0	Hanover	587.00	10.9
Airport	752.00	17.3	Hickory	604.00	9.1
Albermarle	717.00	11.7	Highland	611.00	15.0
Anson	589.00	9.4	Iredell	693.00	11.2
Appomattox	605.00	14.5	Kannapolis	606.83	11.3
Avery	698.00	7.8	Kernersville	609.00	14.9
Bloomsburg	600.00	7.0	Lancaster	761.00	9.1
Botetourt	645.17	14.3	Louisa	624.00	9.2
Buchanan	613.00	11.3	Macon	614.00	6.7
Cabarrus	632.00	7.1	Mecklenburg	648.00	8.4
Caldwell	701.00	7.3	Montgomery	620.00	8.6
Catawba	643.83	12.4	Mooresville	579.00	5.1
Charlotte	612.00	6.8	Nash	597.00	12.1
Charlottesville	655.00	14.5	Nelson	658.00	8.3
Chester	623.00	6.2	Newton	589.00	5.1
Cleveland	607.50	8.9	New Castle	602.00	5.7
Clover	780.00	22.3	Norfolk	630.17	7.6
Conover	634.00	4.4	Northampton	614.00	5.7
Danville	604.00	9.2	Orange	643.00	7.5
Davidson	725.00	9.3	Pamlico	666.00	9.6
Fort Mill	658.00	7.9	Pittsylvania	662.00	7.0
Franklin	661.00	15.3	Rock Hill	569.00	8.6
Frederick	681.00	11.5	Rockingham	636.50	10.3
Gaston	585.00	9.1	Rowan	617.67	7.7
Gates	622.00	8.4	Rutherford	677.17	9.1
Giles	678.00	11.6	Salisbury	577.00	8.6
Gloucester	740.00	11.6	Stokes	624.00	6.7
Granville	622.00	8.9	Surry	624.00	9.2
Grayson	605.00	11.3	Union	582.00	4.1
Greene	746.00	20.1	York	672.00	7.2

Looking for Patterns in the Data

Detecting a relationship between two variables often begins with a graph. In the case of bivariate data, a **scatterplot** (or **scatter diagram**) is the traditional graphical method used to display the relationship between two variables. In a **scatterplot**, measurements are plotted in pairs with one variable plotted on each axis. When examining a scatterplot we are trying to draw conclusions concerning the overall pattern of the data. Does the pattern roughly follow a straight line? Is the pattern upward sloping or downward sloping? Are the data values tightly clustered in the pattern or widely dispersed? Are there significant deviations from the pattern?

A number of different scatterplots are shown in the following figures. In the first two scatterplots (Figures 4.21 and 4.22) the data are strongly related and, in fact, fall on a straight line. In Figure 4.21 the slope of the relationship is positive, that is, as the x-variable increases the y-variable also increases.

Scatterplot 1

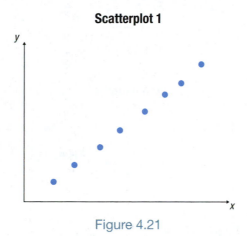

Figure 4.21

In Figure 4.22 the relationship is negative; as the x-variable increases, the y-variable decreases. This is also called an **inverse relationship**.

Scatterplot 2

Figure 4.22

Figures 4.23 and 4.24 show less obvious relationships between the data. Figure 4.23 reveals a very imprecise relationship between x and y, although as x increases, y tends to increase. The relationship between x and y is much more apparent in Figure 4.24 than in Figure 4.23.

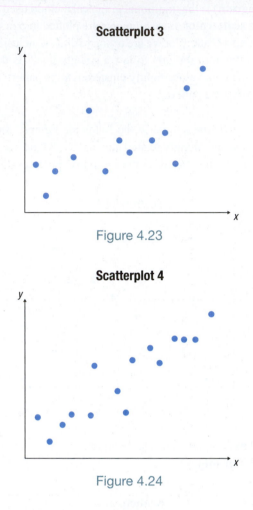

Figure 4.23

Figure 4.24

Figure 4.25 reveals a downward sloping relationship between x and y. That is, as x increases, y tends to decrease. The relationship is not as exact as the relationship in Figure 4.22. In Figure 4.26, there is no apparent relationship between x and y. That is, there is no tendency for y to increase or decrease as x increases.

Figure 4.25

Scatterplot 6

Figure 4.26

Let's explore the data in Table 4.20 with a scatterplot (see Figure 4.27). Examining the scatterplot reveals a slightly upward sloping relationship. That is, as monthly rent increases, the vacancy rate also increases.

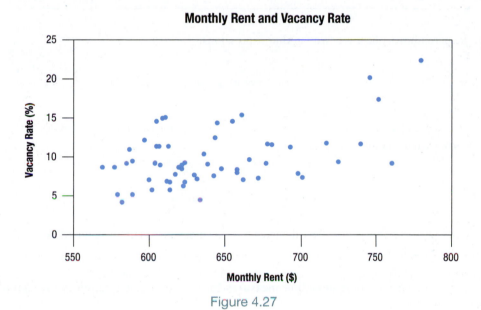

Figure 4.27

While the relationship between monthly rent and vacancy rate is apparent, the relationship is not very strong. Interestingly, the law of supply and demand for rental space suggests that higher rents are a result of an increase in demand, contributing to higher prices. Thus, one would expect to see lower vacancy rates as rents rise, which could result in the development of more apartment units. However, that doesn't appear to be the case with these data. The fact that there appears to be a positive relationship between monthly rent and vacancy rate does not prove that high rent causes high vacancy rates even though such a conclusion may seem reasonable. One of the temptations that must be avoided is equating association with causation. There are many potential confounding variables in this observational data. However, we can quantify the strength of this relationship using the correlation coefficient, which is a measure of the linear association between two variables.

Measuring the Degree of Linear Relationship: The Correlation Coefficient

Francis Galton
1822 – 1911

Galton was the ninth child of a wealthy English family in Birmingham, England. He did extensive traveling early in his life and became interested in heritability of human traits. He was also greatly influenced by his first cousin, Charles Darwin. He believed that physical characteristics such as weight, height, and intelligence, as well as some personality traits were inherited. He collected data on these traits from mothers, fathers, and their children and found some interesting results. Tall fathers, for example, tended to have shorter children than themselves. Short fathers tended to have children that were taller than themselves. He also found this same property in seeds of wheat. He named this phenomena regression toward the mean. This is how the term regression entered statistics. Galton was partially responsible for the development of the correlation coefficient. He needed some method of assessing the strength of the physical relationships between parents and their children.
He enlisted a young English statistician named Karl Pearson to work on a measure that could be used to determine association. It was Pearson that developed the correlation coefficient for this purpose. It is unclear exactly what role Galton played in the development of the correlation coefficient, however in the social science literature Galton is recognized as a co-developer of the statistic.

A scatter diagram is a useful exploratory tool for detecting relationships between two variables. Eventually, however, a researcher will want to know the strength of the relationship between the two variables. Karl Pearson developed a measure in 1896 called the **correlation coefficient**, r, to measure the degree of linear relationship.

> **Formula**
>
> ## Correlation Coefficient
>
> The correlation coefficient is an index number used to summarize the strength of a linear relationship.
>
> $$r = \frac{1}{n-1} \left\{ \sum_{i=1}^{n} \left(\frac{x_i - \bar{x}}{s_x} \right) \left(\frac{y_i - \bar{y}}{s_y} \right) \right\} \qquad -1 \le r \le 1$$

Within the parentheses, there are two familiar expressions:

$$\frac{y_i - \bar{y}}{s_y},$$

which is a z-score that shows how far y deviates from its mean measured in standard deviation units (s_y is the standard deviation of y), and

$$\frac{x_i - \bar{x}}{s_x},$$

which is a z-score that shows how far x deviates from its mean measured in standard deviation units (s_x is the standard deviation of x).

Summing the products of these **deviation measures** for each data pair determines the sign of the correlation coefficient.

Note: The formula presented for the correlation coefficient is difficult to calculate by hand. Often, you will see the correlation coefficient presented as the following equivalent computational formula. However, there are many statistical software programs that can calculate the correlation coefficient, so it is not often you will need to do it by hand.

> **Formula**
>
> ## Computational Formula for the Correlation Coefficient
>
> The computational formula for the correlation coefficient is as follows.
>
> $$r = \frac{n \sum x_i y_i - \left(\sum x_i \right) \left(\sum y_i \right)}{\sqrt{n \sum x_i^2 - \left(\sum x_i \right)^2} \sqrt{n \sum y_i^2 - \left(\sum y_i \right)^2}}$$

Positive Relationships

When r is positive, there is a tendency for y to increase as x increases. If both of the deviations are positive, then each of the observations is above its mean. If both are negative, then each is below its mean. In a positive linear relationship, when one of the variables is above its mean, the other variable tends to be above its mean. Similarly, if one variable is below its mean, the other tends to be below its mean. Such is the case in Figure 4.28.

Positive Relationship between *x* and *y*

Figure 4.28

The points in the group labeled **A** are points whose x-values are greater than the mean of x and whose y-values are greater than the mean of y. Since for each point in group **A**, the deviations $x_i - \bar{x}$ and $y_i - \bar{y}$ will be positive numbers, the expression $\left(\dfrac{x_i - \bar{x}}{s_x} \right)\left(\dfrac{y_i - \bar{y}}{s_y} \right)$ will be the product of two positive numbers, which will be *positive*.

The points in the group labeled **B** are points whose x-values are below the mean of x and whose y-values are below the mean of y. Since $x_i - \bar{x}$ and $y_i - \bar{y}$ will both be negative numbers for members of this group, the expression $\left(\dfrac{x_i - \bar{x}}{s_x} \right)\left(\dfrac{y_i - \bar{y}}{s_y} \right)$ will be the product of two negative numbers, which will be *positive*.

Since all the points fall into either group **A** or **B**, all the products in the summation are positive. Thus, the correlation coefficient (r) will have a positive value for an upward sloping (positive) relationship. Now, let's look at a downward sloping relationship.

Negative Relationships

Negative Relationship between *x* and *y*

Figure 4.29

In downward sloping (negative) relationships, a value above the mean for one variable will tend to be associated with a value below the mean for the other variable. In Figure 4.29, for every point in group **C** the value of y is above y's mean, but the value of x is below x's mean. In group **D** the reverse is true. This kind of relationship is often called an **inverse relationship**. You see this kind of relationship quite often. For example, consider the relationship between price and quantity sold. For virtually all goods, as the price increases, the quantity sold will decrease. Examining the correlation measure for negative relationships reveals that the product in the summation $\left(\dfrac{x_i - \bar{x}}{s_x}\right)\left(\dfrac{y_i - \bar{y}}{s_y}\right)$ will be negative for points in groups **C** and **D**. Since all but one of the points belong to groups **C** and **D**, then the summation will consist of mostly negative values for $\left(\dfrac{x_i - \bar{x}}{s_x}\right)\left(\dfrac{y_i - \bar{y}}{s_y}\right)$; therefore, the value of the correlation coefficient will be negative.

Example 4.20

Table 4.21 contains the age and the annual maintenance costs of a certain model compact vehicle. Calculate the correlation coefficient to determine if a relationship exists between the age and annual maintenance cost of that model.

Table 4.21 – Vehicle Age and Maintenance Cost		
Observation	**Age**	**Annual Maintenance Cost (Dollars)**
1	2	225
2	4	400
3	5	475
4	7	650
5	9	800
6	12	1175

Solution

To calculate the correlation coefficient, we will use the following computational formula.

$$r = \frac{n\sum x_i y_i - \left(\sum x_i\right)\left(\sum y_i\right)}{\sqrt{n\sum x_i^2 - \left(\sum x_i\right)^2}\,\sqrt{n\sum y_i^2 - \left(\sum y_i\right)^2}}$$

Some useful summary statistics needed to calculate r are as follows.

$$n = 6,\ \sum x_i y_i = 30{,}275,\ \sum x_i = 39,\ \sum y_i = 3725,\ \sum x_i^2 = 319,\ \text{and}\ \sum y_i^2 = 2{,}879{,}375$$

Using the summary statistics, we get the following calculation for the correlation coefficient.

$$r = \frac{6(30275) - (39)(3725)}{\sqrt{6(319) - (39)^2}\,\sqrt{6(2879375) - (3725)^2}}$$
$$\approx 0.9950$$

Given the large magnitude of r, one can say that there is a strong, positive, linear relationship between the age of the vehicle and the annual maintenance cost. That is, as the vehicle gets older, one can expect to spend more money on annual maintenance.

Many calculators and virtually all statistics programs will compute the correlation coefficient, r, so it is unlikely that you will need to manually compute the measure. The Discovering Technology section at the end of this chapter will show you how to calculate the correlation coefficient using a TI-84 Plus calculator, Microsoft Excel, and Minitab.

Properties of the Correlation Coefficient

This leads to the following properties of the correlation coefficient.

Properties

The correlation coefficient, r, measures the degree of linear relationship, i.e. how well the data cluster around a line.

- The value of r is always between -1 and $+1$.

- A value of r near -1 or $+1$ means the data are tightly bundled around a line.

- A value of r near -1 or $+1$ means that we have a very strong negative relationship or positive relationship, respectively, such that predictions within the scope of the data are very reliable.

- Positive association is indicated by $r > 0$ and an upward sloping relationship.

- Negative association is indicated by $r < 0$ and a downward sloping relationship.

- A value of r near zero means there is no linear relationship between x and y.

- It does not matter whether you correlate y with x or x with y; you will still get the same value for r.

Correlation Coefficients

$r = 0.95$

$r = -0.86$

$r = -0.05$

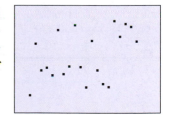

Avoiding Some Correlation Pitfalls

A high correlation does not imply causation. Suppose that a high correlation has been observed between the weekly sales of ice cream and the number of snake bites each week. It seems unlikely that ice cream sales would cause snakes to bite people or that more snake bites would cause higher ice cream sales. Yet when the data are analyzed, you may find an unexpectedly high correlation. If the two variables aren't actually related, what could explain such an observed relationship?

The apparent relationship is an illusion caused by a phenomenon called **common response**. That is, both variables are related to a third variable. In this case the high temperatures in the summer cause increases in both ice cream sales and reptile activity.

Correlating summary measures (such as means) will tend to provide an inflated correlation

199

4

measurement. Ignoring the variation of the individual values magnifies the correlation measure and gives a somewhat distorted view of the underlying relationship.

Baldness and Heart Attacks

Newsweek magazine reported the results of a medical study that identified increased risk of heart attack with male pattern baldness. This kind of information based on correlation can be useful in the sense that men with pattern baldness can be extra vigilant about possible cardiac arrest. On the surface, this increased risk is hard to explain. What do you think is a reasonable explanation?
The medical researchers that studied this problem came to the conclusion that there may be a third variable, often called a lurking variable, which is a male hormone that induces both pattern baldness and propensity to increased heart attack. This finding opened new research questions that centered around finding such a male hormone.

Suppose there is good reason to believe that a causal relationship exists between two variables, but when a correlation is performed the value of the correlation is near zero, indicating no association. Does the lack of correlation between the two variables prove no relationship exists? There are several reasons two related variables might not have a high correlation.

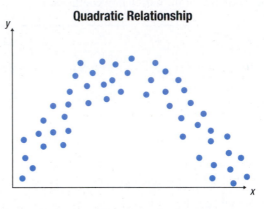

Figure 4.30

A low correlation could mean that no *linear* relationship exists. In Figure 4.30 the relationship between x and y is not a straight line. The correlation measure for these points is going to be very close to zero. Yet, there does appear to be a very strong relationship between x and y. The kind of relationship exhibited by this data is called a **quadratic relationship**.

Another problem that can produce low correlations is **confounding**. Confounding occurs when more than one variable affects the dependent variable, and the effects of the variables cannot be distinguished from each other. Suppose that the variable y is dependent on x. Thus, as x changes, it produces changes in y. Such a relationship should produce a significant correlation measure between the two variables. But suppose there is another variable z, which also affects y. As z changes so does y. It is certainly possible that changes in z will mask the changes caused by x.

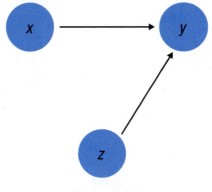

Figure 4.31

The range of x-values selected for correlation can also significantly affect the value of the correlation coefficient. If the range of the x data is large, the correlation will usually be greater than if the range of the x-values is small, as shown in Figure 4.32.

Small Range of x-Values

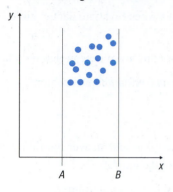

Figure 4.32

If the points below A in Figure 4.33 were removed and the points above B were removed from the data, (as shown in Figure 4.32), the correlation between x and y will be considerably weaker. Thus, when looking at correlation data, you should ask if the data are complete.

Complete Data

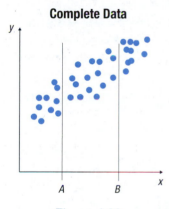

Figure 4.33

Sometimes unrelated variables are highly correlated. When this occurs the variables are said to have a **spurious** correlation. For example, over a short period of time daily car sales and the number of penguins in Antarctica might be related. However, it is doubtful that a significant change in the penguin population will cause a change in car sales, or vice versa.

4.8   Exercises

Basic Concepts

1. Give an example of a business situation in which knowledge of a relationship between two variables is desired.

2. If a relationship can be uncovered, what are the potential benefits?

3. What are bivariate data? How is bivariate data different from univariate data?

4. What graphical tool is often used in the discovery of relationships?

5. What are four common questions you should ask when studying a graphical representation of bivariate data?

6. If bivariate data exhibit an inverse relationship, what does that mean?

7. How do you construct exact relationships between two variables?

8. In what range is the value of r when bivariate data exhibit a positive relationship? A negative relationship?

9. If the value of r is small, does this always mean that no relationship exists? Explain.

10. What is confounding? Why is confounding a problem?

Exercises

11. Consider the following scatterplots and answer the following questions regarding the overall pattern of the data for each of the graphs.

   • Does the pattern roughly follow a straight line?

   • Is the pattern upward sloping or downward sloping?

   • Are the data values tightly clustered in the pattern or widely dispersed?

   • Are there significant deviations from the pattern?

   a.

Scatterplot A

   b.

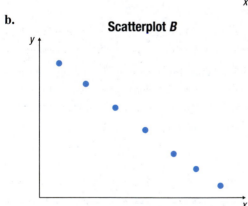

Scatterplot B

12. Consider the following scatterplots and answer the following questions regarding the overall pattern of the data for each of the graphs.

- Does the pattern roughly follow a straight line?

- Is the pattern upward sloping or downward sloping?

- Are the data values tightly clustered in the pattern or widely dispersed?

- Are there significant deviations from the pattern?

a.

b.

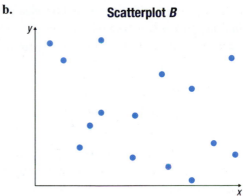

13. A manufacturing company which produces laminate for countertops is interested in studying the relationship between the number of hours of training an employee receives and the number of defects per countertop produced. Ten employees are randomly selected. The number of hours of training which each employee has received is recorded and the number of defects on the most recent countertop produced is determined. The results are as follows.

Table for Exercise 13 – Employee Training	
Hours of Training	Defects per Countertop
1	1
4	4
7	0
3	3
2	5
2	4
5	3
5	2
1	5
6	1

    a. Analyze the data collected for the study by answering the following questions.

        i. Do the variables selected for measurement seem appropriate for answering the question that the manufacturing company is interested in?

        ii. What biases or errors might be present in the data?

        iii. How are the data collected – through observation or controlled experiment?

    b. Plot the data points on a scatterplot.

    c. Based on the scatterplot in part **b.**, answer the following questions regarding the overall pattern of the data.

        i. Does the pattern roughly follow a straight line?

        ii. Is the pattern upward sloping or downward sloping? Are the data values tightly clustered in the pattern or widely dispersed?

        iii. Are there significant deviations from the pattern?

14. Illustrate, using a scatterplot, a data set that would have a correlation coefficient of 1.

15. Illustrate, using a scatterplot, a data set that would have a correlation coefficient of -1.

16. Describe the relationships indicated by the correlation coefficients as tightly clustered in a positive linear fashion, tightly clustered in a negative linear fashion, loosely clustered in a positive linear fashion, loosely clustered in a negative linear fashion, or no linear relationship.

    a. $r = 0.9$ **b.** $r = 0.5$

    c. $r = -0.9$ **d.** $r = -0.5$

    e. $r = 0$

17. Describe the relationships indicated by the correlation coefficients as tightly clustered in a positive linear fashion, tightly clustered in a negative linear fashion, loosely clustered in a positive linear fashion, loosely clustered in a negative linear fashion, or no linear relationship.

    a. $r = 0.8$ **b.** $r = 0.4$

    c. $r = -0.8$ **d.** $r = -0.4$

    e. $r = 0.1$

18. A sample of 10 female swimmers, all 17 years old, is selected from a local swim league. Each swimmer's best time (in seconds) in the 50-yard freestyle and in the 100-yard individual medley are obtained. The 100-yard individual medley consists of swimming 25 yards with each of the four major strokes. The data are given in the following table.

Table for Exercise 18 – Best Times									
Freestyle	27.4	27.0	26.8	30.7	28.5	28.6	29.6	30.8	31.5	29.8
Medley	66.3	66.4	66.7	78.7	69.4	72.0	73.5	81.1	78.6	73.5

   a. Construct a scatterplot of the data.

   b. Does there appear to be a negative or positive relationship between the variables?

   c. Compute the correlation coefficient.

19. A personnel director is interested in studying the relationship (if any) between age and salary. Sixteen employees are randomly selected and their ages and salaries are recorded.

Table for Exercise 19 – Ages and Salaries			
Age	Salary ($)	Age	Salary ($)
25	22,000	49	39,000
55	45,000	37	45,000
27	43,000	62	60,000
30	30,000	40	35,000
22	24,000	35	34,000
33	53,000	29	30,000
19	18,000	58	73,000
45	38,000	52	42,000

   a. Plot the data points on a scatterplot.

   b. Determine the correlation coefficient.

   c. Describe the relationship indicated by the correlation coefficient and the scatterplot.

20. The following variables have high positive linear correlations. Is it reasonable to conclude that an increase in one variable causes an increase in the other variable? Explain what could be causing this apparent relationship.

   a. Height and vocabulary

   b. Absenteeism from school and sale of cough syrup

   c. Sale of turkey and sale of toys

21. The following variables have high positive linear correlations. Is it reasonable to conclude that an increase in one variable causes an increase in the other variable? Explain what could be causing this apparent relationship.

   a. Sale of air conditioners and sale of tomatoes

   b. Sale of greeting cards and sale of chocolates

   c. The number of wrecks on a local highway and absenteeism from work

T   Discovering Technology

Using the TI-84 Plus Calculator

Sample Mean and Standard Deviation

Use the data from Example 4.9 (4, 10, 9, 11, 9, 7) for this exercise.

1. Press **STAT**, and select **1:Edit**. Press **ENTER**. Then input the data into list **L1**.

2. Press **STAT**, and then press the right arrow key so that **CALC** is highlighted. Choose **1:1-Var Stats** and press **ENTER**. Press **2ND** and **1** to choose the data in list L1 and press **ENTER**.

3. Observe the output screen for the sample mean and sample standard deviation.

Figure 4.34

Correlation Coefficient

For this exercise, use the data from Example 4.20.

1. Press **STAT** and select **1:Edit**.

2. Enter the x data values (ages) in **L1** and the y data values (costs) in **L2**.

3. Access the catalog by pressing **2ND** then **0**.

4. Press the down arrow and find **DiagnosticOn**. Press **ENTER**, then **ENTER** again.

5. Press **STAT**, and then press the right arrow key so that **CALC** is highlighted.

6. Highlight **4: LinReg (ax + b)**. Press **ENTER**.

7. Press **ENTER** again. The calculator will display values for a and b in the equation $y = ax + b$ for the data entered in the lists.

8. The last value listed will be the correlation coefficient for the two sets of data, r.

Figure 4.35

Using Excel

Sample Mean and Standard Deviation

Use the data from Example 4.9 (4, 10, 9, 11, 9, 7) for this exercise.

1. Enter the data into Column A.

2. Under the **Data** tab, choose **Data Analysis**, then **Descriptive Statistics**. Select **OK**.

3. Enter the Input Range **$A$1:$A$6**, select the radio button next to **Output Range** and enter **A8**. Check the box next to **Summary Statistics**. Select **OK**.

4. Observe the output screen for the summary statistics.

	A	B
1	4	
2	10	
3	9	
4	11	
5	9	
6	7	
7		
8	*Column1*	
9		
10	Mean	8.333333
11	Standard Error	1.021981
12	Median	9
13	Mode	9
14	Standard Deviation	2.503331
15	Sample Variance	6.266667
16	Kurtosis	1.137392
17	Skewness	-1.13891
18	Range	7
19	Minimum	4
20	Maximum	11
21	Sum	50
22	Count	6

Figure 4.36

Percentiles and Quartiles

Use the data from Example 4.15 for this exercise.

Microsoft Excel has a function (PERCENTILE) to calculate the kth percentile of a data set and a function (QUARTILE) to calculate the quartiles of a data set. The method that Excel uses to calculate percentiles and quartiles is different from that which was shown in this chapter. The method is beyond the scope of this text, but it is important to realize that the percentiles and quartiles calculated by Excel may be different from those calculated by hand. In addition, Excel 2010 has two percentile functions, PERCENTILE.EXC and PERCENTILE.INC and two quartile functions, QUARTILE.EXC and QUARTILE.INC. The .EXC functions return the kth percentile in a range where k is between 0 and 1, exclusive. The .INC functions return the kth percentile in a range where k is between 0 and 1, inclusive. The regular PERCENTILE function is the one found in Excel 2007 and earlier, and calculates the percentiles in the same way that the PERCENTILE.INC function does.

Just as we do in this chapter, notice that Excel equates the first quartile with the 25th percentile, the second quartile with the 50th percentile, and the third quartile with the 75th percentile.

1. In cell A1 type the label **Data**.

2. Enter the data values in Column A, beginning in cell A2.

3. In cell B1 type the label **Percentile**.

4. In Column B, enter the following percentiles, beginning in cell B2. **10, 25, 40, 50, 60, 75, 90**.

5. In cell D1 type the label **Quartile**.

6. In Column D, enter the following quartiles, beginning in cell D2. **1, 2, 3, 4**.

7. The percentile function is **PERCENTILE.INC(array, k)** where *array* corresponds to the data set you are interested in and k is the kth percentile, expressed as a number between 0 and 1.

8. Now calculate the 10th percentile by entering the following function in cell C2.

$$=\text{PERCENTILE.INC(A2:A41,0.10)}$$

As calculated by Excel, the 10th percentile is 28.8. Calculate the 25th, 40th, 50th, 60th, 75th, and 90th percentiles using the appropriate function in cells C3 through C8, respectively.

	A	B	C	D	E
1	Data	Percentile		Quartile	
2	67	10	28.8	1	
3	45	25	42.5	2	
4	63	40	47.6	3	
5	58	50	54	4	
6	35	60	56.4		
7	54	75	64.5		
8	27	90	73.4		
9	66				
10	21				

Figure 4.37

9. The quartile function is **QUARTILE.INC(array, quart)** where *array* corresponds to the data set you are interested in and *quart* is the quartile (1 through 4).

10. Now calculate the first quartile by entering the following function in cell E2.

$$=\text{QUARTILE.INC(A2:A41,1).}$$

As calculated by Excel, the first quartile is 42.5. Notice this value is the same as the 25th percentile. Calculate the second, third, and fourth quartiles using the appropriate function in cells E3 through E5, respectively. Notice that the second and third quartiles are equal to the 50th and 75th percentile, respectively, and that the fourth quartile is the maximum value in the data set (corresponding to the 100th percentile).

	A	B	C	D	E
1	Data	Percentile		Quartile	
2	67	10	28.8	1	42.5
3	45	25	42.5	2	54
4	63	40	47.6	3	64.5
5	58	50	54	4	82
6	35	60	56.4		
7	54	75	64.5		
8	27	90	73.4		
9	66				
10	21				

Figure 4.38

Correlation Coefficient

For this exercise, use the data from Example 4.20.

1. Enter the label **Age** in cell A1, and the x data values in cells A2 through A7.

2. Enter the label **Annual Maintenance Cost (Dollars)** in cell B2, and the y data values in cells B2 through B7.

3. Under the **Data** tab select **Data Analysis**.

4. Select **Correlation** and press **OK**.

5. Enter **$A$2:$B$7** as the Input Range. Make sure that the radio button next to **Columns** is selected.

6. Select the radio button next to **Output Range** and enter **$E$1**. Press **OK**.

7. In the summary output, the correlation coefficient, r, is displayed (0.99501).

	A	B	C	D	E	F	G
1	Age	Annual Maintenance Cost (Dollars)				Column 1	Column 2
2	2	225			Column 1	1	
3	4	400			Column 2	0.99501	1
4	5	475					
5	7	650					
6	9	800					
7	12	1175					

Figure 4.39

Using Minitab

Sample Mean and Standard Deviation

Use the data from Example 4.9 (4, 10, 9, 11, 9, 7) for this exercise.

1. Enter the data into Column C1.

2. Under **Stat**, choose **Basic Statistics**, then **Display Descriptive Statistics**.

3. In the Display Descriptive Statistics dialog box, input **C1** under Variables. Select **OK**.

4. Observe the Output Screen for the summary statistics.

```
Descriptive Statistics: C1

Variable   N   N*   Mean   SE Mean   StDev   Minimum    Q1   Median     Q3   Maximum
C1         6    0   8.33      1.02    2.50      4.00   6.25     9.00  10.25     11.00
```

Figure 4.40

Correlation Coefficient

For this exercise, use the data from Example 4.20.

1. Enter the x data values (ages) in Column C1.

2. Enter the y data values (costs) in Column C2.

3. Under **Stat**, highlight **Basic Statistics**, and select **Correlation**.

4. Enter **C1** and **C2** in the box for the Variables. Press **OK**.

5. The correlation coefficient r is displayed in the summary output.

```
Correlations: C1, C2

Pearson correlation of C1 and C2 = 0.995
P-Value = 0.000
```

Figure 4.41

R | Chapter 4 Review

Key Terms and Ideas

- Numerical Descriptive Statistics
- Parameters
- Statistics
- Inferential Statistics
- Measures of Location (Central Tendency)
- Arithmetic Mean
- Sample Mean
- Population Mean
- Deviation
- Weighted Mean
- Trimmed Mean
- Median
- Resistant Measure
- Mode
- Bimodal
- Multimodal
- Positively Skewed (Skewed to the Left)
- Negatively Skewed (Skewed to the Right)
- Moving Average
- Variability (Dispersion or Spread)
- Deviation from the Mean
- Range
- Mean Absolute Deviation
- Population Variance
- Sample Variance
- Population Standard Deviation
- Sample Standard Deviation

- Empirical Rule
- Chebyshev's Theorem
- P^{th} Percentile
- Percentile
- Quartile
- Interquartile Range
- Outlier
- z-Score
- Shape
- Subsetting
- Coefficient of Variation
- Mean of Grouped Data
- Variance of Grouped Data
- Population Proportion
- Sample Proportion
- Univariate Data
- Bivariate Data
- Scatterplot (Scatter Diagram)
- Linear Relationship
- Positive Relationship
- Negative Relationship (Inverse Relationship)
- Correlation Coefficient
- Quadratic Relationship
- Common Response
- Confounding

4

Key Formulas

	Section

Arithmetic Mean

$$\frac{1}{n}\left(x_1 + x_2 + \cdots + x_N\right)$$

4.1

Population Mean

$$\mu = \frac{1}{N}\left(x_1 + x_2 + \cdots + x_N\right) = \frac{\sum x_i}{N}$$

4.1

Sample Mean

$$\bar{x} = \frac{1}{n}\left(x_1 + x_2 + \cdots + x_n\right) = \frac{\sum x_i}{n}$$

4.1

Weighted Mean

$$\bar{x} = \frac{w_1 x_1 + w_2 x_2 + \cdots + w_n x_n}{w_1 + w_2 + \cdots + w_n} = \frac{\sum\left(w_i x_i\right)}{\sum w_i}$$

4.1

Median Location

$$L(m) = \frac{n+1}{2}$$

4.1

Range

largest value − smallest value

4.2

Mean Absolute Deviation

$$\mathrm{MAD} = \frac{\sum|x_i - \bar{x}|}{n}$$

4.2

Population Variance

$$\sigma^2 = \frac{\sum\left(x_i - \mu\right)^2}{N}$$

4.2

Sample Variance

$$s^2 = \frac{\sum\left(x_i - \bar{x}\right)^2}{n-1}$$

4.2

Key Formulas (cont.)

	Section
Population Standard Deviation $$\sigma = \sqrt{\sigma^2}$$	4.2
Sample Standard Deviation $$s = \sqrt{s^2}$$	4.2
Empirical Rule **One Sigma Rule:** $\mu \pm 1\sigma$ contains about 68% of the data **Two Sigma Rule:** $\mu \pm 2\sigma$ contains about 95% of the data **Three Sigma Rule:** $\mu \pm 3\sigma$ contains about 99.7% of the data	4.2
Chebyshev's Theorem The proportion of any data set lying within k standard deviations of the mean is at least $1 - \dfrac{1}{k^2}$ for $k > 1$	4.2
Location of the Pth Percentile $$\ell = n\left(\frac{P}{100}\right)$$	4.2
Percentile $$\text{percentile of } x = \frac{\text{number of data values less than or equal to } x}{\text{total number of data values}} \cdot 100$$	4.3
Location of Q_1 $$\ell = n\left(\frac{25}{100}\right)$$	4.3
Location of Q_2 $$\ell = n\left(\frac{50}{100}\right)$$	4.3
Location of Q_3 $$\ell = n\left(\frac{75}{100}\right)$$	4.3

Key Formulas (cont.)

	Section

Interquartile Range

$$IQR = Q_3 - Q_1$$

4.3

Outlier

Data value greater than $Q_3 + 1.5 \cdot$ Interquartile Range

or

Data value less than $Q_1 - 1.5 \cdot$ Interquartile Range

4.3

z-Score

$$z = \frac{x - \mu}{\sigma}$$

4.3

Coefficient of Variation

Population data: $CV = \left(\dfrac{\sigma}{\mu} \cdot 100 \right)\%.$

Sample data: $CV = \left(\dfrac{s}{\overline{x}} \cdot 100 \right)\%.$

4.5

Mean of Grouped Data

Population data: $\mu = \dfrac{\Sigma(f_i M_i)}{N}$

Sample data: $\overline{x} = \dfrac{\Sigma(f_i M_i)}{n}$

4.6

Variance of Grouped Data

Population data: $\sigma^2 = \dfrac{\Sigma(M_i - \mu)^2 f_i}{N}$

Sample data: $s^2 = \dfrac{\Sigma(M_i - \overline{x})^2 f_i}{n-1}$

4.6

Computational Formulas for the Variance of Grouped Data

Population Data: $\sigma^2 = \dfrac{\Sigma(f_i M_i^2)}{N} - \left(\dfrac{\Sigma(f_i M_i)}{N} \right)^2$

Sample Data: $s^2 = \dfrac{\Sigma(f_i M_i^2) - \dfrac{(\Sigma(f_i M_i))^2}{n}}{n-1}$

4.6

Key Formulas (cont.)

	Section
Class Midpoint	
$$M_i = \frac{\text{lower class boundary} + \text{upper class boundary}}{2}$$	4.6
Population Proportion	
$$p = \frac{x}{N}$$	4.7
Sample Proportion	
$$\hat{p} = \frac{x}{n}$$	4.7
Correlation Coefficient	
$$r = \frac{1}{n-1}\left\{\sum_{i=1}^{n}\left(\frac{x_i - \bar{x}}{s_x}\right)\left(\frac{y_i - \bar{y}}{s_y}\right)\right\}$$	4.8
Computational Formula for the Correlation Coefficient	
$$r = \frac{n\sum x_i y_i - \left(\sum x_i\right)\left(\sum y_i\right)}{\sqrt{n\sum x_i^2 - \left(\sum x_i\right)^2}\sqrt{n\sum y_i^2 - \left(\sum y_i\right)^2}}$$	4.8

AE   Additional Exercises

1. A carpenter is attempting to repair a porch and needs twenty boards which are eight feet long. The salesman at the hardware store says he has twenty boards that "average" eight feet long. When the carpenter checks what he has bought, there are ten boards at six feet and ten boards at ten feet. Do you feel the salesman accurately represented the lengths? Discuss.

2. The maximum heart rates achieved while performing a particular aerobic exercise routine are measured (in beats per minute) for 9 randomly selected individuals.

Table for Exercise 2 – Maximum Heart Rates (Beats per Minute)								
145	155	130	185	170	165	150	160	125

   a. Calculate the sample variance of the maximum heart rate achieved.

   b. Calculate the sample standard deviation of the maximum heart rate achieved.

   c. Calculate the range of the maximum heart rate achieved.

   d. What are some of the factors which might contribute to the variation in the observations?

3. A sample of teenagers was asked how many times they went to the movies in the past 3 months. The frequency distribution table summarizes the results.

Table for Exercise 3 – Teenager Movie Visits	
Number of Visits	Frequency
0	13
1	18
2	11
3	7
4	4
5	3
6	0
7	3
8	3
9	0
10	2

   a. What proportion of the sample visited the movies at least 3 times in the previous 3 months?

   b. Find the mean and standard deviation of the number of visits using the formulas for grouped data.

   c. Compute the interval one standard deviation about the mean.

   d. Find the percent of data falling in the interval one standard deviation about the mean.

   e. Is the percent of the data falling in the interval one standard deviation about the mean close to what the Empirical Rule predicts? What is the reason for the discrepancy, if any?

4. A high school math teacher summarized the 35 math SAT scores for the students in her calculus class. The mean for the class was 521 and the median was 535. The range of the scores was 235 and the highest score in the entire class was 675. Approximately 40% of the class scored higher than 562. State whether each of the following is true or false.

   a. The 45th percentile exceeds 540.

   b. The lowest score in the class was 440.

   c. The z-score for a score of 510 is a negative number.

   d. The third quartile exceeds 562.

   e. The percentile rank of 562 is 40.

5. Consider the following number of defective circuit boards produced by two different machines on seven randomly selected days.

Table for Exercise 5 – Defective Circuit Boards							
Machine A	2	3	7	4	5	1	0
Machine B	2	3	4	3	4	2	4

   a. Calculate the average number of defective circuit boards produced by each machine.

   b. Calculate the variance of the number of defective circuit boards produced by each machine.

   c. Calculate the standard deviation of the number of defective circuit boards produced by each machine.

   d. Which machine do you think is better? Why?

6. A basketball coach has one remaining scholarship to offer and has narrowed his choice to two players. Listed in the following table are the points scored per game over the last season for each player.

Table for Exercise 6 – Points Scored		
Game Number	Braudrick	Douglas
1	27	35
2	34	21
3	29	50
4	25	28
5	28	missed
6	35	32
7	31	29
8	33	missed
9	33	23
10	25	35
11	28	31
12	32	36
Total	360	320

   a. What level of measurement does the data possess?

   b. What statistical criteria might you use to select the better player? Justify your answer.

   c. Calculate the statistics you proposed in b.

   d. Which player is more consistent? Why?

7.   Consider the literacy data given in the following table.

Table for Exercise 7 – Literacy Rates			
Country	Literacy Rate (%)	Country	Literacy Rate (%)
Australia	99.0	Luxembourg	99.0
Bolivia	90.7	Mexico	92.8
Canada	99.0	Netherlands	99.0
Denmark	99.0	Peru	89.6
France	99.0	Saudi Arabia	85.0
India	74.0	United States of America	99.0
Kenya	73.0	Zimbabwe	91.2

Source: United Nations Development Programme Report, 2009

a.   What is the mean literacy rate for these selected countries?

b.   What is the standard deviation of these literacy rates?

c.   How many countries in this group would we expect to have literacy rates between one standard deviation below the mean and one standard deviation above the mean?

d.   How many countries in this group actually have literacy rates between one standard deviation below the mean and one standard deviation above the mean?

e.   What assumption did you make in answering part **c.** above?

8.   A manufacturer considers her production process to be "in control" if the proportion of defective items is less than 3%. She randomly selects 200 items and determines that 9 of the items are defective.

a.   Calculate the sample proportion of defective items.

b.   Based on the sample, do you think it is reasonable for the manufacturer to conclude that the production process is "out of control"? Why or why not?

9. A pharmacist is interested in studying the relationship between the amount of a particular drug in the bloodstream (in mg) and reaction time (in seconds) of subjects taking the drug. Ten subjects are randomly selected and administered various doses of the drug. The reaction times (in seconds) are measured 15 minutes after the drug is administered with the following results.

Table for Exercise 9 – Reaction Times	
Amount of Drug (mg)	Reaction Time (Seconds)
1	0.5
2	0.7
3	0.6
4	0.7
5	0.8
6	0.8
7	0.9
8	0.6
9	0.9
10	1.0

a. Analyze the data collected for the study by answering the following questions:

   i. Do the variables selected for measurement seem appropriate for answering the question the pharmacist is interested in?

   ii. What biases or errors might be present in the data?

   iii. What level of measurement (nominal, ordinal, interval, ratio) does the data possess?

b. Plot the data points on a scatterplot.

c. Based on the scatterplot in part b., answer the following questions regarding the overall pattern of the data.

   i. Does the pattern roughly follow a straight line?

   ii. Is the pattern upward sloping or downward sloping? Are the data values tightly clustered in the pattern or widely dispersed?

   iii. Are there significant deviations from the pattern?

10. Sometimes the following descriptions are assigned to the correlation coefficient.

$$r = 0 \quad \text{no linear relationship}$$
$$-0.5 < r < 0 \quad \text{weak negative linear relationship}$$
$$0 < r < 0.5 \quad \text{weak positive linear relationship}$$
$$-0.8 < r \le -0.5 \quad \text{moderate negative linear relationship}$$
$$0.5 \le r < 0.8 \quad \text{moderate positive linear relationship}$$
$$-1.0 < r \le -0.8 \quad \text{strong negative linear relationship}$$
$$0.8 \le r < 1.0 \quad \text{strong positive linear relationship}$$
$$r = 1 \quad \text{exact positive linear relationship}$$
$$r = -1 \quad \text{exact negative linear relationship}$$

Describe the relationships indicated by the correlation coefficients below using the descriptions defined above.

a. $r = 0.9$ **b.** $r = 0.5$

c. $r = -0.9$ **d.** $r = -0.5$

e. $r = 0$

11. Describe the relationships indicated by the correlation coefficients below using the descriptions defined in problem 10 above.

a. $r = 0.8$ **b.** $r = 0.4$

c. $r = -0.8$ **d.** $r = -0.4$

e. $r = 0.1$

12. Consider the following data.

Table for Exercise 12						
x	1	2	3	4	5	6	7
y	1	4	9	16	25	36	49

a. Plot the data points on a scatterplot.

b. Determine the correlation coefficient.

c. Describe the relationship between x and y.

13. Consider the following data.

Table for Exercise 13						
x	1	2	3	4	5	6	7
y	1.00	1.41	1.73	2.00	2.24	2.45	2.65

a. Plot the data points on a scatterplot.

b. Determine the correlation coefficient.

c. Describe the relationship between x and y.

P   Discovery Project

Describing Everyday Data

1. Consult the Farmer's Almanac and determine the Almanac's weather prediction for your area during the last three months. Evaluate the accuracy of the Almanac's predictions. How accurate is the Almanac's forecast? What does it mean to be accurate? Is the almanac forecast a better predictor than simply guessing? Describe your findings in a report.

2. Determine the average length of a rock and roll song. *Billboard* magazine would be an excellent source for the data.

3. Suppose there were 20 quizzes during your statistics course. What would be the advantages and disadvantages of using the trimmed mean to compute your quiz average?

4. Answer the following questions regarding the median incomes for the years 2007-2008 and 2009-2010 using the data set titled HOUSEHOLD INCOME from the data sets that are available for download from the following website.

    http://www.hawkeslearning.com/Support/Downloads.htm

    The data set is available in Excel or Minitab format. The data provided are the 2-year median household incomes from 2007 to 2010 for all 50 states in the United States, and were collected by the U.S. Census Bureau.

    a. What is the mean of the median incomes

        i.   for 2007-2008?   **ii.**   for 2009-2010?

    b. What is the median of the median incomes

        i.   for 2007-2008?   **ii.**   for 2009-2010?

    c. What is the mode of the median incomes

        i.   for 2007-2008?   **ii.**   for 2009-2010?

    d. Given the relationship between the mean, median, and mode of the household incomes calculated in parts **a.** through **c.**, describe the shape of the distributions of the median household incomes in both 2007-2008 and 2009-2010.

    e. Calculate the range for the median household incomes.

    f. Calculate the variance of the median household incomes.

    g. Calculate the standard deviation of the median household incomes.

    h. Discuss the measures of variability (in years) that you calculated in parts **e.** through **g.** for both 2007-2008 and 2009-2010.

    i. Construct a box plot of the median household incomes for 2007-2008 and 2009-2010. Compare the two box plots, discussing the spread, the center of the distribution, and any unusual incomes during either of the years.

    j. Discuss the change in median incomes by state. That is, calculate the difference in median income by state and discuss the change and whether you think the change is significant or negligible.

Chapter Five
Probability, Randomness, and Uncertainty

D   Discovering the Real World

The Credit Card Industry

The majority of students on college campuses are approached by vendors representing credit card companies soliciting students to apply for credit cards. The solicitors preach to the students the importance of establishing a credit history before graduating. Having this credit history will make it easier for the students to obtain financing for cars, homes, and even apartment leases upon graduation. Below is a list of facts about credit cards and the credit card industry that may surprise you.

- 84% of the student population overall have credit cards.

  Source: Sallie Mae, "How Undergraduate Students Use Credit Cards," April, 2009

- 50% of college undergraduates had four or more credit cards in 2008.

  Source: Sallie Mae, "How Undergraduate Students Use Credit Cards," April, 2009

- 76% of undergraduates have credit cards with an average debt of $2200.

  Source: Nellie Mae, "Undergraduate Students and Credit Cards in 2004: An Analysis of Usage Rates and Trends"

- Seniors graduated with an average credit card debt of more than $4100. Close to one fifth of seniors carried balances greater than $7000.

  Source: Sallie Mae, "How Undergraduate Students Use Credit Cards," April, 2009

- 609.8 million credit cards are held by U.S. consumers.

  Source: "The Survey of Consumer Payment Choice," Federal Reserve Bank of Boston, January 2010

- The average credit card debt per household with credit card debt: $15,788.

  Source: www.creditcards.com

- There are 576.4 million total credit cards in circulation in U.S. as of year-end 2009.

  Source: Nilson Report, February 2010

- 3.5 is the average number of credit cards held by cardholders as of year-end 2008.

  Source: "The Survey of Consumer Payment Choice," Federal Reserve Bank of Boston, January 2010

- 78 percent of American households (about 91.1 million) had one or more credit cards at the end of 2008.

  Source: Nilson Report, April 2009

- 14.67% is the average APR on a credit card with a balance on it as of February 2010.

  Source: Federal Reserve's G.19 report on consumer credit, May 2010

- $852.6 billion is the amount of total U.S. revolving debt (98 percent of which is made up of credit card debt) as of March 2010.

  Source: Federal Reserve's G.19 report on consumer credit, March 2010

- $2.45 trillion is the amount of total U.S. consumer debt as of March 2010.

  Source: Federal Reserve's G.19 report on consumer credit, May 2010

- The top 10 U.S. credit card issuers held an 87.55 percent market share of $972.73 billion in general purpose card outstandings in 2008. That includes Visa, MasterCard, American Express, and Discover.

  Source: Nilson Report, April 2009

- 13.01% is the U.S. credit card default rate.

  Source: Fitch Ratings, April 2010

- 4.27% is the U.S. credit card 60-day delinquency rate.

  Source: Fitch Ratings, April 2010

Like any other industry, banks take profit maximization as their ultimate goal. Given the statistics just presented, it is clear that over the past several years, competition in the U.S. credit card industry has been fierce because there is a great deal of money to be made. However, maximizing profit doesn't come without risk. This competition has led many banks, including large, full-service institutions, to lose customers to issuers that are aggressively expanding their card portfolios. Some of the more visible winners of this competition are Capital One Bank, Citibank, HSBC, Chase Bank, and Bank of America. The tremendous volume of card loans generated has created an equally immense need for inexpensive, reliable funding. To be competitive and successful in this industry, credit card issuers must have sophisticated risk-based and risk-controlling technology models to determine not only who is issued a credit card but also the rate it can offer to a particular customer based on that customer's risk profile.

The traditional approach of assessing a customer's risk is by examining their credit score. By comparing the applicant's/borrower's credit history to all other borrowers' information, banks check whether the applicant's credit history is the same as those who regularly default and even declare bankruptcy to determine their credit worthiness. There are also other factors that are examined such as income, lengths of continuous employment at current (and past) jobs, number of credit cards, and whether one owns or rents their home. Rather than simply seeking to minimize the loss or zero risk of a card, examining these factors allows the banks to determine an acceptable level of risk. Thus, when issuing a card to a customer, banks attempt to find the best balance or combination of risks and revenue. The capacity and the technology of controlling risks are the key to a bank's profitability.

Once the accounts are acquired (i.e., a customer has been approved for a credit card), banks begin managing the customer's account, which includes authorization of repeated credit purchases, credit line management, fraud detection, sales promotion, cross-selling, and collections. Risk management closely monitors the delinquency and utilization patterns along with charge-offs, bankruptcies and fraud occurrences. Based on the aforementioned items, authorization and credit line policies are modified to reduce loss or improve profitability. Behavioral score cards, vintage performance dashboards, and reports are developed to support the decision process. Marketing analyzes activation and sales patterns and prompts the customers to buy more through rewards and other offers. Cross-selling of financial products like personal loans and mortgages to existing customers is another area marketing analytics focuses on. The collections analytics team identifies the likely defaulters using behavioral score cards and, based on the risk profiles, devises strategies for collections. They use the past experiences to improve the collection efficiency and work closely with the operations team to implement data driven strategies. The operations analytics team, typically aligned with customer service operations, forecasts the call volumes and analyzes call logs to improve the call routing and support processes. Collection efficiency of the analysts in terms of number of calls and amount collected are some of the metrics monitored by this department.

In summary, card issuers use a variety of tools and utilities in consumer lending to perform an array of analytics activities ranging from simple reporting to complex statistical modeling. Assessing customers' risk is the foundation during the whole process of risk control including risk identification, risk estimation, and risk assessment. Only by assessing customers' risk and issuing cards to the applicants can the bank card business begin. Accurately assessing customers' risk is an important basis for credit. It can play a role of early warning. Based on it, banks can develop appropriate risk control policies to improve their revenues and profit levels.

Introduction

We all have to cope with uncertainty. Uncertainty is that uneasy feeling we get when our gas tank is nearly empty a few miles from the closest gas station. You remember in the past that your car continued to run when the gauge was even lower, but you are not quite sure it will this time. In everyday usage, the word *probably* describes an event or circumstance the speaker believes will occur. At the same time the speaker reserves the possibility that it may not occur. In this sense, *probably* reflects a strong but nonspecific degree of belief.

Probability is used to quantify uncertainty. If a person says he believes there is a 0.95 probability that his car will make it to a gas station before it runs out of gas, he has made a precise statement which, no doubt, reflects his past experiences and indicates a strong belief in his chances of finding a gas station. This statement is vastly different from the statement that there is a 0.40 probability the car will make it to a gas station, which casts considerably more doubt on his prospects. The probability statement provides more precise information than phrases like *maybe I'll make it, I might make it*, or *I should make it*.   Therein lies its value.

Uncertainty doesn't necessarily have to be associated with events in the future. There can be plenty of uncertainty about the past. What was the Native American population of the continental United States in the year 1492? Nobody knows, so ascribing a probability to the statement *the population of Native Americans in 1492 was between 1.1 and 1.8 million* attaches someone's degree of belief to the statement.

The word **randomness** suggests a certain haphazardness or unpredictability. Randomness and uncertainty are both vague concepts that deal with variation. And even though these words are not synonyms, their discussion leads to the concept of probability.

A simple example of randomness involves a coin toss. Unless the person tossing the coin possesses magical powers, the outcome of the toss (heads or tails) is uncertain. Since the coin tossing experiment is unpredictable, the outcome is said to exhibit randomness. There are many different kinds of randomness. Some are easily described, like the toss of a coin, and others are extraordinarily complex, like molecular motion or changes in stock prices.

Even though individual flips of a coin are unpredictable, if we flip the coin a large number of times, a pattern will emerge. For most coins, roughly half of the flips will be heads and half will be tails. This long-run regularity of a random event is described with probability. Our discussions of randomness will be limited to phenomena that in the short run are not exactly predictable but do exhibit long-run regularity.

5.1   Probability Concepts

Games of chance, such as tossing a coin, provide a way to demonstrate some of the fundamental laws (rules) of probability. Statistically speaking, the playing of the game is called an **experiment**.

Definition

A **random experiment** is defined as any activity or phenomenon that meets the following conditions.

1.   There is one distinct outcome for each trial of the experiment.

2.   The outcome of the experiment is uncertain.

3.   The set of all distinct outcomes of the experiment can be specified and is called the **sample space**, denoted by S.

Definition

An **outcome** is any member of the sample space.

If you were to toss a single coin there are only 2 outcomes {Head, Tail} in the experiment. The sample space of an experiment contains every potential outcome that could occur in one trial of the experiment. For a coin tossing experiment, the sample space would be $S = \{$Head, Tail$\}$. The sample space may be called the **outcome set**, since it contains all possible outcomes of an experiment. The result of a random experiment must be an outcome.

Definition

An **event** is a set of outcomes.

Examples of random experiments, events, and outcomes follow. These experiments will be used throughout the chapter to illustrate the laws of probability.

Experiment Number 1: Toss a coin and observe the outcome. Have we met the three conditions of a random experiment?

1.   There will only be one outcome for each trial of the experiment since it is not possible to observe both a head and a tail on the same toss.

2.   The outcome is unknown before the toss.

3.   The sample space can be specified and contains two outcomes, $S = \{$Head, Tail$\}$.

Because each of the three conditions is satisfied, for all practical purposes, this experiment meets the conditions of a random experiment.

It is possible for a coin to land on its edge and thus neither be heads nor tails. Any theory, however, involves a certain degree of idealization, which means, in this case, that landing on an edge will not be considered a possible outcome.

5

Experiment Number 2: Toss a coin three times and observe the number of heads. Have we met the three conditions of a random experiment?

1.   There will be only one outcome since, for example, it is not possible to have **exactly** one and **exactly** two heads on the same trial.

2.   The outcome will be unknown before tossing the coin three times.

3.   The sample space can be specified and is composed of eight outcomes:

$S = \{$TTT, TTH, THT, THH, HTT, HTH, HHT, HHH$\}$.

This experiment meets the conditions of a random experiment.

An **event** could be obtaining more than one head, which involves the following set of outcomes: {THH, HTH, HHT, HHH}.

Experiment Number 3: Select a student from a class of size 100 and observe his or her grade point average (GPA). Have we met the conditions of a random experiment?

1.   Although we can only select one student at a time, we are measuring GPA, which is a continuous random variable between zero and four.

2.   The outcome will be unknown before selecting the student.

3.   The sample space cannot be specified with certainty.

Given points **1.** and **3.**, this experiment does not meet the conditions of a random experiment.

Experiment Number 4: Assume we have a deck of playing cards consisting of 13 hearts, 13 clubs, 13 spades, and 13 diamonds. Draw a card from a well shuffled deck and observe the suit of the card. Have we met the three conditions of a random experiment?

1.   There will be only one outcome.

2.   The suit will be unknown since the card will be drawn at random.

3.   The sample space consists of the following set of outcomes:

$S = \{$heart, club, spade, diamond$\}$.

This experiment meets the conditions of a random experiment.

If the random experiment involves drawing a card and observing a spade or a club, then the **event** would be given by the set of outcomes {spade, club}.

Experiment Number 5: Inspect a transistor to determine if it meets quality control standards. Have we met the three conditions of a random experiment?

1.   There will only be one outcome.

2.   The outcome of the experiment will be unknown if the transistor is selected from a manufacturing process that occasionally produces defective parts.

3.   The sample space consists of the set of outcomes, $S = \{$meets standards, does not meet standards$\}$.

This experiment meets the conditions of a random experiment.

5.1   Exercises

Basic Concepts

1. Describe randomness.

2. What is probability?

3. What are the conditions of a random experiment?

4. Consider the random experiment of flipping a fair coin twice. What is the sample space for this experiment?

5. What is an event?

6. Consider the random experiment of rolling a fair die once. Give an example of an event for this experiment and list the outcomes associated with that particular event.

Exercises

7. Consider the following random experiment. A potato chip manufacturer is interested in determining if the brand of potato chip which it manufactures is preferred over three of its major competitors. Several customers are randomly selected and asked which brand of potato chip they prefer: Brand A, Brand B, Brand C, or Brand D.

   a. Determine the sample space for the experiment described.

   b. If the manufacturer makes Brand A, list the outcomes in the event $M =$ {customer does not prefer the manufacturer's brand}.

8. Consider the following random experiment. A doctor is interested in determining whether or not his patients think that he listens attentively to what they are saying. He randomly selects several patients and administers an anonymous survey that asks which of the following categories best describes his attentiveness: Very Attentive, Somewhat Attentive, Not Attentive.

   a. Determine the sample space for the above experiment.

   b. Determine all possible outcomes for the event $A =$ {the doctor is not described as very attentive}.

5

5.2   Interpreting Probability

Origins of Probability

Galileo performed one of the earliest probability analyses. Galileo was approached by an Italian nobleman who had observed that three dice were more likely to obtain a sum of 10 than a sum of 9 when thrown. Galileo became interested in the uncertainties of throwing dice and wrote a short work outlining his findings. Galileo's work set forth some of the theory of probability. The next step in the birth of probability came from the French. A French nobleman, Chevalier de Mere, won quite a few francs by getting unwitting souls to bet him that he would not roll at least one six in a sequence of four tosses of a die. He then lost his profits by wagering that he would get at least one double 6 in a sequence of 24 tosses of two dice. Chevalier de Mere asked two of the leading mathematicians of the time, Pierre de Fermat (1601–1665) and Blaise Pascal (1623 – 1662), to consider his problem. The exchange of letters between Fermat and Pascal led to some of the first analysis of random phenomena, and development of the first principles of probability theory.

What is Probability?

If someone is asked, *What is probability?*, they are likely to respond correctly with *It is the chance of something happening*. A more formal definition of probability is the likelihood of the occurrence of a particular event. However, there are many ideas of what probability is and how it should be calculated. There is **subjective probability**, which is one's personal belief about the occurrence of a specific event. With subjective probability, there is no specific calculation, and the probability simply reflects a person's opinion and/or past experiences. On the other hand, **objective probability** is the likelihood of the occurrence of a particular event that is based on recorded outcomes. Objective probabilities are more accurate than subjective probabilities. In this section we will discuss two types of objective probability, relative frequency and the classical approach, as well as subjective probability.

Relative Frequency

Someone who wanted to determine the probability of getting a head on the toss of a coin could toss a coin a large number of times and observe the number of times that a head appeared. The probability could be computed as the number of times a head was observed divided by the number of times the coin was flipped. This is the relative frequency interpretation of probability.

Formula

Relative Frequency

If an experiment is performed n times, under identical conditions, and the event A happens k times, the **relative frequency** of A is given by the following expression.

$$\text{Relative Frequency of } A = \frac{k}{n}$$

Let's flip a coin 42 times and observe the relative frequency of a head during those tosses.

Relative Frequency of Heads for 42 Flips						
Flip Number						
1	2	3	4	5	6	7
Relative Frequency						
1.00	0.5	0.6667	0.75	0.6	0.5	0.4286
Flip Number						
8	9	10	11	12	13	14
Relative Frequency						
0.375	0.4444	0.4	0.3636	0.4167	0.3846	0.3571
Flip Number						
15	16	17	18	19	20	21
Relative Frequency						
0.3333	0.375	0.4118	0.3889	0.3684	0.35	0.3810
Flip Number						
22	23	24	25	26	27	28
Relative Frequency						
0.3636	0.3478	0.375	0.36	0.3462	0.3333	0.3214
Flip Number						
29	30	31	32	33	34	35
Relative Frequency						
0.3448	0.3333	0.3548	0.375	0.3636	0.3529	0.3429
Flip Number						
36	37	38	39	40	41	42
Relative Frequency						
0.3611	0.3514	0.3421	0.3590	0.35	0.3415	0.3571

5

In Figure 5.1, you can see that the proportion of heads is very unstable during the first 15 flips. The proportion of heads begins to stabilize at around flip 20, although there is still some fluctuation in its value. Looking at the first 42 flips makes you wonder whether this is a "fair" coin, since heads is occurring only $0.3571(100) = 35.71\%$ of the time. In Figure 5.2, which starts at about 200 flips, the percentage of heads becomes very stable. By flip number 296, there are 141 heads and 155 tails which equate to (approximately) a 0.4764 probability (or relative frequency) of heads. Although this is slightly less than expected, such a percentage is reasonable considering the randomness of the coin toss.

Figure 5.1

Figure 5.2

Relative Frequency of Heads (Flips 1350 – 1450)

Figure 5.3

As can be seen in Figure 5.2 and Figure 5.3, the percentage of heads converges on some point which is close to 0.5. Figure 5.3 starts at 1350 flips. As you can see, the percentage of heads remains very stable. At flip number 1450 there are 718 heads and 732 tails which (approximately) equate to a 0.4952 probability (or relative frequency) of heads.

What is the Relative Frequency of a Head on Our Coin?

Our best available guess is 0.4952, since it is the observed relative frequency of heads using all 1450 tosses.

$$\text{Relative Frequency of } A = \frac{k}{n} = \frac{718}{1450} \approx 0.4952$$

How good is this guess for the relative frequency of heads? Since the coin has been tossed a large number of times and the observed frequency is very stable, the guess should be very good.

Will the Observed Relative Frequency Ever Reach 0.5?

No mathematical or physical law requires the observed relative frequency to ever reach some predetermined level. But if the probability of observing a head was really 0.5, the observed relative frequency should closely approach this value after a large number of flips.

> ### A Summary
>
> **The experiment:** Toss a coin and observe which side of the coin appears on top.
>
> **Duration of the experiment:** Toss the coin 1450 times. $n = 1450$
>
> **Observe the event "getting a head":** The event was observed 718 times. $k = 718$
>
> **Relative frequency of the event "getting a head":** $\dfrac{k}{n} = \dfrac{718}{1450} \approx 0.4952$

The relative frequency of a head seems to converge to the expected relative frequency of 0.5. This kind of convergence is sometimes called **statistical regularity**. Although the outcomes of the experiment may vary, in the long run the relative frequency of an outcome tends to some value, its probability.

Problems with the Relative Frequency Idea

The problem with the relative frequency approach to defining probability is that probability only exists for events that can be repeated under the same conditions. Coin, dice, and card experiments can easily be repeated. However, because of the strict requirements of identical and repeatable experiments, many events in which it would be desirable to have relative frequency probabilities do not satisfy the requirement of repetition. Whether the next launch of the rocket will be successful, or whether you will make an A in your statistics course are examples of experiments that are not repeatable under the exact same conditions. Thus, they are not appropriate for the application of the relative frequency idea. This perspective greatly limits the application of the relative frequency interpretation of probability. Despite its limitations, the relative frequency approach is a widely held interpretation of probability.

Example 5.1

Suppose we perform an experiment drawing three numbers, without replacement, from an urn containing 64 numbers. Have we met the conditions of a random experiment?

Solution

No! This does not meet the conditions of a random experiment because it cannot be repeated under the same conditions. That is, as each number is selected, it is not returned to the urn prior to the subsequent selection, thus decreasing the amount of numbers remaining in the urn. For this reason, the experiment is different for each number drawn, and cannot be considered a random experiment.

Classical Approach

The second objective approach commonly used in probability is the classical approach. **Classical probability** can be measured as a simple proportion: the number of outcomes that compose the event divided by the number of outcomes in the sample space, when it can be assumed that all of the outcomes are equally likely.

Formula

Classical Probability

Using the classical approach to probability, the probability of an event A, denoted $P(A)$, is given by

$$P(A) = \frac{\text{number of outcomes in } A}{\text{total number of outcomes in the sample space}}.$$

Example 5.2

In experiment number 2 (from Section 5.1), a coin was tossed three times and the number of heads was observed. The sample space consists of 8 outcomes {TTT, TTH, THT, THH, HTT, HTH, HHT, HHH}. Let A be the event of getting at least one head. What is $P(A)$?

Solution

Since the event A consists of 7 outcomes, {TTH, THT, THH, HTT, HTH, HHT, HHH}, and there are 8 equally likely outcomes in the sample space,

$$P(A) = \frac{7}{8} = 0.875.$$

Example 5.3

In experiment number 4 (from Section 5.1), let A be the event of drawing a heart. What is $P(A)$?

Solution

Since there are 13 outcomes in A (i.e. 13 hearts in a deck of cards) and 52 outcomes in the sample space (i.e. 52 cards in the deck) the probability of event A is as follows.

$$P(A) = \frac{13}{52} = \frac{1}{4} = 0.25$$

If the sample space is composed of equally-likely outcomes, then once the set of outcomes is determined, computing a probability is simply a matter of counting the members in each set and dividing by the total number of outcomes.

It is very important to remember that the classical approach rests on the assumption of equally-likely outcomes. If the assumption is not reasonable, some other method of determining the probability must be used.

Example 5.4

In experiment number 2, a coin was tossed three times and the number of heads was observed. The outcomes in the sample space were as follows.

$$S = \{TTT, TTH, THT, THH, HTT, HTH, HHT, HHH\}$$

Could the sample space in this experiment have been formulated differently? Could, for example, the outcomes be defined as the number of heads in three tosses, $S = \{0, 1, 2, 3\}$?

Solution

Using the classical definition of probability requires that the outcomes be equally likely. Since the sample space $S = \{0, 1, 2, 3\}$ does not contain equally likely events, the classical definition of probability is not applicable.

Subjective Approach

The subjective viewpoint regards the probability of an event as a measure of the degree of belief that the event has occurred or will occur. Someone's degree of belief in some event will depend on his or her life experiences. Different life experiences produce different degrees of belief. Hence, the subjective approach must allow for differences in the degree of belief among reasonable people examining the same evidence. One of the significant advantages of this view is the ability to discuss the probability of events that cannot be repeated. Thus, a subjectivist would be willing to assign the probability of making an A in your statistics course. Someone who adopts the subjective view could use the frequency interpretation to influence the determination of a subjective probability. For example, suppose that a coin had been tossed 20,000 times and had come up heads 63% of the time. It would certainly be reasonable for a subjectivist to use this information in the formulation of a statement of probability about the outcome of the next toss.

Criticism of the Subjective View

If science is defined as finding out what is probably true, there should be a probability criterion on which all reasonable persons could agree. But if probability is subjective, how can it be used as a universally accepted criterion? Two reasonable persons might examine the same data and reach different conclusions about their degree of belief about some proposition.

5.2   Exercises

Basic Concepts

1. What are the two main branches of probability?

2. What are the two approaches to objective probability?

3. What are some of the problems associated with the relative frequency approach?

4. True or false: According to the mathematical law of probability, the observed relative frequency of heads when flipping a coin will eventually reach 0.5 since the probability of heads is 0.5.

5. What is statistical regularity?

6. Describe the classical approach to probability.

7. Using the classical approach, describe how you would determine the probability of event A.

8. What is the subjective approach to probability? Discuss the problems of applying the subjective interpretation.

Exercises

9. A gambler has made a weighted die. In order to decide which of the six sides is most likely to turn up, he tosses the die 33 times and notes the number of dots on the upper-most surface. The results of the experiment are shown in the following table.

Table for Exercise 9 – Rolls of a Weighted Die										
1	2	1	3	1	4	1	5	6	3	1
3	1	5	1	2	1	3	1	2	1	2
2	1	3	5	1	2	1	2	1	4	6

  a. Using the relative frequency approach, what is the probability of observing each side?

  b. Which side do you think the gambler will bet on when the die is tossed?

10. Assume there are two red, two yellow, and two blue buttons in a hat. A button is drawn out of the hat, the color is noted, and the button is returned. This is repeated fifty times. The results are listed in the following table.

Table for Exercise 10 – Button Drawing				
Yellow	Yellow	Red	Yellow	Red
Red	Red	Blue	Red	Blue
Blue	Red	Red	Yellow	Red
Red	Blue	Yellow	Red	Yellow
Yellow	Blue	Red	Blue	Red
Red	Red	Red	Red	Yellow
Blue	Yellow	Yellow	Blue	Red
Yellow	Red	Red	Red	Yellow
Red	Yellow	Yellow	Yellow	Red
Red	Red	Blue	Red	Blue

Using the relative frequency approach, what is the probability of drawing each color?

11. Twenty-five insurance agents are randomly selected and asked if they own a handgun. Twenty-two of those surveyed said that they do own a handgun. If an insurance agent is randomly selected, estimate the probability that the agent will own a handgun.

12. Thirty elementary school teachers are randomly selected and asked if they favor standardized testing of elementary school children. Twenty of those surveyed said that they did favor standardized testing of elementary school children. If an elementary school teacher is randomly selected, estimate the probability that the teacher will favor standardized testing for elementary school children.

13. Fifty chief executive officers (CEOs) of publicly traded companies are randomly selected and their salaries are determined. Forty-five of the CEOs selected have salaries in excess of $500,000. If a CEO from one of the selected publicly traded companies is randomly selected, find the probability that the CEO will have a salary in excess of $500,000.

14. Forty emergency calls to which a local police department responded were randomly selected. Of the forty emergency calls fifteen were categorized as domestic arguments. Estimate the probability that the next emergency call to which the local police department responds will be a domestic argument.

15. For the following situations, decide which probability interpretation is most reasonable to use: relative frequency, subjective, or classical.

    a. Whether or not you will have a wreck on your next trip to the mall.

    b. Whether or not a car coming off the Ford assembly line will have a defect.

    c. The probability that you will graduate from college in four calendar years.

    d. Whether a person will be in an automobile accident during the next year.

    e. The probability that you will be dealt a full house from a well-shuffled deck of cards.

16. For the following situations, decide which probability interpretation is most reasonable to use: relative frequency, subjective, or classical.

    a. Suppose you have purchased a lottery ticket. Describe your chances of winning the lottery.

    b. The probability you will enjoy a vacation trip to Mexico.

    c. The probability your company's sales will exceed seven million dollars this year.

    d. One hundred people receive keys to a new car in a radio contest. Only one key actually fits the car. The probability that key number 25 will open the car door.

    e. The probability that you will get a ticket if you drive 70 mph on the interstate between work and home this coming Tuesday.

    f. The probability that the S&P 500 will increase or decrease by at least 25 points in one day.

17. A couple plans to have two children.

    a. List all possible outcomes for the sexes of the two children.

    b. Find the probability that the couple will have 2 boys.

    c. Find the probability that the couple will have at least 1 girl.

18. Consider a student who is taking a multiple choice examination where there are five possible answers for each question. Since the student has not studied or attended any of the classes, the student decides to randomly guess at each question.

    a. Find the probability that the student will answer the first question correctly.

    b. Find the probability that the student will answer the first question incorrectly.

19. A game show contestant has to choose one of three doors to win a prize. Behind one door the prize is a trip to Hawaii; behind another door, the prize is a color TV; behind the final door, the prize is a bag of potatoes. If a contestant randomly selects a door,

    a. Find the probability that the contestant will win a trip to Hawaii.

    b. Find the probability that the contestant will not win a trip to Hawaii.

5

5.3   Probability, Statistics, and Business

Most of the time, when working with samples, statisticians try to deduce from the samples the population parameters (means, proportions, variances, etc.) of certain variables. This process of making judgments about population parameters is called **statistical inference**. Because samples are random, there is no guarantee that the sample will be representative of the population. If the sample is not representative, then using the sample mean as an estimate (inference) of the population mean would not be very wise. Probability is used to assess the quality of our inference. All statistical conclusions must be endowed with a degree of uncertainty. Because probability is used to assess the reliability of sample inferences, it is the foundation of all inferential statistics.

The probability concept also has many direct applications in business. When a manager wonders whether dropping a bid price by 5% will increase the probability of winning the bid, he or she is thinking about chance. Probability is also used as a criterion in designing and evaluating product reliability, evaluating insurance, inventory management, project management, and in the study of queuing theory (a probabilistic analysis of waiting lines).

Probability theory emerged from the need to better understand a game of chance. Business decisions, like games, have uncertain outcomes. In an effort to make better decisions, businesses spend considerable amounts of money trying to quantify uncertainty. This means trying to turn uncertainty into a probability. Insurance companies have historically done a good job of quantifying uncertainty. In fact, a special kind of statistician called an actuary has emerged to assist in the development of insurance models which quantify uncertainty and aid in business decisions.

For example, the next time you watch a 30-second commercial during the Super Bowl, consider the fact that a company has just spent roughly $3 million for the air time plus a substantial amount of money developing the advertisement. Without knowing the effect of the advertisement in advance, extensive amounts of money are put at risk with an uncertain outcome. The manager making the decision uses subjective probability to assess the risk and reward.

5.3   Exercises

Basic Concepts

1. What is statistical inference?

2. Discuss the relationship between probability and statistics.

3. Give three applications of probability in business.

4. Describe the importance of probability in the insurance industry.

5. What type of probability does the manager of a company use when purchasing a commercial spot during the Super Bowl? Explain why.

5.4   Laws of Probability

Interpreting probability using the classical approach is a good way of thinking about the basic probability principles. In this section we will discuss certain laws that probabilities must obey, regardless of how probability is defined.

Probability Law 1

A probability of zero means the event cannot happen.

For example, the probability of observing three heads in two tosses of a coin is zero.

Probability Law 2

A probability of one means the event must happen.

For example, if we toss a coin, the probability of getting either a head or tail is one.

Probability Law 3

All probabilities must be between zero and one inclusively. That is, $0 \leq P(A) \leq 1$.

The closer the probability is to 1, the more likely the event. The closer the probability is to 0, the less likely the event.

Probability Law 4

The sum of the probabilities of all outcomes in a sample space must equal one. That is, if $P(A_i)$ is the probability of outcome A_i, and there are n such outcomes, then $P(A_1) + P(A_2) + \cdots + P(A_n) = 1$.

There are a number of rules concerning the relationships between events that are useful in determining probabilities.

Definition

A **compound event** is an event that is defined by combining two or more events.

Suppose that the marketing director of *Sports Illustrated* believed that anyone who possessed an income greater than $50,000 and/or subscribed to more than one other sports magazine could potentially be a good prospect for a direct mail marketing campaign.

5

Let the events

$$A = \{\text{annual income is greater than \$50,000}\}$$

and

$$B = \{\text{subscribes to more than one other sports magazine}\}.$$

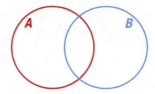

Figure 5.4

There are several different types of compound events. To illustrate the concepts consider the two events A and B in Figure 5.4. The set of outcomes in which either or both of these events occurs is called the **union** of the two sets.

> ### Definition
>
> The **union** of the events A and B is the set of outcomes that are included in A or B or both, and is denoted $A \cup B$.

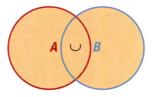

Figure 5.5

Notice in Figure 5.5 that the union includes all outcomes in both A and/or B.

> ### Definition
>
> The **intersection** of the events A and B is the set of all outcomes that are included in both A and B. Symbolically, the intersection of A and B is denoted $A \cap B$ and is read "A intersect B."

Suppose the marketing director was interested in persons who possessed an annual income greater than \$50,000 and subscribed to more than one other sports magazine, that set would be called the intersection of A and B.

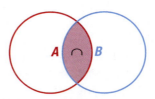

Figure 5.6

Notice in Figure 5.6 that the intersection includes only those outcomes in both A **and** B.

Two other useful concepts are the notions of the **complement** of an event and events which are **mutually exclusive**.

> ### Definition
>
> The **complement** of an event A is the set of all outcomes in the sample space that are not in A.

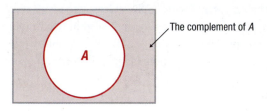

Figure 5.7

The complement of the set A is written as A^c. Notice in Figure 5.7 that the complement of event A includes all outcomes which are not in A. For the event $A = \{$annual income is greater than $\$50,000\}$, the complement of A would be

$$A^c = \{\text{annual income is less than or equal to } \$50,000\}.$$

Also note that $A^c \cup A = S$.

> ### Probability Law 5
>
> The probability of A^c is given by $P(A^c) = 1 - P(A)$.

Sometimes it is much easier to calculate the probability of the complement of an event than the actual event.

Example 5.5

Consider the event $A = \{$annual income is greater than $\$50,000\}$. Suppose the probability of A was 0.08. Determine the probability of observing someone whose income was less than or equal to $\$50,000$.

Solution

$$
\begin{aligned}
P(\text{annual income is less than or equal to } \$50,000) &= P(A^c) \\
&= 1 - P(A) \\
&= 1 - 0.08 \\
&= 0.92
\end{aligned}
$$

Example 5.6

Consider an experiment to see how many tosses of a coin will be required to obtain the first head. The first head could be observed on the first toss or the second toss but there is no upper limit on the number of tosses that could be required. Therefore, the sample space for this experiment is the set of positive integers {1, 2, 3, ...}. Not only is the sample space infinitely large, but there is another problem: the outcomes are not equally likely. This is a potentially ugly environment in which to compute a probability. But let's make matters slightly worse. Suppose we want to know the probability that it will require at least two tosses to get the first head. That is, we want to know the probability of getting the first head in two or more tosses of the coin. This means that we must compute the probability of 2, the probability of 3, and so on up to infinity and add them up in some way. The problem is rather insidious if approached directly.

Solution

The problem becomes rather trivial by determining the complement of the event and computing its probability. The complement of obtaining the first head in two or more tosses is getting a head on the first toss. The probability of getting a head on the first toss is 0.5, assuming the coin is fair. Therefore, we have the following.

$$P(\text{two or more tosses to obtain the first head}) = 1 - P(\text{head on the first toss})$$
$$= 1 - 0.5$$
$$= 0.5$$

As shown in this example, when you see the key words "at least" in a probability problem, you will often want to use the complement to find the appropriate probability.

Mutually Exclusive

Two events are mutually exclusive if they cannot occur at the same time. For example, if you were to select one card from a standard deck, the two outcomes:

- The card is a Jack
- The card is a Seven

are mutually exclusive, because you cannot select a card that is both a Jack *and* a Seven.
However, the two outcomes

- The card is a Jack
- The card is a Club

are not mutually exclusive, because you can select a card that is both a Jack *and* a Club.

Another idea that is helpful in determining probabilities is the notion of **mutual exclusivity**.

Definition

Two events are **mutually exclusive** if they have no outcomes in common.

Mutual exclusivity is also called **disjointedness**. Figure 5.8 represents two disjoint events.

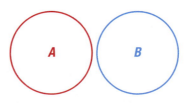

Figure 5.8

Probability Law 6

Union of Mutually Exclusive Events

If the events A and B are mutually exclusive, then

$$P(A \cup B) = P(A) + P(B).$$

Probability Law 7

Intersection of Mutually Exclusive Events

If the events A and B are mutually exclusive, then

$$P(A \cap B) = 0.$$

Example 5.7

Suppose $P(A) = 0.27$ and $P(B) = 0.19$. If A and B are mutually exclusive, what is the probability of $A \cup B$?

Solution

Since these are mutually exclusive events,

$$P(A \cup B) = P(A) + P(B) = 0.27 + 0.19 = 0.46.$$

There is a more generalized rule that eliminates the assumption of mutual exclusivity between the sets.

Probability Law 8

The Addition Rule

For any two events A and B,

$$P(A \cup B) = P(A) + P(B) - P(A \cap B).$$

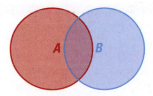

Figure 5.9

The addition rule is identical to Probability Law 6 when the events are mutually exclusive, since A intersect B will be an empty set whose probability will be zero (i.e. $P(A \cap B) = 0$).

Example 5.8

Suppose that the marketing manager mentioned earlier believed that the probability that someone earns more than $50,000 is 0.2 and the probability that someone will subscribe to more than one sports magazine is 0.3. If the probability of finding someone in both categories is 0.08, what is the probability of finding someone who is earning over $50,000 or subscribes to more than one sports magazine, or both?

Solution

The problem involves the union of two events. Using the same event names (A and B) as in previous examples, the desired probability is

$$P(A \cup B) = P(A) + P(B) - P(A \cap B) = 0.2 + 0.3 - 0.08 = 0.42.$$

Therefore, the probability of finding someone who is earning over $50,000 or subscribes to more than one sports magazine, or both is 0.42.

5.4   Exercises

Basic Concepts

1. What laws must probability obey, regardless of the methodology used to derive the probabilities?

2. Suppose you are taking a test next week. Interpret each of the following statements.

    a. P(receiving an A on the test) = 0

    b. P(receiving an A on the test) = 1

    c. P(receiving an A on the test) = 0.3

3. What is a compound event?

4. Define the following set operations: union, intersection, and complement.

5. If you know the probability of two events, what else must you know in order to calculate the probability of *one event or the other*?

6. If events A and B are mutually exclusive, what is $P(A \cap B)$?

Exercises

7. Determine if the following values could be probabilities. If the value cannot be a probability, explain why.

   a. 0

   b. $\dfrac{36}{25}$

   c. $\dfrac{7}{8}$

   d. -0.4

   e. 0.23

8. Determine if the following values could be probabilities. If the value cannot be a probability, explain why.

   a. 1

   b. $\dfrac{15}{16}$

   c. $\dfrac{4}{3}$

   d. 0.99

   e. -0.05

9. Interpret the following probabilities with respect to the occurrence of some event.

   a. $P(\text{event}) = 0$

   b. $P(\text{event}) = 1.0$

   c. $P(\text{event}) = 0.45$

   d. $P(\text{event}) = 65\%$

   e. $P(\text{event}) = -1.0$

10. Find the following probabilities.

    a. The probability of an event that must happen.

    b. The probability of an event that cannot happen.

    c. The probability of having a boy or a girl in a single birth.

    d. The probability of rolling a two and a five in a single toss of a die.

11. The annual premium amounts charged by life insurance companies to their clients are set very carefully. If the amount is too high, the client will take his or her business to another company. If it is too low, the insurance company may not make enough profit to stay in business. In order to properly determine a premium, the company often relies on life tables. These tables allow one to compute the probabilities of death at various ages. They are constructed only after collecting and reviewing extensive data on age at death from a large group of people. A life table is normally constructed assuming that 100,000 people are alive at age 0. This number is simply a reference value used to make comparisons throughout the table. Other numbers could be used. The table then gives the number of people of the original 100,000 that are alive at the beginning of various years of life. In order for the insurance company to optimally set premiums, a separate table should be constructed for the different genders and races. The following abbreviated life table is valid only for females.

Table for Exercise 11 – Life Table

Year	0	1	5	10	15	20	25	30	35
Number Alive	100,000	99,090	98,912	98,815	98,716	98,477	98,204	97,897	97,500
Year	40	45	50	55	60	65	70	75	80
Number Alive	96,958	96,097	94,766	92,623	89,449	84,565	77,772	68,200	55,535

a. What is the probability that a newborn female lives until the age of 40?

b. What is the probability that a newborn female dies before she reaches the age of 50?

12. A mail order company classifies its customers by gender and location of residence. The market research department has gathered data from a random sample of 759 customers.

Table for Exercise 12 – Mail Order Customers

Location	Gender	
	Male	Female
Suburban	196	298
Urban	92	173

a. What is the probability that a customer is male?

b. What is the probability that a customer is female?

c. What is the probability that a customer is a suburban male?

d. What is the probability that a customer is an urban female?

e. What is the probability that a customer is a suburban male or an urban female?

f. What is the probability that a customer is urban?

g. What is the probability that a customer is not a suburban female?

h. What approach to probability did you use to calculate your answers?

i. Are the events {customer is urban} and {customer is suburban} mutually exclusive? Explain.

13. A large life insurance company is interested in studying the insurance policies held by married couples. In particular, the insurance company is interested in the amount of insurance held by the husbands and the wives. The insurance company collects data for all of its 1000 policies where both the husband and the wife are insured. The results are summarized in the following table.

Table for Exercise 13 – Life Insurance Coverage				
	Amount of Life Insurance on Husband ($)			
	0 – 50,000	50,000 – 100,000	100,000 – 150,000	More than 150,000
Amount of Life Insurance on Wife ($) 0 – 50,000	400	200	50	50
50,000 – 100,000	50	50	30	30
100,000 – 150,000	20	10	25	25
More than 150,000	20	10	15	15

a. For a randomly selected policy, what is the probability that the husband will have between $50,000 and $100,000 of insurance?

b. For a randomly selected policy, what is the probability that the wife will have between $100,000 and $150,000 of insurance?

c. For a randomly selected policy, what is the probability that the wife will have more than $150,000 of insurance or the husband will have more than $150,000 of insurance?

d. For a randomly selected policy, what is the probability that the wife will have between $0 and $50,000 of insurance and the husband will have between $0 and $50,000 of insurance?

e. For a randomly selected policy, what is the probability that the wife will not have between $0 and $50,000 of insurance?

f. For a randomly selected policy, what is the probability that the husband will have more than $50,000 of insurance?

g. What approach to probability did you use to calculate your answers?

h. Are the events {the wife has more than $150,000 in insurance} and {the husband has between $50,000 and $100,000 of insurance} mutually exclusive? Explain.

5.5   Conditional Probability

Researchers often want to examine a limited portion of the sample space. For example, consider the question of whether cigarette smoking harms those that are indirectly exposed to the smoke. Suppose that 3 percent of women who do not smoke die of cancer. However, if a nonsmoking woman is married to a smoking husband (not to be confused with a husband who is on fire), the probability of dying of cancer is 0.08. This probability is a **conditional probability,** because the sample space is being limited by some condition – in this case, limited to only wives of smoking husbands. In this instance, the dramatic effect of a smoking husband on cancer rates is readily evident.

$$P(\text{a nonsmoking woman dies of cancer})$$
$$\neq$$
$$P(\text{a nonsmoking woman dies of cancer given that her husband smokes})$$

Similarly, the results from a market survey indicate that 39 percent of the customers surveyed believe a product is of high quality. However, if the analysis is limited to only women, 54 percent of women surveyed believe the product is of high quality. Based on the survey, it appears that women have a much higher regard for the company's product than men. The difference in attitude would probably be something that could affect how the company spends its marketing dollars.

The notion of conditional probability is defined below.

Definition

The probability that one event will occur given that some other event has occurred is a **conditional probability**.

To compute a conditional probability, apply the following rule.

Probability Law 9

The conditional probability of A, given that B has occurred, is

$$P(A \mid B) = \frac{P(A \cap B)}{P(B)}.$$

The notation $P(A \mid B)$ is read as *the probability of A given the occurrence of B*. The vertical bar within a probability statement will always mean *given*.

Note that since we are finding the conditional probability of A, given that B has occurred, $P(B) \neq 0$.

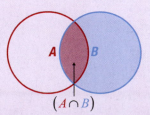

$(A \cap B)$

Figure 5.10

The events A and B can be reversed in the preceding rule to compute $P(B|A)$. It would be very unlikely that $P(A|B)$ equals $P(B|A)$.

Example 5.9

Suppose a marketing research firm has surveyed a panel of consumers to test a new product and produced the following cross tabulation indicating the number of panelists that liked the product, the number that did not like the product, and the number that were undecided.

Table 5.1 – Market Research Survey				
Age	Like	Not Like	Undecided	Total
18-34	213	197	103	513
35-50	193	184	67	444
Over 50	144	219	83	446
Total	550	600	253	1403

If an individual is between 35 and 50 years old, what is the probability he or she will like the product?

Solution

Let the events

$A = \{$like the product$\}$, and

$B = \{$age between 35 and 50$\}$.

Then the desired probability can be formulated as

$$P(A|B) = \frac{P(A \cap B)}{P(B)}.$$

The $P(A \cap B)$ is called a joint probability since it is the probability of the occurrence of more than one event. To compute $P(A \cap B)$ use the empirical approach.

$$P(A \cap B) = \frac{193}{1403} \approx 0.1376$$

Similarly, $P(B)$ can be computed as

$$P(B) = \frac{444}{1403} \approx 0.3165.$$

Consequently, $P(A|B)$ is

$$P(A|B) \approx \frac{0.1376}{0.3165} \approx 0.4348.$$

Note that this answer could have also been obtained by simply dividing 193 by 444.

5.5   Exercises

Basic Concepts

1. Define conditional probability.

2. How do you calculate $P(A\,|\,B)$?

Exercises

3. The following table was given in Section 5.4, Exercise 12.

Table for Exercise 3 – Mail Order Customers		
	Gender	
Location	Male	Female
Suburban	196	298
Urban	92	173

   a. Given that the customer is a male, what is the probability that he is urban?

   b. Given that the customer is urban, what is the probability that he is male?

   c. Given that the customer is female, what is the probability that she is suburban?

   d. Given that the customer is suburban, what is the probability that the customer is a woman?

4. The following table was given in Section 5.4, Exercise 13.

Table for Exercise 4 – Life Insurance Coverage				
	Amount of Life Insurance on Husband ($)			
	0 – 50,000	50,000 – 100,000	100,000 – 150,000	More than 150,000
Amount of Life Insurance on Wife ($) **0 – 50,000**	400	200	50	50
50,000 – 100,000	50	50	30	30
100,000 – 150,000	20	10	25	25
More than 150,000	20	10	15	15

   a. Given the wife has between $100,000 and $150,000 of insurance, what is the probability that the husband has more than $150,000 of insurance?

   b. Given the wife has between $0 and $50,000 of insurance, what is the probability that the husband has between $0 and $150,000 of insurance?

   c. Given that the husband has between $0 and $50,000 of insurance, what is the probability that the wife will have more than $150,000 of insurance?

   d. Given that the husband has more than $150,000 of insurance, what is the probability that the wife will have more than $150,000 of insurance?

5. A computer software company receives hundreds of support calls each day. There are several common installation problems, call them A, B, C, and D. Several of these problems result in the same symptom, *lock up* after initiation. Suppose that the probability of a caller reporting the symptom *lock up* is 0.7 and the probability of a caller having problem A and a *lock up* is 0.6.

    a. Given that the caller reports a lock up, what is the probability that the cause is problem A?

    b. What is the probability that the cause of the malfunction is not problem A given that the caller is experiencing a lock up?

6. A television advertising representative has determined the following probabilities based on past experience. The probability that an individual will watch an ad during the Super Bowl is 0.10. Given that the individual watches the ad, the probability that the individual will buy the product is 0.005. It is also known that the probability that an individual would buy the product is 0.02. Given that an individual buys the product, find the probability that the individual watched the television ad during the Super Bowl.

7. Medical researchers have determined that there is a 2% chance that an individual will have a gene which gives him a predisposition for heart disease. Given that an individual has the gene, the probability that heart disease will develop is 25%. It is also known that the probability that an individual has heart disease is 12%.

    a. Find the probability that an individual will have the gene and develop heart disease.

    b. Given that a person has heart disease, what is the probability that they have the gene?

5.6   Independence

An extremely important concept in statistical analysis is **independence**. It describes a special kind of relationship between two events. Two events are said to be independent if knowledge of one event does not provide information of the other event's occurrence. In other words, the occurrence of one event does not affect the occurrence of another event if the events are independent.

Example 5.10

Experiment: roll a fair die two times. Consider the two events

A = {rolling a six on the first roll of a fair die} and,

B = {rolling a four on the second roll of a fair die}.

Are these two events independent?

Solution

Since knowledge of the outcome of the first roll does not help one make an inference of the outcome of the second roll, events A and B are independent.

Symbolically the idea of independence is expressed in the following definition.

Definition

Two events, A and B, are **independent** if and only if

$$P(A \mid B) = P(A) \text{ and } P(B \mid A) = P(B).$$

In many cases, regarding the independence of two events, intuition and common sense will lead you to the correct determination. However, there are situations in which independence can only be discovered by formal application of the definition.

What does it mean if two events are not independent? The obvious response is to say that they are **dependent**, a term that is just as much a part of statistical vocabulary as independent. If events are dependent, they are related; the nature of the relationship and whether the relationship can be used for predictive purposes are problems often examined by statisticians.

During the course of a business negotiation, both negotiators may exhibit numerous types of idiosyncratic behavior. If they have jewelry, they might manipulate it. If they smoke cigarettes, they might play with their lighters or packs of cigarettes. Are their mannerisms independent of the importance of the issue they are negotiating? Does the negotiator tend to smoke a cigarette or play with jewelry when he or she has a strong position? Good negotiators will pick up dependencies and use the information to their advantage. However, the concept of association implied by dependence must not be confused with the idea of causation. It may be that one of the events does indeed cause the other, but the fact that they are not independent (dependent) is not evidence of causation.

Basketball and Dependence

If you were playing basketball and made a large number of consecutive shots, then there might be a temptation to boast of your skills. Alternatively, someone might suggest that you were on a lucky streak. Two psychologists examined the "lucky streak" phenomenon by analyzing the sequence of made and missed shots by professional basketball players. The selected players made roughly 50 percent of their shots. Their analysis found that there was no evidence of the hot hand – that is, there was no evidence of a dependent relationship between the consecutive shots. They did not find more long streaks of made baskets than would be expected to occur by chance.

Probability Law 10

Multiplication Rule for Independent Events

If two events, A and B, are independent, then

$$P(A \cap B) = P(A)P(B).$$

If n events, $A_1, A_2, \ldots, A_n$, are independent, then

$$P(A_1 \cap A_2 \cap \ldots \cap A_n) = P(A_1)P(A_2) \cdots P(A_n).$$

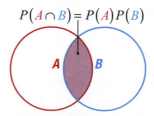

Figure 5.11

The multiplication rule for independent events (Probability Law 10) is sometimes called the **product rule**. The rule simply states that the probability of the joint occurrence of mutually independent events is the product of their probabilities.

Example 5.11

A coin is flipped, a die is rolled, and a card is drawn from a deck of 52 cards. Find the probability of getting tails on the coin, a five on the die, and a Jack of clubs from the deck of cards.

Solution

Since the three events (flipping a coin, rolling a die, and selecting a card) are independent, we can use the product rule.

We know the following.

$$P(\text{tails on coin}) = \frac{1}{2},$$

$$P(\text{five on die}) = \frac{1}{6}, \text{ and}$$

$$P(\text{Jack of clubs}) = \frac{1}{52}.$$

Therefore, we can calculate the probability as follows using the product rule.

$$P(\text{tails on coin} \cap \text{five on die} \cap \text{Jack of clubs})$$
$$= P(\text{tails on coin})P(\text{five on die})P(\text{Jack of clubs}).$$
$$= \left(\frac{1}{2}\right)\left(\frac{1}{6}\right)\left(\frac{1}{52}\right)$$
$$= \frac{1}{624}$$
$$\approx 0.0016.$$

So the probability of getting tails on the coin, rolling a five on the die, and then selecting the Jack of clubs from the deck of cards is approximately 0.0016.

Struck by Lightning

Many people have a greater chance of meeting someone who survived a lightning strike than someone who won the lottery. There are an estimated 1800 new thunderstorms being created around the world every minute and the odds against someone being struck by lightning are 606,944 to one. That is unless you are Roy C. Sullivan. Mr. Sullivan was a former U.S. park ranger who was struck by lightning over seven times in 41 years.

Example 5.12

This is an actual case that stirred up quite a controversy.

People v. Collins (1968)

On June 18, 1964, at about 11:30 AM, Mrs. Juanita Brooks was assaulted and robbed while walking through an alley in the San Pedro area of Los Angeles. Mrs. Brooks described her assailant as a young woman with a blond pony tail. At about the same time John Bass was watering his lawn and witnessed the assault. He described the assailant as a Caucasian woman with dark-blond hair. As she ran from the alley she jumped into a yellow automobile driven by a black man with a mustache and a beard.

Several days later the police arrested two individuals based on the descriptions provided by the assailant and the witness. The two suspects were eventually charged with the crime. During the trial the prosecution called a professor of mathematics to testify. The prosecutor set forth the following probabilities for the characteristics of the assailants:

Table 5.2 – Assailant Characteristics Data	
Characteristic	**Probability**
Yellow automobile	0.10
Man with mustache	0.25
Girl with ponytail	0.10
Girl with blonde hair	0.33
Black man with beard	0.10
Interracial couple in a car	0.001

How did the prosecution use these probabilities to argue its case?

Solution

If the events are assumed to be independent, then the product rule can be used to calculate the likelihood of observing their joint occurrence.

$$P\begin{pmatrix} \text{Yellow automobile} \cap \text{Man with} \\ \text{mustache} \cap \text{Girl with ponytail} \cap \text{Girl} \\ \text{with blond hair} \cap \text{Black man with} \\ \text{beard} \cap \text{Interracial couple in car} \end{pmatrix} = (0.10)(0.25)(0.10)(0.33)(0.10)(0.001)$$
$$= 0.0000000825$$

Based on the product rule, the mathematician testified that there was about a 1 in 12 million chance that a couple selected at random would possess these characteristics. The prosecution added that the probability was the chance that "any other couple possessed the distinctive characteristics of the defendants." The jury convicted the defendants. On appeal, the Supreme Court of California reversed the decision, based on two main points. First there was no proof offered that the probabilities used in the probability calculation were correct and there was no evidence that the events were independent. Second, the prosecution's evidence pertaining to a randomly selected couple was not pertinent to the problem of the existence of any other couple possessing the same characteristics.

Example 5.13

In a production process, a product is assembled by using four independent parts (A, B, C, and D). In order for the product to operate properly, each part must be free of defects. The probability that the parts are defect-free is given by $P(A) = 0.9$, $P(B) = 0.7$, $P(C) = 0.8$, and $P(D) = 0.9$.

  a. What is the probability that all four parts have defects?

  b. What is the probability that the product does not work?

Solution

  a. Since there are four parts with the probability of each part working (i.e., being defect-free) given to be $P(A) = 0.9$, $P(B) = 0.7$, $P(C) = 0.8$, and $P(D) = 0.9$, for all parts to have defects, we need the complement of each of the parts. That is,

$$P(\text{all four parts have defects}) = P(A^c \cap B^c \cap C^c \cap D^c).$$

  Since each part operates independently of the others, then the probability that all four parts have defects is the product of the probabilities of each of the part's complement.

$$P(\text{all four parts have defects}) = P(A^c) \cdot P(B^c) \cdot P(C^c) \cdot P(D^c)$$
$$= (1-0.9)(1-0.7)(1-0.8)(1-0.9) = 0.0006$$

  Thus, the probability that all four parts have defects is 0.0006, or 0.06%.

  b. The probability that the product does not work is the probability that at least one of the parts does not work (since each part must be defect-free for the product to work).

$$P(\text{product does not work}) = P(\text{at least one part does not work})$$
$$= 1 - P(\text{all parts work})$$
$$= 1 - P(A \cap B \cap C \cap D)$$
$$= 1 - (0.9)(0.7)(0.8)(0.9) = 1 - 0.4536 = 0.5464$$

  Thus, given the above probabilities, there is nearly a 55% chance that the product will not work.

Winning the Lottery Twice!

In 1986 a woman won the New Jersey state lottery twice and in 1988 a man in Pennsylvania also won the State lottery twice for a total of 5.4 and 6.8 million dollars respectively. The *New York Times* reported the odds to be 1 in 17 trillion. How is this to be explained? When millions of people play a lottery daily it can be shown that the odds of winning twice is about 1 in 30 in a 4-month period and the odds are even better in a 7 year period. Thus what appeared to be an almost theoretical impossibility, turns out to be quite a probable event. What one can say is that even when the probability of an event is very small, if there are millions of possibilities, then the rare event rarely remains rare. This is what has been called "The Law of Real Large Numbers" by Diaconis and Mosteller.

Source: Diaconis, P. and Mosteller, F. (1989). Methods for Studying Coincidences," Journal of the American Statistical Association, 84, 853-861.

5.6   Exercises

Basic Concepts

1. Explain the difference between dependent and independent events.

2. Are mutually exclusive events dependent or independent? Explain your answer.

3. If events A and B are independent, what is $P(A|B)$ equal to?

4. What is the product rule?

5. In the case *People v. Collins* an appeals court overturned the conviction. What flaws did the appeals court detect in the case against the accused assailants?

Exercises

6. The following table was given in Section 5.4, Exercise 12.

Table for Exercise 6 – Mail Order Customers		
	Gender	
Location	Male	Female
Suburban	196	298
Urban	92	173

Are the events {customer is urban} and {customer is suburban} independent? Explain.

7. The following table was given in Section 5.4, Exercise 13.

Table for Exercise 7 – Life Insurance Coverage				
	Amount of Life Insurance on Husband ($)			
	0 – 50,000	50,000 – 100,000	100,000 – 150,000	More than 150,000
Amount of Life Insurance on Wife ($) 0 – 50,000	400	200	50	50
50,000 – 100,000	50	50	30	30
100,000 – 150,000	20	10	25	25
More than 150,000	20	10	15	15

Are the events {the husband has more than $150,000 in insurance} and {the wife has more than $50,000 in insurance} independent? Explain.

8. Suppose you were flipping a coin. What is the probability that you would observe a head:

    a. on two consecutive flips?   b. on three consecutive flips?

    c. on four consecutive flips?   d. on 100 consecutive flips?

9. Suppose an atomic reactor has two independent cooling systems. The probability that Cooling System A will fail is 0.01 and the probability that Cooling System B will fail is 0.01. What is the probability that both systems will fail simultaneously?

10.   Mandy is 30, and the probability that she will survive until age 65 is 0.90. Ashley is 45, and the probability that she will survive until age 65 is 0.95.

   a.   Find the probability that both Mandy and Ashley will survive until age 65.

   b.   Find the probability that only Mandy will survive until age 65.

   c.   Find the probability that neither Mandy nor Ashley will survive until age 65.

   d.   What assumption about the lives of Mandy and Ashley did you make in answering the above questions?

11.   An insurance company is considering insuring two large oil tankers against spills. The limit of the liability on the coverage is $10,000,000. The company believes that the probability of an oil spill requiring the maximum liability coverage during the policy period is 0.001 per tanker.

   a.   What is the probability that neither tanker would have a spill requiring the maximum liability coverage during the policy period?

   b.   What is the probability that only one tanker would have a spill requiring the maximum liability coverage during the policy period?

   c.   What is the probability that both tankers would have spills requiring the maximum liability coverage during the policy period?

12.   Coin flipping can be used to model other real life phenomena and aid in certain probability calculations. An example of this would be to compute the probability that the World Series ends in some specified number of games. The World Series is a best of seven game series played at the end of the regular baseball season between the champion of the American League and the champion of the National League. The first team to win four games is declared the champion of baseball for that year. If we assume the probability of either team winning a game is approximately 0.5 and the games are independent events, the probability that the series ends in either 4, 5, 6, or 7 games can be computed.

   a.   What is the probability that the series ends in exactly 4 games? Write the sample space consisting of 16 equally likely simple events similar to the sample space resulting from tossing a coin four times.

   b.   What is the probability that the series ends in exactly 5 games?

   c.   Assume the probability that the series ends in exactly 6 games is $\frac{5}{16}$. Use this information together with your answers to the first two parts of this problem to compute the probability that the series ends in exactly 7 games.

13.   Drug usage in the workplace costs employers incredible amounts of money each year. Drug testing potential employees has become so prevalent that drug users are finding it extremely hard to find jobs. Drug tests, however, are not completely reliable. The most common test used to detect drugs is approximately 98% accurate. To decrease the likelihood of making an error, all potential employees are screened through two tests, which are independent, and each has about 98% accuracy.

   a.   If a person were drug free, what is the probability he or she would fail both tests?

   b.   If a person were a drug user, what is the probability he or she would pass both tests?

5.7 Bayes' Theorem

We have completed the discussions about conditional probability and independent events in Section 5.5 and Section 5.6. **Bayes' theorem** (also referred to as **Bayes' rule** or **Bayes' law**) is somewhat of an extension of conditional probability in which we calculate probabilities based on new information. Please note and understand that the additional information is obtained for a subsequent event, and the new information is used to revise the initial probability. Recall the following formulas for conditional probability.

$$P(B \mid A) = \frac{P(A \cap B)}{P(A)} \text{ provided that } P(A) > 0$$

$$P(A \mid B) = \frac{P(A \cap B)}{P(B)} \text{ provided that } P(B) > 0$$

Bayes' theorem is developed from extending the definition of conditional probability. If we rearrange the formula for the probability of A given B and solve for A intersect B, we have the following.

$$P(A \cap B) = P(A \mid B)P(B)$$

Substituting this expression for A intersect B in the formula for the conditional probability of B given A, we have the following.

$$P(B \mid A) = \frac{P(A \cap B)}{P(A)} = \frac{P(A \mid B)P(B)}{P(A)}$$

Now recall the multiplication rule for independent events. That is, we know that if A and B are independent, then $P(A \cap B) = P(A)P(B)$. If this rule holds for two events, A and B, then we can generalize the multiplication rule as follows:

$$P(A) = P(A \cap B_1) + P(A \cap B_2) + \cdots + P(A \cap B_k)$$

where B_1, B_2, ..., B_k are k mutually exclusive and collectively exhaustive events.

Using conditional probabilities,

$$P(A \cap B_i) = P(A \mid B_i)P(B_i).$$

Using this substitution, we can generalize the multiplication rule as follows.

$$P(A) = P(A \mid B_1)P(B_1) + P(A \mid B_2)P(B_2) + \cdots + P(A \mid B_k)P(B_k)$$

Substituting this formula for $P(A)$ into the formula for the conditional probability of B given A, we can derive Bayes' theorem.

Theorem

Bayes' Theorem

$$P(B_i \mid A) = \frac{P(A \mid B_i)P(B_i)}{P(A \mid B_1)P(B_1) + P(A \mid B_2)P(B_2) + \cdots + P(A \mid B_k)P(B_k)}$$

where B_i is the ith event out of k mutually exclusive and collectively exhaustive events.

The following example will illustrate how Bayes' theorem can be applied.

Example 5.14

Of the travelers arriving at a small airport, 60% fly on major airlines, 30% fly on privately owned planes, and the remainder fly on commercially owned planes not belonging to a major airline. Of those traveling on the major airlines, 50% are traveling for business reasons, whereas 60% of those arriving on private planes and 90% of those arriving on the other commercially owned planes are traveling for business reasons. Suppose that we randomly select one person arriving at the airport. What is the probability that the person:

  a. is traveling on business?

  b. is traveling for business on a privately owned plane?

  c. arrived on a privately owned plane given that the person is traveling for business reasons?

Solution

First of all, we should define the events associated with this problem. Once we have defined the events, we can then note the probabilities that are given to us.

Let

$$M = \text{ Major Airline}$$
$$P = \text{ Private Plane}$$
$$C = \text{ Commercial Airline}$$
$$B = \text{ Travel for Business Reasons.}$$

In the context of Bayes' theorem, $k = 3$ for this example, since major airline, private plane, and commercial airline are the only ways in which travelers can arrive at the airport, and a particular passenger cannot arrive on more than one type of airplane (thus M, P, and C are mutually exclusive and exhaustive events). When we apply the theorem, $P(B_1)$ will be $P(M)$, $P(B_2)$ will be $P(P)$, and $P(B_3)$ will be $P(C)$. We know the following.

$$P(M) = 0.6$$
$$P(P) = 0.3$$
$$P(C) = 0.1$$
$$P(B \mid M) = 0.5$$
$$P(B \mid P) = 0.6$$
$$P(B \mid C) = 0.9$$

a. The first question asks what the probability is that a randomly selected person is traveling on business. We know that a person is traveling on business if they are traveling on business via any of the flight methods. Thus we have

$$\begin{aligned}
P(B) &= P(B \cap M) + P(B \cap P) + P(B \cap C) \\
&= P(B \mid M)P(M) + P(B \mid P)P(P) + P(B \mid C)P(C) \\
&= (0.5)(0.6) + (0.6)(0.3) + (0.9)(0.1) \\
&= 0.57.
\end{aligned}$$

Thus, there is a 57% chance that the randomly selected traveler will be traveling for business.

b. To determine if the traveler is traveling for business on a privately owned plane, we want to find

$$P(B \cap P) = P(B \mid P)P(P) = (0.6)(0.3) = 0.18.$$

So, there is an 18% chance that the traveler will be traveling for business on a privately owned plane.

c. This part of the problem wants us to determine the probability that the person arrived on a privately owned plane, given that he or she is traveling for business reasons. We can write this probability statement as $P(P \mid B)$.

By the definition of conditional probability, we can write

$$P(P \mid B) = \frac{P(P \cap B)}{P(B)} = \frac{P(B \mid P)P(P)}{P(B)}.$$

From the items calculated in parts **a.** and **b.**, we know

$$\begin{aligned}
P(P \mid B) &= \frac{P(P \cap B)}{P(B)} = \frac{P(B \mid P)P(P)}{P(B)} \\
&= \frac{P(B \mid P)P(P)}{P(B \mid M)P(M) + P(B \mid P)P(P) + P(B \mid C)P(C)} \\
&= \frac{(0.6)(0.3)}{0.57} \\
&\approx 0.3158.
\end{aligned}$$

Thus, we know that if the passenger is traveling on business, there is about a 32% chance that he or she will be traveling by private plane.

Even though it was fairly subtle (given that we performed the calculations in parts **a.** and **b.**), please note the use of Bayes' theorem in the previous calculation.

5.7   Exercises

Basic Concepts

1. Briefly explain the relationship between conditional probability and Bayes' theorem.

2. Other than conditional probability, which other rule which you previously studied is used in the derivation of Bayes' theorem?

3. What is Bayes' theorem?

4. How is Bayes' theorem used to "revise" a probability based on additional information?

Exercises

5. The issue of Corporate Tax Reform has been cause for much debate in the United States, especially in the House Ways and Means Committee as well as the Senate Finance Committee. Among those in the legislature, 45% are Republicans and 55% are Democrats. It is reported that 30% of the Republicans and 70% of the Democrats favor some type of Corporate Tax Reform to prevent American companies from operating in foreign countries. Suppose a member of Congress is randomly selected and they are found to favor some type of corporate tax reform. What is the probability that this person is a Democrat?

6. Males and females are observed to react differently to sad, emotional movies. It has been observed that 70% of the females say they cry at some point during those types of movies, whereas only 40% of the males admit to crying during those types of movies. A group of 40 people, of whom 25 are female, was shown a sad, emotional movie and the subjects were asked if they cried. A response picked at random from the 40 indicated that they cried. What is the probability that it was a male?

7. As items come to the end of a production line, an inspector chooses which items are to go through a complete inspection. Eight percent of all items produced are defective. Sixty percent of all defective items go through a complete inspection, and 20% of all good items go through a complete inspection. Given that an item is completely inspected, what is the probability that it is defective?

8. Two teaching methods for a business statistics class, online and face-to-face, are available during the course of an academic year. The failure rate (students that receive below a C− and thus, will have to repeat the course) is 4% for the online class and 8% for the face-to-face class. However, the online class is more expensive and hence is offered only 25% of the time. (The face-to-face class is offered the other 75% of the time.) A student takes the statistics class via one of the methods of delivery but failed the course. What is the probability that the student took the online class?

9. A personnel director has two lists of applicants for jobs. List 1 contains names of 15 women and 5 men whereas List 2 contains the names of 5 women and 12 men. A name is randomly selected from List 1 and added to List 2. A name is then randomly selected from the augmented List 2. Given that the name selected is that of a man, what is the probability that a woman's name was originally selected from List 1?

5.8 | Counting

To compute certain probabilities, such as the probability of having winning numbers in the state lottery, requires the ability to count the number of possible outcomes for a given experiment or a sequence of experiments.

However, often it is impractical to list out all the possibilities. Therefore, we will develop some techniques to facilitate our counting.

The Fundamental Counting Principle

> **Theorem**
>
> Fundamental Counting Principle
>
> E_1 is an event with n_1 possible outcomes and E_2 is an event with n_2 possible outcomes. The number of ways the events can occur in sequence is $n_1 \cdot n_2$. This principle can be applied for any number of events occurring in sequence.

Example 5.15

A local bank has three branches. Each branch has four departments and each department has two employees. How many employees does the bank have?

Solution

$$
\underset{\left(\substack{\text{number of}\\\text{branches}}\right)}{3} \quad \cdot \quad \underset{(\text{departments})}{4} \quad \cdot \quad \underset{\left(\substack{\text{employees per}\\\text{department}}\right)}{2} \quad = \quad \underset{\left(\substack{\text{total number}\\\text{of employees}}\right)}{24}
$$

Thus, the bank has 24 total employees.

Example 5.16

Non-personalized license plates in the state of Utah consist of three numbers followed by three letters (excluding I, O, and Q). How many license plates are possible?

Solution

There are ten digits (0-9) possible for each of the first three characters. Likewise, there are 23 letters possible for the last three characters. Therefore, we have the following.

$$
\underset{(\text{digit})}{10} \cdot \underset{(\text{digit})}{10} \cdot \underset{(\text{digit})}{10} \cdot \underset{(\text{letter})}{23} \cdot \underset{(\text{letter})}{23} \cdot \underset{(\text{letter})}{23} = \underset{\left(\substack{\text{possible license}\\\text{plates}}\right)}{12,167,000}
$$

Therefore, there are 12,167,000 possible non-personalized license plates in the state of Utah.

Example 5.17

You have a stack of 5 textbooks: English, History, Statistics, Geology, and Psychology. How many ways can you arrange these textbooks on a shelf?

Solution

Because you can put each book on the bookshelf only once, you have five possible choices for the first book. Similarly, there are four possible choices for the second book, three possible choices for the third book, two possible choices for the fourth book, and only one book left for the fifth book. Using the fundamental counting principle, we have

$$(5)(4)(3)(2)(1) = 120 \text{ possible arrangements.}$$

Factorials

The product $(5)(4)(3)(2)(1)$ in Example 5.17 is a special type of product called a **factorial**. Factorials occur so frequently that they have their own notation as follows.

Formula
Factorial
Suppose n is a positive whole number. Then,
$$n! = n(n-1)(n-2)\cdots(3)(2)(1).$$
Note: $0! = 1$ by definition.

$n!$ is read as *n factorial*. Note from the previous example that $n!$ represents the number of ways to arrange n items.

Using this notation, $5! = (5)(4)(3)(2)(1) = 120$.

Combinations

We have seen how the fundamental counting principle and factorial notation can help us when counting the number of possible arrangements of items. If we are interested in counting the number of arrangements and order is *not* important, we are dealing with a **combination**.

> **Definition**
>
> A **combination** is a collection or grouping of objects where the order is not important.

The number of combinations of n objects taken k at a time can be found as follows.

> **Formula**
>
> ## Combination
>
> The number of combinations of n unique objects taken k at a time is:
>
> $$_nC_k = \frac{n!}{(n-k)!\,k!}.$$
>
> Note that some alternate notations for combinations that you may see are $\binom{n}{k}$, C_k^n, and $C(n,k)$.
>
> All of these denote the number of combinations of n objects taken k at a time.

Example 5.18

In the Mega Millions lottery, a player selects five different numbers from 1 to 56 (inclusive) and then another (a sixth number, called the Mega Ball) from 1 to 46 (inclusive). If the first five numbers match the player's numbers in any order along with the Mega Ball number, the player wins.

  a. What is the total number of winning combinations?

  b. What is the probability of winning?

Solution

  a. First, we have to determine the number of ways of selecting 5 numbers from 56. Then, we have to multiply the number of combinations of choosing 5 numbers from 56 by 46, the number of ways you can select the Mega Ball. Thus, we have the following.

$$_{56}C_5 = \frac{56!}{(56-5)!5!} = \frac{56!}{51!5!} = \frac{(56)(55)(54)(53)(52)\,\cancel{(51!)}}{\cancel{51!}\,5!}$$

$$= \frac{(56)(55)(54)(53)(52)}{(5)(4)(3)(2)(1)} = 3{,}819{,}816$$

So, the total number of winning combinations is
$$46 \cdot 3{,}819{,}816 = 175{,}711{,}536.$$

b. The probability of winning with any one combination is

$$\frac{1}{175,711,536}$$

which is approximately 0. The actual probability is 0.00000000569115.

Permutations

We have just looked at counting where order did not matter. But there are many situations in which the order or arrangement of the objects is important. If we are interested in counting the number of arrangements and order *is* important, we are dealing with **permutations**.

Definition

A **permutation** is a specific order or arrangement of objects in a set. There are $n!$ permutations of n unique objects.

Example 5.19

To complete your holiday shopping, you need to go to the bakery, department store, grocery store, and toy store. If you are going to visit the stores in sequence, how many different sequences exist?

Solution

This is a permutation problem because order matters. By the permutation definition there are $4! = (4)(3)(2)(1) = 24$ sequences.

There are times when not all objects in a set will be used for a permutation problem. Consider the following.

Example 5.20

At a local fast food restaurant, the door to the kitchen is secured by a five button lock, labeled 1, 2, 3, 4, 5. To open the door, the correct three digit code must be pushed but each button can only be pushed once. How many different codes are possible?

Solution

This is a permutation problem of 5 objects, but we are taking only 3 at a time. There are 5 buttons available for the first character in the code, 4 for the second, and 3 for the third. Therefore, there are

$$(5)(4)(3) = 60 \quad \text{possible codes.}$$

In the previous example, we had a permutation of 5 objects taken 3 at a time. In general,

---

Formula

Permutations

The number of permutations of n unique objects taken k at a time is:

$$_nP_k = \frac{n!}{(n-k)!}.$$

Note that some alternate notations for permutations that you may see are P_k^n and $P(n,k)$. All of these denote the number of permutations of n objects taken k at a time.

Example 5.21

Seven bids are placed for a commercial construction job and three will be selected in order of quality. How many ways can the bids be selected in first, second, and third place?

Solution

Because we have seven bids and the order (first, second, third) is important, we need to find the number of permutations of 7 objects taken 3 at a time.

$$_7P_3 = \frac{7!}{(7-3)!} = \frac{7!}{4!} = \frac{(7)(6)(5)\,4!}{4!} = 210$$

There are times when we are interested in finding the number of permutations where some of the objects are duplicates. For instance, consider the word EYE.

If we have interchanged the two E's, the resulting permutation is not distinguishable from the original. To count the number of **distinguishable permutations**, we need the following formula.

Formula

Distinguishable Permutations

If given n objects, with n_1 alike, n_2 alike, ..., n_k alike, then the number of distinguishable permutations of all n objects is $\dfrac{n!}{(n_1!)(n_2!)(n_3!)\cdots(n_k!)}.$

Example 5.22

How many distinguishable permutations can be made from the word *Mississippi*?

Solution

There are 11 letters in the word *Mississippi*, one M, four I's, four S's, and two P's. So there are,

$$\frac{11!}{(1!)(4!)(4!)(2!)} = 34,650$$

distinguishable permutations of the letters in *Mississippi*.

It is important to remember that combinations are used when order is not important, and permutations are used when order is important.

Concept		Formula
Fundamental Counting Principle	If one event has n_1 outcomes and another event has n_2 outcomes, the number of ways the event can occur in sequence is the product of n_1 and n_2.	$n_1 \cdot n_2$
Factorial	$n!$ is the product of each of the positive whole numbers from 1 to n.	$n! = n(n-1)(n-2)\cdots(3)(2)(1)$
Combination	A collection or grouping of objects where order is not important.	The number of combinations of n unique objects taken k at a time is given by the following formula. $${}_nC_k = \frac{n!}{(n-k)!k!}$$
Permutation	A specific order or arrangement of objects.	The number of arrangements for n objects when order is important is given by n factorial: $n!$.
	The number of permutations of n unique objects taken k at a time.	$${}_nP_k = \frac{n!}{(n-k)!}$$
	Given n objects with n_1 alike, n_2 alike, …, n_k alike, then the number of distinguishable permutations is given by the formula to the right.	$$\frac{n!}{(n_1!)(n_2!)(n_3!)\cdots(n_k!)}$$

5.8   Exercises

Basic Concepts

1. What is the fundamental counting principle?

2. What is a factorial?

3. Describe the difference between permutations and combinations.

4. Give an example of a situation in which you would need to calculate the number of distinguishable permutations.

Exercises

5. The blue plate lunch at a local cafeteria consists of an entrée, a side item, and a desert. If there are 6 choices for an entrée, 5 choices for a side item, and 4 choices for a dessert, how many different lunches are available?

6. You are interested in buying a home in a new subdivision. The builder offers 3 basic floor plans, each with 4 possible arrangements for the garage, and siding in 6 different colors. How many different homes can be built?

7. Compute each of the following.

    a. 1!   **b.** 3!

    c. 5!   **d.** 7!

8. Compute each of the following.

    a. 2!   **b.** 4!

    c. 6!   **d.** 8!

9. A DJ needs to select 6 songs from a CD containing 12 songs to compose an event's musical lineup. How many different lineups are possible?

10. In how many ways can 11 kids be picked for the 9 positions on a baseball team?

11. How many distinguishable permutations can be made from the word STATISTICS?

12. How many distinguishable permutations can be made from the word SASSAFRAS?

13. A person tosses a coin 11 times. In how many ways can he get 9 heads?

14. How many 5 card hands can be dealt from a deck of 52 cards?

T   Discovering Technology

Using the TI-84 Plus Calculator

Factorials

Calculate 10! using a TI-84 Plus calculator.

1. To calculate the value of 10! using a TI-84 Plus calculator, first enter $\boxed{1}$ $\boxed{0}$ in the calculator.
2. Press $\boxed{\text{MATH}}$, scroll to **PRB**, and select option **4:!**.
3. Press $\boxed{\text{ENTER}}$ twice. The value of 10! (3628800) is displayed.

Figure 5.12

Combinations

Use the information from Example 5.18 for this exercise.

1. In order to calculate the number of winning combinations for the Mega Millions lottery, we have to determine the number of ways of selecting 5 numbers from 56. This is a combination of 56 objects, taken 5 at a time, or $_{56}C_5$. To find the number of combinations using the calculator, first enter $\boxed{5}$ $\boxed{6}$, press $\boxed{\text{MATH}}$, scroll to **PRB**, and choose option **3:nCr**.
2. Press $\boxed{\text{ENTER}}$. Enter the number $\boxed{5}$ and press $\boxed{\text{ENTER}}$ again.
3. The number of combinations is given as 3819816. We then need to multiply this by 46, which is the number of ways you can select the Mega Ball. Multiply the answer of 3819816 by 46 to get 175711536, which is the total number of winning combinations. Compare these results to those in Example 5.18.

Figure 5.13

Permutations

Use the information from Example 5.21 for this exercise.

1. Seven bids were placed for a commercial construction job, and we want to know in how many ways the bids can be selected in first second and third place. Since order is important, this is equivalent to a permutation of 7 objects, taken 3 at a time, or $_7P_3$. To calculate the number of permutations using the calculator, first enter ⌜7⌟, press **MATH**, scroll to **PRB**, and choose option **2:nPr**.

2. Press **ENTER**. Enter the number ⌜3⌟ and press **ENTER** again. The number of permutations is then displayed. Compare this to our solution to Example 5.21.

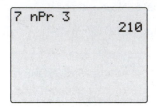

Figure 5.14

Using Excel

Probability and Independence

A coin is flipped, a die is rolled, and a card is drawn from a deck. Find the probability of getting heads on the coin, a 2 on the die, and drawing a spade from the deck of cards.

1. Fill in the data in a worksheet as shown here.

◢	A	B	C	D
1		Coin	Die	Card
2	**Chance of Success**	1	1	13
3	**Possible Outcomes**	2	6	52

Figure 5.15

2. In cell B4 divide B2 by B3 by entering the following formula.

$$=B2/B3$$

3. Copy the formula in cell B4 to C4 and D4.

4. Now we have the probability of success for each event listed in Row 4. In cell E4, multiply cells B4 through D4 using the following formula.

$$=B4*C4*D4$$

5. This gives us the result that the probability of getting heads on the coin, a 2 on the die and drawing a spade from the deck of cards is 0.020833 or about 2.1%.

◢	A	B	C	D	E
1		Coin	Die	Card	
2	**Chance of Success**	1	1	13	
3	**Possible Outcomes**	2	6	52	
4		0.5	0.166667	0.25	0.020833

Figure 5.16

Combinations

For this exercise, use the information from Example 5.18.

1. The first step in this example was to determine the number of ways of selecting 5 numbers from 56. This is equivalent to determining the number of combinations of 56 unique objects taken 5 at a time. In an Excel spreadsheet, enter **n** and **k** in cells A1 and B1, respectively, and **56** and **5** in cells A2 and B2, respectively.

	A	B
1	n	k
2	56	5

Figure 5.17

2. Label cell C1 **nCk** to represent the number of combinations of *n* objects taken *k* at a time. The formula to calculate the number of combinations in Microsoft Excel is

COMBIN(number, number_chosen)

where *number* is the number of items and *number_chosen* is the number of items in each combination. In cell C2, enter the formula to calculate the number of combinations of lottery numbers.

=COMBIN(A2, B2)

3. Observe the number of combinations as calculated by Microsoft Excel. Compare this to our solution in part **a.** of Example 5.18.

	A	B	C
1	n	k	nCk
2	56	5	3819816

Figure 5.18

Permutations

For this exercise, use the information from Example 5.21.

1. Since order is important and we are interested in finding the number of ways the bids can be selected in first, second, and third place, we need to calculate the number of permutations of 7 unique objects taken 3 at a time. In an Excel spreadsheet, enter **n** and **k** in cells A1 and B1, respectively, and **7** and **3** in cells A2 and B2, respectively.

	A	B
1	n	k
2	7	3

Figure 5.19

2. Label cell C1 *nPk* to represent the number of permutations of *n* objects taken *k* at a time. The formula to calculate the number of permutations in Microsoft Excel is

PERMUT(number, number_chosen)

where *number* is the number of objects and *number_chosen* is the number of objects in each permutation. In cell C2, enter the formula to calculate the number of permutations of bid orders.

=PERMUT(A2, B2)

3. Observe the number of permutations as calculated by Microsoft Excel. Compare this to our solution in Example 5.21.

	A	B	C
1	n	k	nPk
2	7	3	210

Figure 5.20

R   Chapter 5 Review

Key Terms and Ideas

- Probability
- Randomness
- Random Experiment
- Sample Space
- Outcome
- Event
- Subjective Probability
- Objective Probability
- Relative Frequency
- Statistical Regularity
- Classical Probability
- Statistical Inference
- Probability Law 1
  (Probability of 0)
- Probability Law 2
  (Probability of 1)
- Probability Law 3
  $\left(0 \le P(A) \le 1\right)$
- Probability Law 4
  $\left(P(A_1) + P(A_2) + \cdots + P(A_n) = 1\right)$
- Compound Event
- Union
- Intersection
- Complement

- Probability Law 5
  $\left(P(A^c) = 1 - P(A)\right)$
- Mutual Exclusivity
- Probability Law 6
  (Union of Mutually Exclusive Events)
- Probability Law 7
  (Intersection of Mutually Exclusive Events)
- Probability Law 8
  (The Addition Rule)
- Conditional Probability
- Probability Law 9
  (Conditional Probability)
- Independent Events
- Dependent Events
- Probability Law 10
  (Multiplication Rule for Independent Events)
- Bayes' Theorem
- Fundamental Counting Principle
- Factorial
- Combination
- Permutation
- Distinguishable Permutation

Key Formulas

	Section
Relative Frequency	
If an experiment is performed n times and event A happens k times, then $$\text{Relative Frequency of } A = \frac{k}{n}.$$	5.2
Classical Probability	
$$P(A) = \frac{\text{number of outcomes in } A}{\text{total number of outcomes in the sample space}}$$	5.2

Key Formulas (cont.)

	Section
Probability Law 3	5.4

$$0 \leq P(A) \leq 1$$

Probability Law 4	5.4

$$P(A_1) + P(A_2) + \cdots + P(A_n) = 1$$

Probability Law 5	5.4

$$P(A^c) = 1 - P(A)$$

Union of Mutually Exclusive Events	5.4

$$P(A \cup B) = P(A) + P(B)$$

Intersection of Mutually Exclusive Events	5.4

$$P(A \cap B) = 0$$

The Addition Rule	5.4

$$P(A \cup B) = P(A) + P(B) - P(A \cap B)$$

Conditional Probability	5.5

$$P(A \mid B) = \frac{P(A \cap B)}{P(B)}$$

Independent Events	5.6

Two events are independent if and only if

$$P(A \mid B) = P(A) \text{ and } P(B \mid A) = P(B).$$

Multiplication Rule for Independent Events	5.6

$$P(A \cap B) = P(A) P(B)$$

Bayes' Theorem	5.7

$$P(B_i \mid A) = \frac{P(A \mid B_i) P(B_i)}{P(A \mid B_1) P(B_1) + P(A \mid B_2) P(B_2) + \cdots + P(A \mid B_k) P(B_k)}$$

where B_i is the ith event out of k mutually exclusive and collectively exhaustive events.

Key Formulas (cont.)

	Section
Fundamental Counting Principle If E_1 is an event with n_1 possible outcomes and E_2 is an event with n_2 possible outcomes, the number of ways the events can occur in sequence is $n_1 \cdot n_2$	5.8
***n* Factorial** $$n! = n(n-1)(n-2)\cdots(3)(2)(1)$$	5.8
Combination $$_nC_k = \frac{n!}{(n-k)!\,k!}$$	5.8
Permutation $$_nP_k = \frac{n!}{(n-k)!}$$	5.8
Number of Distinguishable Permutations $$\frac{n!}{(n_1!)(n_2!)(n_3!)\cdots(n_k!)}$$	5.8

AE   Additional Exercises

1. A couple plans to have three children.

   a. List all possible outcomes for the sexes of the three children.

   b. Find the probability that the couple will have three girls.

   c. Find the probability that the couple will have at least one boy.

2. 671 registered voters were surveyed and asked their political affiliation and whether or not they favor a national healthcare policy. The results of the survey are displayed in the table below.

Table for Additional Exercise 2 – Survey Results			
Position on National Healthcare	Democrat	Independent	Republican
Favor	161	40	130
Do Not Favor	110	40	190

   If one of the surveyed voters is randomly selected, answer the following questions.

   a. What is the probability that the voter will be a Republican?

   b. What is the probability that the voter will not favor a national healthcare policy?

   c. What is the probability that the voter will be a Democrat or an Independent?

   d. What is the probability that the voter will be a Democrat and favor a national healthcare policy?

   e. Given that the voter is a Republican, what is the probability that the voter will favor a national healthcare policy?

   f. If the voter does not favor a national healthcare policy, what is the probability that the voter is an Independent?

   g. Are the events {voter is a Democrat} and {voter favors national healthcare policy} independent? Explain.

3. A roulette wheel has 38 outcomes labeled 1 through 36 plus 0 and 00. The wheels are supposed to be designed so that each outcome is equally likely. The numbers 0 and 00 are often referred to as house numbers because the only way that a player can win when these outcomes are observed is by directly betting on the numbers. A great deal of the money wagered on a roulette wheel is wagered on odd or even numbers, or columns or rows of numbers. The numbers 0 and 00 are not in any row or column, nor are they odd or even.

   a. What is the probability of observing an even number (0 and 00 are neither odd nor even)?

   b. What is the probability of observing a number between 1 and 12, inclusive?

   c. What is the probability of observing 0 or 00?

   d. What is the probability of observing a 4?

   e. What is the probability of not observing 7, 13, or 21?

4. A survey of customers in a particular retail store showed that 10% were dissatisfied with the customer service. Half of the customers who were dissatisfied dealt with Bill, the senior customer service representative. If Bill responds to 40% of all customer service inquiries in the retail store, find the following probabilities.

    a. The probability that a customer will be unhappy, given that the representative was Bill.

    b. The probability that the service representative was not Bill, given that the customer complained.

5. A package of documents needs to be sent to a given destination, and it is important that it arrive within one day. To maximize the chances of on-time delivery, three copies of the documents are sent via three different delivery services. Service A is known to have a 90% on-time delivery record, Service B has an 88% on-time delivery record, and Service C has a 91% on-time delivery record. Assuming that the delivery services and their records are independent, what is the probability that at least one copy of the documents will arrive at its destination on time?

6. A boxcar contains six complex electronic systems. Two of the six are to be randomly selected for thorough testing and then classified as defective or not defective. If two of the six systems are actually defective:

    a. find the probability that at least one of the two systems tested will be defective.

    b. find the probability that both are defective.

7. *Odds in favor of* and *odds against* are often used to express chances of occurrences. For example, if the odds are 5 to 2 that it will rain tomorrow then we would be wise to carry an umbrella with us. How exactly are odds related to probabilities? If the probability of event A occurring is p, then the odds in favor of A occurring are a to b such that $\dfrac{a}{b} = \dfrac{p}{(1-p)}$. The odds against A occurring are b to a.

    a. What are the odds of rolling a six when a single die is thrown?

    b. What are the odds against getting a head when a coin is tossed?

    c. What are the odds against getting 3 consecutive heads when a coin is tossed 3 times?

    d. Suppose the odds in favor of your favorite athletic team winning this weekend are 8 to 3. What is the probability that they will win?

8. Consider a well-shuffled deck of cards with 13 hearts, 13 spades, 13 clubs, and 13 diamonds.

    a. Find the probability that the first card dealt is a heart.

    b. Find the probability that the first card dealt is a spade.

    c. Find the probability that the first card dealt is not a spade.

    d. If you know that the first card dealt will not be a spade, find the probability that it will be a heart.

    e. Suppose you saw the bottom card, and it was the queen of hearts. What is the probability that the first card dealt will be a heart?

9. A box contains eighteen large marbles and ten small marbles. Each marble is either green or white. Twelve of the large marbles are green and four of the small marbles are white. If a marble is randomly selected from the box, what is the probability that it is white or large?

10. User passwords for a certain computer network consist of four letters followed by two numbers. How many different passwords are possible?

11. Hydraulic assemblies for landing coming from an aircraft rework facility are each inspected for defects. Historical records indicate that 8% have defects in shafts only, 6% have defects in bushings only, and 2% have defects in both shafts and bushings. One of the hydraulic assemblies is selected randomly. What is the probability that:

   a. the assembly has a bushing defect?

   b. the assembly has a shaft or bushing defect?

   c. the assembly has exactly one of the two types of defects?

   d. the assembly has neither type of defect?

you will get the job!

Chapter Six

Discrete Probability Distributions: Information about the Future

Introduction

The notions of randomness and uncertainty were introduced in Chapter 5. This chapter extends those ideas by developing concepts to describe a pattern of randomness for an entire set of outcomes produced by some random phenomenon. In the coin tossing experiment the description was rather easy. The totality of outcomes contained only two values, *heads* and *tails*. Probabilities for these events were constructed by assuming a fair coin and by applying the classical definition of probability.

Analyzing more complex random phenomena requires a method of organizing information about random processes and a vocabulary to describe the organizational concepts. Two such descriptive notions will be introduced in this chapter: **random variables** and **probability distributions**.

> **Definition**
>
> A **random variable** is a numerical outcome of a random process.

> **Definition**
>
> A **probability distribution** is a model that describes a specific kind of random process.

Probability distributions are the best descriptors of random processes. However, for "real world" random processes they are often difficult to obtain.

6.1   Types of Random Variables

Quantitative random variables are classified as discrete or continuous. This categorization refers to the types of values that outcomes of the random variable can assume. **Discrete random variables** are analyzed in this chapter, and the next chapter discusses **continuous random variables**.

In defining random variables, there is a naming convention. Capital letters, such as X, will be used to refer to the random variable, while small letters, such as x will refer to specific values of the random variable. Often the specific values will be subscripted, $x_1, x_2, ..., x_n$.

Discrete Random Variables

With a discrete random variable, you can count the number of outcomes that the random variable might possess.

> **Definition**
>
> A **discrete random variable** is a random variable which has a countable number of possible outcomes.

In fact, the values that many discrete random variables assume are the counting numbers from 0 to N, where N depends upon the nature of the variable.

Procedure

Discrete Random Variables

When describing a discrete random variable, you should do the following.

1. State the variable.

2. List all of the possible values of the variable.

3. Determine the probabilities of these values.

Example 6.1

Random Phenomenon: Toss a six-sided die and observe the outcome of the toss.

1. *Identify the random variable:* $X =$ outcome of the toss of a six-sided die.

2. *Range of values:* Integers between 1 and 6, inclusive.

    In this instance, $x_1 = 1$, $x_2 = 2$, ..., $x_6 = 6$.

Table 6.1 – Tossing a Die	
Value of X	**Probability**
1	$\frac{1}{6}$
2	$\frac{1}{6}$
3	$\frac{1}{6}$
4	$\frac{1}{6}$
5	$\frac{1}{6}$
6	$\frac{1}{6}$

3. *Probability distribution:* The outcomes of the toss of a six-sided die and their probabilities are given in Table 6.1. The probabilities are deduced using the classical method and the assumption of a fair die.

Not all discrete random variables have probability distributions that are easy to determine. In Examples 6.2 and 6.3, neither random variable has an easily determined probability distribution.

Example 6.2

Random Phenomenon: Suppose we observe the number of defective integrated circuits received in a batch of 1000.

1. *Identify the random variable:* $X =$ the number of defective integrated circuits in a batch of 1000.

2. *Range of values*: Integers between 0 and 1000, where $N = 1000$. If symbols were chosen to represent the values they could be given as $x_1 = 0$, $x_2 = 1$, ..., $x_{N+1} = 1000$. That is, there could be 0 defective circuits or they could all (1000) be defective.

3. *Probability distribution*: Unknown.

Example 6.3

Random Phenomenon: A stock market analyst is interested in the number of stocks on the New York Stock Exchange (NYSE) that increased in price on the previous day. She realizes that the number that increases is a random variable. She will have to develop a description of the randomness in order to study the stocks.

1. *Identify the random variable*: $X =$ number of stocks that increased in price.

2. *Range of values*: Integers between 0 and the number of stocks trading on the NYSE on the previous day.

3. *Probability distribution*: Unknown, but could be estimated using the relative frequency idea from Section 5.2 in conjunction with historical data of the NYSE.

Continuous Random Variables

Heights, weights, volumes, and time measurements, for example, are usually measured on a continuous scale. These measurements can take on any value in some interval.

> **Definition**
>
> A **continuous random variable** is a random variable whose measurements can assume any one of a countless number of values in an interval.

Example 6.4

Random Phenomenon: Sylars Watch Manufacturer has a process that indicates that all watches will be assembled between 20 to 45 minutes. Obviously, the time to assemble a watch will be a function of the style and other accoutrements that are selected by the customer. Since time is measured on a continuous scale and the variability of watch assembly is not predictable due to different workers, the overall time to assemble a watch is considered to be a continuous random variable.

1. *Identify the random variable*: $X =$ time to assemble a watch.

2. *Range of values*: Between 0 and ∞. Note that X is measured on a continuous scale.

3. *Probability distribution*: Unknown.

6.1   Exercises

Basic Concepts

1. What is a random variable?

2. What is a probability distribution?

3. Do all random variables have easily determined probability distributions? Explain.

4. What are the two types of random variables discussed in the chapter? What distinguishes the two types?

Exercises

5. Classify the following as either a discrete random variable or a continuous random variable.

   a. The number of pages in a standard math textbook.

   b. The amount of electricity used daily in a home.

   c. The number of customers entering a restaurant in one day.

   d. The time spent daily on the phone after supper by a teenager.

   e. Campers at a state park over Labor Day weekend.

6. Classify the following as either a discrete random variable or a continuous random variable.

   a. The speed of a train.

   b. The possible scores on the SAT reasoning test.

   c. The number of pizzas delivered on a college campus each day.

   d. The daily takeoffs at Chicago's O'Hare Airport.

   e. The high temperatures in Maine and Florida tomorrow.

7. Classify the following as either a discrete random variable or a continuous random variable.

   a. The number of emergency phone calls received per day by a local fire department.

   b. The speed of pitches of major league baseball pitchers.

   c. The weight of a lobster caught in Maine.

   d. The number of defective circuits on a computer chip.

   e. The time it takes for a 5-year battery to die.

8. Classify the following as either a discrete random variable or a continuous random variable.

   a. The total points scored per football game for a local high school team.

   b. The daily price of a stock.

   c. The interest rate charged by local banks for 30-year mortgages.

   d. The number of times a backup of the computer network is performed in a month.

   e. The amount of sugar imported by the U.S. in a day.

| 6.2 | **Discrete Probability Distributions** |

So far, the random variable concept is so general that it is not very useful by itself. What would make it useful is to determine what numerical values the random variable could assume and to assess the probability of each of these values. This information defines a probability distribution for a discrete random variable.

Definition

A **discrete probability distribution** consists of all possible values of the discrete random variable along with their associated probabilities.

Discrete probability distributions always have three characteristics.

Properties

Properties of Discrete Probability Distributions

1. The sum of all the probabilities must equal 1. That is,

$$P(X = x_1) + P(X = x_2) + \cdots + P(X = x_n) = 1.$$

2. The probability of any value must be between 0 and 1, inclusively. That is,

$$0 \le P(X = x_i) \le 1.$$

3. The probabilities are additive. That is,

$$P(X = x_i \text{ or } X = x_j) = P(X = x_i) + P(X = x_j).$$

The association of the possible values with their respective probabilities can be expressed in three different forms: in a table, in a graph, and in an equation.

Example 6.5

Consider the random phenomenon of tossing a fair coin three times and counting the number of heads. Since the coin is assumed to be fair, $P(H) = P(T) = 0.5$. What is the probability distribution for the number of heads observed in three tosses of a coin?

Solution

The random variable is $X =$ the number of heads in three tosses of a coin.

Table 6.2 – Tossing a Coin		
Value of X	**P(X = x)**	**Simple Events**
0	$\frac{1}{8}$	TTT
1	$\frac{3}{8}$	HTT THT TTH
2	$\frac{3}{8}$	HHT HTH THH
3	$\frac{1}{8}$	HHH
Total	$\sum P(X = x) = 1.0$	

The probabilities given in Table 6.2 can be deduced using the classical approach to probability.

Example 6.6

K. J. Johnson is a computer salesperson. During the last year he has kept records of his computer sales for the last 200 days. He recognizes that his daily sales constitute a random process and he wishes to determine the probability distribution for daily sales. Then, from the probability distribution, he would like to know what the probability is that he will sell **a.** at least 2 computers each day and **b.** at most two computers each day.

Solution

The random variable is X = the number of computers sold each day.

Table 6.3 – Frequency Distribution	
Sales	**Frequency**
0	40
1	20
2	60
3	40
4	40

The probabilities for this random variable are computed in Table 6.4 based upon 200 days of sales data obtained from Mr. Johnson's records using the relative frequency approach.

Table 6.4 – Probability Distribution	
Sales x	Probability $P(X = x)$
0	$\dfrac{40}{200} = 0.2$
1	$\dfrac{20}{200} = 0.1$
2	$\dfrac{60}{200} = 0.3$
3	$\dfrac{40}{200} = 0.2$
4	$\dfrac{40}{200} = 0.2$
Total	$\sum P(X = x) = 1.0$

a. The probability that Mr. Johnson will sell at least 2 computers each day is calculated as follows.

$$P(X \geq 2) = P(X = 2) + P(X = 3) + P(X = 4) = 0.3 + 0.2 + 0.2 = 0.7$$

b. To find the probability that Mr. Johnson will sell at most 2 computers each day, the calculation is as follows.

$$P(X \leq 2) = P(X = 0) + P(X = 1) + P(X = 2) = 0.2 + 0.1 + 0.3 = 0.6$$

Example 6.7

Phone operators of a TV shopping network usually receive about three calls per minute from interested shoppers. Suppose we are interested in the next three telephone calls and the number of callers that place an order. The probability distribution was created based on historical data. Given the probability distribution, what is the probability that at least one caller will purchase something each minute?

$$X = \text{the number of callers that place an order}$$

Table 6.5 – Caller Data	
x	$P(X = x)$
0	0.2
1	0.5
2	0.1
3	0.2

Solution

The probability that at least one caller will purchase something is:

$$P(X \geq 1) = P(X = 1) + P(X = 2) + P(X = 3) = 0.5 + 0.1 + 0.2 = 0.8$$

or

$$P(X \geq 1) = 1 - P(X = 0) = 1 - 0.2 = 0.8.$$

Thus, the probability that at least one caller will purchase an item is 80%.

Where Do Probability Distributions Come From?

In the previous examples, the probability distributions were given. But in the real world there are very few instances in which the probability distribution is conveniently available. Substantial effort is usually required to obtain the probability distribution. Probabilities are determined using the same techniques described in Chapter 5: classical, relative frequency, or subjective.

Sometimes, however, you get lucky. The random variable you wish to analyze either conforms to or can be approximated by an experiment that has a known probability distribution. Four well known discrete distributions will be discussed in this chapter: uniform, binomial, Poisson, and hypergeometric.

Each of the discrete distributions possesses a **probability distribution function**. These functions assign probabilities to each value of the random variable.

Example 6.8

The following function is a discrete probability distribution function.

$$P(X = x) = \begin{cases} \dfrac{x^2}{30}, & \text{if } x = 1, \ 2, \ 3, \ 4 \\ 0 & \text{otherwise} \end{cases}$$

Summarize the probability distribution for this function.

Solution

To determine the probability for a value, use the value as the argument to the function. For example, to determine the probability that $X = 3$, calculate the following.

$$P(X = 3) = \frac{3^2}{30} = \frac{9}{30}$$

The probability that $X = 4$ can be computed in the same way.

$$P(X = 4) = \frac{4^2}{30} = \frac{16}{30}$$

The resulting probability distribution is summarized in Table 6.6.

Table 6.6 – Probability Distribution	
x	$P(X = x)$
1	$\dfrac{1}{30}$
2	$\dfrac{4}{30}$
3	$\dfrac{9}{30}$
4	$\dfrac{16}{30}$
Total	$\sum P(X = x) = \dfrac{30}{30} = 1$

Note that the distribution possesses the essential properties of all probability distributions; that is, the probabilities sum to one, and all the probabilities are between 0 and 1.

6.2   Exercises

Basic Concepts

1. Discrete probability distributions always have three characteristics. What are they?

2. What is the value of describing a random variable with a probability distribution?

3. What are three different ways to express possible values of a random variable along with their associated probabilities?

4. How is a probability distribution created?

5. Identify four discrete probability distributions.

6. What is a probability distribution function?

Exercises

7. Determine whether or not the following distribution is a probability distribution. If the distribution is not a probability distribution, give the characteristic which is not satisfied by the distribution.

Table for Exercise 7	
x	$P(X = x)$
1	$\dfrac{1}{3}$
2	$\dfrac{2}{3}$
3	$\dfrac{1}{3}$

8. Determine whether or not the following distribution is a probability distribution. If the distribution is not a probability distribution, give the characteristic which is not satisfied by the distribution.

Table for Exercise 8	
x	P(X = x)
−2	0.25
2	0.50
3	0.25

9. Tell whether or not the following distribution is a probability distribution. If the distribution is not a probability distribution, give the characteristic which is not satisfied by the distribution.

Table for Exercise 9	
x	P(X = x)
2	0.30
3	−0.50
4	0.50
5	0.70

10. Determine whether or not the following distribution is a probability distribution. If the distribution is not a probability distribution, give the characteristic which is not satisfied by the distribution.

Table for Exercise 10	
x	P(X = x)
5	0.46
10	0.25
15	0.25

11. Determine whether or not the following distribution is a probability distribution. If the distribution is not a probability distribution, give the characteristic which is not satisfied by the distribution.

Table for Exercise 11	
x	P(X = x)
−10	0.18
−5	0.39
3	0.08
8	0.35

12. Determine whether or not the following distribution is a probability distribution. If the distribution is not a probability distribution, give the characteristic which is not satisfied by the distribution.

Table for Exercise 12	
x	P(X = x)
100	−0.10
200	0.50
300	0.50

13. Determine whether or not the following distribution is a probability distribution. If the distribution is not a probability distribution, give the characteristic which is not satisfied by the distribution.

$$P(X = x) = \frac{x}{16}, \text{ for } x = 1,\ 2,\ 3,\ 4,\ 5$$

14. Determine whether or not the following distribution is a probability distribution. If the distribution is not a probability distribution, give the characteristic which is not satisfied by the distribution.

$$P(X = x) = \frac{x^2}{30}, \text{ for } x = 1,\ 2,\ 3,\ 4$$

6.3   Expected Value and Variance

The notion of **expected value** is one of the most important concepts in the analysis of random phenomena. Expected value is important because it is a summary statistic for a probability distribution that can be used as a criterion for comparing alternative decisions in the presence of uncertainty. Conceptually, expected value is closely allied with the notion of mean or average.

The expected value of a random variable should be very close to the average value of a large number of observations from the random process, and the larger the number of observations collected, the more likely the average of the observations will be close to the expected value. It should not be interpreted as the value of the random variable we *expect* to see. In fact, for discrete random variables the expected value is rarely one of the possible outcomes of the random variable.

Formula

Expected Value

The **expected value** of a discrete random variable X is the mean of the random variable X. It is denoted $E(X)$ and is given by computing the expression

$$\mu = E(X) = \sum \left[xp(x) \right]$$

where $p(x) = P(X = x)$.

Essentially the expected value is a weighted average, in which each possible value of the random variable is weighted by its probability.

Table 6.7 – Calculating K.J. Johnson's Expected Number of Sales per Day		
x	$p(x)$	$xp(x)$
0	0.2	0
1	0.1	0.1
2	0.3	0.6
3	0.2	0.6
4	0.2	0.8
Total		$E(X) = 2.1$

$$E(X) = \sum \left[xp(x) \right] = 0(0.2) + 1(0.1) + 2(0.3) + 3(0.2) + 4(0.2) = 2.1$$

The expected value of the probability distribution given in Example 6.6 is computed in Table 6.7. In the long run, K.J. Johnson will average 2.1 sales per day, which is the expected value of the probability distribution of K.J. Johnson's daily sales. That is, the expected value of a distribution can be considered the long run average of that distribution.

Using Expected Values to Compare Alternatives

Expected value analysis can also be used to make comparisons when choosing between uncertain alternatives.

Example 6.9

Suppose you are confronted with two investment alternatives that possess uncertain outcomes described by the probability distributions given in Table 6.8.

Solution

Table 6.8 – Investment Alternatives			
Option A		**Option B**	
Profit (Dollars)	**Probability**	**Profit (Dollars)**	**Probability**
−2000	0.2	−3000	0.2
0	0.1	−1000	0.1
1000	0.3	2000	0.2
2000	0.3	3000	0.3
4000	0.1	4000	0.2

$$E(X_A) = \sum \left[x_A p(x_A) \right]$$
$$= (-2000)0.2 + (0)0.1 + (1000)0.3$$
$$+ (2000)0.3 + (4000)0.1$$
$$= \$900$$

$$E(X_B) = \sum \left[x_B p(x_B) \right]$$
$$= (-3000)0.2 + (-1000)0.1$$
$$+ (2000)0.2 + (3000)0.3 + (4000)0.2$$
$$= \$1400$$

Because of the randomness of the profit variable, it is difficult to evaluate the investments by merely eyeballing the two distributions. However, by calculating the expected values of the two alternatives the information in each distribution is condensed to a single point. This point characterizes the center of the distribution and facilitates comparison. The expected values of Options A and B are $900 and $1400, respectively. Thus, in the long run Option B would be $500 more profitable. The phrase *in the long run* is a significant qualifier. It means that under repeated investments you would receive an average profit of $1400 from Option B. But on any one investment in Option B, you may lose as much as $3000 or make as much as $4000.

Variance of a Discrete Random Variable

In Example 6.9, while Option B has a greater expected value, it may differ in risk from Option A. Option B offers a greater chance of making a substantial gain, but it also has the potential for significant loss. The expected value of a distribution measures only one dimension of the random variable, namely its central value. To gauge the variability of a random variable we need another measure similar to the variance measure previously constructed, but one which accounts for the difference in probabilities of the variable.

Formula

Variance

The **variance** of a discrete random variable X is given by the following formula.

$$\sigma^2 = V(X) = \sum\left[(x-\mu)^2\, p(x)\right]$$

Note: Sometimes it is easier to calculate the variance of a discrete random variable using the computational formula

$$\sigma^2 = V(X) = \sum\left[x^2 p(x)\right] - \mu^2.$$

Both formulas are equivalent.

Once again, the variance can be considered an average. In this case it is the weighted average of the squared deviations about the mean. This is very similar to the computations for the sample variance (s^2) and population variance (σ^2) given by $s^2 = \dfrac{\sum(x_i - \bar{x})^2}{n-1}$ and $\sigma^2 = \dfrac{\sum(x_i - \mu)^2}{N}$, respectively. The larger the variance, the more variability in the outcomes.

To manually compute the variance of the random variable, it's often a good idea to construct a table.

Example 6.10

The calculation of the variances of the random variables described in Example 6.9 are given in Tables 6.9 and 6.10.

Solution

Table 6.9 – Variance of Option A		
Option A		
Profit (Dollars)	**Probability**	$(x-\mu)^2 \, p(x)$
−2000	0.2	$(-2000-900)^2\,0.2 = 1{,}682{,}000$
0	0.1	$(0-900)^2\,0.1 = 81{,}000$
1000	0.3	$(1000-900)^2\,0.3 = 3000$
2000	0.3	$(2000-900)^2\,0.3 = 363{,}000$
4000	0.1	$(4000-900)^2\,0.1 = 961{,}000$
	Total	$\sigma^2 = V(X) = \$3{,}090{,}000$

The standard deviation is computed by taking the square root of the variance. In this instance, the standard deviation of Option A is given by

$$\sigma = \sqrt{V(X)} = \sqrt{3{,}090{,}000} \approx \$1757.84.$$

Table 6.10 – Variance of Option B

	Option B	
Profit (Dollars)	Probability	$(x - \mu)^2 \, p(x)$
−3000	0.2	$(-3000 - 1400)^2 \, 0.2 = 3,872,000$
−1000	0.1	$(-1000 - 1400)^2 \, 0.1 = 576,000$
2000	0.2	$(2000 - 1400)^2 \, 0.2 = 72,000$
3000	0.3	$(3000 - 1400)^2 \, 0.3 = 768,000$
4000	0.2	$(4000 - 1400)^2 \, 0.2 = 1,352,000$
	Total	$\sigma^2 = V(X) = \$6,640,000$

The standard deviation of Option B is given by

$$\sigma = \sqrt{6,640,000} \approx \$2576.82.$$

A larger standard deviation reflects greater variability in profits and increased risk. When risk is considered, the decision becomes more difficult. The option with the largest expected value (Option B) is also the option with the greatest risk. Hence, the decision maker must subjectively evaluate the trade-off between greater expected return and increased risk.

Example 6.11

EJN Trucking Company is trying to decide which of two trucks to purchase. Each truck costs $125,000 but comes with different features and amenities. EJN's financial advisor estimates that the returns from each truck for the next five years will follow the probability distribution in the following table.

Table 6.11 – Return on Truck Purchase

Return (Dollars)	Truck 1	Truck 2
−5000	0.02	0.15
0	0.03	0.10
5000	0.20	0.10
10,000	0.50	0.10
15,000	0.20	0.30
20,000	0.03	0.20
25,000	0.02	0.05

Calculate the expected return, variance, and standard deviation for each truck and recommend which truck to purchase.

Solution

Table 6.12 – Calculations

Return (Dollars)	Truck 1	Truck 2	Truck 1 $xp(x)$	Truck 2 $xp(x)$	Truck 1 $(x-\mu)^2 p(x)$	Truck 2 $(x-\mu)^2 p(x)$
−5000	0.02	0.15	−100	−750	4,500,000	36,037,500
0	0.03	0.10	0	0	3,000,000	11,025,000
5000	0.20	0.10	1000	500	5,000,000	3,025,000
10,000	0.50	0.10	5000	1000	0	25,000
15,000	0.20	0.30	3000	4500	5,000,000	6,075,000
20,000	0.03	0.20	600	4000	3,000,000	18,050,000
25,000	0.02	0.05	500	1250	4,500,000	10,512,500

$$E(X) = \qquad \$10,000 \qquad \$10,500$$

$$V(X) = \sigma^2 = \qquad \$25,000,000 \qquad \$84,750,000$$

$$\sqrt{V(X)} = \sigma = \qquad \$5000 \qquad \$9206$$

The expected value, variance, and standard deviation of purchasing Truck 1 are $10,000, $25,000,000, and $5000, respectively. Similarly the expected value, variance, and standard deviation of purchasing Truck 2 are $10,500, $84,750,000, and $9206. Given that the risk (standard deviation) of Truck 2 is nearly twice that of Truck 1, it appears that selecting Truck 1 is the best decision, even though the expected return is slightly less.

6.3   Exercises

Basic Concepts

1. Why is the notion of expected value important in the analysis of random phenomena?

2. True or false: the expected value of a random variable is usually one of the possible outcomes of the random variable.

3. Suppose the expected value of a random variable was known to be 6.3. Interpret the meaning of the expected value.

4. Give an example of a situation in which expected value would be useful to compare alternatives.

5. How is the variance (or standard deviation) of a random variable related to risk?

Exercises

6. Find the expected value, the variance, and the standard deviation for a random variable with the following probability distribution.

Table for Exercise 6 – Probability Distribution					
x	−5	−2	0	2	5
p(x)	0.06	0.15	0.58	0.18	0.03

7. Find the expected value, the variance, and the standard deviation for a random variable with the following probability distribution.

Table for Exercise 7 – Probability Distribution					
x	400	420	440	460	480	500
p(x)	0	0.1	0.1	0.2	0.2	0.4

8. A regional hospital is considering the purchase of a helicopter to transport critical patients. The relative frequency of X, the number of times the helicopter is used to transport critical patients each month, is derived for a similarly sized hospital and is given in the following probability distribution.

Table for Exercise 8 – Probability Distribution: Number of Helicopter Transports							
x	0	1	2	3	4	5	6
p(x)	0.15	0.20	0.34	0.19	0.06	0.05	0.01

a. Find the average number of times the helicopter is used to transport critical patients each month.

b. Find the variance of the number of times the helicopter is used to transport critical patients.

c. Find the standard deviation of the number of times the helicopter is used to transport critical patients.

d. Find the probability that the helicopter will not be used at all during a month to transport critical patients.

e. Find the probability that the helicopter will be used at least once to transport critical patients.

f. Find the probability that the helicopter will be used at most twice to transport critical patients.

g. Find the probability that the helicopter will be used more than three times to transport critical patients.

9. Based on past experience, an architect has determined a probability distribution for X, the number of times a drawing must be examined by a client before it is accepted.

Table for Exercise 9 – Probability Distribution: Number of Times Examined					
x	1	2	3	4	5
p(x)	0.1	0.2	0.3	0.2	0.2

a. Find the average number of times a drawing must be examined by a client before it is accepted.

b. Find the variance of the number of times a drawing must be examined by a client before it is accepted.

c. Find the standard deviation of the number of times a drawing must be examined by a client before it is accepted.

d. What is the probability that a drawing must be examined five times before being accepted by the client?

e. Find the probability that the drawing must be examined at least twice before being accepted by the client.

f. Find the probability that a drawing must be examined at most three times before being accepted by the client.

g. Find the probability that a drawing must be examined less than twice before being accepted by the client.

10. The manager of a retail clothing store has determined the following probability distribution for X, the number of customers who will enter the store on Saturday.

Table for Exercise 10 – Probability Distribution: Customers on Saturday					
x	10	20	30	40	50	60
p(x)	0.10	0.20	0.30	0.20	0.10	0.10

a. Find the expected number of customers who will enter the store on Saturday.

b. Find the standard deviation of the number of customers who will enter the store on Saturday.

c. Find the variance of the number of customers who will enter the store on Saturday.

d. Find the probability that more than 30 customers will enter the store on Saturday.

e. Find the probability that at most 20 customers will enter the store on Saturday.

f. Find the probability that at least 40 customers will enter the store on Saturday.

g. What is the probability that exactly 10 customers will enter the store on Saturday?

11. An entrepreneur is considering investing in a new venture. If the venture is successful, he will make $50,000. However, if the venture is not successful, he will lose his investment of $10,000. Based on past experience, he believes that there is a 40% chance that the venture will be successful.

   a. Use the information in the problem to determine the probability distribution of the amount of money to be made (or lost) on the venture.

   b. Determine the expected amount of money to be made on the venture.

   c. Determine the standard deviation of the amount of money to be made on the venture.

12. An investor is considering two alternative investment options with the following payoff distributions.

Table A for Exercise 12 – Option 1	
Payoff	P(Payoff)
–$100,000	$\frac{1}{3}$
$30,000	$\frac{1}{3}$
$100,000	$\frac{1}{3}$

Table B for Exercise 12 – Option 2	
Payoff	P(Payoff)
–$20,000	0.25
$0	0.50
$20,000	0.25

   a. Calculate the expected payoff for each of the investment options.

   b. Calculate the standard deviation of the payoff for each of the investment options.

   c. Which investment option would you choose? Explain.

13. A cereal manufacturer has two new brands of cereal which it would like to produce. Because resources are limited, the cereal manufacturer can only afford to produce one of the new brands. A marketing study produced the following probability distributions for the amount of sales for each of the new brands of cereal.

Table A for Exercise 13 – Cereal A	
Sales	P(Sales)
–$150,000	0.2
$200,000	0.3
$300,000	0.3
$400,000	0.2

Table B for Exercise 13 – Cereal B	
Sales	P(Sales)
–$10,000	0.40
$300,000	0.40
$600,000	0.10
$1,000,000	0.10

a. What are the expected sales of each of the new brands of cereal?

b. What is the standard deviation of the sales for each of the brands of cereal?

c. If both of the brands of cereal cost the same amount to produce, which brand of cereal do you think the cereal manufacturer should produce? Explain.

6.4   The Discrete Uniform Distribution

The **discrete uniform distribution** is one of the simplest probability distributions. Each value of the random variable is assigned an identical probability. There are many situations in which the discrete uniform distribution arises. Some common examples are rolling a fair die or flipping a fair coin.

Formula

Discrete Uniform Probability Distribution Function

Mathematically, the **discrete uniform probability distribution function** is given by

$$P(X = x) = \frac{1}{n}$$

where $n =$ the number of values that the random variable may assume.

Example 6.12

What is the probability distribution for the outcome of the throw of a single six-sided die?

Solution

If the die is fair, then each of the outcomes is equally likely, and thus we have a discrete uniform distribution in which all probabilities equal $\frac{1}{6}$. The probability distribution is given in Table 6.13.

Table 6.13 – Throwing a Die	
x	**P(X = x)**
1	$\frac{1}{6}$
2	$\frac{1}{6}$
3	$\frac{1}{6}$
4	$\frac{1}{6}$
5	$\frac{1}{6}$
6	$\frac{1}{6}$

Example 6.13

Suppose a purchasing agent has just received a pricing and delivery schedule from a new vendor. The delivery schedule was quoted as 1 to 4 weeks. The agency wishes to construct the probability distribution for the time until delivery.

Solution

The probabilities of the random variable X = the number of weeks until delivery are given in Table 6.14.

Table 6.14 – Delivery Distribution	
x	$P(X = x)$
1	$\frac{1}{4}$
2	$\frac{1}{4}$
3	$\frac{1}{4}$
4	$\frac{1}{4}$

Without any prior information, the agent believes any time frame is as likely as any other. Hence, the number of weeks until delivery will be assumed to have a discrete uniform distribution. Over time the purchasing agent will undoubtedly revise the distribution as more information is gathered about the company's delivery schedule.

Example 6.13 illustrates an important principle in the application of the discrete uniform distribution. That is, when there is little or no information concerning the outcome of a random variable, the discrete uniform distribution may be a reasonable initial alternative.

6.4 Exercises

Basic Concepts

1. What is the most significant property of the uniform distribution?

2. What is the discrete uniform probability distribution function?

3. Explain why the uniform distribution is often used when there is little or no information concerning the outcome of a random variable.

Exercises

4. In the casino game of roulette, a wheel is spun and a ball is set in motion, ultimately coming to rest in one of the 38 slots on the wheel. Any slot is as likely as any other to capture the ball. Of the 38 slots, 18 are red, 18 are black, and 2 are green. Suppose the entry fee to play a single game is $1 and the participant bets on red. If the ball comes to rest in one of the red slots, he wins $1 in addition to getting back the original $1 entry fee. If the ball does not end up in a red slot, the $1 entry fee is lost. Let X denote the monetary gain when betting $1 on red, in a single game of roulette. Gain is defined as the amount won minus the fee to play.

    a. What are the possible values of X?

    b. Is X a discrete or continuous random variable? Explain.

    c. Construct the probability distribution of X.

    d. Find the expected value of X and interpret this number.

    e. Do you feel that in any casino games you would have a positive expected gain? Why?

5. An experiment consists of tossing two coins and a die simultaneously.

    a. List the 24 equally likely outcomes.

    b. Define the random variable X as the sum of the number of heads on the two coins and the number of dots on the die. What are the possible values of X?

    c. Construct the probability distribution of X in the form of a table.

    d. Find the expected value of X.

6.5   The Binomial Distribution

The binomial distribution arises from experiments with repeated two-outcome trials, where only one of the outcomes is counted. Experiments of this kind are rather common in the business world. In market research, a survey respondent (a trial) either will or will not recognize a company's brand. The number that recognize the brand is a count that may be modeled as a **binomial random variable**. When a customer (a trial) enters a bank for service, he or she may have to wait. If we are counting customers who have to wait, then the count may conform to the binomial model. Experiments are required to meet several conditions in order to qualify as a binomial experiment.

Definition

A **binomial experiment** is a random experiment which satisfies all of the following conditions.

1. There are only two outcomes in each trial of the experiment. (One of the outcomes is usually referred to as a *success*, and the other as a *failure*.)

2. The experiment consists of n identical trials as described in Condition 1.

3. The probability of success on any one trial is denoted by p and does not change from trial to trial. (Note that the probability of a failure is $(1 - p)$ and also does not change from trial to trial.)

4. The trials are independent.

5. The binomial random variable is the count of the number of successes in n trials.

A binomial random variable is formed by counting the number of successes in n trials of an experiment with two outcomes. One of the simplest of binomial random variables is produced by tossing a coin.

Example 6.14

Toss a coin 4 times and record the number of heads. Is the number of heads in 4 tosses a binomial random variable?

Solution

1. There are only two outcomes, heads or tails.

2. The experiment will consist of 4 tosses of a coin. (Hence, $n = 4$.)

3. The probability of getting a head (success) is $\dfrac{1}{2}$ and does not change from trial to trial. (Hence, $p = \dfrac{1}{2}$.)

4. The outcome of one toss will not affect other tosses.

5. The variable of interest is the count of the number of heads in 4 tosses.

All the conditions of a binomial experiment are met, so the number of heads in 4 tosses of a coin is a binomial random variable.

The probability distribution for this experiment is given in the following table.

Table 6.15 – Tossing a Coin		
Events	**Number of Heads**	**Probability**
TTTT	0	$\frac{1}{16}$
HTTT THTT TTHT TTTH	1	$\frac{4}{16}$
HHTT HTHT HTTH THHT THTH TTHH	2	$\frac{6}{16}$
THHH HTHH HHTH HHHT	3	$\frac{4}{16}$
HHHH	4	$\frac{1}{16}$

The probability distribution can be derived using the classical method described in the previous chapter. To derive the binomial distribution by listing the simple events is unnecessarily tedious (see Table 6.15). Instead of tossing the coin 4 times, suppose the coin is tossed 10 times in the experiment. The number of simple events for an experiment with 10 tosses would be 1024, and an experiment with 20 tosses would require listing a staggering 1,048,576 simple events. Fortunately, there is a far simpler method of obtaining the probability distribution. The binomial probability distribution function provides a relatively simple method of calculating binomial probabilities.

Formula

Binomial Probability Distribution Function

The **binomial probability distribution function** is

$$P(X = x) = {}_nC_x p^x (1-p)^{n-x}$$

where ${}_nC_x$ represents the number of possible combinations of n objects taken x at a time (without replacement) and is given by

$${}_nC_x = \frac{n!}{x!(n-x)!} \quad \text{where } n! = n(n-1)(n-2)\cdots(2)(1) \text{ and } 0! = 1;$$

n = the number of trials,

p = the probability of a success, and

x = the number of successes in n trials.

To calculate a binomial probability, the parameters of the distribution (n and p) as well as the value of the random variable must be specified. For example, to determine the probability of 3 heads in 4 tosses of a coin, you would substitute the values $X = 3$, $n = 4$, and $p = \frac{1}{2}$, into the binomial probability distribution function as follows.

$$P(X = 3) = {_4}C_3 \left(\frac{1}{2}\right)^3 \left(1 - \frac{1}{2}\right)^{4-3}$$

Since ${_4}C_3 = \dfrac{4!}{3!(4-3)!} = \dfrac{(4)\,(3!)}{(3!)\,(1!)} = 4$, then

$$P(X = 3) = 4\left(\frac{1}{2}\right)^3 \left(\frac{1}{2}\right) = \frac{4}{16} = \frac{1}{4} = 0.25.$$

The probability that 2 heads would be tossed would be computed in a similar manner.

$$P(X = 2) = {_4}C_2 \left(\frac{1}{2}\right)^2 \left(1 - \frac{1}{2}\right)^{4-2}$$

$$= \frac{4!}{(2!)(2!)}\left(\frac{1}{2}\right)^2 \left(\frac{1}{2}\right)^2 = \frac{6}{16} = \frac{3}{8} = 0.375.$$

The complete distribution can be computed by substituting the remaining values of the random variable into the probability distribution function.

Table 6.16 – Tossing a Coin	
Number of Heads	**Probability**
0	$\dfrac{1}{16}$
1	$\dfrac{4}{16}$
2	$\dfrac{6}{16}$
3	$\dfrac{4}{16}$
4	$\dfrac{1}{16}$

The coin toss is a classical binomial experiment easily related to the rules of a binomial experiment. In many instances the relationship of an experiment to the binomial definition is not as clear.

Example 6.15

Roll a single six-sided die 4 times and record the number of sixes observed. Does the number of sixes rolled in 4 tosses of a die meet the conditions required of a binomial random variable? Construct the probability distribution for this experiment.

Solution

1. The experiment either produces a six or does not, and thus satisfies the two outcome requirement. (At first glance, there appears to be a problem. There are six sides to a die and there would appear to be six possible outcomes rather than the two required for a single trial of a binomial experiment.)

2. The experiment is repeated 4 times. (Hence, $n = 4$.)

3. The probability remains constant from trial to trial. (One roll of the die does not affect other rolls.)

4. If X is the number of sixes in 4 rolls, then it has a binomial distribution.

To obtain the probability distribution for X, use the binomial probability distribution function with parameters $n = 4$ and $p = \dfrac{1}{6}$.

$$P(X = 0) = {}_4C_0 \left(\frac{1}{6}\right)^0 \left(1 - \frac{1}{6}\right)^{4-0} = (1)(1)\left(\frac{5}{6}\right)^4 \approx 0.4823$$

$$P(X = 1) = {}_4C_1 \left(\frac{1}{6}\right)^1 \left(1 - \frac{1}{6}\right)^{4-1} = (4)\left(\frac{1}{6}\right)\left(\frac{5}{6}\right)^3 \approx 0.3858$$

$$P(X = 2) = {}_4C_2 \left(\frac{1}{6}\right)^2 \left(1 - \frac{1}{6}\right)^{4-2} = (6)\left(\frac{1}{6}\right)^2 \left(\frac{5}{6}\right)^2 \approx 0.1157$$

$$P(X = 3) = {}_4C_3 \left(\frac{1}{6}\right)^3 \left(1 - \frac{1}{6}\right)^{4-3} = (4)\left(\frac{1}{6}\right)^3 \left(\frac{5}{6}\right) \approx 0.0154$$

$$P(X = 4) = {}_4C_4 \left(\frac{1}{6}\right)^4 \left(1 - \frac{1}{6}\right)^{4-4} = (1)\left(\frac{1}{6}\right)^4 (1) \approx 0.0008$$

Table 6.17 – Throwing a Die	
x	**Probability**
0	0.4823
1	0.3858
2	0.1157
3	0.0154
4	0.0008

The calculations required to produce the distribution were reasonably simple in this case. Note that given we have an equation for $P(X = x)$ we can tabulate, i.e. form a table of $P(X = x)$ for all possible x's given p and n. (See Example 6.15.) However, if there had been forty rolls rather than four, the determination of the probabilities would have been rather burdensome, at best. Binomial tables containing a large collection of binomial distributions have been constructed in order to avoid tedious calculations. The tables consist of cumulative probabilities $\big($i.e. $P(X \le x)\big)$ for various values of n, p, and x, and are found in Appendix A, Table E. Probabilities can also be found using technology such as a TI-84 Plus calculator, Microsoft Excel, or Minitab. The Discovering Technology section at the end of this chapter shows how to find binomial probabilities using these types of technology.

Example 6.16

A department store has decided to interview a random sample of ten customers to determine if they should modify their return policy. The company believes that 70% of the customers will approve the change and 30% will not approve the change. Using the probabilities given, what is the probability that at least three of the ten customers will approve the change in return policy?

Solution

Instead of laboriously computing the distribution, use the binomial tables in Appendix A, Table E. The binomial parameters are $n = 10$ and $p = 0.7$. Let X equal the number of customers that will approve the change. The following table shows the cumulative binomial distribution for X.

Table 6.18 – Binomial Table $n = 10$, $p = 0.7$	
x	$P(X \le x)$
0	0.0000
1	0.0001
2	0.0016
3	0.0106
4	0.0473
5	0.1503
6	0.3504
7	0.6172
8	0.8507
9	0.9718
10	1.0000

The probability that at least three customers will approve the change is equal to $1 - P(X \le 2) = 1 - 0.0016 = 0.9984$. Therefore, it's almost certain that at least three customers will approve the change in return policy.

The Shape of a Binomial Distribution

Binomial Distribution for $n = 12$ and $p = 0.1$

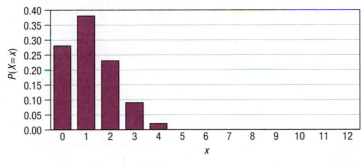

Figure 6.1

Binomial distributions have taken various shapes in the previous problems. The shape of the distribution depends upon the parameters n and p. If p is small the distribution tends to be skewed with a tail on the right (i.e. there are more successes when x is smaller). (See Figure 6.1.)

Figure 6.2

If p is near 0.5 the distribution is symmetrical. The graph in Figure 6.2 displays the probabilities for $n = 9$ and $p = 0.5$.

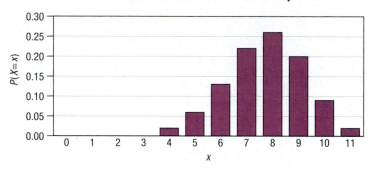

Figure 6.3

If p is large the distribution tends to be skewed with a long tail on the left (i.e. there are more successes when x is larger). The graph in Figure 6.3 displays the binomial probabilities for $n = 11$ and $p = 0.7$.

Expected Value and Variance

To calculate the expected value of a binomial random variable, such as in Example 6.16, would require a substantial number of arithmetic operations. Fortunately, you can avoid this calculation by utilizing a special characteristic of the binomial distribution.

Formula

Expected Value

The **expected value** of a binomial random variable can be computed using the expression

$$\mu = E(X) = np,$$

where n and p are the parameters of the binomial distribution.

It is important to remember that this formula is only valid for binomial random variables. Also, there is a simple method for obtaining the variance for a binomial random variable.

Formula

Variance and Standard Deviation

To find the **variance** of a binomial random variable, use the expression

$$\sigma^2 = V(X) = np(1-p).$$

Thus, the **standard deviation** of a binomial random variable is given by

$$\sigma = \sqrt{V(X)} = \sqrt{np(1-p)}.$$

Example 6.17

Compute the expected value and the variance of the number of customers that will approve the change in return policy in Example 6.16.

Solution

Since the random variable is binomial, we can use the shortcuts $E(X) = np$ and $V(X) = np(1-p)$. Since $n = 10$ and $p = 0.7$, the expected value is given by the following expression.

$$E(X) = np = 10(0.7) = 7$$

The variance is

$$\sigma^2 = V(X) = np(1-p) = 10(0.7)(0.3) = 2.1,$$

which implies that the standard deviation is $\sqrt{2.1} \approx 1.4491$.

Thus, if 10 randomly selected customers are polled, we would expect 7 of the 10 to approve the change in return policy, and the standard deviation would be 1.4491 customers.

6.5 Exercises

Basic Concepts

1. Describe the characteristics of a binomial experiment.

2. What are the parameters of a binomial probability model?

3. Give an example of a binomial experiment in a business context.

4. What is the binomial probability distribution function?

5. Describe the shape of a binomial distribution. Does the shape change? What influences the shape of the distribution?

6. How do you calculate the expected value of a binomial random variable? The variance? The standard deviation?

Exercises

7. Calculate $_nC_x$ for each of the following combinations of x and n.

   a. $n = 5, x = 4$

   b. $n = 10, x = 8$

   c. $n = 15, x = 1$

   d. $n = 20, x = 0$

8. Calculate $_nC_x$ for each of the following combinations of x and n.

   a. $n = 4, x = 2$

   b. $n = 12, x = 8$

   c. $n = 18, x = 15$

   d. $n = 23, x = 20$

9. The random variable X is a binomial random variable with $n = 9$ and $p = 0.1$.

   a. Find the expected value of X.

   b. Find the standard deviation of X.

   c. Find the probability that X equals 2. (Use the formula for $P(X = x)$.)

   d. Find the probability that X is at most 3.

   e. Find the probability that X is at least 2.

   f. Find the probability that X is less than 5.

10. The random variable X is a binomial random variable with $n = 12$ and $p = 0.8$.

   a. Find the expected value of X.

   b. Find the standard deviation of X.

   c. Find the probability that X equals 7. (Use the formula for $P(X = x)$.)

   d. Find the probability that X is at most 4.

   e. Find the probability that X is at least 1.

   f. Find the probability that X is more than 10.

11. A real estate agent has ten properties that she shows. She feels that there is a ten percent chance of selling any one property during a week. The chance of selling any one property is independent of selling another property.

    a. What probability model would be appropriate for describing the number of properties sold each week?

    b. Compute the expected number of properties to be sold in a week.

    c. Compute the standard deviation of the number of properties sold each week.

    d. Compute the probability of selling one property in one week.

    e. Compute the probability of selling five properties in one week.

    f. Compute the probability of selling at least three properties in one week.

12. A small commuter airline is concerned about reservation no-shows and, correspondingly, how much they should overbook flights to compensate. Assume their commuter planes will hold 15 people. Industry research indicates that 20% of the people making a reservation will not show up for a flight. Whether or not one person takes the flight is considered to be independent of other persons holding reservations.

    a. What probability model would be appropriate for the number of passengers that actually take the flight?

    b. If the airlines decide to book 18 people for each flight, how often will there be at least one person who will not get a seat?

    c. If they book 17 people, how often will there be at least one person who will not get a seat?

    d. If they book 16 people, how often will there be at least one person who will not get a seat?

    e. If they book 18 people for each flight, how often will there be one or more empty seats?

    f. If they book 17 people, how often will there be one or more empty seats?

    g. If they book 16 people, how often will there be one or more empty seats?

    h. Based on the results from parts **b.** to **g.** above, which booking policy do you prefer? Explain your answer.

13. Seven plants are operated by a garment manufacturer. They feel there is a ten percent chance for a strike at any one plant and the risk of a strike at one plant is independent of the risk of a strike at another plant. Let X = number of plants of the garment manufacturer that strike.

    a. Determine the probability distribution for X.

    b. Interpret the results for $P(X = 0)$, $P(X = 4)$, and $P(X = 7)$.

    c. Compute the expected value of X.

    d. Compute the standard deviation for X. Is this value large in relation to the expected value? In what units is the standard deviation expressed?

14. A company that makes traffic signal lights buys switches from a supplier. Out of each shipment of 1000 switches, the company will take a random sample of 10 switches. Let X equal the number of defective switches in the sample.

   a. The company has a policy of rejecting a lot if they find any defective switches in the sample. What is the probability that the shipment will be accepted if, in fact, 2% of the switches are actually defective?

   b. What is the probability that the shipment will be accepted if the percent of defective switches is actually 5%?

   c. The company decides to change their policy and will accept the lot if they find no more than one defective switch. Repeat parts **a.** and **b.** for this new policy.

15. Parents have always wondered about the sex of a child before it is born. Suppose that the probability of having a male child was 0.5, and that the sex of one child is independent of the sex of other children.

   a. Determine the probability of having exactly two girls out of four children.

   b. What is the probability of having four boys out of four children?

16. A certain aspirin is advertised as being preferred by 4 out of 5 doctors. If the advertisement is assumed to be true, answer the following questions.

   a. What is the probability that at least half of ten doctors chosen at random will prefer this brand of aspirin?

   b. What is the probability that 9 out of 10 of the doctors will prefer this brand?

17. In manufacturing integrated circuits, the yield of the manufacturing process is the percentage of good chips produced by the process. The probability that an integrated circuit manufactured by the Ace Electronics Company will be defective is $p = 0.05$. If a random sample of 15 circuits is selected for testing, answer the following questions.

   a. What is the probability that no more than one integrated circuit will be defective in the sample?

   b. What is the expected number of defective integrated circuits in the sample?

18. The Alvin Secretarial Service procures temporary office personnel for major corporations. They have found that 90% of their invoices are paid within 10 working days. If a random sample of 12 invoices is checked, answer the following questions.

   a. What is the probability that all of the invoices will be paid within 10 working days?

   b. What is the probability that six or more of the invoices will be paid within 10 working days?

19. An experiment consists of rolling a pair of dice 10 times. On each roll the sum of the dots on the two dice is noted.

a. Find the probability that on any roll of the two dice the sum of the dots is either 7 or 11.

b. Find the probability that in the 10 rolls of the pair of dice, a 7 or 11 occurs 5 times.

c. Find the probability that in the 10 rolls of the pair of dice, a 7 or 11 does not occur at all.

d. Find the mean and variance of the number of times we see a 7 or 11 in the 10 rolls of the dice.

20. *Would you say you eat to live or live to eat?* was asked to each person in a sample of 1001 adults in a Gallup Poll taken in April 1996. Seventy-four percent of the respondents answered eat to live, 23% answered live to eat, and 3% had no opinion. Assuming these percents are accurate, find the probability, in 12 randomly chosen adults, that the number who would answer "eat to live" is:

a. exactly 7.

b. no more than 10.

c. at most 11.

d. at least 3.

6.6 The Poisson Distribution

The binomial random variable requires a fixed number of repetitions of the experiment, where the outcomes are either successes or failures. The **Poisson distribution** is similar to the binomial in that the random variable represents a count of the total number of successes. The major difference between the two distributions is that the Poisson does not have a fixed number of trials. Instead, the Poisson uses a fixed interval of time or space in which the number of successes are recorded. Thus, there is no theoretical upper limit on the number of successes, although large numbers of successes are not very likely. The word *success* in the Poisson context can sometimes take on rather unpleasant connotations. For example, the randomness exhibited by the number of airplane crashes, oil tanker spills, and car accidents in some fixed period of time seem to conform to the randomness described by a Poisson random variable.

Kicked by Horses

A real world example of the Poisson distribution involves the distribution of Prussian cavalry deaths from getting kicked by horses, in the period 1875-1894. The Prussian military kept meticulous records on horse-kick deaths in each of its army corps, and the data are neatly summarized in a 1963 book called *Lady Luck*, by the late Warren Weaver. There were a total of 196 kicking deaths – these being the successes. The trials were each army corps's observations on the number of kicking deaths sustained during the year. With 14 army corps and data for 20 years, there were 280 trials. The Poisson formula predicts, for example, that there will be 34.1 instances of having exactly two deaths in a year. In fact, there were 32 such cases. Pretty good, eh?

In business environments many variables seem to follow a pattern of randomness similar to that described by the Poisson distribution. One of the Poisson's principal areas of use in business is the analysis of waiting lines. Other random phenomena, such as airplane arrivals at an airport, trucks arriving at a loading dock, users logging on to a computer system, or the number of defects in a given surface area, can be modeled with a Poisson distribution. These variables are often of interest in determining personnel requirements, inventories, and quality control.

Definition

In order to qualify as a **Poisson random variable** an experiment must meet two conditions.

1. Successes occur one at a time. (That is, two or more successes cannot occur at exactly the same point in time or exactly at the same point in space.)

2. The occurrence of a success in any interval is independent of the occurrence of a success in any other interval.

If these two conditions are met, it can be proven that the random variable for the number of successes follows the Poisson probability distribution function.

Formula

Poisson Probability Distribution Function

The **Poisson probability distribution function** is given by:

$$P(X = x) = \frac{e^{-\lambda}\lambda^x}{x!}, \text{ for } x = 0, 1, 2, \dots$$

where $e = 2.71828\dots$, and

λ = the mean number of successes.

The Poisson distribution has only one parameter, λ, pronounced *lambda*. One peculiar feature of the distribution is that the variance of the distribution is equal to the mean (lambda). That is, $\mu = \sigma^2 = \lambda$.

Tables for the Poisson distribution are found in Appendix A, Table F. The tables give $P(X = x)$ for particular values of x and λ. Poisson probabilities can also be computed using technology such as a TI-84 Plus calculator, Microsoft Excel, or Minitab. The Discovering Technology section at the end of this chapter shows how to calculate Poisson probabilities using available technology.

The shape of the distribution varies dramatically with the parameter λ. If λ is small, say 0.3, then the corresponding distribution is found in Table 6.19 and displayed in Figure 6.4.

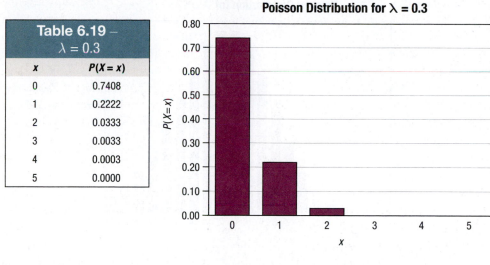

Table 6.19 – $\lambda = 0.3$	
x	P(X = x)
0	0.7408
1	0.2222
2	0.0333
3	0.0033
4	0.0003
5	0.0000

Figure 6.4

As λ increases to 3 (see Table 6.20), the distribution in Figure 6.5 exhibits a mound shape with skewness.

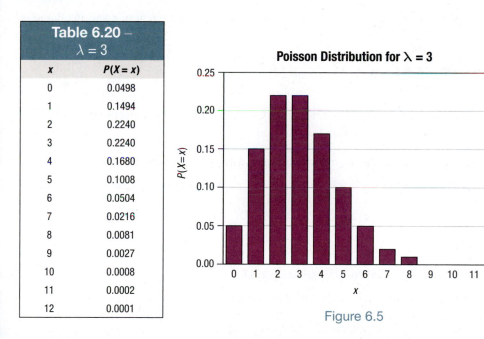

Table 6.20 – $\lambda = 3$	
x	P(X = x)
0	0.0498
1	0.1494
2	0.2240
3	0.2240
4	0.1680
5	0.1008
6	0.0504
7	0.0216
8	0.0081
9	0.0027
10	0.0008
11	0.0002
12	0.0001

Figure 6.5

As λ becomes even larger, say $\lambda = 12$, the distribution begins to closely resemble a bell-shaped distribution as shown in Figure 6.6.

Figure 6.6

Poisson Random Variables for Time

Most Poisson applications relate to the number of occurrences of some event in a specific duration of time.

Example 6.18

Suppose a bank has one automatic teller machine. Customers arrive at the machine at a rate of 20 per hour and according to a Poisson pattern.

a. What is the probability that no one will arrive in a 15-minute interval?

b. What is the probability that in a 15-minute period at least 3 persons will use the automated teller machine?

Solution

a. Let X = number of arrivals in a 15-minute period.

This problem contains one of the standard techniques used in working with Poisson random variables, that is, translating the arrival rate to correspond to the desired time interval. In this problem, the rate is given at 20 per hour which corresponds to a rate of 5 per $\frac{1}{4}$ hour (15 minutes). Thus, $\lambda = 5$ and the desired probability is

$$P(X = 0) = \frac{e^{-5}5^0}{0!}, \text{ or}$$

$P(X = 0) \approx 0.0067$.   (**Note:** 0! is defined to be 1.)

The probability could also have been found by using the table in Appendix A or appropriate technology.

b. To find the probability that in a 15-minute period at least 3 persons will use the automated teller machine, we are interested in $P(X \geq 3)$.

$$P(X \geq 3) = 1 - P(X \leq 2)$$
$$= 1 - \big(P(X = 0) + P(X = 1) + P(X = 2)\big)$$
$$= 1 - \big(0.0067 + 0.0337 + 0.0842\big) \ \big(\text{using the Poisson table}\big)$$
$$= 1 - 0.1246 = 0.8754$$

Poisson Random Variables for Length or Space

Instead of counting the number of successes in a time interval, there are a number of applications of the Poisson distribution that measure the number of successes in some area or length. The average number of successes in the area or length will define the parameter of the Poisson random variable.

Example 6.19

The telephone company is considering purchasing optical cable from Optica, Inc. The company wishes to replace approximately 100,000 feet of conventional cable with optical fiber. Since optical fiber is very difficult to repair, it is important that the number of optical cable defects are minimized. Optica claims that on average there is one defect per 200,000 feet of cable. What is the probability that the replaced cable will contain no defects?

Solution

Let $\lambda =$ the number of defects in 100,000 feet of optical cable.

Based on previous experience, we assume that the number of defects is approximated by a Poisson distribution with Poisson parameter

$$\lambda = \frac{100,000}{200,000} = \frac{1}{2} \ \text{(average number of defects per 100,000 ft of cable)}.$$

Using the tables provided in Appendix A, Table F,

$$P(X = 0) = 0.6065.$$

6.6   Exercises

Basic Concepts

1. How is the Poisson distribution similar to the binomial distribution?

2. What are the two conditions that an experiment must meet in order to be considered a Poisson random variable?

3. What are some uses of the Poisson probability model in business?

4. What is the Poisson probability distribution function?

5. What is the parameter of the Poisson probability model?

6. What is the expected value of a Poisson random variable? The variance? The standard deviation?

Exercises

7. Suppose that, on average, 5 students enrolled in a small liberal arts college have their automobiles stolen during the semester. What is the probability that exactly 2 students will have their automobiles stolen during the current semester?

8. The number of calls received by an office on Monday morning between 8:00 AM and 9:00 AM has a Poisson distribution with λ equal to 4.0.

    a. Determine the probability of getting no calls between eight and nine in the morning.

    b. Calculate the probability of getting exactly five calls between eight and nine in the morning.

    c. What will be the expected number of calls received by the office during this time period? What is the variance?

    d. Graph the probability distribution of the number of calls using values from Appendix A, Table F.

9. The director of a local hospital is studying the occurrence of medication errors. Medication errors are deemed to occur when a patient is given the wrong amount of medication or the wrong medication is given to a patient. Based on past experience, the director believes that medication errors follow a Poisson process with an average rate of 2 per week. (For the following problems, assume that 1 month = 4 weeks.)

    a. What is the probability that there are no medication errors in one week?

    b. What is the probability that there are no medication errors in one month?

    c. Find the average number of medication errors in one week.

    d. Find the average number of medication errors in one month.

    e. Find the standard deviation of the number of medication errors in one month.

    f. How likely is it that at least 4 medication errors will be observed in one month?

10. The number of weaving errors in a twenty foot by ten foot roll of carpet has a Poisson distribution with $\lambda = 0.1$.

   a. Using Appendix A, Table F, construct the probability distribution for the carpet.

   b. What is the probability of observing less than 2 errors in the carpet?

   c. What is the probability of observing more than 5 errors in the carpet?

11. A bank is evaluating their staffing policy to assure they have sufficient staff for their drive up window during the lunch hour. If the number of people who arrive at the window in a 15-minute period has a Poisson distribution with $\lambda = 5$, answer the following questions.

   a. How many people are expected to arrive during the lunch hour?

   b. What is the probability that no one will show up during the lunch hour of 12:00 PM to 1:00 PM?

   c. What is the probability that more than 6 people will show up in any 15-minute period?

12. An aluminum foil manufacturer wants to improve the quality of his product and is trying to develop a probability model for the flaws that occur in a sheet of foil. Assume that X, the number of flaws per square foot, has a Poisson distribution. If flaws occur randomly at an average of one flaw per 50 square feet, what is the probability that a box containing a 200 square foot roll will contain one flaw? More than one flaw?

13. A manufacturing company is concerned about the high rate of accidents that occurred on the production line last week. There were 6 accidents in the last week and this may require a report to be sent to the government agency for safety. Calculate the probability of 6 accidents occurring in a week when the average number of accidents per week has been 3.5. Assume that the number of accidents per week follows a Poisson distribution.

6.7 The Hypergeometric Distribution

The binomial and the hypergeometric random variables are very similar. Both random variables have only two outcomes in each trial of the experiment. They both count the number of successes in n trials of an experiment. The hypergeometric distribution differs from the binomial distribution in the lack of independence between trials, which also implies that the probability of success will vary between trials. In addition, hypergeometric distributions have finite populations in which the total number of successes and failures are known.

Because the binomial and hypergeometric distributions are closely related, a small change in an experiment can switch the distribution of the random variable. A binomial experiment, such as counting the number of red cards drawn in 8 draws from a deck with replacement, can easily be modified to a hypergeometric by not replacing the cards. Since there are 26 red cards (successes) and 26 black cards (failures), the probability of drawing a red card on the first draw is $\frac{26}{52}$ or $\frac{1}{2}$. If a red card is drawn on the first draw and not replaced, the probability of drawing a red card on the next draw is slightly less $\left(\frac{25}{51}\right)$ since there is one less red card in the deck. If the next card drawn is also red, then the probability of a red card on the third draw will be diminished to $\frac{24}{50}$. The probability distribution function of the hypergeometric distribution is given below.

Generally, the hypergeometric distribution can be used when one is sampling n items from a population of size N without replacement and it is known that there are A successes in the population (thus, $N-A$ failures in the population). Using the hypergeometric distribution, you can find the probability of x successes in the sample of size n.

Formula

Hypergeometric Probability Distribution Function

The **hypergeometric probability distribution function** is given by:

$$P(X = x) = \frac{{}_A C_x \, {}_{N-A} C_{n-x}}{{}_N C_n},$$

where

A = the total number of successes possible in the population,

N = the size of the total population,

n = the size of the sample drawn,

x = the number of successes in the sample of size n, and

$$\text{maximum of } (0, n + A - N) \le x \le \text{minimum of } (A, n).$$

There cannot be more successes than there are successes in the population, nor can there be more successes than the total size of the sample. Thus, the maximum value of X is the smaller of A and n.

Example 6.20

In a volatile stock market, suppose a mutual fund contains 15 stocks. Four of the stocks in the fund have positive gains, while 11 of the stocks are losing money. If two of the stocks from the mutual fund are chosen at random (without replacement), what is the probability distribution for the number of stocks in the sample that will have a positive gain?

Solution

The random variable under consideration is given as

$X =$ the number of stocks in the mutual fund that have positive gains.

The parameters of the distribution are

$A = 4$ (a success in this case is a money-making stock)

$N = 15$, and

$n = 2$.

The maximum value of X in this case is 2. Using the hypergeometric distribution function, we have the following.

$$P(X = 0) = \frac{_4C_0 \; _{15-4}C_{2-0}}{_{15}C_2} \approx 0.5238$$

$$P(X = 1) = \frac{_4C_1 \; _{15-4}C_{2-1}}{_{15}C_2} \approx 0.4190$$

$$P(X = 2) = \frac{_4C_2 \; _{15-4}C_{2-2}}{_{15}C_2} \approx 0.0571$$

The probability distribution is summarized in the following table.

Table 6.21 – Probability Distribution	
x	**P(X = x)**
0	0.5238
1	0.4190
2	0.0571

Expected Value and Variance

It is apparent that the hypergeometric distribution can be rather tedious to calculate, based on the calculations in Example 6.20. Since there are no tables for the distribution, determining an expected value using the definition $\left(\sum [xp(x)] \right)$ would consume a considerable amount of time. Fortunately, there is a simpler method.

Formula

Expected Value

The **expected value** of a hypergeometric random variable can be obtained using the following expression.

$$\mu = E(X) = n\left(\frac{A}{N}\right)$$

Formula

Variance

The **variance** of a hypergeometric random variable is given by the following expression.

$$\sigma^2 = V(X) = n\left(\frac{A}{N}\right)\left(1-\frac{A}{N}\right)\frac{(N-n)}{(N-1)}$$

Example 6.21

Compute the expected value and variance for the random variable defined in Example 6.20.

Solution

$$E(X) = 2\left(\frac{4}{15}\right) \approx 0.5333$$

$$\sigma^2 = V(X) = 2\left(\frac{4}{15}\right)\left(1-\frac{4}{15}\right)\frac{(15-2)}{(15-1)} \approx 0.3632$$

Thus, if the experiment were repeated many times, the average number of stocks in the mutual fund that had positive gains would be 0.5333. This leads us to believe that this particular fund is not very promising.

6.7   Exercises

Basic Concepts

1. How does the hypergeometric model differ from the binomial model?

2. What is the hypergeometric probability distribution function?

3. What are the parameters of the hypergeometric model?

4. How do you calculate the expected value of a hypergeometric random variable? The variance?

Exercises

5. Suppose a batch of 50 light bulbs contains 3 light bulbs that are defective. Let $X =$ the number of defective light bulbs in a random sample of 10 light bulbs (where the sample is taken without replacement).

    a. What probability model would be appropriate for describing the number of defective light bulbs in the sample?

    b. Find the expected number of defective bulbs.

    c. Find the standard deviation of the number of defective bulbs.

    d. Find the probability that at least 1 of the bulbs sampled will be defective.

    e. Find the probability that at most 2 of the bulbs sampled will be defective.

    f. Find the probability that more than 3 of the bulbs sampled will be defective.

6. A small electronics firm has 60 employees. Ten of the employees are older than 55. An attorney is investigating a client's claim regarding age discrimination. The attorney randomly selects 15 employees without replacement and records the number of employees over age 55.

    a. What probability model would be appropriate for describing the number of employees over age 55 in a sample of 15 selected without replacement?

    b. Find the average number of employees over age 55 in the sample.

    c. Find the standard deviation of the number of employees over age 55 in the sample.

    d. Find the probability that at least 2 of the employees selected will be over age 55.

    e. Find the probability that less than 2 of the employees selected will be over age 55.

    f. Find the probability that at most 4 of the employees will be over age 55.

7. A bank has to repossess 100 homes. Fifty of the repossessed homes have market values that are less than the outstanding balance of the mortgage. An auditor randomly selects 10 of the repossessed homes (without replacement) and records the number of homes that have market values less than the outstanding balance of the mortgage.

    a. Find the expected number of homes the auditor will find with market values less than the outstanding balance of the mortgage.

    b. Find the standard deviation of the number of homes the auditor will find with market values less than the outstanding balance of the mortgage.

    c. What is the probability that all of the audited homes will have outstanding balances in excess of the mortgage?

    d. What is the probability that none of the audited homes will have outstanding balances in excess of the mortgage?

8. A small liberal arts college in the Northeast has 200 freshmen. Eighty of the freshmen are female. Suppose thirty freshmen are randomly selected (without replacement).

    a. Find the expected number of females in the sample.

    b. Find the standard deviation of the number of females in the sample.

    c. Find the probability that none of the selected students will be female.

    d. Find the probability that all of the selected students will be female.

T   Discovering Technology

Using the TI-84 Plus Calculator

Binomial Distribution

Roll a single six-sided die 4 times and record the number of threes observed. What is the probability that you roll exactly 3 threes in the 4 rolls of the die? What is the probability that you roll less than 3 threes in the 4 rolls of the die?

1. Let X be the number of threes in 4 rolls of the die. Then X has a binomial probability distribution with $n = 4$ and $p = \dfrac{1}{6}$. To find the probability of rolling exactly 3 threes in 4 rolls of the die, we need to determine $P(X = 3)$. To do this, press **2ND**, and **VARS** to access the **DISTR** menu.

2. Since we are first interested in the probability of X being equal to a particular value, namely 3, select option **0:binompdf**. Press **ENTER**.

3. The syntax for entering the values into the calculator is binompdf(n, p, x). Thus, to find the probability of rolling exactly 3 threes in 4 rolls of the die, enter **binompdf(4,1/6,3)** and press **ENTER**.

4. The probability is displayed. Thus, $P(X = 3) \approx 0.0154$.

Figure 6.7

5. To find the probability of rolling less than 3 threes in 4 rolls of the die, we need to find $P(X \leq 2)$. To do this, again press **2ND**, and **VARS** to access the **DISTR** menu.

6. This time, we need the cumulative probability. Thus, select option **A:binomcdf**. Press **ENTER**.

7. The syntax here is the same, binomcdf(n, p, x), except the cdf function returns the cumulative binomial probability up to and including x. Thus, to find the probability of rolling less than 3 threes in 4 rolls of the die, enter **binomcdf(4,1/6,2)** and press **ENTER**.

8. The probability is displayed. Thus, $P(X \leq 2) \approx 0.9838$.

Figure 6.8

Poisson Distribution

For this exercise, use the information from Example 6.18.

1. Let X = the number of arrivals in a 15-minute period, and we know that X has a Poisson distribution with $\lambda = 5$. We want to find the probability that there are no arrivals in a 15-minute period, i.e. $P(X = 0)$. To do this, press **2ND**, and **VARS** to access the **DISTR** menu.

2. Since we are first interested in the probability of X being equal to a particular value, namely 0, select option **B:poissonpdf**. The syntax is poissonpdf(λ, x). Thus, to find the probability that there are no arrivals in a 15-minute interval, enter **poissonpdf(5,0)**. Press **ENTER**.

3. The probability is displayed. Thus, $P(X = 0) \approx 0.0067$. Compare this result with that in Example 6.18.

Figure 6.9

4. Next we want to find the probability that at least three people will use the machine in a fifteen minute interval, i.e. $P(X \geq 3)$. This is equivalent to finding $1 - P(X \leq 2)$. To find this probability, enter **1** and **−**, and then press **2ND**, and **VARS** to access the **DISTR** menu.

5. This time, we need the cumulative probability. Thus, select option **C:poissoncdf**. Press **ENTER**.

6. The syntax is the same, namely poissonpdf(λ, x), except the cdf function returns the cumulative poisson probability up to and including x. Enter **poissoncdf(5,2)**. Press **ENTER**.

7. The probability is displayed. Thus, $P(X \geq 3) \approx 0.8753$. Note that the answer obtained on the calculator is slightly different from that obtained in Example 6.18. This is due to rounding error when doing the calculations by hand using values from the table in Appendix A. Answers obtained using technology will be more accurate than those obtained using the tables in the Appendix.

Figure 6.10

Using Excel

Binomial Distribution

For this exercise, use the data from Example 6.16.

$$n = 10, p = 0.7$$

1. Set up the worksheet as shown in Figure 6.11.

	A	B	C	D	E
1	x	P(X=x)		x	P(X<=x)
2	0			0	
3	1			1	
4	2			2	
5	3			3	
6	4			4	
7	5			5	
8	6			6	
9	7			7	
10	8			8	
11	9			9	
12	10			10	

Figure 6.11

2. Using the binomial distribution function in Microsoft Excel, you can calculate both individual and cumulative probabilities for each number of successes. The function in Microsoft Excel that calculates binomial probabilities is

BINOM.DIST(number_s, trials, probability_s, cumulative)

where **number_s** = the number of successes,

trials = the number of trials,

probability_s = the probability of success, and

cumulative = a logical value which is either TRUE (Excel returns the cumulative probability) or FALSE (Excel returns the probability there are exactly **number_s** successes).

3. Begin with the individual probabilities. In cell B2, enter the formula to return the probability that there are 0 successes in 10 trials as follows.

=BINOM.DIST(0,10,0.7,FALSE)

Press **Enter**.

4. Continue in Column B by entering the appropriate parameters into the BINOM.DIST function for each individual number of successes. Excel will return the probabilities as shown in Figure 6.12.

	A	B	C	D	E
1	x	P(X=x)		x	P(X<=x)
2	0	5.9049E-06		0	
3	1	0.00013778		1	
4	2	0.0014467		2	
5	3	0.00900169		3	
6	4	0.03675691		4	
7	5	0.10291935		5	
8	6	0.20012095		6	
9	7	0.26682793		7	
10	8	0.23347444		8	
11	9	0.12106082		9	
12	10	0.02824752		10	

Figure 6.12

5. Next, calculate the cumulative probabilities. In cell E2, enter the formula to return the cumulative probability that there are no more than 0 successes in 10 trials.

$$=\textbf{BINOM.DIST(0,10,0.7,TRUE)}$$

Press **Enter**.

6. Continue in Column E by entering the appropriate parameters into the BINOM.DIST function for the cumulative probabilities for each number of successes. Excel will return the cumulative probabilities as shown in Figure 6.13. Compare these to the values we found in Example 6.16.

	A	B	C	D	E
1	x	P(X=x)		x	P(X<=x)
2	0	5.9049E-06		0	5.9049E-06
3	1	0.00013778		1	0.00014369
4	2	0.0014467		2	0.00159039
5	3	0.00900169		3	0.01059208
6	4	0.03675691		4	0.04734899
7	5	0.10291935		5	0.15026833
8	6	0.20012095		6	0.35038928
9	7	0.26682793		7	0.61721721
10	8	0.23347444		8	0.85069165
11	9	0.12106082		9	0.97175248
12	10	0.02824752		10	1

Figure 6.13

Poisson Distribution

Use Microsoft Excel to create the Poisson probability distribution in Table 6.19 ($\lambda = 0.3$).

1. Enter the labels x and $P(X = x)$ in cells A1 and B1, respectively.

2. In Column A, enter the values of x in Table 6.19, i.e. 0, 1, 2, 3, 4, and 5.

3. The Excel function for the Poisson distribution is:

POISSON.DIST(x, mean, cumulative)

where **x** = the number of successes in the sample,

mean = the mean number of successes, and

cumulative = a logical value which is either TRUE (Excel returns the cumulative probability) or FALSE (Excel returns the probability there are **x** successes).

4. In cell B2, enter the formula to find the probability of 0 successes when the mean number of successes is 0.3.

=POISSON.DIST(0,0.3,FALSE)

5. The probability is displayed as 0.7408 rounded to 4 decimal places. Continue to enter the POISSON.DIST formula in Column B for the remaining values of x. You will obtain the values displayed in Table 6.19 (see Figure 6.14).

	A	B
1	x	P(X=x)
2	0	0.740818221
3	1	0.222245466
4	2	0.03333682
5	3	0.003333682
6	4	0.000250026
7	5	1.50016E-05

Figure 6.14

Hypergeometric Distribution

$$A = 2, n = 16, N = 30$$

1. Set up the worksheet as shown in Figure 6.15.

	A	B
1	x	P(X = x)
2	0	
3	1	
4	2	

Figure 6.15

2. The Excel function for the hypergeometric distribution is:

HYPGEOM.DIST (sample_s, number_sample, population_s, number_pop, cumulative)
where **sample_s** = the number of successes in the sample,

number_sample = the size of the sample,

population_s = the number of successes in the population,

number_pop = the population size, and

cumulative = a logical value which is either TRUE (Excel returns the cumulative probability) or FALSE (Excel returns the probability there are **sample_s** successes).

3. Navigate your cursor to cell B2 and enter the formula as follows.

=HYPGEOM.DIST(0,16,2,30,FALSE)

Press **Enter**.

4. Cell B2 should now read 0.209195 which is the corresponding probability for 0 successes in this distribution. Apply the same formula to cells B3 and B4, replacing "sample_s" with **1** and **2**, respectively. The values for B3 and B4 should read 0.514943 and 0.275862.

	A	B
1	x	$P(X = x)$
2	0	0.209195
3	1	0.514943
4	2	0.275862

Figure 6.16

Using Minitab

Binomial Distribution

$n = 12, p = 0.1, x = 0, 1, 2, 3, 4$

1. Set up the worksheet in C1 and C2 as shown in Figure 6.17.

↓	C1	C2
	x	p(x)
1	0	
2	1	
3	2	
4	3	
5	4	

Figure 6.17

2. Press **Calc**, **Probability Distributions**, and **Binomial**.

3. Designate **Probability**. (Alternately, **Cumulative Probability**).

4. Complete the dialog box by entering **12** for Number of trials, **0.1** for Event probability, **C1** for Input column, and **C2** for Optional storage.

5. Press **OK** and read the output in C2.

↓	C1	C2
	x	p(x)
1	0	0.282430
2	1	0.376573
3	2	0.230128
4	3	0.085233
5	4	0.021308

Figure 6.18

Poisson Distribution

$$\lambda = 5,\ x = 0,\ 1,\ 2,\ 3,\ 4$$

1. Set up the worksheet in C1 and C2 as shown in Figure 6.19.

↓	C1	C2
	x	p(x)
1	0	
2	1	
3	2	
4	3	
5	4	

Figure 6.19

2. Press **Calc**, **Probability Distributions**, and **Poisson**.

3. Designate **Probability**. (Alternately, **Cumulative Probability**).

4. Complete the dialog box by entering **5** for Mean, **C1** for Input column, and **C2** for Optional storage.

5. Press **OK** and read the output in C2.

↓	C1	C2
	x	p(x)
1	0	0.006738
2	1	0.033690
3	2	0.084224
4	3	0.140374
5	4	0.175467

Figure 6.20

R | Chapter 6 Review

Key Terms and Ideas

- Random Variable
- Probability Distribution
- Discrete Random Variable
- Continuous Random Variable
- Discrete Probability Distribution
- Probability Distribution Function
- Expected Value of a Discrete Random Variable
- Variance of a Discrete Random Variable
- Probability Distribution Function
- Discrete Uniform Distribution
- Discrete Uniform Distribution Function
- Binomial Distribution
- Binomial Experiment
- Binomial Random Variable

- Binomial Probability Distribution Function
- Binomial Tables
- Expected Value of a Binomial Random Variable
- Variance of a Binomial Random Variable
- Standard Deviation of a Binomial Random Variable
- Poisson Distribution
- Poisson Random Variable
- Lambda
- Poisson Probability Distribution Function
- Poisson Tables
- Hypergeometric Distribution
- Hypergeometric Probability Distribution Function

Key Formulas

	Section
Expected Value of a Discrete Random Variable X $$\mu = E(X) = \sum \left[xp(x) \right]$$ where $p(x) = P(X = x)$	6.3
Variance of a Discrete Random Variable X $$\sigma^2 = V(X) = \sum \left[(x - \mu)^2 \, p(x) \right]$$	6.3
Computational Formula for Variance of a Discrete Random Variable X $$\sigma^2 = V(X) = \sum \left[x^2 p(x) \right] - \mu^2$$	6.3

Key Formulas (cont.)

	Section

Discrete Uniform Probability Distribution Function

$$P(X = x) = \frac{1}{n}$$

6.4

where n = the number of values that the random variable may assume.

Binomial Probability Distribution Function

$$P(X = x) = {}_nC_x p^x (1-p)^{n-x}$$

6.5

where ${}_nC_x = \dfrac{n!}{x!(n-x)!}$, n = the number of trials, and p = the probability of a success.

Expected Value of a Binomial Random Variable

$$\mu = E(X) = np$$

6.5

Variance and Standard Deviation of a Binomial Random Variable

$$\sigma^2 = V(X) = np(1-p)$$

$$\sigma = \sqrt{V(X)} = \sqrt{np(1-p)}$$

6.5

Poisson Probability Distribution Function

$$P(X = x) = \frac{e^{-\lambda}\lambda^x}{x!}$$

6.6

where $e = 2.71828...$ and λ = mean number of successes.

Expected Value of a Poisson Random Variable

$$\mu = E(X) = \lambda$$

6.6

Variance and Standard Deviation of a Poisson Random Variable

$$\sigma^2 = V(X) = \lambda$$

$$\sigma = \sqrt{V(X)} = \sqrt{\lambda}$$

6.6

Hypergeometric Probability Distribution Function

$$P(X = x) = \frac{{}_AC_x \,{}_{N-A}C_{n-x}}{{}_NC_n}$$

6.7

where A = the total number of successes possible, N = the size of the total population, and n = the size of the sample drawn.

Expected Value of a Hypergeometric Random Variable

$$\mu = E(X) = n\left(\frac{A}{N}\right)$$

6.7

Variance of a Hypergeometric Random Variable

$$\sigma^2 = V(X) = n\left(\frac{A}{N}\right)\left(1-\frac{A}{N}\right)\frac{(N-n)}{(N-1)}$$

6.7

AE   Additional Exercises

1. A statistics professor has determined the following probability distribution for *X*, the grade which a student will earn in a business statistics class.

Table for Exercise 1 – Grade Distribution		
Grade	*x*	*P(X = x)*
A	4.0	0.15
B	3.0	0.35
C	2.0	0.25
D	1.0	0.15
F	0.0	0.10

   a. What is the average grade that a student will earn in a business statistics class?

   b. Find the variance of the grades that students will earn in a business statistics class.

   c. Find the standard deviation of the grades which students will earn in a business statistics class.

   d. What is the probability that a student will earn a grade of 4.0?

   e. Find the probability that a student will earn a grade of at least 2.0.

   f. Find the probability that a student will earn a grade of at most 1.0.

   g. Find the probability that a student will earn a grade of more than 3.0.

2. The U.S. Department of Labor has issued a new set of guidelines governing certain work practices for employees. It estimates that only 20% of all firms will be subject to the new guidelines. To validate the estimate of the number of firms that will be affected by the new guidelines, the department randomly selects a sample of twenty firms for a study. Assuming their initial estimate of 20% is correct, answer the following questions.

   a. What is the probability that 1 or fewer of the sampled firms will be subject to the new rules?

   b. What is the probability that between 15 and 25 percent of the sampled firms will be subject to the rules?

   c. One of the directors in the department remarked he thought that ten firms out of the sample would be subject to the rules. If the initial estimate is correct, what is the chance of this occurring?

3. Historically, the probability that a library book will be returned in one week is *p* = 0.50. The head librarian for the University Staff Hospital library is monitoring a random sample of 10 books to determine if the historical proportion of the books returned within one week, 0.50, has changed. Assuming the historical return rate is still the same, answer the following questions.

   a. What is the probability that between four and six books will be returned in one week?

   b. What is the chance that eight or more books will be returned in one week?

   c. What is the probability that only one book will be returned in one week?

4. The number of fatalities resulting from automobile accidents for a 10-mile stretch of an interstate highway averages 1 per 100,000 automobiles. During a particular holiday weekend, 500,000 automobiles traveled over the 10-mile segment. Using a Poisson distribution, find the probability of each of the following.

   a. No fatalities

   b. 3 fatalities

   c. At least one fatality

5. Compute the mean and variance for the following random variables.

   a. The number of sixes obtained in 10 rolls of a single die.

   b. The number of hearts in a 13 card bridge hand. (Draw 13 cards from a standard deck without replacement.)

   c. The number of free throws made by a professional basketball player in his next 10 attempts. (Assume the player makes 88% of his free throws in the long run.)

   d. The number of cracked eggs selected when randomly selecting 5 eggs from a 12-egg carton containing 2 cracked eggs.

   e. The number of dots on the upper face when a single die is thrown.

6. A manufacturer of digital cameras knows that a shipment of 30 cameras sent to a large discount store contains eight defective cameras. The manufacturer also knows that the store will choose two of the cameras at random, test them, and accept the shipment if neither one is defective.

   a. Find the probability that at least one is defective.

   b. What is the probability that the shipment is accepted?

7. In a certain shipment of sixteen radios, four are defective. Eight of the radios are selected at random without replacement. What is the probability that at least one of the eight radios is defective?

8. According to the American Hotel and Lodging Association (AH&LA), women accounted for 31% of business travelers in the year 2009. Suppose that to attract these women business travelers, the AH&LA found that 80% of hotels offer hair dryers in the bathrooms. Consider a random and independent sample of 15 hotels.

   Source: American Hotel & Lodging Association

   a. Based on the information given, how many of the 15 hotels are expected to offer hair dryers in the bathrooms?

   b. Find the probability that all of the hotels in the sample offer hair dryers in the bathrooms.

   c. Find the probability that more than 5 but less than 9 of the hotels in the sample offer hair dryers in the bathrooms.

9. A carnival has a game of chance: a fair coin is tossed. If it lands heads, you win $1, and if it lands tails, you lose $0.50. How much should a ticket cost to play this game if the carnival wants to break even?

10. You are working on a multiple choice test which consists of 15 problems. Each of the problems has five answers, only one of which is correct. If you are totally unprepared for the test and are guessing, what is the probability that your first correct answer is within the first fifteen problems?

11. An automobile manufacturer is always trying to improve the quality of its vehicles. Assume that the number of defects per vehicle follows a Poisson distribution. If these defects occur randomly at an average rate of five per vehicle, what is the probability that a randomly selected vehicle will have at least one defect?

12. When proofreading a statistics textbook, one can expect to find a number of errors, whether they are typographical, symbolic, or even incorrect mathematical calculations. On average, a statistics textbook will contain 30 errors. What is the probability that when proofreading a text, one finds at least three errors? Assume that the number of errors found follows a Poisson distribution.

13. While on a shopping spree, you randomly select five portable music players from an electronics store that sells 20 portable music players. Of these 20 music players, 12 will last beyond the 1-year limited warranty and will not need to be replaced or repaired. What is the probability that at least three of the five portable music players selected will not last beyond the limited warranty period without needing to be replaced or repaired?

14. A jeweler was given a collection of twelve diamonds, of which three were synthetic (fake). If the jeweler selected two of these diamonds at random (without replacement), what is the probability that neither jewel is found to be synthetic?

15. L-Mart Inspections is a building inspection company. There were ten new commercial construction buildings completed in the last month and the sites are now available for inspection. L-Mart plans to inspect some of the new constructions for code violations and believes that half of the buildings will have violations.

    a. What probability model would be appropriate for describing the number of buildings in the sample that have code violations? Explain your answer.

    b. If L-Mart randomly selects four buildings to inspect, what is the probability that three of the buildings will have violations?

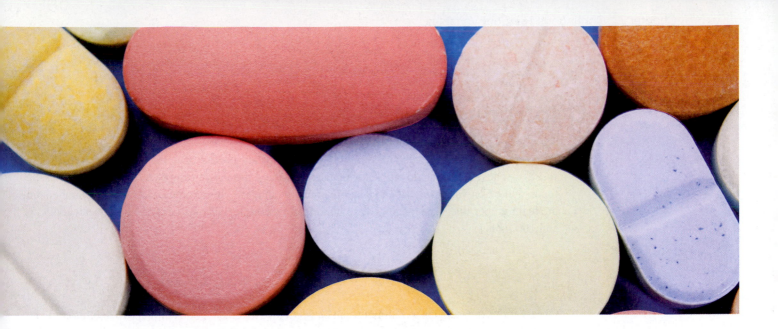

Chapter Seven
Continuous Random Variables

Introduction

In the previous chapter random variables were primarily counts of some phenomenon called a *success*. These counts could only take on discrete values, usually starting at zero. In this chapter the focus will be on variables that can take on any value in some continuum, i.e. a range of numbers on the real number line. The range of adult heights, for example, lies on a continuum between about 26 and 100 inches.

Observations measured in a continuum can be very close together. For example, two heights could be

68.234175245987149399 inches and

68.234175245987149398 inches.

As a practical matter, it is hard to imagine a situation that would require the knowledge of a person's height to 18 decimal places. Variables like heights and weights are **continuous random variables** even though heights are usually given to the nearest inch or centimeter and weights to the nearest pound or kilogram. Variables measured in these units give the appearance of being discrete, yet they are continuous.

One of the striking differences between discrete and continuous variables concerns the way probability is defined. In a probability distribution for a discrete random variable, each possible outcome of the random variable is assigned its own probability. However, for continuous random variables, there are infinitely many outcomes, and each has *no* probability. Outcomes of a continuous random variable do not have probability assigned to any one point, because there are simply too many points. If we attempted to assign each value even an infinitesimal probability, the sum of all probabilities would exceed one. Thus, for continuous random variables, probability is only assigned to intervals.

7.1   The Uniform Distribution

There are two types of uniform distributions, discrete and continuous. You already studied the discrete uniform distribution in Chapter 6. Recall that for the discrete uniform distribution, we assign the same probability to each possible value of the random variable. Conceptually, both distributions distribute probability evenly across a sample space. For the continuous uniform distribution, the probability is spread out over some range from a to b, as shown in Figure 7.1.

Continuous Uniform Distribution

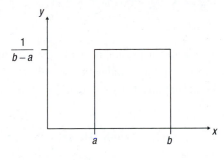

Figure 7.1

Continuous random variables do not have probability distribution functions. Instead they have **probability density functions**, which are denoted by $f(x)$. The probability density function is used to calculate probabilities for continuous random variables. The probability density function for the uniform random variable, its expected value (mean), and its standard deviation are given in the following box. The parameters of the probability density function are the minimum and maximum values of the uniform random variable, and are referred to as a and b, respectively.

Formula

Uniform Probability Density Function

The **uniform probability density function** is as follows.

$$f(x) = \begin{cases} \dfrac{1}{b-a} & \text{for } a \leq x \leq b \\ 0 & \text{otherwise} \end{cases}$$

The mean and standard deviation are given by the following expressions.

$$\mu = \frac{a+b}{2} \quad \text{and} \quad \sigma = \frac{b-a}{\sqrt{12}}$$

When the uniform probability density function is graphed, it produces a rectangle or square. The probability of observing a random variable in some interval is expressed as the area under the density function associated with the interval. Since the density function for the uniform distribution produces a rectangle, calculating the probability of an interval is as simple as calculating the area of a rectangle, which does not require complicated geometry.

Example 7.1

The Blacksburg Bus Company is interested in the wait time of its passengers between 8:00 AM and 8:00 PM. Suppose that the wait time follows a uniform distribution with the minimum wait time being one minute and the maximum wait time of 15 minutes.

a. What is the probability that a passenger will wait between 10 and 15 minutes?

b. What is the expected wait time between 8:00 AM and 8:00 PM?

c. What is the standard deviation of the wait time between 8:00 AM and 8:00 PM?

Solution

a. Let X = the number of minutes a passenger waits for the bus. The probability of a random variable is given by the area under the probability density function. For the uniform distribution, the probability is simply calculated using the formula for the area of a rectangle.

$$\text{Area} = \text{Width} \cdot \text{Height}$$

According to the uniform probability density function $f(x) = \dfrac{1}{b-a}$. Therefore,

$$\text{Height} = \frac{1}{b-a} = \frac{1}{15-1} = \frac{1}{14}.$$

We are interested in the probability that a passenger will wait between 10 and 15 minutes for the bus, so

$$\text{Width} = 15 - 10 = 5, \text{ and}$$

$$\text{Area} = \text{Width} \cdot \text{Height} = 5\left(\frac{1}{14}\right) = \frac{5}{14}.$$

Since the probability of observing a uniform random variable is the area under the density function associated with the interval, we calculate the probability of a passenger waiting between 10 and 15 minutes as follows (see Figure 7.2).

$$P(10 \leq X \leq 15) = \frac{5}{14}$$

Continuous Uniform Distribution

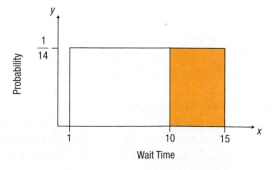

Figure 7.2

b. The expected value of a continuous uniform random variable is given by

$$E(X) = \mu = \frac{a+b}{2} = \frac{(1+15)}{2} = 8.$$

Therefore, the expected wait time for a passenger is 8 minutes.

c. The standard deviation of a continuous uniform random variable is given by

$$\sigma = \frac{b-a}{\sqrt{12}} = \frac{15-1}{\sqrt{12}} \approx 4.0415.$$

Therefore, the standard deviation for the wait time of a passenger for a bus operated by the Blacksburg Bus Company is 4.0415 minutes.

What Do Data From a Uniform Random Variable Look Like?

While the density function for the uniform distribution has a flat top, a histogram of data from a uniformly distributed random process will not be perfectly flat. Suppose that we generated 100 observations from a random process that was uniformly distributed between 2 and 15. Clearly, the frequency of each category is not identical (see Figure 7.3).

Figure 7.3

However, if we were to generate 1000 observations, the distribution would begin to level (see Figure 7.4).

Figure 7.4

343

7.1   Exercises

Basic Concepts

1. Probability is defined differently for discrete and continuous random variables. Describe this difference.

2. What is a probability density function?

3. How is the continuous uniform distribution different from the discrete uniform distribution?

4. What is the uniform probability density function?

5. Describe the shape of the density function for a uniform distribution.

Exercises

6. Suppose a continuous random variable is uniformly distributed between 10 and 70.

   a. What is the mean of the distribution?

   b. What is the standard deviation of the distribution?

   c. What is the probability that a randomly selected value will be above 45?

   d. What is the probability that a randomly selected value will be less than 30?

   e. What is the probability that a randomly selected value will be between 25 and 50?

   f. Find the probability that a randomly selected value will exactly equal 35.

7. Polar Bear Frozen Foods manufactures frozen French fries for sale to grocery store chains. The final package weight is thought to be a uniformly distributed random variable. Assume X, the weight of French fries has a uniform distribution between 57 ounces and 63 ounces.

   a. What is the mean weight for a package?

   b. What is the standard deviation for the weight of a package?

   c. What is the probability that a store will receive a package weighing less than 59 ounces?

   d. What is the probability that a package will contain between 60 and 63 ounces?

   e. What is the probability that a package will contain more than 62 ounces?

   f. Find the probability that a package will contain exactly 60 ounces.

8. The annual increase in height of cedar trees is believed to be distributed uniformly between six and eleven inches.

    a. Draw a picture of the distribution of growth in height of cedar trees.

    b. What is the mean growth per year?

    c. What is the standard deviation of the growth per year?

    d. What is the probability that a randomly selected cedar tree will grow between 9 and 10 inches in a given year?

    e. Find the probability that a randomly selected cedar tree will grow less than 8 inches in a given year.

    f. Find the probability that a randomly selected cedar tree will grow more than 9 inches in a given year.

    g. Find the probability that a randomly selected cedar tree will grow exactly 7 inches in a given year.

9. A particular employee arrives to work sometime between 8:00 AM and 8:30 AM. Based on past experience the company has determined that the employee is equally likely to arrive at any time between 8:00 AM and 8:30 AM.

    a. On average, what time does the employee arrive?

    b. What is the standard deviation of the time at which the employee arrives?

    c. If a call comes in for the employee at 8:10 AM, find the probability that the employee will be there to take the call.

    d. Find the probability that the employee will arrive between 8:20 AM and 8:25 AM.

    e. Find the probability that the employee will arrive after 8:15 AM.

    f. Find the probability that the employee will arrive at exactly 8:10 AM.

7.2   The Normal Distribution

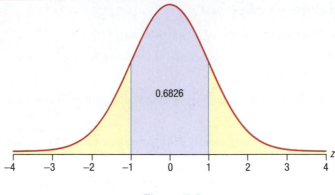

Figure 7.5

The **normal distribution**, originally called the Gaussian distribution, was named after Karl Gauss who published a work in 1833 describing the mathematical definition of the distribution. Gauss developed this distribution to describe the error in predicting the orbits of planets.

Normal distributions are all bell-shaped, but the bells come in various shapes and sizes. Since all normal distributions are symmetric, the mean, median, and mode are all equal.

Although normally distributed random variables can range in value from negative infinity to positive infinity, values that are a great distance from the mean rarely occur. You may recall that when we discussed box plots, these values were called outliers.

Properties

Properties of the Normal Distribution

1. The normal distribution is symmetric. That is, the curve's shape to the left of the mean is the mirror image of the curve's shape to the right of the mean.

2. The highest point on the normal curve is located at the mean, which is also the location of the median and the mode of the distribution.

3. The area under the curve of the normal distribution equals 1.

4. Due to symmetry, the area to the right of the mean equals the area to the left of the mean, and each of these areas equals 0.5.

5. The shape of the normal distribution is defined by its two parameters, the mean (μ) and the standard deviation (σ).

In Figure 7.5 the shaded area represents the probability of being within $\pm 1\sigma$ of the mean. Regardless of the values of the mean and standard deviation of the normal distribution, the area under the curve and the probability of being within one standard deviation ($\pm 1\sigma$) of the mean equals 0.6826.

Figure 7.6 illustrates the area under the curve within two standard deviations of the mean. The probability of being within ±2σ of the mean equals 0.9544 for every normal distribution.

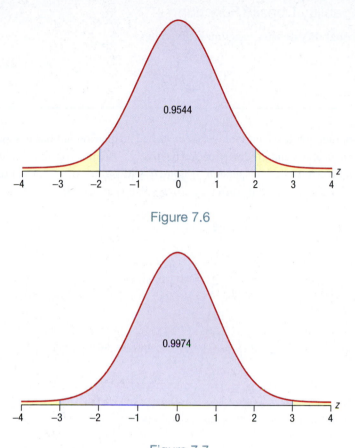

0.9544

Figure 7.6

0.9974

Figure 7.7

Figure 7.7 illustrates the area under the curve within three standard deviations of the mean. As you can see, virtually all of the area under the curve is within three standard deviations of the mean. The probability of being within ±3σ of the mean equals 0.9974.

Since the empirical rule given in Chapter 4 is based on the normal distribution, these results are identical to the empirical rule frequencies.

Although a normal distribution has a bell shape, a bell shape does not imply a normal distribution. A number of non-normal distributions with bell shapes will be introduced in later chapters. In fact the word *normal* can be somewhat misleading. The name suggests that the distribution is a fact of nature. It is not. However, many variables seem to possess a shape that resembles the normal distribution.

The normal distribution is the preeminent distribution used in the statistical theory we will examine. Many statistical inference procedures, either directly or indirectly, have been developed based on the normal distribution. These procedures usually assume that the population from which a random sample is drawn is normally distributed.

Like other theoretical distributions, a normal distribution is completely defined by its probability density function.

Astronomy and the Normal Distribution

Although DeMoivre derived the equation of the normal distribution as a consequence of problems involving the binomial distribution, Gauss and LaPlace were inspired by their desire to predict the location of planets, stars, meteors, and comets. Their predictions of the location of celestial bodies produced errors and they needed to describe those errors. In the 19th century the normal distribution was referred to as the astronomers' error law. Much of the early scientific use of the normal distribution was the analysis of errors.

Formula

Normal Probability Density Function

The **normal probability density function** is given by:

$$f(x) = \frac{1}{\sigma\sqrt{2\pi}} e^{-\frac{(x-\mu)^2}{2\sigma^2}}$$

At first glance, this function does not look very *normal*! The distribution has two parameters, μ and σ, which are the mean and standard deviation, respectively. The mean defines the location, and the standard deviation determines the dispersion. Figure 7.8 illustrates three normal distributions with identical standard deviations. The only difference in the distributions is the central location, the mean.

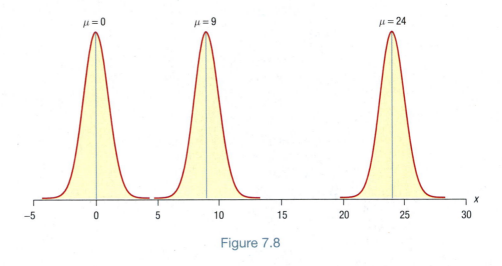

Figure 7.8

In Figure 7.9, there are two distributions with identical means, but with different standard deviations. Changing the standard deviation parameter can have rather significant effects on the shape of the distribution.

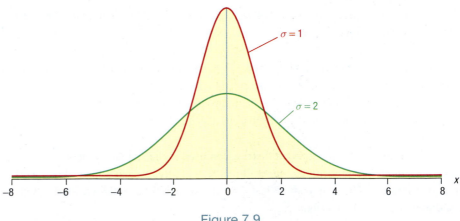

Figure 7.9

Looking at Data from Normal Distributions

The normal distribution is a theoretical construct with a bell shape. Therefore it would not be unreasonable to expect data drawn from a normal population to exhibit the bell-shaped characteristic. In two small samples taken from a normal population, shown in Figures 7.10 and 7.11, the data do show a faint resemblance of a bell shape. But as you can see, the shapes of the histograms developed from these small samples are somewhat unpredictable, even though the bell-shaped pattern is to some extent apparent.

Figure 7.10

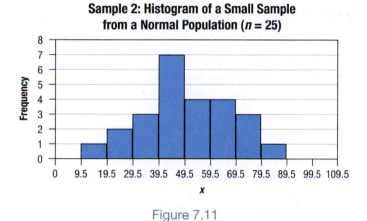

Figure 7.11

For large samples, the representation of the bell curve is usually more visible. While the large sample ($n = 200$) certainly is not a perfect bell curve, it is recognizable (see Figure 7.12).

Figure 7.12

Bringing the Normal Distribution Out of the Closet: Adolphe Quetelet 1796 – 1874

Quetelet was born in Belgium and received his doctorate from the University of Ghent in 1819. In 1824 he went to Paris for three months to study astronomy and was exposed to the theory of probability by LaPlace and Fourier. Quetelet, however, became deeply interested in social science and believed the astronomers' error law and other laws of physics could be applied to phenomena in the social world. In 1844 Quetelet announced that the astronomers' error law (the normal distribution) could be applied to human features such as height, weight, and girth. He believed that models that astronomers were using could eventually be used in a new science of "social mechanics." Quetelet was instrumental in popularizing the normal distribution in disciplines other than astronomy.

Using the probability density function to determine the probability of some interval would be complicated. Fortunately, there is an easier way. A special normal distribution, called the standard normal, can be used to determine probabilities for any normal random variable. The standard normal distribution will be discussed in Section 7.3.

7.2   Exercises

Basic Concepts

1. How was the normal distribution developed?

2. Are the normal and uniform distributions probability models?

3. List the properties of the normal distribution.

4. What is the shape of the normal distribution?

5. What are the parameters of the normal distribution?

6. If the variance of a normal distribution is constant, what affect will changes in the mean have on the distribution?

7. If the mean of a normal distribution is constant, what effect will changes in the standard deviation have on the distribution?

Exercises

8. Sketch a normal curve and mark each of the following on the x-axis.

    a. μ

    b. $\mu + \sigma$

    c. $\mu - \sigma$

9. Sketch a normal curve and use labels to illustrate the empirical rule.

10. Sketch three normal curves on a single axis that have the same standard deviation but different means.

11. Sketch three normal curves on a single axis that have the same mean, but different standard deviations.

7.3   The Standard Normal Distribution

Given that the normal distribution is a function of two continuous parameters μ and σ, there are an infinite number of combinations for μ and σ, and thus, an infinite number of normal distributions. The **standard normal distribution** in Figure 7.13 is a special version of the normal distribution.

Definition

The **standard normal distribution** is a normal distribution with a mean of zero and a standard deviation of one.

$$\mu = 0 \text{ and } \sigma = 1$$

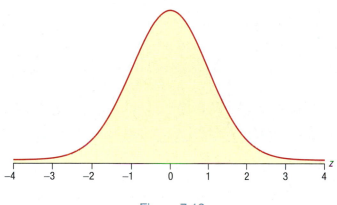

Figure 7.13

The standard normal distribution, also called the z-distribution, provides a basis for computing probabilities for all normal distributions. The technique used to convert any normal random variable into a standard normal random variable is called "standardizing" the random variable and was discussed earlier in Chapter 4.

Appendix A, Tables A, B, and C contain probability calculations for various areas under the standard normal curve. Specifically, Appendix A, Tables A and B provide the probability that a standard normal random variable will be less than a specified value. For example, to compute the probability that a standard normal random variable will be less than 1 (see Figure 7.14), look up the value 1.00 in Table B. The table value of 0.8413 is the area under the curve between negative infinity and 1, which is also the probability that the random variable will assume a value in that interval.

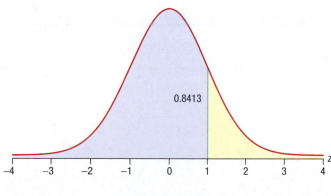

Figure 7.14

Note: We will primarily use Tables A and B in the following examples to determine areas and probabilities under the standard normal curve, but note that Table C (areas between 0 and z) could also be used.

The z-distribution will be used throughout the remainder of this text. Thus, it is important to comprehend the use of the z-tables in the determination of probabilities for a variety of problems. Because of the manner in which the tables are constructed, some types of problems will require manipulation of the table values.

Example 7.2

Compute the probability that a standard normal random variable is less than 1.27.

Solution

Drawing a picture, even when the problem is rather simple, is a good idea. Remember that the probability is represented by the area under the standard normal curve.

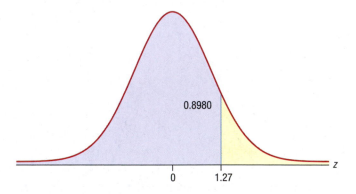

0.8980

0   1.27 z

Figure 7.15

Determining the area under standard normal curve to the left of a particular value requires little effort since the tables are constructed to give the cumulative probabilities. That is, the table gives probabilities that the random variable z is less than (or less than or equal to) some value (i.e. $P(z < z_0)$ where z_0 is the number of standard deviations above or below the mean). In this case, the construction of Table B exactly matches the kind of interval we are examining. Thus, merely looking up the value corresponding to 1.27 in the table is sufficient to obtain the probability.

z	0.00	0.01	...	0.06	0.07
0.0	0.5000	0.5040		0.5239	0.5279
0.1	0.5398	0.5438		0.5636	0.5675
⋮					
1.1	0.8643	0.8665		0.8770	0.8790
1.2	0.8849	0.8869		0.8962	0.8980
1.3	0.9032	0.9049		0.9131	0.9147

$$P(z < 1.27) = 0.8980$$

Example 7.3

Determine the probability that a standard normal random variable is between −1.08 and 0.

Solution

First, draw a picture.

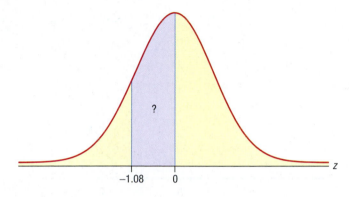

Figure 7.16

In this case, we cannot simply look up the value in the table to obtain the probability of interest. Understanding that the table gives us cumulative probabilities, if we find the probability that z is less than 0 and then subtract the probability that z is less than −1.08, we will get the probability that z is between −1.08 and 0.

Because the standard normal distribution is symmetric and the total area under the curve is equal to 1, we know that the probability that z is less than 0 is equal to 0.5 (since 0 is the mean or the central value). We can find $P(z < -1.08)$ to be 0.1401 from looking up −1.08 in Appendix A, Table A. Thus, to find the area between −1.08 and 0 we subtract the area to the left of −1.08 from 0.5 as follows.

$$P(-1.08 < z < 0) = P(z < 0) - P(z < -1.08) = 0.5 - 0.1401 = 0.3599$$

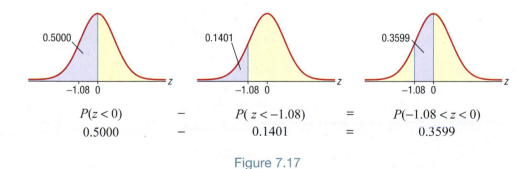

Figure 7.17

Example 7.4

What is the probability that a standard normal random variable will be between 1 and 2?

Solution

Again, first draw a picture.

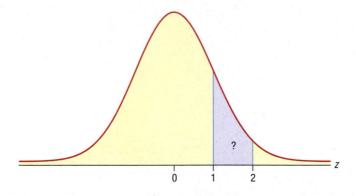

Figure 7.18

Because the Table B gives us cumulative probabilities, we can find the probability that z is less than 2 and then subtract the probability that z is less than 1, yielding the probability that z is between 1 and 2. Thus, we have the following. Figure 7.19 illustrates the areas involved in this calculation.

$$P(1 < z < 2) = P(z < 2) - P(z < 1) = 0.9772 - 0.8413 = 0.1359$$

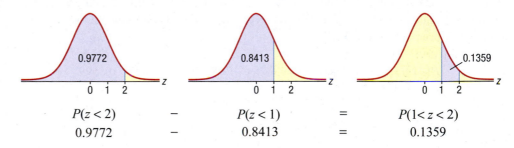

Figure 7.19

Again, drawing the picture proves to be invaluable in this example.

Example 7.5

Given that z is a standard normal random variable, find the value of z for each situation.

a. The area to the left of z is 0.9147.

b. The area between 0 and z is 0.3665.

c. The area to the left of z is 0.1469.

d. The area to the right of z is 0.7967.

Solution

a. First, draw a picture.

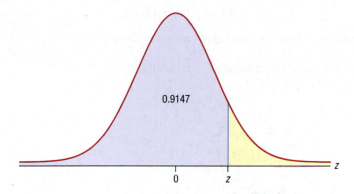

Figure 7.20

Note that this problem is slightly different from the previous one. In Example 7.4, you were asked to find a probability, given that you know the value of z. In this example, you are given a probability and asked to find the corresponding value of z. Recall that Appendix A, Table B gives you the cumulative probability of the area less than some value of z.

In order to find the value of z, look in the body of Appendix A, Table B and find the probability value 0.9147. Once you've found the value (the probability), determine the corresponding value of z. In this case, the value of z is 1.37. So,

$$P(z < 1.37) = 0.9147$$

and the value of z is 1.37 with the area to the left of it being 0.9147.

b. Again, draw a picture.

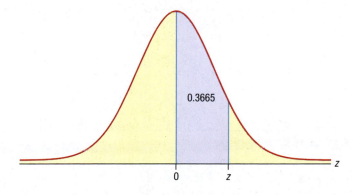

Figure 7.21

Recall that the table gives us cumulative probabilities. Since the area between 0 and z is 0.3665, if we write this as a cumulative probability, we need to add the area to the left of 0. Thus, the area to the left of z is 0.8665. This is the probability (less than some value of z) that is provided in Table B. Therefore, find 0.8665 in the body of Table B and locate the corresponding value of z. In this case,

$$P(z < 1.11) = 0.8665.$$

So, the value of z with the area 0.3665 between 0 and z is 1.11.

c.   Just as in parts **a.** and **b.**, a picture can be helpful.

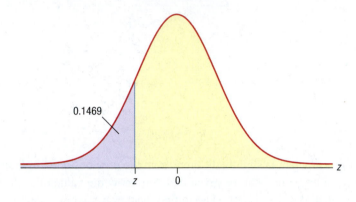

Figure 7.22

Please note that the value of z is to the left of 0. Thus, the value of z is going to be negative. Note that the area to the left of z represents the cumulative probability (in Figure 7.22). So, to find the value of z, we only need to find 0.1469 in the body of Appendix A, Table A. The value of z with the area 0.1469 to the left of it is −1.05. That is, $P(z < -1.05) = 0.1469$.

d.   Once again, a picture can be very helpful.

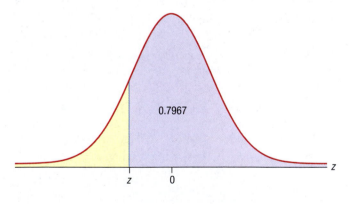

Figure 7.23

Note that from the picture, we have the area to the right of z. However, we know that the total area under the curve is 1. Thus, if the area to the right of z is 0.7967, then the area to the left of z is $1 - 0.7967 = 0.2033$. From the picture, it is clear that if we find 0.2033 in the body of Appendix A, Table A, the corresponding value of z is the value we are interested in. This value of z is −0.83. Therefore, the value of z with the area 0.7967 to the right is −0.83.

Standardizing Normal Random Variables

> ## Formula
>
> ## Standardizing a Normal Random Variable
>
> The following formula can transform any normal random variable into a **standard normal random variable**, z.
>
> $$z = \frac{x - \mu}{\sigma}$$
>
> where x is a normal random variable with mean μ and standard deviation σ.

If we look at the individual pieces, exactly how the transformation works is not very mysterious. First, the numerator, $x - \mu$, centers the z-distribution around zero. By subtracting the mean of the random variable from each data value, the mean of the resulting random variable will be zero. A short example illustrates this point. Suppose that a population contained the following data values shown in Table 7.1.

Table 7.1	
Data Set A	**Data Set A − 6**
1	$1 - 6 = -5$
5	$5 - 6 = -1$
6	$6 - 6 = 0$
12	$12 - 6 = 6$
Mean = 6	Mean = 0

Data Set A has a mean of 6. If 6 is subtracted from each of the data values, the resulting deviations are shown in the second column of the table. The deviations have a mean of zero. Essentially the location of the data set has been shifted to zero. The interrelationship of the data points to one another has not changed. Try this experiment on larger sets of data to convince yourself that subtracting the mean from each value of a data set will produce a data set that always has a mean of zero.

The standard deviation of Data Set A is approximately 3.9370. Let's standardize each data value in Data Set A (see Table 7.2). The resulting z-values indicate how far the data values in Table 7.1 are from the mean, measured in standard deviation units. The first value in Table 7.2 indicates that 1 is -1.27 standard deviation units from the mean. The mean and standard deviation of the transformed values in Table 7.2 are zero and one, respectively. You can verify that the mean and standard deviation of the z-scores are actually zero and one. Compute the population standard deviation rather than the sample standard deviation.

Table 7.2 – Standardizing Data Set A		
Data	**Formula**	**Value of z**
1	$\dfrac{1-6}{3.9370} \approx$	−1.27
5	$\dfrac{5-6}{3.9370} \approx$	−0.25
6	$\dfrac{6-6}{3.9370} \approx$	0
12	$\dfrac{12-6}{3.9370} \approx$	1.52

Now that we have seen how to convert any normal random variable into a standard normal random variable, let's apply it.

Example 7.6

Suppose that the times it takes for an accountant to process payroll are normally distributed with a mean of 10 minutes and a standard deviation of 20 minutes. Find the probability that the payroll processing time on a randomly selected payroll date will take between 10 minutes and 40 minutes.

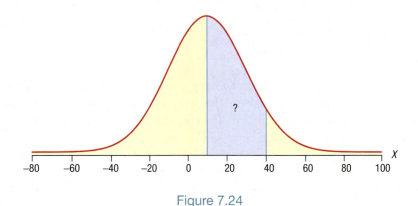

Figure 7.24

Solution

Suppose X, the payroll processing time, is a normally distributed random variable with a mean of 10 minutes and a standard deviation of 20 minutes.

Standardizing the random variable X yields the following.

$$P(10 < X < 40) = P\left(\frac{10-10}{20} < \frac{x-\mu}{\sigma} < \frac{40-10}{20}\right) = P(0 < z < 1.5)$$

Note that for each argument in the above probability statement, we subtracted the mean and divided by the standard deviation.

Once the problem has been converted to a problem involving z (see Figure 7.25), the appropriate probability can be determined from the standard normal table in Appendix A, Table B.

We find the probability that z is less than 1.5 and subtract 0.5 (the probability that z is less than 0) as follows.

$$P(0 < z < 1.5) = P(z < 1.5) - P(z < 0) = 0.9332 - 0.5 = 0.4332$$

Thus, the probability that the payroll processing time on a randomly selected payroll date will take between 10 and 40 minutes is 0.4332.

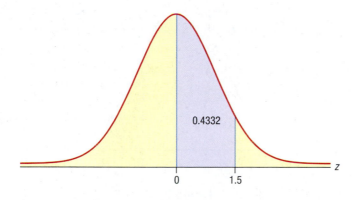

0.4332

0   1.5

Figure 7.25

Example 7.7

Continuing Example 7.6, find the probability that a randomly selected payroll date will take more than 30 minutes to process by the accountant.

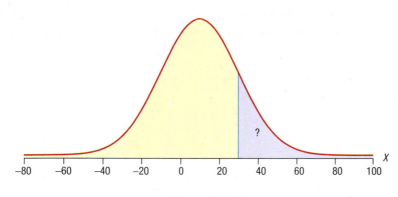

?

−80   −60   −40   −20   0   20   40   60   80   100

Figure 7.26

Solution

Suppose X, the payroll processing time, is a random variable with a mean of 10 minutes and a standard deviation of 20 minutes. Standardizing the random variable, we have the following.

$$P(X > 30) = P\left(z > \frac{30 - 10}{20}\right)$$
$$= P(z > 1)$$

To find the probability that the accountant will take more than 30 minutes to process the payroll, we need to find the probability that the accountant will take less than 30 minutes, and subtract this from 1. Using Appendix A, Table B, we have the following.

$$P(z > 1) = 1 - P(z < 1)$$
$$= 1 - 0.8413$$
$$= 0.1587$$

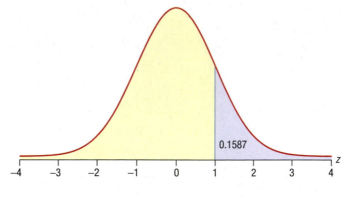

Figure 7.27

Note that the *X*-value of 30 transformed into the *z*-value of 1. In other words, 30 is one standard deviation away from the mean.

Example 7.8

Suppose that a national testing service gives a test in which the results are normally distributed with a mean of 400 and a standard deviation of 100. If you score a 644 on the test, what percentage of the students taking the test exceeded your score?

Solution

Let X = a student's score on the test.

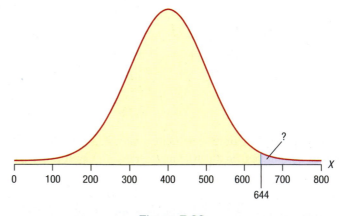

Figure 7.28

The first step is to standardize the random variable. Then, using Appendix A, Table B, find the appropriate probability.

$$P(X > 644) = P\left(z > \frac{644 - 400}{100}\right)$$

$$= P(z > 2.44)$$

$$= 1 - P(z < 2.44)$$

$$= 1 - 0.9927 = 0.0073$$

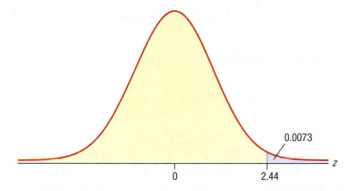

Figure 7.29

Thus, only 0.73% of the students scored higher than your score of 644.

Example 7.9

Using the information provided in Example 7.8, what score must a student get to be in the 90th percentile?

Solution

Recall that X = a student's score on the test follows a normal distribution with a mean of 400 and a standard deviation of 100. As with the examples before, a picture can be very helpful.

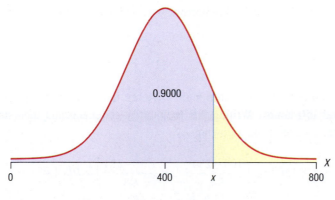

Figure 7.30

From the picture above, it can be seen that we want to find the value of X (the student's score) that represents the 90th percentile (i.e., the student scored the same or better than 90% of other students taking the test). We would write $P(X \leq x) = 0.90$.

If we standardize the probability statement, we get the following.

$$P\left(z \le \frac{x-\mu}{\sigma}\right) = 0.90$$

Next, substitute the values for the mean and standard deviation to get the following.

$$P\left(z \le \frac{x-400}{100}\right) = 0.90$$

Note that we can rewrite the above probability statement as

$$P(z \le z_0) = 0.90 \text{ where } z_0 = \frac{x-400}{100}.$$

Therefore, we need to find the value of z with the area 0.90 to the left of it. To do this, we look in the body of the standard normal table (Table B in Appendix A) for the value 0.9000. It's not often that we can find the exact probability in the body of the table. When this is the case, we find the closest probability along with the corresponding value of z. In this case, the value of $z = 1.28$.

Thus,

$$1.28 = \frac{x-400}{100}.$$

Solving for x, we get $x = 528$.

Therefore, a student who scores 528 on the test will be in the 90th percentile.

Example 7.10

Suppose that for 132 space shuttle missions, the flight duration follows a normal distribution with a mean of 234 hours and a standard deviation of 94 hours. If you just read in the newspaper that a space shuttle will be launched tomorrow, what is the probability that the duration of the flight will be between 100 and 150 hours?

Solution

Let X = the duration of a space shuttle flight. We are interested in the probability that X is between 100 and 150 hours. Writing this probability statement and then standardizing the random variable X, we have the following.

$$P(100 < X < 150) = P\left(\frac{100-234}{94} < z < \frac{150-234}{94}\right) = P(-1.43 < z < -0.89)$$

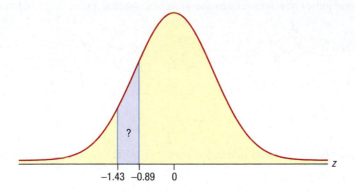

Figure 7.31

To find the probability that z is between -1.43 and -0.89, we will need to find the probability that z is less than -0.89 and subtract the probability that z is less than -1.43. Using Table A in Appendix A, we have the following.

$$P(-1.43 < z < -0.89) = P(z < -0.89) - P(z < -1.43) = 0.1867 - 0.0764 = 0.1103$$

Thus, there is approximately an 11% chance that the flight will last between 100 and 150 hours.

7.3   Exercises

Basic Concepts

1. What is the standard normal distribution? What are the parameters of the distribution?

2. Why is the standard normal distribution important?

3. Describe the connection between the z-transformation and the standard normal random variable.

Exercises

4. What proportion of the area under the standard normal curve falls between the following z-values?

    a. 0 and 0.67

    b. 0 and 1.645

    c. 0 and 1.96

    d. 0 and 2.575

5. What proportion of the area under the standard normal curve falls between the following *z*-values?

   a. −0.67 and 0

   b. −1.645 and 0

   c. −1.96 and 0

   d. −2.575 and 0

6. What proportion of the area under the standard normal curve falls between the following *z*-values?

   a. −0.85 and 0.85

   b. −0.55 and 0.55

   c. −1.56 and 1.98

   d. −2.23 and 2.96

7. What proportion of the area under the standard normal curve falls between the following *z*-values?

   a. −0.97 and 0.97

   b. −0.54 and 1.82

   c. −1.95 and 2.28

   d. −2.89 and 1.59

8. Using the standard normal tables in Appendix A, determine the following probabilities. Sketch the associated areas.

   a. $z \leq 0$

   b. $z \geq 0$

   c. $z \leq -1$

   d. $z \leq 1$

   e. $z \geq -1$

   f. $z \geq 1$

9. Using the standard normal tables in Appendix A, determine the following probabilities. Sketch the associated areas.

   a. $z \leq -0.44$

   b. $z \geq 0.44$

   c. $-0.44 \leq z \leq 0.44$

   d. $z \leq -0.67$

   e. $z \geq 0.67$

   f. $-0.67 \leq z \leq 0.67$

10. Using the standard normal tables in Appendix A, determine the following probabilities. Sketch the associated areas.

    a. $z \leq -1.28$

    b. $z \geq 1.28$

    c. $-1.28 \leq z \leq 1.28$

    d. $z \leq -1.96$

    e. $z \geq 1.96$

    f. $-1.96 \leq z \leq 1.96$

11. Using the standard normal tables in Appendix A, determine the following probabilities. Sketch the associated areas.

    a. $P(0 \leq z \leq 0.79)$

    b. $P(-1.57 \leq z \leq 2.33)$

    c. $P(z \geq 1.89)$

    d. $P(z \leq -2.77)$

12. Using the standard normal tables in Appendix A, determine the following probabilities. Sketch the associated areas.

    a. $P(0 \leq z \leq 1.24)$

    b. $P(-2.64 \leq z \leq 3.32)$

    c. $P(z \geq 3.22)$

    d. $P(z \leq -3.39)$

13. Find the value of z such that 0.05 of the area under the curve lies to the right of z.

14. Find the value of z such that 0.01 of the area under the curve lies to the right of z.

15. Find the value of z such that 0.10 of the area under the curve lies to the right of z.

16. Find the value of z such that 0.05 of the area under the curve lies to the left of z.

17. Find the value of z such that 0.01 of the area under the curve lies to the left of z.

18. Find the value of z such that 0.10 of the area under the curve lies to the left of z.

19. Find the value of z such that 0.7458 of the area under the curve lies between $-z$ and z.

20. Find the value of z such that 0.9505 of the area under the curve lies between $-z$ and z.

21. Find the value of z such that 0.90 of the area under the curve lies between $-z$ and z.

22. The random variable X has a normal distribution with a mean of 30 and a standard deviation of 5.

    a. Find the probability that X is between 25 and 35.

    b. Find the probability that X is greater than 40.

    c. Find the probability that X is less than 20.

23. The random variable X has a normal distribution with a mean of 200 and a standard deviation of 25.

    a. Find the probability that X is between 160 and 220.

    b. Find the probability that X is greater than 240.

    c. Find the probability that X is less than 150.

24. The Arc Electronic Company had an income of $200,000 last year. Suppose the mean income of firms in the industry for the year is $1,000,000 with a standard deviation of $500,000. If incomes for the industry are normally distributed, what proportion of the firms in the industry earned less than Arc?

25. A certain component for the newly developed electronic diesel engine is considered to be defective if its diameter is less than 8.0 mm or greater than 10.5 mm. The distribution of the diameters of these parts is known to be normal with a mean of 9.0 mm and a standard deviation of 1.5 mm. If a component is randomly selected, what is the probability that it will be defective?

26. A television manufacturer is studying television remote control unit usage. One of the criteria they are measuring is the distance at which people attempt to activate the television set with the remote unit. They have discovered that activation distances are normally distributed with an average activation distance of six feet with a standard deviation of three feet. If a remote unit's maximum range is ten feet, what fraction of the time will users attempt to operate the remote outside of the operating limit?

27. According to the Bureau of Labor Statistics, the mean weekly earnings for people working in a sales related profession in 2010 was $631. Assume that the weekly earnings are approximately normally distributed with a standard deviation of $90.
    Source: Bureau of Labor Statistics

    a. What are the mean weekly earnings for people working in a sales related profession in 2010?

    b. If a salesperson was randomly selected, find the probability that his or her weekly earnings exceed $700.

    c. If a salesperson was randomly selected, find the probability that his or her weekly earnings are at most $525.

    d. If a salesperson was randomly selected, find the probability that his or her weekly earnings are between $400 and $615.

    e. Do you feel that it is reasonable to assume that the weekly earnings have a normal distribution? Why or why not?

28. The repair time for air conditioning units is believed to have a normal distribution with a mean of 38 minutes.

    a. What is the standard deviation of repair time if 40% of the units are repaired between 33 and 43 minutes?

    b. Using the value of the standard deviation that you calculated in **a.**, what is the probability that a repair will be longer than an hour?

    c. Using the value of the standard deviation that you calculated in **a.**, what is the probability that the repair time for an air conditioning unit will be less than 25 minutes?

29. VGA monitors manufactured by TSI Electronics have life spans which have a normal distribution with an average life span of 15,000 hours and a standard deviation of 2000 hours. If a VGA monitor is selected at random, find the following probabilities.

    a. The probability that the life span of the monitor will be less than 12,000 hours.

    b. The probability that the life span of the monitor will be more than 18,000 hours.

    c. The probability that the life span of the monitor will be between 13,000 hours and 17,000 hours.

30. A beer distributor believes the amount of beer in a 12-ounce can of beer has a normal distribution with a mean of 12 ounces and a standard deviation of 1 ounce. If a 12-ounce beer can is randomly selected, find the following probabilities.

    a. The probability that the 12-ounce can of beer will actually contain less than 11 ounces of beer.

    b. The probability that the 12-ounce can of beer will actually contain more than 12.5 ounces of beer.

    c. The probability that the 12-ounce can of beer will actually contain between 10.5 and 11.5 ounces of beer.

31. A statistics teacher believes that the final exam grades for her business statistics class have a normal distribution with a mean of 82 and a standard deviation of 8.

    a. Find the score which separates the top 10% of the scores from the lowest 90% of the scores.

    b. The teacher plans to give all students who score in the top 10% of scores an A. Will a student who scored a 90 on the exam receive an A? Explain.

    c. Find the score which separates the lowest 20% of the scores from the highest 80% of the scores.

    d. The teacher plans to give all students who score in the lowest 10% of scores an F. Will a student who scored a 65 on the exam receive an F? Explain.

32. An investor believes that the yields of his mutual funds have a normal distribution with an average yield of 10% and a standard deviation of 2%. The investor would like to identify the stocks which yield the highest 5% to keep in his portfolio.

    a. Calculate the yield which separates the highest 5% of yields from the lowest 95% of yields.

    b. If a stock yielded 14% would it be kept? Explain.

    c. If a stock yielded 13% would it be kept? Explain.

33.   In order for you to become a member of Mensa, a worldwide organization with approximately 100,000 members, your IQ score must be in the top 2%. The word *mensa* is Latin for "table," and was chosen to denote a group or round table of people with equal ability. In 1996, Mensa, which was founded by two British barristers, celebrated its 50th birthday. American Mensa Ltd., which was founded in 1960 has almost 50,000 members. Marilyn vos Savant, who is reputed to have the highest recorded IQ, is a member. Assuming that IQ scores have an approximately normal distribution with a mean and standard deviation of 100 and 15, respectively, answer the following questions.

a.   What IQ must one have in order to become a member of Mensa?

b.   What percent of all Americans have an IQ of at least 145?

c.   What percent of all members of Mensa have an IQ of at least 145?

d.   If Mensa decided to become more exclusive, and accepted only the top 1% instead of the top 2% as members, what IQ would one need in order to become a member of Mensa?

7.4   Approximations to Other Distributions

To approximate other distributions, the normal distribution can be very useful. Although it is a continuous distribution, it is used to approximate discrete distributions, specifically the binomial and the Poisson.

The Binomial Distribution

Calculating binomial probabilities can be quite time consuming if n is large. For example, suppose that you intend to sample 2000 subjects for a marketing research survey. If 50 percent of the population believes your product is superior to the competition's, what is the probability of obtaining 600 or fewer subjects who believe your company's product is superior?

$$P(X \leq 600) = P(X = 0) + P(X = 1) + P(X = 2) + \cdots + P(X = 599) + P(X = 600)$$

Determining the appropriate probability using the binomial distribution would require the calculation of 601 individual probabilities, many of which would have extremely large combinations such as the following.

$$_{2000}C_{400}\,0.5^{400}\,(1-0.5)^{1600}$$

Computing this and the other 600 similar calculations would be a formidable task. The normal distribution is useful in approximating binomial probabilities. The larger the binomial parameter, n, the more accurate the approximation. Determining the probability described above using the normal approximation is trivial in comparison to calculating the exact probability using the binomial.

Recall that the normal distribution is a function of two parameters, the mean and the standard deviation. Thus, if the normal distribution is used to approximate the binomial distribution, it seems reasonable that the mean and standard deviation of the normal should be the same as the mean and standard deviation of the binomial that is being approximated. Specifically, let

$$\mu = E(X) = np, \text{ and}$$

$$\sigma = \sqrt{V(X)} = \sqrt{np(1-p)}.$$

To approximate a binomial with $n = 20$ and $p = 0.5$ would require a normal distribution with

$$\mu = (20)(0.5) = 10$$

$$\sigma = \sqrt{(20)(0.5)(1-0.5)} = \sqrt{5} \approx 2.2361.$$

In this example, the shapes of the distributions are quite similar and consequently the approximation will be good.

Binomial Distribution with *n* = 20, *p* = 0.5

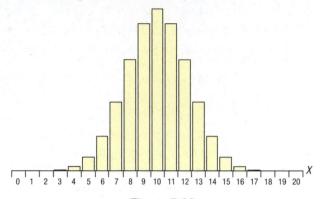

Figure 7.32

Binomial Distribution with *n* = 20, *p* = 0.5

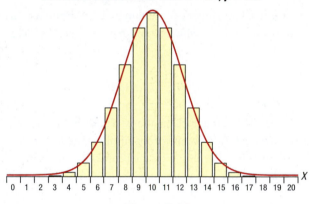

Figure 7.33

So, when should the normal distribution be used to approximate the binomial distribution? Generally, the approximation is reasonable when the mean of the binomial, *np*, is greater than or equal to 5 and $n(1-p)$ is greater than or equal to 5. The approximation becomes quite good when *np* is greater than or equal to 10 and $n(1-p)$ is greater than or equal to 10.

The normal approximation to the binomial can be improved by using **continuity correction**.

Definition

Continuity correction is used when a discrete distribution is approximated using a continuous distribution. To apply continuity correction, subtract or add 0.5 (depending on the question at hand) to a selected value in order to find the desired probability.

Suppose that you wished to determine the probability that a binomial random variable ($n = 20$ and $p = 0.5$) is equal to 5. Recall that for a continuous random variable X, the probability that X is equal to some specific value is equal to zero since there is no area under the curve for a single point, say $X = 5$. Therefore, to approximate the probability using the normal would be equivalent to approximating the area of the shaded region given in Figure 7.34. To approximate the area of the region using the normal would require finding the area under the curve between 4.5 and 5.5.

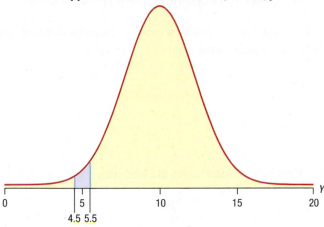

Normal Approximation to the Binomial, $n = 20$, $p = 0.5$

Figure 7.34

The continuity correction should be used whenever the normal distribution is used to approximate the binomial distribution. The following examples will illustrate how to approximate the binomial distribution using the normal distribution with continuity correction.

Example 7.11

   a. Assuming $n = 20$ and $p = 0.5$, use the normal distribution to approximate the probability that a binomial random variable is 5 or less.

   b. Find the probability that the same random variable from part **a.** is greater than 4.

Solution

   a. This implies finding the area of the rectangles for 0, 1, 2, 3, 4, and 5.

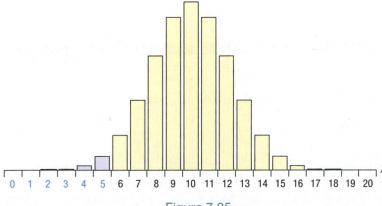

Binomial Distribution, $n = 20$, $p = 0.5$

Figure 7.35

Instead of using the normal approximation $P(Y \le 5)$, use the continuity correction $P(Y \le 5.5)$ in order to accumulate all of the probabilities under the normal curve that correspond to the region associated with the point 5.

To use the normal approximation the mean and standard deviation of the binomial must be calculated.

$$\mu = E(X) = np = (20)(0.5) = 10$$

$$\sigma = \sqrt{np(1-p)} = \sqrt{(20)(0.5)(1-0.5)} = \sqrt{5} \approx 2.2361$$

Using the normal distribution, Y, with a mean of 10 and a standard deviation of 2.2361 to approximate the binomial using continuity correction,

$$P(Y \le 5.5) = P\left(z \le \frac{5.5 - 10}{2.2361}\right)$$

$$= P(z \le -2.01)$$

$$= 0.0222.$$

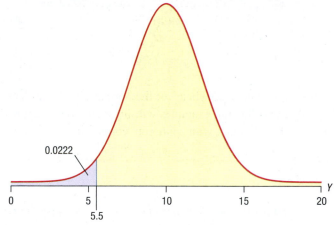

Normal Approximation to the Binomial, *n* = 20, *p* = 0.5

Figure 7.36

Thus, the probability that the random variable is 5 or less is 0.0222.

b.

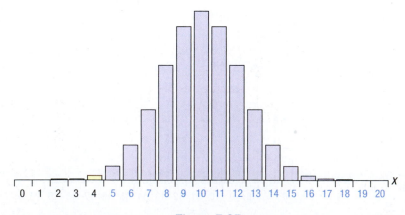

Binomial Distribution, *n* = 20, *p* = 0.5

Figure 7.37

We are interested in the probability that a binomial random variable, X, is greater than 4. Since this is a discrete distribution, the probability that X is greater than 4 is equal to the probability that X is greater than or equal to 5. Thus, when using the normal approximation, we need to apply continuity correction and consider the probability that the normal random variable is greater than or equal to 4.5.

Using the normal distribution, Y, with a mean of 10 and a standard deviation of 2.2361 to approximate the binomial using continuity correction,

$$P(Y \geq 4.5) = P\left(z \geq \frac{4.5 - 10}{2.2361}\right)$$
$$= P(z \geq -2.46)$$
$$= 1 - P(z \leq -2.46)$$
$$= 1 - 0.0069$$
$$= 0.9931.$$

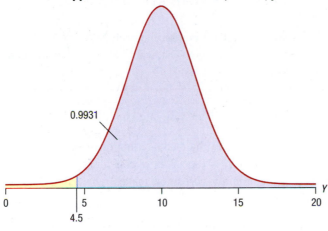

Normal Approximation to the Binomial, $n = 20$, $p = 0.5$

Figure 7.38

Thus, the probability that the random variable is greater than 4 is 0.9931.

Example 7.12

An advertising agency hired on behalf of Tech's development office conducted an ad campaign aimed at making alumni aware of their new capital campaign. Upon completion of the new campaign, the agency claimed that 20% of alumni in the state of Virginia were aware of the new campaign. To validate the claim of the agency, the development office surveyed 1000 alumni in the state and found that 150 were aware of the campaign. Assuming that the ad agency's claim is true, what is the probability that no more than 150 of the alumni in the random sample were aware of the new campaign?

Solution

Let X = the number of alumni that were aware of the campaign.

X is a binomial random variable with $n = 1000$ and $p = 0.20$.

So, $np = 200$ and $n(1-p) = 800$. Therefore the normal distribution is appropriate to use as an approximation to the binomial distribution.

The mean is $\mu = np = 200$ and the standard deviation is

$$\sigma = \sqrt{np(1-p)} = \sqrt{160} \approx 12.6491.$$

We are interested in the probability that no more than 150 of the alumni in the sample were aware of the campaign, or $P(X \leq 150)$. However, since we using the normal distribution to approximate the binomial, continuity correction must be applied.

Let Y be a normally distributed random variable with a mean of 200 and a standard deviation of 12.6491. Applying continuity correction, we are interested in the following probability.

$$P(Y \leq 150.5) = P\left(z \leq \frac{150.5 - 200}{12.6491}\right) = P(z \leq -3.91) \approx 0.$$

Thus, if the marketing agency's claim is true, the probability that 150 or fewer alumni are aware of the campaign is practically zero. This would lead the development office to believe that the agency's claim is false.

Note: the z-value -3.91 is not listed in the tables given in Appendix A. However, using technology such as a calculator or computer software, it can be calculated that the actual probability is approximately 0.000046.

The smaller the sample size, the more the binomial distribution deviates from the normal distribution. For this reason, continuity correction is especially useful for small sample sizes. Example 7.13 illustrates the difference that continuity correction makes when using the normal distribution to approximate the binomial.

Example 7.13

A popular restaurant near Tech's campus accepts 200 reservations on Saturdays, the day of a Tech football game. Given that many of the reservations are made weeks in advance of game day, the restaurant expects that about eight percent will be no-shows. What is the probability that the restaurant will have no more than 20 no-shows on the next Saturday of a football weekend?

Solution

Let $X =$ the number of no-shows.

X is a binomial random variable with $n = 200$ and $p = 0.08$.

Since $np = 16$ and $n(1-p) = 184$ are both greater than 5, the normal distribution can be used to approximate the binomial probability.

For the binomial, $\mu = np = 16$ and

$$\sigma = \sqrt{np(1-p)} = \sqrt{(200)(0.08)(1-0.08)} = \sqrt{14.72} \approx 3.8367.$$

Using the normal distribution, Y, with a mean of 16 and a standard deviation of 3.8367, to approximate the binomial without continuity correction results in

$$P(Y \le 20) = P\left(z \le \frac{20-16}{3.8367}\right) = P(z \le 1.04) = 0.8508.$$

Using continuity correction,

$$P(Y \le 20.5) = P\left(z \le \frac{20.5-16}{3.8367}\right) = P(z \le 1.17) = 0.8790.$$

Thus, using the normal approximation and continuity correction, the probability that the restaurant will have no more than 20 no-shows is 0.8790. Notice that the continuity correction has a significant impact on the accuracy of the approximation. Using the binomial distribution, the exact probability is 0.8775.

The Poisson Distribution

Approximating the Poisson distribution with the normal distribution is similar to approximating the binomial distribution. To use this approximation, the mean and standard deviation of the normal should be set to the mean and standard deviation of the Poisson. Since the mean and variance of the Poisson are both λ, the appropriate mean, variance, and standard deviation for the normal would be as follows.

$$\mu = \lambda, \sigma^2 = \lambda, \ \sigma = \sqrt{\lambda}$$

Just as with the binomial distribution, when using the normal distribution to approximate the Poisson distribution, continuity correction should be applied for the best possible approximation. This is because we are using a continuous distribution to approximate a discrete distribution. The approximation becomes increasingly accurate as the mean becomes larger. However, it is not very accurate when the mean of the Poisson distribution is less than 5.

Example 7.14

The manager of a sporting goods store wants to determine the best method of staffing employees without being wasteful. The problem that is often encountered is having too many employees and too few customers, and vice versa. The manager realized that during the initial kick-off of certain sports seasons such as football, baseball, and basketball, the store attracts, on average, 30 customers per hour. On some days, the number is higher, on others, the number is lower. The manager would like to determine the probability of at least 40 customers arriving in a given hour. Use the normal approximation to the Poisson distribution to find the probability.

Solution

Let X = the number of customers that enter the store in a given hour.

Notice that since we are counting the number of customers in a specific time interval, X has a Poisson distribution with

$$\mu = 30 \text{ and } \sigma = \sqrt{30} \approx 5.4772.$$

Figure 7.39

If Y represents the normal random variable with a mean of 30 and a standard deviation of 5.4772, it should be a good approximation to the Poisson. We are interested in the probability that at least 40 customers arrive in a given hour, or $P(X \geq 40)$. Because we are using the normal distribution to approximate the Poisson distribution, we must use continuity correction. Thus, we are interested in the following probability.

$$P(X \geq 40) = P(Y \geq 39.5) = P\left(z \geq \frac{39.5 - 30}{5.4772}\right) = P(z \geq 1.73) = 1 - P(z < 1.73) = 1 - 0.9582 = 0.0418$$

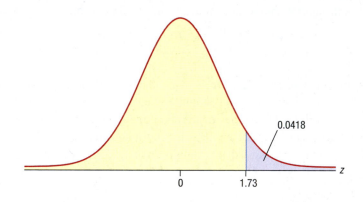

Figure 7.40

Thus, using the normal approximation to the Poisson, the probability that at least 40 customers arrive in a given hour is 0.0418.

7.4   Exercises

Basic Concepts

1.   Why would you want to use the normal distribution to approximate a binomial distribution or a Poisson distribution?

2.   What are the parameters of a normal distribution used to approximate a binomial distribution?

3.   What are the parameters of a normal distribution used to approximate a Poisson distribution?

4.   What is continuity correction? How does it improve the normal approximation to the binomial or Poisson?

Exercises

5.   Management at a small engineering company is considering the addition of a company cafeteria area. A random sample of 50 persons out of the total number of persons employed by the firm will be surveyed to see if they are in favor of the addition. Assume that the true percentage of persons that favor the addition is 90%.

   a.   Find the expected number of employees in the sample who will favor the addition of the cafeteria area.

   b.   Find the standard deviation of the number of employees in the sample who will favor the addition of the cafeteria area.

   c.   What is the probability that between 35 and 37 employees (inclusive) in the sample will favor the cafeteria?

   d.   What is the probability that more than 40 of the employees in the sample will favor the cafeteria?

   e.   What is the probability that at most 38 of the employees in the sample will favor the cafeteria?

6.   The accounting department of a large corporation checks the addition of expense reports submitted by executives before paying them. Historically, they have found that 15% of the reports contain addition errors. An auditor randomly selects 60 expense reports and audits them for addition errors.

   a.   Find the expected number of reports in the sample that will have addition errors.

   b.   Find the standard deviation of the number of reports sampled that will have addition errors.

   c.   Find the probability that fewer than 10 of the sampled expense reports will have addition errors.

   d.   Find the probability that at least 30 of the sampled expense reports will have addition errors.

   e.   Find the probability that between 5 and 15 (inclusive) of the sampled expense reports will have addition errors.

7. A local electronics store purchased a market research study which suggests that 60 percent of all homes have DVD recorders/players. A sample of 200 homes is selected to confirm the study's findings. If the marketing study is correct, answer the following questions.

   a. Find the expected number of homes sampled which will have DVD recorders/players.

   b. Find the standard deviation of the number of homes in the sample which will have video recorders/players.

   c. What is the probability that at most 80 of the sampled homes will have DVD recorders/players?

   d. What is the probability that between 100 and 120 (inclusive) homes sampled will have DVD recorders/players?

   e. What is the probability that at least 130 of the sampled homes will have DVD recorders/players?

8. Suppose a virus is believed to infect two percent of the population. If a sample of 3000 randomly selected subjects are tested, answer the following questions.

   a. Find the expected number of subjects sampled that will be infected.

   b. Find the standard deviation of the number of subjects sampled that will be infected.

   c. What is the probability that fewer than 30 of the subjects in the sample will be infected?

   d. What is the probability that between 40 and 80 (inclusive) of the subjects in the sample will be infected?

   e. Find the probability that at least 70 of the subjects in the sample will be infected.

9. A company manufacturing metal sheets believes that the number of defects on a 10' by 10' sheet of metal follows a Poisson distribution with an average defect rate of 5 per sheet.

   a. Find the standard deviation of the number of defects per sheet.

   b. Using the Poisson table in Appendix A, Table F, find the probability of observing at least 10 defects per sheet.

   c. Using the normal approximation to the Poisson, find the probability of observing at least 10 defects per sheet.

   d. How do the answers in parts b. and c. compare?

10. Service calls arriving at an electric company follow a Poisson distribution with an average arrival rate of 60 per hour.

   a. Find the average number of service calls in a 30-minute period.

   b. Find the standard deviation of the number of service calls in a 30-minute period.

   c. Using the normal approximation to the Poisson, find the probability that the electric company receives at least 40 service calls in a 30-minute period.

   d. Using the normal approximation to the Poisson, find the probability that the electric company receives at most 20 service calls in a 30-minute period.

   e. Using the normal approximation to the Poisson, find the probability that the electric company receives between 25 and 50 (inclusive) service calls in a 30-minute period.

11. Patients arriving at the emergency room of a local hospital follow a Poisson distribution with an average arrival rate of 15 per half hour.

   a. Find the average number of patients that arrive at the emergency room in one hour.

   b. Find the standard deviation of the number of patients that arrive at the emergency room in one hour.

   c. Find the probability that at least 15 patients will arrive at the emergency room in one hour.

   d. Find the probability that between 30 and 50 patients (inclusive) will arrive at the emergency room in one hour.

   e. Find the probability that at most 35 patients will arrive at the emergency room in one hour.

T   **Discovering Technology**

Using the TI-84 Plus Calculator

Normal Distribution: Finding the Area between Two Values

Use the information from Example 7.6 for this exercise. Suppose that the times it takes for an accountant to process payroll is normally distributed with a mean of 10 minutes and a standard deviation of 20 minutes. Find the probability that a randomly selected payroll date will take between 10 minutes and 40 minutes to process.

1. To use a TI-84 Plus calculator to find this probability, you will use the normalcdf function. The syntax for the normalcdf function on the calculator is as follows.

    normalcdf(lower bound, upper bound, mean, standard deviation)

    Press **2ND** and then **VARS** to access the **DISTR** menu.

2. Select option **2:normalcdf**. Press **ENTER**.

3. We are interested in the probability that the payroll will take between 10 and 40 minutes to process, assuming the times are normally distributed with a mean of 10 minutes and a standard deviation of 20 minutes. Thus, the function you want to enter is

    normalcdf(10,40,10,20). Press **ENTER** and observe the results.

Figure 7.41

Normal Distribution: Finding the Area to the Right of Some Value

Use the information from Example 7.8 for this exercise. Suppose that a national testing service gives a test in which the results are normally distributed with a mean of 400 and a standard deviation of 100. If you score a 644 on the test, what percentage of the students taking the test exceeded your score?

1. Press **2ND** and then **VARS** to access the **DISTR** menu. Select option **2:normalcdf**.

2. We are interested in the probability of scoring greater than 644 on a test where the results are normally distributed with a mean of 400 and a standard deviation of 100. To enter this on the calculator, we enter a very large positive number as the upper bound in the normalcdf function. Enter **normalcdf(644,1E99,400,100)** into the calculator. Here, 1E99 represents infinity, and is entered by pressing **1**, **2ND**, **,**, **9**, **9**.

3. Press ENTER and observe the results.

Figure 7.42

Normal Distribution: Finding the Area to the Left of Some Value

Find the probability that a normal random variable with a mean of 10 and a standard deviation of 20 will lie below −10.

1. Press 2ND and then VARS to access the **DISTR** menu. Select option **2:normalcdf**.

2. We are interested in the probability that a normal random variable with a mean of 10 and a standard deviation of 20 will lie below −10. To enter this on the calculator, we enter a very large negative number as the lower bound in the normalcdf function to represent negative infinity. Enter **normalcdf(-1E99,-10,10,20)** into the calculator. Here, -1E99 represents negative infinity and is entered by pressing (−), 1, 2ND, , , 9, 9.

3. Press ENTER and observe the results.

Figure 7.43

Normal Distribution: Finding the Value of z Given an Area

Use the information from Example 7.5 for this exercise.

1. To find the value of z given a particular area, use the **invNorm** function within the **DISTR** menu. The syntax for the invNorm function is as follows.

invNorm(area to the left of z)

Part **a.** of Example 7.5 asks to find the value of z if the area to the left of z is 0.9147. Thus, to find this probability, press 2ND and then VARS to access the **DISTR** menu and select option **3:invNorm**.

2. Since we are given the area to the left of z, simply enter **invNorm(0.9147)** into the calculator. Press ENTER and observe the results. The value of z is approximately 1.37.

3. Part **b.** of Example 7.5 asks to find the value of z if the area between 0 and z is 0.3665. Since the function in the calculator uses the area to the left of z, we need to add 0.5 to 0.3665 to get the area from negative infinity to z. Thus, press 2ND and then VARS to access the **DISTR** menu and select option **3:invNorm**. Enter **invNorm(0.8665)** into the calculator and press ENTER to observe the results. The value of z is approximately 1.11.

4. Part **d.** asks for the value of z if the area to the right of z is 0.7967. Since the invNorm function uses the area to the left of z, we will need to enter $1- 0.7967$ as the area. Enter **invNorm(1-0.7967)** and press ENTER to observe the results. The value of z is approximately -0.83.

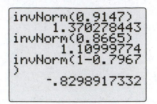

Figure 7.44

Using Excel

Normal Distribution

Consider a normally distributed random variable, X, with a mean of 15 and a standard deviation of 5. Use Microsoft Excel to find **a.** the area to the left of 8.4, **b.** the area between 9 and 11, and **c.** the area to the right of 17.

1. The function in Microsoft Excel that gives the area under a normal curve to the left of some value is as follows.

    NORM.DIST(x, mean, standard_dev, cumulative)

    where x is the value of the random variable, **mean** is the mean of the random variable, **standard_dev** is the standard deviation of the random variable, and **cumulative** = TRUE if we are interested in the area to the left of x and FALSE if we are interested in the value of the probability density function at the particular value of x (this is usually not very useful). To find the area to the left of 8.4, enter the following formula into Excel.

    =NORM.DIST(8.4,15,5,TRUE)

    The result is the area to the left of 8.4, which is approximately 0.0934.

2. To find the area between 9 and 11, we will need to find the area to the left of 11 and subtract the area to the left of 9, since the formula in Excel only calculates areas to the left of x. To find the area between 9 and 11, enter the following formula into Excel.

    =NORM.DIST(11,15,5,TRUE)-NORM.DIST(9,15,5,TRUE)

    The result is the area between 9 and 11, which is approximately 0.0968.

3. To find the area to the right of 17, we will need to use Excel to find the area to the left of 17, and then subtract this area from 1. To find the area to the right of 17, enter the following formula into Excel.

    =1-NORM.DIST(17,15,5,TRUE)

    The result is the area to the right of 17, which is approximately 0.3446.

Normal Distribution: Finding the Value of z Given an Area

Use the information from Example 7.5 for this exercise.

1. The function in Microsoft Excel that returns the value of z given the area to the left of z is as follows.

    =NORM.INV(probability, mean, standard_dev)

    where **probability** is the area to the left of z, **mean** is the mean of the normal distribution, and **standard_dev** is the standard deviation of the normal distribution.   Part **a.** of Example 7.5 asks for the value of z if the area to the left of z is 0.9147.   To find the value of z in this case, enter the following formula in Microsoft Excel.

    =NORM.INV(0.9147,0,1)

    The result is displayed, and the value of z is approximately 1.37.

2. Part **b.** of Example 7.5 asks to find the value of z if the area between 0 and z is 0.3665. Since the NORM.INV function in Excel uses the area to the left of z, we need to add 0.5 to 0.3665 to get the area from negative infinity to z.   Enter the following formula into Excel to find the value of z.

    =NORM.INV(0.8665,0,1)

    The result is displayed, and the value of z is approximately 1.11.

3. Part **d.** asks for the value of z if the area to the right of z is 0.7967.   Since the NORM. INV function uses the area to the left of z, we will need to enter $1 - 0.7967$ as the area. Enter the following formula into Excel to find the value of z.

    =NORM.INV(1-0.7967,0,1)

    The result is displayed, and the value of z is approximately −0.83.

Using Minitab

Normal Distribution

Consider a normally distributed random variable, X, with a mean of 400 and a standard deviation of 100. Use Minitab to find **a.** the area to the left of 246, **b.** the area between 550 and 640, and **c.** the area to the right of 644.

1. Minitab finds the cumulative normal probabilities to the left of a given z-value. Set up a Minitab worksheet similar to the one shown in Figure 7.45.

↓	C1	C2
	x	
1	246	
2	640	
3	550	
4	644	

Figure 7.45

2. Go to **Calc, Probability Distributions**, and **Normal**. In the Normal Distribution dialog box, insure that **Cumulative probability** is selected. Enter **400** for the mean and **100** for the Standard deviation. Enter **C1** as the Input column, and click **OK**.

3. The results are displayed in the Session window. Keep in mind that these probabilities are the probabilities that the normal random variable with a mean of 400 and a standard deviation of 100 is less than or equal to x. Therefore, in order to find the probabilities indicated that are not the area to the left of x, we need to work with the probabilities algebraically.

```
Cumulative Distribution Function

Normal with mean = 400 and standard deviation = 100

  x   P( X <= x )
246      0.061780
640      0.991802
550      0.933193
644      0.992656
```

Figure 7.46

4. To find the area to the left of 246, it is simply given in the Session window as approximately 0.0618. To find the area between 550 and 640, we need to subtract the cumulative probability for 550 from the cumulative probability for 640. Thus, the area between 550 and 640 is approximately $0.9918 - 0.9332 = 0.0586$. To find the area to the right of 644, we need to subtract the area to the left of 644 from 1. Thus, the area to the right of 644 is $1 - 0.9927 = 0.0073$.

Normal Distribution: Finding the Value of z Given an Area

Given that z is a standard normal random variable, find the value of z if **a.** the area to the left of z is 0.1052, **b.** the area between 0 and z is 0.2532, and **c.** the area to the right of z is 0.6045.

1. Minitab will return the value of z given an area to the left of z. So, we need to enter the areas to the left of z for parts **a.**, **b.**, and **c.**, into Column C1 of Minitab.

↓	C1	C2
	Area to the left of z	
1	0.1052	
2	0.7532	
3	0.3955	

Figure 7.47

2. Go to **Calc**, **Probability Distributions**, and **Normal**. In the Normal Distribution dialog box, insure that **Inverse cumulative probability** is selected, and that the Mean is set to **0** and the Standard deviation is set to **1**. Enter **C1** as the Input column, and click **OK**.

3. Observe the results in the Session window. The values of x for parts **a.**, **b.**, and **c.** are given as approximately −1.25, 0.68, and −0.27, respectively.

```
Inverse Cumulative Distribution Function

Normal with mean = 0 and standard deviation = 1

P( X <= x )        x
    0.1052   -1.25247
    0.7532    0.68459
    0.3955   -0.26501
```

Figure 7.48

Binomial Distribution

For a binomial random variable with $n = 20$ and $p = 0.3$, calculate the probability that X is less than or equal to 4.

1. Enter 0, 1, 2, 3, 4 in the first five rows of C1.

2. Select **Calc**, choose **Probability Distributions**, and then choose **Binomial**.

3. Make sure **Cumulative probability** is selected, then enter **20** as the Number of trials and **0.3** as the Event probability.

4. In the Input column, select **C1** and press **OK**.

5. The cumulative distribution function is displayed in the Session window. When $x = 4$ the probability that $X \leq 4$ is approximately 0.2375.

```
Cumulative Distribution Function

Binomial with n = 20 and p = 0.3

x   P( X <= x )
0      0.000798
1      0.007637
2      0.035483
3      0.107087
4      0.237508
```

Figure 7.49

| **R** | **Chapter 7 Review** |

Key Terms and Ideas

- Continuous Random Variables
- Continuous Uniform Distribution
- Probability Density Function
- Uniform Probability Density Function
- Normal Distribution
- Normal Probability Density Function
- Standard Normal Distribution

- z-Distribution
- z-Score
- Standard Normal Random Variable
- Normal Approximation to the Binomial Distribution
- Continuity Correction
- Normal Approximation to the Poisson Distribution

Key Formulas

	Section
Uniform Probability Density Function	
$$f(x) = \begin{cases} \dfrac{1}{b-a} & \text{for } a \le x \le b \\ 0 & \text{otherwise} \end{cases}$$	7.1
Expected Value for a Continuous Uniform Random Variable	
$$\mu = E(x) = \frac{a+b}{2}$$	7.1
Standard Deviation for a Continuous Uniform Random Variable	
$$\sigma = \frac{b-a}{\sqrt{12}}$$	7.1
Normal Probability Density Function	
$$f(x) = \frac{1}{\sigma\sqrt{2\pi}} e^{-\frac{(x-\mu)^2}{2\sigma^2}}$$	7.2
Standardizing a Normal Random Variable	
$$z = \frac{x-\mu}{\sigma}$$	7.3

Key Formulas (cont.)

	Section
Normal Approximation for the Binomial Distribution	
If X is a binomial random variable where $np \geq 5$ and $n(1-p) \geq 5$, then X can be approximated by a normal distribution with $\mu = np$ and $\sigma = \sqrt{np(1-p)}$.	7.4
Normal Approximation for the Poisson Distribution	
If X has a Poisson distribution, X can be approximated by a normal distribution with $\mu = \lambda$ and $\sigma = \sqrt{\lambda}$.	7.4

AE   Additional Exercises

1. Joe is always between 5 and 35 minutes late for work. Assuming that X, the number of minutes that Joe is late for work, has a continuous uniform distribution, answer the following questions.

  a. Find the mean of X.

  b. Find the standard deviation of X.

  c. Find the probability that on any randomly selected day, Joe is at least a half hour late for work.

  d. What percent of the time will X fall within one standard deviation of its mean?

  e. Compare this percent to that given by the Empirical Rule of Chapter 4. Why the large discrepancy?

2. Using the standard normal tables, determine the following probabilities. Sketch the associated areas.

  a. $P(0 \le z \le 0.85)$

  b. $P(-1.25 \le z \le 2.25)$

  c. $P(z \ge 1.75)$

  d. $P(z \le -2.75)$

3. Using the standard normal tables, determine the following probabilities. Sketch the associated areas.

  a. $P(0 \le z \le 1.00)$

  b. $P(-2.50 \le z \le 3.01)$

  c. $P(z \ge 3.25)$

  d. $P(z \le -2.50)$

4. Find the value of z such that 0.99 of the area under the curve lies between $-z$ and z.

5. Find the value of z such that 0.80 of the area under the curve lies between $-z$ and z.

6. The weights of newborn baby boys born at a local hospital are believed to have a normal distribution with an average weight of 7.25 lb and a standard deviation of 1 lb. If a newborn baby boy, born at the local hospital, is randomly selected, answer the following questions.

  a. Find the probability that the weight of the newborn baby boy will be more than 8 lb.

  b. Find the probability that the weight of the newborn baby boy will be less than 6 lb.

  c. Find the probability that the weight of the newborn baby boy will be between 6.5 lb and 8.5 lb.

  d. Find the weight that separates the lowest 10% of the weights from the highest 90% of the weights.

  e. If babies in the lowest 10 percent of weights are kept for observation, would a baby that weighed 5 lb be kept for observation?

7. Medication errors in a hospital can be dangerous and expensive. Medication errors are defined as giving a patient a non-prescribed medication in any quantity or the improper dosage of a prescribed medication. Suppose the national average for medication errors is one out of every 1000 patients. A hospital believes that their medication error rate is comparable to the national average. If the hospital randomly selects 5000 patients, answer the following questions.

   a. Find the expected number of patients in the sample that will have had a medication error.

   b. What is the standard deviation of the number of patients in the sample that will have had a medication error?

   c. What is the probability of observing one or more patients who have had medication errors in the sample?

   d. What is the probability of observing two or more patients who have had medication errors in the sample?

   e. Do you have any concerns about the accuracy of the probabilities you determined in parts **c.** and **d.**?

8. The number of violent crimes committed in a large city follows a Poisson distribution with an average rate of 10 per month.

   a. Find the expected number of violent crimes committed in a 3-month period.

   b. Find the standard deviation of the number of violent crimes committed in a 3-month period.

   c. Find the probability that at least 45 violent crimes will be committed in a 3-month period.

   d. Find the probability that between 25 and 40 (inclusive) violent crimes will be committed in a 3-month period.

   e. Find the probability that less than 20 violent crimes will be committed in a 3-month period.

9. According to the 2011 Statistical Abstract of the United States, 20.5% of the scores on the critical reading portion of the SAT Reasoning Test exceeded 600. Approximately 17.4% of the scores were less than 400, according to the same reference. Assuming that scores on the critical reading portion of the SAT are approximately normally distributed, what are the mean and the standard deviation of the scores on the verbal portion of the SAT Reasoning Test?
   Source: U.S. Census Bureau

10. The annual average per capita consumption of red meat in the United States in 2008 was 108.3 pounds, according to the 2011 Statistical Abstract of the United States. This figure was down from a per capita average of 126.4 pounds in 1980. Assume that both in 2008 and 1980 the per capita amount of red meat consumed was a normal random variable with a standard deviation of 15 pounds.
   Source: U.S. Census Bureau

    a. In 1980, what percent of the population consumed at least 100 pounds of red meat?

    b. In 2008, what percent of the population consumed at least 100 pounds of red meat?

    c. In 1980, what percent of the population consumed at most 130 pounds of red meat?

    d. In 2008, what percent of the population consumed at most 130 pounds of red meat?

    e. Do you feel that it is reasonable to assume that the per capita amount of red meat consumed has a normal distribution? Why or why not?

11. A cell phone manufacturer has developed a new type of battery for its phones. Extensive testing indicates that the population battery life (in days) obtained by all batteries of this new type is normally distributed with a mean of 700 days and a standard deviation of 100 days. The manufacturer wishes to offer a guarantee providing a discount on batteries if the original battery purchased does not exceed the days stated in the guarantee. What should the guaranteed battery life be (in days) if the manufacturer desires that no more than 5% of the batteries will fail to meet the guaranteed number of days?

12. The manager of a retail store wants to determine the best method of staffing employees without being wasteful. The problem that is often encountered is having too many employees and too few customers, and vice versa. The manager realized that during the holiday season they attract, on average, 90 customers per hour. On some days, the number is higher, on others, the number is lower. The manager would like to determine the probability of at least 2 customers arriving in a given minute during holiday season. Use the normal approximation to find the probability.

13. A machine used to regulate the amount of dye dispensed for mixing shades of paint can be set so that it discharges an average of μ milliliters of dye per can of paint. The amount of dye discharged is known to have a normal distribution with a variance equal to 0.0160. If more than 6 milliliters of dye are discharged when making a particular shade of blue paint, the shade is unacceptable. Determine the setting of μ so that no more than 1% of the cans of paint will be unacceptable.

14. The length of time required to complete a college achievement test is found to be normally distributed with a mean of 75 minutes and a standard deviation of 15 minutes. When should the test be terminated if we wish to allow sufficient time for 95% of the students to complete the test?

15. A manufacturing plant utilizes 3000 electric light bulbs that have a length of life that is normally distributed with a mean of 500 hours and a standard deviation of 50 hours. To minimize the number of bulbs that burn out during operation hours, all the bulbs are replaced after a given period of operation. How often should the bulbs be replaced if we want not more than 2% of the bulbs to burn out between replacement periods?

16. Howe's Finance Corporation provides financing for customers at an automotive dealership. The average loan amount is $24,000 with a standard deviation of $8000. Assuming that the loan amount is normally distributed, what is the probability that a randomly selected consumer buying a car will want to finance at least $20,000?

17. Suppose that the income of families in a large community follows a normal distribution. Two families are randomly selected and their incomes are $55,000 and $85,000, respectively. The two incomes correspond to z-scores of −0.5 and 2.0 respectively. Calculate the mean and standard deviation of the income of families in the neighborhood.

18. Suppose that the 30th percentile of a normal distribution is equal to 756 and that the 90th percentile of this normal distribution is 996. Find the mean and standard deviation of the normal distribution.

Chapter Eight

Samples and Sampling Distributions

D   **Discovering the Real World**

Holiday Spending

Gift Cards More Popular Than Ever, According to NRF

Consumers to spend a total of $26.3 billion on gift cards this holiday

As gift cards become increasingly convenient, more personalized and more sought after, retailers can expect to see an increase in gift card sales this holiday season. The fifth annual National Retail Federation (NRF) Gift Card Survey, conducted by BIGresearch, found that gift card sales will total $26.3 billion this holiday season, compared to $24.8 billion in 2006. Additionally, the average consumer will spend more on gift cards than they did last year ($122.59 vs. $116.51 in 2006).

"It is no secret why gift cards are the most popular item on consumers' shopping lists," said NRF President and CEO Tracy Mullin. "Many consumers pulled back on spending for themselves this year and are hoping to receive a gift card so they can purchase something they may have had their eye on for months."

According to a recent NRF survey, gift cards will be the one of the most popular gifts this holiday season as 56.6 percent of consumers plan to buy them, compared to other top-selling gift items like clothing, toys, and books. And with popularity rising among consumers of all ages, 87.7 percent of shoppers said they will purchase two or more gift cards this holiday season.

"As consumers become more strapped for time, they are constantly looking for ways to make holiday shopping easier and less stressful," said Phil Rist, Vice President of Strategy for BIGresearch. "Gift cards allow for a faster and more convenient way to buy gifts for family and friends."

About the Survey

The NRF 2007 Holiday Consumer Intentions and Actions Survey was designed to gauge consumer behavior and shopping trends related to the winter holidays. The survey, which polled 7982 consumers, was conducted for NRF by BIGresearch from October 31–November 7, 2007. The consumer poll has a margin of error of plus or minus 1.0 percent.

Source: Grannis, Kathy, and Scott Krugman. "Gift Cards More Popular Than Ever, According to NRF." **National Retail Federation**. N.p., 13 Nov. 2007.

Questions to Consider

1. What sampling issues are expressed in this article?

2. What questions would you have regarding the manner in which the survey data were collected?

Introduction

When buying or even looking for a new car, one of the features that a customer is most interested in is the miles per gallon (mpg) rating. Have you ever wondered how these mpg ratings are obtained for vehicles? The miles per gallon ratings are obtained via **sampling** and **sampling distributions**. That is, auto manufacturers randomly sample vehicles, drive them in various conditions (city or highway) and note the mileages obtained under such conditions. This sample provides convincing evidence of the vehicle's average miles per gallon in the city and/or on the highway, and thus the manufacturer is able to establish (and eventually publish) these estimates with confidence.

The quality of all statistical analyses and procedures depends on the quality of the sample data. If the sample data are not representative of the population, analyzing the data and drawing conclusions will be unreliable. This is a simple idea, and an important one. It suggests the beginning of any statistical investigation should be focused on the development of representative sample data. Interestingly, obtaining a representative sample requires the introduction of randomness in the sampling procedure.

This chapter begins a bridge from the study of probability to the study of statistical inference. One of the critical links between statistical inference and probability is the sampling distribution, which is the probability distribution of a statistic. In the ensuing chapter, the language of probability will be used to help define the level of confidence associated with the statistical inferences that are made.

8.1   Random Samples

People do a lot of sampling every day. If someone buys a cup of coffee, he or she does not have to drink the whole cup to determine whether the coffee is too hot. He or she takes a sample, a sip or two, then makes a decision. The sampling procedure produces as much information as sampling the entire cup and, if the coffee is too hot, with much less pain. Sampling is a fundamental component of existence. When we cross an intersection, we sample traffic in both directions before making the decision to cross. When driving a car, we visually sample traffic conditions, speed, and engine gauges. Sampling is a good idea. It saves time and is less costly than performing a census on the entire population. If the sample is representative of the population, the information it provides can be almost as good as performing a census. The relevant question is, how do you select a sample that is representative of the population?

Drawing samples that are representative of a population is often quite difficult. One "sampling method" that is often seen in the media is **voluntary sampling**. Usually some question is posed to a large audience and people "volunteer" to participate in the sample. The Internet has made this kind of sampling popular. However, this sampling method can often produce misleading results.

For example, the advice columnist Ann Landers once asked her readers, *If you had to do it over again, would you have children?* Over 10,000 readers wrote in; 70% said "No," and many described the misery their children had brought them. Voluntary response of this kind over-represents people with strong feelings, especially strong negative feelings. A national random sample on the same issue found that over 90% of parents would choose to have children again. Voluntary response can easily produce 70% "No" in a sample when the truth about the population is close to 90% "Yes." Yet the media continue to conduct voluntary response polls and discuss them as if they provide useful information.

Ann Landers' sample is an extreme example of **selection bias**. Selection bias occurs when the sample that has been selected overrepresents or underrepresents some segments of the population it is intended to represent. In the case of Ann Landers' sample, the sampling bias was the result of parents with strong negative opinions feeling compelled to respond to the survey. For this reason, the majority of responses were negative. Voluntary samples typically are not representative samples of a population.

> **Definition**
>
> A sample is **biased** if it overrepresents or underrepresents some segment(s) of the population.

Literary Digest

For the 1936 presidential election, *Literary Digest* conducted a poll to determine the winner. Over 10 million questionnaires were sent to those who owned automobiles and/or telephones. Over 2.4 million questionnaires were returned, and *Literary Digest* predicted that Alf Landon would defeat Franklin D. Roosevelt with 57% of the vote. George Gallup also conducted a poll of 50,000 random voters and predicted Roosevelt as the winner. Many people laughed at Gallup because *Literary Digest* had been correctly predicting the outcome of the presidential election since 1916 and based its predictions on such a large sample. Gallup was correct and Roosevelt won with 62% of the vote. Where did *Literary Digest* go wrong? First, only those who were rich enough to afford the luxury of a car or phone were sampled. Also, those who were not happy with Roosevelt were more likely to respond. In the end, *Literary Digest* went bankrupt and Gallup started his own company, which still predicts the elections.

Choosing a Representative Sample

In order to select a sample, the population must be clearly defined. Otherwise, it is not clear what the sample represents. A significant portion of the effort required to produce good samples goes into the creation of a **sampling frame**, which is a list containing all the members of the population. Once this list is available, it is possible to construct a **simple random sample**.

> **Definition**
>
> A **sampling frame** is a list which identifies all members of the population.

Using Randomness to Remove Sample Bias

One of the more powerful methods of coping with bias is the use of randomness in the sampling process.

> **Definition**
>
> A **simple random sample** from a finite population is one in which every possible sample of the same size n has the same probability of being selected.

This definition implies that every possible sample has an equal chance of being selected. There are a number of ways of selecting a simple random sample. If the sampling frame is small, each member of the list can be assigned a unique number. These numbers can be placed in a hat and drawn randomly. Another technique is to use a **random number table**.

Random number tables are composed of random digits, arranged in groups of 4 or 5 to improve readability. To use the random number table, randomly pick a row and column in the table as a starting point and select the appropriate number of digits. The appropriate number of digits is based on the size of the sampling frame. If the size of the sampling frame is from 1 to 99, then two digits should be selected. If the size of the sampling frame is from 100 to 999, then three digits should be selected. It makes no difference whether the random numbers are selected across the column or down the row. Random numbers which have the correct number of digits, but which are larger than the frame, should be ignored. Once the appropriate random numbers have been selected, sample the items in the frame which correspond to the random numbers. The resulting sample will constitute a simple random sample. Of course, with a computer readily available, software programs are used to generate sets of random numbers. Techniques on using software to generate random numbers for samples are given in the Discovering Technology section later in this chapter.

Example 8.1

Using a random number table, select a simple random sample of 20 customer accounts from a sampling frame that contains 897 accounts.

Solution

If each account is given a number between 1 and 897, then three digits from the table can be used to indicate a specific account. Any consecutive three digits that are greater than 897 should be ignored. We can select the initial starting position as row three and column two, and select three digits across the row. (Note that the choices of the starting position and whether to select digits across the row or column are arbitrary.) Those accounts colored in red are selected (see Table 8.1).

Table 8.1 – Random Number Table									
985	201	776	714	905	686	072	210	940	558
609	709	343	350	500	739	981	180	505	431
398	082	773	250	725	682	482	940	524	201
527	756	785	183	452	996	340	628	898	083
137	467	007	818	475	406	106	871	177	817
886	854	020	086	507	584	013	676	667	951
903	476	493	296	091	106	299	594	673	488
751	764	969	918	260	892	893	785	613	682
347	834	113	862	481	176	741	746	850	950
580	477	697	473	039	571	864	021	816	544

Thus, for our random sample of 20 accounts we will consider the following account numbers: 82, 773, 250, 725, 682, 482, 524, 201, 527, 756, 785, 183, 452, 340, 628, 83, 137, 467, 7, and 818.

Random samples form the basis of almost all inferential statistics. But random sampling is not a haphazard, purposeless selection technique. On the contrary, removing potential selection bias is difficult, and randomness is the only tool for the job.

Suppose you want to obtain a random sample of students at your college. If 8000 students attend the college, deriving the sampling frame would be a formidable task. Fortunately, a college's student population is well-defined, and with the aid of a computer and the registrar's database, a random sample could be selected.

Well-defined populations are the exception, not the rule. Consider the problem of defining the sampling frame for the customer base of a retail shopping mall. Developing a clear definition of the mall's market area would be difficult. Assuming this could be done, developing the list of all potential consumers in the market area would be almost impossible. There would be numerous sources that would have a large segment of the names (utility companies, phone companies, etc.), but however complete these lists may be, there will be people who will be omitted. The principal liability of the simple random sampling technique is the necessity of forming the sampling frame. There are sampling techniques that are designed to overcome this problem, one of which is called cluster sampling. This and other sampling techniques are discussed in Section 8.5.

8.1   Exercises

Basic Concepts

1. Why is the quality of sample data so important?

2. Why is randomness useful in sampling?

3. What is wrong with a voluntary survey?

4. What is a biased sample?

5. What is a sampling frame? Why is this concept important?

6. Discuss how you would draw a simple random sample of the students at your college.

7. What makes drawing a simple random sample from a geographic area a difficult task?

Exercises

8. A magazine reported the results of a survey in which readers were asked to send in their responses to several questions regarding good eating. Consider the reported results to the question, *How often do you eat chocolate?*

Table for Exercise 8 – Survey Responses	
Category	**% of Responses**
Frequently	13
Occasionally	45
Seldom	37
Never	5

   a. Were the responses to this survey obtained using voluntary sampling techniques? Explain your answer.

   b. What types of biases may be present in the responses?

   c. Is 13% a reasonable estimate of the proportion of all Americans who eat chocolate frequently? Explain.

9. A magazine reported the results of a survey in which readers were asked to send in their responses to several questions regarding anger. Consider the reported results to the question, *How long do you usually stay angry?*

Table for Exercise 9 – Survey Responses	
Category	**% of Responses**
A few hours or less	48
A day	12
Several days	9
A month	1
I hold a grudge indefinitely	22
It depends on the situation	8

   a. Were the responses to this survey obtained using voluntary sampling techniques? Explain your answer.

   b. What types of biases may be present in the responses?

   c. Is 22% a reasonable estimate of the proportion of all Americans who hold a grudge indefinitely? Explain.

10. Students in a marketing class have been asked to conduct a survey to determine whether or not there is a demand for an insurance program at a local college. The students decide to randomly select students from the local college and mail them a questionnaire regarding the insurance program. Of the 150 surveys that were mailed, 50 students responded to the following survey item: *Pick the category which best describes your interest in an insurance program.*

Table for Exercise 10 – Survey Responses	
Category	**% of Responses**
Very Interested	50
Somewhat Interested	15
Interested	10
Not Very Interested	5
Not At All Interested	20

    a. What types of biases may be present in the responses?

    b. Is 50% a reasonable estimate of the proportion of all students who would be very interested in an insurance program at the local college? Explain.

    c. Is 50% a reasonable estimate of the proportion of all business majors who would be very interested in an insurance program at the local college? Explain.

    d. What strategies do you think the marketing students could have used to get a less biased response to their survey?

    e. Suppose the program was created and only a few people registered. How could the survey question have been reworded to better predict actual enrollment?

11. Television news programs often conduct opinion surveys by announcing some question on the air and advising viewers to call different numbers for a *yes* or *no* response. National television programs do the same thing except they use 900 numbers and the respondent must pay for the call. Suppose that a national news program asks its viewers to phone in a response to the following: *Women should be permitted to assume combat roles in the military.* The results of the particular survey were 34% *yes* and 66% *no*. Is it reasonable to believe that the results of the survey reflect the attitudes of the nation on this issue? What biases exist in this sampling method?

12. A local politician wants to know what the residents of his community think about an increase in the local property tax to pay for improvements to the highway. He decides to conduct a survey.

    a. What is the population of interest to the politician?

    b. Can you think of any good sources for a sampling frame?

    c. What are the shortcomings (if any) of the sources you picked for the sampling frame?

8.2   Introduction to Sampling Distributions

The notion of a statistic, such as the sample mean, as a random variable may seem odd. After all, when you studied descriptive statistics in Chapter 4, statistics were numbers, not variables. The focus of Chapter 4, however, was to demonstrate techniques for summarizing data after they had been collected. When we consider a sampling distribution, the "after the data have been collected" condition is dropped. A sampling distribution for the sample mean would describe the means of all possible samples of a particular sample size from a specified population. Since the value of the sample mean for any particular sample depends on the sample we draw, the sample mean is a random variable.

As we studied in Chapter 6, the best way to summarize outcomes of some random phenomenon (in this case the sample mean) is with a probability distribution. The probability distribution which describes the distribution of the sample means is called the **sampling distribution of the sample mean**.

Although the discussion has focused on the sample mean, the same arguments could be applied to the sample variance, sample proportion, and sample standard deviation. All of these sample statistics have sampling distributions as well.

Statistics as Random Variables

Statistical inference is the focus of a large portion of the remainder of this text. As we previously discussed, statistical inference uses statistics calculated from samples as the basis of an inference about the population. If the sample is representative, then the sample statistics (which are random variables that depend on which random sample is selected) ought to be close to their population counterparts. In other words, the sample mean ought to be close to the population mean, the sample proportion ought to be close to the population proportion, and the sample standard deviation should be close to the population standard deviation. Since sample statistics will be used as the basis of the statistical inference, we must know how those statistics vary from one sample to another. Once the variability of the sample statistic is understood, we will be able to make probability statements regarding our inferences.

Why Calculate the Sample Mean?

When analyzing ratio data, the first piece of summary information that an analyst wants to determine is the mean. For most populations, performing a census to determine the population mean is impractical. The only alternative is to use sample information. It seems reasonable that the sample mean, $\bar{x}$, would contain an enormous amount of information about the population mean, μ, and would thus be a sensible estimate of the population mean. Generally, if you wish to estimate a population value – be it the mean, standard deviation, or proportion – the corresponding sample value will be a good **estimator**.

> **Definition**
>
> A **point estimator** is a single-valued estimate, calculated from the sample data, which is intended to be close to the true population value.

Can you be sure that the sample mean will always be close to the population mean? When dealing with random variables, nothing is certain, but there are methods of reducing the probable error. To understand how this is achieved, we must examine how the sample mean varies. The next section will introduce the distribution of the sample mean and how this distribution can be used to make statistical inferences.

8.2   Exercises

Basic Concepts

1. Is the sample mean always close to the population mean?

2. Under what conditions is the sample mean considered a random variable?

3. What is the sampling distribution of the sample mean?

4. Describe how statistics as random variables are crucial to statistical inference.

5. What is an estimator? Give an example.

8.3   The Distribution of the Sample Mean

Sample means vary because sample data vary from sample to sample. As an illustration, suppose that an automobile manufacturer wished to determine the average miles per gallon (mpg) of a specific vehicle model that it manufactures. Since determining the mpg of each vehicle is very time consuming, the manufacturer has decided to select two vehicles from a batch of six. Suppose that the actual mpg of the six vehicles are given in Table 8.2.

Table 8.2 – Miles per Gallon	
Car	MPG
A	25
B	27
C	40
D	29
E	28
F	30
Mean	29.8
Variance	23.14
Standard Deviation	4.81

It is important to realize that we are assuming the above set of data constitutes a population. The mean mpg rating of the population in Table 8.2 is approximately 29.8 and the population standard deviation is approximately 4.81.

$$\mu \approx 29.8$$
$$\sigma \approx 4.81$$

Both of these measures are considered population parameters. The mpg ratings given in Table 8.2 are not known by the manufacturer when the shipment arrives. The manufacturer's job is to estimate the population mean using a sample estimate, in this case using the sample mean from a sample of size two.

How many different samples of size two can be drawn? Assuming no replacement, there would be 15 possible samples of size two if order does not matter. A list of all possible samples and the resulting sample means is given in Table 8.3.

Sample Number	Car 1	Car 2	First Observation	Second Observation	Mean $\bar{x}$
1	A	B	25	27	26.0
2	A	C	25	40	32.5
3	A	D	25	29	27.0
4	A	E	25	28	26.5
5	A	F	25	30	27.5
6	B	C	27	40	33.5
7	B	D	27	29	28.0
8	B	E	27	28	27.5
9	B	F	27	30	28.5
10	C	D	40	29	34.5
11	C	E	40	28	34.0
12	C	F	40	30	35.0
13	D	E	29	28	28.5
14	D	F	29	30	29.5
15	E	F	28	30	29.0

Table 8.3 – MPG Sample Measurements

The sample means in Table 8.3 vary and when something varies there are at least three questions to ask:

1.   What is the central value of the variable?

2.   What is the variability of the variable?

3.   Is there a pattern (distribution) to the variability?

What is the Central Value of $\bar{x}$?

Intuitively, you would expect the sample mean to be larger than μ some of the time and smaller than μ some of the time. For large samples, the distribution of $\bar{x}$ will be relatively symmetrical, and consequently, $\bar{x}$ should be larger than μ about 50% of the time and smaller than μ about 50% of the time. But for small samples, the distribution of the sample mean may not be symmetrical. This is the case for the sample means in Table 8.3. 10 of the 15 means are below 29.8, which is the population mean for the population in Table 8.2. Generally, however, the sample means should be near the population mean, or symbolically, $\bar{x}$ should be near μ. It can be shown theoretically that the mean of the $\bar{x}$'s equals μ. In the example, the mean of the sample means is approximately 29.8, which equals the population mean. This is not a coincidence.

Definition

If the average value of an estimator equals the population parameter being estimated, the estimator is said to be **unbiased**.

Estimators are similar to marksmen. When you shoot, you want to hit what you are shooting at.

When you estimate, you want to get as close as possible to the population characteristic you are estimating. But, bullets do not always land exactly where the marksman aims. If the gun sights are properly adjusted, then the shots will be dispersed around the middle of the target area.

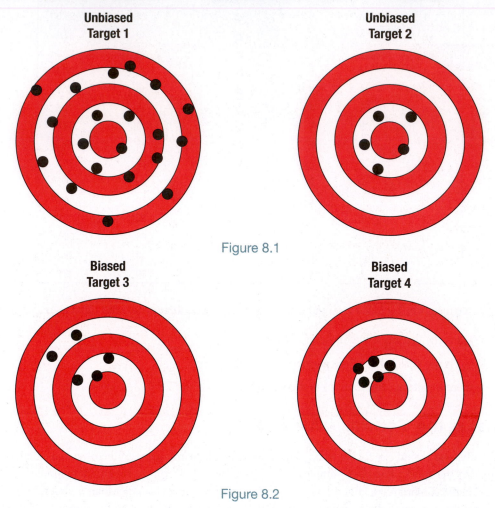

Unbiased Target 1

Unbiased Target 2

Figure 8.1

Biased Target 3

Biased Target 4

Figure 8.2

An estimator which produces estimates centered around the true value is said to be unbiased.

Unbiasedness is desirable property for an estimator to possess. Just as the ideal target rifle is one that would hit in exactly the same place every shot, the ideal estimator is one that is unbiased. Since the mean of $\bar{x}$ is always equal to μ, $\bar{x}$ is an unbiased estimator of μ.

> **Definition**
>
> The sample mean, $\bar{x}$, is an unbiased estimator of μ.
>
> The sample proportion, $\hat{p}$, is an unbiased estimator of p.
>
> The sample variance, s^2, is an unbiased estimator of σ^2.

If an estimator is unbiased, its variability determines its reliability. If an unbiased estimator is extremely variable, then the individual estimates it produces may not be as close to the parameter being estimated as estimates produced by a biased estimator with little variability. This is why it is important to not only consider the central value, but also the variability of a random variable, in this case the sample mean.

What is the Variability of $\bar{x}$?

The variability of an estimator reveals a great deal about the quality of that estimator. In order to assess how well the sample mean estimates the population mean, the standard deviation of the sample means must be determined.

Formula
Standard Deviation of the Sample Mean: Infinite Population
It can be shown that for a population of infinite size, the standard deviation of $\bar{x}$, denoted as $\sigma_{\bar{x}}$, is
$$\sigma_{\bar{x}} = \frac{\sigma}{\sqrt{n}},$$
where σ is the population standard deviation and n is the sample size.

We refer to $\sigma_{\bar{x}}$ as the **standard error of the mean**, generally called the **standard error**, which is the standard deviation of a point estimator. We will use the standard error of the mean to indicate how far the sample mean is from the population mean.

Suppose, for example, you were drawing a sample from a population whose standard deviation is 4.81. The standard deviation of the sample means for samples of size $n = 2$ would be

$$\sigma_{\bar{x}} = \frac{\sigma}{\sqrt{n}} = \frac{4.81}{\sqrt{2}} \approx 3.40.$$

If the population is finite, as in Table 8.2, then the finite population correction factor $\left(\sqrt{\frac{N-n}{N-1}}\right)$ must be applied to the calculation of the standard deviation of the sample mean.

Formula
Standard Deviation of the Sample Mean: Finite Population
For a finite population the standard deviation of $\bar{x}$ is
$$\sigma_{\bar{x}} = \sqrt{\frac{N-n}{N-1}} \cdot \frac{\sigma}{\sqrt{n}},$$
where $N =$ the size of the population and $n =$ the size of the sample.

Note: For most of the populations you will encounter, the assumption is made that the population is infinite.

Correcting for the finite population represented in Table 8.2, the standard deviation of the sample means for samples of size 2 would be

$$\sigma_{\bar{x}} = \sqrt{\frac{N-n}{N-1}} \cdot \frac{\sigma}{\sqrt{n}} = \sqrt{\frac{6-2}{6-1}} \cdot \frac{4.81}{\sqrt{2}} \approx 3.04.$$

Examining the formula for $\sigma_{\bar{x}}$, we notice that as the size of the sample, n, increases, the variability of the sample mean decreases. The possibility of changing the variability of an estimator means the accuracy of the estimator can be manipulated. For example, suppose that instead of using a sample of size two, the manufacturer decides to use a sample of size three. Six items chosen three at a time could produce 20 samples of size three. The samples and means of each sample are shown in Table 8.4. The standard deviation of the sample means for $n = 3$ in Table 8.4 is approximately 2.15, using the finite population correction factor.

For both samples ($n = 2$ and $n = 3$), the mean of the sample means is approximately 29.8, which equals the population mean. But the standard deviation of the sample means for samples of size three is 2.15, which is smaller than the standard deviation for samples of size two (3.04).

$$\text{For } n = 2, \ \sigma_{\bar{x}} \approx 3.04$$

$$\text{For } n = 3, \ \sigma_{\bar{x}} \approx 2.15$$

Table 8.4 – MPG Sample Measurements						
Sample No.	Car 1	Car 2	Car 3	First Observation	Second Observation	Third Observation	Mean $\bar{x}$
1	A	B	C	25	27	40	30.67
2	A	B	D	25	27	29	27.00
3	A	B	E	25	27	28	26.67
4	A	B	F	25	27	30	27.33
5	A	C	D	25	40	29	31.33
6	A	C	E	25	40	28	31.00
7	A	C	F	25	40	30	31.67
8	A	D	E	25	29	28	27.33
9	A	D	F	25	29	30	28.00
10	A	E	F	25	28	30	27.67
11	B	C	D	27	40	29	32.00
12	B	C	E	27	40	28	31.67
13	B	C	F	27	40	30	32.33
14	B	D	E	27	29	28	28.00
15	B	D	F	27	29	30	28.67
16	B	E	F	27	28	30	28.33
17	C	D	E	40	29	28	32.33
18	C	D	F	40	29	30	33.00
19	C	E	F	40	28	30	32.67
20	D	E	F	29	28	30	29.00

Since both estimators are unbiased and are centered around the population mean, the standard deviation of the estimator is a measure of how close the estimator (in this case the sample mean) is to the population mean. In other words, the lower $\sigma_{\bar{x}}$ is, the better the estimator is.

As in many problems, there is an accuracy versus dollars trade off. Greater accuracy can be achieved by taking a larger sample, but larger samples normally cost more to collect and examine.

Suppose a population had a mean of 43,660 and a standard deviation of 2500. The distributions in Figure 8.3 are the distributions of the sample mean for samples of size 25, 100, and 200, respectively, from this population. Based on the graphs in Figure 8.3, it seems that the estimator shown for $n = 200$ would be preferred. Because it has less variability, the estimates of μ ($\bar{x}$'s) from samples of $n = 200$ should be closer (on average) to μ.

Sampling Distribution of the Sample Mean, σ = 2500

$n = 200$
$n = 100$
$n = 25$

$\mu = 43{,}660$

Figure 8.3

Properties

Characteristics of the Sample Mean

If an estimate of the population mean is required, the sample mean possesses two desirable characteristics.

1. The mean of the sample means is the population mean. Another way of expressing this concept is to say that the expected value of $\bar{x}$ is equal to the population mean. Symbolically, this can be expressed as

$$E(\bar{x}) = \mu \text{ (unbiasedness)}.$$

2. If the sample size is increased, the standard error of the sample mean decreases. This implies that the quality of the estimator tends to improve as the sample size increases.

$$\sigma_{\bar{x}} = \frac{\sigma}{\sqrt{n}}$$

The second characteristic has an important consequence for estimating a population parameter. By choosing a sufficiently large sample size, an estimate can be obtained with some specified level of accuracy. Being able to predetermine the accuracy of an estimate is an important topic in statistics and will be discussed in the next chapter.

Is There a Familiar Pattern to the Variability?

The **central limit theorem** is a very important theorem that summarizes the distribution of the sample mean. The distribution of the sample mean becomes closer to a normal distribution as the sample size becomes larger, regardless of the distribution of the population from which the sample is drawn.

The most important feature of the central limit theorem is that it can be applied to any population. Because the theorem does not have any distributional assumptions, it is widely applicable and is one of the cornerstones of statistical inference. Many of the statistical techniques discussed in subsequent chapters will have their theoretical basis in this theorem.

Theorem

Central Limit Theorem

If a sufficiently large random sample (i.e. $n \geq 30$) is drawn from a population with mean μ and standard deviation σ, the distribution of the sample mean will have the following characteristics.

1. An approximately normal distribution regardless of the distribution of the underlying population.

2. $\mu_{\bar{x}} = E(\bar{x}) = \mu$ (The mean of the sample means equals the population mean.)

3. $\sigma_{\bar{x}} = \dfrac{\sigma}{\sqrt{n}}$ (The standard deviation of the sample means equals the standard deviation of the population divided by the square root of the sample size.)

The only restrictive feature of the theorem is that the sample size must be sufficiently large for the theorem to be applicable. Even if the distribution of the population deviates substantially from the normal distribution, a sample size of at least 30 will be sufficiently large to produce a sampling distribution for $\bar{x}$ that is approximately normal. Additionally, points 2 and 3 are true even for small samples when it can be shown that the samples are drawn from a population that is normally distributed. That is, if the population is known to be normally distributed, then the sampling distribution of $\bar{x}$ will be normally distributed for any sample size.

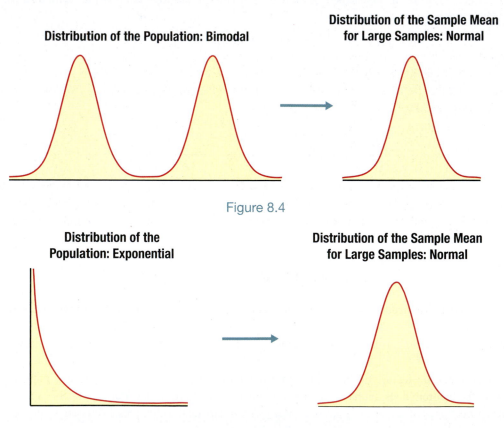

Distribution of the Population: Bimodal

Distribution of the Sample Mean for Large Samples: Normal

Figure 8.4

Distribution of the Population: Exponential

Distribution of the Sample Mean for Large Samples: Normal

Figure 8.5

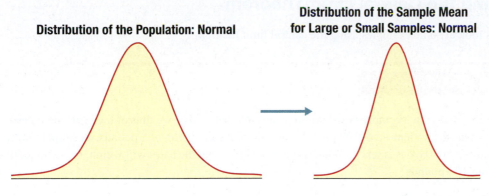

Distribution of the Population: Normal

Distribution of the Sample Mean for Large or Small Samples: Normal

Figure 8.6

Distribution of the Population: Uniform

Distribution of the Sample Mean for Large Samples: Normal

Figure 8.7

If we are dealing with samples that are sufficiently large, once we have determined that the central limit theorem applies and that the sampling distribution of the sample means is approximately normally distributed, we can standardize any sample mean to find probabilities that we are interested in. Thus, we can adapt the *z*-score formula as follows.

$$z = \frac{\bar{x} - \mu_{\bar{x}}}{\sigma_{\bar{x}}} = \frac{\bar{x} - \mu}{\sigma / \sqrt{n}}$$

If the population standard deviation is unknown and the sample size is sufficiently large ($n \geq 30$), we can estimate σ with s, and the formula for the *z*-score becomes the following.

$$z = \frac{\bar{x} - \mu_{\bar{x}}}{s_{\bar{x}}} = \frac{\bar{x} - \mu}{s / \sqrt{n}}$$

The History of the Central Limit Theorem

Pierre-Simon Laplace is credited with the initial statement of the central limit theorem in 1776. He developed the theorem while working on the probability distribution of the sum of meteor inclination angles.

Although the theorem is stated with respect to the sample mean, it is a more general theorem regarding the sum of random variables. If you add up a sufficiently large number of random variables the sum will be normally distributed.

Using the Central Limit Theorem

Let's look at some examples using the central limit theorem.

Example 8.2

A recycling plant compresses aluminum cans into bales. The weights of the resulting bales are known to have a mean of 100 pounds and a standard deviation of 8 pounds. A simple random sample of 50 bales is taken. What is the probability that the bales will weigh, on average, less than 104 pounds?

Solution

The information provided in the problem is as follows:

$$\mu = 100 \text{ pounds, } \sigma = 8 \text{ pounds, and}$$
$$n = 50 \text{ bales.}$$

Let X be a random variable that represents the weight of a bale.

By the central limit theorem (since $n \geq 30$), the distribution of $\bar{x}$ will be approximately normal with a mean equal to the population mean, 100, and a standard error given by $\sigma_{\bar{x}} = \dfrac{\sigma}{\sqrt{n}} = \dfrac{8}{\sqrt{50}}$.

Thus, to find the probability that the bales weigh, on average, less than 104 pounds, we are interested in the following probability.

$$P(\bar{x} < 104) = ?$$

Figure 8.8

Since $\bar{x}$ is a normal random variable, the probability that $\bar{x}$ is less than 104 pounds is determined using a z-transformation.

$$P(\bar{x} < 104) = P\left(z < \frac{\bar{x} - \mu}{\sigma / \sqrt{n}}\right) = P\left(z < \frac{104 - 100}{8 / \sqrt{50}}\right) = P(z < 3.54) \approx 1$$

Since Table B's largest value is 3.49, yielding a probability of 0.9998, we say that the probability is approximately 1.

The concept of using a statistic to estimate a population parameter introduces the concept of the **error of estimation** (also referred to as **sampling error** or **estimation error**).

> ### Definition
>
> In general, the **error of estimation** is the difference between a sample statistic and the parameter that it is estimating.

The following example will further illustrate the error of estimation and how it is used.

Example 8.3

Suppose that a population has an unknown mean and a standard deviation of 5. A sample of size 100 is drawn from the population. If the sample mean is used as an estimate of the population mean, what is the probability that the sample mean will be within one unit of the true mean?

Solution

Since the error of estimation is defined to be the difference between the sample statistic and the parameter that it is estimating, for this problem the error of estimation is one. In the problem at hand, the error is given by $\mu - \bar{x}$ since we will be using $\bar{x}$ to estimate μ.

However, we are just as interested in negative errors as positive errors. Thus, to satisfy the condition that an error of less than one unit has been made, we are interested in the following probability.

$$P\left(\mu - 1 < \bar{x} < \mu + 1\right)$$

To find the probability that the sample mean is within one unit of the population mean, we must standardize the equation above so that we can use the standard normal distribution. Standardizing, we obtain the following.

$$P\left(\frac{(\mu-1)-\mu}{5/\sqrt{100}} < z < \frac{(\mu+1)-\mu}{5/\sqrt{100}}\right) = P(-2 < z < 2)$$

$$P(-2 < z < 2) = P(z < 2) - P(z < -2) = 0.9772 - 0.0228 = 0.9544$$

Sampling Distribution of $\bar{x}$

The probability
that the error of
estimation of $\bar{x}$
is less than 1

$\mu - 2$ $\mu - 1$ $\bar{x}$ $\mu + 1$ $\mu + 2$

Figure 8.9

Standard Normal Distribution

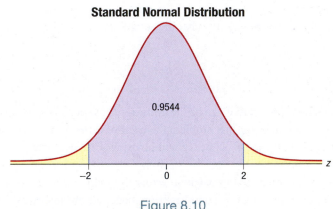

0.9544

-2 0   2 z

Figure 8.10

Thus, if a sample size of 100 is drawn from the population given, the probability that the sample mean will be within one unit of the population mean is 0.9544.

In this problem, we were able to determine the probability of making an error of less than one unit without knowing the mean of the population. This probability (0.9544) tells us a lot about the quality of the estimate $\bar{x}$. The estimate of μ obtained from a sample of size 100 is more than just an estimate now, it is an estimate with a "level of confidence" in its quality.

Example 8.4

The lifetime of a certain type of micro transistor is normally distributed with a mean of 156 hours. It is known that 1.5 percent of the transistors have a lifetime greater than 167 hours.

a. What is the standard deviation of the distribution of the lifetimes of the transistors?

b. Using the standard deviation calculated in part **a.**, what is the probability that the average lifetime of a sample of 25 transistors is at least 155 hours?

Solution

a. Let X = lifetime of a micro transistor.

X is normally distributed with a mean of 156 hours. (Note that even though $n < 30$ we can still use the methods presented because we are told the underlying population is normal.) We also know that 1.5% of the transistors have a lifetime greater than 167 hours. This can be written as

$$P(X > 167) = 0.015 \qquad\qquad (1)$$

Since σ is unknown, we can find σ using (1) such that

$$P\left(z > \frac{167 - 156}{\sigma}\right) = 0.015.$$

Figure 8.11

$$P(z > z_0) = 0.0150 \text{ and } P(z < z_0) = 0.9850 \text{ where } z_0 = \frac{167 - 156}{\sigma}$$

Finding the value 0.9850 in Table B, we see that $z_0 = 2.17$.

Thus, $\dfrac{167 - 156}{\sigma} = 2.17$, and solving for σ, we get

$$\sigma \approx 5.0691.$$

b. $P(\bar{x} \geq 155) = ?$

Since X is normally distributed, then $\bar{x}$ is also normally distributed. Thus,

$$P(\bar{x} \geq 155) = P\left(z \geq \frac{155-156}{5.0691 \Big/ \sqrt{25}} \right)$$
$$= P(z \geq -0.99)$$
$$= 1 - P(z < -0.99)$$
$$= 1 - 0.1611$$
$$= 0.8389$$

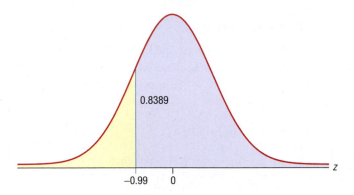

Figure 8.12

Thus, the probability that the average lifetime of a sample of 25 transistors is at least 155 hours is 0.8389.

8.3   Exercises

Basic Concepts

1. What key three questions should be asked when considering a random variable?

2. Explain the difference between a biased estimator and an unbiased estimator.

3. Give three examples of estimators that are unbiased.

4. Is an unbiased estimator always closer to the parameter being estimated than a biased estimator? Explain.

5. What is the standard error of the mean? What does it indicate?

6. What are two desirable characteristics of the sample mean?

7. Explain the central limit theorem.

8. What effect does increasing the sample size have on the accuracy of an estimate?

9. What is the error of estimation?

Exercises

10. Suppose the random variable X has a mean of 20 and a standard deviation of 5. Calculate the mean and the standard deviation of the sample mean for each of the following sample sizes (assume the population is infinite).

    a. $n = 35$

    b. $n = 50$

    c. $n = 75$

    d. What happens to the size of the standard deviation of the sample mean as the sample size increases?

11. Suppose the random variable X has a mean of 50 and a standard deviation of 10. Calculate the mean and standard error for each of the following sample sizes (assume the population is infinite).

    a. $n = 40$

    b. $n = 55$

    c. $n = 100$

    d. What happens to the size of the standard error as the sample size increases?

12. If there is a normally distributed random variable with a mean of 75 and a standard deviation of 22, what is the probability that the mean of a sample of size 19 will be greater than 80?

13. If a sample of size 40 is drawn from a population that has a mean of 276 and a variance of 81, what is the probability that the mean of the sample will be less than 273?

14. Suppose there is a normally distributed population with a mean of 250 and a standard deviation of 50. If $\bar{x}$ is the average of a sample of 36, find the following probabilities.

    a. $P(\bar{x} \leq 240)$ b. $P(\bar{x} \geq 255)$

    c. $P(246 \leq \bar{x} \leq 260)$ d. $P(234 \leq \bar{x} \leq 245)$

15. Suppose there is a normally distributed population with a mean of 100 and a standard deviation of 10. If $\bar{x}$ is the average of a sample of 50, find the following probabilities.

    a. $P(\bar{x} \leq 110)$ b. $P(\bar{x} \geq 90)$

    c. $P(95 \leq \bar{x} \leq 115)$ d. $P(85 \leq \bar{x} \leq 98)$

16. A company fills bags with fertilizer for retail sale. The weights of the bags of fertilizer have a normal distribution with a mean weight of 15 lb and standard deviation of 1.70 lb.

    a. What is the probability that a randomly selected bag of fertilizer will weigh between 14 and 16 pounds?

    b. If 35 bags of fertilizer are randomly selected, find the probability that the average weight of the 35 bags will be between 14 and 16 pounds.

17. A travel agency conducted a survey of the prices charged by ocean cruise ship lines and determined they were approximately normally distributed with a mean of $110 per day and a standard deviation of $20 per day.

   a. If an ocean cruise ship line is chosen at random, find the probability that it will charge less than $99 per day.

   b. What is the probability that the average charge for a randomly selected sample of 35 ocean cruise ship lines will be less than $99 per day?

18. The turkeys found in a particular county have an average weight of 15.6 pounds with a standard deviation of 4.00 pounds. Forty-five turkeys are randomly selected for a county fair.

   a. Find the probability that the average weight of the turkeys will be less than 14.5 pounds.

   b. What is the probability that the average weight of the turkeys will be more than 17 pounds?

   c. Find the probability that the average weight of the turkeys will be between 13 and 18 pounds.

19. The average score for a water safety instructor (WSI) exam is 75 with a standard deviation of 12. Fifty scores for the WSI exam are randomly selected.

   a. Find the probability that the average of the fifty scores is at least 80.

   b. Find the probability that the average of the fifty scores is at most 70.

   c. Find the probability that the average of the fifty scores is between 72 and 78.

20. A college food service buys frozen fish in boxes labeled 10 pounds. The true average weight of the boxes is 8 pounds with a standard deviation of 2 pounds. The food service director suspects that the boxes do not contain as much fish as advertised. He decides to inspect 40 boxes from the next shipment. If the average weight is less than 10 pounds he will reject the entire shipment. Find the probability that the food service director will not reject the shipment.

21. The AQI, or the Air Quality Index, is an index used to determine the ozone level in a city. Depending upon the AQI reading, it may not be safe to jog or even to go outside. Readings in the 0-50 range mean that the air quality conditions are considered "good," 51-100 are "moderate," 101-150 means "unhealthy for sensitive groups," 151-200 means "unhealthy," 201-300 means "very unhealthy," and 301-500 means "hazardous." Suppose that an industrial region has an average AQI reading of 102 with a standard deviation of 40. Find the probability that for a random sample of 50 days, the average AQI reading is:

   Source: airnow.gov

   a. at least 105.

   b. at most 90.

   c. between 100 and 115.

8.4   The Distribution of the Sample Proportion

There are many instances in which the variable of interest is a proportion. A manufacturer might be interested in knowing what fraction of his manufacturing components are defective. A market researcher might be interested in knowing what proportion of persons on a mailing list will buy the company's product. A college might be concerned with the fraction of freshmen that may be struggling academically after the first year. Population proportions must be estimated just like population means.

The symbols used to represent the population and sample proportions are

p, the **population proportion**, and

$\hat{p}$, the **sample proportion**.

Note: $\hat{p}$ is pronounced *p-hat*.

Determining the Sample Proportion, $\hat{p}$

Suppose you are trying to determine the fraction of manufacturing components that are defective. If you select 120 components at random and 38 components are defective, then

$$\hat{p} = \text{proportion in the sample that are defective} = \frac{38}{120} \approx 0.3167.$$

In general, when calculating a proportion, the number in the sample that possess the characteristic of interest goes in the numerator, and the size of the sample is placed in the denominator.

Formula

Sample Proportion

The **sample proportion** is given by

$$\hat{p} = \frac{x}{n},$$

where x is the number of observations in the sample possessing the characteristic of interest and n is the total number of observations in the sample.

Just as the sample mean was a good estimate of the population mean, the sample proportion is a good estimate of the population proportion. Also, recall that the sample mean varied depending on the sample selected. The sample proportion varies in the same manner.

Since $\hat{p}$ varies, three familiar questions must be examined.

1. What is the central value?

2. What is the variability?

3. Is there a pattern to the variability?

What is the Central Value of $\hat{p}$?

The expected value of the sample proportion, $\hat{p}$, is the population proportion, p. Symbolically, this is expressed as

$$E\left(\hat{p}\right) = p.$$

Since the expected value of the estimator $\hat{p}$ is equal to p, $\hat{p}$ is an unbiased estimator of p.

What is the Variability of $\hat{p}$?

Formula

Standard Deviation of the Sample Proportion

The standard deviation of $\hat{p}$ is given by

$$\sigma_{\hat{p}} = \sqrt{\frac{p(1-p)}{n}},$$

where p is the population proportion and n is the sample size.

Note that $\sigma_{\hat{p}}$ is affected by the values of p and n.

The standard deviation of $\hat{p}$ decreases as n becomes larger. Also, the numerator reaches a maximum when $p = 0.5$ and declines as you move away from that figure. So for a fixed value of n, $\hat{p}$ has its greatest standard deviation when the population proportion equals 0.5. If the population proportion is unknown (which is usually the case), p can be estimated by $\hat{p}$, and the standard deviation of the sample proportion is estimated as

$$\sigma_{\hat{p}} \approx \sqrt{\frac{\hat{p}(1-\hat{p})}{n}}.$$

The sampling distribution of $\hat{p}$ approaches normality as n becomes sufficiently large. The sample size is generally considered "sufficiently large" if $np \geq 5$ and $n(1-p) \geq 5$.

Figure 8.13

The sampling distribution of $\hat{p}$ is summarized below, for both finite and infinite populations.

Properties

Sampling Distribution of the Sample Proportion

If the population is infinite and the sample is sufficiently large $\left(np \ge 5 \text{ and } n(1-p) \ge 5\right)$, the distribution of $\hat{p}$ has the following characteristics.

1. An approximately normal distribution.

2. $\mu_{\hat{p}} = E\left(\hat{p}\right) = p.$ (The mean of the sample proportions equals the population proportion.)

3. $\sigma_{\hat{p}} = \sqrt{\dfrac{p(1-p)}{n}} \approx \sqrt{\dfrac{\hat{p}(1-\hat{p})}{n}}.$

If the population is finite and the sample is sufficiently large, the distribution of $\hat{p}$ has the following characteristics.

1. An approximately normal distribution.

2. $\mu_{\hat{p}} = E\left(\hat{p}\right) = p.$

3. $\sigma_{\hat{p}} = \sqrt{\dfrac{N-n}{N-1}} \cdot \sqrt{\dfrac{p(1-p)}{n}} \approx \sqrt{\dfrac{N-n}{N-1}} \cdot \sqrt{\dfrac{\hat{p}(1-\hat{p})}{n}}$, where N is the size of the population.

Since $\hat{p}$ is a good estimator of p, one of the natural questions to ask is, can limits be established for the error of estimation? Since the sampling distribution of $\hat{p}$ is known, probabilities for various errors of estimation can be determined.

Example 8.5

A series of tests conducted by the company Switches, We Got'em indicates that a particular type of switch manufactured by the company will operate correctly 95% of the time. If 400 switches are selected at random, what is the probability that less than 90% of the switches will operate correctly?

Solution

The information given in the problem is summarized below.

$$p = P \text{ (switch operates correctly)} = 0.95$$

$$n = 400$$

We want to determine the probability that less than 90% of the switches in the sample will operate correctly. This translates to $P(\hat{p} < 0.90)$. Since $np = 400(0.95) = 380 \geq 5$ and $n(1-p) = 400(0.05) = 20 \geq 5$, the sample size is sufficiently large, and we can assume that the sample proportion is approximately normally distributed with $\mu = p = 0.95$ and

$$\sigma = \sqrt{\frac{p(1-p)}{n}} = \sqrt{\frac{0.95(1-0.95)}{400}}.$$

Therefore, we can standardize the distribution of the sample proportion as follows.

$$P(\hat{p} < 0.90) = P\left(z < \frac{\hat{p} - p}{\sqrt{\frac{(p)(1-p)}{n}}} \right) = P\left(z < \frac{0.90 - 0.95}{\sqrt{\frac{(0.95)(1-0.95)}{400}}} \right)$$

$$= P(z < -4.59)$$

$$\approx 0$$

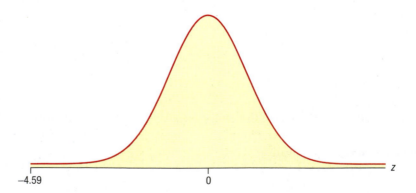

Figure 8.14

Just as in Example 8.2, almost the entire curve is to the right of −4.59 and thus, the probability that $\hat{p} < 0.90$ is approximately 0.

With the information we have developed thus far, we can begin to draw conclusions (i.e. make inferences about the population proportion). If it is true that this particular type of switch works 95% of the time, then it is extremely unlikely that we would observe a sample proportion of less than 90% in a sample of 400 switches.

Example 8.6

Suppose a sample of 400 people is used to perform a taste test. If the true proportion of the population that prefers Pepsi is 0.5, what is the probability that less than 44% of the persons in the sample will prefer Pepsi?

Solution

Assume the population from which the sample is drawn is extremely large and the finite population correction factor is not applicable. The distribution of $\hat{p}$ would then be normal with $\mu = E\left(\hat{p}\right) = p = 0.5$ and

$$\sigma_{\hat{p}} = \sqrt{\frac{(0.5)(1-0.5)}{400}} = 0.025.$$

The probability that the sample proportion is less than 0.44 is given by

$$P\left(\hat{p} < 0.44\right) = P\left(z < \frac{0.44 - 0.5}{0.025}\right)$$
$$= P(z < -2.40) = 0.0082.$$

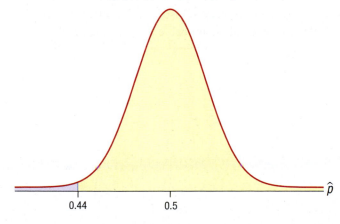

Sampling Distribution of $\hat{p}$

Figure 8.15

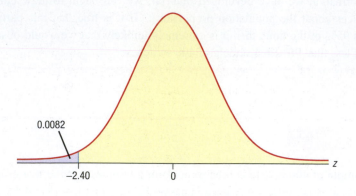

Figure 8.16

If the true proportion of people in the population who prefer Pepsi is 0.5, it is extremely unlikely (0.0082 is less than 1 in 100) to observe a sample proportion as low as 0.44. Suppose that you had to make a decision as to whether cola drinkers were indifferent between Pepsi and Coke. If they were indifferent, the proportion who prefer Pepsi should be around 0.5. If you used a sample of 400 people and observed a sample proportion of 0.438 that preferred Pepsi, which of the conclusions would you believe?

Conclusion A: Cola drinkers are indifferent between Pepsi and Coke. Stated another way, the proportion of persons that favor Pepsi is about 0.5.

Conclusion B: Cola drinkers are not indifferent between Pepsi and Coke. In other words, people prefer one brand of cola over the other.

The likelihood of observing a sample with $\hat{p}$ less than 0.44 is very rare (0.0082) given that the true proportion who prefer Pepsi is $p = 0.5$. This leads us to doubt Conclusion A, and select Conclusion B. The decision-making problem given above is really a statistical inference problem. Although the procedure for analyzing inference problems will be presented in subsequent chapters, the problem illustrates the connection between probability and inference. To reach a decision (make the inference) we used the fact that if the true proportion is really 0.5, a proportion below 0.44 is highly improbable for a sample of 400.

Example 8.7

Suppose a sample of 400 is used to estimate the proportion of U.S. citizens over 18 that favors expanding government regulation of major financial institutions. If the proportion that favors expanding government regulation is really 0.6, what is the probability that the error of estimation will be less than five percent?

Solution

Since the true value of the population proportion is 0.6, the value of $\hat{p}$ must fall between 0.55 and 0.65 in order for the error of estimation to be less than 0.05.

Figure 8.17

In order to determine the probability that $\hat{p}$ will fall in this interval, its distribution must be determined. Since the distribution of $\hat{p}$ is approximately normal for large samples $\left(np \geq 5 \text{ and } n(1-p) \geq 5\right)$, the distribution of $\hat{p}$ will be approximately normal with

$$\mu_{\hat{p}} = p = 0.6$$

$$\sigma_{\hat{p}} = \sqrt{\frac{p(1-p)}{n}} = \sqrt{\frac{0.6(1-0.6)}{400}} \approx 0.0245.$$

To find the probability that $\hat{p}$ is within 0.05 of the true proportion, we must find

$$P\left(p - 0.05 < \hat{p} < p + 0.05\right) = P\left(0.55 < \hat{p} < 0.65\right).$$

Sampling Distribution of $\hat{p}$

Figure 8.18

Using the z-transformation,

$$= P\left(\frac{0.55-0.6}{0.0245} < z < \frac{0.65-0.6}{0.0245}\right)$$

$$= P(-2.04 < z < 2.04)$$

$$= P(z < 2.04) - P(z < -2.04)$$

$$= 0.9793 - 0.0207$$

$$= 0.9586.$$

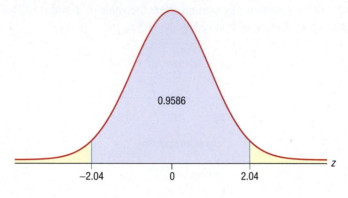

Figure 8.19

Therefore, for a sample of 400 U.S. citizens over 18 years of age, it is highly likely (0.9586) that the error of estimation will be less than five percent.

8.4   Exercises

Basic Concepts

1. What does the symbol $\hat{p}$ represent?

2. What is the connection between $\hat{p}$ and p?

3. Is $\hat{p}$ an unbiased estimator? If so, of what?

4. What are the conditions that make the sample size n "sufficiently large" for a sample proportion?

5. Describe the sampling distribution of $\hat{p}$ if n is sufficiently large.

Exercises

6. A random sample of 40 electronic components has 5 defective components.

    a. Find the sample proportion of components that are defective.

    b. Find the sample proportion of components that are not defective.

7. A random sample of 100 employees of a large steel company has 30 females and 70 males.

    a. Find the sample proportion of female employees.

    b. Find the sample proportion of male employees.

8. Suppose that the true proportion of registered voters who favor the Republican presidential candidate is 0.45. Find the mean and standard deviation of the sample proportion for samples of the following sizes.

    a. $n = 30$

    b. $n = 45$

    c. $n = 65$

    d. What happens to the size of the standard deviation of the sample proportion as the sample size increases?

9. Suppose that the true proportion of Americans over 25 years old that have a 4-year college degree is 0.35. Find the mean and the standard deviation of the sample proportion for samples of the following sizes.

    a. $n = 38$

    b. $n = 52$

    c. $n = 75$

    d. What happens to the size of the standard deviation of the sample proportion as the sample size increases?

10. Suppose the true population proportion is $p = 0.50$. What is the probability that the sample proportion of a sample of size 20 will be greater than 0.60?

11. Suppose the true population proportion, $p = 0.30$. What is the probability that the sample proportion of a sample of size 30 will be less than 0.20?

12. Suppose that the true proportion of Americans who save at least 10% of their income is 0.15.

    If $\hat{p}$ is the sample proportion of Americans surveyed who save at least 10% of their income from a sample of size 38, find the following probabilities.

    a. $P\left(\hat{p} > 0.25\right)$

    b. $P\left(\hat{p} < 0.09\right)$

    c. $P\left(0.10 < \hat{p} < 0.20\right)$

    d. $P\left(0.18 < \hat{p} < 0.25\right)$

13. Suppose that the true proportion of airline pilots between the ages of 35 and 45 is 0.60. If $\hat{p}$ is the sample proportion of airline pilots between the ages of 35 and 45 from a sample of size 100, find the following probabilities.

   a. $P\left(\hat{p} > 0.55\right)$

   b. $P\left(\hat{p} < 0.45\right)$

   c. $P\left(0.50 < \hat{p} < 0.60\right)$

   d. $P\left(0.60 < \hat{p} < 0.75\right)$

14. The director of a radio station in a large metropolitan area believes that the proportion of young professionals (his target market) in the area who prefer rock 'n' roll music has increased from 25% to 35%. The director randomly decides to select 50 young professionals and ask them if they prefer rock 'n' roll to any other type of music. If the sample proportion is greater than 0.35, he will switch to a new format emphasizing rock 'n' roll.

   a. If the true proportion of young professionals who prefer rock 'n' roll has not changed, find the probability that the radio director will switch to the new format.

   b. If the true proportion of young professionals who prefer rock 'n' roll has changed as the director suspects, find the probability that the radio director will switch to the new format.

15. The property manager of a large office building would like to make the building smoke free; however, he does not want to upset too many of his customers. He decides to randomly select 50 of the workers in the building and ask them whether or not they smoke. If the sample proportion of workers who smoke is less than 0.30, the property manager will make the building smoke free.

   a. Find the probability that the property manager will make the building smoke free when the true proportion of smokers is 0.5.

   b. Find the probability that the property manager will not make the building smoke free when the true proportion of smokers is 0.2.

16. Eighty percent of the flights arriving in Atlanta for a large U.S. airline are on time. If the FAA randomly selects 50 of the airline's flights, find the probability that:

   a. at least 85% of the sampled flights will be on time.

   b. at most 70% of the sampled flights will be on time.

   c. between 75% and 85% of the sampled flights will be on time.

17. Approximately 7% of the nation's public school children in grades 2 through 5 took medication in 2003 for attention deficit hyperactivity disorder (ADHD), a developmental disorder characterized by impulsiveness or difficulty concentrating or sitting still. The main treatment prescribed for ADHD is Ritalin, a relatively safe drug with few side effects. Assume that a suburban elementary school had an enrollment of 286 students in 2003.

   a. Find the probability that at least 4% of the school children took medication for ADHD.

   b. Find the probability that between 5% and 8% of the school children took medication for ADHD.

| 8.5 | **Other Forms of Sampling** |

Random sampling is an effective means of obtaining a sample that is representative of the population. As we discussed previously, acquiring an exact sampling frame for the population under consideration is a requirement for simple random sampling, a requirement which can be time-consuming and expensive. In addition, it is sometimes not even possible to list all the members of a population. There are other sampling strategies that are designed to reduce the cost of sampling or add control to the sampling procedure. These techniques can be categorized as probability samples or non-probability samples.

Probability samples enable an analyst to determine the probable errors that an estimator might generate. Essentially, they allow the analyst a known degree of confidence in his or her estimation. All of statistical inference relies on probability sampling. **Non-probability samples** are convenient means of obtaining sample data. If data from a non-probability sample are used to estimate a population parameter, there is no statistical theory that helps define the potential error of the estimate, and hence no statement about the estimate's reliability can be made.

Non-Probability Samples

Non-probability samples come in several forms. A **judgment sample** is a sample in which sample values are selected by an expert in the field. One of the common uses of judgment samples is in auditing. When auditing a company's accounts receivable, an auditor may use a judgment sample to verify the accounts outstanding. The quality of the sample data is related to the competence of the expert. If the expert is good at what he or she does, this type of sampling can produce reasonable representations of the population and correspondingly good estimates of the population parameters.

Another type of non-probability sample is the **convenience sample**. As the name implies, a convenience sample is nothing more than a convenient group of observations. For example, the students in your statistics class would be a convenience sample of students at your college. Although convenience samples could be representative of the population, they tend to possess more bias than other forms of sampling. Consider your statistics class. It is likely that the class is dominated by some particular group of majors and has a disproportionate number of sophomores and juniors. Despite their shortcomings, convenience samples are convenient, and certainly a convenient sample is often better than no sample at all. There are two disadvantages to using a convenience sample. First, it's difficult to determine how representative the sample is of the population, or if it is representative at all. Secondly, even though the convenience sample may not be representative of the population, statistical methods are often applied. Since it's not a probability sample, the results will lack validity and are questionable.

Convenience Sampling

Figure 8.20

The worst forms of non-probability samples are voluntary or self-selected samples. Those samples were discussed in Section 8.1.

Systematic Sampling

One type of sampling technique, the **systematic sample**, does not belong to the probability or non-probability sample categories.

> **Definition**
>
> **Systematic sampling** involves including every k^{th} member of the population in the sample.

For example, suppose a sample of 1000 names is to be selected from a mailing list that contains 80,000 names. If every 80th person in the mailing list is selected for inclusion in the sample, the result will be a sample of 1000 names. If the names are in random order, then the systematic sample will produce a random sample. Unfortunately, it is difficult to be sure that the list does not possess some pattern that would bias the sample. Still systematic samples are regularly used in sampling mailing lists. They are also used in controlling production quality. Production-oriented quality control plans frequently call for the inspection (sample) of every k^{th} item on an assembly line. Conceptually, inspection of every k^{th} item is no different from sampling every k^{th} item from a mailing list. A sample using this technique is often called an "almost random sample."

Systematic Sampling

Every 4th bottle

Figure 8.21

Systematic samples are generally good samples. But if there is some pattern in the sampling frame that corresponds to the sampling pattern, an unrepresentative sample can result. For example, suppose you wished to sample a list containing total daily sales. If you sampled every 7th item, the result would be a sample that contained sales from only one day of the week.

Cluster Sampling

Although simple random sampling may be feasible, the traveling cost required to obtain the sample information may be prohibitive. Suppose McDonald's wishes to perform quality control inspection on its franchises. If a random sample of 500 is drawn from the 31,000 McDonald's restaurants scattered throughout the world, the quality inspection team could be in for quite a bit of traveling. Instead of selecting the stores individually, suppose we divide the world into 200 regions. Then we could select 20 of these regions and inspect every store in the region. This technique is called **cluster sampling**. Using this technique, an inspector would be required to travel much less and a substantial reduction in cost would result.

Clusters are not always geographic. Suppose you were interested in soliciting student opinion regarding book store prices. If everyone in the college had a 9:00 AM class, one reasonable method of clustering would be to use each class as a cluster. Randomly select a set of clusters (classes) and interview each member of the selected cluster. The selection of clusters depends greatly on the population and variables of interest.

Cluster Sampling

Figure 8.22

Definition

Cluster Sampling involves dividing the population into clusters, and randomly selecting a sample of clusters to represent the population.

Cluster sampling can be as effective as simple random sampling if the clusters are as heterogeneous as the population. Unfortunately, clusters are almost never as diverse as the population. If the clusters are geographic, people in the same geographic area tend to be less diverse than the population. If the clusters are not heterogeneous, cluster sampling is less efficient than simple random sampling for a given sample size. Generally, smaller cluster sizes will result in more representative samples.

In simple random sampling, constructing the sampling frame is frequently cumbersome. Cluster sampling simplifies the task, since the initial frame is composed only of clusters. Only those clusters selected as part of the sample must be completely enumerated and sampled.

Cluster sampling is a good alternative to simple random sampling when the population's geographic area is spread out over a large area or when the sampling frame for a simple random sample is difficult to construct.

Stratified Sampling

The fundamental goal of sampling is to obtain a sample that is representative of the population. *Representative* means that characteristics of the population are proportionally represented in the sample. For example, if a population contained 60% females and 30% with blood type A, then it would be desirable for the sample to have 60% females and 30% with blood type A. Simple random samples make no guarantees regarding the constituency of the sample. If a random sample could be taken such that it was assured that population characteristics were properly represented, the resulting samples would tend to be more representative of the population. This is precisely the objective of **stratified sampling**.

Definition

In **stratified sampling**, the population is divided into strata, which are sub-populations. Subjects are then chosen from each stratum in order to accurately represent the population. A stratum can be any identifiable characteristic that can be used to classify the population. If the population consists of people, then strata could be sex, income, political party, religion, education, race, and location.

Stratified Sampling

Figure 8.23

Suppose you wished to estimate the average profit of companies on the New York Stock Exchange (NYSE). Given that all companies are not the same size, it would be best to stratify the companies by some characteristic such as market capitalization or number of employees. For example, if companies were classified in intervals based on market cap, stratification would mean that a fixed percentage of the companies would come from each interval (stratum), thus, giving us a representative sample of the population of companies on the NYSE. The mean and standard deviation from each stratum would be calculated separately and combined to form an estimate of the population parameters. The standard deviation of each subgroup should be smaller than the standard deviation of the population. When these standard deviations are combined to form an estimate of the population standard deviation, the resulting standard deviation will be smaller than if a simple random sample had been used.

In summary, stratified sampling can provide greater accuracy if the population is heterogeneous, and sub-populations of the population can be identified that are relatively homogeneous.

8.5   Exercises

Basic Concepts

1. What are advantages and disadvantages of non-probability samples?

2. What is a judgment sample? Give an example not in the text of when a judgment sample would be appropriate.

3. What is a convenience sample? Are these samples usually representative of the population?

4. What are the worst forms of non-probability samples?

5. Explain the idea of systematic sampling. What are the advantages and disadvantages of this sampling procedure?

6. Explain the idea of cluster sampling. What are the advantages and disadvantages of this sampling procedure?

7. Explain the idea of stratified sampling. What are the advantages and disadvantages of this sampling procedure?

Exercises

8. An employee-owned company has 6000 female employees and 2000 male employees. The human resources department decides to develop a survey on several different benefit plans, including child care and retirement benefits, that they may offer employees in the future. The results of the survey are to be presented to the board of directors for consideration. Because the human resources department wants to be sure of equal representation of the sexes, it has decided to randomly select 500 females and 500 males. What kind of sampling method is the human resources department using? If the sample is used to make inferences regarding the desirability of various benefits packages for all employees, discuss any deficiencies in the sampling procedure.

9. Explain why a systematic sample is not a random sample.

10. Suppose you were instructed to draw a simple random sample from a metropolitan area in order to gain information on the citizens' view on a proposed amendment to the state constitution. To create a simple random sample, you must create a sampling frame. You have decided to use the telephone directory as your frame for the metropolitan area.

    a. Identify the population under consideration.

    b. What kinds of people will be omitted from your frame?

    c. What kinds of biases will be introduced in your sample as a consequence of the omission you described in part **b.**? Can you think of ways of compensating for the bias?

11. A social researcher in Florida wants to determine the average number of children per family in the state.

   a. What is the population of interest?

   b. What variable will be measured?

   c. What level of measurement is the variable of interest?

   d. Discuss the steps that would be necessary for each of the following sampling methods.

      i.   Simple random sampling

      ii. Cluster sampling

      iii. Stratified sampling

   e. What sampling method do you believe would be the most cost-effective? Justify your answer.

12. A stock analyst wants to estimate the average yearly earnings of stocks on the New York Stock Exchange.

   a. What is the population of interest?

   b. Discuss the steps necessary to apply each of the following sampling methods.

      i.   Simple random sampling

      ii. Cluster sampling

      iii. Stratified sampling

13. A news reporter in Orlando, Florida wants to conduct a survey to determine how local residents feel about the institution of a state income tax. Since there will be a lot of people from which to choose, he goes to Disney World and randomly selects individuals entering the complex. He asks the selected people whether or not they favor a state income tax in Florida. The responses to the survey are as follows.

Table for Exercise 13 – Survey Responses	
Category	% of Responses
Favor a Florida State Income Tax	50
Do Not Favor a Florida State Income Tax	50

   a. What sampling technique was used for this survey?

   b. What biases may be present in the responses?

   c. Is 50% a reasonable point estimate of the proportion of Orlando residents who favor the state income tax? Explain.

T   Discovering Technology

Using the TI-84 Plus Calculator

Random Samples

Select a simple random sample of 20 customer accounts from a sampling frame that contains 897 accounts.

1. Choose **MATH**, then select **PRB** and option **5:randInt.**

2. Press **ENTER**, and then input the (minimum number, maximum number, random numbers desired), in this case (**1,897,20**).

3. Press **ENTER**, and the random numbers are listed on the screen.

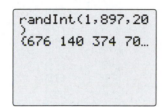

randInt(1,897,20
)
{676 140 374 70...

Figure 8.24

Using Excel

Random Samples

Select a simple random sample of 20 customer accounts from a sampling frame that contains 897 accounts.

1. The function in Microsoft Excel for generating a random number in a certain range is

    RANDBETWEEN(bottom, top)

    where *bottom* is the lower limit of the sampling frame and *top* is the upper limit of the sampling frame. In cell A1, enter the function with the parameters for this problem.

    =RANDBETWEEN(1,897)

    Press **Enter**. A random customer account has been chosen.

2. Place the cursor over the bottom right-hand corner of cell A1, click and drag the box down to cell A20. Twenty random numbers are shown in those cells.

	A
1	345
2	28
3	345
4	698
5	775
6	603
7	838
8	398
9	462
10	554
11	70
12	164
13	201
14	794
15	784
16	25
17	582
18	611
19	377
20	275

Figure 8.25

Using Minitab

Random Samples

Select a random sample of 20 customer accounts from a sampling frame that contains 897 accounts.

1. Press **Calc**, select **Random Data**, and then choose **Integer**.

2. Complete the dialog box (Figure 8.26) with **20** for the number of rows of data to generate, **C1** for the column in which to store the data, **1** for Minimum value, and **897** for Maximum value. Press **OK**.

Figure 8.26

3. Observe the 20 random numbers in column C1.

↓	C1
1	595
2	240
3	5
4	841
5	889
6	456
7	574
8	255
9	183
10	328
11	886
12	548
13	756
14	77
15	400
16	516
17	500
18	603
19	657
20	790

Figure 8.27

| R | Chapter 8 Review |

Key Terms and Ideas

- Sampling
- Sampling Distribution
- Voluntary Sampling
- Selection Bias
- Biased Sample
- Sampling Frame
- Simple Random Sample
- Random Number Table
- Sampling Distribution of the Sample Mean
- Point Estimator
- Unbiased Estimator
- Standard Error of the Mean

- Central Limit Theorem
- Population Proportion
- Sample Proportion
- Sampling Distribution of the Sample Proportion
- Error of Estimation
- Probability Samples
- Non-Probability Samples
- Judgment Sample
- Convenience Sample
- Systematic Sampling
- Cluster Sampling
- Stratified Sampling

Key Formulas

	Section

Standard Deviation of the Sample Mean: Infinite Population

$$\sigma_{\bar{x}} = \frac{\sigma}{\sqrt{n}}$$

8.3

where σ is the population standard deviation and n is the sample size.

Standard Deviation of the Sample Mean: Finite Population

$$\sigma_{\bar{x}} = \sqrt{\frac{N-n}{N-1}} \cdot \frac{\sigma}{\sqrt{n}}$$

8.3

where N = the size of the population and n = the size of the sample.

Characteristics of the Sample Mean

1. $E(\bar{x}) = \mu$

2. $\sigma_{\bar{x}} = \frac{\sigma}{\sqrt{n}}$

8.3

Key Formulas (cont.)

	Section

Central Limit Theorem

If a sufficiently large random sample $(n \geq 30)$ is drawn from a population with mean μ and standard deviation σ, the distribution of the sample mean will have the following characteristics.

1. An approximately normal distribution regardless of the distribution of the underlying population.

    8.3

2. $\mu_{\bar{x}} = E(\bar{x}) = \mu$

3. $\sigma_{\bar{x}} = \dfrac{\sigma}{\sqrt{n}}$

Sample Proportion

$$\hat{p} = \frac{x}{n}$$

8.4

where x is the number of observations in the sample possessing the characteristic of interest and n is the total number of observations in the sample.

Standard Deviation of the Sample Proportion: Infinite Population

If the population is infinite and the sample is sufficiently large $\left(np \geq 5 \text{ and } n(1-p) \geq 5\right)$,

$$\sigma_{\hat{p}} = \sqrt{\frac{p(1-p)}{n}} \approx \sqrt{\frac{\hat{p}(1-\hat{p})}{n}}$$

8.4

where p is the population proportion and n is the sample size.

Standard Deviation of the Sample Proportion: Finite Population

If the population is infinite and the sample is sufficiently large $\left(np \geq 5 \text{ and } n(1-p) \geq 5\right)$,

$$\sigma_{\hat{p}} = \sqrt{\frac{N-n}{N-1}} \cdot \sqrt{\frac{p(1-p)}{n}} \approx \sqrt{\frac{N-n}{N-1}} \cdot \sqrt{\frac{\hat{p}(1-\hat{p})}{n}}$$

8.4

where N is the size of the population.

AE   Additional Exercises

1. A national news network is interested in the opinion which Americans have regarding a national healthcare policy. During the evening news, they display a 900 telephone number and ask their viewers to call in and respond to the question: *Do you favor a national healthcare policy in the U.S.?*

| Table for Exercise 1 – Survey Responses ||
Category	% of Responses
Yes	45
No	45
Do not have enough information to decide	10

   a. What sampling technique was used for this survey?

   b. What biases may be present in the responses?

   c. Is 45% a reasonable estimate of the proportion of all Americans who favor a national healthcare policy? Explain.

2. An entrepreneur wants to open a new Indian restaurant in a resort community. To determine if there is a market for the new restaurant, the entrepreneur decides to conduct a survey.

   a. What is the population of interest to the entrepreneur?

   b. Can you think of any good sources for a sampling frame?

   c. What are the shortcomings (if any) of the sources you picked for the sampling frame?

3. A manufacturer is developing a new type of paint. Test panels were exposed to various corrosive conditions to measure the protective ability of the paint. Based on the results of the test, the manufacturer has concluded that the mean life before corrosive failure for the new paint is 168 hours with a standard deviation of 30 hours. If the manufacturer's conclusions are correct, find the probability that the paint on a sample of 60 test panels will have a mean life before corrosive failure of less than 150 hours.

4. Seventy-five percent of the students graduating from high school in a small Iowa farm town attend college. The town's Chamber of Commerce randomly selects 30 recent graduates and inquires whether or not they will attend college.

   a. Find the probability that at least 80% of the surveyed students will be attending college.

   b. Find the probability that at most 70% of the surveyed students will be attending college.

   c. Find the probability that between 65% and 85% of the surveyed students will be attending college.

   d. Why might the data gathered from this sample misestimate the proportion of students who will actually be attending college?

5. A biology professor is interested in the proportion of students at his college who are pre-med majors. In his next class, he asks for the students who are pre-med majors to raise their hands. Fifty percent of the students raised their hands.

    a. What type of sampling technique was used for this survey?

    b. What types of biases may be present in the responses?

    c. Is 50% a reasonable point estimate of the proportion of students at the college who are pre-med majors? Explain.

6. A report released by the U.S. Census Bureau in November of 2001 stated that fewer families fit the traditional family makeup in 2000 as compared with 1970. The report stated that in 1970, 5.6 million families were headed by women with no husband present, and 16.6% of the households consisted of people living alone. In 2000, these figures increased to 16.2 million and 26.7% respectively. The report also stated that in 1970, 40% of the households consisted of married couples with no children while in the year 2000 this decreased to 25%. Suppose that a random sample of 150 households is chosen in 2000, and the percent of these that consist of people living alone is determined.

    a. Find the probability that at least 30% of those households sampled consist of people living alone.

    b. Find the probability that at most 28% of those sampled households consist of people living alone.

    c. Find the probability that between 26% and 36% of those households consist of people living alone.

7. It is known that the percentage return for a group of stocks in the technology sector is normally distributed with a mean of 15 percent and a standard deviation of 22 percent. Suppose you selected a random sample of 10 stocks from this sector.

    a. What are the mean and standard deviation of $\bar{x}$?

    b. Find the interval containing 68.26% of all possible sample mean returns.

8. A restaurant wants to determine the average time to prepare meals for its customers. To aid in this process, the restaurant randomly selects the meal preparation time of 150 of its customers and finds that the average preparation time is 18 minutes with a standard deviation of eight minutes. Describe the distribution of the sample mean of preparation time for its customers.

9. With such a large number of people using text messages as a means of communication, a company is interested in determining the number of work hours lost due to text messaging. Based on a survey of 30 randomly selected employees (anonymously, of course), the company has determined that the average amount of time spent texting over a one-month period is 180 minutes with a standard deviation of 60 minutes.

   a. What is the probability that the average amount of time spent using text messages is more than 210 minutes in this one-month period?

   b. Thinking that it's practically impossible for her employees to spend, on average, three hours a month texting while at work, the manager conducts another survey. She randomly samples 45 employees and finds that the average amount of time spent texting while at work over a one-month period is less than 180 minutes. Is it reasonable to conclude that the average amount of time spent using text messaging has decreased since the initial survey? Justify your answer.

   c. How might the data gathered from this sample not accurately depict the loss of productivity from text messaging?

10. Suppose that a random sample of size 64 was selected and the researcher found that the mean was 30 and the standard deviation was 4.

    a. What is the probability that the sample mean is more than 31.25?

    b. What assumptions were made in part a.?

11. A town is considering building a high school football stadium approximately one-half mile from a well-established housing development. The residents of the development opposed the stadium construction due to the noise coming from the stadium during games. In presenting their argument, the residents indicated that any noise more than 103 decibels would be unacceptable. Using a sample of 35 games previously played in the old arena, the town found that the average decibels were 100 with a standard deviation of 8 decibels.

    a. What is the probability that a randomly selected game will generate noise in excess of 103 decibels at the stadium?

    b. What is the probability that a randomly selected game will generate a noise level of exactly 103 decibels?

    c. Suppose a compromise was made that required the noise level to be lower than 103 decibels 95% of the time. Will the mean level of the noise have to be lowered to comply with the new regulation? If so, by how much? Assume that the standard deviation remains at 8 decibels.

12. The town manager believes that 60% of the residents will approve the construction of the proposed high school football stadium. A random sample of 100 residents will be used to estimate the proportion of residents that will approve the construction.

    a. Assuming that the town manager is correct and that $p = 0.6$, describe the sampling distribution of $\hat{p}$.

    b. What is the probability that between 50% and 70% of the residents will approve the stadium construction?

13. It is believed that 90% of all adults and 85% of all kids between the ages of 12 and 17 have cellular phones. Suppose a sample of 500 adults and 400 kids was taken.

    a. Describe the sampling distribution of the proportion of adults that have cellular phones. Assume that the stated probabilities above are true.

    b. Describe the sampling distribution of the proportion of kids that have cellular phones. Assume that the stated probabilities above are true.

    c. What is the probability that the sample proportion of adults having cell phones will be within 2% of the true proportion?

    d. What is the probability that the sample proportion of kids having cell phones will be within 4% of the true proportion?

14. A survey of college students was conducted to learn about their attitudes toward alcohol abuse on college campuses. Sixty-two percent of student respondents indicated that they believe there was a high rate of alcohol abuse on college campuses. Suppose that a sample of 250 college students was taken. What is the probability that more than seventy percent believed that there was a high rate of alcohol abuse on college campuses?

15. A credit card issuer believes that 75% of college students between the ages of 18 and 22 have more than $5000 of credit card debt. The credit card issuer conducted a survey of 500 college students between the ages of 18 and 22.

    a. What is the probability that at least 70% of college students between the ages of 18 and 22 have credit card debt in excess of $5000?

    b. Assuming that the credit card issuer is correct, what is the probability that the proportion will be within three percent of the population proportion?

    c. What is the probability that the proportion will not be within three percent of the population proportion?

16. A marketing firm conducts a survey by mail with a 20% response rate. If the firm mailed 1000 surveys for a new study, what is the probability that at least 220 individuals will respond?

17. Suppose that it has been reported by a group of researchers that the average number of hours of TV viewing per household per week in the United States is 50.4 hours. Suppose the standard deviation is 11.8 hours, and a random sample of 42 U.S. households is taken.

    a. What is the probability that the sample average is more than 35 hours? If the sample average is actually more than 35 hours, what would it mean in terms of the figures presented by the researchers?

    b. Suppose the population standard deviation is unknown. If 71% of all sample means are greater than 49 hours and the population mean is still 50.4 hours, what is the value of the population standard deviation? Use a sample size of 42.

P   Discovery Project

Industry Statistics

Pick an industry (software companies, computer manufacturers, oil, etc.) and research the financial information about companies in that industry. The Internet, your library, the prospectus from the companies, etc. may be useful resources to help you with your research. Also, select a statistic to track or follow about the companies such as sales, profit, revenue, etc. Using the information that you gather about the companies, answer the following questions.

1. Identify the population of interest.

2. Select a random sample of 10 companies in the industry that you've selected and record the variable of interest for each of the companies.

3. Compute the average and standard error of the variable of interest.

4. What is the error of estimation using the 10 companies selected?

5. Discuss what you have learned about the statistics you studied and the companies in your population.

Chapter Nine

Estimating Means and Proportions: Single Samples

| D | **Discovering the Real World** |

The Travel Industry

Each year, the U.S. Bureau of Labor Statistics (BLS) makes forecasts on travel. They forecast the times when people will travel the most, the lengths of the vacations, the method of transportation, and many other factors that interest travelers and the travel industry. The travel industry depends on travelers to support its businesses, and travelers rely on the travel industry to provide them with services such as transportation, lodging, food and beverage, and entertainment. Businesses rely on forecasts from the BLS to give them insight on travel-related expenditures, employment expectations within the industry, and earnings of various travel-related occupations. Of course, regardless of which branch of the government (or other entity in the travel industry) is making the predictions, these forecasts are evaluated based on their validity and reliability.

In 2008, U.S. households spent, on average, a total of $1415 on transportation, lodging, food and beverage, entertainment, and gifts while on vacation. This dollar amount represents, on average, three percent of total household expenditures. The largest travel expense was transportation, which represented 44 percent of the total amount spent. Households spent the least amount on the most discretionary travel-related expense, entertainment, which represented 10 percent of all travel-related expenses.

Source: U.S. Bureau of Labor Statistics

The estimates provided by the BLS are viewed as the "best possible" information available. That is, they don't know the true population mean amount that households spend on travel. However, they took a random sample of households that vacationed and estimated the mean from that sample. These forecasts (or estimates) can be later validated when the actual outcomes become available, giving the analysts an idea of the accuracy of their predictions.

Many estimates are *subjective*; that is, a person with experience in the field estimates an unknown population value. If the expert is experienced with the population or process, then formal statistical methods may add very little quality to the resulting forecast. These types of predictions are called **judgment estimates**.

The problem with judgment estimates is that their degree of accuracy or reliability cannot be determined. If a decision has important consequences, standard statistical methods should be applied, especially if there is no expert to rely upon. Even if an expert is available, statistics can offer estimates with known reliability. In this chapter, we will discuss estimation techniques to estimate the population mean, standard deviation, and proportion.

Introduction

The previous chapter concentrated on sampling and how the practice of drawing samples produces a statistic (e.g. the sample mean) that is a random variable. This chapter marks the beginning of statistical inference. It discusses the properties of sample summary statistics (sample means, sample proportions, etc.) as well as methods for assessing the reliability of those estimates.

The importance of the statistical measures to a decision-making process depends on the confidence the decision maker has in the estimated values. How much faith can be placed in the accuracy of the estimate? Statistical inference permits the estimation of statistical measures (means, proportions, etc.) with a known degree of confidence. This ability to assess the quality of estimates is one of the significant benefits statistics brings to decision making and problem solving.

Applications of statistics are usually concerned with learning about populations or processes. Populations are described using summary measures called parameters, such as the mean, the standard deviation, and the proportion. Processes are also described with summary measures and with various models. Unfortunately, determining the exact mean and standard deviation of a population is seldom an easy task and often is not feasible. Statistical methods rely upon samples of the population to obtain information about the population's parameters. Whether the sample data will produce reliable information about a process or population depends on the methods used to collect the data.

> ### Definition
>
> Using properly drawn sample data to draw conclusions about a population is called **statistical inference**.

All forms of statistical inference are tarnished with a degree of uncertainty. If samples are selected randomly, the uncertainty of the inference is measurable. Drawing random samples allows us to measure the level of confidence associated with an estimate. If samples are not selected randomly, there will be no known relationship between the sample estimates and the population parameters they are supposed to estimate. Therefore, if a statistical technique requires a sample, assume the sample will be a random sample from a population or process.

This chapter is devoted to the one-sample problem. That is, a sample consisting of n measurements, x_1, x_2, ..., x_n, of some population will be analyzed with the objective of making inferences about the population. Inference will be introduced with the one-sample problem, not because it is the most common, but because the procedures and basic principles of inference are easier to understand.

9.1   What is an Estimator?

What is meant by the terms **estimator** and **estimate**?

> **Definition**
>
> An **estimator** is a strategy or rule that is used to estimate a population parameter. If the rule is applied to a specific set of data, the result is an **estimate**.

The sample mean is an *estimator* of the population mean. A specific sample mean, $\bar{x}$, such as 103.4, is an *estimate* of the population mean (μ).

In this chapter, we will study two different kinds of estimators, point and interval. A **point estimator** uses a single point to estimate a population parameter. For example, $\bar{x}$ is a *point estimator* of a population mean ($\bar{x} = 12.7$ is a *point estimate* of a population mean).

Table 9.1 – Point Estimators		
Point Estimator	**Parameter Being Estimated**	**Point Estimate**
$\bar{x}$	μ	$\bar{x} = 12.7$
$\hat{p}$	p	$\hat{p} = 0.37$
s	σ	$s = 6.4$

An **interval estimate** defines an upper and lower boundary for an interval that will hopefully contain the population parameter. Often times, the interval estimate for a parameter is a function of the point estimate of that parameter.

The value of drawing random samples resides in the ability to assess the reliability of sample inference. Reliability is expressed in probability. That is why so much of the text has been devoted to probabilistic ideas.

9.1   Exercises

Basic Concepts

1. What is statistical inference?

2. What is an estimator?

3. What is a judgment estimate? What are some drawbacks of judgment estimates?

4. Explain, in your own words, the difference between the terms *estimator* and *estimate*.

5. What is the difference between a point estimate and an interval estimate?

6. Give three examples of point estimators. Identify the parameters being estimated by these estimators.

7. Describe the primary advantages of *random* sampling procedures.

9.2 Point Estimation of the Population Mean

Like other statistical inference methods, estimation begins with the collection of data. Two important questions come to mind.

1. How should the data be used to estimate the population mean?

2. How can you tell a good estimator from a bad one?

Good estimators conform to the rules of horseshoes: the closer, the better. If the objective is to estimate a population mean, closeness is measured in terms of the distance the estimate is from the actual population mean.

One of the more puzzling questions is, *how can you judge the accuracy of your estimate without knowing the true value of the population parameter?* It's like shooting an arrow at a bull's-eye without being able to see it. If you can't see the bull's-eye, how do you know how close you are?

> ## Definition
>
> An estimator's average squared distance from the true parameter is referred to as its **mean squared error** (MSE). The mean squared error for the sample mean is given by
>
> $$\text{MSE}(\overline{x}) = E(\overline{x} - \mu)^2 .$$

A perfect estimator would have a mean squared error of zero, but there is no such thing as a perfect estimator. Since statistical estimators depend on data that is randomly drawn, estimates are random variables and will seldom be equal to the true population characteristic. The goal is to find an estimator whose average squared error is the smallest. Unfortunately, there are a number of different estimators, and without restricting the kinds of estimators that will be considered, very little progress can be made. One desirable restriction is **unbiasedness**. As was discussed in Chapter 8, to be an **unbiased estimator**, the expected value of the estimator must be equal to the parameter that is being estimated. For example, $\overline{x}$ is an unbiased estimator of the population mean since

$$E(\overline{x}) = \mu.$$

Unfortunately, there are a number of estimators that are unbiased estimators of the population mean including the sample mean, sample median, or any single sample value. **Among unbiased estimators of the population mean, the sample mean has the smallest mean squared error**. There is no other unbiased estimator that can consistently do a better job of estimating the population mean.

The fact that the sample mean is a good estimator of the population mean should not be surprising, since you would expect most sample statistics to be reasonably good estimators of their population counterparts. One of the exceptions is the sample range, which is a poor estimate of the population range. In short, the best estimate is the one that is unbiased with the smallest mean squared error.

9.2 Exercises

Basic Concepts

1. What are two important questions to consider when estimating a population mean?

2. What is mean squared error?

3. What is an unbiased estimator? Give an example.

4. Why is the sample mean considered the best point estimate of the population mean?

5. Are all estimators unbiased? Explain.

6. Generally, we expect most sample statistics to be good estimators of their population counterparts. Which statistic is the exception to this idea?

7. What are two characteristics of the best available estimate for a parameter?

9.3 Interval Estimation of the Population Mean

Rarely will a point estimate of the population mean result in a value that exactly matches the population mean, μ. In fact, the probability that a point estimate for the population mean exactly equals the population mean is zero for data drawn from continuous distributions. Yet, if an estimate is used for decision making, it is desirable that there be some indication of its potential error. One of the significant limitations of simply reporting a point estimate is the lack of information concerning the estimator's accuracy.

Interval estimates, however, are constructed to provide additional information about the precision of the estimate. An **interval estimator** is made by developing an upper and lower boundary for an interval that will hopefully contain the population parameter. It would be easy to construct an interval estimator that would always contain the population parameter: for example, the interval from negative infinity to positive infinity. But this particular estimator would not contain any useful information about the location of the population parameter. In interval estimation, the narrower the width of the interval for a given level of confidence, the better the estimator.

The foundations of a good interval estimator for the population mean can be found in Chapter 8, which defined how $\bar{x}$ should vary. Recall that if the sample size is reasonably large ($n \geq 30$), the central limit theorem ensures that $\bar{x}$ has an approximately normal distribution with mean, μ, and standard deviation, $\dfrac{\sigma}{\sqrt{n}}$.

Sampling Distribution of $\bar{x}$

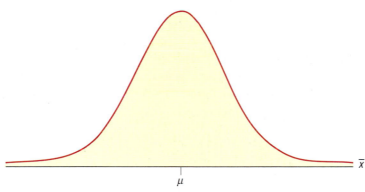

Figure 9.1

The sampling distribution can be used to develop an interval estimator. For a standard normal random variable, we know that

$$P(-1.96 < z < 1.96) = 0.95.$$

Since $\bar{x}$ can be transformed into a standard normal random variable by using the z-transformation, $z = \dfrac{\bar{x} - \mu}{\sigma_{\bar{x}}}$, then by substitution,

$$P\left(-1.96 < \frac{\bar{x} - \mu}{\sigma_{\bar{x}}} < 1.96\right) = 0.95,$$

The Father of Confidence Intervals: Jerzy Neyman 1894 – 1981

Jerzy Neyman grew up in Poland. However, a significant part of Poland was under Russian control during his youth and Neyman received his training in mathematics in Russia. In her book *Neyman – from Life*, Constance Reid attributes the development of the confidence interval to Jerzy Neyman. "During the years 1934-38 Neyman made four fundamental contributions to the science of Statistics. Each of them would have been sufficient to establish an international reputation, both for their immediate effect and for the impetus which the new ideas and methods had on the thinking of young and old alike. He put forward the theory of confidence intervals, the importance of which in statistical theory and analysis of data cannot be overemphasized. His contribution to the theory of contagious distributions is still of great utility in the interpretation of biological data. His paper on sampling stratified populations paved the way for a statistical theory which, among other things, gave us the Gallup poll. [His] work, and that of Fisher, each with a different model for randomized experiments, led to the whole new field of experimentation so much used in agriculture, biology, medicine, and physical sciences."

and with some algebraic manipulation we obtain

$$P\left(\bar{x}-1.96\sigma_{\bar{x}}<\mu<\bar{x}+1.96\sigma_{\bar{x}}\right)=0.95.$$

If the sample size is greater than or equal to 30, then by the central limit theorem there is a 0.95 probability that the sample mean will be within 1.96 standard deviations of the population mean **before a particular sample is selected**.

Sampling Distribution of $\bar{x}$

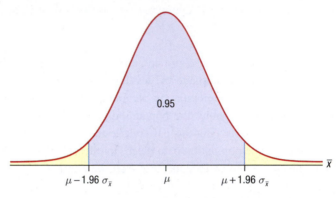

Figure 9.2

Since we discussed the sampling distribution of the sample mean in Chapter 8, the idea of defining the probability that the sample mean should fall within some specified distance of the population mean is not a new one. The expression does suggest a specific form for the interval since it provides an interval and the associated probability that the population mean will fall within the interval:

$$\bar{x}\pm1.96\sigma_{\bar{x}}.$$

However, the provision *before a particular sample is selected* modifies the interpretation of the interval for a specific sample. After the sample is selected, the sample mean is no longer a random variable. Suppose a sample has been drawn from a population with a standard deviation of 200, and the following characteristics have been observed:

$$\sigma=200,\qquad\text{(given)}$$

$$n=100,\qquad\text{(chosen by researcher)}$$

$$\bar{x}=150,\qquad\text{(obtained from the sample)}$$

Remember,

$$\sigma_{\bar{x}}=\frac{\sigma}{\sqrt{n}}=\frac{200}{\sqrt{100}}=\frac{200}{10}=20.$$

The resulting interval would be

$$150\pm1.96(20).$$

That is,

$$150 - 1.96\,(20) \approx 111 \qquad\qquad 150 \qquad\qquad 150 + 1.96\,(20) \approx 189$$

Is the population mean, μ, inside this interval? If not, what fraction of the time will μ be inside the interval? Even though the interval is calculated using a technique that captures the population mean 95% of the time, it would not be appropriate from a relative frequency point of view, to state that

$$P(111 < \mu < 189) = 0.95 \qquad \text{Wrong Interpretation}$$

since the population mean is an unknown but constant quantity. Either μ will always be inside the interval or will always be outside the interval. If this is true, then what information do we have about the interval? We say that this interval was constructed with a **confidence level** of 95% (i.e., $100(1-\alpha)\%$) or a **confidence coefficient** of 0.95 (i.e. $1 - \alpha$). Hence, the term **confidence interval** is used to describe the method of construction rather than a particular interval.

In summary, a 95% confidence interval can be interpreted to mean that if all possible samples of a given size are taken from a population, 95% of the samples would produce intervals that captured the true population mean and 5% would not. In Figure 9.3 we have plotted, as line segments, the confidence intervals from 20 random samples. All but one of the intervals, number 5, capture the mean, μ.

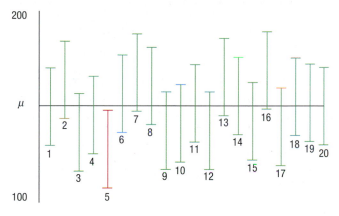

Confidence Intervals from 20 Random Samples

Figure 9.3

So far we have examined only the 95% confidence interval, yet the idea is a general one and can be extended to any specified degree of confidence.

In a practical sense, the selection of the degree of confidence depends upon the importance of the decision for which the confidence interval will be utilized. If we are launching a space vehicle, we would want to be very certain that the vehicle would have sufficient fuel to return safely. For other decisions, we might be willing to accept an 80% confidence of correctly estimating the population mean, especially if the cost of gathering additional data is large.

The expression $\bar{x} \pm z_{\alpha/2} \dfrac{\sigma}{\sqrt{n}}$ creates the "generalized" confidence interval.

Formula

Confidence Interval for the Population Mean

A $100(1-\alpha)\%$ **confidence interval** for the population mean is given by

$$\bar{x} \pm z_{\alpha/2} \frac{\sigma}{\sqrt{n}}$$

if either of the following conditions are true:

1. $n \geq 30$ (use s as an approximation for σ), or

2.   if σ is known and the population being studied is normally distributed.

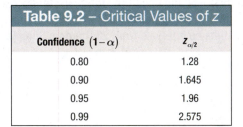

The term $z_{\alpha/2}$ represents the z-value required to obtain an area of $\alpha/2$ in the right tail under the standard normal curve. The z-values for obtaining various $(1-\alpha)$ areas centered under the standard normal curve are given in Table 9.2 and graphed in Figure 9.4.

Table 9.2 – Critical Values of z	
Confidence $(1-\alpha)$	$z_{\alpha/2}$
0.80	1.28
0.90	1.645
0.95	1.96
0.99	2.575

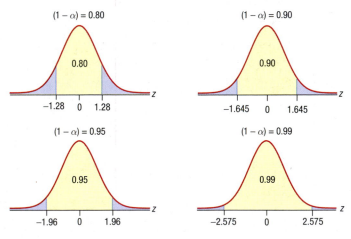

Figure 9.4

Example 9.1

Construct 80%, 90%, 95%, and 99% confidence intervals for the population mean if the standard deviation of the population is 900. Use the following sample data.

$$n = 100$$
$$\bar{x} = 425$$

Solution

80% Confidence Interval:

$$425 \pm 1.28 \cdot \frac{900}{\sqrt{100}} \quad \text{or 310 to 540}$$

310   425   540

90% Confidence Interval:

$$425 \pm 1.645 \cdot \frac{900}{\sqrt{100}} \quad \text{or 277 to 573}$$

277   425   573

95% Confidence Interval:

$$425 \pm 1.96 \cdot \frac{900}{\sqrt{100}} \quad \text{or 249 to 601}$$

249   425   601

99% Confidence Interval:

$$425 \pm 2.575 \cdot \frac{900}{\sqrt{100}} \quad \text{or 193 to 657}$$

193   425   657

The intervals in Example 9.1 illustrate that to achieve more confidence we must pay a price. For a given sample size, the only way to achieve greater confidence is to widen the interval. However, the resulting information provides a less precise location of the population mean.

Example 9.2

Given the rapid rate of technological innovation, SWN Management Company wants to study the number of new products introduced by top technology firms in the United States. Suppose a random sample of 150 companies is selected and each company is asked to report the number of new products that it has introduced in the last 12 months. It is also known from past experience that the population standard deviation of the new products introduced is 7.85. The sample mean is found to be 8.56 products. Calculate a 98 percent confidence interval for the population mean number of new products introduced in the last 12 months.

Solution

From the information given, we know that

$$n = 150, \ \bar{x} = 8.56, \text{ and } \sigma = 7.85.$$

Since we want a 98% confidence interval, we need to find the value of $z_{\alpha/2}$. Remember that $z_{\alpha/2}$ represents the z-value required to obtain an area of $(1-\alpha)$ centered under the standard normal curve. Therefore, for a 98% confidence interval, $\alpha = 0.02$.

Similar to finding the values in Table 9.2 we can use Table A to find the corresponding value of z. That is, if we are to use Table A, we look up 0.01 $(\alpha/2)$ in the body of the standard normal table. Since the exact value of 0.01 is not in the table, we use the value that is closest. In this case, we can see that 0.0099 corresponds to a z-value of -2.33. Since the standard normal distribution is symmetrical, $z_{\alpha/2}$ capturing an area of 0.01 in the upper tail of the distribution will be 2.33.

$$\text{For } \alpha = 0.02, \; z_{\alpha/2} = z_{0.01} = 2.33.$$

A 98% confidence interval is then calculated as follows.

$$\bar{x} \pm z_{\alpha/2}\frac{\sigma}{\sqrt{n}}$$

$$8.56 \pm 2.33\frac{7.85}{\sqrt{150}}$$

$$7.07 \text{ to } 10.05$$

| 7.07 | 8.56 | 10.05 |

Thus, we are 98% confident that the true mean number of new products introduced in the last 12 months will be contained in the above interval.

So far, the confidence interval has been discussed as a way of placing bounds on the location of a parameter with a specific degree of confidence. But we can also think about the confidence interval as a means of describing the quality of a point estimate. Let's look at the expression for the confidence interval for the population mean.

Confidence Interval for μ

$$\underbrace{\bar{x}}_{\text{point estimate}} \quad \pm \quad \underbrace{z_{\alpha/2}\frac{\sigma}{\sqrt{n}}}_{\substack{\text{margin of error with a} \\ \text{specific level of} \\ \text{confidence}}}$$

Another interpretation of the confidence interval is given below the expression of the confidence interval for μ. The part of the expression that is added and subtracted to the point estimate, $z_{\alpha/2}\frac{\sigma}{\sqrt{n}}$, can be thought of as the **margin of error** (also known as the **maximum error of estimation**) using the point estimate $\bar{x}$ with a specified level of confidence. For example, the 95% confidence interval in Example 9.1 was given as

$$425 \pm 1.96 \cdot \frac{900}{\sqrt{100}}$$

$$425 \pm 176.4.$$

We could say that we are 95% confident that the point estimate of μ, $\bar{x} = 425$, has a margin of error of 176.4 or an error of estimation no larger than 176.4. Being able to assess the error of an estimate is one of the most useful applications of statistical methods.

9.3   Exercises

Basic Concepts

1. What is an interval estimator?

2. What is the distinction between probability and confidence?

3. What is the role of the z-value in the confidence interval expression?

4. Describe in words the ideas behind the construction of a confidence interval.

5. Consider the following statement: *If the sample size is greater than or equal to 30, then by the central limit theorem there is a 0.95 probability that the sample mean will be within 1.96 standard deviations of the population mean before a particular sample is selected.* Explain why the phrase "before a particular sample is selected" is important here.

6. Explain what is wrong with the following expression: $P(111 < \mu < 189) = 0.95$.

7. Define the following terms: confidence level, confidence coefficient, confidence interval.

8. What are the conditions required in order to construct a $100(1-\alpha)\%$ confidence interval using the expression $\bar{x} \pm z_{\alpha/2} \dfrac{\sigma}{\sqrt{n}}$?

9. Describe the effect on the width of a confidence interval as each of the following increases:

    a. n **b.** $1 - \alpha$

    c. α **d.** $\bar{x}$

10. What expression indicates the margin of error? Is this the same as the maximum error of estimation?

Exercises

11. Find $z_{\alpha/2}$ for the following levels of α.

    a. $\alpha = 0.05$ **b.** $\alpha = 0.01$ **c.** $\alpha = 0.10$

12. Find $z_{\alpha/2}$ for the following levels of α.

    a. $\alpha = 0.04$ **b.** $\alpha = 0.02$ **c.** $\alpha = 0.08$

13. Find $z_{\alpha/2}$ for the following confidence levels.

    a. 98%   **b.** 94%   **c.** 92%

14. Find $z_{\alpha/2}$ for the following confidence levels.

    a. 96%   **b.** 88%   **c.** 85%

15. Consider a normally distributed population with a standard deviation of 64. If a random sample of size 90 from the population produces a sample mean of 250, construct a 95% confidence interval for the true mean of the population.

16. Construct a 90% confidence interval for the true mean of a normal population if a random sample of size 40 from the population yields a sample mean of 75 and the population has a standard deviation of 5.

17. A psychologist is studying learning in rats. The psychologist wants to determine the average time required for rats to learn to traverse a maze. She randomly selects 40 rats and records the time it takes for the rats to traverse the maze in minutes. The sample average time required for the rats to traverse the maze is 5 minutes with a standard deviation of 1 minute. Estimate the average time required for rats to learn to traverse the maze with a 90% confidence interval.

18. A paint manufacturer is developing a new type of paint. Thirty panels were exposed to various corrosive conditions to measure the protective ability of the paint. The mean life for the samples was 168 hours before corrosive failure. The life of paint samples is assumed to be normally distributed with a population standard deviation of 30 hours. Find the 95% confidence interval for the mean life of the paint.

19. The chief purchaser for the State Education Commission is reviewing test data for a metal link chain which will be used on children's swing sets in elementary school playgrounds. The average breaking strength for a sample of 50 pieces of chain is 5000 pounds. Based on past experience, the breaking strength of metal chains is known to be normally distributed with a standard deviation of 100 pounds. Estimate the actual mean breaking strength of the metal link chain with 99% confidence.

20. Tomatoes are grown in Florida for shipment to other parts of the country by the Anderson Produce Company. A random sample of 40 boxes is selected at one warehouse for weighing. The average weight for the sample is 33.5 pounds per box with a standard deviation of 2.1 pounds. Find a 90% confidence interval for the true average weight of the boxes of tomatoes.

21. Thirty-five strands of piano wire were selected at random from a recent shipment by the quality control department at Elkins Piano Company. The strands of piano wire were tested to failure in tests of tensile strength. The mean tensile strength of the sample was 30,000 pounds per square inch (psi) with a sample standard deviation of 1950 psi. Find the 98% confidence interval for the true mean tensile strength.

22. When preparing a standardized test to be given to all the sixth graders, the Standard Test Company gave a version of the test to a random sample of 45 sixth graders and timed how long it took them to finish the test. The average time required to finish the test for the sample was 2 hours and 15 minutes with a standard deviation of 30 minutes. Estimate the true average time required to finish the test with 95% confidence.

23. According to the 2009 College Senior Survey administered by the Higher Education Research Institute at UCLA, 56.4% of college seniors spend 10 hours or less studying or doing homework in a typical week. Suppose a random sample of 50 college seniors was selected from all the college seniors in the Southeast region to determine the homework habits of college seniors in the Southeast region of the United States. Each student in the sample is asked approximately how many hours per week he or she spends studying or doing homework. If the mean and standard deviation of the sample are 9.6 hours and 3.1 hours, respectively, construct a 99% confidence interval for the mean number of hours a week that a college senior in the region spends studying or doing homework per week.

Source: Cooperative Institutional Research Program at the Higher Education Research Institute at UCLA

9.4   Interval Estimation of the Population Mean: Small Samples, σ Unknown

In the previous section, we assumed that the population standard deviation was known. In practice this assumption is not very realistic, since the standard deviation describes variability about the mean. If the population standard deviation is known, the mean is usually also known, and there is no need to create an interval estimate for it. Why estimate something we already know?

If σ is not known and $n < 30$, the derivation of the confidence interval must be changed slightly. Provided the population from which the sample is drawn is normally distributed, the distribution of the quantity

$$t = \frac{\bar{x} - \mu}{s / \sqrt{n}} \quad \text{where } s \text{ is the standard deviation of the sample,}$$

has a **Student's t-distribution**. The t-distribution was discovered by W.S. Gossett in 1908. Gossett published the result under the pen name of Student while working as an employee of the Guinness Brewery in Dublin.

The t-distribution is very much like the normal distribution (see Figure 9.5). It is a symmetrical, bell-shaped distribution with slightly thicker tails than a normal distribution. The t-distribution has one parameter, **degrees of freedom**. The shape of the t-distribution approaches the normal distribution as the sample size, n, and thus the degrees of freedom become larger.

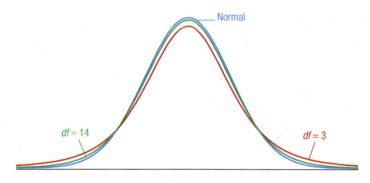

Figure 9.5

Formula

Degrees of Freedom

The **degrees of freedom** for any t-distribution are computed in the following manner.

$$df = \text{number of sample observations} - 1 = n - 1$$

Formula

Confidence Interval for the Population Mean: Small Samples, σ Unknown

If σ is unknown and the sample is drawn from a normal population, a $100(1-\alpha)\%$ confidence interval for the population mean is given by

$$\bar{x} \pm t_{\alpha/2,df} \frac{s}{\sqrt{n}},$$

where $t_{\alpha/2,df}$ is the critical value for a t-distribution with $n-1$ degrees of freedom which captures an area of $\alpha/2$ in the right tail of the distribution.

The form of the confidence interval is identical to the previous confidence interval for the mean of a population except that the z has been replaced by a t and σ has been replaced by s. The interpretation of the confidence interval is identical to the previous interval.

Example 9.3

Given the following data drawn from a normal population with unknown mean and variance, construct a 95% confidence interval for the population mean. Seven values have been selected randomly from the population.

$$25, 19, 37, 29, 40, 28, 31$$

Solution

The sample mean and standard deviation are $\bar{x} \approx 29.8571$ and $s \approx 7.0811$, respectively. The degrees of freedom associated with the problem is

$$df = n - 1 = 7 - 1 = 6.$$

The t-value corresponding to 6 degrees of freedom and 95% confidence $(\alpha = 0.05)$ is given in Table D of Appendix A as $t_{0.025,6} = 2.447$. We also know that the general form of the confidence interval for the population mean when σ is unknown and $n < 30$ is given by

$$\bar{x} \pm t_{\alpha/2,df} \frac{s}{\sqrt{n}}$$

$$29.8571 \pm 2.447\left(\frac{7.0811}{\sqrt{7}}\right)$$

$$29.8571 \pm 6.5492.$$

Thus, we are 95 percent confident that the interval

$$23.3079 \text{ to } 36.4063$$

will contain the population mean.

An alternate interpretation would be that we are 95% confident that the point estimate, 29.8571, has a maximum error of estimation of 6.5492.

Example 9.4

A manufacturing company is interested in the amount of time it takes to complete a certain stage of the production process. The project manager randomly samples 10 products as they come from the production line and notes the time of completion. The average completion time is 23.45 minutes with a sample standard deviation of 4.32 minutes. Based on this sample, construct a 95% confidence interval for the average completion time for that stage in the production process. Assume that the population distribution of the completion times is approximately normal.

Solution

Since the company wants to calculate a 95% confidence interval for the average completion time, μ, we know that $\alpha = 0.05$. We also have a sample size of $n = 10$ and thus, $n - 1 = 10 - 1 = 9$ degrees of freedom. Therefore, we use $t_{\alpha/2,df} = t_{0.025,9} = 2.262$ from Table D in the Appendix.

We also know that the general form of the confidence interval for the population mean when σ is unknown and $n < 30$ is given by

$$\bar{x} \pm t_{\alpha/2,df} \frac{s}{\sqrt{n}}$$

$$23.45 \pm 2.262 \frac{4.32}{\sqrt{10}}$$

$$23.45 \pm 3.0901$$

$$20.36 \text{ to } 26.54.$$

Thus, we are 95% confident that the true average completion time of that stage of the process is between 20.36 minutes and 26.54 minutes.

Interval Estimation of the Population Mean: A Summary

In Sections 9.3 and 9.4 we have outlined two procedures for determining an interval estimate for the population mean. In Section 9.3 we used the interval $\bar{x} \pm z_{\alpha/2} \frac{\sigma}{\sqrt{n}}$ for normally distributed populations in which the population standard deviation, σ, is known. We also stated that if the population standard deviation is unknown and the sample size is sufficiently large $(n \geq 30)$, we may use the sample standard deviation, s, as an approximation for σ. For these sufficiently large sample sizes, the t-distribution very closely resembles the z-distribution, so using the critical value $z_{\alpha/2}$ when constructing the confidence interval is acceptable. In Section 9.4 we introduced the t-distribution with $n - 1$ degrees of freedom and used the interval $\bar{x} \pm t_{\alpha/2,df} \frac{s}{\sqrt{n}}$ to find an interval estimate for the population mean for a normally distributed population where σ is unknown and $n < 30$. The following flow chart is useful when deciding how to construct an interval estimate for the population mean.

Finding a Confidence Interval for the Population Mean

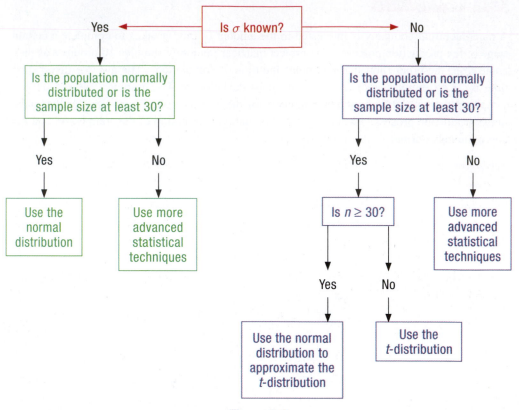

Figure 9.6

9.4   Exercises

Basic Concepts

1. Why is the assumption that the population standard deviation is known when estimating the population mean not very realistic?

2. What effect does knowing the standard deviation of the population have on the construction of the confidence interval?

3. What is the Student's *t*-distribution?

4. What are the conditions in which the *t*-distribution is used in interval estimation of the population mean?

5. What is the parameter of the *t*-distribution? How is it calculated?

Exercises

6. Find the *t*-value such that 0.025 of the area under the curve is to the right of the *t*-value. Assume the degrees of freedom equal 13.

7. Find the *t*-value such that 0.01 of the area under the curve is to the right of the *t*-value. Assume the degrees of freedom equal 21.

8. Find $t_{\alpha/2,df}$ for the following combinations of α and n.

   a. $\alpha = 0.05,\ n = 15$ **b.** $\alpha = 0.01,\ n = 20$ **c.** $\alpha = 0.10,\ n = 8$

9. Find $t_{\alpha/2,df}$ for the following combinations of α and n.

   a. $\alpha = 0.05,\ n = 12$ **b.** $\alpha = 0.01,\ n = 18$ **c.** $\alpha = 0.10,\ n = 22$

10. A random sample, consisting of the values listed below, was taken from a normally distributed population. Assuming the standard deviation of the population is unknown, construct a 99% confidence interval for the population mean.

Table for Exercise 10			
27.4	26.5	25.7	31.4
28.2	21.9	16.3	22.7
18.8	34.4	29.2	20.5

11. Construct an 80% confidence interval for the mean of a normal population assuming that the values listed below comprise a random sample taken from the population. The population standard deviation is unknown.

Table for Exercise 11				
83.9	87.4	65.2	86.0	73.1
80.3	92.7	87.5	69.3	77.5
91.9	71.1	79.1	72.4	88.2

12. An FDA representative randomly selects 8 packages of ground chuck from a grocery store and measures the fat content (as a percent) of each package. The resulting measurements are given below.

Table for Exercise 12 – Fat Contents			
13%	12%	14%	17%
15%	16%	18%	15%

   a. Calculate the sample mean and the sample standard deviation of the fat contents.

   b. Construct a 90% confidence interval for the true mean fat content of all the packages of ground beef.

   c. What assumption did you make about the fat content in constructing your interval?

13. A hospital would like to determine the mean length of stay for its patients having abdominal surgery. A sample of 15 patients revealed a sample mean of 6.4 days and a sample standard deviation of 1.4 days.

   a. Find a 95% confidence interval for the mean length of stay for patients with abdominal surgery.

   b. Interpret this interval and state any assumptions that were made in the construction of the interval.

14. An independent group of food service personnel conducted a survey on tipping practices in a large metropolitan area. They collected information on the percentage of the bill left as a tip for 25 randomly selected bills. The average tip was 12.3% of the bill with a standard deviation of 2.7%.

   a. Construct an interval to estimate the true average tip (as a percent of the bill) with 99% confidence.

   b. Interpret the interval, and state any assumptions that were made in the construction of the interval.

15. A travel agent is interested in the average price of a hotel room during the summer in a resort community. The agent randomly selects 15 hotels from the community and determines the price of a regular room with a king size bed. The average price of the room for the sample was $115 with a standard deviation of $30.

   a. Construct an interval to estimate the true average price of a regular room with a king size bed in the resort community with 90% confidence.

   b. Interpret the interval, and state any assumptions that were made in the construction of the interval.

16. In 2010 the median home price in all regions of the United States was $221,800. It is commonly thought that better schools are found in wealthier areas. In *Forbes* magazine's list of the "Best Schools for your Real Estate Buck," the top 10 cities in America were identified where your housing dollar will go the furthest in getting your children a great education. 17,589 towns and cities were analyzed using results from the most recent National Assessment for Educational Progress data, and the top 10 school districts were identified. The list counteracted the idea that more money equals better schools, as Falmouth, Maine topped the list beating out high-dollar school districts like Manhattan Beach, California. The top 10 cities are given below, along with the median home price for each city.

Table for Exercise 16 – Best Schools for your Real Estate Buck		
Education Rank	City	Median Home Price ($)
1	Falmouth, Maine	351,550
2	Mercer Island, Washington	708,740
3	Pella, Iowa	148,200
4	Barrington, Rhode Island	296,010
5	Bedford, New Hampshire	293,730
6	Manhattan Beach, California	1,278,980
7	Moraga, California	722,010
8	Parkland, Florida	426,390
9	St. Johns, Florida	181,700
10	Southlake, Texas	476,880

Source: Forbes magazine

   a. Construct a 90% confidence interval for the median home price of cities on the top 10 list.

   b. Is the average median price for these cities higher than the median price for the U.S. as a whole?

   c. What population assumption needs to be made here?

   d. How would your solutions to **a.** and **b.** change if these were mean rather than median values?

9.5	**Precision and Sample Size: Means**

The more accurate an estimate, the greater its potential value in decision making. The only way to accurately determine an unknown population parameter is to perform a census, though this is usually impractical because of cost or time considerations. Realistically, the best interval estimate a decision maker could hope for would be an interval with a small width possessing a large amount of confidence. The width of the confidence interval defines the precision with which the population mean is estimated; the smaller the interval, the greater the precision. If the width of a confidence interval could be controlled, we could achieve estimates with a level of accuracy that is appropriate for the decision at hand.

Determining the Sample Size: σ Known

There are three components that affect the width of the confidence interval for the population mean:

$z_{\alpha/2}$ Represents the distance the confidence interval boundary is from the sample mean $\bar{x}$ in standard deviation units. The distance is related to the specific level of confidence.

σ Represents the population standard deviation.

n Represents the sample size.

As discussed in Section 9.3, the level of confidence will affect confidence interval width. That is, the higher the level of confidence, the wider the confidence interval. The population standard deviation, σ, is a constant, and does not change. The sample size, however, is selected by the decision maker. The larger the sample, the smaller the width of the resulting confidence interval for some given level of confidence. Since the sample size can be increased, which reduces the width of the confidence interval, how large should the sample be? Taking too large a sample wastes money, while taking too small a sample produces an estimator that does not possess sufficient reliability.

The sample size should be selected in relation to the size of the margin of error the decision maker is willing to accept. This can be achieved by setting the error equal to one-half the confidence interval width.

$$E = \text{margin of error} = z_{\alpha/2} \frac{\sigma}{\sqrt{n}}$$

The preceding equation can be solved for the sample size, n.

$$E = z_{\alpha/2} \frac{\sigma}{\sqrt{n}}$$

$$\sqrt{n} = z_{\alpha/2} \frac{\sigma}{E}$$

$$n = \left(\frac{z_{\alpha/2}\sigma}{E} \right)^2$$

By selecting a level of confidence and the maximum error, the relationship can be used to determine the sample size necessary to estimate the population mean with the desired accuracy. **In order to assure the desired level of confidence, always round the value obtained for the sample size up to the next integer.**

Six Degrees of Separation: A Law of Small Worlds

What is the number of people a randomly chosen person in Omaha, Nebraska needs to contact before she can find a connection with a randomly chosen housewife in New England? How many intervening people do you think separates you from the President of the United States? Unsuspecting readers might guess very large numbers but the actual numbers are quite small. The answer to both of these questions may very well be less than 6! Psychologists have done ingenuous experiments and have actually calculated this *degree of separation*, on average, to be six. What is amazing about this degree of separation is that it is equally true for the President of the United States and a sweet vendor in Bangladesh. The degree of separation for the number of clicks that you will need to make to get to a website that interests you, as well as the analysis of terrorist networks turn out to be of similar nature. The new science of networks can shed useful light and help to derive general laws applicable to many of these types of questions.

Example 9.5

Consider a population having a standard deviation of 15. We want to estimate the mean of the population. How large of a sample is needed to construct a 95% confidence interval for the mean of this population if the margin of error is equal to 1.5?

Solution

$$n = \left(\frac{z_{\alpha/2}\sigma}{E}\right)^2 = \left(\frac{1.96 \cdot 15}{1.5}\right)^2 = 384.16$$

Rounding up, we have to have a sample size of 385 to ensure that we get at least a 95% confidence interval with a margin of error equal to 1.5.

Example 9.6

Suppose that a quality control manager at Argon Chemical Company wishes to measure the average amount of cleaning fluid the company is placing in their 12-ounce bottles. The manager is concerned that if the bottles are overfilled, then there could be a chance of an explosion. At the same time, if they are underfilled customers would be unhappy and begin to buy the competitor's product. From previous samples, they believe that the standard deviation is 0.3 ounce. How large a sample must be taken in order to be 95% confident of estimating the mean amount of cleaning fluid in a 12-ounce bottle to within 0.05 ounce?

Solution

$$n = \left(\frac{z_{\alpha/2}\sigma}{E}\right)^2 = \left(\frac{1.96 \cdot 0.3}{0.05}\right)^2 = 138.2976$$

To be assured of finding the desired level of confidence, always round up. Thus, we are 95% confident that a sample of $n = 139$ observations would produce an estimate of the mean amount of cleaning fluid in a 12-ounce bottle to within 0.05 ounce. Being able to know the accuracy of your estimate is one of the significant benefits of inferential statistics. In this case, if 139 bottles are measured, we will be 95% confident that the resulting sample mean is within five one-hundredths of an ounce of the true mean. That's close.

Determining the Sample Size: σ Unknown

In the previous discussion of determining the sample size necessary to estimate a population mean with a desired accuracy, σ was assumed to be known. This assumption is usually unreasonable in most problem-solving environments.

The most obvious method for obtaining an estimate of σ is to take a small sample and use the sample standard deviation as an estimate of the population standard deviation. Replacing σ with s in the sample size determination relationship will provide an initial estimate of the required sample size.

Another alternative is to use the value of the sample standard deviation obtained in a previous study, sometimes called a **pilot study**. Keep in mind that in order to construct a confidence interval for a population mean when the population standard deviation is unknown, one of the following conditions must be met:

- The sample size is greater than or equal to 30 (use the z-distribution)
- The underlying population is normally distributed (use the t-distribution)

Example 9.7

An airline's maintenance manager desires to estimate the average time (in hours) required to replace a jet engine in a Boeing 767. How large a sample would be necessary if the manager wishes to be 95% confident of estimating the population mean to within one-quarter of an hour ($E = 0.25$)? Assume a preliminary sample of size $n = 30$ has a mean replacement time of 16.7 hours with a standard deviation of 4.3 hours.

Solution

Using the results from the initial sample,

$$n = \left(\frac{z_{\alpha/2}s}{E}\right)^2 = \left(\frac{1.96 \cdot 4.3}{0.25}\right)^2 = 1136.4989$$

$n = 1137$. (Always round up to assure required confidence.)

Notice that while the sample data values are being collected they can be used to improve the estimate of the population standard deviation. For example, suppose the sample standard deviation after sampling the first 1000 observations was 4.1. Using this estimate of s instead of 4.3 results in a sample size of 1034 compared to the original specification of 1137. The notion of modifying the sample size estimate as additional data are observed can be applied at regular intervals during the sampling process until the estimate of the standard deviation stabilizes.

9.5   Exercises

Basic Concepts

1. What is the value of having a confidence interval with a small width?

2. Can a confidence interval be constructed with a width of your choice? Explain.

3. What are the three components that affect the width of the confidence interval for the population mean? Describe how changes in these three components affect the width of the confidence interval.

4. What is the margin of error? What is the connection between the expression for the margin of error and the equation to determine the sample size?

5. What is the rounding rule regarding the determination of the sample size?

6. What is the difference between the method of determining the sample size when σ is known versus when σ is unknown?

7. What is a pilot study?

8. Note that E and s can be viewed as measures of variation. Compare and contrast the meanings of E and s in layman's terms.

Exercises

9. A technician working for the Chase-National Food Additive Company would like to estimate the preserving ability of a new additive. This additive will be used for Auntie's brand preserves. Based on past tests, it is believed that the time to spoilage for this additive has a standard deviation of 6 days. To be 90% confident of the true mean time to spoilage, what sample size will be needed to estimate the mean time to spoilage with an accuracy of one day?

10. A computer software company would like to estimate how long it will take a beginner to become proficient at creating a graph using their new spreadsheet package. Past experience has indicated that the time required for a beginner to become proficient with a particular function of the new software product has an approximately normal distribution with a standard deviation of 15 minutes. Find the sample size necessary to estimate the true average time required for a beginner to become proficient at creating a graph with the new spreadsheet package to within 5 minutes with 95% confidence.

11. A hot-dog vendor is evaluating a downtown location by counting the number of people who walk past the prospective location on a particular day during lunch time (i.e. 11:00 AM to 2:00 PM). A preliminary study has indicated a standard deviation of about 30 people per lunch period. How many lunch periods will be needed to estimate the average number of people who walk past the prospective location during the lunch period to within 9 people with 90% confidence?

9.6   Estimating the Population Proportion

An attribute is a characteristic that members of a population either possess or do not possess. Attributes are almost always measured as the **proportion** of the population that possesses the characteristic.

Many decisions require a measure of a population attribute. Television and radio stations base their advertising charges on ratings reflecting the *percentage of television viewers who are watching a particular program*. A political analyst wants to know the *fraction of voters who favor a particular candidate*. A social researcher needs the *fraction of teachers who believe group learning is a beneficial instructional method*. An insurance company is interested in estimating the *fraction of their policies that will result in claims*. A quality control engineer requires the *percentage defective in a lot of goods*. A marketing researcher demands the *fraction of persons on a mailing list that will purchase the product as a result of a direct mail marketing campaign*. The items in italics are measures of attributes of some population. Researchers estimate the proportion of population members possessing those characteristics.

Estimating the proportion of the population that possesses an attribute is straightforward. A random sample is selected and the sample proportion is computed as follows:

x = number in the sample that possesses the attribute,

n = sample size, and

$$\hat{p} = \frac{x}{n}.$$

The symbol above the p indicates an estimate of the quantity specified. Since $\hat{p}$ is computed from a random sample, $\hat{p}$ is a random variable whose value depends on which random sample is selected.

Example 9.8

Estimate the fraction of defective transistors in a lot containing 100,000 transistors. Suppose a sample of size 800 is drawn from the lot, and 5 transistors were found to be defective.

Solution

x = number in the sample that possesses the attribute = 5

n = sample size = 800

Then,

$$\hat{p} = \frac{5}{800} \approx 0.0063,$$

which is an estimate of the proportion of defective transistors in the lot of 100,000.

A natural question to ask is, *How good is the estimate of the fraction of defective transistors?* The answer to this question naturally arises in the discussion of interval estimation for proportions.

Interval Estimation of the Population Proportion

The concept of confidence intervals, used to apprise a decision maker of the reliability of estimates of a population mean, can also be applied to estimating proportions. In order to develop the confidence interval for a population proportion, the sampling distribution of the point estimate must be developed. (See Section 8.4 for review.)

The random variable, $\hat{p}$, has a binomial distribution which is approximated with a normal random variable.

Thus, the sample proportion, $\hat{p}$, is normally distributed with mean, p, and standard deviation,

$$\sigma_{\hat{p}} = \sqrt{\frac{p(1-p)}{n}}.$$

If the true population proportion is unknown, the standard deviation of the sample proportion, $\hat{p}$, is denoted symbolically as $\sigma_{\hat{p}}$ and is given by

$$\sigma_{\hat{p}} = \sqrt{\frac{p(1-p)}{n}} \approx \sqrt{\frac{\hat{p}(1-\hat{p})}{n}},$$

where $\hat{p}$ is used as an estimate of p.

As before,

$$P(-1.96 < z < 1.96) = 0.95.$$

Substituting

$$z = \frac{\hat{p} - p}{\sigma_{\hat{p}}}$$

results in

$$P\left(-1.96 < \frac{\hat{p} - p}{\sigma_{\hat{p}}} < 1.96\right) = 0.95.$$

Manipulating the inequalities results in

$$P\left(\hat{p} - 1.96\sigma_{\hat{p}} < p < \hat{p} + 1.96\sigma_{\hat{p}}\right) = 0.95,$$

which suggests that the interval

$$\hat{p} \pm 1.96\sigma_{\hat{p}}$$

would be a good choice for a 95% confidence interval for the population proportion. As before, the probability that the interval will contain the true population proportion is 0.95 *before a specific sample is drawn*. After a specific sample is drawn, the only available information about the interval is that the technique which generated it will bound the true proportion 95% of the time.

Formula

Confidence Interval for the Population Proportion

If the sample size is sufficiently large, i.e. $np \geq 5$ and $n(1-p) \geq 5$, the $100(1-\alpha)\%$ confidence interval for the population proportion is given by the expression

$$\hat{p} \pm z_{\alpha/2} \sigma_{\hat{p}},$$

where $z_{\alpha/2}$ is the value of z which captures an area of $\alpha/2$ in the right tail of the standard normal distribution, and $\sigma_{\hat{p}}$ is the standard deviation of $\hat{p}$.

Example 9.9

Suppose a sample of 410 randomly selected radio listeners revealed that 48 listened to WJLN.

$$\hat{p} = \frac{48}{410} \approx 0.1171$$

This is a point estimate of the proportion that listen to WJLN.

To obtain an interval estimate, the amount of confidence to be placed in the interval must be specified. Suppose we desire 95% confidence.

Solution

$$z_{\alpha/2} = z_{0.05/2} = z_{0.025} = 1.96, \text{ and } \sigma_{\hat{p}} \approx \sqrt{\frac{\hat{p}(1-\hat{p})}{n}} = \sqrt{\frac{0.1171(1-0.1171)}{410}} \approx 0.0159.$$

Note that the sample proportion $\hat{p}$ is used in place of p in the computation of $\sigma_{\hat{p}}$. For any realistic problem, this will always be the case. Fortunately, unless $\hat{p}$ and p are far apart, the value of $\sigma_{\hat{p}}$ will not be greatly affected.

Computing the confidence interval $\hat{p} \pm z_{\alpha/2} \sigma_{\hat{p}}$ results in

$$0.1171 \pm 1.96(0.0159)$$
$$0.1171 \pm 0.0312$$
$$0.0859 \text{ to } 0.1482.$$

We are 95% confident in the procedure that created this interval. Another interpretation would be that we are 95% confident that the point estimate, 0.1171, has a maximum error of estimation of 0.0312. A maximum error of only 0.0312 with 95% confidence suggests a rather high level of accuracy in the estimation of the proportion.

9.6 Exercises

Basic Concepts

1. What is a proportion? What type of information does it give us about the population?

2. How is the sample proportion found?

3. Describe, in layman's terms, how a confidence interval is constructed for a population proportion.

Exercises

4. Acid rain accumulations in lakes and streams in the northeastern part of the United States are a major environmental concern. A researcher wants to know what fraction of lakes contain hazardous pollution levels. He randomly selects 200 lakes and determines that 45 of the selected lakes have an unsafe concentration of acid rain pollution.

    a. Calculate the best point estimate of the population proportion of lakes that have unsafe concentrations of acid rain pollution.

    b. Determine a 95% confidence interval for the population proportion.

    c. If a local politician states that only 20% of the lakes are contaminated, does the study provide overwhelming evidence at the 95% level to contradict his views?

5. *The Richland Gazette*, a local newspaper, conducted a poll of 1000 randomly selected readers to determine their views concerning the city's handling of snow removal. The paper found that 650 people in the sample felt the city did a good job.

    a. Compute the best point estimate for the percentage of readers who believe the city is doing a good job of snow removal.

    b. Construct a 90% confidence interval for this percentage.

6. The clinical testing of drugs involves many factors. For example, patients that have been given placebos, which are harmless compounds that have no effect on the patient, often will still report that they feel better. Assume that in a study of 500 random subjects conducted by the Poppins Sucre Drug Company, the percentage of patients reporting improvement when given a placebo was 37%.

    a. What would be a 95% confidence interval for the true proportion of patients who exhibit the placebo effect? Interpret this interval in terms of the problem.

    b. What would the 99% confidence interval be?

    c. To gain the additional 4% of confidence how much wider did the interval become?

7. The Peacock Cable Television Company thinks that 40% of their customers have more outlets wired than they are paying for. A random sample of 400 houses reveals that 110 of the houses have excessive outlets.

    a. Construct a 99% confidence interval for the true proportion of houses having too many outlets.

    b. Do you feel the company is accurate in its belief about the proportion of customers who have more outlets wired than they are paying for? Justify your answer.

8. Running continues to be a very popular sport in America. At a major race, like the Peachtree Road Race in Atlanta, there may be over 10,000 people entered to run. The race promoters for a road race in the Pacific Northwest took a random sample of 750 runners out of the 5000 runners entered to estimate the number of runners who will need hotel accommodations. Five hundred runners indicated they would need hotel accommodations.

   a. Construct a 90% confidence interval for the true proportion of runners who will need hotel accommodations.

   b. Is the confidence interval obtained sufficiently narrow to be of help in planning the number of hotel rooms which will be necessary to accommodate the runners? Justify your answer.

9. In the fourth quarter of 2010 the home ownership rate was 66.5%. This rate is 2.7 percentage points lower than the 2004 peak of 69.2%, and the lowest rate since 1998. Home ownership fell at an alarming pace in the fourth quarter of the year, despite the fact that home prices fell, affordability was much improved, and inventories of new and existing homes were running quite high. Suppose that a random sample of 120 households was selected from an area in the Midwest that is particularly economically depressed. Suppose that 57 of the households sampled were owned by the residents of the homes.

   Source: U.S. Census Bureau

   a. Construct a 95% confidence interval for the proportion of households in the area sampled that are owned by the residents of the homes.

   b. Is there evidence at the 95% level that the proportion of the households in the area sampled that are owned by residents is less than the national rate?

10. In the *Gallup Poll Monthly*, it was reported that 31% of the people surveyed in a recent poll claimed that vegetables were their least favorite food. Surprisingly, only 14% responded with liver, and 10% of those surveyed did not submit a response because they claimed that they liked everything. The poll was based upon a sample of 1001 people. Assuming that a random sample was chosen, construct a 90% confidence interval for the percentage of all Americans who say that vegetables are their least favorite food.

11. The Federal Trade Commission (FTC) conducted a study investigating the accuracy of bar-code scanners. It was concluded that these computer scanners, used mainly at grocery, department store, and drugstore checkout counters, ring up the wrong price about 5% of the time. In most instances, however, the error was in favor of the shopper, according to the FTC. Suppose that your local grocery store conducts a study to determine the accuracy of its scanners. Assume 13 shoppers are randomly chosen and their bills, as indicated by the scanner, are checked against the correct bill computed by conventional means. Suppose that of the 200 items scanned, 21 of the items were charged incorrectly by the scanner.

   a. Construct a 95% confidence interval for the proportion of items that were rung up incorrectly by the scanner.

   b. Does it appear that the local grocery store has a larger error rate than 5%?

9.7 Precision and Sample Size: Proportions

Just as for the population mean, a specific level of accuracy in estimating a population proportion is desirable. Suppose, for example, that a direct-mail marketer would like to estimate the fraction of a mailing list that will purchase the company's product. To be profitable, a purchase response of at least 0.008 is required. Because the proportion to be estimated is of such a small magnitude, a high degree of precision in estimating the proportion is necessary. How large a sample would be required if the population proportion (the actual proportion of persons on the mailing list that will buy the product) is to be estimated with an accuracy of 0.002? We are saying that we want our maximum error to be less than two one-thousandths. That would seem to be a highly precise estimate. But, the quantity we are trying to estimate (the proportion of people on the list that will buy the product) could easily be near 0.008. The maximum error is about 25% as large as the value we are trying to estimate. When we estimate extremely small quantities, highly precise estimates are necessary.

The technique for deriving the sample size parallels the discussion of precision and sample size for the sample mean (Section 9.5). Setting one-half the entire width of the confidence interval equal to the maximum allowable error yields

$$E = \text{margin of error} = z_{\alpha/2}\sigma_{\hat{p}} = z_{\alpha/2}\sqrt{\frac{p(1-p)}{n}}.$$

Solving for n yields

$$n = \frac{z_{\alpha/2}^2 p(1-p)}{E^2}.$$

Generally the population proportion is unknown and is estimated from a pilot study. In this case the sample size necessary to estimate the population proportion to within a particular error with a certain level of confidence is given by

$$n \approx \frac{z_{\alpha/2}^2 \hat{p}(1-\hat{p})}{E^2}$$

where $\hat{p}$ is the estimate of the population proportion obtained from the pilot study.

If an estimate of the population proportion is not available, then the population proportion is set equal to 0.5. The value 0.5 maximizes the quantity $p(1-p)$ and thus provides the most conservative estimate of the sample size possible. Hence if no estimate of the population proportion is available, the sample size necessary to estimate the population proportion to within a particular error with a certain level of confidence is given by

$$n = \frac{z_{\alpha/2}^2 (0.5)(1-0.5)}{E^2} = \frac{z_{\alpha/2}^2 (0.25)}{E^2}.$$

By selecting a level of confidence and an error, a sample size can be determined that will likely (at the level of confidence) produce an estimate with at least the desired accuracy. Remember to always round the sample size to the next largest integer to assure the desired level of accuracy.

Example 9.10

How large a sample would be required to estimate the proportion of buyers on a mailing list that will buy the product with an accuracy of 0.002 with a 95% degree of confidence if the true proportion is approximately 0.008?

Solution

$$p \approx 0.008,$$
$$z_{\alpha/2} = 1.96 \text{ for 95\% confidence, and}$$
$$E = 0.002$$

Using the sample size determination expression yields

$$n = \frac{z_{\alpha/2}^2 p(1-p)}{E^2}$$
$$= \frac{1.96^2 (0.008)(1-0.008)}{0.002^2}$$
$$= 7621.7344 \approx 7622 \text{ (always round up).}$$

Thus, to be 95% confident that the proportion is estimated with an error of at most 0.002 requires a sample size of at least 7622.

Example 9.11

Using Example 9.9, suppose that the radio station WJLN desires to estimate the proportion of the market they hold with a maximum error of 0.01 and a confidence coefficient of 0.95. How large a sample would be required to estimate the fraction of listeners to within the desired level of accuracy? Since we don't know the true population proportion, let's assume the previous point estimate of 0.1171 is the true proportion.

Solution

$$p \approx 0.1171,$$
$$z_{\alpha/2} = 1.96 \text{ for 95\% confidence, and}$$
$$E = 0.01$$

Using the sample size determination expression yields

$$n = \frac{z_{\alpha/2}^2 p(1-p)}{E^2} = \frac{1.96^2 (0.1171)(1-0.1171)}{0.01^2} \approx 3971.7377 \approx 3972 \text{ (always round up).}$$

Thus to be 95% confident that the proportion of listeners is estimated with an error of at most 0.01 would require a sample size of at least 3972.

Suppose we did not have a previous estimate of the population proportion in Example 9.11. In this case we would estimate p with 0.5. The sample size necessary to estimate the true proportion of listeners to within 1% with 95% confidence is given by

$$n = \frac{z_{\alpha/2}^2 (0.5)(1-0.5)}{E^2} = \frac{1.96^2 (0.5)(1-0.5)}{0.01^2} = 9604.$$

Notice that the required sample size is significantly larger when an estimate of the population proportion is not available.

9.7 Exercises

Basic Concepts

1. It seems that estimating proportions produces estimates which are much more precise than those for means. Explain why this is the case.

2. The population proportion is often unknown. How is this issue dealt with when determining sample size?

3. What is the guideline to follow when there is no estimate available for the population proportion? Why is this done?

4. How do the resulting required sample sizes differ when there is an estimate available versus when there is no estimate available for the population proportion?

Exercises

5. The Big Green Poster Company wants to estimate the fraction of poster sites controlled by their competition, Bird's Billboard Service. What sample size would be necessary to estimate this fraction to within 3% with 95% confidence? (They think Bird's controls about 33 percent of the boards.)

6. Researchers working in a remote area of Africa feel that 40% of families in the area are without adequate drinking water either through contamination or unavailability. What sample size will be necessary to estimate the percentage without adequate water to within 5% with 99% confidence?

7. Companies that provide environmental cleanup for hazardous waste and toxic chemicals are growing rapidly. W.R. Gross is thinking about entering this field with a subsidiary called Saf-t-Soil. They wish to estimate the true proportion of U.S. corporations that produce hazardous waste as a by-product of their manufacturing process to within 10% with 80% confidence. What sample size will be needed?

8. The public relations manager for a political candidate would like to determine if the registered voters in the candidate's district agree with the politician's view on a particular issue. Find the sample size necessary for the public relations manager to estimate the true proportion to within 5% with 85% confidence.

T   Discovering Technology

Using the TI-84 Plus Calculator

Confidence Intervals for the Population Mean

Construct the 90% confidence interval for the population mean of a normal population if the sample standard deviation is 900, the sample mean is 425, and the sample size is 100.

1. Choose **STAT** , then select **TESTS** and choose option **7:ZInterval**.

2. Press **ENTER** and choose **Stats**. Input **900** for σ (or s in this case), **425** for $\bar{x}$, **100** for n, and **0.90** for the C-Level. Highlight **Calculate**, and press **ENTER**.

3. The confidence interval, (276.96, 573.04) is listed on the screen.

Figure 9.7

Figure 9.8

Confidence Intervals for the Population Mean (Small Samples, σ Unknown)

Construct the 95% confidence interval for the population mean of a normal population if the sample standard deviation is 6.5, the sample mean is 125, and the sample size is 15.

1. Choose **STAT** , then select **TESTS** and choose option **8:TInterval**.

2. Press **ENTER** and choose **Stats**. Input **125** for $\bar{x}$, **6.5** for Sx, **15** for n, and **0.95** for the C-Level. Highlight **Calculate**, and press **ENTER**.

3. The confidence interval, (121.4, 128.6) is listed on the screen.

Figure 9.9

Figure 9.10

9

Confidence Intervals for the Population Proportion

Suppose a sample of 410 randomly selected radio listeners revealed that 48 listened to WJLN. Construct a 95% confidence interval for the population proportion of radio listeners that listen to WJLN.

1.   Choose **STAT**, then select **TESTS** and choose option **A:1-PropZInt**.

2.   Enter **48** for x (the number in the sample that listened to WJLN), **410** for n (the sample size), and **0.95** for C-Level. Highlight **Calculate**, and press **ENTER**.

3.   The confidence interval, (0.08595, 0.14819), is listed on the screen, along with the sample proportion (0.1171), and the sample size (410).

Figure 9.11

Figure 9.12

Using Excel

Confidence Intervals for the Population Mean

Construct a 90% confidence interval for the population mean of a normal population if the sample standard deviation is 900, the sample mean is 425, and the sample size is 100.

1.   Label cells A1 and B1 with **Lower Limit** and **Upper Limit**, respectively.

2.   The function that returns the margin of error for a confidence interval in Microsoft Excel is

CONFIDENCE.NORM(alpha, standard_dev, size)

where *alpha* is the level of significance, *standard_dev* is the population standard deviation (estimated by s if the sample size is greater than or equal to 30), and *size* is the sample size. First we will find the lower limit of the confidence interval. In cell B1, enter the following formula.

=425-CONFIDENCE.NORM(0.10,900,100)

The lower limit is computed to be 276.9632.

3.   Now, enter the following formula in cell B2 to find the upper limit of the interval.

=425+CONFIDENCE.NORM(0.10,900,100)

The upper limit is computed to be 573.0368. Thus, we are 90% confident that the true population mean is between 276.9632 and 573.0368.

	A	B	C
1	Lower Limit	276.9632	
2	Upper Limit	573.0368	
3			

Figure 9.13

Confidence Intervals for the Population Mean (Small Samples, σ Unknown)

Construct the 95% confidence interval for the population mean of a normal population if the sample standard deviation is 6.5, the sample mean is 125, and the sample size is 15.

1. Label cells A1 and B1 with **Lower Limit** and **Upper Limit**, respectively.

2. The function that returns the margin of error for a confidence interval in Microsoft Excel when sigma is unknown, the sample size is less than 30, and the data are drawn from a normally distributed population is

   CONFIDENCE.T(alpha, standard_dev, size)

   where *alpha* is the level of significance, *standard_dev* is the sample standard deviation, and *size* is the sample size. First we will find the lower limit of the confidence interval. In cell B1, enter the following formula.

   =125−CONFIDENCE.T(0.05,6.5,15)

   The lower limit is computed to be 121.4004.

3. Now, enter the following formula in cell B2 to find the upper limit of the interval.

   =125+CONFIDENCE.T(0.05,6.5,15)

   The upper limit is computed to be 128.5996. Thus, we are 95% confident that the true population mean is between 121.4004 and 128.5996.

⊿	A	B	C
1	Lower Limit	121.4004	
2	Upper Limit	128.5996	
3			

Figure 9.14

Using Minitab

Confidence Intervals for the Population Mean

Construct the 90% confidence interval for the population mean of a normal population if the sample standard deviation is 900, the sample mean is 425, and the sample size is 100.

1. Choose **Stat**, select **Basic Statistics**, and then choose **1-Sample Z**.

2. Select the radio button for **Summarized data**, and enter **100** for the Sample size, **425** for the Mean, and **900** for the Standard deviation.

3. Click on the **Options** button. Set the Confidence level to **90.0** and be sure that **not equal** is selected as the Alternative. Click **OK**.

4. Click **OK** again. The 90% confidence interval is given in the Session window as (227.0, 573.0).

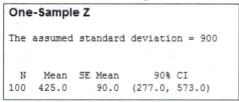

```
One-Sample Z

The assumed standard deviation = 900


  N    Mean   SE Mean      90% CI
100    425.0     90.0   (277.0, 573.0)
```

Figure 9.15

Confidence Intervals for the Population Mean (Small Samples, σ Unknown)

Construct the 95% confidence interval for the population mean of a normal population if the sample standard deviation is 6.5, the sample mean is 125, and the sample size is 15.

1. Choose **Stat**, select **Basic Statistics**, and then choose **1-Sample t**.

2. Select the radio button for **Summarized data**, and enter **15** for the Sample size, **125** for the Mean, and **6.5** for the Standard deviation.

3. Click on the **Options** button. Set the Confidence level to **95.0** and be sure that **not equal** is selected as the Alternative. Click **OK**.

4. Click **OK** again. The 95% confidence interval is given in the Session window as (121.40, 128.60).

```
One-Sample T

 N    Mean  StDev  SE Mean        95% CI
15  125.00   6.50     1.68  (121.40, 128.60)
```

Figure 9.16

Confidence Intervals for the Population Proportion

Use the information in Example 9.9 to construct the 95% confidence interval.

1. Choose **Stat**, select **Basic Statistics**, and then choose **1 Proportion**.

2. Select **Summarized data** and enter **48** as the Number of events and **410** as the Number of trials.

3. Click on the **Options** button. Set the Confidence level to **95.0** and that **not equal** is selected as the Alternative. Check the box for **Use test and interval based on normal distribution**.

4. Press **OK**, and then **OK** again.

5. The 95% confidence interval is given in the Session window as (0.085953, 0.148194).

```
Test and CI for One Proportion

Sample   X    N  Sample p        95% CI
1       48  410  0.117073  (0.085953, 0.148194)

Using the normal approximation.
```

Figure 9.17

R   Chapter 9 Review

Key Terms and Ideas

- Judgment Estimates
- Statistical Inference
- Estimator
- Estimate
- Point Estimator
- Point Estimate
- Interval Estimate
- Mean Squared Error
- Unbiasedness
- Unbiased Estimator
- Interval Estimator
- Confidence Interval

- Confidence Level
- Confidence Coefficient
- Confidence Interval for the Population Mean
- Margin of Error (Maximum Error of Estimation)
- Student's t-Distribution
- Degrees of Freedom
- Pilot Study
- Proportion
- Sample Proportion
- Confidence Interval for the Population Proportion

Key Formulas

	Section
Mean Squared Error for the Sample Mean $$\text{MSE}\left(\overline{x}\right) = E\left(\overline{x} - \mu\right)^2$$	9.2
$100\left(1-\alpha\right)\%$ Confidence Interval for the Population Mean $$\overline{x} \pm z_{\alpha/2}\frac{\sigma}{\sqrt{n}}$$ where $\overline{x}$ is the sample mean, $z_{\alpha/2}$ is the critical value for a z-distribution which captures an area of $\alpha/2$ in the right tail of the distribution, σ is the population standard deviation (approximated by s if σ is unknown and $n \geq 30$), and n is the sample size.	9.3
Student's t-Distribution $$t = \frac{\overline{x} - \mu}{s/\sqrt{n}}$$ where $\overline{x}$ is the sample mean, s is the sample standard deviation, and n is the sample size.	9.4

Key Formulas (cont.)

	Section

Degrees of Freedom

$$df = \text{number of sample observations} - 1 = n - 1$$

9.4

$100(1-\alpha)\%$ Confidence Interval for the Population Mean: Small Samples, σ Unknown

$$\bar{x} \pm t_{\alpha/2, df} \frac{s}{\sqrt{n}}$$

9.4

where $t_{\alpha/2, df}$ is the critical value for a t-distribution with $n - 1$ degrees of freedom which captures an area of $\alpha/2$ in the right tail of the distribution.

Margin of Error for the Population Mean (Maximum Error of Estimation)

$$E = z_{\alpha/2} \frac{\sigma}{\sqrt{n}}$$

9.5

Determining the Sample Size for the Population Mean: σ Known

$$n = \left(\frac{z_{\alpha/2}\sigma}{E} \right)^2$$

9.5

Determining the Sample Size for the Population Mean: σ Unknown

$$n = \left(\frac{z_{\alpha/2}s}{E} \right)^2$$

9.5

Point Estimate of the Population Proportion

$$\hat{p} = \frac{x}{n}$$

9.6

where x is the number in the sample that possesses the attribute and n is the sample size.

$100(1-\alpha)\%$ Confidence Interval for the Population Proportion

$$\hat{p} \pm z_{\alpha/2}\sigma_{\hat{p}}$$

9.6

where $z_{\alpha/2}$ is the critical value for a z-distribution which captures an area of $\alpha/2$ in the right tail of the distribution, and $\sigma_{\hat{p}}$ is the standard deviation of $\hat{p}$.

Key Formulas (cont.)

	Section

Margin of Error for the Population Proportion (Maximum Error of Estimation)

$$E = z_{\alpha/2}\sigma_{\hat{p}} = z_{\alpha/2}\sqrt{\frac{p(1-p)}{n}}$$

9.7

where $z_{\alpha/2}$ is the critical value for a z-distribution which captures an area of $\alpha/2$ in the right tail of the distribution, p is the population proportion, and n is the sample size.

Determining the Sample Size for the Population Proportion

$$n = \frac{z_{\alpha/2}^2 p(1-p)}{E^2} \approx \frac{z_{\alpha/2}^2 \hat{p}(1-\hat{p})}{E^2}$$

9.7

where $z_{\alpha/2}$ is the critical value for a z-distribution which captures an area of $\alpha/2$ in the right tail of the distribution, p is the population proportion, and $\hat{p}$ is the sample proportion.

Determining the Sample Size for the Population Proportion: No Estimate $\left(\hat{p}\right)$ Available

9.7

$$n = \frac{z_{\alpha/2}^2(0.5)(1-0.5)}{E^2} \approx \frac{z_{\alpha/2}^2(0.25)}{E^2}$$

AE   Additional Exercises

1. The owner of Sloppy Jack's bar is thinking about installing some video game machines. To estimate the profitability of the machines, he measures the number of times a competitor's machines are played over a randomly selected sample of days. The preliminary sample showed that the standard deviation of the number of times the machines are played is 10 times per day. Find the sample size (in days) necessary to estimate the average number of times the machines will be played in a day to within 5 plays with 99% confidence.

2. A random sample of fifteen eleven-year-old boys is selected in order to estimate the mean height for boys belonging to that age group. The resulting measurements in inches are given in the table below.

Table for Exercise 2 – Heights (Inches)				
55	58	52	58	54
57	56	54	58	56
52	59	55	61	57

   a. Calculate the sample mean and the sample standard deviation of the heights.

   b. Construct a 95% confidence interval for the mean height of all eleven-year-old boys.

   c. What assumption did you make about the heights in constructing your interval?

3. R. Cramden, chief development officer for Fontana Area Transport bus company, is concerned about the declining use of the bus system. He wishes to estimate the percentage of Fontana residents who consider safety a significant factor in their decision about whether or not to ride a bus. This will be a preliminary study so he is willing to develop an estimate with an error of 10% at a confidence level of 90%.

   a. What sample size will be needed?

   b. If 150 residents in a random sample of 500 Fontana residents say that they consider safety a significant factor in their decision about whether or not to ride a bus, estimate the true proportion of Fontana residents who think safety is a significant factor in their decision about whether or not to ride a bus with 95% confidence.

4. According to a 2001 study conducted by the American Stock Exchange, 87% of 500 young Americans surveyed said that they can't count on Social Security as a source of income when they retire. Construct a 90% confidence interval for the proportion of young Americans who feel they can't count on Social Security as a source of income when they retire.

5. The State Bureau of Standards must inspect gasoline station pumps on a regular basis to be sure they are operating properly. A recent survey of a randomly selected group of 61 pumps produced a sample mean of 9.75 gallons dispensed for a pump reading ten gallons. If the sample had a standard deviation of 1.12 gallons, find the 80% confidence interval for the mean amount of gas dispensed when a gas pump reads ten gallons.

6. In a population of non-unionized employees, 55% are sympathetic toward unionization. The American Federation of Labor has drawn a random sample of 250 persons selected from this population to investigate union interest. Construct a 90% confidence interval for the proportion of the sample that will be sympathetic toward unionization.

7. Suppose a study designed to collect data on smokers and nonsmokers uses a preliminary estimate of the proportion that smoke of 22%. How large a sample should be taken to estimate the proportion of smokers in the population with a margin of error of 0.02 with 88% confidence?

8. As part of an annual review of its accounts, a discount brokerage firm selects a random sample of 15 customers. Their accounts are reviewed for a total account valuation, which showed a mean of $32,000 with a sample standard deviation of $8200.

   a. What is a 99% confidence interval for the mean account valuation of the population of customers? Interpret the interval in terms of the problem.

   b. What assumption about the account distribution is necessary to solve this problem?

9. Direct Music has 250 retail outlets throughout the United States. The firm is evaluating a potential location for a new outlet, based in part, on the mean annual income of the individuals in the marketing area of the new location. A sample of size 36 was taken; the sample mean income is $31,100. The population standard deviation is estimated to be $4500. Construct a confidence interval using a confidence coefficient of 0.95.

10. Suppose we want to determine the sample size required to give us a 95% confidence interval that estimates, to within $500, the average salary of a Virginia Tech employee. Also, suppose that from a previous experiment, we know that $s = \$6300$. What is the minimum sample size required?

11. A reporter for a student newspaper is writing an article on the cost of off-campus housing. A sample of 16 efficiency apartments within a half-mile of campus resulted in a sample mean of $650 per month and a sample standard deviation of $55. Construct a 95% confidence interval estimate of the mean rent per month for the population of efficiency apartments within a half-mile of campus. We will assume that this population is normally distributed.

12. A stock market analyst wants to estimate the average return on a certain stock. A random sample of 15 days yields an average return of 10.37% and a standard deviation of 3.5%. Give a 90% confidence interval for the true average return on the stock. Assume that the stock returns are normally distributed.

13. Voting, Inc. specializes in voter polls and surveys designed to keep political office seekers informed of their position in a race. Using telephone surveys, interviewers ask registered voters who they would vote for if the election were held that day. In a current election campaign, Voting, Inc. has found that 220 registered voters, out of 500 contacted, favor a particular candidate. Find a 95% confidence interval estimate for the proportion of the population of registered voters that favor the candidate.

14. Before beginning a pension program for its workers, a corporation wishes to estimate the proportion of its workers who have been employed at the company for at least 20 years. A random sample of 138 employees yielded 16 who have been with the corporation for at least 20 years. Construct a 92.2% confidence interval for the proportion of employees who have worked for this corporation for at least 20 years. Interpret your results.

P   Discovery Project

Existing-Home Sales

The article below was taken from the monthly newsletter of the National Association of Realtors. The article presents common examples of the use of statistics in the media. Answer the questions at the end of the article.

Source: National Association of Realtors – realtor.org

Existing-Home Sales Slow in June but Remain Above Year-Ago Levels

WASHINGTON – July 22, 2010

With the scheduled closing deadline for the home buyer tax credits, existing-home sales slowed in June but remained at relatively elevated levels, according to the National Association of Realtors.

Existing-home sales, which are completed transactions that include single-family, townhomes, condominiums, and co-ops, fell 5.1 percent to a seasonally adjusted annual rate of 5.37 million units in June from 5.66 million in May, but are 9.8 percent higher than the 4.89 million-unit pace in June 2009.

Lawrence Yun, NAR chief economist, said the market shows uncharacteristic yet understandable swings as buyers respond to the tax credits. "June home sales still reflect a tax credit impact with some sales not closed due to delays, which will show up in the next two months," he said.

"Broadly speaking, sales closed after the home buyer tax credit will be significantly lower compared to the credit-induced spring surge. Only when jobs are created at a sufficient pace will home sales return to sustainable healthy levels."

According to Freddie Mac, the national average commitment rate for a 30-year, conventional, fixed-rate mortgage fell to a record low 4.74 percent in June from 4.89 percent in May; the rate was 5.42 percent in June 2009.

The national median existing-home price for all housing types was $183,700 in June, which is 1.0 percent higher than a year ago. Distressed homes were at 32 percent of sales last month, compared with 31 percent in May; it was also 31 percent in June 2009.

NAR president Vicki Cox Golder, owner of Vicki L. Cox & Associates in Tucson, Ariz., said softer home sales expected this summer don't tell the whole story. "Despite these market swings, total annual home sales are rising above 2009 and we're looking for overall gains again this year as well as in 2011," she said. "Conditions have become more balanced in much of the country, which is good for both buyers and sellers. However, consumers find it even more challenging to navigate the transaction process, especially for distressed properties, which only underscores the value Realtors bring to buyers and sellers in this market."

A parallel NAR practitioner survey shows first-time buyers purchased 43 percent of homes in June, down from 46 percent in May. Investors accounted for 13 percent of sales in June, little changed from 14 percent in May; the remaining purchases were by repeat buyers. All-cash sales were at 24 percent in June compared with 25 percent in May.

Total housing inventory at the end of June rose 2.5 percent to 3.99 million existing homes available for sale, which represents an 8.9-month supply at the current sales pace, up from an 8.3-month supply in May.

"The supply of homes on the market is higher than we'd like to see. But home prices are still holding their ground because prices had already over-corrected in many local markets," Yun said.   Raw unsold inventory remains 12.7 percent below the record of 4.58 million in July 2008.

Single-family home sales fell 5.6 percent to a seasonally adjusted annual rate of 4.70 million in June from a level of 4.98 million in May, but are 8.5 percent above the 4.33 million pace in June 2009. The median existing single-family home price was $184,200 in June, up 1.3 percent from a year ago.

Single-family median existing-home prices were higher in 10 out of 19 metropolitan statistical areas reported in June in comparison with June 2009. In addition, existing single-family home sales rose in 12 of the 19 areas from a year ago while two were unchanged.

Existing condominium and co-op sales slipped 1.5 percent to a seasonally adjusted annual rate of 670,000 in June from 680,000 in May, but are 20.5 percent higher than the 556,000-unit pace in June 2009. The median existing condo price was $180,000 in June which is 1.4 percent below a year ago.

Regionally, existing-home sales in the Northeast rose 7.9 percent to an annual level of 960,000 in June and are 17.1 percent above June 2009. The median price in the Northeast was $244,300, down 1.2 percent from a year ago.

Existing-home sales in the Midwest dropped 7.5 percent in June to a pace of 1.23 million but are 11.8 percent higher than a year ago. The median price in the Midwest was $155,900, down 0.1 percent from June 2009.

In the South, existing-home sales fell 6.5 percent to an annual level of 2.01 million in June but are 11.0 percent above June 2009. The median price in the South was $163,600, unchanged from a year ago.

Existing-home sales in the West dropped 9.3 percent to an annual pace of 1.17 million in June but are 0.9 percent higher than a year ago. The median price in the West was $221,800, up 1.5 percent from June 2009.

Questions to Consider:

1. What type of statistical measure is used throughout the article?

2. Are point and interval estimates used in the article?

3. What is the population from which the sample is drawn?

4. What sampling method is used to obtain the data?

5. Are margins of error presented?

6. Overall, do you believe the statistics used in this article are reasonably good estimates of the population? Justify your answer.

Chapter Ten

Hypothesis Testing: Single Samples

Introduction

Given the recent economic crisis – with rising gas/oil prices, the decline of the banking industry and financial markets, as well as many other companies and corporations – one would think that the travel industry would also face a significant decrease in the amount of money spent on vacations (or the number of vacations taken). As part of the domino effect, one would also expect the number of jobs available in the travel industry to be limited. If we wanted to determine if the previous assumptions were valid, we would be interested in making inferences about the average amount of money spent on vacations in the last year versus the amount of money spent on vacations five years ago. Similarly, we would also want to make inferences about the number of jobs available in the travel industry last year versus five years ago. In both cases, we are interested in testing how the value of a parameter relates to (i.e. whether it is less than, equal to, or greater than) some specific numerical value. This type of inference is called **hypothesis testing**, which is the subject of this chapter.

Like the above example, we perform some type of hypothesis test on a daily basis. Whether it's trying to determine if the shower temperature is reasonable (acceptable, too hot, or too cold) or if we want to confirm if the day's weather forecast is accurate (will it be too hot to wear a sweater, adequate, or if you'll be underdressed), this type of subjective decision making is always a part of our lives. We rely on personally accumulated data to help us make informed decisions. What distinguishes statistical hypothesis testing from the everyday variety is the use of statistical measures in the statement of the hypotheses, the collection of the sample data, and the use of the sample data in a well-defined decision-making process.

10.1   Developing a Hypothesis

It should come as no surprise that the first step in all statistical hypothesis testing is a statement of hypothesis, regardless of the nature of the problem. Further, if the hypothesis test is conducted using statistical methods, the hypotheses must be stated in terms of statistical measures such as the population mean, population proportion, or population variance.

Example 10.1

Tech Travel, which is a travel agency managed by C. Dubya, is interested in determining how much money people are spending on travel and entertainment during the economic downturn. From previous data, the agency knows that in 2008, U.S. households spent, on average, a total of $1415 on transportation, food, lodging, and pleasure trips. However, the agency believes that in spite of economic woes, people will spend more in 2010 on travel than they did two years ago. The agency's question can be investigated with two hypotheses.

- Travel expenses have not increased since 2008.
- Travel expenses have increased since 2008.

To apply statistical methodology to answer the question, the preceding hypotheses must be translated into a problem statement concerning a statistical measure (e.g. mean, proportion, or variance). The statistical measure will be used in the definition of a criterion to decide the issue. There is a great deal of variability in the amount of money one spends on vacations, depending on the location, whether they are flying or driving, etc. Comparing the mean amount spent per family with that which was spent previously is a reasonable method of evaluating the differences. The average amount spent in 2008 is known to be $1415. However, the average amount spent in 2010 is currently unknown, and that uncertainty is what makes the problem difficult. If μ is defined as

$$\mu = \text{average amount spent on travel expenses in 2010,}$$

then the two claims can be written in the following manner.

$$H_0: \mu \leq \$1415 \quad \left(\begin{array}{l} \text{Money spent on 2010 travel expenses is} \\ \text{less than or equal to that spent in 2008.} \end{array} \right)$$

$$H_a: \mu > \$1415 \quad \left(\begin{array}{l} \text{More money is spent on travel expenses} \\ \text{in 2010 than in 2008.} \end{array} \right)$$

H_0 is called the **null hypothesis**, and contends that the mean amount spent on travel expenses in 2010 is less than or equal to that spent in 2008. This would contradict the agency's assertion. H_a is called the **alternative hypothesis** (sometimes denoted by H_1) and declares that the mean amount spent in 2010 is more than that spent in 2008. Once the problem statement is formulated in terms of the population parameter (in this case, the parameter is μ), sample data can be developed to help determine which hypothesis is more reasonable.

This example illustrates another important component of a hypothesis test, namely, deciding whether H_a should be one-sided or two-sided. In Example 10.1, H_a involves a **one-sided alternative**, since the agency is only interested in whether travelers spent *more* money. If the agency was interested in whether travelers spent *less* money in 2010, then H_a would also have been one-sided, but the inequality would have been in the other direction $\left(H_a: \mu < \$1415 \right)$. A **two-sided alternative** $\left(H_a: \mu \neq \$1415 \right)$ would indicate that the agency is concerned with expenditures that are *above or below* the average in 2008. One-sided alternatives require a **one-tailed hypothesis test** and two-sided alternatives require a **two-tailed hypothesis test**.

The hypothesis testing procedure is a method for choosing between two competing hypotheses. Although the procedure will change for different kinds of hypotheses, there are common elements among all tests.

Elements of Hypothesis Testing

Common Elements and Comments about Hypothesis Testing

- The null hypothesis is presumed to be true unless sample data produces overwhelming evidence to the contrary. That is, the test statistic is calculated under the assumption that the null hypothesis is true.

- Sample data are used to calculate a test statistic. The form of the test statistic will change depending upon the statistical measure used in the hypothesis statement, μ, p, or σ^2, as well as with the assumptions about knowledge of the population. The **test statistic** is a component of the criteria used to evaluate the hypothesis. If the test statistic falls into a **rejection region**, the null hypothesis, H_0, will be rejected in favor of the alternative hypothesis, H_a.

- The test statistic is usually designed so that if the null hypothesis is true, the value of the test statistic will "probably" be close to zero. (The notion of *close* is arbitrary. Developing a formal test of a hypothesis will draw upon the theory of sampling distributions and the language of probability to more precisely define the meaning of close.)

- The conclusion of the hypothesis test results in a decision to either **reject** or **fail to reject** the null hypothesis. Note that we do not accept the null hypothesis; we only conclude that there is insufficient evidence to support the alternative.

- There is no way to be absolutely certain your decision is correct, no matter which hypothesis is selected.

Roles of the Null and Alternative Hypotheses

In the previous example, we treated the null and alternative hypotheses equally. In the practice of statistical inference, this is not the case. In a statistical test of a hypothesis, the null hypothesis, H_0, is a statement that is presumed to be true unless there is overwhelming evidence in favor of the alternative. In other words, the null hypothesis is given the benefit of the doubt.

A familiar example of the disparate treatment of H_0 and H_a is found in our judicial system.

H_0: Defendant is not guilty.   (null hypothesis)

H_a: Defendant is guilty.   (alternative hypothesis)

A jury must believe that the evidence (data) demonstrates guilt "beyond a reasonable doubt" in order to convict a defendant (H_a). However, the defendant only has to demonstrate there is insufficient evidence of guilt in order to be acquitted (H_0). In other words, the null hypothesis $(H_0:$ Defendant is not guilty.$)$ is presumed to be true unless there is overwhelming evidence to the contrary. The defendant is innocent until proven guilty.

In Example 10.1 the hypotheses were formulated as follows.

	Correct		Incorrect
	$H_0: \mu \leq \$1415$	instead of	~~$H_0: \mu > \$1415$~~
	$H_a: \mu > \$1415$		~~$H_a: \mu \leq \$1415$~~

The <u>Incorrect</u> Formulation: H_0: $\mu > \$1415$

The statement H_0: $\mu > \$1415$ says that the average travel expenditures, μ, are greater than they were in 2008. The essence of the problem at hand is whether the statement $\mu > \$1415$ is true or untrue. By placing this statement in H_0, the statement is presumed to be true unless there is overwhelming evidence to the contrary. This does not make sense! Why would the agency try to prove that travel expenditures are greater in 2010 starting out by assuming that expenditures in 2010 are greater?

The <u>Correct</u> Formulation: H_0: $\mu \leq \$1415$

The agency must demonstrate overwhelming evidence that travel expenditures are greater in 2010 than in 2008 in order to be correct. To achieve this goal, the statement $\mu > \$1415$ must be placed in the alternative hypothesis. In the correct version, the null hypothesis, H_0: $\mu \leq \$1415$, states that we believe that travel expenditures are no greater in 2010 than they were in 2008. Because the null hypothesis is presumed to be true, this hypothesis is not rejected unless there is overwhelming evidence to the contrary. If H_0: $\mu \leq \$1415$ is rejected in favor of H_a: $\mu > \$1415$, then there is overwhelming evidence that households spent more on travel in 2010 than in 2008.

To summarize, the agency wants to know if the travel expenditures are greater in 2010 than in 2008. If the agency is correct, the null hypothesis will be rejected in favor of the alternative. Rejecting the null in favor of the alternative means the sample data overwhelmingly support the idea that expenditures in 2010 are greater than those in 2008. Failing to reject the null hypothesis would mean that there is insufficient evidence to warrant concluding that people spent more on travel in 2010 than they did in 2008.

This type of formulation is typical of a test of a hypothesis in a research environment (testing a new theory or idea against an established one). In many research hypothesis tests, the investigator hopes to reject the null hypothesis in order to demonstrate the significance of a new idea. It is often the case that if there is a **standard value** for a population parameter, the null hypothesis states that the population parameter is equal to the standard value.

When formulating hypotheses, remember that the equality statement will always be contained in the null hypothesis. That is, the null hypothesis will contain $\leq, \geq$, or $=$, whereas the alternative hypothesis will contain $<, >$, or $\neq$. This is because the only way to support an equality statement in the alternative hypothesis would be to perform a census, which defeats the purpose of sampling.

Example 10.2

Suppose a potato chip manufacturer is concerned that the bagging equipment is not functioning properly when filling 10 ounce bags. He wants to test a hypothesis that will help determine if there is a problem with the bagging equipment. What are the correct hypotheses?

$$H_0: \mu = 10 \text{ oz} \qquad H_0: \mu \neq 10 \text{ oz}$$
$$\text{or}$$
$$H_a: \mu \neq 10 \text{ oz} \qquad H_a: \mu = 10 \text{ oz}$$

Solution

Since the standard value in this problem is 10 ounces, the null hypothesis is H_0: $\mu = 10$ oz. The bagging equipment ordinarily functions properly; thus the manufacturer requires overwhelming evidence that the machine is overfilling or underfilling the bags before shutting down the equipment.

When the alternative hypothesis allows values above and below the standard value, it is called a **two-sided alternative**. In this case, the manufacturer hopes that he *fails to reject* the null hypothesis, since that would suggest that his equipment is putting the right amount in the bags. This type of hypothesis is quite common in quality control and other kinds of system monitoring.

Example 10.3

Analysts at JBN Technologies want to assess the use of Macintosh and Windows computers in their line of work. With the rise in popularity of Apple and the Mac OS X operating system, analysts believe that businesses are more likely to allow their employees to deploy Macintoshes as their enterprise desktops. How should JBN Technologies analysts formulate an appropriate hypothesis to determine if more businesses will deploy Macintoshes as their enterprise desktops?

Solution

The correct formulation of the hypotheses is as follows, where p is the fraction of enterprises that would support the use of Macintoshes as their desktops.

$$H_0: p \leq 0.5 \quad \left(\begin{array}{l} \text{There is not sufficient evidence to believe that more} \\ \text{businesses deploy Macs as their enterprise desktops.} \end{array} \right)$$

$$H_a: p > 0.5 \quad \left(\begin{array}{l} \text{There is sufficient evidence to believe that more} \\ \text{businesses deploy Macs as their enterprise desktops.} \end{array} \right)$$

The problem's formulation requires the use of a different statistical measure, p, the population proportion. The crucial value of the proportion is 0.5, since in order to believe that more businesses will support Macs, this proportion has to be more than 50%. The null hypothesis contends that there is not sufficient evidence that more businesses will support the use of Macs. If the null hypothesis is rejected in favor of the alternative, $H_a: p > 0.5$, then there is overwhelming evidence that businesses are prepared to adopt Mac computers.

The alternative is one-sided, since the intent of the study is to learn if Macs are more popular in the workplace. (A two-sided alternative would imply interest in finding out if more companies or fewer companies are willing to accept Macs in the workplace, a rather useless idea.) The analysts hope the conclusion of the test will be to *reject* the null hypothesis, since that would support the analysts' claim.

Procedure

Formulating Hypothesis Testing Problems

To be successful at formulating hypothesis testing problems you must be able to:

- determine the appropriate statistical measure to test the desired hypothesis (population mean, proportion, or variance).

- determine the appropriate value to use in the null hypothesis. (This may be stated in the problem as a standard value or may need to be deduced from the information at hand.)

- decide whether the alternative should be one-sided or two-sided.

10.1 Exercises

Basic Concepts

1. What is a hypothesis?

2. What is the first step in the test of a hypothesis?

3. Describe the common elements present in all hypothesis tests.

4. Summarize the difference between the null and alternative hypotheses.

5. Define and give an example of a one-sided alternative. How does this differ from a two-sided alternative?

6. What is the connection between one and two-sided alternatives and one and two-tailed tests?

7. Is there a way to be absolutely certain your decision is correct when performing a hypothesis test? Explain.

8. What are the three important things you must be able to do in order to be successful at formulating hypothesis testing problems?

10.2   Reaching a Conclusion

The point of a hypothesis test is to select one of the competing hypotheses $(H_0$ or $H_a)$ as the *correct* decision. Once you have reached a conclusion, there will remain a possibility that the inference (decision) is incorrect. The idea of performing the correct statistical test, exactly following the procedure, and still possibly making an incorrect decision may be bothersome. Uncertainty, however, is a part of every statistical inference.

How Uncertain Are Our Conclusions?

Errors can happen in two ways. In statistical terminology, we call these Type I errors and Type II errors.

> **Definition**
>
> Rejecting H_0 when in fact H_0 is the correct choice is called a **Type I error**.

> **Definition**
>
> Failing to reject H_0 when in fact H_a is the correct choice is called a **Type II error**.

In statistical tests of hypotheses, we can only control Type I errors. This fact often influences how we construct hypotheses.

Example 10.4

Carmen is the President and CEO of a regional healthcare organization. She has set a target that the average waiting time for non-emergency patients will not exceed 20 minutes. In a random sample (taken over numerous days), she finds the waiting time for 25 patients. Assuming that the population standard deviation of waiting times is eight minutes, how should Carmen formulate the hypotheses as she examines the waiting times?

Solution

Formulation A

H_0: Average waiting time does not exceed 20 minutes.

H_a: Average waiting time is more than 20 minutes.

Type I error using Formulation A

To have an error, you must make a mistake. Suppose this is what happened.

Truth: H_0: Average waiting time does not exceed 20 minutes.

Carmen's Decision: H_a: Average waiting time is more than 20 minutes.

To have a Type I error, we must reject the null hypothesis when the null hypothesis is true. In this case, Carmen believes that the average waiting time is more than 20 minutes when it is not.

Consequence: Given that the average wait time is not more than 20 minutes, Carmen would investigate ways to change policies and practices that would even further reduce the average waiting time, when reducing the waiting time is not necessary.

Type II error using Formulation A

Carmen's Decision: H_0: Average waiting time does not exceed 20 minutes.

Truth: H_a: Average waiting time is more than 20 minutes.

A Type II error means that we fail to reject the null hypothesis when the alternative hypothesis is true.

Consequence: In this case, Carmen believes that the average wait time is not more than 20 minutes, but it is actually more than 20 minutes. In this situation, Carmen doesn't make changes to any policies and practices to make things better, leaving patients unhappy.

Let's examine another formulation.

Formulation B

H_0: Average waiting time is 20 minutes or more.

H_a: Average waiting time is less than 20 minutes.

Type I error using Formulation B

Truth: H_0: Average waiting time is 20 minutes or more.

Carmen's Decision: H_a: Average waiting time is less than 20 minutes.

Carmen believes that the average wait time is less than 20 minutes when, in fact, the average wait time is 20 minutes or more.

Consequence: Carmen believes that the average wait time is adequate (less than 20 minutes) when in fact the average wait time is 20 minutes or more. In this situation, Carmen doesn't make any changes to improve the services being provided, risking the potential loss of patients.

Type II error using Formulation B

Carmen's Decision: H_0: Average waiting time is 20 minutes or more.

Truth: H_a: Average waiting time is less than 20 minutes.

Carmen believes that the average wait time is 20 minutes or more when in fact the average wait time is less than 20 minutes.

Consequence: Since Carmen believes that the average wait time is more than 20 minutes, she will likely change policies and procedures to reduce the waiting time. However, since the true average waiting time is less than 20 minutes, Carmen will change policies and procedures to further reduce the waiting time when a reduction in waiting time is unnecessary.

For Carmen, the most serious consequence is not making changes to the process when changes are needed in order to reduce the average waiting time to below 20 minutes. She has decided that she would rather improve a process that doesn't necessarily need improving rather than risk losing patients due to a long waiting time. Therefore, given the two formulations, A and B, Formulation B is the correct formulation of the problem, since the most serious consequence is the result of a Type I error. That's good because the probability of a Type I error can be controlled in the hypothesis testing procedure.

The last example demonstrates an important consideration when formulating hypotheses that is worth repeating. **Formulate the hypotheses so that the most serious consequence of a mistaken decision is associated with a Type I error.**

| | **True State of Nature** | |
	H_0 is true	H_0 is false
Reject H_0	Type I Error (Rejecting a true H_0) α	Correct Decision
Fail to Reject H_0	Correct Decision	Type II Error (Failing to reject a false H_0) β

(Decision)

Definition

The probability of a Type I error is denoted by α (the Greek letter alpha), and is referred to as the **level of the test**, or **significance level of the test**.

It is up to the researcher to select alpha prior to the start of the hypothesis test.

Definition

The probability of a Type II error is denoted by β (the Greek letter beta). Unfortunately, the value of β depends on the actual value of the population parameter and thus cannot be determined unless the actual value of the population parameter is known, which is rarely the case.

In examining the relationship between α and β, a natural question which might arise is:

Why not make both α and β as small as possible?

Unfortunately, the probability of making a Type I error, α, and the probability of making a Type II error, β, are inversely related. Thus, the smaller we make the probability of making a Type I error, the more likely we are to make a Type II error. In general, choose the largest level of α which is tolerable to avoid unnecessarily increasing the probability of making a Type II error.

10.2   Exercises

Basic Concepts

1. Describe a Type I error.

2. Describe a Type II error.

3. Explain how Type I and Type II errors influence the construction of a hypothesis.

4. Can both Type I and Type II errors be controlled in the hypothesis testing procedure? Explain.

5. What is the level of the test?

6. Why is a Type II error difficult to express numerically?

Exercises

7. For the following situations, develop the appropriate H_0 and H_a and state what the consequences would be for Type I and Type II errors.

   a. The Standard Tire Company has introduced a new tire in Europe that will be guaranteed to last at least 30,000 kilometers. Standard Tire has hired an independent agency to determine if there is overwhelming evidence that their tires will last through the warranty period.

   b. Mrs. Russell, head product tester for Hathaway Tool Corporation, is testing a newly designed series of bar hooks. The hooks have been designed to give way if they get too hot. The previous design gave way at 240 degrees. Develop a test to determine if the newly designed hooks give way at a higher temperature than the previous design.

   c. The regional manager of a large grocery store chain believes that a checkout person using a new scanner can process a full basket of groceries in less than seven minutes. Develop a test to determine if there is overwhelming evidence to support the manager's belief.

8. For the following situations, develop the appropriate H_0 and H_a and state what the consequences would be for Type I and Type II errors.

   a. A company that manufactures one-half inch bolts selects a random sample of bolts to determine if the diameter of the bolts differs significantly from the required one-half inch.

   b. A company that manufactures safety flares randomly selects 100 flares to determine if the flares last at least three hours on average.

   c. A consumer group believes that a new sports coupe gets significantly fewer miles to the gallon than advertised on the sales sticker. To confirm this belief, they randomly select several of the new coupes and measure the miles per gallon.

10.3 A Procedure for Testing a Hypothesis

A procedure for testing a hypothesis is given below. In examining this procedure, you will notice we have already discussed the first five steps. The next two steps define a **decision rule**, which is a criterion used to determine whether the null or the alternative hypothesis will be chosen. As you look over these steps, do not be overly concerned if you do not understand everything. You will learn by following the examples.

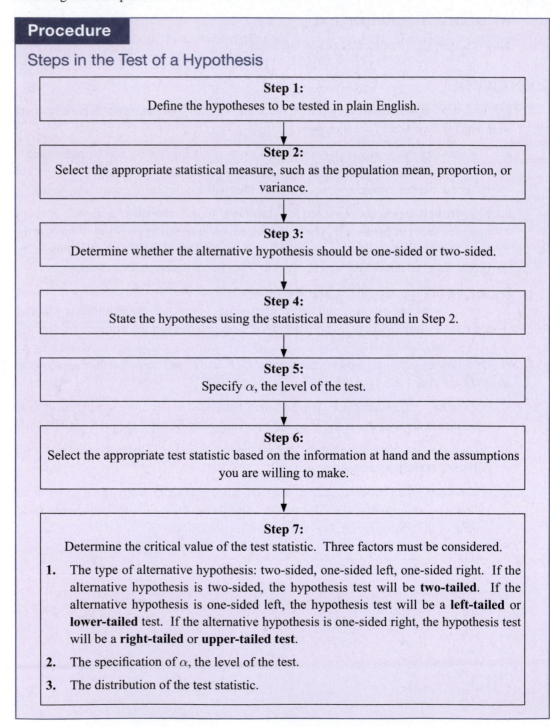

Procedure

Steps in the Test of a Hypothesis

Step 1:
Define the hypotheses to be tested in plain English.

Step 2:
Select the appropriate statistical measure, such as the population mean, proportion, or variance.

Step 3:
Determine whether the alternative hypothesis should be one-sided or two-sided.

Step 4:
State the hypotheses using the statistical measure found in Step 2.

Step 5:
Specify α, the level of the test.

Step 6:
Select the appropriate test statistic based on the information at hand and the assumptions you are willing to make.

Step 7:
Determine the critical value of the test statistic. Three factors must be considered.

1. The type of alternative hypothesis: two-sided, one-sided left, one-sided right. If the alternative hypothesis is two-sided, the hypothesis test will be **two-tailed**. If the alternative hypothesis is one-sided left, the hypothesis test will be a **left-tailed** or **lower-tailed** test. If the alternative hypothesis is one-sided right, the hypothesis test will be a **right-tailed** or **upper-tailed test**.

2. The specification of α, the level of the test.

3. The distribution of the test statistic.

Procedure

Steps in the Test of a Hypothesis (cont.)

Step 8:
Collect sample data and compute the value of the test statistic.

↓

Step 9:
Make the decision.

- If the value of the test statistic is in the rejection region, reject the null hypothesis in favor of the alternative.

- If the value of the test statistic is not in the rejection region, fail to reject the null hypothesis.

↓

Step 10:
State the conclusion in terms of the original question.

10.3 Exercises

Basic Concepts

1. State the 10 steps of a hypothesis test.

10.4 Testing a Hypothesis about a Population Mean

The following example will be used to illustrate the hypothesis testing procedure. The example goes through all the steps and contains detailed explanations of several new concepts.

Example 10.5

Suppose that the average amount of money a student spends on textbooks per semester at college campuses in the U.S. is $500 with a standard deviation of $100. A local university wants to know if its students are spending that amount on textbooks. The level of the test is to be set at 0.05. A random sample of 75 college students has been selected and the resulting average is $540 spent per semester on textbooks.

Solution

Step 1: Define the hypotheses in plain English.

The hypotheses are fairly straightforward.

- **Null Hypothesis:** The local university's students are spending an average of $500 per semester on textbooks.

- **Alternative Hypothesis:** The students are not spending an average of $500 per semester on textbooks.

Step 2: Select the appropriate statistical measure.

Since the problem states that the university wants to know how much it's students spend on textbooks per semester in relation to the national average, the hypotheses will concern the population mean. Let

$$\mu = \text{average amount spent on textbooks per semester by college students}$$
$$\text{at the local university.}$$

Step 3: Determine whether the hypothesis should be one-sided or two-sided.

Nothing is mentioned in the problem to indicate that the university is interested in learning if its students are spending more than the national average per semester on textbooks, nor is there anything mentioned that the university is interested in learning if its students are spending less than the national average per semester on textbooks. Thus, it is assumed that the university is interested in determining if there is any departure from the national average. Consequently, the problem needs to be formulated as a two-sided alternative.

Step 4: State the hypotheses using the appropriate statistical measure.

$$H_0: \mu = \$500$$
$$H_a: \mu \neq \$500$$

The null hypothesis states that the average amount spent per semester on textbooks at the local university is equal to the national average (the standard value). The alternative states that it is not. This is an application of testing against the standard value.

Step 5: Specify the level of the test.

The level of the test was given in the problem statement as the $\alpha = 0.05$ level. If it were not specified in the problem statement, then it would be necessary to select the value. Typical values for the level of the test are 0.10, 0.05, and 0.01. It is important to remember that the level of the test specifies the probability of a Type I error.

Step 6: Select the appropriate test statistic.

What we know about the population will affect the methods used to perform the test of the hypothesis. There are three key questions:

1. Is the standard deviation of the population known?

2. Is the distribution of the population normal?

3. Is the sample size sufficiently large?

Is the standard deviation of the population known?

In most instances, the standard deviation of the population will not be known.

Is the distribution of the population normal?

If we have a large sample size $(n \geq 30)$, we will not have to be concerned about the distribution of the population. If the sample size is small $(n < 30)$, then the underlying population will need to be normally distributed. When the population is normal, the histogram of the sample data should be bell-shaped. However, should the frequency distribution show substantial deviation from normality, a nonparametric procedure may be warranted. These methods are discussed in Chapter 16.

Is the sample size sufficiently large?

Fortunately, the central limit theorem enables us to define the following rule to make this decision.

We will continue with steps 7 through 10 after we discuss the test statistic.

Formula

z-Test Statistic

If $n \geq 30$ or if the standard deviation of the population, σ, is known and the sample is drawn from a normal population, then by the central limit theorem the test statistic is approximately given by

$$z = \frac{\bar{x} - \mu_0}{\sigma_{\bar{x}}}, \text{ where } \sigma_{\bar{x}} = \frac{\sigma}{\sqrt{n}}.$$

z has a standard normal distribution, and if σ is unknown and n is greater than or equal to 30, s can be used as an approximation of σ.

The symbol for the hypothesized mean is μ_0. In Example 10.5, $\mu_0 = \$500$. The difference between the symbols μ and μ_0 is that μ_0 is a specific claim about the value of μ. The actual value of μ is unknown since it is a population parameter.

In this example, the symbol μ refers to the mean amount spent per semester on textbooks at the local university. If this value were known, then it would be easy to decide which of the two claims

$$H_0: \mu = \$500 \qquad \text{(null hypothesis)}$$
$$H_a: \mu \neq \$500 \qquad \text{(alternative hypothesis)}$$

is correct.

What is the Rationale for the Test Statistic?

Although the true population mean is unknown, the central limit theorem implies that the sample mean, $\bar{x}$, is usually close to the population mean, μ, if the sample data are representative of the population. **If the null hypothesis is true**, the value of the population mean is specified in the null hypothesis, $H_0: \mu = \$500$. If the null is true, the sample mean $\bar{x}$ should be close to 500 (μ_0).

Another way of expressing this idea of closeness is to measure the difference between the sample mean and the hypothesized mean, $\bar{x} - \mu_0$. The difference should be close to zero, if the null is true.

Conversely, a large distance between $\bar{x}$ and μ_0

would indicate either an improbable value of $\bar{x}$ has been observed given that the null hypothesis is true; or that the null hypothesis is false and μ_0 is not the actual value of the population mean.

Although this approach has intuitive appeal, the notion of *close* is too ambiguous to have practical value. Changing the unit of measure from inches to miles may affect one's perception of closeness. What is needed is a statistic that will be the same regardless of the scale of measurement the sample data possesses.

Measuring Closeness

Converting the distance, $\bar{x} - \mu_0$, into standard deviation units using the z-transformation eliminates the problems caused by differing measurement scales, since the value of the z-transformation will be the same regardless of the unit of measure.

Recall that the z-transformation is $z = \dfrac{x - \mu}{\sigma}$. The variable x represents the random variable that will be transformed. In this case, the random variable to be transformed is $\bar{x}$. The symbol μ in the z-transformation is the mean of the random variable being transformed $(\bar{x})$. In Example 10.5, the null hypothesis states that $\mu_0 = \$500$. If the null hypothesis is true, μ_0 is the mean of the population and thus the mean of the sample means.

Using the data provided in Example 10.5 and substituting $\bar{x}$, σ, n, and μ_0 into the formula for the z-transformation yields the following.

Modern Hypothesis Testing

Many of the concepts in modern hypothesis testing were first proposed by Jerzy Neyman and Egon Pearson. They were the first to recognize the need for an alternative hypothesis. They developed and named the notions of Type I and Type II errors.
They proposed a term, "critical region," to denote a set of sample statistical values leading to the rejection of the hypothesis being tested.

$$z = \frac{\bar{x} - \mu_0}{\sigma_{\bar{x}}} = \frac{540 - 500}{100 \Big/ \sqrt{75}} \approx 3.46$$

$$\text{where } \sigma_{\bar{x}} = \frac{\sigma}{\sqrt{n}} = \frac{100}{\sqrt{75}} \approx 11.5470$$

The resulting z-value, 3.46, is the number of standard deviations that the sample mean (540) is from the hypothesized value of the mean, 500 (μ_0). If $\bar{x}$ is close to the hypothesized value, then the z-value is close to zero. If $\bar{x}$ is far away from the hypothesized value, the z-value will be large in absolute value. The larger z grows in absolute value, the more difficult it becomes to believe H_0 is true.

In this example, $\bar{x}$ is 3.46 standard deviation units from the hypothesized mean. How far does the hypothesized value of the null hypothesis, μ_0, need to be from $\bar{x}$ before we would be willing to believe that the deviation was not caused by ordinary sampling variability? If $\bar{x}$ is an improbable event given H_0 is true, then we have two choices. We could believe H_0 is true, and we have witnessed a rare event caused by sampling variability, or we say that we doubt that we have witnessed a rare event and believe H_0 is not true.

Is an $\bar{x}$ that is 3.46 standard deviation units from the mean a rare occurrence? When we study P-values later in this chapter this question will be answered quite specifically. But to answer that question, an understanding of the sampling variability of the z-value will be essential in judging the likelihood that any difference between $\bar{x}$ and μ_0 is due to ordinary sampling variability.

The original motivation for using the z-transformation was that it provided a reasonable method of measuring how close the sample mean was to the hypothesized mean. While this is true, it suggests concern only with the results of one sample. However, to develop a theory which can be applied to all possible samples, we must broaden the discussion to all potential values of $\bar{x}$ and z. In other words, we must concern ourselves with the sampling distribution of $\bar{x}$.

Fortunately, the central limit theorem provides a great deal of information about the variability of $\bar{x}$. Namely, that it is normally distributed and has the same mean as the population from which the sample is drawn. The standard deviation of $\bar{x}$ is known to be $\frac{\sigma}{\sqrt{n}}$, the standard deviation of the population being sampled divided by the square root of the sample size.

The test statistic, z, is a random variable because $\bar{x}$ is a random variable. It will be normally distributed because $\bar{x}$ has a normal distribution.

In addition to measuring the distance between $\bar{x}$ and μ_0, the numerator of the z-transformation $(\bar{x} - \mu_0)$ also centers the distribution of the transformed random variable $\bar{x}$ around zero. If the null hypothesis is true, then the mean of $\bar{x}$ will be correctly specified in the null hypothesis, and the centering will work. If the null hypothesis is false, the mean of z will not be centered at 0. By using the z-transformation, we have shifted the focus of the problem. Instead of focusing our attention directly on $\bar{x}$ and its relation to the hypothesized value of μ, the crux of the hypothesis test will be to decide whether the z-value is near zero, or not. In our case the question is whether 3.46 is close to 0 or not. This question will be addressed when the decision rule is constructed.

If the null hypothesis is true, the z-test statistic has a normal distribution with a mean of 0 and a standard deviation of 1.

The Standard Normal Random Variable

$$\mu = 0 \quad \sigma = 1$$

Figure 10.1

A Decision Strategy

Suppose we claim that any value of the z-test statistic less than 1.96 in absolute value represents ordinary sampling variability of $\bar{x}$.

Any value of $|z|$ greater than or equal to 1.96 is attributed to a false null hypothesis.

If the null is true, the distribution of the z-test statistic is known (normal with mean 0 and standard deviation 1), and a plan can be developed for defining ordinary variability of the test statistic. Since the test statistic has a z-distribution, we know that 95% of the time the value of z should be within ± 1.96 standard deviations of the mean.

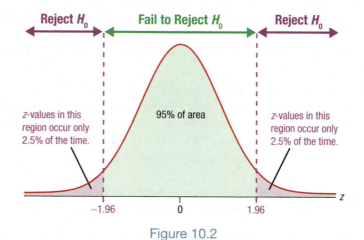

Figure 10.2

It is important to recognize that this strategy attempts to define the notion of close $\left(\text{for } \alpha = 0.05\right)$ as being within 1.96 standard deviation units. Using this definition of closeness, if $\bar{x}$ is within 1.96 standard deviations of the hypothesized value, it is close enough to the hypothesized value in the null to have occurred from ordinary sampling variability; otherwise it is too far to have happened by chance, and we reject the null. In our example, 1.96 is a **critical value** of z.

Continuing Example 10.5

Step 7: Determine the critical value.

In determining the critical value of the test statistic, we must take into account whether the alternative hypothesis is one-sided or two-sided, the level of the test, and the distribution of the test statistic.

Table 10.1 – Critical Values of the z-Test Statistic for Two-Sided Alternatives		
Level of the Test	**Definition of Ordinary Variability**	$z_{\alpha/2}$
0.20	80% interval around hypothesized mean	1.28
0.10	90% interval around hypothesized mean	1.645
0.05	95% interval around hypothesized mean	1.96
0.01	99% interval around hypothesized mean	2.575

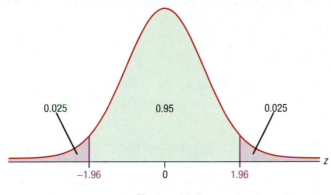

Figure 10.3

Notice that the **rejection region** is divided into two parts. The right-hand rejection region indicates above average amounts of money spent on textbooks per semester and, conversely, the left-hand rejection region indicates below average amounts of money spent on textbooks per semester. Also observe that the probability associated with the level of the test is divided equally between the two rejection regions, each region receiving 0.025. If the probabilities in these two regions are added together, $0.025 + 0.025 = 0.05$, the sum equals the level of the test (α).

The level of the test can be thought of as a tolerance for rareness. The level of the test defines a rejection region which can be thought of as an intolerance zone. If the test statistic falls in this "intolerance" region, then the data have produced a sample mean that has in turn produced a test statistic that is too rare to have occurred by chance if the null were true.

Essentially, we lose faith in the null hypothesis; the data have cast too much doubt. Consequently, the null hypothesis is rejected. The "fail to reject" zone can be interpreted as a zone of ordinary sampling variation given the null is true. For two-tailed tests with the level of the test set to $\alpha = 0.05$, the investigator is stating that any value of the test statistic which falls in a 95% interval around the hypothesized mean represents ordinary sampling variability. For the z-test statistic, this corresponds to a critical value of 1.96 standard deviation units (see Table 10.1). If the test statistic falls into the ordinary sampling variation zone, the null hypothesis will not be rejected.

Step 8: Compute the test statistic.

A random sample of 75 students revealed a mean of $540 per semester spent on textbooks at the local university. As discussed earlier, the z-statistic is given by

$$z = \frac{\bar{x} - \mu_0}{\sigma_{\bar{x}}} = \frac{540 - 500}{100/\sqrt{75}} \approx 3.46.$$

The sample mean is 3.46 standard deviations from the hypothesized value. Is the sample mean too far away from the value of the mean specified in the null hypothesis for us to believe that the null is true?

Step 9: Make the decision.

The critical values of the test statistic are ±1.96. If the null were true, observing a value of z larger in absolute value than 1.96 would occur only 5% of the time.

The test statistic $z = 3.46$ implies $\bar{x}$ is 3.46 standard deviation units from the mean which is substantially more than 1.96 standard deviations from the hypothesized value. The decision must be to reject H_0. The sample mean is too far from the hypothesized value for us to believe the difference is caused by ordinary sampling variation. Essentially, $\bar{x}$ exceeds the tolerance for rareness that we have imposed by setting $\alpha = 0.05$.

Step 10: State the conclusion in terms of the original question.

There is significant evidence at the 0.05 level that the students at the local university do not spend, on average, $500 per semester on textbooks. It would be tempting to conclude that the students at the local university spend more since the sample mean is greater than the national average. However, we did not test a hypothesis for spending more; we tested whether the students at the local university were spending *more or less* than the national average, and the conclusion must be consistent with the hypothesis tested.

Stating a Conclusion

There is not significant evidence at the [stated α] level to reject H_0.   ← No —   Do we reject H_0?   — Yes →   There is significant evidence at the [stated α] level that we reject H_0 in favor of H_a.

One temptation might be to reformulate the hypothesis and test to see if the students are spending more on textbooks with the same data. This would be practically unethical. If you already know the z-value for a given set of data, you should not formulate a hypothesis and use the data to "support" that hypothesis.

Part of every hypothesis test is arbitrary. The level of the test, α, is arbitrarily defined by the researcher. As you have seen (e.g. Table 10.1) in the last example, the level of the test affects the decision rule.

Since the level of the test is arbitrarily set, the decision rule inherits the arbitrariness.

Example 10.6

The manufacturer of a portable music player (PMP) has shown that the average life of the product is 72 months with a standard deviation of 12 months. The manufacturer is considering using a new parts supplier for the PMPs and wants to test that the new hard drives will increase the life of the PMP. Before manufacturing the PMPs on a large scale, the manufacturer sampled 200 PMPs and found the average life to be 78 months. Test the hypothesis using $\alpha = 0.01$ that the new hard drives will increase the life of the PMPs. Assume that the standard deviation of the new PMPs is the same as the standard deviation of the older model.

Solution

Step 1: Define the hypotheses in plain English.

The hypotheses are fairly straightforward.

- **Null Hypothesis:** The new hard drives do not increase the life of the PMPs.
- **Alternative Hypothesis:** The new hard drives increase the life of the PMPs.

Step 2: Select the appropriate statistical measure.

Once again, the population mean is the parameter of interest. If the average life of the new PMP is greater than the average life of the older model, all other things being equal, the new PMP will be better. Let

$$\mu = \text{mean life of the new PMP.}$$

Step 3: Determine whether the hypothesis should be one-sided or two-sided.

The goal is to determine if there is sufficient evidence to conclude that the new PMPs will last longer, on average, than the older model. The alternative hypothesis will be one-sided, and this will be a one-tailed test.

Step 4: State the hypotheses using the appropriate statistical measure.

In the formulation of the hypotheses, we will be comparing the life of the new PMP model to that of the older model. Comparing a new model or design against an older model (or "standard value") is very common in hypothesis testing.

$$H_0: \mu \leq 72$$
$$H_a: \mu > 72$$

Remember, if the purpose is to demonstrate that there is overwhelming evidence in support of a hypothesis, that hypothesis should be the alternative hypothesis. In this example, the aim is to determine if there is sufficient evidence to conclude that the new PMP model will last longer.

Step 5: Specify the level of the test.

The level of the test is given in the problem to be $\alpha = 0.01$. If it were not given, the investigator would choose the level.

Step 6: Select the appropriate test statistic.

The problem states that the standard deviation of the new model is equivalent to the standard deviation of the older model, which is known. Therefore, the standard deviation of the new PMP's life is assumed to be known. Since the sample is relatively large, the central limit theorem asserts the distribution of $\bar{x}$ should be approximately normal. Because the standard deviation is known and the sample size is greater than or equal to 30, the z-test statistic can be used.

Step 7: Determine the critical value.

Since the alternative hypothesis is one-sided, we want to know if there is evidence that μ (the mean life of the new PMP model) is greater than the hypothesized value, μ_0 (72 months – the standard life of the existing PMP model). If $\bar{x}$ (the mean life of the 200 PMP sample) is much larger than μ_0, that would suggest that H_a is a more reasonable choice than H_0. Of course, through ordinary sampling variation, $\bar{x}$ could be larger than μ_0. How much greater than μ_0 does $\bar{x}$ have to be in order for us to believe that the mean life of the new PMP is greater than 72 months (H_a)? To locate the critical value of z for a one-sided "greater than" alternative, find the value of z that cuts off α worth of probability in the right-hand tail of the distribution. The critical values for "greater than" alternative hypotheses are given in Table 10.2 for typical values of α.

Table 10.2 – Critical Values of the z-Test Statistic for One-Sided (Greater Than) Alternatives		
Level of the Test	**Definition of Ordinary Variability**	**z_α**
0.20	Lower 80% of the distribution	0.84
0.10	Lower 90% of the distribution	1.28
0.05	Lower 95% of the distribution	1.645
0.01	Lower 99% of the distribution	2.33

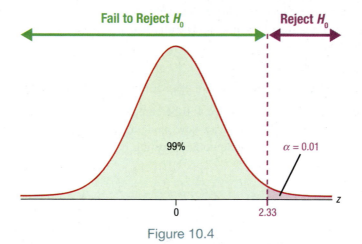

Figure 10.4

The z-test statistic has a standard normal distribution; so critical values are determined from the probability distribution of the standard normal distribution. The alternative hypothesis is one-sided $(H_a: \mu > 72)$ and $\alpha = 0.01$. From Table 10.2, we find that the appropriate critical value for the test is 2.33. This critical value means that if $\bar{x}$ is 2.33 standard deviations larger than μ_0, then H_0 should be rejected in favor of H_a. However, if H_0 is true, ordinary variation would cause $\bar{x}$ to be greater than 2.33 about 1% of the time. Thus, 1% of the time H_0 will be rejected when it is true, a Type I error.

Step 8: Compute the test statistic.

Suppose a random sample of 200 PMPs revealed a mean life of 78 months. The resulting z-test statistic is

$$z = \frac{\bar{x} - \mu_0}{\sigma_{\bar{x}}} = \frac{78 - 72}{12/\sqrt{200}} \approx 7.07$$

indicating that the sample mean life of the new PMP model is more than 7 standard deviations larger than the life of the current PMP model. It is quite unlikely that ordinary sampling variation could have caused the sample mean to be more than 7 standard deviations from the hypothesized value.

Step 9 Make the decision.

Since the test statistic is larger than the critical value, the conclusion is to reject H_0 in favor of H_a.

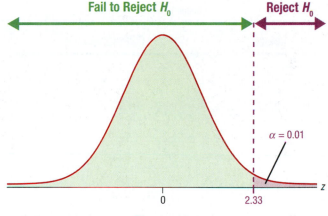

Figure 10.5

Step 10: State the conclusion in terms of the original question.

There is sufficient evidence at the $\alpha = 0.01$ level to conclude that the life of the new PMP model is superior to the life of the older model.

In the previous two examples, we have considered a two-sided alternative and a one-sided "greater than" alternative. When considering a one-sided "less than" alternative, the procedure is very similar to that of a one-sided "greater than" alternative. The null hypothesis will be rejected if the calculated value of the test statistic, z, is less than or equal to the critical value, $-z_\alpha$, for the specified level of significance.

Procedure

One-sided "Less Than" Alternatives

For tests that are based on a test statistic which has a standard normal distribution and a "less than" alternative, find the value of z that cuts off α worth of probability in the left-hand tail of the distribution. The critical values for "less than" alternative hypotheses are given in Table 10.3 for typical values of α.

Table 10.3 – Critical Values of the z-Test Statistic for One-Sided (Less Than) Alternatives		
Level of the Test	**Definition of Ordinary Variability**	$-z_\alpha$
0.20	Upper 80% of the distribution	−0.84
0.10	Upper 90% of the distribution	−1.28
0.05	Upper 95% of the distribution	−1.645
0.01	Upper 99% of the distribution	−2.33

Figure 10.6

The figure above shows the rejection region for a test with a one-sided "less than" alternative hypothesis, and a significance level of 0.20. For this test, the null hypothesis will be rejected if the calculated value of the test statistic is less than or equal to −0.84.

Small Sample Hypothesis Tests, σ Unknown

The previous hypothesis testing strategy assumes that the sample size is greater than or equal to 30 or the standard deviation of the population is known and the population is normal. However, in most instances, the population standard deviation is just as unknown as the population mean, and sometimes a sample size of at least 30 is not feasible from either a cost or a time perspective. Despite the added uncertainty, the general approach to testing a hypothesis is the same provided that it is reasonable to assume that the population from which you are sampling is normal. Some of the technical details concerning the distribution of the test statistic change, since the sample standard deviation, s, will be used in place of the population standard deviation, σ, and $n < 30$. This modification will cause a change in the distribution of the test statistic.

Formula

t-Test Statistic

If the sample size, n, is less than 30 and the standard deviation of the population is unknown, but the distribution of the population is normal, then the test statistic is given by

$$t = \frac{\bar{x} - \mu_0}{s_{\bar{x}}}, \text{ where } s_{\bar{x}} = \frac{s}{\sqrt{n}}.$$

The test statistic has a t-distribution with $n - 1$ degrees of freedom.

If we assume the population we are sampling from is normally distributed, the test statistic has a t-distribution. Consequently, the test statistic will be called a t-test statistic. The fundamental notion embodied in the test statistic has not changed. It measures how far the sample mean is from the hypothesized mean in standard deviation units.

In the previous chapter, we discussed the t-distribution in conjunction with estimating the population mean. Steps 6 and 7 in our hypothesis testing procedure must be modified to reflect the change in distribution of the test statistic.

Example 10.7

A website developer has indicated to potential clients that for the sites he has developed, visitors spend an average of 45 minutes per day on the sites. One of his potential clients conducted a survey of 20 visitors to several of the sites created by the website developer and found that the average time spent on the sites was 35 minutes with a standard deviation of 7 minutes. Determine if there is sufficient evidence to conclude that the average time spent on his sites is different from what he has indicated. Conduct the test at the 0.05 level.

Solution

Step 1: Define the hypotheses in plain English.

- **Null Hypothesis:** The average time spent on the sites is 45 minutes.

- **Alternative Hypothesis:** The average time spent on the sites is not 45 minutes.

Step 2: Select the appropriate statistical measure.

In this case, the population consists of the time spent per visitor on all of the sites created by the website developer. Since the test is specified as having an average amount of time spent on the sites of 45 minutes, the parameter of interest will be the population mean. Let

μ = mean time spent on sites created by the website developer.

Step 3: Determine whether the hypothesis should be one-sided or two-sided.

From the information given, we must assume that the client is interested in learning if the time spent on the sites is either significantly longer than 45 minutes or significantly less than 45 minutes. Thus the alternative hypothesis will be two-sided, and this will be a two-tailed test.

Step 4: State the hypotheses using the appropriate statistical measure.

The resulting hypotheses would be as follows.

$$H_0: \mu = 45$$
$$H_a: \mu \neq 45$$

Step 5: Specify the level of the test.

The level of the test is specified in the problem as 0.05.

Step 6: Select the appropriate test statistic.

Since the standard deviation of the population is not known and the sample size is less than 30, the t-test statistic is utilized. However, to know that the test statistic possesses a t-distribution, we must be willing to make the assumption that the population possesses a reasonably normal distribution. At this state, we will have to assume normality of the population since there is not enough data available to provide much assistance.

Step 7: Determine the critical value.

The t-distribution has a parameter called degrees of freedom, which must be defined in order to utilize the distribution. Since 20 visitors were sampled, the degrees of freedom are

$$df = n - 1 = 20 - 1 = 19.$$

t-Distribution, df = 19

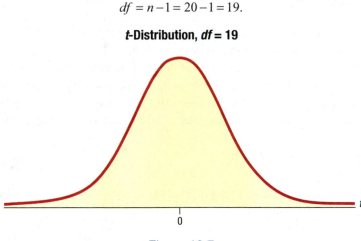

Figure 10.7

Since the alternative hypothesis (H_a) is two-sided, two tails of the distribution must be cut off as rejection regions. Each tail will receive half of the allotted level of the test. The value of t which begins the right-hand side rejection region is $t_{0.025, 19}$, which corresponds to a critical value of 2.093. Since the t-distribution is symmetrical, the left-hand side rejection region begins at −2.093.

Figure 10.8

What the picture says is that it is unlikely (only occurs 5% of the time) that ordinary sampling variability will cause $\bar{x}$ to differ as much as 2.093 standard deviation units from the hypothesized mean of 45.

Step 8: Compute the test statistic.

Since 20 visitors were sampled and the mean time spent on the sites was 35 minutes with a standard deviation of 7 minutes, the computed value of the test statistic is

$$t = \frac{\bar{x} - \mu_0}{s_{\bar{x}}} = \frac{35 - 45}{7 / \sqrt{20}} \approx -6.389.$$

The observed sample mean is more than 6 standard deviations less than the hypothesized mean. The value of the test statistic signifies a striking difference between $\bar{x}$ and the hypothesized value, μ_0. It is very unlikely that the cause of such a difference would be ordinary sampling variation.

Step 9: Make the decision.

Since the test statistic falls in the rejection region, we must reject H_0 in favor of H_a.

Step 10: State the conclusion in terms of the original question.

There is sufficient evidence at the $\alpha = 0.05$ level to conclude that the average time spent on the developer's sites is different from 45 minutes. Since the sample mean is much less than the indicated 45 minutes, it is tempting to conclude that the time spent on the sites is less than 45 minutes. However, we did not test the one-sided hypothesis. If the potential client would like to demonstrate that visitors spend significantly less time on the developer's sites, then they must start over by testing a one-sided hypothesis and obtaining new sample data.

Example 10.8

The Alexander Bolt Company produces half-inch A-class stainless steel bolts. The specified quality standard is that they have a mean tensile strength of more than 4000 pounds per square inch (psi). These bolts are primarily used in the manufacturing of farm implements. The company is very concerned about quality and wants to be sure that its A-class product does not fall below the standard. A sample of 25 bolts is to be randomly selected from stock and tested for tensile strength. Design a test of hypothesis to determine if there is overwhelming evidence that the bolts meet the specified quality standard.

Solution

Step 1: State the hypotheses in plain English.

- **Null Hypothesis:** The half-inch A-class stainless steel bolts have a mean tensile strength of no more than 4000 psi.

- **Alternative Hypothesis:** The half-inch A-class stainless steel bolts have a mean tensile strength of more than 4000 psi.

Step 2: Select the appropriate statistical measure.

Although there is nothing in the problem that specifically defines the population parameter of interest, there is mention of a "standard value." By comparing the mean breaking strength of the population to the **standard value**, 4000 psi, a conclusion about the quality can be drawn. Let

$$\mu = \text{mean tensile strength of the half-inch A-class stainless steel bolts.}$$

Step 3: Determine whether the hypothesis should be one-sided or two-sided.

If the population mean is significantly above the standard, then there would be evidence that the desired quality is being produced. Thus, the alternative hypothesis will be one-sided, and this will be a one-tailed test.

Step 4: State the hypotheses using the appropriate statistical measure.

The resulting hypotheses would be as follows.

$H_0: \mu \leq 4000$ (Mean tensile strength is below the quality standard of 4000 psi. The bolt production system is flawed.)

$H_a: \mu > 4000$ (Mean tensile strength is above 4000 psi. The bolt production system is working well.)

Step 5: Specify the level of the test.

The level of the test is not stated in the problem, and thus the selection is left to the researcher. This is a typical predicament and, unfortunately, leaves us with what appears to be an arbitrary decision regarding the level of the test. What is a reasonable value for the level of the test?

In this problem, making a Type I error (rejecting the null hypothesis when it is true) means that management believes that the bolt production system is not flawed, when it is. If we set α too high, management will too frequently believe the quality level is being met, when it is not and poor quality bolts will not be readily detected. After all is said and done, the decision remains arbitrary. A Type I error could be calamitous, so a small level of alpha would be desirable. Let's use $\alpha = 0.01$.

Step 6: Select the appropriate test statistic.

The population standard deviation is unknown and will be estimated with the sample standard deviation, and the sample size is less than 30. Assuming the distribution of the tensile strengths is reasonably normal, the *t*-statistic is appropriate.

Step 7: Determine the critical value.

Since the alternative hypothesis implies that we are interested in discovering if the mean is above the standard value, the rejection region will only be on the right-hand side of the sampling distribution. Because the level of the test has been chosen to be 0.01, the interval encompassing 0.01 of the area on the right-hand side of the curve will define the rejection region. Remember that the test statistic has a *t*-distribution.

To obtain the critical value, first determine the degrees of freedom.

$$df = n - 1 = 25 - 1 = 24$$

Using Table D in Appendix A, $t_{0.01,24} = 2.492$. If the observed *t*-value is greater than or equal to 2.492 then the observed mean is presumed to be too far from the hypothesized value to have occurred from ordinary sample variation. If $t \geq 2.492$, the presumption will be that the mean tensile strength is greater than 4000 psi.

t-Distribution, df = 24

Figure 10.9

Step 8: Compute the test statistic.

Suppose that after randomly selecting and testing 25 bolts, the average tensile strength is 4014 psi, with a standard deviation of 20 psi. Symbolically this is expressed as

$$\bar{x} = 4014, \ s = 20, \text{ and } n = 25.$$

The resulting test statistic is

$$t = \frac{\bar{x} - \mu_0}{s_{\bar{x}}} = \frac{4014 - 4000}{20 / \sqrt{25}} = 3.5.$$

The sample mean is 3.5 standard deviations more than the hypothesized value of the mean. Is this difference caused by ordinary sampling variation, or is this evidence of a false null hypothesis? Keep in mind that rejecting the null hypothesis (H_0) will be a desirable outcome since the null states that the bolts are below quality standards.

Step 9: Make the decision.

Since the test statistic falls in the rejection region, it indicates that the sample mean is too far from the hypothesized value to believe it is due to ordinary sampling variation. (In other words, $\bar{x} = 4014$ is a rare observation and has exceeded our tolerance for rareness.) The null hypothesis must be rejected in favor of the alternative hypothesis.

Step 10: State the conclusion in terms of the original question.

There is sufficient evidence to conclude that the bolt tensile strength is greater than 4000 pounds, and there is at most a 1% chance that this conclusion is incorrect.

This conclusion is stated in a rather absolute manner. However, there is always a degree of subjectivity in any statistical conclusion. In this instance, we arbitrarily selected the value of α, and we assumed that the distribution of the population was normal.

P-Values

We have been examining the classical approach to hypothesis testing. This approach results in a conclusion to reject or fail to reject the null hypothesis. What is missing from this type of conclusion is the degree to which the data confirm the inference. For example, suppose we are testing a hypothesis with the following rejection region.

If the test statistic were $z = 2.03$, it would fall into the rejection region (but just barely), and the null hypothesis would be rejected.

Compare this to a test statistic of $z = 3.09$, which falls far into the rejection region and results in the same conclusion, rejecting the null hypothesis. Clearly, the sample data that produced a test statistic of 3.09 provide much stronger evidence against the null hypothesis than the sample that produced a test statistic of 2.03. Expressing the difference in the strength of our conclusion is accomplished using **P-values**.

> ## Definition
>
> A **P-value** is the probability of observing a value of a test statistic as extreme or more extreme than the one observed, assuming the null hypothesis is true.

For a left-tailed (or lower-tailed) test, the P-value is given by $P(z \le z_0)$, where z_0 is the observed value of the test statistic. For a right-tailed (or upper-tailed) test, the P-value is given by $P(z \ge z_0)$, where z_0 is the observed value of the test statistic. For two-tailed tests, the P-value is given by $2P(z \ge |z_0|)$, where z_0 is the observed value of the test statistic. P-values can be determined using the tables in Appendix A or technology such as a TI-84 Plus calculator, Microsoft Excel, or Minitab. The Discovering Technology section at the end of this chapter shows how to calculate P-values for z and t-test statistics.

The following graph displays the application of the definition to the test statistics $z = 2.03$ and $z = 3.09$.

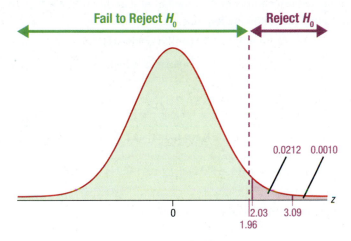

Figure 10.10

If the null hypothesis is true, the probability of observing a test statistic greater than or equal to 2.03 is 0.0212, which is the P-value. One interpretation of the P-value is that if 10,000 researchers were to draw similar size samples, assuming the null hypothesis is true, ordinary sampling variation would cause roughly 212 researchers to obtain samples that would produce test statistics as large or larger than 2.03.

Compare this to the test statistic 3.09. If the null hypothesis is true, the chance of sampling variation causing a test statistic greater than or equal to 3.09 is 0.0010, which is the P-value. If 10,000 researchers drew similar size samples, ordinary sampling variation would cause only about 10 of the researchers to observe test statistics as large or larger than 3.09. The further the test statistic penetrates the rejection region, the more confidence the researcher can place in his or her conclusions.

Many statistical computer programs provide P-values in their output. It is a fairly easy task to utilize these values to perform a classical hypothesis test.

> ## Procedure
>
> ### Performing a Hypothesis Test Using *P*-Values
>
> - If the computed *P*-value is less than or equal to α, reject the null hypothesis in favor of the alternative.
>
> - If the computed *P*-value is greater than α, fail to reject the null hypothesis.

The rationale for this approach is that if the *P*-value is less than or equal to alpha, we have observed a test statistic that is more unusual than the level of the test, which defines how rare an event we must observe in order to reject the null hypothesis. Conversely, if the *P*-value is larger than α, the test statistic is not sufficiently rare (it could have been caused by ordinary sampling variation) to reject the null hypothesis.

P-Values for a Two-Sided Hypothesis

So far, we have only discussed calculating a *P*-value for a one-tailed hypothesis test. To compute the *P*-value for a two-tailed test, simply double the tail probability of the test statistic. For example, suppose that we are testing the following hypotheses.

$$H_0: \mu = 5 \qquad \text{(null hypothesis)}$$
$$H_a: \mu \neq 5 \qquad \text{(alternative hypothesis)}$$

Further, suppose the computed test statistic is $z = 2.31$.

For an upper one-tailed test (i.e., $H_a: \mu > 5$), we would calculate the *P*-value as

$$P\text{-value} = P(z \geq 2.31) = P(z \leq -2.31) = 0.0104.$$

However, for a two-tailed test, the *P*-value is calculated as

$$P\text{-value} = 2P(z \geq |z_0|), \text{ where } z_0 \text{ is the observed value of the test statistic.}$$

Thus, to compute the *P*-value for a two-tailed test, we double the tail area.

$$P\text{-value} = 0.0104 + 0.0104 = 0.0208$$

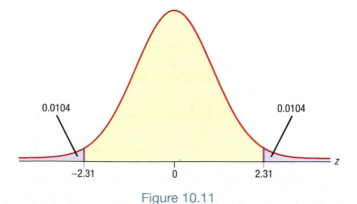

Figure 10.11

The *P*-value of 0.0208 is the likelihood of observing a value of the test statistic greater than or equal to 2.31 or less than or equal to −2.31 given the null hypothesis is true.

P-Values for *t*-Test Statistics

Because there are numerous *t*-distributions, one for each different degree of freedom, the *t*-tables are constructed differently from the *z*-tables. The *t*-table only provides *t*-values for frequently used tail probabilities. Note in the *t*-table (Appendix A, Table D) that we only have six tail areas. Because of this limitation, in most instances the exact value of the *t*-test statistic will not be in the table. When this circumstance arises, find the closest *t*-values with the appropriate degrees of freedom that surround the test statistic. For example, suppose the value of the test statistic was

$$t = 2.40 \text{ with 17 degrees of freedom.}$$

t-Distribution, df =17

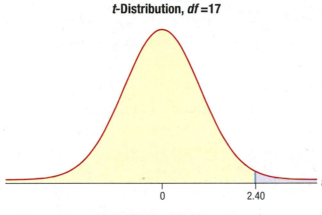

Figure 10.12

The value of the test statistic falls between 2.110, which corresponds to $\alpha = 0.025$, and 2.567, which corresponds to $\alpha = 0.01$.

df	$t_{0.100}$	$t_{0.050}$	$t_{0.025}$	$t_{0.010}$
1	3.078	6.314	12.706	31.821
2	1.886	2.920	4.303	6.965
3	1.638	2.353	3.182	4.541
⋮				
17	1.333	1.740	2.110	2.567
18	1.330	1.734	2.101	2.552
⋮				

Thus, we report a bound for the *P*-value. In this case, the *P*-value is $0.01 < P\text{-value} < 0.025$. That is, the *P*-value falls between 0.01 and 0.025. If the level of the test, α, is greater than 0.025, then our test is significant at the 0.025 level. Thus, the *P*-value would be reported as being significant at the 0.025 level, but not significant at the 0.01 level. Exact *P*-values for *t*-test statistics can be found using technology such as a TI-84 Plus calculator, Microsoft Excel, and Minitab. Methods for finding exact *P*-values can be found in the Discovering Technology section at the end of this chapter.

Because of the arbitrariness of assigning the level of the test, researchers will often report the significance of their findings as *P*-values. Determining the significance of the results is left to the researcher. There are few universal standards when it comes to evaluating the significance of a *P*-value. However, most researchers would view a *P*-value of less than 0.01 as significant and a *P*-value of greater than 0.10 as insignificant. Interpreting *P*-values between 0.01 and 0.10 will depend on the circumstances. However, once the *P*-value gets above 0.05, the conclusion starts to become unconvincing.

10.4   Exercises

Basic Concepts

1. What is the rationale for the z-statistic?

2. What are the three key questions to be asked in the hypothesis testing procedure in order to determine which test statistic is appropriate?

3. Describe the distribution of the z-test statistic.

4. If the variance for a population is not known, how is the test statistic affected?

5. What are critical values? How do critical values influence the decision rule in the hypothesis testing procedure?

6. Suppose a null hypothesis were rejected at $\alpha = 0.05$. Would it be rejected at 0.10? Explain.

7. Suppose a null hypothesis were rejected at $\alpha = 0.05$. Would it be rejected at 0.01? Explain.

8. What is a P-value?

9. Discuss how P-values are used in the test of a hypothesis.

Exercises

10. Determine the critical value(s) of the test statistic for each of the following large sample tests for the population mean.

    a. Left-tailed test, $\alpha = 0.01$

    b. Right-tailed test, $\alpha = 0.10$

    c. Two-tailed test, $\alpha = 0.05$

11. Determine the critical value(s) of the test statistic for each of the following large sample tests for the population mean.

    a. Left-tailed test, $\alpha = 0.05$

    b. Right-tailed test, $\alpha = 0.02$

    c. Two-tailed test, $\alpha = 0.08$

12. A random sample of 1000 observations produces a sample mean of 53.5 with a standard deviation of 5.3. Test the hypothesis that the mean is not equal to 55 at $\alpha = 0.05$.

13. A random sample of 200 observations indicate a sample mean of 4117 with a standard deviation of 300. Test the hypothesis that the mean is greater than 4100 at $\alpha = 0.01$.

14. The head of the Veterans Administration has been receiving complaints from a Vietnam veterans' organization concerning disability checks. The organization claims that checks are continually late. The checks are supposed to arrive no later than the tenth of each month. The administrator randomly selects 100 disabled veterans and measures the arrival time in relation to the tenth of the month for each check. If the check arrives early, it receives a negative value. For example, if the check arrives on the eighth of the month, it is measured as –2. If the check arrives on the twelfth of the month, it is measured as + 2.

   a. What statistical measure should you use in your statement of hypothesis?

   b. Formulate hypotheses to test the veterans' organization's claim.

   c. Suppose in the sample of 100 disabled veterans receiving checks, the average number of days late was 1.2 with a standard deviation of 1.4. Calculate the test statistic for your hypothesis.

   d. If the test is conducted at the 0.05 level, construct the decision rule for the test statistic.

   e. Is there overwhelming evidence at the 0.05 level that the checks arrive late?

   f. If you are the head of the Veterans Administration, what is your conclusion?

15. Hurricane Andrew swept through southern Florida causing billions of dollars of damage. Because of the severity of the storm and the type of residential construction used in this semitropical area, there was some concern that the average claim size would be greater than the historical average hurricane claim of $24,000. Several insurance companies collaborated in a data gathering experiment. They randomly selected 84 homes and sent adjusters to settle the claims. In the sample of 84 homes, the average claim was $27,500 with a standard deviation of $2400.

   a. What is the population being studied?

   b. What statistical measure should you use in your hypothesis?

   c. State your hypotheses.

   d. Test the hypothesis at the 0.01 level.

   e. Is there overwhelming evidence (at the 0.01 level) that home damage is greater than the historical average? Write your conclusion in the context of the original problem.

16. A retail computer store is considering offering a two-year service warranty, instead of its current one-year plan. In order to do this, they must determine the average service costs for their systems in the second year of operation. A committee of technicians, sales, and management staff believe that the average repair cost in the second year should be approximately $50. Seventy-five customers who purchased machines between two and four years earlier are randomly selected. Tracking the service needs of these customers reveals an average service cost of $38 with a standard deviation of $10.

   a. What is the population being studied?

   b. What variable is being measured in this problem?

   c. What level of measurement does the variable possess?

   d. Test the committee's claim at the 0.10 level.

   e. What concerns might you have about the data that were collected?

17. In preparation for upcoming wage negotiations with the union, the managers for the Bevel Hardware Company want to establish the time required to assemble a kitchen cabinet. A first line supervisor believes that the job should take 45 minutes on average to complete. A random sample of 125 cabinets has an average assembly time of 47 minutes with a standard deviation of 10 minutes.

   a. Is there overwhelming evidence to contradict the first line supervisor's belief at a 0.05 significance level? Make your conclusion using the P-value approach.

   b. What is the lowest average assembly time that would allow the union to conclude that the supervisor is incorrect?

18. The Better Business Bureau has received several complaints that a flour company is underfilling its five pound bags of flour. The Bureau randomly selects 750 bags of flour and determines the weight of each bag. The sample average weight of the bags is 4.80 pounds with a standard deviation of 0.15 pounds.

   a. Is there overwhelming evidence at the 0.01 level that the bags are underfilled?

   b. What is the lowest average bag weight that would allow the Bureau to conclude that the bags are underfilled?

19. A horticulturist working for a large plant nursery is conducting experiments on the growth rate of a new shrub. Based on previous research, the horticulturist feels the average daily growth rate of the new shrub is 1 cm per day. A random sample of 45 shrubs has an average growth of 0.90 cm per day with a standard deviation of 0.30 cm. Will a test of hypothesis at the 0.05 significance level support the claim that the growth rate is less than 1 cm per day?

20. Del Valley Foods requires that corn supplied for canning must weigh more than 5 ounces per ear. South Valley Farms claims that the corn they supply meets the required specifications. 200 ears of corn are selected at random from a delivery. The sample has a mean of 5.01 ounces and a standard deviation of 0.30 ounce. Will a test of hypothesis at $\alpha = 0.10$ support South Valley Farms' claim?

21. Government regulations restrict the amount of pollutants that can be released to the atmosphere through industrial smokestacks. To demonstrate that their smokestacks are releasing pollutants below the mandated limit of 5 parts per billion pollutants, REM Industries collects a random sample of 300 readings. The mean pollutant level for the sample is 4.85 parts per billion with a standard deviation of 0.30 parts per billion. Do the data support the claim that the average pollutants produced by REM Industries are below the mandated level at a 0.01 significance level?

22. The director of the IRS has been flooded with complaints that people must wait more than 45 minutes before seeing an IRS representative. To determine the validity of these complaints, the IRS randomly selects 400 people entering IRS offices across the country and records the times that they must wait before seeing an IRS representative. The average waiting time for the sample is 55 minutes with a standard deviation of 15 minutes.

   a. What is the population being studied?

   b. Are the complaints substantiated by the data at $\alpha = 0.10$?

23. The manufacturer of Brand X floor polish is developing a new polish that it hopes will dry faster than the competition's polish. The competition's polish is advertised to have an average drying time of 10 minutes. A random sample of 1000 Brand X polishes has an average drying time of 9.3 minutes with a standard deviation of 0.5 minute. Based on the data, can the manufacturer conclude that the drying time for Brand X is faster than the competition's brand at a 0.05 significance level?

24. Determine the critical value(s) of the test statistic for each of the following small sample tests for the population mean where the assumption of normality is satisfied.

   a. Left-tailed test, $\alpha = 0.01$, $n = 15$

   b. Right-tailed test, $\alpha = 0.10$, $n = 20$

   c. Two-tailed test, $\alpha = 0.05$, $n = 8$

25. Determine the critical value(s) of the test statistic for each of the following small sample tests for the population mean where the assumption of normality is satisfied.

   a. Left-tailed test, $\alpha = 0.005$, $n = 12$

   b. Right-tailed test, $\alpha = 0.025$, $n = 5$

   c. Two-tailed test, $\alpha = 0.10$, $n = 25$

26. Consider the following random sample of size six from a normal population. Based on the sample, perform a hypothesis test to test the claim that the mean of the population is not equal to 10 at $\alpha = 0.05$.

Table for Exercise 26					
10	15	12	9	11	10

27. Consider the following random sample of size eight from a normal population. Based on the sample, perform a hypothesis test to test the claim that the mean of the population is greater than 100 at $\alpha = 0.05$. Calculate the P-value for this hypothesis test.

Table for Exercise 27							
100	150	120	90	95	110	100	80

28. Consider the following random sample of size seven from a normal population. Based on the sample, perform a hypothesis test to test the claim that the mean of the population is less than 0.5 at $\alpha = 0.10$. Calculate the P-value for this hypothesis test.

Table for Exercise 28						
0.3	0.5	0.4	0.6	0.5	0.4	0.4

29. NarStor, a computer disk drive manufacturer, claims that the average time to failure for its hard drives is 14,400 hours. You work for a consumer group that has decided to examine this claim. Technicians ran 16 drives continuously for three years. Recently the last drive failed. The time to failure (in hours) are given below.

Table for Exercise 29 – Time Until Failure (Hours)							
330	620	1870	2410	4620	6396	7822	8102
8309	12,882	14,419	16,092	18,384	20,916	23,812	25,814

a. What is the population being studied?

b. What is the variable being measured?

c. What level of measurement does the variable possess?

d. Conduct a hypothesis test to determine whether there is overwhelming evidence that the average time to failure is less than the manufacturer's claim. Use $\alpha = 0.01$.

e. What assumption did you make in performing the test in part **d.**?

30. The admitting office at Sisters of Mercy Hospital wants to be able to inform patients of the average level of expenses they can expect per day. Historically, the average has been approximately $1240. The office would like to know if there is evidence of an increase in the average daily billing. Twenty randomly selected patients have an average daily charge of $1491 with a standard deviation of $342.

a. What is the population being studied?

b. Conduct a hypothesis test to determine whether there is evidence that average daily charges have increased at $\alpha = 0.10$.

c. What assumption did you make in performing the test in part **b.**?

31. A supplier has agreed to provide the manager of a large hospital with light bulbs that he claims will last more than 1000 hours. Twenty-five bulbs are randomly selected and tested by the hospital's maintenance department. The sample has an average life of 1099 hours with a standard deviation of 99 hours.

a. Perform a hypothesis test to determine whether the data support the supplier's claim at $\alpha = 0.05$.

b. What assumption did you make in performing the test in part **a.**?

c. What is the P-value for the hypothesis test performed in part **a.**?

32. The managers of a large department store wish to test reactions of shoppers to a new in-store video screen which will broadcast continuous information about the store and the items currently on sale. The video production company claims that the average shopper will watch for five or more minutes. The managers randomly select 17 shoppers and determine how long they watch the video. The average time is 4.5 minutes with a standard deviation of 2.5 minutes.

a. Perform a hypothesis test to determine whether there is overwhelming evidence to refute the claim of the video production company. Use $\alpha = 0.01$.

b. What assumption did you make in performing the test in part **a.**?

33. A group of local businessmen is thinking about developing land into a shopping mall. To evaluate the desirability of the location, they count the number of shoppers who visit the neighboring shopping center each day. A random sample of 25 days reveals a daily average of 107 shoppers with a standard deviation of 23 shoppers. The businessmen will develop the land if the average number of shoppers per day is more than 100.

   a. Based on the sample data, should the businessmen develop the land? Perform a hypothesis test and use $\alpha = 0.10$.

   b. What assumption did you make in performing the test in part **a.**?

34. The Dodge Reports are used by many companies in the construction field to estimate the time required to complete various jobs. The company has received several complaints that the time required to install 130 square feet of bathroom tile is greater than the eight hours reported in the current manual. A researcher for Dodge randomly selects 10 construction workers and determines the time required to install 130 square feet of bath tile. The average time required to install the tile for the sample is 8.5 hours with a standard deviation of 1 hour.

   a. Use a hypothesis test to determine whether the customers' complaints are substantiated by the data. Use $\alpha = 0.05$.

   b. What assumption did you make in performing the test in part **a.**?

35. Officials in charge of televising an international chess competition in South America want to determine if the average time per move for the top players has remained under five minutes over the last two years. Video tapes of matches which have been played over the two-year period are reviewed and a random sample of 50 moves are timed. The sample mean is 3.5 minutes with a standard deviation of 1.5 minutes.

   a. What is the population under study?

   b. Can the officials conclude at $\alpha = 0.05$ that the time per move is still under five minutes?

36. Buckshot Heaven is developing a new shotgun shell that they hope will have a significantly tighter pellet pattern than their competition. Twenty-five shells are tested at fifty yards. The average pellet pattern of the sample was 8.7 inches in diameter with a standard deviation of 2.0 inches. Their competitor advertises that the average pellet pattern of their shells is nine inches.

   a. Does the test completed by Buckshot Heaven support the claim that their shell pattern is tighter than the competition at a level of significance of 0.10?

   b. What assumption did you make in performing the test in part **a.**?

37. For each of the following combinations of the P-value and α, decide whether you would reject or fail to reject the null hypothesis.

   a. P-value $= 0.0935$, $\alpha = 0.10$

   b. P-value $= 0.0311$, $\alpha = 0.05$

   c. P-value $= 0.0545$, $\alpha = 0.01$

   d. P-value $= 0.0489$, $\alpha = 0.05$

38. For each of the following combinations of the P-value and α, decide whether you would reject or fail to reject the null hypothesis.

    a. *P*-value = 0.0839, $\alpha = 0.05$

    b. *P*-value = 0.0174, $\alpha = 0.02$

    c. *P*-value = 0.0444, $\alpha = 0.10$

    d. *P*-value = 0.0374, $\alpha = 0.01$

39. Consider the following large sample hypothesis tests for the population mean. Compute the P-value for each test and decide whether you would reject or fail to reject the null hypothesis at $\alpha = 0.05$.

    a. $H_0: \mu \leq 15$, $H_a: \mu > 15$, $z = 1.58$

    b. $H_0: \mu \geq 1.9$, $H_a: \mu < 1.9$, $z = -2.25$

    c. $H_0: \mu = 100$, $H_a: \mu \neq 100$, $z = 1.90$

40. Consider the following large sample hypothesis tests for the population mean. Compute the P-value for each test and decide whether you would reject or fail to reject the null hypothesis at $\alpha = 0.01$.

    a. $H_0: \mu \leq 10$, $H_a: \mu > 10$, $z = 2.00$

    b. $H_0: \mu \geq 82$, $H_a: \mu < 82$, $z = -2.45$

    c. $H_0: \mu = 100$, $H_a: \mu \neq 100$, $z = 2.70$

41. Consider the following small sample hypothesis tests for the population mean. Compute the P-value for each test and decide whether you would reject or fail to reject the null hypothesis at $\alpha = 0.01$. See the Discovering Technology section at the end of this chapter for instructions on finding exact P-values for *t*-statistics.

    a. $H_0: \mu \leq 25$, $H_a: \mu > 25$, $t = 2.7$, $n = 15$

    b. $H_0: \mu \geq 0.85$, $H_a: \mu < 0.85$, $t = -2.5$, $n = 7$

    c. $H_0: \mu = 1000$, $H_a: \mu \neq 1000$, $t = 2.0$, $n = 15$

42. Consider the following small sample hypothesis tests for the population mean. Compute the P-value for each test and decide whether you would reject or fail to reject the null hypothesis at $\alpha = 0.05$. See the Discovering Technology section at the end of this chapter for instructions on finding exact P-values for *t*-statistics.

    a. $H_0: \mu \leq 120$, $H_a: \mu > 120$, $t = 1.5$, $n = 20$

    b. $H_0: \mu \geq 0.2$, $H_a: \mu < 0.2$, $t = -2.75$, $n = 18$

    c. $H_0: \mu = 50$, $H_a: \mu \neq 50$, $t = 2.4$, $n = 5$

43.   A.C. Bone has developed a duck hunting boot which it claims can remain immersed for more than 12 hours without leaking. Five hundred pairs of the boots are tested and the time until first leakage is measured. The average time until first leakage for the sample is 12.25 hours with a standard deviation of 3.0 hours.

    a.   Find the *P*-value to test the claim that the average time until first leakage for the hunting boot is more than 12 hours.

    b.   Does this sample support A.C. Bone's claim at $\alpha = 0.10$?

10.5   Practical Significance versus Statistical Significance

A hypothesis is rarely exactly true. As the sample size becomes larger, the likelihood of rejecting the null becomes greater. As a result, sometimes even though the results of a hypothesis test are statistically significant, they may not be practically significant.

Example 10.9

Suppose that Fife Police Department was concerned about speeds in the town school zone, especially during the beginning and end of the school day. The posted speed limit is 25 mph and the deputy officer, Benny, believes that the speed is being exceeded. To test his belief, Benny randomly samples (via radar) 2000 vehicles in the school zone over a 30-day period and finds that the average speed is 25.01 mph with a standard deviation of 0.10 mph.

Solution

To test Benny's claim, the hypothesis test is carried out with the following hypotheses.

$$H_0: \mu \leq 25 \text{ mph}$$
$$H_a: \mu > 25 \text{ mph}$$

The resulting test statistic is

$$z = \frac{\bar{x} - \mu_0}{s / \sqrt{n}} = \frac{25.01 - 25}{0.10 / \sqrt{2000}} \approx 4.47.$$

The P-value for this test is 0.000004 and suggests that the test statistic is extremely rare, if the null hypothesis is true. Considering that a test statistic this large would result from ordinary sampling variation in only about 4 in a million samples, we should reject the null hypothesis $\left(H_0: \mu \leq 25 \text{ mph}\right)$ in favor of the alternative. With Benny being a "strictly by the book" officer, he would conclude that the speeds are significantly higher than 25 mph in the school zone and the department should allocate extra resources in that area to reduce speeds. However, from a practical perspective, there isn't much difference (other than random variation) between 25 mph and 25.01 mph, and it isn't likely to be detected by radar. So, despite the "statistical significance" of the test, the practical significance is negligible.

It is important to keep the practical significance of the hypothesis test in mind when making conclusions. This is one of the reasons why in the 10-step procedure for hypothesis testing, Step 10 is to state the conclusion in terms of the original problem. This step may help shed light on the practical significance of the hypothesis test.

10.5   Exercises

Basic Concepts

1. Describe the difference between statistical significance and practical significance.

2. Give an example of a situation in which results could be statistically significant but not practically significant.

Exercises

3. Consider Exercise 17 in Section 10.4:

   In preparation for upcoming wage negotiations with the union, the managers for the Bevel Hardware Company want to establish the time required to assemble a kitchen cabinet. A first line supervisor believes that the job should take 45 minutes on average to complete. A random sample of 125 cabinets has an average assembly time of 47 minutes with a standard deviation of 10 minutes. Is there overwhelming evidence to contradict the first line supervisor's belief at a 0.05 significance level?

   Discuss the statistical and practical significance for this problem.

4. Consider Exercise 19 in Section 10.4:

   A horticulturist working for a large plant nursery is conducting experiments on the growth rate of a new shrub. Based on previous research, the horticulturist feels the average daily growth rate of the new shrub is 1 cm per day. A random sample of 45 shrubs has an average growth of 0.90 cm per day with a standard deviation of 0.30 cm. Will a test of hypothesis at the 0.05 significance level support the claim that the growth rate is less than 1 cm per day?

   Discuss the statistical and practical significance for this problem.

5. Consider Exercise 22 in Section 10.4:

   The director of the IRS has been flooded with complaints that people must wait more than 45 minutes before seeing an IRS representative. To determine the validity of these complaints, the IRS randomly selects 400 people entering IRS offices across the country and records the times which they must wait before seeing an IRS representative. The average waiting time for the sample is 55 minutes with a standard deviation of 15 minutes. Are the complaints substantiated by the data at $\alpha = 0.10$?

   Discuss the statistical and practical significance for this problem.

6. Consider Exercise 32 in Section 10.4:

   The managers of a large department store wish to test reactions of shoppers to a new in-store video screen which will broadcast continuous information about the store and the items currently on sale. The video production company claims that the average shopper will watch for five or more minutes. The managers randomly select 17 shoppers and determine how long they watch the video. The average time is 4.5 minutes with a standard deviation of 2.5 minutes. Perform a hypothesis test to determine whether there is overwhelming evidence to refute the claim of the video production company. Use $\alpha = 0.01$.

   Discuss the statistical and practical significance for this problem.

10.6   The Relationship between Confidence Interval Estimation and Hypothesis Testing

Previously, we discussed interval estimation for the population mean and the population proportion. We know that when estimating the population mean, a $100(1-\alpha)\%$ confidence interval for μ is given by

$$\bar{x} \pm z_{\alpha/2} \frac{\sigma}{\sqrt{n}}.$$

In this chapter, we have shown that the two-sided hypothesis test about the population mean μ is

$$H_0 : \mu = \mu_0$$
$$H_a : \mu \neq \mu_0$$

where μ_0 is some specific value of the population mean.

From the previous chapter, we know that $100(1-\alpha)\%$ of the confidence intervals will contain μ. Therefore, if we reject the null hypothesis when the confidence interval does not contain the value of μ_0, we will reject the null hypothesis when it is actually true with probability of α. You may recall that α represents the probability of committing a Type I error (i.e., we reject the null hypothesis when the null hypothesis is true). So, when we construct a $100(1-\alpha)\%$ confidence interval and reject the null hypothesis when the interval does not contain μ_0, this is equivalent to performing a two-tailed hypothesis test using α as the level of the test.

Procedure

Using a Confidence Interval to Test a Hypothesis

Step 1: If the hypothesis test is of the form

$$H_0 : \mu = \mu_0$$
$$H_a : \mu \neq \mu_0$$

Step 2: Calculate the $100(1-\alpha)\%$ confidence interval.

If $n \geq 30$ or if the standard deviation of the population, σ, is known and the sample is drawn from a normal population, then the confidence interval is given by

$$\bar{x} \pm z_{\alpha/2} \frac{\sigma}{\sqrt{n}}.$$

If $n < 30$ and the standard deviation of the population is unknown, but the distribution of the population is normal, then the confidence interval is given by

$$\bar{x} \pm t_{\alpha/2} \frac{s}{\sqrt{n}}.$$

Step 3: If μ_0 falls within the interval, then we fail to reject the null hypothesis; however, if μ_0 falls outside the calculated interval, then reject the null hypothesis in favor of the alternative.

Let's look at Example 10.5. Recall that the hypotheses were

$$H_0: \mu = 500$$
$$H_a: \mu \neq 500.$$

The test statistic for this example was $z = 3.46$ and the null hypothesis was rejected at the 0.05 level.

Note that the value $z_{\alpha/2} = z_{0.025} = 1.96$. Calculating a 95% confidence interval for μ, we get

$$\bar{x} \pm z_{\alpha/2} \frac{\sigma}{\sqrt{n}}$$
$$540 \pm 1.96 \frac{100}{\sqrt{75}}$$
$$517.37 \text{ to } 562.63.$$

Because the hypothesized value ($\mu = 500$) does not fall in this interval, we reject the null hypothesis, which is consistent with the earlier conclusion. Although we presented this demonstration using the confidence interval for the population mean, the same relationship exists for other population parameters.

10.6 Exercises

Basic Concepts

1. How can a confidence interval be used to test a hypothesis?

Exercises

2. AAA Controls makes a switch that is advertised to activate a warning light if the power supplied to a machine reaches 100 volts. A random sample of 250 switches is tested and the mean voltage at which the warning light occurs is 98 volts with a sample standard deviation of 3 volts. Using the confidence interval approach, test the hypothesis that the mean voltage activation is different from AAA Controls' claim at the 0.05 level.

3. Researchers studying the effects of diet on growth would like to know if a vegetarian diet affects the height of a child. The researchers randomly selected 12 vegetarian children that were six years old. The average height of the children is 42.5 inches with a standard deviation of 3.8 inches. The average height for all six-year-old children is 45.75 inches.

    a. Using confidence intervals, test to determine whether there is overwhelming evidence at $\alpha = 0.05$ that six-year-old vegetarian children are not the same height as other six-year-old children.

    b. What assumption did you make in performing the test?

4. High-power experimental engines are being developed by the Stevens Motor Company for use in its new sports coupe. The engineers have calculated the maximum horsepower for the engine to be 600 HP. Sixteen engines are randomly selected for horsepower testing. The sample has an average maximum HP of 620 with a standard deviation of 50 HP.

   a. Use the confidence interval approach to determine whether the data suggest that the average maximum HP for the experimental engine is significantly different than the maximum horsepower calculated by the engineers. Use a significance level of $\alpha = 0.01$.

   b. What assumption did you make in performing the test?

5. The nutrition label for Oriental Spice Sauce states that one package of sauce has 1190 milligrams of sodium. To determine if the label is accurate, the FDA randomly selects two hundred packages of Oriental Spice Sauce and determines the sodium content. The sample has an average of 1167.34 milligrams of sodium per package with a sample standard deviation of 252.94 milligrams.

   a. Calculate a 99% confidence interval for the mean sodium content in Oriental Spice Sauce.

   b. Using the confidence interval approach, is there evidence that the sodium content is different than the nutrition label states?

10.7 Testing a Hypothesis about a Population Proportion

The topic that we will develop in this section will be a hypothesis testing approach for categorical values (nominal data). The inferences that we will make with these data will concern one population. We will use the information in the sample proportion $\left(\hat{p}\right)$ to test hypotheses about the population proportion, p.

Testing hypotheses about a population proportion could involve a variety of problems.

- What fraction of a student's grades will be A's?
- What fraction of graduating seniors obtain jobs with starting salaries in excess of $38,000?
- What fraction of products that a company produces are defective?
- What fraction of the voters favor the incumbent in the next election?
- What fraction of the customers who purchase a Ford Focus are extremely satisfied?
- What fraction of the time will a baseball player get a hit?
- What fraction of the time will a drug be successful in treating a specific disorder?

Developing the Test

Testing a hypothesis concerning a population proportion is nearly identical to testing a hypothesis about a population mean. The major changes in the procedure include the use of the population proportion $\left(p\right)$ in the formulation of the hypotheses rather than the population mean $\left(\mu\right)$, and the calculation of the test statistic. Let's try an example.

Example 10.10

Sir is a novice at trading stocks and feels that if the Dow Jones Industrial Average (DJIA) is up on a given day, then it's likely that his portfolio will also be up, regardless of the stocks in his portfolio. In a random sample of 520 stocks on the New York Stock Exchange (NYSE), 350 increased on days that the DJIA increased. Sir would like to know if there is sufficient evidence to conclude that more than 60% of stocks in the NYSE increase on days that the DJIA increases.

Solution

Step 1: State the hypotheses in plain English.

- **Null Hypothesis:** At most 60% of the stocks on the NYSE increase on days when the DJIA increases.
- **Alternative Hypothesis:** More than 60% of the stocks increase on days when the DJIA increases.

Step 2: Select the appropriate statistical measure.

Since the problem concerns the fraction of stocks on the NYSE that increase on days that the DJIA increases, the appropriate statistical measure will be a proportion.

Let p = fraction of the stocks on the NYSE that increase on days when the DJIA increases.

Step 3: Determine whether the hypothesis should be one-sided or two-sided.

Since the problem is concerned with finding evidence that the percentage is greater than 60%, the formation of the hypothesis will be one-sided.

Step 4: Specify the hypotheses using the appropriate statistical measure.

$H_0: p \leq 0.60$ (At most 60% of stocks on the NYSE increase when the DJIA increases.)

$H_a: p > 0.60$ (More than 60% of the stocks on the NYSE increase when the DJIA increases.)

Step 5: Specify the level of the test.

Since the level of the test is not specified in the problem, let's use $\alpha = 0.05$.

Step 6: Select the appropriate test statistic.

If $np_0 \geq 5$ and $n(1 - p_0) \geq 5$, the appropriate test statistic is given by

$$z = \frac{\hat{p} - p_0}{\sigma_{\hat{p}}}, \text{ where } \sigma_{\hat{p}} = \sqrt{\frac{p_0(1 - p_0)}{n}}.$$

The test statistic measures how far the sample proportion, $\hat{p}$, is from the hypothesized value of the population proportion, p_0, measured in standard deviation units. This is exactly the same concept used in testing hypotheses about a population mean.

Step 7: Determine the critical value.

The test statistic, z, has a normal distribution with a mean of zero and a standard deviation of one. So designating the decision rule is exactly like what has been done in previous problems. The test is a one-tailed test, and in **Step 5** we decided to set the level of the test at 0.05.

The decision rule will be to reject $H_0: p \leq 0.60$ if the value of z is greater than or equal to 1.645. In essence, we are saying that ordinary sampling variation might account for a $\hat{p}$ up to 1.645 standard deviations larger than the hypothesized value. But if a z-value is observed that is larger than 1.645, then we will lose faith in the null hypothesis because we have observed something that is too rare by the standards we have set for the level of the test.

Figure 10.13

Step 8: Compute the test statistic.

In a random sample of 520 stocks on the NYSE, 350 increased on days that the DJIA increased. The value of $\hat{p}$ is given by

$$\hat{p} = \frac{350}{520} \approx 0.6731.$$

Because $np_0 = 520(0.60) = 312 \geq 5$ and $n(1-p_0) = 520(1-0.60) = 208 \geq 5$, we can use the z-test statistic. Thus, the test statistic is calculated as follows.

$$z = \frac{\hat{p} - p_0}{\sqrt{\dfrac{p_0(1-p_0)}{n}}} = \frac{0.6731 - 0.60}{\sqrt{\dfrac{0.60(1-0.60)}{520}}} \approx 3.40.$$

In this case, the sample proportion favoring Sir's assumption is 3.40 standard deviation units larger than the hypothesized proportion. If H_0 is true, would we expect a sample proportion that is 3.40 standard deviation units larger than the hypothesized proportion? Or, is this test statistic $(z = 3.40)$ overwhelming evidence in favor of the alternative hypothesis?

Step 9: Make the decision.

The value of the test statistic, $z = 3.40$, falls in the rejection region. According to the standard of rareness established in **Step 5**, the z-value is too rare to believe the null hypothesis $(H_0: p \leq 0.60)$ is true. The sample evidence contradicts the null hypothesis, and it will be rejected in favor of the alternative hypothesis,

$$H_a: p > 0.60.$$

Step 10: State the conclusion in terms of the original question.

At a significance level of 0.05, there is overwhelming evidence that over 60% of the stocks on the NYSE increase when the DJIA increases.

Example 10.11

A consulting firm believes that at least 39% of online shoppers have bad experiences on websites and do not purchase anything because sites are too complicated and do not possess good user interfaces. Out of a random sample of 60 online shoppers, only 15 reported frustration with the sites and thus, didn't make a purchase. Do the data provide sufficient evidence to refute the claim made by the consulting firm? Test using a significance level of 0.01.

Solution

Step 1: State the hypotheses in plain English.

- **Null Hypothesis:** At least 39% of online shoppers do not purchase anything due to a bad website experience.

- **Alternative Hypothesis:** Less than 39% of online shoppers do not purchase anything due to a bad website experience.

Step 2: Select the appropriate statistical measure.

Since the problem concerns the fraction of online shoppers that had a bad experience, the appropriate statistical measure will be a proportion. Let

$$p = \text{fraction of the online shoppers that did not purchase}$$
$$\text{anything due to a bad experience on the website.}$$

Step 3: Determine whether the hypothesis should be one-sided or two-sided.

Since the problem is concerned with finding evidence that the percentage of online shoppers that have a bad experience is less than 39%, it should be a one-sided alternative.

Step 4: State the hypotheses using the appropriate statistical measure.

$$H_0: p \geq 0.39 \quad \left(\begin{array}{l} \text{At least 39\% of online shoppers have a} \\ \text{bad experience and do not purchase anything.} \end{array} \right)$$

$$H_a: p < 0.39 \quad \left(\begin{array}{l} \text{Less than 39\% of online shoppers have a} \\ \text{bad experience and do not purchase anything.} \end{array} \right)$$

Step 5: Specify the level of the test.

The significance level of 0.01 is given in the problem.

Step 6: Select the appropriate test statistic.

Because $np_0 = 23.4$ and $n(1 - p_0) = 36.6$ are both greater than or equal to 5, the z-test statistic is appropriate.

$$z = \frac{\hat{p} - p_0}{\sigma_{\hat{p}}}, \text{ where } \sigma_{\hat{p}} = \sqrt{\frac{p_0 (1 - p_0)}{n}}$$

Step 7: Determine the critical value.

The null hypothesis should only be rejected if there is overwhelming evidence that the population proportion is less than 0.39. Since the test is one-sided, the entire probability associated with the level of the test $(\alpha = 0.01)$ will be placed in the left-hand tail of the distribution.

Figure 10.14

From Table 10.3, we know that the critical value of the test statistic is −2.33 (because it's a lower one-sided alternative hypothesis). The rejection region is the shaded area in the above graph. The decision will be to reject the null hypothesis if the value of the test statistic is less than or equal to −2.33.

Step 8: Compute the test statistic.

In a random sample of 60 online shoppers, 15 indicated that they did not have a good experience with the sites and thus, did not purchase anything. The value of $\hat{p}$ is given by

$$\hat{p} = \frac{15}{60} = 0.25, \text{ and}$$

$$z = \frac{\hat{p} - p_0}{\sqrt{\dfrac{p_0(1-p_0)}{n}}} = \frac{0.25 - 0.39}{\sqrt{\dfrac{0.39(1-0.39)}{60}}} \approx -2.22.$$

Remember:

In order to assure approximate normality of the test statistic we need to verify whether $np_0 \geq 5$ and $n(1 - p_0) \geq 5$.

The z-value indicates that the sample proportion is 2.22 standard deviation units below the hypothesized proportion. Could this have happened by ordinary sampling variation, or is this overwhelming evidence that the null hypothesis is not true and the alternative should be selected?

Step 9: Make the decision.

Note that the value of z does not fall in the rejection region. Thus, we fail to reject $H_0: p \geq 0.39$. There is not sufficient evidence (at the 0.01 level) to reject the null hypothesis. Another way of thinking about this is that the sample proportion was not sufficiently rare to reject $H_0: p \geq 0.39$.

Step 10: State the conclusion in terms of the original question.

At the 0.01 level of significance, there is not overwhelming evidence to conclude that shoppers' experiences during online shopping deterred them from making a purchase.

Calculating a *P*-Value for a Proportion

We can use exactly the same idea in determining a *P*-value for a test about a population proportion as we did in calculating *P*-values for the test statistic of a hypothesis concerning a population mean. Let's calculate the *P*-value using the test statistic we computed in **Step 8** of the previous example. If the null is true, how rare is the sample proportion that produced a *z*-value of −2.22?

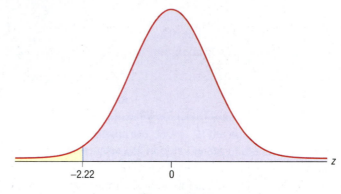

Figure 10.15

Assuming the null is true, $\hat{p}$ has a normal distribution that is centered around 0.39. If the null is really true, a *z*-value of −2.22 is uncommon. The probability of observing a value as small or smaller than −2.22 is the *P*-value. Using Table A in the Appendix,

$$P\text{-value} = P(z \le -2.22) = 0.0132.$$

Table 10.4 – If *P*-value = 0.0132	
Level of the Test	**Reject or Fail to Reject H_0**
0.10	Reject
0.05	Reject
0.01	Fail to Reject
0.005	Fail to Reject

If the null is true, the *P*-value measures the rareness of the test statistic under ordinary sampling variation. In other words, how often would we see a test statistic as small or smaller than the test statistic we have observed. Presuming the null is true, ordinary sampling variation produces a test statistic smaller than −2.22 about 1 time out of every 100. Should H_0 be rejected? Have we observed a test statistic that is too "rare" for H_0 to be true? The level of the test defines an unacceptable level of rareness for the test statistic. In the previous example $\alpha = 0.01$. Setting $\alpha = 0.01$ implies we are only willing to make a Type I error (reject the null hypothesis when it is true) once in every 100 trials of the experiment. Since the *P*-value of our test statistic, 0.0132 is greater than 0.01, the null hypothesis was not rejected.

If the level of the test had been 0.05, then to reject H_0 in favor of H_a requires a test statistic whose rareness under ordinary sampling variation is less than 0.05. For $\alpha = 0.05$, the null hypothesis is rejected in favor of the alternative, since the test statistic has a *P*-value (0.0132) less than α. In general, if the *P*-value is smaller than the level of the test, α, H_0 is rejected in favor of H_a. If the level of the test is less than the *P*-value, then H_0 is not rejected.

Note that if H_a had required a two-tailed test we would double the single tail area. Thus, for a test statistic of −2.22 and a two-tailed H_a the resulting *P*-value would be $2(0.0132) = 0.0264$.

10.7 Exercises

Basic Concepts

1. How does testing a hypothesis about a proportion differ from testing a hypothesis about a mean?

2. What is the appropriate test statistic to be used in hypothesis testing of a population proportion?

3. What conditions must be met in order to perform a hypothesis test about a population proportion?

4. How are P-values determined for a proportion?

Exercises

5. Determine the critical value(s) of the test statistic for each of the following large sample tests for the population proportion.

   a. Left-tailed test, $\alpha = 0.05$

   b. Right-tailed test, $\alpha = 0.01$

   c. Two-tailed test, $\alpha = 0.10$

6. Determine the critical value(s) of the test statistic for each of the following large sample tests for the population proportion.

   a. Left-tailed test, $\alpha = 0.07$

   b. Right-tailed test, $\alpha = 0.04$

   c. Two-tailed test, $\alpha = 0.09$

7. A commercial airline is concerned about the increase in usage of carry-on luggage. For years, the percentage of passengers with one or more pieces of carry-on luggage has been stable at approximately 38%. The airline recently selected 300 passengers at random and determined that 148 possessed carry-on luggage. Is there overwhelming evidence of an increase in carry-on luggage at a significance level of 0.01?

8. Ordinarily, when a company recruits a technical staff member, about 25% of the applicants are qualified. However, based on the information in 120 recently received resumes, 18 appear to be technically qualified.

   a. Is there overwhelming evidence that the percentage of qualified applicants is less than 25%? Test at the 0.05 level.

   b. What concerns might you have about the data in this problem?

9. The National Center for Drug Abuse is conducting a study to determine if heroin usage among teenagers has changed. Historically, about 1.3 percent of teenagers between the ages of 15 and 19 have used heroin one or more times. In a recent survey of 1824 teenagers, 37 indicated they had used heroin one or more times.

   a. Is there overwhelming evidence of a change in heroin usage among teenagers? Test at the 0.05 level.

   b. What concerns might you have about the data in this problem?

10

10. Paper International, Inc. has a large staff of salespeople nationwide. Top officials of the company believe that 75% of their salespeople have met their monthly sales goals by the end of the third week of each month. To investigate this, they randomly select 250 salespeople and examine their sales records at the end of the third week of the current month. One-hundred seventy-five of the 250 salespeople surveyed had already met their monthly sales goals.

   a. Does this sample support the belief of the top officials at the company at $\alpha = 0.10$?

   b. What concerns might you have about the manner in which the data were collected?

11. Ships arriving in U.S. ports are inspected by customs officials for contaminated cargo. Assume, for a certain port, that 20% of the ships arriving in the previous year contained cargo that was contaminated. A random selection of 50 ships in the current year included five that had contaminated cargo.

   a. Do the data suggest that the proportion of ships arriving in the port with contaminated cargoes has decreased in the current year at $\alpha = 0.01$?

   b. Do you have any concerns about the sample size? Explain.

12. Grain elevators store hundreds of thousands of bushels of grain each year that are waiting to be processed. It is critical to control the amount of moisture in the grain so that it does not spoil. A large storage facility is deemed to be "in control" if less than 1% of the grain elevators have a moisture content that exceeds 10%. One-hundred fifty grain elevators are randomly selected and the moisture content is measured. Two of the grain elevators sampled have a moisture content in excess of 10%.

   a. Is there sufficient evidence for the manager to conclude that the storage facility is "out of control?" Use $\alpha = 0.05$.

   b. Do you have any concerns about the sample size? Explain.

13. Electronic circuit boards are randomly selected each day to determine if any of the boards are defective. A random sample of 100 boards from one day's production has four boards that are defective.

   a. Based on the data, is there overwhelming evidence that more than 5% of the circuit boards are defective? Test at the $\alpha = 0.10$ level.

   b. Do you have any concerns about the sample size? Explain.

14. Loch Ness Fish Farm breeds fish for commercial sale. The fish are kept in breeder tanks until at least 70% of the fish are five inches long at which time they are transferred to outdoor ponds. To determine if it is the appropriate time to transfer the fish, 50 fish are randomly selected and measured. If 33 of the fish are found to be over five inches long, does the sample data suggest that it is the appropriate time to transfer the fish at $\alpha = 0.05$?

15. Digger and Digger, a precious metals mining company, is considering the development of a new mining area. They have a lease on an area which they believe contains gypsum. The area will be profitable to mine if more than 15% of the rocks contain more than trace amounts of the mineral. Eighty rocks are randomly selected and the amount of gypsum is measured. Thirteen rocks in the sample are observed to have more than trace amounts of the mineral. Based on the sample data, should Digger and Digger conclude that the area will be profitable to mine? Use $\alpha = 0.01$.

16. A socially conscious corporation wants to relocate their headquarters to another part of town. One concern expressed by workers is that their commuting distance will increase. The corporation has decided that if more than 50% of the employees will have to drive farther to the proposed new location, they will cancel the move. In a random sample of 398 employees, 201 indicated that their commuting distance to the new office will be longer. Based on the sample data, should the corporation cancel the move? Use a significance level of 0.01.

17. A production process will normally produce defective parts 0.2% of the time. In a random sample of 1400 parts, three defectives are observed.

   a. Is this overwhelming evidence at the 0.05 level to indicate that the defective rate of the process has increased?

   b. Compute the P-value for the test statistic.

   c. Based on the P-value, would the decision change at $\alpha = 0.01$?

18. Bombay Charlie's, a fast food Indian restaurant, is thinking about adding a certain spice to their chicken curry dish to attract more customers. The restaurant manager has decided to add the spice if more than 80% of his customers prefer the taste of the chicken curry with the spice added. Sixty-five customers are randomly selected to participate in a blind taste test. Fifty-four of these customers prefer the chicken curry with the added spice.

   a. Find the P-value for the hypothesis test that the manager will perform to decide if more than 80% of the customers prefer the taste of the chicken curry with the added spice.

   b. Do the data suggest that more than 80% of the customers prefer the curry with the new spice at $\alpha = 0.05$?

19. The news program for KOPE, the local television station, claims to have 40% of the market. A random sample of 500 viewers conducted by an independent testing agency found 192 who claim to watch the KOPE news program on a regular basis.

   a. Find the P-value for testing the hypothesis that the news program for KOPE does not have at least 40% of the market as it claims.

   b. Is there sufficient evidence to reject the hypothesis that KOPE does not have at least 40% of the market at a significance level of 0.05?

20. The length of time that a storm window will last before beginning to leak is of interest to a window manufacturer who wishes to guarantee his windows. He believes that more than 50% of the windows will last at least four years. To research this, 931 windows, which were installed at least four years ago, are randomly selected and checked for leakage. Five hundred of the windows are found to still be leak-free.

   a. Find the P-value for testing the hypothesis that more than 50% of the windows will be leak-free in four years.

   b. Does the sample support the hypothesis that more than 50% of the windows will be leak-free in four years at $\alpha = 0.05$?

21. In order to discourage soldiers from smoking, the Pentagon raised the price of cigarettes by $4 a carton in October of 1996. This increased the average price of a carton of brand-name cigarettes to $17.50, an increase of about 30%. Prior to the price increase, about 32% of military personnel smoked, as opposed to 25% of all adult Americans. Suppose that following the price increase, a random sample of military personnel is selected to determine smoking habits. With $\alpha = 0.05$, can we conclude that the price increase was effective in decreasing the percentage of smokers if 50 of the 200 military personnel sampled smoke?

22. Wearing bright or fluorescent orange colored clothing clearly reduces the risk of being shot or killed by hunting. According to an October 1996 article appearing in *The Augusta Chronicle* (Georgia), about two-thirds of the hunters shot in Georgia and South Carolina during the preceding five years were not wearing bright clothes. Of the 52 that were killed, only 19 wore orange. Suppose that a random sample of 100 hunters in Georgia are surveyed and it is determined that of the 100, 62 routinely wear fluorescent orange colored clothing while hunting. With $\alpha = 0.10$, can it be concluded that over half the hunters in Georgia routinely wear fluorescent orange colored clothing while hunting?

23. According to the Federal Communications Commission, about 49% of the households in the United States had cable television in 1985. Suppose that a sample of 200 households is selected in 2003 and it is determined that 125 of them have cable television.

    a. With $\alpha = 0.05$, can it be concluded that a higher proportion of households in 2003 have cable television as compared with 1985?

    b. In the sample of 200, what is the fewest number of people who have cable television that would allow the conclusion that a higher proportion of households in 2003 have cable television as compared to 1985?

24. Selling autographed sports memorabilia has become a multimillion dollar industry in the United States. But just how does the purchaser of an autographed football jersey or an autographed baseball know that the autograph is indeed authentic? Unfortunately, the sports memorabilia market is teeming with con artists who prey upon the trusting nature of sports fans. In 1996, the FBI said that 70% of all autographed sports memorabilia is fraudulent. Assume that 50 pieces of autographed sports memorabilia are sampled at a large memorabilia show and that 40 of them are determined to be fraudulent.

    a. With $\alpha = 0.05$, can it be concluded that the proportion of fraudulent autographed sports memorabilia at the show differs from the FBI claim?

    b. What is the greatest number of fraudulent items in the sample of 50 that would not allow a conclusion that sports memorabilia at the show differs from the FBI claim?

10.8   Testing a Hypothesis about a Population Variance

In this section we want to adapt the hypothesis testing procedure to test a hypothesis concerning a population variance. Before we can perform the test about the population variance, we need to review a few topics that were discussed earlier in the text.

Recall that the sample variance is

$$s^2 = \frac{\sum (x_i - \bar{x})^2}{n-1}$$

and it serves as the point estimate of the population variance, σ^2. As with the other tests developed for the population mean and the population proportion, we first need to develop a sampling distribution for

$$\frac{(n-1)s^2}{\sigma^2}$$

that will allow us to calculate a confidence interval as well as perform hypothesis testing about the population variance.

Note that if we have a random sample of size n, taken from a normal population, then the sampling distribution of $\frac{(n-1)s^2}{\sigma^2}$ follows a **chi-square distribution** with $n-1$ degrees of freedom.

The chi-square distribution is a positively skewed (or skewed to the right) distribution. Like the t-distribution, the shape of the distribution is a function of its degrees of freedom. See Figure 10.16, which illustrates the chi-square distributions with 4 and 10 degrees of freedom.

χ^2 Distribution with 4 and 10 Degrees of Freedom

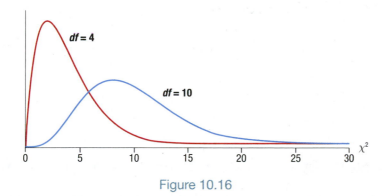

Figure 10.16

To use the chi-square distribution, we need a chi-square point, denoted by χ^2_α (the Greek letter chi, pronounced *Ki*). We'll later call this our critical value for the chi-square distribution. As is shown in Figure 10.17, χ^2_α is the point on the horizontal axis under the curve with an area of α to the right of it. The value of χ^2_α depends on the right-hand tail area, α, and the number of degrees of freedom of the chi-square distribution. The values are tabulated in Table G in Appendix A. Looking at the chi-square table, the rows correspond to the appropriate number of degrees of freedom (the first column listed down the left side of the table), while the columns represent the right-hand tail area.

χ^2 Distribution

Figure 10.17

Using Table G in Appendix A, suppose we want to find the chi-square value that gives us a right-hand tail area of 0.05 with 5 degrees of freedom. To do this, we would look down the left-most column for 5 degrees of freedom and then the column labeled $\chi^2_{0.05}$. Doing so, we find that $\chi^2_{0.05}$ is 11.070.

df	$\chi^2_{0.100}$	$\chi^2_{0.050}$	$\chi^2_{0.025}$	$\chi^2_{0.010}$	$\chi^2_{0.005}$
1	2.706	3.841	5.024	6.635	7.879
2	4.605	5.991	7.378	9.210	10.597
3	6.251	7.815	9.348	11.345	12.838
4	7.779	9.488	11.143	13.277	14.860
5	9.236	11.070	12.833	15.086	16.750
6	10.645	12.592	14.449	16.812	18.548

Now that we've established the sampling distribution associated with the sample variance, we can make inferences about the population variance. Suppose we have a random sample of size n taken from a normal population and that s^2 is the estimate of the population variance, σ^2.

Formula

$100(1 - \alpha)\%$ Confidence Interval for σ^2

A $100(1-\alpha)\%$ confidence interval for σ^2 is given by

$$\frac{(n-1)s^2}{\chi^2_{\alpha/2}} \leq \sigma^2 \leq \frac{(n-1)s^2}{\chi^2_{1-\alpha/2}}$$

where $\chi^2_{\alpha/2}$ and $\chi^2_{1-\alpha/2}$ are points under the curve of the chi-square distribution with $n-1$ degrees of freedom.

Example 10.12

The quality control supervisor of a bottling plant is concerned about the variance of fill per bottle. Regulatory agencies specify that the standard deviation of the amount of fill should be less than 0.1 ounce. To determine whether the process is meeting this specification, the supervisor randomly selects ten bottles, weighs the contents of each, and finds that the sample standard deviation of these measurements is 0.04. Compute a 95% confidence interval for the standard deviation of ounces of fill for the bottling plant.

Solution

We want to find a 95% confidence interval for the variance. We are given that

$$n = 10, \; s = 0.04, \text{ and } \alpha = 0.05.$$

To calculate a 95% confidence interval, we use the following formula.

$$\frac{(n-1)s^2}{\chi^2_{\alpha/2}} \leq \sigma^2 \leq \frac{(n-1)s^2}{\chi^2_{1-\alpha/2}}$$

Thus, we need to find the values of $\chi^2_{0.025}$ and $\chi^2_{0.975}$ for $n - 1 = 10 - 1 = 9$ degrees of freedom.

Using Table G in Appendix A, at 9 degrees of freedom,

$$\chi^2_{0.025} = 19.023$$
$$\chi^2_{0.975} = 2.700.$$

Substituting the values in the formula above, we have

$$\frac{(10-1)(0.04)^2}{19.023} \leq \sigma^2 \leq \frac{(10-1)(0.04)^2}{2.700}$$
$$0.000757 \leq \sigma^2 \leq 0.00533$$

So, a 95% confidence interval for the variance of fill of the bottles is between 0.000757 and 0.00533 ounce. However, the problem mentions the tolerance for the standard deviation of fill. So, to ensure that we make our interpretation in terms of the problem, to find a 95% confidence interval for the standard deviation, we take the square root of the confidence interval for the variance, yielding

$$0.0275 \leq \sigma \leq 0.0730.$$

The 95% confidence interval for the standard deviation of fill for the bottles is between 0.0275 and 0.0730 ounce, indicating that the process is meeting the specifications of being less than 0.1 ounce.

To adapt the hypothesis testing procedure to test a hypothesis concerning a population variance, let's look at the following example.

Example 10.13

A drug manufacturer believes that the manufacturing process is in control if the standard deviation of the dosage in each tablet is at most 0.10 milligram. The quality control manager is willing to shut down the manufacturing process if there is overwhelming evidence that the process has excessive variation. A sample of 30 tablets is evaluated and the sample standard deviation is found to be 0.14 milligram. Should the quality control manager shut down the manufacturing process? Use $\alpha = 0.01$.

Solution

Step 1: State the hypotheses in plain English.

- **Null Hypothesis:** The manufacturing process does not have excessive variation.
- **Alternative Hypothesis:** The manufacturing process has excessive variation.

Step 2: Select the appropriate statistical measure.

Since the issue in this example is variation, the hypothesis can be stated in terms of the variance.

Step 3: Determine whether the hypothesis should be one-sided or two-sided.

Most hypothesis tests concerning a variance will be one-sided. Small variation is desirable, while too much variation is undesirable. Because we are interested in determining if there is evidence that the process has excessive variation, the test will be one-tailed.

Step 4: State the hypotheses using the appropriate statistical measure.

$$H_0: \sigma^2 \leq 0.01 \quad \text{(The variation of the process is acceptable.)}$$
$$H_a: \sigma^2 > 0.01 \quad \text{(The variation of the process is excessive.)}$$

Step 5: Specify the level of the test.

The level of the test is specified in the problem as $\alpha = 0.01$.

Step 6: Select the appropriate test statistic.

The test statistic will change rather dramatically. In the previous test statistics, the goal was to measure how far the sample statistic, $\bar{x}$ or $\hat{p}$, was from the hypothesized value. The test statistic for a population variance is given by

$$\chi^2 = \frac{(n-1)s^2}{\sigma_0^2} \quad \text{with } n - 1 \text{ degrees of freedom.}$$

Let's look at the pieces of this statistic. The term σ_0^2 refers to the hypothesized value of the variance. In this sample, $\sigma_0 = 0.10$, which implies $\sigma_0^2 = 0.01$. The actual variance of the population is unknown. But the sample variance, s^2, should be reasonably close to the unknown population variance, σ^2. If the null is true, then $\sigma_0^2 = 0.01$ and the ratio $\frac{s^2}{\sigma_0^2}$ should be near 1, since s^2 should be close to σ_0^2. Assuming the null is true, multiplying this ratio by $(n-1)$ should produce a result near $(n-1)$. If the χ^2 expression is a great deal larger than $(n-1)$, then s^2 will be a great deal larger than σ_0^2. Such an event would cast doubt on the validity of the null hypothesis.

Step 7: Determine the critical value.

The role of the critical value in this test is no different from other hypothesis tests discussed earlier. It defines a range of values for the test statistic, the rejection region, that will be too rare to have likely occurred from ordinary sampling variability. From a probabilistic standpoint, the level of the test defines the size of the rejection region. Should the value of the test statistic fall in this region, the null hypothesis will be rejected. Determining the critical value will require knowledge of the sample size. The company studied a sample of 30 tablets. Since the level of the test is 0.01 and there are $df = 30 - 1 = 29$ degrees of freedom, the critical value is 49.588 (see Table G). The test statistic will exceed this value because of ordinary variation only 1% of the time.

df	...	$\chi^2_{0.025}$	$\chi^2_{0.010}$	$\chi^2_{0.005}$
1		5.024	6.635	7.879
2		7.378	9.210	10.597
3		9.348	11.345	12.838
⋮				
29		45.722	49.588	52.336
30		46.979	50.892	53.672
⋮				

χ^2 **Distribution, df = 29**

Fail to Reject H_0   **Reject H_0**

$\alpha = 0.01$

χ^2

0   49.588

Figure 10.18

Step 8: Compute the test statistic.

A sample of 30 tablets is evaluated, and the sample standard deviation is found to be 0.14 milligram.

$$\chi^2 = \frac{(n-1)s^2}{\sigma_0^2} = \frac{(30-1)(0.14)^2}{(0.10)^2} = 56.84$$

Step 9: Make the decision.

Since the test statistic, $\chi^2 = 56.84$, exceeds the critical value, 49.588, we will conclude the test statistic is too rare to have been caused by ordinary sampling variation. The null hypothesis is rejected in favor of the alternative hypothesis.

Fail to Reject H_0   **Reject H_0**

χ^2

0   49.588 $\chi^2 = 56.84$

Step 10: State the conclusion in terms of the original question.

There is overwhelming evidence that the process variation exceeds the desired level. The quality control manager will likely shut down the manufacturing process.

10.8   Exercises

Basic Concepts

1. How does testing a hypothesis about a variance differ from testing a hypothesis about a mean?

2. What is the sampling distribution for $\dfrac{(n-1)s^2}{\sigma^2}$?

3. What assumption must hold to use the chi-square distribution to make inferences about the population variance?

4. True or false: the chi-square distribution is skewed to the right.

5. What is the symbol for a critical value for the chi-square distribution? Describe the meaning of this critical value.

6. Give an example where we would want to calculate a confidence interval for σ^2.

Exercises

7. Determine the critical value(s) of the test statistic for each of the following tests for a population variance where the assumption of normality is satisfied.

    a. Right-tailed test, $\alpha = 0.01$, $n = 20$

    b. Right-tailed test, $\alpha = 0.05$, $n = 24$

    c. Right-tailed test, $\alpha = 0.005$, $n = 5$

8. Determine the critical value(s) of the test statistic for each of the following tests for a population variance where the assumption of normality is satisfied.

    a. Right-tailed test, $\alpha = 0.025$, $n = 18$

    b. Right-tailed test, $\alpha = 0.10$, $n = 24$

    c. Right-tailed test, $\alpha = 0.05$, $n = 41$

9. A bolt manufacturer is very concerned about the consistency with which his machines produce bolts that are $\dfrac{3}{4}$ inch in diameter. When the manufacturing process is working normally the standard deviation of the bolt diameter is 0.05 inch. A random sample of 30 bolts has an average diameter of 0.25 inch with a standard deviation of 0.07 inch.

    a. Construct a 95% confidence interval for the standard deviation of the bolt diameter. Interpret the interval.

    b. Can the manufacturer conclude that the standard deviation of bolt diameters is greater than 0.05 inches at $\alpha = 0.05$?

    c. What assumption did you make about the diameter of the bolts in constructing the confidence interval in part **a.** and performing the test in part **b.**?

10. A drug that is used for treating cancer has potentially dangerous side effects if it is taken in doses that are larger than the required dosage for the treatment. The pharmaceutical company that manufactures the drug must be certain that the standard deviation of the drug content in the tablet is not more than 0.1 mg. Twenty-five tablets are randomly selected and the amount of drug in each tablet is measured. The sample has a mean of 20 mg and a variance of 0.015 mg.

    a. Do the data suggest at $\alpha = 0.01$ that the standard deviation of drug content in the tablets is greater than 0.1 mg?

    b. Construct a 99% confidence interval for the variance of the amount of drug in each tablet. Interpret the interval.

    c. What assumption did you make about the amount of drug contained in the tablets in performing the test in part **a.** and constructing the confidence interval in part **b.**?

11. A conservative investor would like to invest some money in a bond fund. The investor is concerned about the safety of her principal (the original money invested). Colonial Funds claims to have a bond fund which has maintained a consistent share price of $7. They claim that this share price has not varied by more than $0.25 on average since its inception. To test this claim, the investor randomly selects 25 days during the last year and determines the share price for the bond fund. The average share price of the sample is $7 with a standard deviation of $0.35.

    a. Can the investor conclude that the standard deviation of share price of the bond fund is greater than 0.25? Test at the 0.01 level.

    b. Construct a 90% confidence interval for the standard deviation of the share price of the bond fund. Interpret the interval.

    c. What assumption did you make about the share price of the bond fund in your test in part **a.** and constructing the confidence interval in part **b.**?

T **Discovering Technology**

Using the TI-84 Plus Calculator

Hypothesis Test for a Population Mean: σ Known or $n \geq 30$

$$z = \frac{\bar{x} - \mu_0}{\sigma_{\bar{x}}}, \quad \text{where } \sigma_{\bar{x}} = \frac{\sigma}{\sqrt{n}}$$

$H_0: \mu \leq 16,000$ $\qquad n = 1000$ $\qquad \bar{x} = 16,500$

$H_a: \mu > 16,000$ $\qquad \sigma = 2500$

1. Choose **STAT**, then select **TESTS** and **1:Z-Test**.

2. Press **ENTER**, and choose **Stats**. Input **16000** for μ_0, **2500** for σ, **16500** for $\bar{x}$, **1000** for n, and select $>\mu_0$. Press **Calculate**.

Figure 10.19

Figure 10.20

3. The results of the hypothesis test are displayed. The test statistic is $z \approx 6.32$, and the P-value is extremely small. Thus, there is sufficient evidence to reject the null hypothesis. The sample mean and sample size are also repeated for convenience.

Hypothesis Test for a Population Mean: σ Unknown, $n < 30$

$$t = \frac{\bar{x} - \mu_0}{s_{\bar{x}}}, \quad \text{where } s_{\bar{x}} = \frac{s}{\sqrt{n}}$$

$H_0: \mu = 35$ $\qquad n = 20$ $\qquad \bar{x} = 27$

$H_a: \mu \neq 35$ $\qquad s = 4$

1. Choose **STAT**, then select **TESTS** and **2:T-Test**.

2. Press **ENTER**, and choose **Stats**. Input **35** for μ_0, **27** for $\bar{x}$, **4** for s (designated on calculator as Sx), **20** for n, and select $\neq \mu_0$. Press **Calculate**.

Figure 10.21

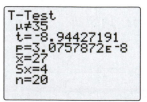

Figure 10.22

3. The results of the hypothesis test are displayed. The test statistic is $t \approx -8.944$, and the P-value is extremely small. Thus, there is sufficient evidence to reject the null hypothesis. The sample mean, sample standard deviation, and sample size are also repeated for convenience.

P-Values for t-Test Statistics

Calculate the P-value for a test statistic of $t = 2.451$ with 17 degrees of freedom. Assume that the hypothesis test was an upper-tailed test.

1. Access the **DISTR** menu by pressing ⬛2ND⬛ and then ⬛VARS⬛.

2. Select option **5:tcdf**. The syntax to find the area under the t-distribution between two t-values is **tcdf(lower bound, upper bound, degrees of freedom)**. To find the P-value for a test statistic of $t = 2.451$ with 17 degrees of freedom, assuming we are performing a one-sided test, enter **tcdf(2.451,1E99,17)**. Note that 1E99 represents a large positive number, and is entered into the calculator by pressing ⬚1⬚, ⬛2ND⬛, and ⬛,⬛.

3. The P-value is given as approximately 0.0127. This is the area to the right of $t = 2.451$ with 17 degrees of freedom. Note that for a two-tailed test, the value displayed would need to be multiplied by 2 to get the P-value in both tails.

Figure 10.23

Hypothesis Test for a Population Proportion

$$z = \frac{\hat{p} - p_0}{\sigma_{\hat{p}}}, \quad \text{where} \quad \sigma_{\hat{p}} = \sqrt{\frac{p_0(1 - p_0)}{n}}$$

$$H_0: p \leq 0.60 \qquad\qquad n = 520$$
$$H_a: p > 0.60 \qquad\qquad x = 350$$

1. Choose ⬛STAT⬛, then select **TESTS** and **5:1-PropZTest**.

2. Input **0.60** for p_0, **350** for x, **520** for n, and select **>p_0**. Press **Calculate**.

Figure 10.24

Figure 10.25

3. The results of the hypothesis test are displayed. The test statistic is $z \approx 3.40$, and the P-value is approximately 0.0003. Thus, there is sufficient evidence to reject the null hypothesis. The sample proportion and sample size are also displayed.

Using Excel

Hypothesis Test for a Population Mean: Sigma Known or $n \geq 30$

Finding the z-test statistic can be one of the more laborious steps in hypothesis testing. Use Excel to compute the z-statistic for Example 10.9.

1. Create a worksheet with these labels.

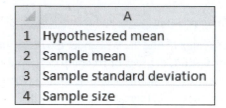

	A
1	Hypothesized mean
2	Sample mean
3	Sample standard deviation
4	Sample size

Figure 10.26

2. Referring to the example, we see that the null hypothesized mean is 25. Enter **25** in cell B1. Enter the sample mean, **25.01** in cell B2. Enter the sample standard deviation, **0.1** in cell B3. Enter the sample size, **2000** in cell B4.

3. Given these values, we have enough data to compute the z-test statistic. In cell B5 construct the following formula for the z-test statistic.

=(B2 −B1)/(B3/(SQRT(B4))

4. Press **Enter**. The resulting z-test statistic is approximately 4.47. Compare this to our calculations in Example 10.9

	A	B
1	Hypothesized mean	25
2	Sample mean	25.01
3	Sample standard deviation	0.1
4	Sample size	2000
5		4.472136

Figure 10.27

You can use this worksheet to compute z for other hypothesis testing questions by changing the values for the hypothesized mean and sample statistics. The formula can also be used in a similar manner to calculate t-statistics when sigma is unknown and $n < 30$, but the sample is drawn from a normal population.

P-Values for t-Test Statistics

Calculate the P-value if the observed value of the test statistic is $t = 2.378$ with 15 degrees of freedom. Assume the hypothesis test is a one-sided, upper-tailed test.

1. The function in Microsoft Excel that returns the probability in the lower (or left) tail of a t-distribution is

    =T.DIST(x, deg_freedom, cumulative)

    where x is the t-statistic, *deg_freedom* is the degrees of freedom for the t-distribution, and *cumulative* is a logical value that if TRUE returns the cumulative distribution function, and if FALSE returns the probability density function. When finding the P-value, cumulative will always be TRUE. To find the P-value for a one-sided, upper-tailed test when the observed value of the test statistic is $t = 2.378$ with 15 degrees of freedom, enter the following formula into Excel.

    =1-T.DIST(2.378,15,TRUE)

2. The P-value is displayed as approximately 0.0156. Remember that the function returns the area to the left of the test statistic. This is why we had to subtract the function from 1 in order to find the area in the upper tail.

Using Minitab

Hypothesis Test for a Population Mean: σ Known or $n \geq 30$

Minitab will compute the z-test statistic for a set of sample data.

Consider the following sample data.

95	110	60	75	85	90	110	90	115	100	90	100
90	50	60	110	90	90	55	100	40	115	50	100
70	70	80	95	90	70	95	80	105	90	55	65

$$H_0: \mu \leq 80 \qquad \alpha = 0.05$$
$$H_a: \mu > 80 \qquad \sigma = 20.15$$

1. Enter the sample data into column C1.

2. Choose **Stat**, **Basic Statistics**, and **1-Sample Z**. Select **Options** to enter the alternative of **greater than** and press **OK**.

3. Enter **C1** for Samples in columns, enter **20.15** for Standard deviation, check the box next to Perform hypothesis test, and enter **80** for Hypothesized mean. Press **OK**.

4. Observe the output screen for the results. The z-test statistic for these sample data is 1.28, and the P-value is 0.100.

```
One-Sample Z: C1

Test of mu = 80 vs > 80
The assumed standard deviation = 20.15


                                      95% Lower
Variable    N    Mean   StDev  SE Mean    Bound      Z      P
C1         36   84.31   20.15     3.36    78.78   1.28  0.100
```

Figure 10.28

Hypothesis Test for a Population Mean: σ Unknown, $n < 30$

Consider the following sample data.

24.9	26.8	27.2	34.1	28.9	25.9	25.1	23.9	23.0	30.9
31.9	35.1	27.4	25.9	19.1	23.7	22.0	25.9	29.0	29.3

$$H_0: \mu = 35 \qquad\qquad \alpha = 0.05$$
$$H_a: \mu \neq 35$$

1. Enter the sample data into column C1.

2. Choose **Stat**, **Basic Statistics**, and **1-Sample t**. Select **Options** to enter the alternative of **not equal** and press **OK**.

3. Enter **C1** for Samples in columns, check the box next to Perform hypothesis test, and enter **35** for Hypothesized mean. Press **OK**.

4. Observe the output screen for the results. The t-test statistic for these sample data is -8.95, and the P-value is 0.000.

```
One-Sample T: C1

Test of mu = 35 vs not = 35


Variable    N     Mean   StDev   SE Mean       95% CI          T       P
C3         20   27.000   4.000    0.894   (25.128, 28.872)  -8.95   0.000
```

Figure 10.29

Note: Performing a hypothesis test for a population proportion using Minitab is very similar to the steps taken to perform a test about a population mean. The tool is found under **Stat**, **Basic Statistics**, and **1 Proportion**. Minitab can either perform the test using raw data values, or summarized data. This is also true for the tests about population means. Summarized data may be used in place of the raw data.

R   Chapter 10 Review

Key Terms and Ideas

- Hypothesis Testing
- Null Hypothesis (H_0)
- Alternative Hypothesis (H_a)
- One-Sided Alternative
- Two-Sided Alternative
- One-Tailed Test
- Two-Tailed Test
- Test Statistic
- Rejection Region
- Reject H_0
- Fail To Reject H_0
- Standard Value
- Type I Error
- Type II Error
- Level of the Test (Significance Level of the Test)

- Decision Rule
- Left-Tailed (Lower-Tailed) Test
- Right-Tailed (Upper-Tailed) Test
- z-Test Statistic
- Critical Value
- t-Test Statistic
- P-Value
- Practical vs. Statistical Significance
- The Relationship between Confidence Interval Estimation and Hypothesis Testing
- Using a Confidence Interval to Test a Hypothesis
- Chi-Square Distribution
- Confidence Interval for σ^2

Key Formulas

	Section
z-Test Statistic for a Population Mean $$z = \frac{\bar{x} - \mu_0}{\sigma_{\bar{x}}}, \text{ where } \sigma_{\bar{x}} = \frac{\sigma}{\sqrt{n}}$$	10.4
t-Test Statistic $$t = \frac{\bar{x} - \mu_0}{s_{\bar{x}}}, \text{ where } s_{\bar{x}} = \frac{s}{\sqrt{n}}$$	10.4
z-Test Statistic for a Population Proportion If $np_0 \geq 5$ and $n(1 - p_0) \geq 5$, the appropriate test statistic is given by $$z = \frac{\hat{p} - p_0}{\sigma_{\hat{p}}}, \text{ where } \sigma_{\hat{p}} = \sqrt{\frac{p_0(1 - p_0)}{n}}.$$	10.7

Key Formulas (cont.)

	Section

$100(1-\alpha)\%$ Confidence Interval for the Population Variance

A $100(1-\alpha)\%$ confidence interval for σ^2 is given by

$$\frac{(n-1)s^2}{\chi^2_{\alpha/2}} \leq \sigma^2 \leq \frac{(n-1)s^2}{\chi^2_{1-\alpha/2}}$$

10.8

where $\chi^2_{\alpha/2}$ and $\chi^2_{1-\alpha/2}$ are points under the curve of the chi-square distribution with $n-1$ degrees of freedom.

Test Statistic for a Population Variance

10.8

$$\chi^2 = \frac{(n-1)s^2}{\sigma^2_0} \quad \text{with } n-1 \text{ degrees of freedom}$$

AE   Additional Exercises

1. A tire company has found that the mean time required for a mechanic to replace a set of four tires is 18 minutes. After instituting a new installation procedure, the company unfortunately believes that the expected time required to replace the set of four tires remains unchanged. A test of the company's belief will be performed.

    a. What are the null and alternative hypotheses for the test of the company's belief?

    b. Describe, in terms of the problem, how a Type I error could occur.

    c. Describe, in terms of the problem, how a Type II error could occur.

2. Tech Transit wishes to test whether the mean number of passenger miles on a particular route exceeds 66,000 passenger miles, the number of passenger miles the company needs on that route to cover all allocated costs. A random sample of 25 trips on the route yields a mean of 70,250 miles and a standard deviation of 9000 miles. It is desired to control the significance level at 1%.

    a. State the appropriate hypotheses for this problem.

    b. Describe, in terms of the problem, how a Type I error could occur.

    c. Describe, in terms of the problem, how a Type II error could occur.

3. A pain reliever currently being used in a hospital is known to bring relief to patients in a mean time of 3.5 minutes. To compare a new pain reliever with the one currently being used, the new drug is administered to a random sample of 50 patients. The mean time to relief for the sample of patients is 2.8 minutes and the standard deviation is 1.14 minutes. Do the data provide sufficient evidence to conclude that the new drug was effective in reducing the mean time until a patient receives relief from pain? Test using $\alpha = 0.10$.

4. Tech uses thousands of fluorescent light bulbs each year. The brand of bulb it currently uses has a mean life of 900 hours. A manufacturer claims that its new brand of bulbs, which cost the same as the brand the university currently uses, has a mean life of more than 900 hours. The university has decided to purchase the new brand if, when tested, the test evidence supports the manufacturer's claim at the 0.05 significance level. Suppose 64 bulbs were tested and they were found to have an average life of 920 hours with a standard deviation of 80 hours. Will the university purchase the new brand of fluorescent bulbs?

5. The daily wages in a particular industry are normally distributed with a mean of $13.20 and a standard deviation of $2.50. If a company in this industry employing 40 workers pays these workers, on average, $12.20, can this company be accused of paying inferior wages? Use the P-value approach with a significance level of 1%.

6. A coin-operated soft drink machine was designed to discharge, on average, 12 ounces of beverage per cup. In a test of the machine, ten cupfuls of beverage were drawn from the machine and measured. The mean and standard deviation of the ten measurements were 12.1 ounces and 0.12 ounce, respectively. Do these data present sufficient evidence to indicate that the mean discharge differs from 12 ounces? Test using $\alpha = 0.10$.

7. Techside Real Estate, Inc. is a research firm that tracks the cost of apartment rentals in Southwest Virginia. In mid-2002, the regional average apartment rental rate was $895 per month. Assume that, based on the historical quarterly surveys, it is reasonable to assume that the population standard deviation is $225. In a current study of apartment rental rates, a sample of 180 apartments in the region provided the apartment rental rates. Do the sample data enable Techside Real Estate, Inc. to conclude that the population mean apartment rental rate now exceeds the level reported in 2002? The sample mean is $915 and the sample standard deviation is $227.50. Make your decision based on $\alpha = 0.10$.

8. Suppose that the national average price for used cars is $10,192. A manager of a local used car dealership reviewed a sample of 25 recent used car sales at the dealership in an attempt to determine whether the population mean price for the used cars at this particular dealership differed from the national mean. The prices for the sample of 25 cars are given in the data with a mean of $9750 and standard deviation of $1400. Test using $\alpha = 0.05$ whether a difference exists in the mean price for used cars at the dealership.

9. Suppose you are responsible for auditing invoices. Historically, about 0.003 of the invoices possessed material errors. During the last audit cycle a number of suggestions were made and implemented, and you hope that the next audit will provide evidence of improvement in the error rate. An audit of 6000 recent invoices reveals 12 material errors.

    a. Do the data suggest an improvement in the invoice error rate at $\alpha = 0.05$?

    b. Compute the P-value of the test statistic.

    c. Based on the P-value, would the decision change at $\alpha = 0.10$?

10. After completing Chemistry 101, Tommy Walker decides to conduct an experiment on his favorite brand of whiskey to determine if the proof rating on the bottle is accurate. He selects eight small eighty-proof bottles from different stores around town and measures the percent of alcohol in each bottle. (**Note:** 80-proof alcohol contains 40% alcohol.)

    The resulting measurements are as follows.

Table for Exercise 10 – Percent of Alcohol per Bottle							
38%	40%	42%	41%	39%	38%	40%	38%

    a. What is the population being studied?

    b. What is the variable being measured?

    c. What level of measurement does the data possess?

    d. Can Tommy conclude that the actual proof of whiskey is not equal to 80 at $\alpha = 0.05$?

    e. What assumption did Tommy make in performing the test in part **d.**?

11. You have decided to become a professional gambler specializing in roulette. If the roulette wheel is fair (each number has a $\dfrac{1}{38}$ chance) then you will lose in the long run. However, you plan to locate wheels that are not balanced properly. An unbalanced wheel will produce some numbers more often than expected. You believe that you have found such a wheel and have started keeping track of the number 29. After 420 spins of the wheel, the number 29 has been observed 14 times. Is this overwhelming evidence at the $\alpha = 0.05$ level that you should start betting heavily on the number 29?

12. A commercial airline is concerned over the increase in weight of a carry-on luggage. In the past, the airline has estimated that the average piece of carry-on luggage will weigh 12 pounds. A random selection of 148 pieces of carry-on luggage has an average weight of 14.2 pounds with a standard deviation of 3.4 pounds. Do you think that the airline's concern is justified? Use $\alpha = 0.01$.

13. Consider the following large sample hypothesis tests for the population mean. Compute the P-value for each test and decide whether you would reject or fail to reject the null hypothesis at $\alpha = 0.01$.

   a. $H_0: \mu \le 15$, $H_a: \mu > 15$, $z = 2.50$

   b. $H_0: \mu \ge 80$, $H_a: \mu < 80$, $z = -1.95$

   c. $H_0: \mu = 1200$, $H_a: \mu \ne 1200$, $z = 3.70$

14. Deli Delivery delivers sandwiches to neighboring office buildings during lunch time in New York City. The deli claims that the sandwiches will be delivered within 20 minutes from receiving the order. Given the hectic schedules of their customers, consistent delivery time is a must. The owner has decided that the standard deviation of delivery times should be at most 4 minutes. To determine how consistently the sandwiches are being delivered, the manager randomly selects 27 orders and measures the time from receiving the order to delivery of the sandwich. The average time to delivery of the sample was 20 minutes with a standard deviation of 4.5 minutes.

   a. Will the manager conclude at $\alpha = 0.10$ that the delivery times vary more than the owner desires?

   b. What assumption did you make about the delivery times in performing the test in part **a.**?

15. Consider the following small sample hypothesis tests for the population mean. Compute the P-value for each of the tests and decide whether you would reject or fail to reject the null hypothesis at $\alpha = 0.05$.

   a. $H_0: \mu = 12$, $H_a: \mu > 12$, $t = 1.75$, $n = 25$

   b. $H_0: \mu = 0.12$, $H_a: \mu < 0.12$, $t = -2.95$, $n = 16$

   c. $H_0: \mu = 55$, $H_a: \mu \ne 55$, $t = 2.35$, $n = 8$

16. In each of the following experimental situations, give the appropriate null and alternative hypotheses to be tested. Define all terms that appear in these hypotheses.

   a. A random sample of 100 customers in a bank are selected and their times to be served are noted. The bank has recently retrained its tellers to be more efficient with the hope of decreasing its average time in servicing its customers, which has been 4 minutes in the past.

   b. A local driver training school claims that at least 75% of its pupils pass the driving test on their first attempt. A sample of 60 students from the school are selected, and their performances on the driving test are noted. Based upon the data collected, we would like to refute the claim of the school.

   c. A spokesperson for a popular diet claims that the average weight lost for someone on the diet will be at least 15 pounds over a two-month period. The amount of weight lost for each person in a sample of 10 people on the diet is determined in order to try to refute the claim of the diet spokesperson.

   d. A tire company tests 68 of its new premium tires to determine if the average lifespan of the tire is more than the average lifespan of its major competitor's best tire. The average lifespan of the competitor's tire is 63,000 miles.

   e. An elementary statistics student conducts an experiment in order to show that a coin from a magic kit is biased. The student flips the coin 500 times.

17. It is essential in the manufacture of machinery to utilize parts that conform to specifications. In the past, diameters of the ball bearings produced by a certain manufacturer had a variance of 0.00156. To cut costs, the manufacturer instituted a less expensive production method. The variance of the diameters of 101 randomly sampled bearings produced by the new process was 0.0021. Do the data provide sufficient evidence to indicate that the diameters of ball bearings produced by the new process are more variable than those produced by the old process? Test using $\alpha = 0.10$.

18. A national news magazine is interested in the proportion of counties in which the cost of living has decreased in the past 24 months. The news magazine believes that the true proportion is less than 30%. In a random sample of 100 counties, 20 counties had cost of living decreases.

   a. Test the news magazine's claim at $\alpha = 0.08$.

   b. Find the P-value for this test.

19. An increasing number of businesses are offering child-care benefits for their workers. However, one union claims that more than 90% of firms in the manufacturing sector still do not offer any child-care benefits to their workers. A random sample of 350 manufacturing firms is selected, and only 28 of them offer child-care benefits.

   a. Does this sample result support the claim of the union? Test using $\alpha = 0.10$.

   b. Calculate the P-value associated with this test.

20. During the holiday season, law enforcement officials estimated that 500 people would be killed and 25,000 injured on the nation's roads. They claimed that more than 50% of the accidents would be caused by drunk driving. A sample of 120 accidents showed that 67 were caused by drunk driving. Use these data to test their claim with $\alpha = 0.05$.

21. The quality control supervisor of a cannery is concerned about the variance of fill per can. Regulatory agencies specify that the standard deviation of the amount of fill should be less than 0.1 ounce. To determine whether the process is meeting this specification, the supervisor randomly selects ten cans, weighs the contents of each, and finds that the sample standard deviation of these measurements is 0.04.

a. Do these data provide sufficient evidence to indicate that the variability is as small as desired? Test using $\alpha = 0.05$.

b. Compute a 95% confidence interval for the variance of ounces of fill for the cannery.

Chapter Eleven

Inferences about Means and Proportions: Two Samples

11

Introduction

Basic concepts of hypothesis testing were introduced in Chapter 10. In this chapter we will continue to follow the same basic hypothesis testing procedure developed in the previous chapter. The inferential methods that were previously developed concerned one population and the value of its mean, proportion, or variance. We use the information in the sample statistics $\bar{x}$, $\hat{p}$, and s^2 to test hypotheses about the population parameters.

There are many instances in which we wish to compare two population means or two population proportions. For example, in Chapter 2 we discussed controlled experiments in which we tried to develop data that would be used to efficiently test hypotheses about two or more proportions. In this chapter, we will begin to explore methods for testing population parameters of this type.

In particular, we will develop methods for comparing two population means, and try to answer questions such as:

- Is $\mu_1 > \mu_2$?   - Is $\mu_1 < \mu_2$?   - Is $\mu_1 \neq \mu_2$?

Similarly, we will develop methods for comparing two population proportions, and try to answer questions such as:

- Is $p_1 > p_2$?   - Is $p_1 < p_2$?   - Is $p_1 \neq p_2$?

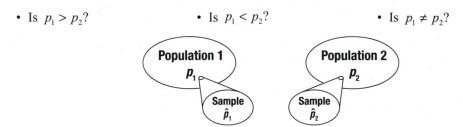

11.1   Comparing Two Population Means

We developed procedures for testing a hypothesis about a single population mean in Section 10.4. In this section, a procedure is developed for comparing two population means when large independent samples are available, or when the samples are drawn from normal populations that have known population standard deviations. Section 11.2 will develop a procedure that is used when the independent samples are small (but have been drawn from normal populations).

There are many situations where our interest is in comparing two population means or the average response of experimental units to two different treatments. For example, a marketing professor may be interested in students' grades when using PowerPoint slides as a delivery method during lecture as opposed to when using the blackboard. A fleet manager may be interested in comparing the average gas mileage for two different makes of cars. An economist may be interested in the difference between costs of living in industrialized and non-industrialized countries.

In each of the above situations, an experiment can be designed for sampling from two separate populations (two different makes of cars or groups of students), or the experiment can be designed for randomly assigning experimental units to two different treatments (students receiving lecture via PowerPoint slides and students receiving lecture via the blackboard).

When observations are randomly selected from two independent populations or experimental units are randomly assigned to two different treatments where the variation between experimental units is small, the sampling design is called an **independent experimental design** or a **completely randomized design**.

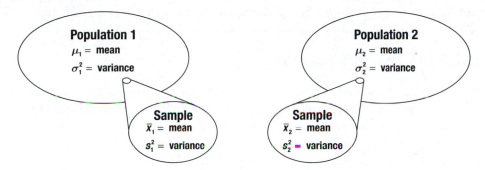

We will use the sample means $\bar{x}_1$ and $\bar{x}_2$ to compare the means of two populations. The sampling distribution for the difference between two sample means, $\bar{x}_1 - \bar{x}_2$, has an approximately normal distribution (when an independent experimental design is used to make comparisons between two population means and the samples drawn from both populations, n_1 and n_2, are "large").

The properties of the sampling distribution for the difference between two sample means are given in the following box.

Properties

Properties of the Sampling Distribution of $\bar{x}_1 - \bar{x}_2$

1. If $n_1 \geq 30$ and $n_2 \geq 30$, the sampling distribution of $\bar{x}_1 - \bar{x}_2$ has an approximately normal distribution.

2. $\mu_{\bar{x}_1 - \bar{x}_2} = \mu_1 - \mu_2$

3. $\sigma_{\bar{x}_1 - \bar{x}_2} = \sqrt{\dfrac{\sigma_1^2}{n_1} + \dfrac{\sigma_2^2}{n_2}}$ if the two samples are independent. σ_1^2 and σ_2^2 can be approximated

   with s_1^2 and s_2^2, respectively, if the population standard deviations are unknown and $n_1 \geq 30$ and $n_2 \geq 30$.

The sampling distribution of $\bar{x}_1 - \bar{x}_2$ will be used in the development of the confidence interval and test statistic for comparing two population means.

Interval Estimation of $\mu_1 - \mu_2$

Given the sampling distribution of $\bar{x}_1 - \bar{x}_2$, we can develop the confidence interval estimate of the difference between two population means.

Formula

$100(1 - \alpha)\%$ Confidence Interval for $\mu_1 - \mu_2$

The $100(1 - \alpha)\%$ confidence interval estimate for the difference in the population means of two independent populations is given by

$$(\bar{x}_1 - \bar{x}_2) \pm z_{\alpha/2} \sqrt{\frac{\sigma_1^2}{n_1} + \frac{\sigma_2^2}{n_2}}$$

where $z_{\alpha/2}$ is the critical value of the standard normal distribution with an area of $\alpha/2$ in the upper tail. σ_1^2 and σ_2^2 can be approximated with s_1^2 and s_2^2, respectively, if the population standard deviations are unknown and the independent samples are large, i.e. $n_1 \geq 30$ and $n_2 \geq 30$.

Hypothesis Testing about $\mu_1 - \mu_2$

You may recall that when comparing the means of two independent populations, you take a random sample of size n_1 from the first population and a random sample of size n_2 from the second population. If both n_1 and n_2 are greater than or equal to 30, then the following test statistic is valid. If one or both of the sample sizes are less than 30, the test is also valid if we assume that the standard deviations of both populations $(\sigma_1$ and $\sigma_2)$ are known, and both populations are normally distributed.

Formula

Test Statistic for a Hypothesis Test about $\mu_1 - \mu_2$

The test statistic for a hypothesis test about $\mu_1 - \mu_2$ is given by

$$z = \frac{(\bar{x}_1 - \bar{x}_2) - (\mu_1 - \mu_2)}{\sqrt{\frac{\sigma_1^2}{n_1} + \frac{\sigma_2^2}{n_2}}}$$

where $\bar{x}_1$ is the sample mean for Population 1, n_1 is the sample size for Population 1, σ_1 is the standard deviation of Population 1, $\bar{x}_2$ is the sample mean for Population 2, n_2 is the sample size for Population 2, and σ_2 is the standard deviation of Population 2. If n_1 and n_2 are greater than or equal to 30, s_1 and s_2 can be used as approximations for σ_1 and σ_2 if the population standard deviations are unknown.

Note that $(\mu_1 - \mu_2)$ represents the hypothesized difference between the two population means and that the z-statistic follows the standard normal distribution.

Often, the goal of a hypothesis test for the difference between two population means is to determine if the population means are different. In this case, the hypothesized difference between the population means will be 0. When the hypothesized difference between the population means is zero, the formula for the test statistic is simplified to

$$z = \frac{\bar{x}_1 - \bar{x}_2}{\sqrt{\dfrac{\sigma_1^2}{n_1} + \dfrac{\sigma_2^2}{n_2}}}.$$

Constructing a confidence interval and performing a hypothesis test for two population means will be developed in Example 11.1.

Example 11.1

A telecommunications analyst is interested in knowing if there is a significant difference in the average quality of service (QOS) between cable television subscribers and satellite television subscribers. She randomly selects 50 cable subscribers and 50 satellite subscribers for the study. She gives each subscriber a survey and asks them to complete it. She tallies the results of the survey and calculates a mean QOS for each service. The QOS is ranked on a scale of 1–10, with the higher value implying that subscribers are more satisfied with their service. The results of the study are shown in Table 11.1.

Table 11.1 – QOS Scores for Subscribers

	n	$\bar{x}$	s
Cable	50	8.8	3
Satellite	50	9.5	2

a. Calculate a 95% confidence interval for the mean difference in average QOS between cable and satellite subscribers.

b. Is there persuasive evidence for the analyst to conclude at $\alpha = 0.05$ that there is a difference in mean QOS between cable and satellite subscribers?

Solution

a. From the information given in the problem, we know the following.

$$n_1 = 50, \ n_2 = 50, \ s_1 = 3, \ s_2 = 2, \ \bar{x}_1 = 8.8, \text{ and } \bar{x}_2 = 9.5$$

Since both sample sizes are at least 30, we can invoke the central limit theorem and assume that the sample means are approximately normally distributed. In addition, we can use the sample standard deviations to estimate the population standard deviations.

Since we want a 95% confidence interval, we need to find the value of $z_{\alpha/2}$. Remember that $z_{\alpha/2}$ represents the z-value required to obtain an area of $(1-\alpha)$ centered under the standard normal curve. Therefore, for a 95% confidence interval, $\alpha = 0.05$.

For $\alpha = 0.05$, $z_{\alpha/2} = z_{0.025} = 1.96$.

A 95% confidence interval for the difference in the average QOS between cable and satellite providers is given by

$$\left(\bar{x}_1 - \bar{x}_2\right) \pm z_{\alpha/2}\sqrt{\frac{s_1^2}{n_1} + \frac{s_2^2}{n_2}}.$$

since both n_1 and n_2 are greater than or equal to 30.

This gives,

$$(8.8 - 9.5) \pm 1.96\sqrt{\frac{3^2}{50} + \frac{2^2}{50}}$$

$$-0.7 \pm 1.96\sqrt{\frac{13}{50}}$$

$$-1.7 \text{ to } 0.3.$$

Thus, we are 95% confident that the true mean difference in QOS between cable and satellite subscribers is between −1.7 and 0.3. Another way to interpret this interval in terms of the problem is that the average QOS for satellite subscribers is between 1.7 rating points higher and 0.3 rating points lower than cable subscribers. Since the confidence interval includes zero, this indicates that the difference in QOS between cable and satellite subscribers is not statistically significant at the 0.05 level.

b.   Step 1: State the hypotheses in plain English.

- **Null Hypothesis:** There is no difference in average QOS between cable and satellite subscribers.

- **Alternative Hypothesis:** There is a difference in average QOS between cable and satellite subscribers.

Step 2: Select the appropriate statistical measure.

Since the analyst is interested in comparing the average QOS between cable and satellite subscribers, the appropriate statistical measures are the following.

$$\mu_1 = \text{the true mean QOS for cable subscribers}$$
$$\mu_2 = \text{the true mean QOS for satellite subscribers}$$

Step 3: Determine whether the hypothesis should be one-sided or two-sided.

The analyst is interested in whether or not there is a difference in the QOS between cable and satellite subscribers. Thus, the alternative hypothesis is two-sided and the test is two-tailed.

Step 4: State the hypotheses using the appropriate statistical measure.

Since the analyst is interested in comparing the two population means, a natural way to make this comparison is to look at the difference between the two population means. Hence the null and alternative hypotheses are formulated as follows.

$$H_0: \mu_1 - \mu_2 = 0 \quad \text{There is no difference in average QOS.}$$
$$H_a: \mu_1 - \mu_2 \neq 0 \quad \text{There is a difference in average QOS.}$$

In this situation, the hypothesized difference in average QOS is 0. However, it is possible to be interested in nonzero differences. For example, if the analyst had been interested in testing if the difference in average QOS was one unit, the hypothesized difference would have been one.

Step 5: Specify the level of the test.

The level of the test is specified in the problem as $\alpha = 0.05$.

Step 6: Select the appropriate test statistic.

The difference in sample means, $\bar{x}_1 - \bar{x}_2$, is an unbiased point estimate of the difference in population means, $\mu_1 - \mu_2$. The sampling distribution for the point estimate, $\bar{x}_1 - \bar{x}_2$, provides essential information for determining the test statistic. If $n_1 \geq 30$ and $n_2 \geq 30$ and an independent experimental design is used, the sampling distribution of $\bar{x}_1 - \bar{x}_2$ has an approximately normal distribution with mean, $\mu_{\bar{x}_1-\bar{x}_2} = \mu_1 - \mu_2$, and standard deviation,

$$\sigma_{\bar{x}_1-\bar{x}_2} = \sqrt{\frac{\sigma_1^2}{n_1} + \frac{\sigma_2^2}{n_2}}.$$

Thus, the test statistic is a standard normal random variable which is calculated by using the familiar z-transformation. The test statistic is given by

$$z = \frac{(\bar{x}_1 - \bar{x}_2) - \mu_{\bar{x}_1-\bar{x}_2}}{\sigma_{\bar{x}_1-\bar{x}_2}} = \frac{(\bar{x}_1 - \bar{x}_2) - \overbrace{(\mu_1 - \mu_2)}^{\substack{\text{Hypothesized} \\ \text{Difference} \\ \text{in Means}}}}{\sqrt{\frac{\sigma_1^2}{n_1} + \frac{\sigma_2^2}{n_2}}},$$

where σ_1 and σ_2 can be approximated by s_1 and s_2 when $n_1 \geq 30$ and $n_2 \geq 30$.

If the null hypothesis is true, z has an approximately normal distribution. If the observed value of $\bar{x}_1 - \bar{x}_2$ is significantly larger or smaller than $\mu_1 - \mu_2$, the hypothesized value of the difference, this will produce a large or small value of the test statistic, causing us to doubt whether or not the null hypothesis is in fact true. How large is *large*? This is answered by the critical value of the test statistic specified in Step 7.

Step 7: Determine the critical value.

The role of the critical value(s) in this test is exactly the same as for all of the hypothesis tests discussed earlier. It defines a range of values for the test statistic, the rejection region, that will be so rare that it is unlikely that the test statistic occurred from ordinary sampling variability, assuming the null is true. The level of the test defines the size of the rejection region. Should the computed value of the test statistic fall within the rejection region, the null hypothesis will be rejected.

If the null hypothesis is true, the test statistic has an approximately normal distribution. Thus the critical value is determined in the same way as for the other tests of hypothesis in which the test statistic had an approximately normal distribution. The rejection region for a two-tailed test with level of significance, $\alpha = 0.05$, is displayed in Figure 11.1.

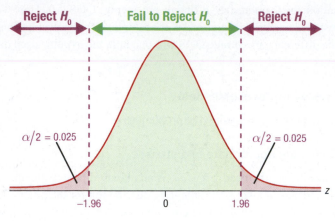

Figure 11.1

Hypothesis Testing Is Not Loved by All Statisticians

The view is expressed in an article by Marks R. Nester entitled "An Applied Statistician's Creed," *Applied Statistics*, 45: 401-410.

"I contend that the general acceptance of statistical hypothesis testing is one of the most unfortunate aspects of 20th century applied science. Tests for the identity of population distributions, for equality of treatment means, for presence of interactions, for the nullity of a correlation coefficient, and so on, have been responsible for much bad science, much lazy science, and much silly science. A good scientist can manage with, and will not be misled by, parameter estimates and their associated standard errors or confidence limits. A theory dealing with the statistical behavior of populations should be supported by rational argument as well as data. In such cases, accurate statistical evaluation of the data is hindered by null hypothesis testing. The scientist must always give due thought to the statistical analysis, but must never let statistical analysis be a substitute for thinking!"

The null hypothesis will be rejected if the computed value of the test statistic is greater than or equal to 1.96 or less than or equal to -1.96. In other words, we will reject the null hypothesis if the observed difference in the average QOS scores is at least 1.96 standard deviations above or below the hypothesized value of 0 (no difference in QOS between cable and satellite).

Step 8: Compute the test statistic.

Based on the data in Table 11.1, the computed value of the test statistic is given by

$$z = \frac{(8.8 - 9.5) - 0}{\sqrt{\dfrac{3^2}{50} + \dfrac{2^2}{50}}} \approx -1.37.$$

Step 9: Make the decision.

Figure 11.2

As shown in Figure 11.2, the calculated value of the test statistic does not fall in the rejection region because -1.37 falls between the critical values -1.96 and 1.96. There is insufficient evidence to conclude that the difference between the observed value and the hypothesized value is due to anything other than ordinary sampling variation. Thus, we fail to reject the null hypothesis at $\alpha = 0.05$.

Step 10: State the conclusion in terms of the original question.

There is not sufficient evidence at the 0.05 level to conclude that the average QOS is significantly different between cable and satellite television subscribers.

11.1   Exercises

Basic Concepts

1. What questions are we interested in answering when comparing two population means?

2. What is an independent experimental design?

3. Which sampling distribution do we use in the formulation of the test statistic when comparing two population means with large samples? What are the properties of this distribution?

4. Does the determination of the critical value(s) for two-sample hypothesis tests differ from one-sample hypothesis tests?

5. What conditions are necessary to perform a large sample test for the difference between two population means?

Exercises

6. Determine the critical value(s) of the test statistic for each of the following large sample tests for the comparison of two population means.

    a. Left-tailed test, $\alpha = 0.05$

    b. Right-tailed test, $\alpha = 0.10$

    c. Two-tailed test, $\alpha = 0.01$

7. Determine the critical value(s) of the test statistic for each of the following large sample tests for the comparison of two population means.

    a. Left-tailed test, $\alpha = 0.04$

    b. Right-tailed test, $\alpha = 0.08$

    c. Two-tailed test, $\alpha = 0.02$

8. A luxury car dealer is considering two possible locations for a new auto mall. The rent on the south side of town is cheaper. However, the dealer believes that the average household income is significantly higher on the north side of town. The dealer has decided that he will locate the new auto mall on the north side of town if the results of a study that he commissioned show that the average household income is significantly higher on the north side of town. The results of the study are as follows.

Table for Exercise 8 – Income (Thousands of Dollars)			
	n	$\bar{x}$	s
North Side	35	50	10
South Side	40	43	5

    a. Calculate a 90% confidence interval for the difference in average income between the north and south sides of town. Interpret the interval.

    b. Based on the study, will the auto dealer decide to locate the new auto mall on the north side of town? Use $\alpha = 0.05$.

9. An internal auditor for Tiger Enterprises has been asked to determine if there is a difference in the average amount charged for daily expenses by two top salesmen, Mr. Ellis and Mr. Ford. The auditor randomly selects 45 days and determines the daily expenses for each of the salesmen.

Table for Exercise 9 – Expenses (Dollars)			
	n	$\bar{x}$	s
Mr. Ellis	45	$55	$8
Mr. Ford	45	$60	$3

a. Calculate a 95% confidence interval for the difference in the average amounts charged for daily expenses between Mr. Ellis and Mr. Ford. Interpret the interval.

b. Based on the survey, can the auditor conclude that there is a difference in the average amounts charged for daily expenses by the two top salesmen? Use $\alpha = 0.05$.

c. Explain how the 95% confidence interval in part **a.** would lead you to make the same decision that was made in part **b.**

10. The military has two different programs for training aircraft personnel. A government regulatory agency has been commissioned to evaluate any differences that may exist between the two programs. The agency administers standardized tests to randomly selected groups of students from the two programs. The results of the tests for the students in each of the programs are as follows.

Table for Exercise 10 – Military Training Programs			
	n	$\bar{x}$	s
Program A	50	85	10
Program B	55	87	9

a. Calculate a 99% confidence interval for the difference between the average scores of the two military programs. Interpret the interval.

b. Can the agency conclude that there is a difference in the average test scores of students in the two programs? Use $\alpha = 0.01$.

11. Tom Sealack, a supply clerk with the Navy, has been asked to determine if a new battery that has been offered to the Navy (at a reduced price) has a shorter average life than the battery they are currently using. He randomly selects batteries of each type and allows them to run continuously so that he can measure the time until failure for each battery. The results of the test are as follows.

Table for Exercise 11 – Battery Life (Hours)			
	n	$\bar{x}$	s
New Battery	35	700	30
Old Battery	35	710	35

a. Do the data suggest at $\alpha = 0.10$ that the time until failure for the new battery is significantly less than the time until failure for the old battery?

b. Calculate the P-value for the test in **a.**

c. Based on the P-value, would the decision change at $\alpha = 0.05$?

12. The City Bank believes that checking account balances are significantly larger for customers who are aged 40 to 49 than those who are aged 30 to 39. To investigate this belief, they randomly select customers from each age group and determine the average daily account balance for each customer for the current month. The results of the study are as follows.

Table for Exercise 12 – Checking Account Balances			
Age Group	n	$\bar{x}$	s
30 – 39	200	$2500	$550
40 – 49	150	$3500	$950

a. Do the data suggest at $\alpha = 0.05$ that the average daily account balances are significantly higher for the 40 to 49 age group than the 30 to 39 age group?

b. Calculate the P-value for the test in **a**.

c. Based on the P-value, would the decision change at $\alpha = 0.10$?

11.2 Comparing Two Population Means: Small Samples, σ_1 and σ_2 Unknown

Is it still possible to make comparisons between two population means if the population standard deviations are unknown and we do not have "large samples"? There are many situations where it is not practical from either a monetary or time perspective to take such samples.

It is still possible to compare the two population means even if large samples are not available. But in order for the confidence interval or hypothesis test to be valid, there are several assumptions that must be met. These assumptions are outlined below.

Assumptions

Assumptions for Small Sample Inferences about $\mu_1 - \mu_2$

1. An independent experimental design is used.

2. Both populations of interest are approximately normal.

3. Both of the populations have approximately equal (but unknown) variances, $\sigma_1^2 = \sigma_2^2 = \sigma^2$.

In order to determine if both populations of interest are approximately normal, it is helpful to draw histograms of the sample observations from each population. If these histograms appear to be approximately normal, then it is reasonable to infer this assumption is satisfied. With limited data, it is sometimes difficult to determine if the sample data are from a normal population. In these situations, you may have to assume normality and recognize that your inferences are predicated on the validity of the assumption. Figure 11.3 shows three histograms of sample data drawn from normal populations.

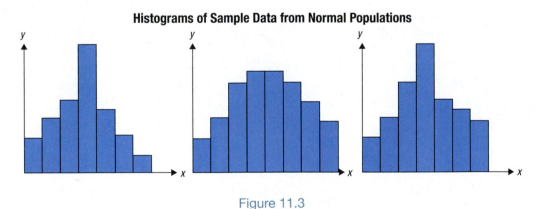

Histograms of Sample Data from Normal Populations

Figure 11.3

To examine the equal variance assumption, we can compare the variances of the samples drawn from the populations. In addition to comparing the sample variances we can draw box plots of the data from each of the distributions and compare the spread of the box plots. If the box plots look approximately the same, it is reasonable to assume that the variances of the two distributions are approximately equal. Figure 11.4 displays two box plots of sample data from populations with approximately equal variances, but with slightly different means.

Figure 11.4

Small Sample Interval Estimation of $\mu_1 - \mu_2$ (σ_1 and σ_2 Unknown)

We can use an interval estimate to determine the difference in two independent population means. In Section 11.1 we used the following interval estimate for the case when samples were drawn from normal populations and σ_1 and σ_2 were known or when the sample sizes n_1 and n_2 were greater than or equal to 30.

$$\left(\bar{x}_1 - \bar{x}_2\right) \pm z_{\alpha/2} \sqrt{\frac{\sigma_1^2}{n_1} + \frac{\sigma_2^2}{n_2}}$$

Assuming that the standard deviations, σ_1 and σ_2 are unknown, we will use the sample standard deviations, s_1 and s_2 to estimate the population standard deviations. However, since we are assuming that the variances are equal, we can estimate the standard error by

$$\sqrt{s_p^2 \left(\frac{1}{n_1} + \frac{1}{n_2}\right)}$$

where $s_p^2 = \dfrac{(n_1 - 1)s_1^2 + (n_2 - 1)s_2^2}{n_1 + n_2 - 2}$, and is called the **pooled variance**. In addition, we will assume that the data follow a normal distribution and thus, replace $z_{\alpha/2}$ with $t_{\alpha/2, df}$ since the samples are not sufficiently large.

Formula

Small Sample $100(1 - \alpha)\%$ Confidence Interval for $\mu_1 - \mu_2$ Assuming Equal Variances

Assuming equal variances, the $100(1 - \alpha)\%$ interval estimate for the difference between two population means if sample sizes are small and σ_1 and σ_2 are unknown is given by

$$\left(\bar{x}_1 - \bar{x}_2\right) \pm t_{\alpha/2, df} \sqrt{s_p^2 \left(\frac{1}{n_1} + \frac{1}{n_2}\right)}$$

where $s_p^2 = \dfrac{(n_1 - 1)s_1^2 + (n_2 - 1)s_2^2}{n_1 + n_2 - 2}$ is the pooled variance and $t_{\alpha/2, df}$ is the critical value of the t-distribution with $n_1 + n_2 - 2$ degrees of freedom, capturing an area of $\alpha/2$ in the upper tail.

11

Small Sample Hypothesis Test about $\mu_1 - \mu_2$ (σ_1 and σ_2 Unknown)

Providing the assumptions previously outlined for small sample inferences have been met, the test procedure developed in Example 11.2 may be used for making comparisons between two population means.

Example 11.2

For a consumer product, the mean dollar sales per retail outlet last year in a sample of 15 stores was \$3425 with a standard deviation of \$200. For a second product, the mean dollar sales per outlet in a sample of 16 stores was \$3250 with a standard deviation of \$175. The sales amounts per outlet are assumed to be approximately normally distributed for both products.

Table 11.2 – Retail Sales (Dollars)			
	n	$\bar{x}$	s
Product 1	15	\$3425	200
Product 2	16	\$3250	175

a. Calculate a 95% confidence interval for the difference in average dollar sales between the two products.

b. Test to see if Product 1 has a higher mean dollar sales record than Product 2. Use $\alpha = 0.01$.

Solution

a. From the information given in the problem, we know that

$$n_1 = 15, \; n_2 = 16, \; s_1 = 200, \; s_2 = 175, \; \bar{x}_1 = \$3425, \text{ and } \bar{x}_2 = \$3250.$$

With σ_1 and σ_2 unknown and small samples (both n_1 and n_2 are less than 30), we assume that the data follow a normal distribution and that the population variances are equal, and estimate the population standard deviation with the sample standard deviations. Since we want a 95% confidence interval, we need to find the value of $t_{\alpha/2,df}$. Note that $df = n_1 + n_2 - 2 = 15 + 16 - 2 = 29$. Thus, $t_{0.025,29} = 2.045$.

A 95% confidence interval for the difference in average dollar sales between the two products is given by

$$\left(\bar{x}_1 - \bar{x}_2\right) \pm t_{\alpha/2,df} \sqrt{s_p^2\left(\frac{1}{n_1} + \frac{1}{n_2}\right)} \text{ where } s_p^2 = \frac{\left(n_1 - 1\right)s_1^2 + \left(n_2 - 1\right)s_2^2}{n_1 + n_2 - 2}.$$

First we need to calculate s_p^2.

$$s_p^2 = \frac{\left(15 - 1\right)200^2 + \left(16 - 1\right)175^2}{15 + 16 - 2} \approx 35150.8621$$

The confidence interval is then calculated as follows.

$$(3425 - 3250) \pm 2.045 \sqrt{35150.8621\left(\frac{1}{15} + \frac{1}{16}\right)}$$

$$175 \pm 2.045 \sqrt{35150.8621\left(\frac{1}{15} + \frac{1}{16}\right)}$$

$$37.20 \text{ to } 312.80$$

Thus, we are 95% confident that the true mean difference in dollar sales between Product 1 and Product 2 is between $37.20 and $312.80. That is, on average, Product 1 generates between $37.20 and $312.80 more dollar sales than Product 2.

b. **Step 1:** Define the hypotheses in plain English.

- **Null Hypothesis:** The average retail sales for Product 1 are less than or equal to the average retail sales for Product 2.

- **Alternative Hypothesis:** The average retail sales for Product 1 are greater than the average retail sales for Product 2.

Step 2: Select the appropriate statistical measure.

Since the retailer is interested in comparing the average retail sales between the two products, the appropriate statistical measures are:

$$\mu_1 = \text{the true mean retail sales of Product 1}$$
$$\mu_2 = \text{the true mean retail sales of Product 2.}$$

Step 3: Determine whether the hypothesis should be one-sided or two-sided.

The retailer is interested in whether the average sales of Product 1 are higher than the average sales of Product 2. Thus, the alternative hypothesis is one-sided and the test is one-tailed.

Step 4: State the hypotheses using the appropriate statistical measure.

The retailer is interested in determining if the average sales for Product 1 are higher than that of Product 2. One way to state this in terms of the statistical measures in Step 2 is $\mu_1 > \mu_2$. But in order to perform the hypothesis test, this statement must be rewritten in a form for which we have a point estimate, or $\mu_1 - \mu_2 > 0$. The resulting hypotheses are as follows.

$$H_0: \mu_1 - \mu_2 \leq 0 \quad \left(\begin{array}{l}\text{Average sales for Product 1 are less than} \\ \text{or equal to the average sales for Product 2.}\end{array}\right)$$

$$H_a: \mu_1 - \mu_2 > 0 \quad \left(\begin{array}{l}\text{Average sales for Product 1 are greater} \\ \text{than the average sales for Product 2.}\end{array}\right)$$

Step 5: Specify the level of the test.

The level of the test is specified in the problem as $\alpha = 0.01$.

Step 6: Select the appropriate test statistic.

To develop the appropriate test statistic, a random variable whose value will be used to make the decision to reject or fail to reject H_0 must be found. The sampling distribution $\bar{x}_1 - \bar{x}_2$ provides essential information in assessing the rareness of the test statistic. If

Note:
$\mu_1 - \mu_2$ in the t-test statistic is the hypothesized difference in means. In most problems this difference is hypothesized to be 0.

$n_1 < 30$ or $n_2 < 30$ and the assumptions previously outlined are satisfied, the sampling distribution of $\bar{x}_1 - \bar{x}_2$ has a t-distribution. The t-test statistic is given by

$$t = \frac{(\bar{x}_1 - \bar{x}_2) - (\mu_1 - \mu_2)}{\sqrt{s_p^2 \left(\dfrac{1}{n_1} + \dfrac{1}{n_2} \right)}},$$

where $s_p^2 = \dfrac{(n_1 - 1)s_1^2 + (n_2 - 1)s_2^2}{n_1 + n_2 - 2}$, and is the **pooled variance**.

In this hypothesis test, the population variances are assumed to be equal. Thus s_p^2, which is the weighted average of the two sample variances, is used to approximate the unknown population variance.

If the null hypothesis is true, t has a t-distribution with $n_1 + n_2 - 2$ degrees of freedom. If the observed value of $\bar{x}_1 - \bar{x}_2$ is significantly larger than the hypothesized value of the difference $(H_0 : \mu_1 - \mu_2 \leq 0)$, this will produce a large value of the test statistic, causing us to doubt whether the null hypothesis is in fact true. How large is *large*? The critical value of the test statistic specified in Step 7 will implicitly define the notion of "large."

Step 7: Determine the critical value.

The role of the critical value in this test is exactly the same as for all of the hypothesis tests discussed earlier. It defines a range of values for the test statistic, the rejection region, that will be so rare that it is unlikely that it occurred from ordinary sampling variability if the null hypothesis is true. The level of the test defines the size of the rejection region. Should the computed value of the test statistic fall in the rejection region, we will presume that the value of the test statistic is too large to have occurred from ordinary sampling variation, and the null hypothesis will be rejected.

If the null hypothesis is true, the test statistic has a t-distribution. Thus, the critical value is determined in the same way as for the other tests of hypothesis where the test statistic had a t-distribution, except that the degrees of freedom are $n_1 + n_2 - 2$, or $(15 + 16 - 2) = 29$. The rejection region corresponding to the alternative hypothesis, $H_a : \mu_1 - \mu_2 > 0$, with $\alpha = 0.01$ for a t-test statistic with 29 degrees of freedom is given in Figure 11.5.

t-Distribution, $df = 29$

Figure 11.5

We will reject the null hypothesis if the computed value of the test statistic is greater than or equal to 2.462.

Step 8: Compute the test statistic.

Based on the data in Table 11.2, the computed value of the test statistic is given by

$$t = \frac{(3425 - 3250) - 0}{\sqrt{(35150.8621)\left(\frac{1}{15} + \frac{1}{16}\right)}} \approx 2.597$$

$$\text{where } s_p^2 = \frac{(14)(200)^2 + (15)(175)^2}{15 + 16 - 2} \approx 35150.8621.$$

Step 9: Make the decision.

Figure 11.6

As displayed in Figure 11.6, the value of the test statistic does fall in the rejection region. In fact, a test statistic of 2.597 says that the observed difference between the means is about 2.6 standard deviations higher than the hypothesized difference of zero. It is highly unlikely that the difference between the observed value and the hypothesized value is due to ordinary sampling variation. Thus, we reject the null hypothesis at $\alpha = 0.01$.

Step 10: State the conclusion in terms of the original question.

There is sufficient evidence at the 0.01 level to conclude that the average sales for Product 1 are significantly higher than the average sales of Product 2.

t-Test about $\mu_1 - \mu_2$ Assuming Unequal Variances

Recall that one of the assumptions when performing the two-sample *t*-test is that the population variances are unknown but equal $\left(\text{that is, } \sigma_1^2 = \sigma_2^2 = \sigma^2\right)$. If you cannot make the assumption that the variances are equal, then it is not appropriate to pool the two sample variances into one common variance, s_p^2. Instead, you must use a separate variance test developed by Satterthwaite (Satterthwaite, F.E., "An Approximate Distribution for Estimates of Variance Components," *Biometrics Bulletin*, 2 (1946): 110-114.). Satterthwaite's test procedure uses a series of computations that involve using the two separate sample variances to calculate the degrees of freedom for the test statistic. The computations for the $100(1-\alpha)\%$ confidence interval and test statistic are as follows.

Formula

Inferences about $\mu_1 - \mu_2$ Assuming Unequal Variances

Confidence Interval:

$$\left(\bar{x}_1 - \bar{x}_2\right) \pm t_{\alpha/2, df} \sqrt{\frac{s_1^2}{n_1} + \frac{s_2^2}{n_2}}$$

Test Statistic:

$$t = \frac{\left(\bar{x}_1 - \bar{x}_2\right) - \left(\mu_1 - \mu_2\right)}{\sqrt{\dfrac{s_1^2}{n_1} + \dfrac{s_2^2}{n_2}}},$$

which follows a *t*-distribution with degrees of freedom equal to:

$$df = \frac{\left(\dfrac{s_1^2}{n_1} + \dfrac{s_2^2}{n_2}\right)^2}{\dfrac{1}{n_1 - 1}\left(\dfrac{s_1^2}{n_1}\right)^2 + \dfrac{1}{n_2 - 1}\left(\dfrac{s_2^2}{n_2}\right)^2}.$$

The calculation for *df* should be rounded down to the nearest integer. Remember, these formulas should only be used if σ_1 and σ_2 are unknown, the sample sizes n_1 and n_2 are less than 30 and drawn from normal populations, and the population variances are assumed to be unequal.

11.2   Exercises

Basic Concepts

1. Why might large samples not be available when attempting to make inferences about two population means?

2. What assumptions are necessary to perform a small sample test for the difference between two population means?

3. What constitutes a "small sample"?

4. What is the test statistic for a small sample hypothesis test about two population means? How does this statistic differ from the test statistic used for large samples?

5. What is a pooled variance? Why is it used?

Exercises

6. Determine the critical value(s) of the test statistic for each of the following small sample tests for the comparison of two population means where the assumptions of normality and equal variance have been satisfied.

   a. Left-tailed test, $\alpha = 0.05$, $n_1 = 10$, $n_2 = 15$

   b. Right-tailed test, $\alpha = 0.10$, $n_1 = 8$, $n_2 = 12$

   c. Two-tailed test, $\alpha = 0.01$, $n_1 = 5$, $n_2 = 7$

7. Determine the critical value(s) of the test statistic for each of the following small sample tests for the comparison of two population means where the assumptions of normality and equal variance have been satisfied.

   a. Left-tailed test, $\alpha = 0.025$, $n_1 = 13$, $n_2 = 25$

   b. Right-tailed test, $\alpha = 0.005$, $n_1 = 7$, $n_2 = 18$

   c. Two-tailed test, $\alpha = 0.10$, $n_1 = 15$, $n_2 = 15$

8. *Popular Science* (Vol. 242, No. 3) reported the results of a comparison of several popular minivans. One of the features that they compared was the time required to accelerate from 0 to 60 miles per hour in seconds. The Dodge Grand Caravan ES was able to accelerate from 0 to 60 mph in 11.3 seconds, on average. The Volkswagen Eurovan took 16.5 seconds on average to accelerate from 0 to 60 mph. Suppose that 15 minivans of each type were tested and that the sample standard deviation of the times required to accelerate from 0 to 60 for each minivan was 4 seconds. Assume that the population variances are approximately equal.

   a. Calculate a 95% confidence interval for the difference in average acceleration time between the two types of minivans. Interpret the interval.

   b. Do the data suggest that there is a significant difference in the time required to accelerate from 0 to 60 between the two types of minivans at $\alpha = 0.05$?

   c. What assumptions did you make about the time required to accelerate from 0 to 60 mph in calculating the confidence interval in part **a.** and for performing the test in part **b.**?

9. A cereal manufacturer has advertised that its product, Fiber Oat Flakes, has a lower fat content than its competitor, Bran Flakes Plus. Because of complaints from the manufacturers of Bran Flakes Plus, the FDA has decided to test the claim that Fiber Oat Flakes has a lower average fat content than Bran Flakes Plus. Several boxes of each cereal are selected and the fat content per serving is measured. The results of the study are as follows. Assume that the population variances are approximately equal.

Table for Exercise 9 – Fat Content (Grams)			
	n	$\bar{x}$	s
Fiber Oat Flakes	16	5	1
Bran Flakes Plus	15	6	2

a. Calculate a 90% confidence interval for the difference in average fat content between Fiber Oat Flakes and Bran Flakes Plus. Interpret the interval.

b. Does the study performed by the FDA substantiate the claim made by the manufacturer of Fiber Oat Flakes at $\alpha = 0.10$?

c. What assumptions must be made in order to calculate the confidence interval in part **a.** and perform the hypothesis test in part **b.**?

10. A large construction company would like to expand its operations into a new geographic area. The company has narrowed the choice of locations down to two cities. A major consideration in deciding between the two cities will be the average hourly wage they must pay for general laborers. The company randomly selects laborers from each city and determines their hourly wage with the following results. Assume that the population variances are approximately equal.

Table for Exercise 10 – Hourly Wages (Dollars)			
	n	$\bar{x}$	s
City A	20	$7	$3
City B	20	$8	$2

a. Calculate a 99% confidence interval for the difference in average hourly wage between City A and City B. Interpret the interval.

b. Do the data indicate that there is a significant difference in hourly wages at $\alpha = 0.05$?

c. Calculate the P-value for the test performed in part **b.**

d. What assumptions must be made in order to calculate the confidence interval in part **a.** and perform the hypothesis test in part **b.**?

11. A Hollywood studio believes that a movie that is considered a drama will draw a larger crowd on average than a movie that is a comedy. To test this theory, the studio randomly selects several movies that are classified as dramas and several movies that are classified as comedies and determines the box office revenue for each movie. The results of the survey are as follows. Assume that the population variances are approximately equal.

Table for Exercise 11 – Box Office Revenues (Millions of Dollars)			
	n	$\bar{x}$	s
Drama	15	180	50
Comedy	13	150	30

a. Calculate a 95% confidence interval for the difference in average revenue at the box office for drama and comedy movies. Interpret the interval.

b. Do the data substantiate the studio's belief that dramas will draw a larger crowd on average than comedies at $\alpha = 0.01$?

c. Calculate the P-value for the test you conducted in part **b.**

d. What assumptions must be made in order to calculate the confidence interval in part **a.** and to perform the hypothesis test in part **b.**?

12. *Consumer Magazine* is reviewing the top of the line amplifiers produced by two major stereo manufacturers. One of the most important qualities of the amplifiers is the maximum power output. Brand A has redone their internal design and claims to have a higher maximum power level than Brand B. To test this claim, *Consumer Magazine* randomly selects amplifiers from each brand and determines the maximum power output. The results of the test are as follows. Assume that the population variances are approximately equal.

Table for Exercise 12 – Amplifier Power Output (Watts)			
	n	$\bar{x}$	s
Brand A	12	800	25
Brand B	10	780	25

a. What assumptions must be made in order to perform the hypothesis test?

b. Do the data substantiate the claim that the Brand A amplifier has a higher average maximum power output than Brand B at $\alpha = 0.05$?

13. The State Environmental Board wants to compare pollution levels in two of its major cities. Sunshine City thrives on the tourist industry and Service City thrives on the service industry. The environmental board randomly selects several areas within the cities and measures the pollution levels in parts per million with the following results. Assume that the population variances are approximately equal.

Table for Exercise 13 – Pollution Levels (ppm)			
	n	$\bar{x}$	s
Sunshine City	15	8.5	0.57
Service City	10	7.9	0.50

a. What assumptions must be made in order to perform a hypothesis test for the difference between these two population means?

b. Will the State Environmental Board conclude at $\alpha = 0.01$ that Service City has a lower pollution level on average than Sunshine City?

c. Repeat part **b.**, assuming that the population variances are not equal.

d. Compare the results of part **b.** and part **c.**

14. In 2009 U.S. charitable giving fell 3.6 percent to $303.75 billion for the year. Total charitable contributions from American individuals, corporations, and foundations fell to $303.75 billion from $315.08 billion for 2008. The largest share of contributions went to religious organizations, representing 33 percent of total giving. The next largest shares went to educational organizations, receiving an estimated 13 percent of the total, and foundations, which received 10 percent of the total. Suppose a sample of 6 employees is randomly chosen from a large corporation and their charitable contributions in 2008 and 2009 are determined. The following table gives these amounts (in dollars). Assume that the population variances are approximately equal.

Table for Exercise 14 – Charitable Contributions ($)	
Giving in 2008	Giving in 2009
232	215
150	125
50	50
400	350
325	210
175	150

Source: Giving USA Foundation, the Center on Philanthropy at Indiana University.

a. Can we conclude with $\alpha = 0.01$ that the average contribution to charity has decreased in this corporation from 2008 to 2009?

b. Give the assumptions for your test.

c. Repeat part **a.**, assuming that the population variances are not equal.

d. Compare the results of part **a.** and part **c.**

11.3   Paired Difference

Suppose we are interested in comparing the durability of the soles of two brands of tennis shoes, Spikes and Kickers. One approach to making this comparison is the independent experimental design discussed in Section 11.1. Using this design one may randomly select 10 people to wear the Spikes brand of shoes for six months and then randomly select 10 other people to wear the Kickers brand of shoes for six months. After the six-month period, the average wear for the 10 pairs of Spikes shoes is measured and compared to the average wear for the 10 pairs of Kickers tennis shoes.

While this is a perfectly reasonable approach, it does have shortcomings. For example, what if one of the people selected to wear the Spikes brand of shoes is a cross-country runner? Certainly the wear on the runner's pair of shoes will be much greater than the wear on the shoes for someone who just wears the shoes in the evenings after work. Certainly many factors could have a large effect on the observed wear of the tennis shoes.

What can we do to help reduce the effect of factors that cloud the issue of tennis shoe durability? One approach is to have each person wear a Spikes brand of shoe on one foot and a Kickers brand of shoe on the other foot, where the feet are randomly selected. After six months, the wear on the Spikes shoe would be compared to the wear on the Kickers shoe for each person. By designing the experiment in this fashion, the external effects of weight, amount of use, and so forth have been significantly reduced when comparing the two brands of tennis shoes. This type of design is an example of a **paired difference experimental design**.

In a paired design, experimental units are paired such that units within a pair are much more alike than in the population as a whole. If we have been successful at significantly reducing the variation among the sample observations by pairing, we will create very potent data to make decisions.

The paired difference design is frequently used in experiments where we are interested in the average response of an experimental unit before and after some treatment. For example a physician might be interested in comparing a patient's heart rate before and after treatment with some drug. A state legislator may be interested in comparing a person's reaction time before and after drinking one ounce of 100-proof alcohol. An instructor may be interested in comparing a student's math skills before and after being taught a particular course. A company may be interested in a customer's response to a product before and after a particular marketing campaign.

Generally, the sample size for a paired difference experimental design is small, and we must make several assumptions for the test to be valid.

Assumptions

Assumptions for the Paired Difference Experimental Design

1. Experimental units can be paired such that they are more alike within the pair than within the population as a whole.

2. The differences have an approximately normal distribution.

To test the mean difference between two related population means, we treat the difference scores, denoted by d_i, as values from a single sample. The general data setup is shown in Table 11.3.

Table 11.3 – General Setup for the Difference between Paired Samples

Observation	Sample 1	Sample 2	Difference
1	x_{11}	x_{21}	$d_1 = x_{11} - x_{21}$
2	x_{12}	x_{22}	$d_2 = x_{12} - x_{22}$
3	x_{13}	x_{23}	$d_3 = x_{13} - x_{23}$
$\vdots$	$\vdots$	$\vdots$	$\vdots$
$\vdots$	$\vdots$	$\vdots$	$\vdots$
i	x_{1i}	x_{2i}	$d_i = x_{1i} - x_{2i}$
$\vdots$	$\vdots$	$\vdots$	$\vdots$
$\vdots$	$\vdots$	$\vdots$	$\vdots$
n	x_{1n}	x_{2n}	$d_n = x_{1n} - x_{2n}$

Note that d_i is the difference between the ith observations in each sample.

We can use an interval estimate to determine the difference in two related population means. Given the differences, we can use the sample standard deviation of the difference to estimate the population standard deviation of the difference. We can construct a $100(1-\alpha)\%$ confidence interval for the mean difference μ_d using the following formula.

Formula

$100(1 - \alpha)\%$ Confidence Interval for μ_d

A $100(1 - \alpha)\%$ confidence interval for the mean difference is given by

$$\bar{x}_d \pm t_{\alpha/2, df}\, \frac{s_d}{\sqrt{n_d}}$$

where $\bar{x}_d$ is the sample mean of the differences, n_d is the number of differences, $\bar{x}_d = \dfrac{\sum d_i}{n_d}$, $s_d = \sqrt{\dfrac{\sum(d_i - \bar{x}_d)^2}{n_d - 1}}$, and $t_{\alpha/2, df}$ is the critical value of the t-distribution with an area of $\alpha/2$ in the upper tail with $n_d - 1$ degrees of freedom (df).

The methodology for the paired difference test of hypothesis and the calculation of the confidence interval for the paired difference will be presented in Example 11.3.

Example 11.3

Bull & Bones Brewhaus is a microbrewery that has two restaurants located within 15 miles of each other. The owner of the microbrewery wants to compare the average daily food sales of the two restaurants. To do so, the owner randomly selects 10 days over a five-month period (college football season) and records the daily food sales. The results are given in Table 11.4.

    a. Calculate a 95% confidence interval for the mean difference in restaurant sales.

b. The owner wants to know if there is evidence of a difference between the average daily food sales of the two restaurants. Test using $\alpha = 0.01$.

Day	Restaurant 1	Restaurant 2	Difference = Restaurant 1 – Restaurant 2
	Table 11.4 – Daily Food Sales for Two Restaurants ($)		
1	5828	7894	−2066
2	9836	11,573	−1737
3	3984	5319	−1335
4	5845	6389	−544
5	5210	6055	−845
6	9668	10,631	−963
7	6768	7866	−1098
8	6726	7976	−1250
9	4399	5652	−1253
10	6692	8083	−1391

Solution

a. From the information given, we know that $n = 10$ and $\alpha = 0.05$. We now need to calculate the mean and standard deviation of the differences, $\bar{x}_d$ and s_d, respectively.

$$\bar{x}_d = \frac{\sum d_i}{n_d} = \frac{(-2066)+(-1737)+\cdots+(-1253)+(-1391)}{10} = -1248.20$$

$$s_d = \sqrt{\frac{\sum(d_i - \bar{x}_d)^2}{n_d - 1}}$$

$$= \sqrt{\frac{(-2066-(-1248.20))^2 + (-1737-(-1248.20))^2 + \cdots}{+ (-1253-(-1248.20))^2 + (-1391-(-1248.20))^2}{10-1}}$$

$$\approx 434.3631$$

Since we want a 95% confidence interval, we need to find the value of $t_{\alpha/2, df}$. Note that $df = n_d - 1 = 9$. Thus, $t_{\alpha/2, df} = t_{0.025, 9} = 2.262$.

A 95% confidence interval for the average difference in sales record between the two restaurants is calculated as follows.

$$\bar{x}_d \pm t_{\alpha/2, df} \frac{s_d}{\sqrt{n_d}}$$

$$-1248.20 \pm 2.262 \left(\frac{434.3631}{\sqrt{10}} \right)$$

$$-1558.90 \text{ to } -937.50$$

We are 95% confident that the true mean difference in sales between Restaurant 1 and Restaurant 2 is between −$1558.90 and −$937.50. That is, on average, Restaurant 2 averages between $937.50 and $1558.90 more in sales.

b. Before performing the hypothesis test, we need to ensure that the assumption that the differences have a normal distribution is reasonable for the sales data. A histogram of differences is provided in Figure 11.7. Based on this histogram, it appears that the differences are approximately normal and it is safe to proceed with the paired difference test.

Figure 11.7

Step 1: Define the hypotheses in plain English.

- **Null Hypothesis:** There is no difference in the average sales between the two restaurants on a given day.

- **Alternative Hypothesis:** There is a difference in the average sales between the two restaurants on a given day.

Step 2: Select the appropriate statistical measure.

In a paired difference experimental design, the population parameter of interest is the population mean of the differences. Thus, the appropriate statistical measure is given by

μ_d = the average of the differences in daily sales between the two restaurants.

Step 3: Determine whether the hypothesis should be one-sided or two-sided.

Since the owner is interested in whether or not the average daily sales are different between the two restaurants, the alternative hypothesis will be two-sided and this will be a two-tailed test.

Step 4: State the hypotheses using the appropriate statistical measure.

The statistical measure being used is the average difference in daily sales between the two restaurants. Since we have a two-tailed test, the hypotheses will be as follows.

$H_0: \mu_d = 0$ (There is no difference in average daily sales between the two restaurants.)

$H_a: \mu_d \neq 0$ (There is a difference in average daily sales between the two restaurants.)

Step 5: Specify the level of the test.

The level of the test is specified in the problem to be $\alpha = 0.01$.

Step 6: Select the appropriate test statistic.

To develop the appropriate test statistic, a random variable whose value will be used to make the decision to reject or fail to reject H_0 must be developed. It is interesting to notice that by evaluating the differences of the paired data, we have effectively reduced the two-sample problem into a small, one-sample t-test for a population mean. The point estimate for the population mean of the differences, μ_d, is $\bar{x}_d$, the sample mean of the differences. Hence, the sampling distribution for $\bar{x}_d$ provides essential information for testing the hypothesis. If the differences are normally distributed and the null hypothesis is assumed to be true, the sampling distribution of $\bar{x}_d$ has a t-distribution with $n_d - 1$ degrees of freedom. Note that n_d represents the number of differences. The t-test statistic is given by

$$t = \frac{\bar{x}_d - \mu_d}{s_d / \sqrt{n_d}},$$

where s_d is the sample standard deviation of the differences.

If the observed value of $\bar{x}_d$ is significantly smaller or larger than μ_d, this will produce a large negative (or positive) value of the test statistic, causing us to question whether the null hypothesis is in fact true. How large is *large*? This is answered by the critical value of the test statistic specified in Step 7.

Step 7: Determine the critical value.

The role of the critical value in this test is exactly the same as for all of the hypothesis tests discussed earlier. It defines a range of values for the test statistic, the rejection region, that will be so rare that it is unlikely that it occurred from ordinary sampling variability. The level of the test defines the size of the rejection region. Should the computed value of the test statistic fall in the rejection region, its value will be presumed to be too rare to have occurred because of ordinary sampling variation, and the null hypothesis will be rejected.

If the null hypothesis is true, the test statistic has a t-distribution with $n_d - 1$ degrees of freedom. Thus, the critical value is determined in the same way as for the other tests of hypothesis where the test statistic had a t-distribution, except that the degrees of freedom are $n_d - 1 = 10 - 1 = 9$. The rejection region for the alternative hypothesis, $H_a: \mu_d \neq 0$, with $\alpha = 0.01$ and 9 degrees of freedom is displayed in Figure 11.8. Because we have a two-sided hypothesis, we reject H_0 if the test statistic is less than or equal to -3.250 or if it is greater than or equal to 3.250.

Figure 11.8

Step 8: Compute the test statistic.

Based on the data in Table 11.4, $\bar{x}_d = -1248.20$, and the computed value of the test statistic is as follows.

$$t = \frac{\bar{x}_d - \mu_d}{s_d / \sqrt{n_d}} = \frac{-1248.20 - 0}{434.3631 / \sqrt{10}} \approx -9.087$$

Step 9: Make the decision.

As shown in Figure 11.9, the value of the test statistic falls in the rejection region to the left. The test statistic indicates that the observed average daily sales are more than 9 standard deviations below the hypothesized value of 0. It is highly unlikely that the difference between the observed value and the hypothesized value is due to ordinary sampling variation. Thus, the null hypothesis is rejected at $\alpha = 0.01$.

Reject H_0   **Fail to Reject H_0**

−9.087
(Test Statistic)

−3.250
(Critical Value)

Figure 11.9

Step 10: State the conclusion in terms of the original question.

There is sufficient evidence for the owner to conclude at the $\alpha = 0.01$ level that the average daily sales between the two restaurants are significantly different.

11.3 Exercises

Basic Concepts

1. Describe the differences between an independent experimental design and a paired design.

2. What are the assumptions for a paired difference experimental design?

3. What is the appropriate statistical measure to use when performing a hypothesis test about a paired difference experiment?

4. How does the hypothesis testing procedure for a paired difference experiment differ from that of a two-sample t-test?

5. What is the test statistic used in a paired difference hypothesis test?

Exercises

6. Determine the critical value(s) of the test statistic for each of the following small sample paired difference tests (assume the differences have an approximately normal distribution).

   a. Left-tailed test, $\alpha = 0.01$, $n_d = 15$

   b. Right-tailed test, $\alpha = 0.10$, $n_d = 20$

   c. Two-tailed test, $\alpha = 0.05$, $n_d = 8$

7. Determine the critical value(s) of the test statistic for each of the following small sample paired difference tests (assume the differences have an approximately normal distribution).

    a. Left-tailed test, $\alpha = 0.005$, $n_d = 12$

    b. Right-tailed test, $\alpha = 0.025$, $n_d = 5$

    c. Two-tailed test, $\alpha = 0.10$, $n_d = 25$

8. Given that most textbooks can now be purchased online, one wonders if students can save money by comparison shopping for textbooks at online retailers and at their local bookstores. To investigate, students at Tech University randomly sampled 25 textbooks on the shelves of their local bookstores. The students then found the "best" available price for the same textbooks via online retailers. The prices for the textbooks are listed in the following table.

Table for Exercise 8 – Textbook Prices

Textbook	Bookstore	Online Retailer
	Price ($)	
1	70	60
2	38	36
3	88	89
4	165	149
5	80	136
6	103	95
7	42	50
8	98	111
9	89	65
10	97	86
11	140	130
12	40	30
13	175	150
14	85	75
15	100	85
16	68	62
17	67	69
18	140	142
19	49	40
20	149	127
21	126	130
22	92	93
23	144	129
24	98	84
25	40	52

    a. Is a paired design appropriate for the above study? Explain.

    b. What assumption must be made in order to perform the test of hypothesis?

    c. Do the data appear to satisfy the assumption described in part **b.**? Why or why not?

    d. Based on the data, is it less expensive for the students to purchase textbooks from the online retailers than from local bookstores? Use $\alpha = 0.01$.

    e. Calculate a 99% confidence interval for the mean difference in cost between the bookstores and the online retailers. Interpret the interval.

9. The management for a large grocery store chain would like to determine if a new cash register will enable cashiers to process a larger number of items on average than the cash register they are currently using. Seven cashiers are randomly selected, and the number of grocery items they can process in three minutes is measured for both the old cash register and the new cash register. The results of the test are as follows.

Table for Exercise 9 – Number of Grocery Items Processed in Three Minutes							
Cashier	1	2	3	4	5	6	7
Old Cash Register	60	70	55	75	62	52	58
New Cash Register	65	71	55	75	65	57	57

a. Is a paired design appropriate for the above experiment? Explain.

b. What assumption must be made in order to perform the test of hypothesis?

c. Do the data appear to satisfy the assumption described in part **b.**? Why or why not?

d. Calculate a 95% confidence interval for the mean difference between the number of items processed using the old cash register and the new cash register. Interpret this interval.

e. Can the management conclude that the new cash register will allow cashiers to process a significantly larger number of items on average than the old cash register at $\alpha = 0.05$?

10. An auto dealer is marketing two different models of a high-end sedan. Since customers are particularly interested in the safety features of the sedans, the dealer would like to determine if there is a difference in the braking distance (the number of feet required to go from 60 mph to 0 mph) of the two sedans. Six drivers are randomly selected and asked to participate in a test to measure the braking distance for both models. Each driver is asked to drive both models and brake once they have reached exactly 60 mph. The distance required to come to a complete halt is then measured in feet. The results of the test are as follows.

Table for Exercise 10 – Braking Distance of High-End Sedans (Feet)						
Driver	1	2	3	4	5	6
Model A	150	145	160	155	152	153
Model B	152	146	160	157	154	155

a. Is a paired design appropriate for the above experiment? Explain.

b. What assumption must be made in order to perform the test of hypothesis?

c. Do the data appear to satisfy the assumption described in part **b.**? Why or why not?

d. Calculate a 90% confidence interval for the average difference between braking distances for Model A and Model B. Interpret the interval.

e. Can the auto dealer conclude that there is a significant difference in the braking distances of the two models of high-end sedans? Use $\alpha = 0.10$.

11.4   Comparing Two Population Proportions

Techniques are developed in this section for comparing two population proportions. A methodology for comparing two population proportions is particularly useful because proportions are among the few measures that can be used for summarizing categorical data. For a more extensive treatment of comparisons for categorical data, see Chapter 15.

There are many situations where comparing two population proportions may be of interest. For example, a sociologist may be interested in comparing the proportion of females who believe it is okay to cry in public to the proportion of males who think it is okay to cry in public. A marketing manager may be interested in comparing the proportion of customers who favor Product A to the proportion of customers who favor Product B.

In order to perform a comparison of two population proportions, the assumptions outlined below must be met.

Assumptions

Assumptions for Comparing Two Population Proportions

1. An independent experimental design is used.

2. The samples are large enough such that $n_1 \hat{p}_1 \geq 5$, $n_1 \left(1 - \hat{p}_1\right) \geq 5$, $n_2 \hat{p}_2 \geq 5$, and $n_2 \left(1 - \hat{p}_2\right) \geq 5$ where $\hat{p}_1$ and n_1 are the sample proportion and sample size, respectively, from the first population and $\hat{p}_2$ and n_2 are the sample proportion and sample size, respectively, from the second population.

Interval Estimation of $p_1 - p_2$

Formula

100(1 − α)% Confidence Interval for $p_1 - p_2$

We can construct a $100(1 - \alpha)\%$ confidence interval estimate for the difference between two population proportions using the following:

$$\left(\hat{p}_1 - \hat{p}_2\right) \pm z_{\alpha/2} \sqrt{\frac{\hat{p}_1\left(1 - \hat{p}_1\right)}{n_1} + \frac{\hat{p}_2\left(1 - \hat{p}_2\right)}{n_2}}$$

where $\hat{p}_1$ and n_1 are the sample proportion and sample size, respectively, from the first population, $\hat{p}_2$ and n_2 are the sample proportion and sample size, respectively, from the second population, $n_1 \hat{p}_1 \geq 5$, $n_1 \left(1 - \hat{p}_1\right) \geq 5$, $n_2 \hat{p}_2 \geq 5$, and $n_2 \left(1 - \hat{p}_2\right) \geq 5$, and $z_{\alpha/2}$ is the critical value for the z-distribution that captures an area of $\alpha/2$ in the upper tail.

Hypothesis Testing about $p_1 - p_2$

The hypothesis testing procedure for comparing two population proportions is developed in Example 11.4.

Example 11.4

A telephone executive has recently been bombarded with complaints from his customers about defective phones. He has two plants that produce the telephones, and he is not sure where the defective phones are coming from. In the past, the plants have had good control over the number of defective phones produced. Because of the recent flurry of complaints, he thinks that one of the plants may have lost control over its production process. To test this theory, he randomly selects 200 phones from each of the plants and counts the number of defective phones. The results of the survey are displayed in Table 11.5.

Table 11.5 – Telephone Survey Data		
	Number Sampled	**Number of Defectives**
Plant A	200	8
Plant B	200	10

a. Are the sample sizes large enough to assume that the sample proportions are approximately normally distributed? Why is this necessary?

b. Calculate a 95% confidence interval for the difference between the proportions of defective phones from Plant A and Plant B.

c. Is there sufficient evidence for the telephone executive to conclude that there is a difference in the proportion of defective telephones produced by the two plants at $\alpha = 0.10$?

Solution

a. Let

$\hat{p}_1 = $ the sample proportion of defective phones produced in Plant A,

$\hat{p}_2 = $ the sample proportion of defective phones produced in Plant B,

$x_1 = $ the sample number of defective phones produced in Plant A,

$x_2 = $ the sample number of defective phones produced in Plant B,

$n_1 = $ the number of phones sampled from Plant A, and

$n_2 = $ the number of phones sampled from Plant B.

Before we can determine a confidence interval or perform a hypothesis test for the difference between the population proportions, we need to be sure that the sample sizes are large enough. You may recall that when dealing with single samples of proportions, the sample size, n, must be large enough such that the sampling distribution of the sample proportion will be approximately normal. To determine whether the sample sizes are large enough, we need to show that $n_1 \hat{p}_1$, $n_1\left(1-\hat{p}_1\right)$, $n_2 \hat{p}_2$, and $n_2\left(1-\hat{p}_2\right)$ are all greater than or equal to 5. If both samples are large enough, then we know that the sampling distribution of the difference between the sample proportions is approximately normal, and we can proceed with constructing the confidence interval and performing the hypothesis test. First, we need to calculate the sample proportions for the defective phones produced in Plant A and Plant B.

$$\hat{p}_1 = \frac{x_1}{n_1} = \frac{8}{200} = 0.04$$

$$\hat{p}_2 = \frac{x_2}{n_2} = \frac{10}{200} = 0.05$$

Now, we can use the sample proportions along with the sample sizes to verify that the samples are large enough such that the sampling distribution of the difference between the sample proportions is approximately normal.

$$n_1\hat{p}_1 = 200(0.04) = 8$$

$$n_1\left(1-\hat{p}_1\right) = 200(1-0.04) = 192$$

$$n_2\hat{p}_2 = 200(0.05) = 10$$

$$n_2\left(1-\hat{p}_2\right) = 200(1-0.05) = 190$$

Since $n_1\hat{p}_1$, $n_1\left(1-\hat{p}_1\right)$, $n_2\hat{p}_2$, and $n_2\left(1-\hat{p}_2\right)$ are all greater than or equal to 5, we can conclude that $\hat{p}_1 - \hat{p}_2$ has an approximately normal distribution. Now that we have verified the conditions required to proceed with the calculation of a confidence interval, we can find the 95% confidence interval for the difference between the population proportions.

b. We know that $n_1 = n_2 = 200$, $x_1 = 8$, and $x_2 = 10$. With the information given, we can calculate the sample proportions.

$$\hat{p}_1 = \frac{x_1}{n_1} = \frac{8}{200} = 0.04$$

$$\hat{p}_2 = \frac{x_2}{n_2} = \frac{10}{200} = 0.05$$

For a 95% confidence interval, $\alpha = 0.05$ and $z_{\alpha/2} = z_{0.025} = 1.96$.

Therefore, the 95% confidence interval is calculated as follows.

$$\left(\hat{p}_1 - \hat{p}_2\right) \pm z_{\alpha/2}\sqrt{\frac{\hat{p}_1\left(1-\hat{p}_1\right)}{n_1} + \frac{\hat{p}_2\left(1-\hat{p}_2\right)}{n_2}}$$

$$(0.04 - 0.05) \pm 1.96\sqrt{\frac{0.04(1-0.04)}{200} + \frac{0.05(1-0.05)}{200}}$$

$$-0.01 \pm 1.96\sqrt{\frac{0.04(1-0.04)}{200} + \frac{0.05(1-0.05)}{200}}$$

$$-0.0506 \text{ to } 0.0306$$

Thus, we are 95% confident that the true difference in the proportion of defectives between Plant A and Plant B is between −0.0506 and 0.0306.

c.   Step 1: Define the hypotheses in plain English.

- **Null Hypothesis:** There is no difference in the proportion of defective telephones produced at the two plants.

- **Alternative Hypothesis:** There is a difference in the proportion of defective telephones produced at the two plants.

Step 2: Select the appropriate statistical measure.

Since the executive is interested in comparing the proportion of defective phones produced at Plant A to the proportion of defective phones produced at Plant B, the appropriate statistical measures are as follows.

p_1 = the true proportion of defective phones produced at Plant A

p_2 = the true proportion of defective phones produced at Plant B

Step 3: Determine whether the hypothesis should be one-sided or two-sided.

The executive's interest is in whether or not there is a difference in the proportion of defective phones produced by the two plants. Thus, the alternative hypothesis is two-sided and this is a two-tailed test.

Step 4: State the hypotheses using the appropriate statistical measure.

Since the executive is interested in comparing the two population proportions, a natural way to perform this comparison is to look at the difference between the two population proportions. Hence, the null and alternative hypotheses are as follows.

$$H_0: p_1 - p_2 = 0 \quad \left(\begin{array}{l} \text{There is no difference in the proportion of} \\ \text{defective phones produced at the two plants.} \end{array} \right)$$

$$H_a: p_1 - p_2 \neq 0 \quad \left(\begin{array}{l} \text{There is a difference in the proportion of} \\ \text{defective phones produced at the two plants.} \end{array} \right)$$

Note: In order for the following hypothesis testing procedure to be valid, the hypothesized difference in the null hypothesis between the two population proportions must be 0.

Step 5: Specify the level of the test.

The level of the test is specified in the problem to be $\alpha = 0.10$.

Step 6: Select the appropriate test statistic.

To develop the appropriate test statistic, a random variable whose value will be used to help make the decision to reject or fail to reject H_0 must be found. The point estimate of $p_1 - p_2$ is $\hat{p}_1 - \hat{p}_2$. The sampling distribution of $\hat{p}_1 - \hat{p}_2$ will be used in determining the critical values of the test statistic. If the assumptions previously outlined are met, and we assume the null hypothesis is true, the sampling distribution of $\hat{p}_1 - \hat{p}_2$ has an approximately normal distribution with mean, 0, and standard deviation,

$$\sigma_{\hat{p}_1 - \hat{p}_2} = \sqrt{ \bar{p}(1 - \bar{p}) \left(\frac{1}{n_1} + \frac{1}{n_2} \right) },$$

$$\text{where } \bar{p} = \frac{x_1 + x_2}{n_1 + n_2}.$$

Note that $\bar{p}$ is the weighted average of the two sample proportion estimates, $\hat{p}_1$ and $\hat{p}_2$. In hypothesis tests comparing two population proportions, we will always assume that the hypothesized difference between the two proportions is zero.

If the null hypothesis is assumed to be true, then $p_1 - p_2 = 0$, which implies that $p_1 = p_2$. Thus $\hat{p}_1$ and $\hat{p}_2$ are estimating the same quantity. Therefore, $\hat{p}_1$ and $\hat{p}_2$ are pooled to derive a better estimate of the population proportion. The test statistic is a standard normal random variable given by

$$z = \frac{\left(\hat{p}_1 - \hat{p}_2\right) - \left(p_1 - p_2\right)}{\sqrt{\bar{p}(1-\bar{p})\left(\dfrac{1}{n_1} + \dfrac{1}{n_2}\right)}}.$$

If the null hypothesis is true, z has an approximately normal distribution. If the observed value of $\hat{p}_1 - \hat{p}_2$ is significantly larger or smaller than 0, this will produce a large or small value of the test statistic, causing us to question whether the null hypothesis is true. How large is *large*? This is answered by the critical value of the test statistic specified in Step 7.

Step 7: Determine the critical value.

The role of the critical value in this test is exactly the same as for all of the hypothesis tests discussed earlier. It defines a range of values for the test statistic, the rejection region, that will be so rare that it is unlikely that it occurred from ordinary sampling variability, assuming H_0 is true. The level of the test defines the size of the rejection region. Should the computed value of the test statistic fall in the rejection region, the null hypothesis will be rejected.

If the null hypothesis is true, the test statistic has a standard normal distribution. Thus the critical value is determined in the same way as for other tests of hypothesis where the test statistic had a standard normal distribution. The rejection region for a two-tailed test with $\alpha = 0.10$ is displayed in Figure 11.10. We will reject the null hypothesis if the computed value of the test statistic is larger than 1.645 or smaller than -1.645.

Figure 11.10

Step 8: Compute the test statistic.

Based on the data in Table 11.5, the computed value of the test statistic is given by

$$z = \frac{\dfrac{8}{200} - \dfrac{10}{200} - 0}{\sqrt{0.045(1-0.045)\left(\dfrac{1}{200} + \dfrac{1}{200}\right)}} \approx -0.48$$

$$\text{where } \bar{p} = \frac{8+10}{200+200} = 0.045.$$

Step 9: Make the decision.

Figure 11.11

As displayed in Figure 11.11, the value of the test statistic does not fall in the rejection region because $-1.645 < -0.48 < 1.645$. Thus, the difference between the observed value and the hypothesized value is likely due to ordinary sampling variation. We fail to reject the null hypothesis at $\alpha = 0.10$.

Step 10: State the conclusion in terms of the original question.

There is insufficient evidence at $\alpha = 0.10$ for the telephone executive to conclude that the proportion of defective phones produced differs between the two plants.

11.4 Exercises

Basic Concepts

1. Why is comparing two population proportions particularly useful?

2. Give two examples of situations in which someone would be interested in comparing population proportions.

3. What assumptions are necessary to perform a hypothesis test for the difference between two population proportions?

4. Which sampling distribution is used in a two-sample test of hypothesis about population proportions? What are the characteristics of this sampling distribution?

5. What is the test statistic that is used when comparing two population proportions?

6. True or false: in order to use the specified test statistic, the hypothesized difference in the null hypothesis between the two population proportions must be zero.

Exercises

7. Determine the critical value(s) of the test statistic for each of the following large sample tests for the comparison of two population proportions.

    a. Left-tailed test, $\alpha = 0.01$

    b. Right-tailed test, $\alpha = 0.05$

    c. Two-tailed test, $\alpha = 0.10$

8. Determine the critical value(s) of the test statistic for each of the following large sample tests for the comparison of two population proportions.

    a. Left-tailed test, $\alpha = 0.025$

    b. Right-tailed test, $\alpha = 0.02$

    c. Two-tailed test, $\alpha = 0.04$

9. A fund-raiser believes that women are more likely to say "Yes" when asked to donate to a worthy cause than men. To test this theory, she randomly selects 100 men and 95 women and asks for donations to the same cause. The results of the survey are as follows.

Table for Exercise 9 – Fund-Raiser Survey		
	Number Surveyed	**# of "Yes" Responses**
Men	100	6
Women	95	9

    a. Are the sample sizes large enough such that a hypothesis test for the difference between two population proportions may be performed? If so, do the data substantiate the fund-raiser's theory at $\alpha = 0.10$?

    b. Calculate the P-value for the test and interpret its meaning.

    c. Calculate a 95% confidence interval for the difference in the proportion of men and women who would most likely donate to a worthy cause. Interpret the interval.

10. A poll is conducted to determine if U.S. citizens think that there should be a national health care system in the U.S. 69% of the 300 women surveyed and 63% of the 250 men surveyed think that there should be a national health care system in the U.S. Are the sample sizes large enough such that a hypothesis test for the difference between two population proportions may be performed? If so, is there sufficient evidence to conclude at $\alpha = 0.05$ that men and women feel differently about this issue?

11. Major television networks have never seemed to have issues showing commercials for beer and other alcoholic beverages. Even though adult viewers tend to enjoy the commercials, most adults seem to think that the commercials target teenagers and young adults (those under 21 years old). To study this belief, the networks conducted a joint poll of viewers and asked them if they felt that beer and other alcoholic beverage commercials targeted teenagers and young adults. The results of the survey are as follows.

Table for Exercise 11 – Network Advertising Survey		
Age Group	Number Surveyed	Number of "Yes" Responses
30 or Younger	1000	450
Older than 30	1000	655

a. Are the sample sizes large enough such that inferences about the difference between two population proportions can be made? If so, calculate a 99% confidence interval for the difference in the proportions of those older than 30 and those 30 or younger that believe alcoholic beverage commercials targeted teenagers and young adults. Interpret the interval.

b. Based on the data, can the networks conclude that the percentage of viewers who believe beer and alcoholic beverage commercials target teenagers and young adults is significantly higher in the over 30 age group than in the 30 or younger age group at $\alpha = 0.01$?

12. A manufacturer is comparing shipments of machine parts from two suppliers. The parts from Supplier A are less expensive; however, the manufacturer is concerned that the parts may be of a lower quality than those from Supplier B. The manufacturer has decided that he will purchase his supplies from Supplier A unless he can show that the proportion of defective parts is significantly higher for Supplier A than for Supplier B. He randomly selects parts from each supplier and inspects them for defects. The results are as follows. Determine whether the sample sizes are large enough such that inferences about the difference between the population proportions can be made. If so, which supplier will the manufacturer choose at $\alpha = 0.05$? Explain.

Table for Exercise 12 – Number of Defective Parts		
	Number Surveyed	Number of Defective Parts
Supplier A	400	8
Supplier B	300	5

T   Discovering Technology

Using the TI-84 Plus Calculator

Comparing Two Population Means (Large Samples or σ_1 and σ_2 Known)

Use the information from Example 11.1 for this exercise.

$$n_1 = 50, \ n_2 = 50, \ s_1 = 3, \ s_2 = 2, \ \bar{x}_1 = 8.8, \ \text{and } \bar{x}_2 = 9.5$$

1. Part **a.** of Example 11.1 asks for the 95% confidence interval for the difference in the population mean QOS for cable and satellite subscribers. To calculate the interval using a TI-84 Plus calculator, press STAT , scroll to **TESTS**, and select option **9:2-SampZInt**.

2. Select **Stats**, and enter **3** for σ_1, **2** for σ_2, **8.8** for $\bar{x}_1$, **50** for n_1, **9.5** for $\bar{x}_2$, **50** for n_2, and **0.95** for C-Level. Press **Calculate**. (Note that since both sample sizes are greater than or equal to 30 we can use s_1 and s_2 as approximations for σ_1 and σ_2.)

3. The 95% confidence interval is displayed as $(-1.699, 0.29939)$. Compare this to the confidence interval we calculated in Example 11.1.

Figure 11.12

4. Part **b.** of Example 11.1 asks if there is sufficient evidence for the analyst to conclude at $\alpha = 0.05$ that there is a difference in average QOS between cable and satellite subscribers. To perform this test using a TI-84 Plus calculator, press STAT , scroll to **TESTS**, and select option **3:2-SampZTest**.

5. Select **Stats**, and enter **3** for σ_1, **2** for σ_2, **8.8** for $\bar{x}_1$, **50** for n_1, **9.5** for $\bar{x}_2$, **50** for n_2, and $\neq \mu_2$ as the alternative hypothesis. Press **Calculate**. (Note that since both sample sizes are greater than or equal to 30 we can use s_1 and s_2 as approximations for σ_1 and σ_2.)

6. The results of the hypothesis test are displayed. The z-statistic is calculated to be -1.37 and the P-value is given as 0.1698. Thus, there is not sufficient evidence at the 0.05 level to conclude that the mean QOS is significantly different between cable and satellite subscribers.

Figure 11.13

Comparing Two Population Means (Small Samples, σ_1 and σ_2 Unknown)

Use the information from Example 11.2 for this exercise.

$$n_1 = 15,\ n_2 = 16,\ s_1 = 200,\ s_2 = 175,\ \bar{x}_1 = 3425,\ \text{and}\ \bar{x}_2 = 3250$$

1. Part **a.** of Example 11.1 asks for the 95% confidence interval for the difference in the population mean dollar sales between the two products. To calculate the interval using a TI-84 Plus calculator, press **STAT**, scroll to **TESTS**, and select option **0:2-SampTInt**.

2. Select **Stats**, and enter **3425** for $\bar{x}_1$, **200** for s_1, **15** for n_1, **3250** for $\bar{x}_2$, **175** for s_2, **16** for n_2, **0.95** for C-Level, and **Yes** for Pooled. Press **Calculate**.

3. The 95% confidence interval is displayed as (37.189, 312.81). Compare this to the confidence interval we calculated in Example 11.2. The slight difference in the values of the interval endpoints is due to rounding error that results from rounding the pooled variance in the hand calculation.

Figure 11.14

4. Part **b.** of Example 11.2 asks to perform a hypothesis test to determine if Product 1 has higher mean dollar sales than Product 2 using $\alpha = 0.01$. To perform this test using a TI-84 Plus calculator, press **STAT**, scroll to **TESTS**, and select option **4:2-SampTTest**.

5. Select **Stats**, and enter **3425** for $\bar{x}_1$, **200** for s_1, **15** for n_1, **3250** for $\bar{x}_2$, **175** for s_2, **16** for n_2, $> \mu_2$ for the alternative hypothesis, and **Yes** for Pooled. Press **Calculate**.

6. The results of the hypothesis test are displayed. The t-statistic is calculated to be 2.597 and the P-value is given as 0.0073. Thus, there is sufficient evidence at the 0.01 level to conclude that the mean dollar sales for Product 1 are significantly higher than the mean dollar sales for Product 2.

Figure 11.15

Comparing Two Population Proportions

Use the information from Example 11.4 for this exercise.

$$x_1 = 8,\ n_1 = 200,\ x_2 = 10,\ n_2 = 200$$

1. Part **b.** of Example 11.4 asks for the 95% confidence interval for the difference between the population proportions of defective phones from Plant A and Plant B. To calculate the interval using a TI-84 Plus calculator, press **STAT**, scroll to **TESTS**, and select option **B:2-PropZInt**.

2. Enter **8** for x_1, **200** for n_1, **10** for x_2, **200** for n_2, and **0.95** for C-Level. Press **Calculate**.

3. The 95% confidence interval is displayed as (−0.0506, 0.03062). Compare this to the confidence interval we calculated in Example 11.4.

Figure 11.16

4. Part **c.** of Example 11.4 asks to perform a hypothesis test to determine if there is sufficient evidence for the telephone executive to conclude that there is a difference in the proportion of defective telephones produced by the two plants at $\alpha = 0.10$. To perform this test using a TI-84 Plus calculator, press `STAT`, scroll to **TESTS**, and select option **6:2-PropZTest**.

5. Enter **8** for x_1, **200** for n_1, **10** for x_2, **200** for n_2, and $\neq p_2$ for the alternative hypothesis. Press **Calculate**.

6. The results of the hypothesis test are displayed. The z-statistic is calculated to be −0.48 and the P-value is given as 0.6295. Thus, there is not sufficient evidence at the 0.10 level for the telephone executive to conclude that the proportion of defective phones produced differs between the two plants.

```
2-PropZTest
 P1≠P2
 z=-.4823830702
 p=.6295338516
 p̂1=.04
 p̂2=.05
 ↓p̂=.045
```

Figure 11.17

Using Excel

When using these data analysis tools in Excel, the raw data must be available and you will not be able to plug in summary statistics. You will be asked to select a range of data to compute the answer. The data analysis tools that are available in Excel are displayed in the following table.

Table 11.6 – Excel Data Analysis Tools for this Chapter		
Section	**Topic**	**Excel Data Analysis Tool**
11.1	Comparing Two Population Means: Large Samples or σ_1 and σ_2 Known	z-Test: Two Sample for Means
11.2	Comparing Two Population Means: Small Samples (σ_1 and σ_2 Unknown and $\sigma_1 = \sigma_2$)	t-Test: Two-Sample Assuming Equal Variances
11.2	Comparing Two Population Means: Small Samples (σ_1 and σ_2 Unknown and $\sigma_1 \neq \sigma_2$)	t-Test: Two-Sample Assuming Unequal Variances
11.3	Paired Difference	t-Test: Paired Two Sample for Means

Paired Difference

For this exercise, use the data from Example 11.3.

1. Enter the daily food sales for each restaurant found in Table 11.4 into Column A and Column B in an Excel worksheet.

	A	B
1	5828	7894
2	9836	11573
3	3984	5319
4	5845	6389
5	5210	6055
6	9668	10631
7	6768	7866
8	6726	7976
9	4399	5652
10	6692	8083

Figure 11.18

2. The data analysis tool in Excel which performs the hypothesis test for paired difference experiments is *t*-Test: Paired Two Sample for Means. Select this option from the data analysis tools window.

3. Input the range $A$1:$A$10 as the Variable 1 Range or click and drag in the worksheet to select the sales for Restaurant 1 as the first variable. Input the range $B$1:$B$10 as the Variable 2 Range or click and drag in the worksheet to select the sales for Restaurant 2 as the second variable.

4. Enter 0 for the Hypothesized Mean Difference.

5. Specify the level of the test by entering 0.01 for Alpha.

6. Press OK. The results of the test are displayed. Notice that the test statistic (t Stat) and the critical values for one tailed (t Critical one-tail) and two tailed (t Critical two-tail) tests are displayed. Compare the test statistic to that which we obtained in Example 11.3.

	A	B	C
1	t-Test: Paired Two Sample for Means		
2			
3		Variable 1	Variable 2
4	Mean	6495.6	7743.8
5	Variance	3845984.044	4238663.733
6	Observations	10	10
7	Pearson Correlation	0.977817104	
8	Hypothesized Mean Difference	0	
9	df	9	
10	t Stat	-9.087224765	
11	P(T<=t) one-tail	3.94487E-06	
12	t Critical one-tail	2.821437921	
13	P(T<=t) two-tail	7.88974E-06	
14	t Critical two-tail	3.249835541	

Figure 11.19

Using Minitab

Using Minitab you can make inferences about the difference between two population means or proportions using either summarized data or raw data. The Minitab functions that are available to perform hypothesis tests and construct confidence intervals for two populations are given in the following table.

Table 11.7 – Minitab Functions for this Chapter		
Section	Topic	Minitab Function
11.1 – 11.2	Comparing Two Population Means	2-Sample t
11.3	Paired Difference	Paired t
11.4	Comparing Two Population Proportions	2 Proportions

Comparing Two Population Means

For this exercise, use the data from Example 11.2.

$$n_1 = 15, \; n_1 = 16, \; \bar{x}_1 = 3425, \; \bar{x}_2 = 3250, \; s_1 = 200, \; s_2 = 175$$

1. The Minitab function for a two sample hypothesis test is **2-Sample t**. Select **Stat**, then **Basic Statistics**, then **2-Sample t**.

2. Select the button next to **Summarized data**.

3. Enter the parameters from Table 11.2. In the "First" row, enter **15** for Sample size, **3425** for Mean and **200** for Standard deviation. In the "Second" row, enter **16** for Sample size, **3250** for Mean, and **175** for Standard deviation.

4. Check the box next to **Assume equal variances**.

5. Press **OK**. The results are displayed in the Session window. Notice that the confidence interval is displayed along with the results of the hypothesis test. Compare the value of the test statistic to the one we determined in Example 11.2.

```
Two-Sample T-Test and CI

                      SE
Sample  N  Mean  StDev  Mean
1       15 3425   200    52
2       16 3250   175    44


Difference = mu (1) - mu (2)
Estimate for difference:  175.0
95% CI for difference:  (37.2, 312.8)
T-Test of difference = 0 (vs not =): T-Value = 2.60  P-Value = 0.015  DF = 29
Both use Pooled StDev = 187.4856
```

Figure 11.20

Paired Difference

For this exercise, use the data from Example 11.3.

1. Enter the sales data for Restaurant 1 in Column C1 and the sales data for Restaurant 2 in Column C2.

2. Go to **Stat**, **Basic Statistics**, and **Paired t**.

3. Enter **C1** for the First sample, and **C2** for the Second sample. Note that Minitab calculates the paired differences by subtracting the second sample from the first sample.

4. Choose **Options** and set the confidence level to **95.0**. Ensure that the Test mean is set to **0** and the Alternative is **not equal**. Click **OK**.

5. Click **OK** and observe the results in the Session window. The 95% confidence interval is given as $(-1559, -937)$. The calculated value of the t-statistic is given as -9.09 and the P-value is given as 0.000. Thus, there is sufficient evidence to conclude at the 0.01 level that the average daily sales are significantly different between the two restaurants.

```
Paired T-Test and CI: Restaurant 1, Restaurant 2

Paired T for Restaurant 1 - Restaurant 2

               N    Mean   StDev   SE Mean
Restaurant 1   10   6496    1961       620
Restaurant 2   10   7744    2059       651
Difference     10  -1248     434       137


95% CI for mean difference: (-1559, -937)
T-Test of mean difference = 0 (vs not = 0): T-Value = -9.09   P-Value = 0.000
```

Figure 11.21

Comparing Two Population Proportions

For this exercise, use the data from Example 11.4.

1. The Minitab function for a hypothesis test about two population proportions is **2 Proportions**. Select **Stat**, **Basic Statistics**, and **2 Proportions**.

2. Choose the button next to **Summarized data**. Enter the data from Table 11.5. In the "First" row, enter **8** for Events and **200** for Trials. In the "Second" row, enter **10** for Events and **200** for Trials.

3. Click **OK**. The results of the hypothesis test are displayed in the Session window. Compare the z-statistic to the one that we found in Example 11.4.

```
Test and CI for Two Proportions

Sample   X     N   Sample p
1        8    200  0.040000
2       10    200  0.050000


Difference = p (1) - p (2)
Estimate for difference:   -0.01
95% CI for difference:   (-0.0506190, 0.0306190)
Test for difference = 0 (vs not = 0):   Z = -0.48   P-Value = 0.629

Fisher's exact test: P-Value = 0.810
```

Figure 11.22

R   Chapter 11 Review

Key Terms and Ideas

- Independent Experimental Design
- Completely Randomized Design
- Sampling Distribution of $\bar{x}_1 - \bar{x}_2$
- Interval Estimation of $\mu_1 - \mu_2$
- Hypothesized Difference
- Hypothesis Test about $\mu_1 - \mu_2$
- Assumptions for Small Sample Inferences about $\mu_1 - \mu_2$
- Small Sample Interval Estimation of $\mu_1 - \mu_2$ (σ_1 and σ_2 Unknown)
- Small Sample Hypothesis Test about $\mu_1 - \mu_2$ (σ_1 and σ_2 Unknown)

- Pooled Variance
- t-Test about $\mu_1 - \mu_2$ Assuming Unequal Variances
- Paired Difference Experimental Design
- Assumptions for the Paired Difference Experimental Design
- Assumptions for Comparing Two Population Proportions
- Interval Estimation of $p_1 - p_2$
- Hypothesis Test about $p_1 - p_2$

Key Formulas

	Section

Sampling Distribution of $\bar{x}_1 - \bar{x}_2$

$$\mu_{\bar{x}_1 - \bar{x}_2} = \mu_1 - \mu_2$$

$$\sigma_{\bar{x}_1 - \bar{x}_2} = \sqrt{\frac{\sigma_1^2}{n_1} + \frac{\sigma_2^2}{n_2}}$$

11.1

σ_1^2 and σ_2^2 can be approximated with s_1^2 and s_2^2, respectively, if $n_1 \geq 30$ and $n_2 \geq 30$.

$100(1 - \alpha)$% Confidence Interval for $\mu_1 - \mu_2$

$$\left(\bar{x}_1 - \bar{x}_2\right) \pm z_{\alpha/2} \sqrt{\frac{\sigma_1^2}{n_1} + \frac{\sigma_2^2}{n_2}}$$

11.1

where n_1 and n_2 are greater than or equal to 30 or σ_1 and σ_2 are known and the samples are drawn from a normal population.

Test Statistic for a Hypothesis Test about $\mu_1 - \mu_2$

$$z = \frac{\left(\bar{x}_1 - \bar{x}_2\right) - \mu_{\bar{x}_1 - \bar{x}_2}}{\sigma_{\bar{x}_1 - \bar{x}_2}} = \frac{\left(\bar{x}_1 - \bar{x}_2\right) - \left(\mu_1 - \mu_2\right)}{\sqrt{\frac{\sigma_1^2}{n_1} + \frac{\sigma_2^2}{n_2}}}$$

11.1

where n_1 and n_2 are greater than or equal to 30 or σ_1 and σ_2 are known and the samples are drawn from a normal population.

Key Formulas (cont.)

	Section

Small Sample $100(1 - \alpha)\%$ Confidence Interval for $\mu_1 - \mu_2$
(σ_1 and σ_2 Unknown and $\sigma_1 = \sigma_2$)

$$\left(\bar{x}_1 - \bar{x}_2 \right) \pm t_{\alpha/2, df} \sqrt{s_p^2 \left(\frac{1}{n_1} + \frac{1}{n_2} \right)}$$

11.2

where $s_p^2 = \dfrac{\left(n_1 - 1 \right) s_1^2 + \left(n_2 - 1 \right) s_2^2}{n_1 + n_2 - 2}$, and $t_{\alpha/2, df}$ is the critical value of the t-distribution

with $n_1 + n_2 - 2$ degrees of freedom capturing an area of $\alpha/2$ in the upper tail.

Test Statistic for a Small Sample Hypothesis Test about $\mu_1 - \mu_2$
(σ_1 and σ_2 Unknown and $\sigma_1 = \sigma_2$)

$$t = \frac{\left(\bar{x}_1 - \bar{x}_2 \right) - \left(\mu_1 - \mu_2 \right)}{\sqrt{s_p^2 \left(\dfrac{1}{n_1} + \dfrac{1}{n_2} \right)}}$$

11.2

Small Sample $100(1 - \alpha)\%$ Confidence Interval for $\mu_1 - \mu_2$
(σ_1 and σ_2 Unknown and $\sigma_1 \neq \sigma_2$)

$$\left(\bar{x}_1 - \bar{x}_2 \right) \pm t_{\alpha/2, df} \sqrt{\frac{s_1^2}{n_1} + \frac{s_2^2}{n_2}},$$

where $t_{\alpha/2, df}$ is the critical value capturing an area of $\alpha/2$ in the upper tail of t-distribution with degrees of freedom equal to:

11.2

$$df = \frac{\left(\dfrac{s_1^2}{n_1} + \dfrac{s_2^2}{n_2} \right)^2}{\dfrac{1}{n_1 - 1} \left(\dfrac{s_1^2}{n_1} \right)^2 + \dfrac{1}{n_2 - 1} \left(\dfrac{s_2^2}{n_2} \right)^2}.$$

Key Formulas (cont.)

	Section

Test Statistic for a Small Sample Hypothesis Test for $\mu_1 - \mu_2$
(σ_1 **and** σ_2 **Unknown and** $\sigma_1 \neq \sigma_2$ **)**

$$t = \frac{(\bar{x}_1 - \bar{x}_2) - (\mu_1 - \mu_2)}{\sqrt{\dfrac{s_1^2}{n_1} + \dfrac{s_2^2}{n_2}}},$$

which follows a t-distribution with degrees of freedom equal to:

$$df = \frac{\left(\dfrac{s_1^2}{n_1} + \dfrac{s_2^2}{n_2}\right)^2}{\dfrac{1}{n_1 - 1}\left(\dfrac{s_1^2}{n_1}\right)^2 + \dfrac{1}{n_2 - 1}\left(\dfrac{s_2^2}{n_2}\right)^2}.$$

11.2

$100(1 - \alpha)\%$ Confidence Interval for μ_d

$$\bar{x}_d \pm t_{\alpha/2,df}\,\frac{s_d}{\sqrt{n_d}},$$

where $\bar{x}_d$ is the sample mean of the differences, s_d is the sample standard deviation of the differences, n_d is the number of differences, and $t_{\alpha/2,df}$ is the critical value of the t-distribution with an area of $\alpha/2$ in the upper tail with $n_d - 1$ degrees of freedom (df).

11.3

Test Statistic for a Paired Difference Hypothesis Test

$$t = \frac{\bar{x}_d - \mu_d}{s_d \Big/ \sqrt{n_d}},$$

where $\bar{x}_d$ is the mean of the sample differences, μ_d is the hypothesized mean of the differences, s_d is the standard deviation of the sample differences, and n_d is the number of differences.

11.3

$100(1 - \alpha)\%$ Confidence Interval for $p_1 - p_2$

$$\left(\hat{p}_1 - \hat{p}_2\right) \pm z_{\alpha/2}\sqrt{\frac{\hat{p}_1\left(1 - \hat{p}_1\right)}{n_1} + \frac{\hat{p}_2\left(1 - \hat{p}_2\right)}{n_2}}$$

where $n_1\hat{p}_1 \geq 5,\, n_1\left(1 - \hat{p}_1\right) \geq 5,\, n_2\hat{p}_2 \geq 5,$ and $n_2\left(1 - \hat{p}_2\right) \geq 5$ and $z_{\alpha/2}$ is the critical value for the z-distribution that captures an area of $\alpha/2$ in the upper tail.

11.4

Key Formulas (cont.)	
	Section
Sampling Distribution of $\hat{p}_1 - \hat{p}_2$	
$$\mu_{\hat{p}_1 - \hat{p}_2} = p_1 - p_2$$ $$\sigma_{\hat{p}_1 - \hat{p}_2} = \sqrt{\bar{p}(1-\bar{p})\left(\frac{1}{n_1} + \frac{1}{n_2}\right)}$$	11.4
where $\bar{p} = \dfrac{x_1 + x_2}{n_1 + n_2}$	
Test Statistic for a Large Sample Hypothesis Test about $p_1 - p_2$	
$$z = \frac{\left(\hat{p}_1 - \hat{p}_2\right) - \left(p_1 - p_2\right)}{\sqrt{\bar{p}(1-\bar{p})\left(\frac{1}{n_1} + \frac{1}{n_2}\right)}}$$	11.4

AE   Additional Exercises

1. Black Bark, a Colorado based company, makes wood burning stoves. They are interested in comparing two designs to determine which design will produce a stove with a greater average burning time. Several prototypes of each design are tested and the time required to burn 15 pounds of wood was measured (the burning time is measured in hours). The results of the test are as follows.

Table for Exercise 1 – Burning Time for Stoves (Hours)			
	n	$\bar{x}$	s
Stove A	32	9.35	0.50
Stove B	35	9.75	0.75

Is there sufficient evidence at $\alpha = 0.05$ for Black Bark to conclude that the mean burning time for Stove B is greater than for Stove A?

2. In each of the following experimental situations give the appropriate null and alternative hypotheses to be tested. Define all terms that appear in these hypotheses.

a. Independent random samples of 50 male nurses and 50 female nurses are selected from the hospitals in a Southern state. Each nurse is asked whether he or she is satisfied with the working conditions in the hospital. It is of interest to see if there is a difference between male nurses and female nurses on satisfaction with working conditions.

b. A group of 45 high school seniors take the SAT reasoning test both before and after a 3-month training course, which is designed to improve SAT scores. We wish to determine if the training course is effective.

c. Starting salaries are determined for 40 female and 40 male electrical engineers. It is of interest to determine if female electrical engineers tend to have higher starting salaries than their male counterparts.

d. Random and independent samples of younger (age $\leq$ 30) and older (age > 30) automobile drivers are chosen and asked whether they have had a speeding ticket in the past 12 months. It is intended to show that younger drivers are more likely than older drivers to have had a speeding ticket in the past 12 months.

e. Do women have a shorter reaction time than men when exposed to a certain stimulus? Random and independent samples of 10 men and 10 women are included in an experiment that measures reaction time to the stimulus.

3. A nutritionist is interested in determining the decrease in cholesterol level which a person can achieve by following a particular diet that is low in fat and high in fiber. Seven subjects are randomly selected to try the diet for six months, and their cholesterol levels are measured both before and after the diet. The results of the study are as follows.

Table for Exercise 3 – Cholesterol Levels							
Subject	1	2	3	4	5	6	7
Before Diet	155	170	145	200	162	180	160
After Diet	152	168	148	195	162	178	157

a. Is a paired design appropriate for the above experiment? Explain.

b. What assumption must be made in order to perform the test of hypothesis?

c. Do the data appear to satisfy the assumption described in part **b.**? Why or why not?

d. Can the nutritionist conclude that there is a significant decrease in average cholesterol level when the diet is used? Use $\alpha = 0.01$.

4. The design group for a monofilament cord manufacturer is testing two possible compositions of the cord for tensile strength. Composition A is more difficult to manufacture than Composition B, so the design group has decided that it will recommend Composition A only if the mean tensile strength for Composition A is shown to be significantly greater than the mean tensile strength for Composition B. Several monofilament cords of each sample are tested and the tensile strengths are measured in pounds per square inch. Assume that the population variances are approximately equal.

Table for Exercise 4 – Tensile Strength (Pounds per Square Inch)			
	n	$\bar{x}$	s
Composition A	20	52,907	2575
Composition B	20	50,219	1210

a. What assumptions must be made in order to perform the hypothesis test?

b. Will the design group recommend Composition A or Composition B for the monofilament cord at $\alpha = 0.10$?

5. Consider Example 11.3. If you were to perform a two-sample t-test, you would find that you would fail to reject the null hypothesis and conclude that there is no difference in average daily sales between the two restaurants. Of course, it would be difficult to believe such a test given that if we examine the data in Table 11.4, we see that each of the daily sales figures from Restaurant 2 is more than that from Restaurant 1. From this observation, it is clear that the average daily sales between the restaurants are different. Why, then, would the t-test be unable to detect this difference? The answer: **an independent samples t-test is not a valid procedure to use with paired data**. The t-test is inappropriate because the assumption of independent samples is invalid since the dependence between restaurants is a function of the days. Perform the two-sample t-test to verify that the independent samples test would lead the owner to fail to reject the null hypothesis and conclude that there is not a significant difference between average daily sales. The data from Table 11.4 are replicated below for your convenience. Use $\alpha = 0.01$.

Table 11.4 – Daily Food Sales for Two Restaurants ($)			
Day	Restaurant 1	Restaurant 2	Difference = Restaurant 1 – Restaurant 2
1	5828	7894	−2066
2	9836	11,573	−1737
3	3984	5319	−1335
4	5845	6389	−544
5	5210	6055	−845
6	9668	10,631	−963
7	6768	7866	−1098
8	6726	7976	−1250
9	4399	5652	−1253
10	6692	8083	−1391

6. Two independent random samples have been selected, 100 from Population 1 and 150 from Population 2. The sample mean and standard deviation from the first population is 1025 and 10, respectively. For the second sample, the mean is 1039 with a standard deviation of 12.

   a. Test that there is no difference between the groups using $\alpha = 0.01$.

   b. Construct a 95% confidence interval for the true mean difference between Population 1 and Population 2.

7. The manufacturer of Brand 1 cigarettes claims that his cigarettes are no more harmful to health than Brand 2 (filtered) cigarettes. Assuming harmfulness is to be associated with nicotine content, the FDA took random samples of 125 cigarettes from Brand 1 and 180 cigarettes from Brand 2. The average nicotine content in the sample of Brand 1 was 24.6 mg with a standard deviation of 1.4 mg; the average nicotine content in the sample of Brand 2 was 24.3 mg with a standard deviation of 1.1 mg.

   a. Is there evidence to refute the manufacturer's claim at $\alpha = 0.05$?

   b. Construct an 85% confidence interval for the true mean difference in nicotine content between Brand 1 and Brand 2.

8. A team of organizational behavior managers investigated the effects of an orientation program on "first day of work" anxiety levels of new employees. 72 new employees were randomly assigned to receive or not to receive a two-day company orientation program prior to their first day at work. Two hours after beginning work, each employee was given a test to measure his or her level of anxiety. The mean score was 1002 for the 37 receiving orientation and 1018 for the 35 who did not receive the orientation. Scores of employees who attended similar orientation programs in the past have had a standard deviation of 142. Scores of employees that did not attend the orientation program had this same standard deviation.

   a. Test to see if there is evidence of a difference in the mean test scores between those who participate in an orientation program and those who do not. Use a level of significance equal to 0.05.

   b. Calculate the observed significance level (P-value) of this test.

9. A property manager of thousands of apartments wants to test the difference in the mean net annual income between two types of leasing arrangements. Arrangement A is to charge a lower rent but to require the tenants to make repairs. Arrangement B is to charge a higher rent and to state that the landlord will make the repairs. A sample of 25 apartments using Arrangement A had a mean net annual income of $1532.50 with a standard deviation of $400. A sample of 22 apartments using Arrangement B had a mean net annual income of $1489.20 with a standard deviation of $100. Test at $\alpha = 0.025$ that Arrangement B will have a lower mean net annual income by **at least $10** than Arrangement A. Assume that the incomes are normally distributed and that the population variances are equal.

10. For a consumer product, the mean dollar sales per retail outlet last year in a sample of 25 stores were $3425 with a standard deviation of $400. For a second product, the mean dollar sales per outlet in a sample of 16 stores were $3250 with a standard deviation of $175. The sales amounts per outlet are assumed to be approximately normally distributed for both products. Test to see if the first product has a better mean dollar sales record than the second product. Use the *P*-value approach and base your decision on a significance level of 0.01. Assume that the population variances are not equal.

11. A random sample of 10 filled sports drink bottles is taken in one bottling plant, and the mean weight of the bottles is found to be 22 ounces with a variance of 0.09 ounces squared. At another plant, 10 randomly selected bottles have a mean weight of 21 ounces with a variance of 0.04 ounces squared. Assuming the weights in both populations are normally distributed and the population variances are equal, test whether there is a difference between the average weights of the bottles being filled at the two plants. Use $\alpha = 0.05$.

12. Employers implementing workplace wellness programs report reduced healthcare costs, mostly because of fitness center memberships and free preventative screenings for their employees, according to the findings of a survey. Survey results show that employers who invest in wellness programs see increased employee retention, attendance, and productivity. The numbers show that medical costs improve and absenteeism costs improve. The following data have been summarized from the 30 employers that implemented wellness programs in 2011 along with their costs in 2010. The standard deviation of the difference in cost between the two years is $1322. Assume that independent, random samples were used in the study.

Table for Exercise 12 – Average Healthcare Cost per Employee (Dollars)		
	2010	**2011**
Mean	$11,151	$10,344
Sample Size	30	30

a. Calculate and interpret a 95% confidence interval for the true average difference in healthcare costs between 2010 and 2011.

b. Using the rejection region approach, test that there is a significant decrease in average healthcare costs between 2010 and 2011. Make your decision based on a significance level of 0.05.

c. Calculate the *P*-value for the test conducted in part **b**.

13.   Travelers are increasingly using tablets and e-readers on planes, trains and buses instead of more basic electronic devices such as cell phones and iPods. According to a study, the use of tablets and e-readers such as iPads, Kindles and Nooks rose more than 50% in 2011 across all modes of transportation. A team of researchers observed 7781 passengers on various airplane, bus and train departures in December of 2011. Their findings were compared with data about technology use among travelers for the fourth quarters of 2009 and 2010. The following table represents the number of travelers surveyed that used tablets or e-readers.

Table for Exercise 13 – Travelers That Used Tablets or e-Readers			
	2009	**2010**	**2011**
Number of Travelers	7545	7185	7781
Number using Tablets or e-Readers	330	570	792

   a.   Perform a hypothesis test using a significance level of 0.05 to determine if there is sufficient evidence to refute the claim that at least 10% of travelers used tablets or e-readers in each of the years surveyed.

   b.   Calculate and interpret a 95% confidence interval for the difference in the percentage of travelers using tablets or e-readers for 2010 and 2011.

   c.   Test that the proportion of travelers using tablets or e-readers has increased from 2010 to 2011. Make your decision based on a significance level of 0.05.

14.   The University of Michigan Transportation Research Institute believes that a teenage rite of passage may be losing its allure among young people. They believe that teens are in no hurry to get their driver's licenses. In a national study conducted in 1983 with a sample of 10,000 16-year-olds in the U.S., 46% had their licenses. In a similar study in 2008, surveying a sample of 10,000 16-year-olds in the U.S., only 31% had their licenses. Using the information provided, test to determine if there is a significant difference between the proportion of 16-year-olds that had their licenses in 1983 and 2008. Use a significance level of 0.05.

15. The results of a recent study show that people blessed with intelligence are not necessarily free from financial woes. Research by an economist at Ohio State University's Center for Human Research, detailed in an upcoming issue of the journal *Intelligence*, indicated that the smart may still suffer monetarily, making their own financial lives difficult by doing things like maxing out their credit cards and missing bill payments. While past studies have shown a link between being smarter and earning more money, that doesn't mean that those who are intelligent and wealthy don't suffer from debt. Generally, respondents with higher IQ scores earn higher income, with each additional IQ point adding $202 to $616 in income annually. However, the study demonstrated that there was no link between total wealth and a person's IQ. With research showing that individuals with higher IQs are not significantly better off financially than those with average or below average intelligence, the suggestion is that those who are smarter may not be as good about saving money as others. In fact, the researchers discovered that more intelligent people were actually slightly more prone to financial problems, with more than 12 percent of those with an IQ of 90 having maxed out their credit cards, versus less than 8 percent of those individuals with an IQ of 75 and below. Among the highly intelligent, or those respondents with an IQ greater than 125, 6 percent had maxed out credit cards and 11 percent occasionally missed payments. Additionally, there appeared to be a financial sweet spot near the average IQ score, for which individuals had the lowest financial distress. The survey looked at information provided by 7400 respondents who took part in the National Longitudinal Survey of Youth, which considered baby boomers nationwide. The latest study is based on data from 2004, when participants were 40 to 47 years old. Respondents were polled about income and total wealth, as well as three measures of financial distress: whether they had maxed out credit cards, if they missed paying bills during the past five years, and if they had ever declared bankruptcy. The Armed Forces Qualification Test, long used to measure intelligence, was used to generate IQ scores. As a takeaway, the study's author, economist Jay Zagorsky, stated that people with lower intelligence should not feel like they are at a financial disadvantage, while those people with a high IQ should not feel they are at an advantage. In other words, every person has the ability to make smart financial decisions like avoiding debt, regardless of his or her intelligence or income. The following table contains some of the data from the study.

Table for Exercise 15 – Income and Intelligence				
IQ	Average Income ($)	Average Total Wealth ($)	n	Percentage with Maxed out Credit Cards
<75	55,000	500,000	2000	7.8
90	70,000	750,000	3000	12.0
>125	95,000	950,000	2400	6.0

Source: "Smarter doesn't mean richer," May 14, 2007, www.creditcards.com

a. Determine if there is a significant difference in the proportion with maxed out credit cards of those with an IQ of 90 versus those with an IQ less then 75. Make your decision based on a significance level of 0.05.

b. Calculate a 95% confidence interval for the difference in the percentage of people with maxed out credit cards between those with an IQ of 90 and those with an IQ more than 125. Interpret the interval.

c. Did you make any assumptions to carry out the test in **a.** or to calculate the confidence interval in **b.**? If so, what were they?

Chapter Twelve
Analysis of Variance (ANOVA)

Introduction

Chapter 11 described methods for comparing two population means. In that chapter we also began to explore some experimental design issues. For example, if variation among sample observations could be significantly reduced by pairing experimental units, the paired difference test was used. But we paid a price for pairing experimental units by giving up degrees of freedom. If variation could not be significantly reduced by pairing, then experimental units should be randomly selected from each of the populations of interest. Often we will be interested in comparing more than two population means. In the sections that follow, we will extend the concepts presented in Chapter 11 to situations where there are more than two populations of interest.

There are many situations where someone might be interested in comparing several population means. A cellular phone manufacturer may wish to compare the average talk times for several different models of cellular phone. A marketing director may want to compare the average numbers of tickets sold for various promotional campaigns. A stock analyst may be interested in the average yields of various types of equities (small, medium, and large cap). A manufacturer may wish to compare the average outputs of operators during various work shifts. An accounting firm may want to compare auditor proficiency resulting from three different training methods. Comparisons of several population means can be made using the *F*-test, which will be discussed in Section 12.4.

Additionally, there are situations where we are interested in the average response of a variable that depends on two factors. For example, a manager might be interested in the number of defects in a plant with respect to different machines used by different operators. This situation can be analyzed using a **two-way ANOVA**, which will be discussed in sections 12.5 and 12.6.

12.1 Are Means from Samples Significantly Different?

In comparing population means, a natural question arises: *is there a significant difference between the means?* A preliminary answer to this question can be obtained by drawing a random sample from each of the populations of interest and computing the sample means. The larger the differences between the sample means, the more likely it would seem that there is a difference among the population means. How large is large enough to conclude that a difference is significant? Since a sample has been drawn from each of the populations, we must decide whether or not the observed differences between the sample means are simply due to random variation among the sample observations or to a real difference in the population means. Often, we wish to perform a hypothesis test to determine if a difference among population means can be attributed to some **treatment**.

> **Definition**
>
> Individuals or objects on which the experiment is performed are called **experimental units**.

Definition

An experimental condition applied to the units is called a **treatment**.

Suppose a marketing firm wants to know if there are differences between the sales of a certain brand of vehicles resulting from three different advertising strategies. The firm experimented with the marketing campaigns for a 12-month period and noted the sales for the three strategies. The marketing firm selected dealerships that have been using the strategies and computed the summary measures for sales that result in the box plots in Figure 12.1. The firm finds that the sample mean of the sales for Strategy 1 is approximately $8.1 million, the sample mean of the sales for Strategy 2 is approximately $6.4 million, and the sample mean of the sales for Strategy 3 is approximately $7.7 million. Is there a significant difference between the sales resulting from the three advertising strategies? Note that in this example the sales are the experimental units and the advertising strategies are the treatments.

Table 12.1 – Sales by Strategy (Millions of Dollars)		
Strategy 1	**Strategy 2**	**Strategy 3**
3	2	4
6	5	2
7	5	5
4	3	6
6	7	6
7	8	7
10	6	9
6	4	8
15	10	14
8	6	8
9	9	7
16	12	16

Box Plots – Sales by Strategy

Figure 12.1

As another example, suppose that an investor is interested in comparing the average portfolio rates of return achieved by two different local investment brokers, Mr. Morgan and Mr. Stanley. To help him make this comparison, the investor randomly selects returns for each of these investment brokers over a 25-year period. From these samples he calculates the summary measures, which result in the box plots shown in Figure 12.2. The sample median portfolio rate of return for Mr. Morgan is 10.9%, which is somewhat higher than the sample median portfolio rate of return for Mr. Stanley, which is 10.1%. Should the investor conclude that Mr. Morgan's average portfolio rate of return is significantly higher than Mr. Stanley's? Note that in this example, the portfolio rates of return are the experimental units and the investment broker is the treatment.

Figure 12.2

The investor might be tempted to simply compare the sample median rates of return for Mr. Morgan and Mr. Stanley before making a decision. However, this would not be a completely informed decision. Examining the box plots will give him a much better understanding of the situation. The box plots show that the majority of the observed rates of return for Mr. Morgan and Mr. Stanley are similar. The center of the box plot of observed rates of return for Mr. Morgan is slightly higher than for Mr. Stanley, but it is very possible that the difference can be attributed to sampling variation. In other words, the difference in the observed sample means is small when compared to the variation in the observed rates of return for the two samples.

Although box plots can provide a sense of whether or not there is a difference among the population means, they cannot help us evaluate how likely it is that the observed difference is due to ordinary sampling variation. The following sections will develop measures that will help us assess how likely it is that the observed differences among the sample means are due to ordinary sampling variation.

12.1 Exercises

Basic Concepts

1. What is the price that is paid when pairing experimental units in a paired difference experiment?

2. Give two examples of business situations in which a manager would be interested in comparing several population means.

3. Give an example of a business situation in which a manager would be interested in the average response of a variable that depends on more than one factor.

4. What are experimental units?

5. What is a treatment?

6. Why does simply comparing the sample means for multiple populations not suffice when determining if there is a significant difference in the population means?

7. Explain how box plots can be useful in analyzing data when comparing population means.

Exercises

8. Consider the following box plots for data collected to compare the average fat contents (in grams) per serving (2 tablespoons) of three popular brands of peanut butter.

Box Plots – Fat Contents

a. Based on the box plots, do you think that there may be a significant difference in the average fat contents per serving of Brand A and Brand B? Explain.

b. Based on the box plots, do you think that there may be a significant difference in the average fat contents per serving of Brand B and Brand C? Explain.

c. Based on the box plots, do you think that there may be a significant difference in the average fat content per serving of Brand A and Brand C? Explain.

9. Consider the following box plots for data collected to compare the average scores achieved on a standardized aptitude test by freshmen, sophomores, juniors, and seniors at a large university.

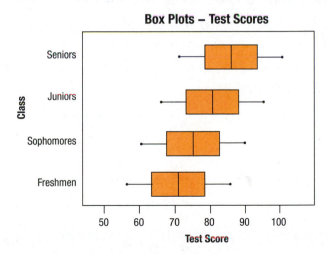

Box Plots – Test Scores

a. Based on the box plots, do you think that there may be a significant difference in the average scores achieved by freshmen and sophomores on the standardized test? Explain.

b. Based on the box plots, do you think that there may be a significant difference in the average scores achieved by freshmen and juniors on the standardized test? Explain.

c. Based on the box plots, do you think that there may be a significant difference in the average scores achieved by juniors and seniors on the standardized test? Explain.

10. Consider the following table containing yields for mutual funds in different asset classes (small, mid, and large cap).

Table for Exercise 10 – Fund Yields by Asset Class					
Small Cap		**Mid Cap**		**Large Cap**	
Fund	**Yield (%)**	**Fund**	**Yield (%)**	**Fund**	**Yield (%)**
Explorer Value	1.13	Capital Value	0.96	Equity Income	3.24
Small-Cap Value Index Admiral	2.46	Mid-Cap Value Index Admiral	2.47	High Dividend Yield Index	3.50
Small-Cap Index Admiral Shares	1.49	Extended Market Index Admiral Shares	1.22	500 Index Admiral Shares	2.35
Strategic Small-Cap Equity	1.10	Mid-Cap Index Admiral Shares	1.52	Diversified Equity	1.23
Explorer	0.17	Mid-Cap Growth	0.10	FTSE Social Index	1.42
Small-Cap Growth Index Admiral	0.21	Mid-Cap Growth Index Admiral	0.32	Growth Equity	0.60

Source: The Vanguard Group, Inc.

a. Identify the experimental units and the treatment in the context of this problem.

b. Compute the mean and median yields for each asset class.

c. Compute the values of the minimum, maximum, first, and third quartiles for each asset class.

d. Construct side-by-side box plots for the three asset classes.

e. Based on the box plots, do you think that there may be a significant difference in the average yields of small-cap and mid-cap funds? Explain.

f. Based on the box plots, do you think that there may be a significant difference in the average yields of mid-cap and large-cap funds? Explain.

g. Based on the box plots, do you think that there may be a significant difference in the average yields of small-cap and large-cap funds? Explain.

h. Based on your analysis, which asset class contains mutual funds with the largest yields, on average? Explain your answer.

11. Consider the following table containing daily production data from a particular week for three different employee shifts.

Table for Exercise 11 – Items Produced		
	First Shift **(7 AM-3 PM)**	**Second Shift** **(3 PM-11 PM)**	**Third Shift** **(11 PM-7 AM)**
Monday	140	168	77
Tuesday	181	224	123
Wednesday	127	162	77
Thursday	172	182	101
Friday	161	219	147
Saturday	152	171	145
Sunday	173	217	111

a. Identify the experimental units and the treatment in the context of this problem.

b. Compute the mean and median numbers of items produced for each shift.

c. Compute the values of the minimum, maximum, first, and third quartiles for each shift.

d. Construct side-by-side box plots for the three shifts.

e. Based on the box plots, do you think that there may be a significant difference in the average numbers of items produced during the first and second shifts? Explain.

f. Based on the box plots, do you think that there may be a significant difference in the average numbers of items produced during the second and third shifts? Explain.

g. Based on the box plots, do you think that there may be a significant difference in the average numbers of items produced during the first and third shifts? Explain.

h. Based on your analysis, which shift would you say is the most productive, on average? Explain your answer.

12.2   Analysis of Variance (ANOVA)

Analysis of variance (ANOVA) deals with differences between or among sample means. You may recall that you were restricted to testing the difference between two population means when using the t-test in Chapter 11. Unlike the t-test, analysis of variance does not restrict the number of means. Instead of asking whether two means differ, we ask if three, four, five, etc., means differ. ANOVA allows us to deal with two or more independent variables simultaneously, asking not only about the effects of each variable separately, but also about how two or more variables interact. In studying ANOVA, we will analyze the differences in means by breaking apart the total variation. The total variation will be separated into two pieces: sum of squares for treatments, SST (the variation attributed to the treatments) and sum of squares for error, SSE (the variation not explained by the treatments).

The variation among all the sample observations, without regard to which treatment or population they are from, can be summarized by the **sample variance**, which is given by the following formula.

Formula

Sample Variance

$$s^2 = \frac{\sum\limits_{j=1}^{k}\sum\limits_{i=1}^{n_j}\left(x_{ij} - \overline{\overline{x}}\right)^2}{n_T - 1}$$

where

$\overline{\overline{x}}$ is the sample mean of all of the observations (the grand mean),

n_j is the number of observations in the jth treatment,

k is the number of treatments, and

n_T is the total number of observations in all samples.

The numerator of s^2 is called the **total sum of squares**, or **TSS**, since it describes the total variation among all of the sample observations. The denominator of s^2 gives the degrees of freedom associated with TSS, $n_T - 1$.

The total variation in the sample measurements can be partially attributed to the treatments from which they came and partially attributed to random variation inherent in sampling. Thus, the total sum of squares can be divided into two components. The first component measures the variation that can be attributed to the treatments and is called the **sum of squares for treatments**, or **SST**.

Formula

Sum of Squares for Treatments

The mathematical expression for the sum of squares for treatments is given by

$$\text{SST} = \sum_{j=1}^{k} n_j \left(\bar{x}_j - \bar{\bar{x}} \right)^2$$

where

n_j is the number of observations in the jth treatment,

$\bar{x}_j$ is the sample mean of the observations in the jth treatment, and

$\bar{\bar{x}}$ is the grand mean.

SST is a summary measure of how much each treatment or population mean differs from the grand mean, $\bar{\bar{x}}$, which is the mean of all of the sample observations. The larger the differences between the treatment means $\left(\bar{x}_j \right)$ and the grand mean $\left(\bar{\bar{x}} \right)$, the more likely the variation in sample observations is due to the treatments rather than to ordinary sampling variation. Thus, SST measures the variation between the treatments. If using statistical software to perform ANOVA, the variation between treatments may also be referred to as "Among variation" or "Between variation." In Microsoft Excel, the source of variation associated with treatments is called "Between Groups" variation. In Minitab, the output yields an ANOVA table that refers to treatment variation as "Factor" variation.

The degrees of freedom associated with SST is $k-1$, where k is the number of treatments. The **mean square for treatments, MST**, is the sum of squares for treatments divided by its degrees of freedom.

$$\text{MST} = \frac{\text{SST}}{k-1}$$

The MST represents the average weighted squared deviation of the sample treatment means from the grand mean. The MST is basically the variance of the sample means.

The second component measures the random variation attributable to sampling and is called the **sum of squares for error**, or **SSE**.

Formula

Sum of Squares for Error

The mathematical expression for the sum of squares for error is given by the following formula.

$$\text{SSE} = \underbrace{\sum_{i=1}^{n_1} \left(x_{i1} - \bar{x}_1 \right)^2}_{\substack{\text{Note: This is the} \\ \text{numerator of} \\ \text{the sample variance} \\ \text{for Sample 1.}}} + \sum_{i=1}^{n_2} \left(x_{i2} - \bar{x}_2 \right)^2 + \cdots + \sum_{i=1}^{n_k} \left(x_{ik} - \bar{x}_k \right)^2$$

The SSE is a summary measure of how much of the total variation in the sample data is not explained by SST. The expression is a direct calculation of SSE which shows that the SSE summarizes how much the sample observations within each treatment vary from the treatment mean. In other words,

SSE is a measure of the variation within the treatments. In practice, it is much easier to calculate SSE as the total variation in the sample, TSS, minus the variation explained by the treatments, SST. The variation associated with the error term is often times called "Within Groups" variation or simply "Error" variation.

The degrees of freedom associated with SSE are $n_T - k$. The **mean square for error**, **MSE**, is the SSE divided by its degrees of freedom. The MSE represents the total variance (or common variance) of the model. That is, the estimate of σ^2 is s^2, which is given by the MSE of the ANOVA model.

$$MSE = \frac{SSE}{n_T - k}$$

Thus we have the following results.

Sum of Squares

$$TSS = SST + SSE$$

$$\sum_{j=1}^{k} \sum_{i=1}^{n_j} \left(x_{ij} - \overline{\overline{x}} \right)^2 = \sum_{j=1}^{k} n_j \left(\overline{x}_j - \overline{\overline{x}} \right)^2 + \left\{ \sum_{i=1}^{n_1} \left(x_{i1} - \overline{x}_1 \right)^2 + \sum_{i=1}^{n_2} \left(x_{i2} - \overline{x}_2 \right)^2 + \cdots + \sum_{i=1}^{n_k} \left(x_{ik} - \overline{x}_k \right)^2 \right\}$$

Degrees of Freedom

$$Total = Treatment + Error$$
$$n_T - 1 = (k-1) + (n_T - k)$$

As you might expect, these measures of variability are the fundamental pieces which will be used to develop a hypothesis test for determining whether or not there is a significant difference among the population means. This is precisely the reason that the test which we will use to make this decision is often called an **analysis of variance**, or **ANOVA**. The expressions are complicated. Fortunately, most statistical analysis programs will compute these quantities.

In the example from Section 12.1 when the marketing firm wanted to determine if there was a difference in the sales resulting from the three advertising strategies, there were two overriding factors which contributed to the analysis. First, the firm considered if there were significant differences between the average sales for the three strategies, which is summarized by SST. The second factor considered was whether or not these differences were larger than the differences among the sample observations for each of the strategies, which is summarized by SSE. These sources of variation (SST and SSE) will be the basis for answering the question of whether or not we can conclude that there is a significant difference among population means.

Is $\mu_1 = \mu_2 = \mu_3$?

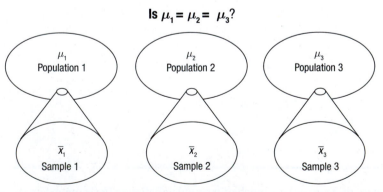

The null hypothesis will state that the population means are equal.

Formulas

ANOVA Formulas

Grand Mean:

$$\bar{\bar{x}} = \frac{\sum\limits_{j=1}^{k}\sum\limits_{i=1}^{n_j} x_{ij}}{n_T}$$

Sum of Squares for Treatments:

$$SST = \sum\limits_{j=1}^{k} n_j \left(\bar{x}_j - \bar{\bar{x}}\right)^2$$

Mean Square for Treatments:

$$MST = \frac{SST}{k-1}$$

Sum of Squares for Error:

$$SSE = \sum\limits_{i=1}^{n_1} \left(x_{i1} - \bar{x}_1\right)^2 + \sum\limits_{i=1}^{n_2} \left(x_{i2} - \bar{x}_2\right)^2 + \cdots + \sum\limits_{i=1}^{n_k} \left(x_{ik} - \bar{x}_k\right)^2$$

Mean Square for Error:

$$MSE = \frac{SSE}{n_T - k}$$

Total Sum of Squares:

$$TSS = SST + SSE = \sum\limits_{j=1}^{k}\sum\limits_{i=1}^{n_j} \left(x_{ij} - \bar{\bar{x}}\right)^2$$

Having calculated the sum of squares for each element for analysis of variance, the data can be summarized in the following table. This is the table often used when performing the F-test, which is discussed in Section 12.4.

Table 12.2 – ANOVA Table

Source of Variation	Sum of Squares	Degrees of Freedom	Mean Square	F-Statistic
Between Groups	SST	$k-1$	$SST/(k-1)$	MST/MSE
Within Groups	SSE	$n_T - k$	$SSE/(n_T - k)$	
Total	TSS	$n_T - 1$	Notes: $TSS = SST + SSE$ Reject H_0 if $F \geq F_\alpha$	

12

12.2   Exercises

Basic Concepts

1. How is the total variation in the dependent variable broken down in analysis of variance?

2. What does the total sum of squares describe? What are its degrees of freedom?

3. What is the mathematical expression for the sum of squares for treatments?

4. What is the grand mean? How is it calculated?

5. What is the mean square for treatments?

6. What is the relationship between TSS, SST, and SSE? Explain why this relationship makes sense.

Exercises

7. The sales by strategy data given in Table 12.1 yield the following statistics.

Table for Exercise 7 – Sales by Strategy (Millions of Dollars)		
Strategy 1	Strategy 2	Strategy 3
3	2	4
6	5	2
7	5	5
4	3	6
6	7	6
7	8	7
10	6	9
6	4	8
15	10	14
8	6	8
9	9	7
16	12	16

$$SST \approx 18.0556$$
$$SSE = 438.5$$

a. What are the degrees of freedom associated with the total sum of squares?

b. What are the degrees of freedom associated with the sum of squares for treatments?

c. Find the mean square for treatments, MST.

d. Find the mean square for error, MSE.

8.   The fund yield data given in Exercise 10 in Section 12.1 give the following summary statistics.

Table for Exercise 8 – Fund Yields by Asset Class					
Small Cap		**Mid Cap**		**Large Cap**	
Fund	**Yield (%)**	**Fund**	**Yield (%)**	**Fund**	**Yield (%)**
Explorer Value	1.13	Capital Value	0.96	Equity Income	3.24
Small-Cap Value Index Admiral	2.46	Mid-Cap Value Index Admiral	2.47	High Dividend Yield Index	3.50
Small-Cap Index Admiral Shares	1.49	Extended Market Index Admiral Shares	1.22	500 Index Admiral Shares	2.35
Strategic Small-Cap Equity	1.10	Mid-Cap Index Admiral Shares	1.52	Diversified Equity	1.23
Explorer	0.17	Mid-Cap Growth	0.10	FTSE Social Index	1.42
Small-Cap Growth Index Admiral	0.21	Mid-Cap Growth Index Admiral	0.32	Growth Equity	0.60

Source: The Vanguard Group, Inc.

$$\text{MST} \approx 1.8464$$
$$\text{MSE} \approx 0.9423$$

a.   Interpret the value of MST.

b.   What are the degrees of freedom associated with the sum of squares for treatments?

c.   Find the sum of squares for treatments.

d.   What are the degrees of freedom for the sum of squares for error?

e.   Find the sum of squares for error.

9.   Consider the production data given in Section 12.1, Exercise 11.

Table for Exercise 9 – Items Produced			
	First Shift (7 AM-3 PM)	**Second Shift (3 PM-11 PM)**	**Third Shift (11 PM-7 AM)**
Monday	140	168	77
Tuesday	181	224	123
Wednesday	127	162	77
Thursday	172	182	101
Friday	161	219	147
Saturday	152	171	145
Sunday	173	217	111

a.   What is the value of the grand mean, $\overline{\overline{x}}$?

b.   What is the value of n_j?

c.   What is the value of k?

d.   What is the value of n_T?

e.   For these data, identify the degrees of freedom associated with the total sum of squares, the degrees of freedom associated with the sum of squares for treatments, and the degrees of freedom associated with the sum of squares for error. Verify that the relationship between the degrees of freedom $(\text{Total} = \text{Treatment} + \text{Error})$ holds.

12.3   Assumptions of the Test

Before proceeding with a description of the hypothesis testing procedure for determining whether or not there is a significant difference among several population means, it is important to point out the assumptions upon which the test is based. If the data do not appear to conform to these assumptions, the hypothesis testing procedures described in the subsequent sections will not be valid inferential techniques.

The first assumption is that the distributions of all k populations of interest are approximately normal. The best way to determine whether or not this assumption is satisfied is to construct a histogram of the sample data for each of the k populations of interest.

If the histograms appear to be approximately normal, then it is reasonable to proceed. Figure 12.3 gives examples of histograms of sample data drawn from normal populations.

Figure 12.3

The second assumption is that the variances of the k populations of interest are equal. We can determine if this assumption is reasonable by drawing box plots of the k samples of data and comparing the spread of the data for each sample. If the spread or variability of the data is approximately the same for each of the k samples, it is reasonable to proceed. For example, consider the box plots for three samples from three populations of interest in Figure 12.4. Although the spread of the box plot is not exactly the same for each of the samples, it is similar enough that it is likely that the observed difference in spread is due to sampling variation, and it is safe to proceed. There is a hypothesis test which can be used to determine if there is a significant difference among the variances, but that testing procedure is beyond the scope of this text.

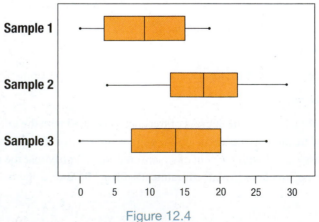

Figure 12.4

Note: There is a simple "rule of thumb" that you can use to check the variance assumption. If the largest standard deviation is no more than twice the smallest, then the presumption is that the assumption holds.

The third assumption that must be met is that each of the k samples must be selected independently from each other and in a random fashion from each of the respective populations.

Assumptions

ANOVA Assumptions

In order for an analysis of variance to be valid, the following three assumptions must hold.

1. The distributions of all k populations of interest are approximately normal.

2. The variances of all k populations are equal.

3. The sample observations are randomly selected and independent.

12.3 Exercises

Basic Concepts

1. Why is it important to validate the assumptions upon which a hypothesis test is based?

2. What is the first assumption on which ANOVA is based?

3. How can we test to see if the data reasonably satisfy the first assumption?

4. What is the second assumption on which ANOVA is based?

5. How can we determine if the second assumption is reasonable for the data we are interested in?

6. What is a simple "rule of thumb" that may be used to check the second assumption?

7. What is the third assumption that must be met before performing ANOVA?

Exercises

8. For each of the following histograms of sample data, decide whether or not you think it is reasonable to assume that the data were drawn from a population that has an approximately normal distribution.

9. For each of the following histograms of sample data, decide whether or not you think it is reasonable to assume that the data were drawn from a population that has an approximately normal distribution.

10. Consider the following box plots.

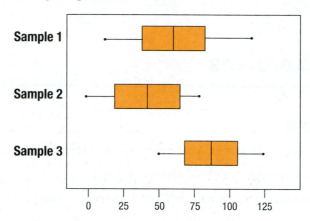

Do you think it is reasonable to assume that the three populations represented by the sample data in these box plots have equal variances? Explain.

11. Consider the following box plots.

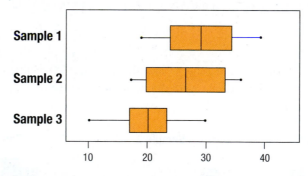

Do you think it is reasonable to assume that the three populations represented by the sample data in these box plots have equal variances? Explain.

12.4   The *F*-Test

In Section 12.2, we introduced most of the formulas that are used to analyze data from more than two samples in order to determine whether or not there is a significant difference among population means. We developed MST, a measure for summarizing the variability among the sample means, and MSE, a measure for summarizing the variability within the samples themselves. We determined that if the variability among the sample means is much larger than the variability within the sample observations, we will doubt the hypothesis that the population means are the same. Alternatively, if the variability among the sample means is small when compared to the variability within the sample observations, it is not likely that the population means are significantly different.

Consider the ratio of the MST (mean square for treatments), the summary measure of the variability among the sample means, to the MSE (mean square for error), the summary measure of the variability within the samples.

$$\frac{\text{MST}}{\text{MSE}}$$

The **F-distribution**, named after the English statistician Sir Ronald Fisher, is a continuous distribution. It will be used in this and subsequent chapters to analyze variation in test statistics formed as ratios of two random variables. The *F*-distribution is not symmetrical, rather, it is skewed to the right. Like the *t*-distribution, its parameters are degrees of freedom. *F*-distributions are associated with test statistics that are quotients. What distinguishes the *F*-distribution is that it has a pair of values for its degrees of freedom. The number of degrees of freedom associated with the numerator is df_{num} and the number of degrees of freedom associated with the denominator is df_{den}.

The shape of the *F*-distribution is a function of its degrees of freedom. Below, the *F*-distributions for $df_{num} = 6$, $df_{den} = 40$ and $df_{num} = 10$, $df_{den} = 4$ are given.

F-Distributions

Like the *t*-distribution, tables have been compiled (see Appendix A) that show critical values of the *F*-distribution at common levels of significance. The notation F_{α} will indicate the value of *F* such that α of the area under the curve lies to the right of this value.

For example, to find the critical value at $\alpha = 0.05$ for an F-distribution with 3 numerator degrees of freedom and 7 denominator degrees of freedom, consult the table in Appendix A, Table H, with $\alpha = 0.05$. The leftmost column corresponds to the denominator degrees of freedom, and the top row corresponds to the numerator degrees of freedom.

Numerator Degrees of Freedom

	1	2	3	4
1	161.4476	199.5000	215.7073	224.5832
2	18.5128	19.0000	19.1643	19.2468
3	10.1280	9.5521	9.2766	9.1172
4	7.7086	6.9443	6.5914	6.3882
5	6.6079	5.7861	5.4095	5.1922
6	5.9874	5.1433	4.7571	4.5337
7	5.5914	4.7374	4.3468	4.1203
8	5.3177	4.4590	4.0662	3.8379

(Row labels are **Denominator Degrees of Freedom**)

From the table, we see that $F_{0.05}$ for 3 numerator degrees of freedom and 7 denominator degrees of freedom is 4.3468. Thus, for this F-distribution, an area of 0.05 lies to the right of 4.3468.

If the assumptions discussed in Section 12.3 are met and we assume that the population means are equal, this quantity will have an F-distribution with $k - 1$ numerator degrees of freedom (associated with MST) and $n_T - k$ denominator degrees of freedom (associated with MSE). If the variability among the sample means is close to the variability within the sample observations, F will be close to 1. However, as the variability among the sample means increases relative to the variability within the sample observations, the value of F will become large, causing us to doubt the assumption that the population means are equal.

Thus, $F = \dfrac{MST}{MSE}$ is a natural test statistic to use in determining whether or not a difference exists among the population means. We will reject the null hypothesis that the population means are equal for large values of the F-test statistic. This is why the analysis of variance (or ANOVA) test for differences among population means is often referred to as the **F-test**.

Procedure

F-Test

A summary of the *F*-test is given below.

Hypotheses:

$$H_0: \mu_1 = \mu_2 = \cdots = \mu_k \qquad \text{(The } k \text{ population means are equal.)}$$

$$H_a: \text{At least one } \mu_i \text{ is different.}$$

Test Statistic:

$$F = \frac{\text{MST}}{\text{MSE}} = \frac{\dfrac{\sum_{j=1}^{k} n_j \left(\bar{x}_j - \bar{\bar{x}} \right)^2}{k-1}}{\dfrac{\sum_{i=1}^{n_1} \left(x_{i1} - \bar{x}_1 \right)^2 + \sum_{i=1}^{n_2} \left(x_{i2} - \bar{x}_2 \right)^2 + \cdots + \sum_{i=1}^{n_k} \left(x_{ik} - \bar{x}_k \right)^2}{n_T - k}}$$

Assumptions:

The k populations of interest are normally distributed with equal variances, and independent, random samples are drawn from each population.

Rejection Region:

H_0 will be rejected for large values of $F = \dfrac{\text{MST}}{\text{MSE}}$. In particular, we will reject H_0 if $F \geq F_\alpha$ with $(k-1)$ numerator degrees of freedom and $(n_T - k)$ denominator degrees of freedom.

Unfortunately, the formulas for MST and MSE given in Section 12.2 are difficult to use for actual calculations. So we present computational formulas for these summary measures.

Formula

Computational Formulas for MST and MSE

Let k be the total number of treatments and $n_T = n_1 + n_2 + \cdots + n_k$ be the total number of observations. Then the computational formulas are as follows.

$$\text{MST} = \frac{\left[\dfrac{\left(\sum_{i=1}^{n_1} x_{i1} \right)^2}{n_1} + \dfrac{\left(\sum_{i=1}^{n_2} x_{i2} \right)^2}{n_2} + \cdots + \dfrac{\left(\sum_{i=1}^{n_k} x_{ik} \right)^2}{n_k} \right] - \dfrac{\left(\sum_{j=1}^{k} \sum_{i=1}^{nj} x_{ij} \right)^2}{n_T}}{k-1}$$

$$\text{MSE} = \frac{\sum_{j=1}^{k} \sum_{i=1}^{n_j} x_{ij}^2 - \left[\dfrac{\left(\sum_{i=1}^{n_1} x_{i1} \right)^2}{n_1} + \dfrac{\left(\sum_{i=1}^{n_2} x_{i2} \right)^2}{n_2} + \cdots + \dfrac{\left(\sum_{i=1}^{n_k} x_{ik} \right)^2}{n_k} \right]}{n_T - k}$$

Hopefully, you will never have to use these formulas in practice, for there are many statistical packages available for this purpose.

Example 12.1

You have just been promoted to sales manager of a company that manufactures robots used to assemble automobiles. Although your sales force is given a suggested price at which to sell the robots, they have considerable leeway in negotiating the final price. Past sales records indicate that sometimes there is a large difference in the selling prices that different sales reps are able to negotiate. You are interested in knowing if this difference is significant, possibly because of a more effective negotiating strategy or exceptional interpersonal skills, or whether this observed difference in selling prices is just due to random variation. You decide to randomly select four sales over the last year for each of your three sales representatives and observe the actual selling prices of the robot. Table 12.3 shows the amounts at which the robots sold in thousands of dollars.

Table 12.3 – Selling Prices (Thousands of Dollars)		
Salesperson 1	**Salesperson 2**	**Salesperson 3**
10	11	11
14	16	13
13	14	12
12	15	15
Total 49	56	51

Based on the results of your survey, can you conclude that there is a significant difference among the average selling prices that the three sales reps have been able to negotiate? Use $\alpha = 0.05$.

Solution

Step 1: State the hypotheses in plain English.

The hypotheses are fairly straightforward.

- **Null Hypothesis:** There is no difference among the average selling prices for the three sales reps.

- **Alternative Hypothesis:** There is a difference among the average selling prices for the three sales reps.

Step 2: Select the appropriate statistical measure.

Since you are interested in comparing the average selling prices for the three sales reps, the population parameters of interest are the true mean selling prices for the three sales reps.

$$\mu_1 = \text{true mean selling price for sales rep \#1}$$
$$\mu_2 = \text{true mean selling price for sales rep \#2}$$
$$\mu_3 = \text{true mean selling price for sales rep \#3}$$

Step 3: Determine whether the hypothesis should be one-sided or two-sided.

Based on the way the test statistic is constructed, we will reject the null hypothesis for large values of the test statistic, meaning that the variability among the sample means is much larger than the variability within the sample observations. The *F*-test is always a one-tailed test.

Step 4: Specify the hypotheses using the appropriate statistical measure.

H_0: $\mu_1 = \mu_2 = \mu_3$ (The mean selling prices are the same for all three sales reps.)

H_a: At least one μ_i is different.

Step 5: Specify the level of the test.

The level of the test is specified in the problem as $\alpha = 0.05$.

Step 6: Select the appropriate test statistic.

There are three key questions we must ask.

1. Are the selling prices for each sales rep normally distributed?

2. Do the selling prices for each sales rep have approximately equal variances?

3. Were the sample selling prices for each sales rep collected in an independent and random fashion?

Based on prior studies, you have reason to believe that the selling prices of the robots for each of the sales reps have an approximately normal distribution and that the variances of the three distributions are approximately equal. Also, you used random sampling to collect your data. Thus, the appropriate test statistic is the *F*-statistic.

$$F = \frac{\text{MST}}{\text{MSE}}$$

Step 7: Determine the critical value.

The level of the test is $\alpha = 0.05$, and there are $(k-1) = (3-1) = 2$ numerator degrees of freedom and $(n_T - k) = (12-3) = 9$ denominator degrees of freedom. Thus the critical value is $F_{0.05} = 4.2565$. We will reject H_0 if the computed value of the test statistic is greater than or equal to 4.2565. Figure 12.5 displays the rejection region.

F-Distribution
$df_{num} = 2, df_{den} = 9$

Fail to Reject H_0 Reject H_0

$\alpha = 0.05$

0   4.2565   F

Figure 12.5

Step 8: Compute the test statistic.

Using the computational formulas presented previously, we have the following.

$$MST = \cfrac{\left[\cfrac{\left(\sum\limits_{i=1}^{n_1} x_{i1}\right)^2}{n_1} + \cfrac{\left(\sum\limits_{i=1}^{n_2} x_{i2}\right)^2}{n_2} + \cdots + \cfrac{\left(\sum\limits_{i=1}^{n_k} x_{ik}\right)^2}{n_k}\right] - \cfrac{\left(\sum\limits_{j=1}^{k}\sum\limits_{i=1}^{nj} x_{ij}\right)^2}{n_T}}{k-1}$$

$$= \cfrac{\left[\cfrac{(49)^2}{4} + \cfrac{(56)^2}{4} + \cfrac{(51)^2}{4}\right] - \cfrac{(156)^2}{12}}{3-1}$$

$$= 3.25$$

$$MSE = \cfrac{\sum\limits_{j=1}^{k}\sum\limits_{i=1}^{n_j} x_{ij}^2 - \left[\cfrac{\left(\sum\limits_{i=1}^{n_1} x_{i1}\right)^2}{n_1} + \cfrac{\left(\sum\limits_{i=1}^{n_2} x_{i2}\right)^2}{n_2} + \cdots + \cfrac{\left(\sum\limits_{i=1}^{n_k} x_{ik}\right)^2}{n_k}\right]}{n_T - k}$$

$$= \cfrac{(10)^2 + (14)^2 + \cdots + (15)^2 - \left[\cfrac{(49)^2}{4} + \cfrac{(56)^2}{4} + \cfrac{(51)^2}{4}\right]}{12-3}$$

$$= 3.5$$

The resulting calculated value of the test statistic is as follows.

$$F = \frac{MST}{MSE} = \frac{3.25}{3.5} \approx 0.9286$$

Step 9: Make the decision.

Since the resulting value of the test statistic, 0.9286, is less than the critical value of 4.2565, we fail to reject the null hypothesis (see Figure 12.6).

Figure 12.6

Step 10: State the conclusion in terms of the original question.

There is not sufficient evidence at $\alpha = 0.05$ to reject the null hypothesis. Thus, we cannot conclude that there is a difference among the average selling prices for the three sales reps.

It is often more convenient to use statistical software (rather than the computational formulas) to perform an analysis of variance such as the one in this example. The summary output from Microsoft Excel is given in Figure 12.7. (See the Discovering Technology section at the end of this chapter for directions on performing ANOVA using Microsoft Excel.)

ANOVA						
Source of Variation	SS	df	MS	F	P-value	F crit
Between Groups	6.5	2	3.25	0.928571429	0.429903823	4.256494729
Within Groups	31.5	9	3.5			
Total	38	11				

Figure 12.7

Note that the *F*-statistic is approximately 0.9286 and the *P*-value associated with the test is approximately 0.4299. Since 0.4299 > 0.05, there is not sufficient evidence to conclude that there is a difference between the selling prices negotiated by the three salespersons.

Note that if the null hypothesis that all population (or group) means are equal is rejected, the analysis of variance procedure does not specifically tell you *which* population means are not equal. There are procedures however, called **multiple comparison procedures**, that allow us to assess differences among specified means. Some of the more popular multiple comparison procedures are the Bonferroni method (1981), Scheffe's method (1953), which develops a general procedure for comparing all possible linear combinations of treatment means, called contrasts, and the Tukey method (1949), which is a procedure specifically for pairwise comparisons when the sample sizes of the treatments are equal. These procedures are beyond the scope of this text, however, and will not be discussed in detail.

12.4   Exercises

Basic Concepts

1. If you found that MST is much larger than MSE, would you tend to think that the population means were similar or different? Explain how this ratio brings you to this conclusion.

2. What kind of distribution does the ratio $\dfrac{MST}{MSE}$ have?

3. What are the degrees of freedom associated with $\dfrac{MST}{MSE}$?

4. If the variability among the sample means is very similar to the variability among the sample observations, what value will *F* be close to? Explain why.

5. Is the null hypothesis generally rejected for large or small values of the *F*-statistic? Explain why this is the case.

6. What are the null and alternative hypotheses for the *F*-test?

7. What are the assumptions of the *F*-test?

8. What is the rejection region for the *F*-test?

9. Can *P*-values be used to make a decision for the *F*-test? What is the decision rule?

Exercises

10. The results of a comparison of four popular minivans are reported in the following table. One of the features the researchers compared was the distance (in feet) required for the minivan to come to a complete stop when traveling at a speed of 60 miles per hour (braking distance). Suppose the braking distances were measured for five minivans of each type with the following results.

Table for Exercise 10 – Braking Distances (Feet)			
Minivan A	Minivan B	Minivan C	Minivan D
150	153	155	167
152	150	150	164
151	156	157	169
149	151	158	162
153	155	155	173

a. Can the researchers conclude at $\alpha = 0.10$ that there is a difference among average braking distances for the four minivan models?

b. What assumptions did the researchers make in performing the test procedure in part **a.**? Do the data appear to satisfy these assumptions? Explain.

11. A steel company is considering the relocation of one of its manufacturing plants. The company's executives have selected four areas that they believe are suitable locations. However, they want to determine if the average wages are significantly different in any of the locations, since this could have a major impact on the cost of production. A survey of hourly wages of similar workers in each of the four areas is performed with the following results.

Table for Exercise 11 – Hourly Wages ($)			
Area 1	Area 2	Area 3	Area 4
10	15	13	20
12	16	14	16
11	18	15	18
13	17	15	17
10	14	12	16

a. Do the data indicate a significant difference among the average hourly wages in the four areas at $\alpha = 0.05$?

b. What assumptions were made in performing the test in part **a.**? Do the data appear to satisfy these assumptions? Explain.

12. A director of training at a large temporary services company has learned of three different methods for teaching a person to type. He is interested in determining if there is a difference in the average typing speeds for employees who are taught to type using each of the three methods. He randomly selects 15 new employees and then randomly assigns five employees to learn to type by each of the training methods. At the end of the course, he measures the number of correct words per minute for each employee. The results are as follows.

Table for Exercise 12 – Typing Speeds (Correct Words per Minute)		
Method 1	Method 2	Method 3
45	50	60
50	55	63
40	49	55
43	52	52
47	53	58

a. Can the director of training conclude that there is a difference among the average typing speeds of the employees for the three methods at $\alpha = 0.10$?

b. What assumptions did the director of training make in performing the test in part a.? Do the data appear to satisfy these assumptions? Explain.

13. A physical trainer has four workouts that he recommends for his clients. The workouts have been designed so that the average maximum heart rate achieved is the same for each workout. To test this design he randomly selects 12 people and randomly assigns three of them to use each of the workouts. During each workout, he measures the maximum heart rate in beats per minute with the following results.

Table for Exercise 13 – Maximum Heart Rates (Beats per Minute)			
Workout #1	Workout #2	Workout #3	Workout #4
180	160	175	185
185	170	180	190
170	175	170	180

a. Can the physical trainer conclude at $\alpha = 0.05$ that there is a difference among the average maximum heart rates which are achieved during the four workouts?

b. What assumptions did the physical trainer make in performing the test procedure in part a.? Do the data appear to satisfy these assumptions? Explain.

14. The results of a survey comparing the costs of staying one night in a full-service hotel (including food, beverages, and telephone calls, but not taxes or gratuities) for several major cities are given in the following table.

Table for Exercise 14 – Hotel Costs per Night ($)				
New York	Los Angeles	Atlanta	Houston	Phoenix
300	240	190	195	238
320	250	198	190	240
325	230	185	200	236
350	245	195	192	248
275	235	182	198	228

a. Do the data suggest that there is a significant difference among the average costs of one night in a full-service hotel for the five major cities at $\alpha = 0.05$?

b. What assumptions were made in performing the test procedure in part **a.**? Do the data appear to satisfy these assumptions? Explain.

c. Based on the analysis you performed in part **b.**, which cities, if any, do you think have significantly different average costs for a one-night stay in a full-service hotel? Explain.

15. Consider the following information regarding the dividends paid per share by companies in the banking, transportation, and energy industries.

Table for Exercise 15 – Dividends per Share ($)		
Banking	Transportation	Energy
1.52	1.00	2.08
3.12	1.20	2.68
1.32	0.20	0.70
0.60	0.40	2.00
1.20	1.09	1.91
1.00	0.61	1.60
1.19	0.35	1.28

a. Do the data provide sufficient evidence to conclude that there is a significant difference among the average dividends paid per share for the three different industries? Use $\alpha = 0.10$.

b. What assumptions were made in performing the test procedure in part **a.**? Do the data appear to satisfy these assumptions? Explain.

c. Based on the analysis you performed in part **b.**, which industries, if any, do you think pay significantly different average dividends per share? Explain.

12.5   Two-Way ANOVA: Randomized Block Design

An accounting firm wants to study auditor proficiency using three training methods for auditing: in-home training, on-site training, and off-site training. One approach in designing this study is to randomly select a number of auditors, say 30, and then randomly assign 10 auditors to each training method. At the end of the training period, the firm could then assign them to a case and measure their proficiency. We could then compare the average proficiency for each of the training methods using the F-test described in Section 12.4. This type of design is called a **completely randomized design**. Although this may be a perfectly reasonable approach, it does have some shortcomings. For example, what if all of the people selected for the off-site training method had more than 20 years of experience, whereas all of the people selected for in-home training were with the company less than two years? Certainly, a person's experience will have an effect on his or her proficiency to evaluate a case, regardless of the training method.

What can be done to reduce the effect of this factor that affects proficiency, but is not directly related to the training methods? One approach would be to divide our study participants into several experience bands called **blocks**: those people with one year of experience, those people with two years of experience, and so on. Then randomly select three people from each experience level (block) and assign each a different training method. Once the data have been collected, compare the average proficiency of each of the training methods for individuals in each of the experience bands (blocks).

This design is an extension of the paired difference design for comparing two means that we discussed in Chapter 11. Although the effect that current proficiency has on the proficiency after training has not been eliminated, it has probably been greatly reduced by this blocking. As with the paired difference design, we give up degrees of freedom when we create blocks. Thus, we must be sure that the reduction in variation among sample observations is enough to compensate for the loss of degrees of freedom. This type of design is an example of a **randomized block design**.

In a randomized block design, we use blocks such that the experimental units within the blocks are as much alike as possible. Then we randomly assign the experimental units within each block to the k populations or treatments of interest. Finally, we compare the response of the experimental units to each of the treatments of interest within each of the blocks. In this way, we eliminate possible variation due to some of the extraneous factors that are unrelated to the treatments. Table 12.4 gives the data collected by the accounting firm for the randomized block design. Note that the data in the body of the table are proficiency scores (out of 100).

Table 12.4 – Randomized Block Design			
Block	**Treatment**		
Years of Experience	In-Home Training	On-Site Training	Off-Site Training
1	91	72	79
2	95	75	77
3	63	63	84
4	84	89	77
5	72	92	73
6	85	94	68
7	72	89	76
8	79	92	92
9	87	71	91
10	76	70	68

Our interest is still in comparing the population (or treatment) means. For the accounting firm, the treatments of interest are the training methods. Thus the null and alternative hypotheses are the same as they were in the F-test.

H_0: $\mu_1 = \mu_2 = \mu_3$ (The average proficiency resulting from each of the training methods is the same.)
H_a: At least one μ_i is different.

The test statistic appears to be exactly the same as that used for the F-test.

$$F = \frac{\text{MST}}{\text{MSE}}$$

However, the denominator of the F-statistic is calculated differently because it takes into account the fact that we were able to eliminate some of the variation among our sample responses by blocking the experimental units. The total sum of squares (TSS) and the degrees of freedom that we discussed in Section 12.2 can be broken down further as follows. Note that in the equations below **SST** is the **sum of squares for treatments**, **SSBL** is the **sum of squares for blocks**, and **SSE** is the **sum of squares for error**.

Sum of Squares:	TSS	=	SST	+	SSBL	+	SSE
Degrees of Freedom:	$n-1$	=	$k-1$	+	$b-1$	+	$(b-1)(k-1)$

where n is the total number of observations,

k is the number of treatments, and

b is the number of blocks.

If any of the observed variation in the sample observations has been reduced by blocking, then the SSE is reduced. But it is important to note that the degrees of freedom that are used by the SSBL are taken from the SSE; thus the reduction in variation achieved by blocking must offset this loss in degrees of freedom. If we are successful at significantly reducing variation by blocking and there is a difference among the sample means, we will be more likely to detect it. The test procedure for the randomized block design is outlined in the following box.

Procedure

Two-Way ANOVA: Randomized Block Design

Hypotheses:

$$H_0: \mu_1 = \mu_2 = \cdots = \mu_k \qquad \text{(The } k \text{ population means are equal.)}$$
$$H_a: \text{At least one } \mu_i \text{ is different.}$$

Test Statistic:

$$F = \frac{\text{MST}}{\text{MSE}} = \frac{\dfrac{\text{SST}}{k-1}}{\dfrac{\text{SSE}}{(k-1)(b-1)}}$$

where SSE = TSS − SST − SSBL.

Assumptions:

The differences in observed responses to treatments for blocked units are normally distributed with equal variances.

Rejection Region:

H_0 will be rejected for large values of $F = \dfrac{\text{MST}}{\text{MSE}}$. In particular, we will reject H_0 if $F \geq F_\alpha$ with $(k-1)$ numerator degrees of freedom and $(k-1)(b-1)$ denominator degrees of freedom.

The calculation formulas for the F-test statistic in the randomized block design are beyond the scope of this text and, therefore, will not be presented in this chapter. We will assume that the reader has access to a statistical package that will produce the results of the F-test for a randomized block design. For reference, the general form of an ANOVA table for a randomized block design is given in Table 12.5.

Table 12.5 – ANOVA Summary Table for a Randomized Block Design				
Source of Variation	**SS**	**df**	**MS**	**F**
Block	SSBL	$b-1$	MSBL	$\dfrac{\text{MSBL}}{\text{MSE}}$
Treatment	SST	$k-1$	MST	$\dfrac{\text{MST}}{\text{MSE}}$
Error	SSE	$(b-1)(k-1)$	MSE	
Total	TSS	$n-1$		

The results of the accounting firm's test are given in Figure 12.8.

12

	A	B	C	D	E	F	G
1	Anova: Two-Factor Without Replication						
2							
3	SUMMARY	Count	Sum	Average	Variance		
4	1	3	242	80.66667	92.33333		
5	2	3	247	82.33333	121.3333		
6	3	3	210	70	147		
7	4	3	250	83.33333	36.33333		
8	5	3	237	79	127		
9	6	3	247	82.33333	174.3333		
10	7	3	237	79	79		
11	8	3	263	87.66667	56.33333		
12	9	3	249	83	112		
13	10	3	214	71.33333	17.33333		
14							
15	In-Home	10	804	80.4	96.48889		
16	On-Site	10	807	80.7	133.3444		
17	Off-Site	10	785	78.5	70.05556		
18							
19							
20	ANOVA						
21	Source of Variation	SS	df	MS	F	P-value	F crit
22	Rows	801.4667	9	89.05185	0.844746	0.586668	2.456281
23	Columns	28.46667	2	14.23333	0.135017	0.874577	3.554557
24	Error	1897.533	18	105.4185			
25							
26	Total	2727.467	29				

Figure 12.8

Note: For the computer output shown in Figure 12.8 "Rows" corresponds to the blocks since in Table 12.4 each row represented a block. "Columns" corresponds to the treatments since the treatments were organized in columns. When performing an ANOVA in Excel or any other statistical software, it is important to be aware of whether the blocks and treatments are organized in rows or columns to accurately interpret the results.

The F-statistic has numerator degrees of freedom equal to 2 and denominator degrees of freedom equal to 18. Thus for $\alpha = 0.05$, the F-critical value is approximately 3.5546. The calculated value of the test statistic is given by

$$F = \frac{MST}{MSE} \approx \frac{14.2333}{105.4185} \approx 0.1350.$$

Identify these values in Figure 12.8. MST and MSE are given in the MS column, the F-statistic is given in the F column, and the critical value is given in the F $crit$ column.

F-Distribution
$df_{num} = 2, df_{den} = 18$

Fail to Reject H_0 Reject H_0

$\alpha = 0.05$

0 0.1350   3.5546 F

Figure 12.9

As shown in Figure 12.9, the value of our test statistic, 0.1350, is less than the F-critical value of 3.5546. Therefore, the accounting firm will fail to reject the null hypothesis. Also note that the P-value is given in the output as approximately 0.8746. Since the P-value is greater than the level of significance, 0.05, the accounting firm will fail to reject the null hypothesis.

There is not sufficient evidence to conclude that there is a significant difference in average proficiency among the three training programs. Because the accounting firm failed to reject the null hypothesis, they may be concerned that they were not able to significantly reduce the variation by blocking.

To determine if the blocking was successful at reducing variation among the sample observations, we use the test statistic $F = \dfrac{MSBL}{MSE}$. Under the null hypothesis that the block means are the same (we were not successful in reducing variation by blocking because the block means were not significantly different), the F-test statistic has an F-distribution with $(b-1)$ numerator degrees of freedom and $(k-1)(b-1)$ denominator degrees of freedom.

At $\alpha = 0.05$ the accounting firm will reject the null hypothesis that block means are all equal (blocking was unsuccessful) if the calculated value of the test statistic is larger than the F-critical value with 9 numerator degrees of freedom and 18 denominator degrees of freedom, which is approximately 2.4563. The calculated value of the test statistic is given by

$$F = \frac{MSBL}{MSE} \approx \frac{89.0519}{105.4185} \approx 0.8447.$$

Again, MSBL and MSE are given in the MS column, the F-statistic is given in the F column, and the critical value is given in the F crit column in Figure 12.8.

F-Distribution
$df_{num} = 9, df_{den} = 18$

Fail to Reject H_0 Reject H_0

$\alpha = 0.05$

0   0.8447   2.4563 F

Figure 12.10

Figure 12.10 displays the rejection region and the calculated value of the test statistic. Since 0.8447 is less than 2.4563, we fail to reject the null hypothesis that the block means are all equal. This decision can also be made using the reported P-value, 0.5867, which is greater than the level of significance, 0.05. The firm was unable to reduce a significant amount of variation among sample observations by blocking.

The hypothesis test just presented is one example of a **two-way analysis of variance (two-way ANOVA)**. This means there are two independent factors considered in the analysis, namely the blocks (the different experience levels) and the treatments (training methods), and treatment was applied once within each block. Note that the researcher's primary interest was to determine if there was a significant difference between training methods. However, the researcher wanted to reduce variation in the model by adding the blocking factor (experience). The researcher was not interested in studying the effect of experience on audit proficiency. Section 12.6 will discuss situations in which the researcher is interested in the effects of both factors.

12.5   Exercises

Basic Concepts

1. What is a completely randomized design? Give an example.

2. Identify a shortcoming that could arise from using a completely randomized design for the example you gave in Exercise 1.

3. What are blocks? What is their purpose?

4. What is a randomized block design? How is it different from a completely randomized design?

5. What are the null and alternative hypotheses when comparing means using a randomized block design?

6. What is the test statistic for the hypothesis test described in Exercise 5?

7. What is the breakdown of the sum of squares for a randomized block design? Does this breakdown make sense? Explain.

8. How are the corresponding degrees of freedom for TSS, SST, SSBL, and SSE related? Verify that this relationship is true.

9. If blocking is successful, how does the value of SSE change?

10. What are the assumptions when performing a two-way ANOVA for a randomized block design?

11. What is the rejection region for the test when performing a two-way ANOVA for a randomized block design?

12. What are the degrees of freedom associated with the test statistic for the two-way ANOVA described in Exercise 11?

Exercises

13. A car dealer is interested in comparing the average gas mileages of four different car models. The dealer believes that the average gas mileage of a particular car will vary depending on the person who is driving the car due to different driving styles. Because of this, he decides to use a randomized block design. He randomly selects six drivers and asks them to drive each of the cars. He then determines the average gas mileage for each car and each driver. The results of the study are as follows.

Table for Exercise 13 – Gas Mileage (MPG)				
	Car A	Car B	Car C	Car D
Driver 1	33	29	27	37
Driver 2	36	32	30	40
Driver 3	34	30	28	38
Driver 4	31	27	25	35
Driver 5	33	29	27	37
Driver 6	35	33	31	41

a. Do you think a randomized block design is appropriate for the car dealer's study? Explain.

b. The results of the two-way ANOVA for the dealer's survey of the average gas mileages of the different car models are given in the following table.

ANOVA			
Source of Variation	SS	df	MS
Rows	84.8333	5	16.9667
Columns	348.5000	3	116.1667
Error	2.5000	15	0.1667
Total	435.8333	23	

Can the dealer conclude that there is a significant difference in average gas mileages of the four car models? Use $\alpha = 0.05$.

c. Was the dealer able to significantly reduce variation among the observed gas mileages by blocking? Use $\alpha = 0.05$.

14. A banana grower has three fertilizers from which to choose. He would like to determine which fertilizer produces banana trees with the largest yield (measured in pounds of bananas produced). The banana grower has noticed that there is a difference in the average yields of the banana trees depending on which side of the farm they are planted (South Side, North Side, West Side, or East Side). Because of the variation in yields among the areas on the farm, the farmer has decided to randomly select three trees within each area and then randomly assign the fertilizers to the trees. After harvesting the bananas, he calculates the yields of the trees within each of the areas. The results are as follows.

Table for Exercise 14 – Banana Yields (Pounds)			
	Fertilizer A	Fertilizer B	Fertilizer C
South Side	53	51	58
North Side	48	47	53
West Side	50	48	56
East Side	50	47	54

a. Do you think a randomized block design is appropriate for the banana grower's study? Explain.

b. The results of the two-way ANOVA for the banana grower's study are given in the following table.

ANOVA			
Source of Variation	SS	df	MS
Rows	36.2500	3	12.0833
Columns	104.0000	2	52.0000
Error	2.0000	6	0.3333
Total	142.2500	11	

Can the banana grower conclude that there is a significant difference among the average yields of the banana trees for the three fertilizers? Use $\alpha = 0.10$.

c. Was the banana grower able to significantly reduce variation among the observed yields by blocking? Use $\alpha = 0.10$.

15. The FAA is interested in knowing if there is a difference in the average numbers of on-time arrivals for four of the major airlines. The FAA believes that the number of on-time arrivals varies by airport. To control for this variation, they randomly select 100 flights for each of the major airlines at each of four randomly selected airports and record the number of on-time flights. The results of the study are as follows.

Table for Exercise 15 – On-Time Flights				
	Airline A	Airline B	Airline C	Airline D
Airport A	87	82	79	81
Airport B	88	84	81	82
Airport C	89	84	83	82
Airport D	90	86	85	83

a. Do you think a randomized block design is appropriate for the FAA's study? Explain.

b. The results of the two-way ANOVA for the FAA's study are given in the following table.

ANOVA			
Source of Variation	SS	df	MS
Rows	29.2500	3	9.7500
Columns	112.7500	3	37.5833
Error	5.7500	9	0.6389
Total	147.7500	15	

Can the FAA conclude that there is a significant difference among the average number of on-time arrivals for the four major airlines? Use $\alpha = 0.01$.

c. Was the FAA able to significantly reduce variation among the observed number of on-time arrivals by blocking? Use $\alpha = 0.01$.

16. A psychologist is interested in determining if there is a difference in the average numbers of suicides for several age groups. The psychologist believes that there may be some variation in the numbers of suicides depending on the region of the country (Northeast, Northwest, Southeast, or Southwest). The psychologist randomly selects 100,000 deaths from each region of the country for each of the age groups of interest and determines the number of suicides. The results of the study are as follows.

Table for Exercise 16 – Suicides							
	Age 15–24	Age 25–34	Age 35–44	Age 45–54	Age 55–64	Age 65–74	Age 75–84
Northeast	15	17	16	17	55	22	27
Northwest	13	16	16	16	49	19	26
Southeast	12	14	15	15	47	17	24
Southwest	13	15	15	16	53	20	25

a. Do you think a randomized block design is appropriate for the psychologist's study? Explain.

b. The results of the two-way ANOVA for the psychologist's study are given in the following table.

ANOVA			
Source of Variation	SS	df	MS
Rows	44.9643	3	14.9881
Columns	4223.3571	6	703.8929
Error	25.7857	18	1.4325
Total	4294.1071	27	

Can the psychologist conclude that there is a significant difference among the average number of suicides for the different age groups? Use $\alpha = 0.10$.

c. Was the psychologist able to significantly reduce variation among the observed number of suicides by blocking? Use $\alpha = 0.05$.

17.   In an experiment designed to compare automated blood pressure devices with those of the standard cuff method, each man in a sample of six patients has his systolic blood pressure determined by three different automated devices and by the standard cuff method. The data are given in the following table.

Table for Exercise 17 – Blood Pressure (mmHg)			
	Device 1	Device 2	Device 3	Standard Cuff
Patient 1	126	128	132	131
Patient 2	134	138	137	140
Patient 3	145	144	150	152
Patient 4	129	134	132	136
Patient 5	154	160	162	160
Patient 6	144	144	148	145

a.   Why was a randomized block design used in this experiment?

b.   From the data, SST and SSE were computed to be 106.4583 and 53.2917, respectively. With $\alpha = 0.05$, can we conclude that the four different methods of determining systolic blood pressure have different mean readings?

c.   SSBL was computed to be 2412.8750. With $\alpha = 0.05$, can we conclude that using people as blocks significantly reduced variation in this study?

12.6 Two-Way ANOVA: The Factorial Design

The techniques that we discussed for a randomized block design can be extended to the situation where there are two factors of interest. A director of personnel might be interested in relating average salary to two factors: age and experience. A supervisor at a manufacturing plant might be interested in relating the average number of defective products to two factors: operator and machine. A doctor might be interested in relating the increase in average patient heart rate to two factors: medication and age.

In order for the director of personnel to evaluate the relationship between average salary, age, and experience, he chooses the following experimental design. He selects four different age groups and three different experience levels and observes two salaries for each of the possible combinations of age and experience. The resulting data are displayed in Table 12.6.

Table 12.6 – Salaries (Thousands of Dollars)				
Years of Experience	**Age**			
	25 – 34	**35 – 44**	**45 – 54**	**55 – 64**
0 – 4	22 27	25 35	34 36	37 43
5 – 9	34 36	35 45	42 48	49 51
10 – 14	39 41	40 50	53 57	51 59

Again, this type of design involves a **two-way analysis of variance** because there are two classifications. It is called a **complete factorial experiment** since there is at least one observation for every possible combination of age and salary. Factorial experiments provide valuable information by enabling the interaction between the two variables to be estimated.

Interaction between the two variables means that the average salary is affected by the combination of age and experience. An example of two variables that interact is shown in Figure 12.11. This is called a **profile plot** (or interacting plot) which plots the means by factors. If there is no interaction at all between two variables, the lines in the profile plot will be perfectly parallel. An example of two variables that do not interact is shown in Figure 12.12. A similar graph for the salary data is shown in Figure 12.13. Based on this graph, there appears to be slight interaction between age and experience when age is between 45 and 65 years and experience is between 10 and 14 years.

Figure 12.11

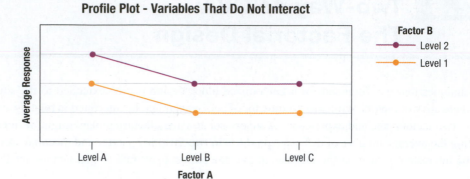

Figure 12.12

There are three effects on average salary that will interest the personnel director. The first is the effect that the interaction between age and experience has on average salary, called the **main effect for interaction**. The second is the effect that experience has on average salary, called the **main effect for experience (Factor A)**. The third is the effect that age has on average salary, called the **main effect for age (Factor B)**.

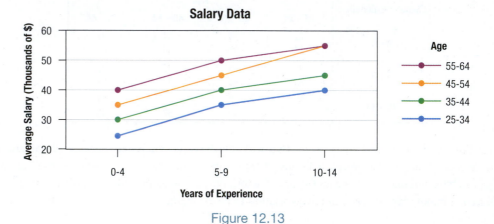

Figure 12.13

The test procedure is somewhat different than those we have discussed previously because of the potential presence of interaction between the two variables. Of primary importance is determining whether or not there is *any* interaction between the two variables. If there is interaction, we will not be able to separate out the effects that age and experience have on average salary, and the hypothesis testing procedure is halted. If there is not interaction, then it is possible to proceed with the hypothesis testing procedure and determine the effect that age has on average salary and the effect that experience has on average salary.

To develop the test procedure for the two-way ANOVA, we again rely on the concepts presented in Section 12.2. The variation among the sample observations, represented by the total sum of squares, can be broken down into four pieces: the **sum of squares for Factor A (SSA)**, the **sum of squares for Factor B (SSB)**, the **sum of squares for interaction (SSAB)**, and the **sum of squares for error**.

Sum of Squares: $\text{TSS} = \text{SSA} + \text{SSB} + \text{SSAB} + \text{SSE}$

Degrees of Freedom: $n-1 = a-1 + b-1 + (a-1)(b-1) + ab(r-1)$

where n is the total number of observations,

a is the number of levels of Factor A,

b is the number of levels of Factor B, and

r is the number of observations in the combinations of levels of Factor A and Factor B (i.e.,the number of observations in each cell).

The test statistics for a two-way ANOVA for a factorial design are derived by dividing the sums of squares by the appropriate degrees of freedom to produce mean squares, and then dividing each of the respective mean squares by the mean square for error. The test procedures are outlined in the following boxes.

Procedure

Test for Interaction between Factors

Hypotheses:

H_0: There is no interaction between Factor A and Factor B.

H_a: There is interaction between Factor A and Factor B.

Test Statistic:

$$F = \frac{\dfrac{\text{SSAB}}{(a-1)(b-1)}}{\dfrac{\text{SSE}}{ab(r-1)}} = \frac{\text{MSAB}}{\text{MSE}}$$

Rejection Region:

Reject the null hypothesis if $F \geq F_\alpha$ with $(a-1)(b-1)$ numerator degrees of freedom and $ab(r-1)$ denominator degrees of freedom.

Note: If the null hypothesis is rejected, then interaction exists. Do not proceed with the main effects tests for Factor A and Factor B if this is the case.

Procedure

Test for Main Effects for Factor A

Hypotheses:

H_0: Factor A has no effect on average response.

H_a: Factor A has an effect on average response.

Test Statistic:

$$F = \frac{\dfrac{\text{SSA}}{(a-1)}}{\dfrac{\text{SSE}}{ab(r-1)}} = \frac{\text{MSA}}{\text{MSE}}$$

Rejection Region:

Reject the null hypothesis if $F \geq F_\alpha$ with $(a-1)$ numerator degrees of freedom and $ab(r-1)$ denominator degrees of freedom.

Procedure

Test for Main Effects for Factor B

Hypotheses:

H_0: Factor B has no effect on average response.

H_a: Factor B has an effect on average response.

Test Statistic:

$$F = \frac{\dfrac{\text{SSB}}{(b-1)}}{\dfrac{\text{SSE}}{ab(r-1)}} = \frac{\text{MSB}}{\text{MSE}}$$

Rejection Region:

Reject the null hypothesis if $F \geq F_\alpha$ with $(b-1)$ numerator degrees of freedom and $ab(r-1)$ denominator degrees of freedom.

The calculations of the F-statistics for a two-way analysis of variance are beyond the scope of this text. We will assume that the reader has access to a statistical software package that will produce the results of a two-way ANOVA. For reference, the general form of the ANOVA table for a factorial experiment is given in Table 12.7.

Table 12.7 – ANOVA Summary Table for a Factorial Experiment

Source of Variation	SS	df	MS	F
Factor A	SSA	$a-1$	MSA	$\dfrac{\text{MSA}}{\text{MSE}}$
Factor B	SSB	$b-1$	MSB	$\dfrac{\text{MSB}}{\text{MSE}}$
Interaction	SSAB	$(a-1)(b-1)$	MSAB	$\dfrac{\text{MSAB}}{\text{MSE}}$
Error	SSE	$ab(r-1)$	MSE	
Total	TSS	$n-1$		

The results of the two-way ANOVA for the personnel director's salary data are given in Figure 12.14.

29	ANOVA						
30	*Source of Variation*	*SS*	*df*	*MS*	*F*	*P-value*	*F crit*
31	Sample	1092.583	2	546.2917	26.59432	3.89E-05	3.885294
32	Columns	828.4583	3	276.1528	13.44354	0.000382	3.490295
33	Interaction	24.41667	6	4.069444	0.198107	0.971006	2.99612
34	Within	246.5	12	20.54167			
35							
36	Total	2191.958	23				

Figure 12.14

Note: In the ANOVA table given in Figure 12.14, "Sample" corresponds to Experience (Factor A) since the experience levels were organized by row. "Columns" corresponds to Age (Factor B) since the age groups are organized by column. "Interaction" corresponds to the interaction between experience and age, and "Within" corresponds to the variation within the sample observations, or the error. When using Excel or other statistical software programs to perform a two-way ANOVA, it is important to pay close attention to how the data are organized so that you can identify Factor A and Factor B and accurately interpret the results.

The personnel director must first decide if there is interaction between age and experience. The appropriate test statistic for testing for interaction, which is the mean square of the interaction term divided by the mean square for error, is given by

$$F = \frac{\text{MSAB}}{\text{MSE}}.$$

Under the null hypothesis that there is no interaction, the F-test statistic has an F-distribution with 6 numerator degrees of freedom and 12 denominator degrees of freedom. At $\alpha = 0.05$, the null hypothesis will be rejected if the calculated value of the test statistic is greater than or equal to 2.9961.

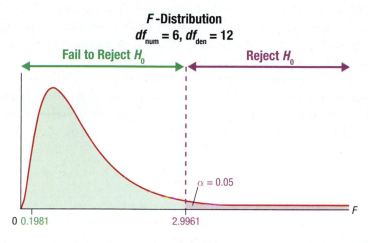

Figure 12.15

From Figure 12.14, the calculated value of the test statistic is approximately 0.1981. Figure 12.15 shows the rejection region and the calculated value of the test statistic. Since 0.1981 is less than 2.9961, we fail to reject the null hypothesis that there is no interaction between experience and age. This is what the personnel director expected based on Figure 12.13. Because there is not evidence of significant interaction between age and experience, it is safe to use the F-test statistic to test whether or not each of the factors, age and experience, has an effect on average salary.

To test whether or not experience has an effect on average salary, we let the null hypothesis be that the average salary is equal for each of the experience levels. The appropriate test statistic, which is the mean square of the experience factor divided by the mean square for error, is given by

$$F = \frac{\text{MSA}}{\text{MSE}}.$$

Under the null hypothesis that level of experience has no effect on average salary, the F-test statistic has an F-distribution with 2 numerator degrees of freedom and 12 denominator degrees of freedom. At $\alpha = 0.05$, the null hypothesis will be rejected if the calculated value of the test statistic is greater than or equal to 3.8853.

F-Distribution
$df_{num} = 2$, $df_{den} = 12$

Fail to Reject H_0   **Reject H_0**

$\alpha = 0.05$

0   3.8853 F

Figure 12.16

From Figure 12.14, the calculated value of the test statistic is approximately 26.5943. Figure 12.16 shows the rejection region. Since 26.5943 is larger than 3.8853, we reject the null hypothesis that the level of experience has no effect on average salary. There is persuasive evidence at $\alpha = 0.05$ that the mean salaries are significantly different for at least two experience levels.

Note: For each of these tests, we could also use the P-values to determine whether to reject or fail to reject the null hypothesis. P-values for Factor A, Factor B, and the interaction (AB) are given in the P-value column of the Excel output. We can compare these P-values to the significance levels rather than comparing the test statistics to the critical values, if we wish. That is, for any given test, if the P-value is less than or equal to α we would reject the null hypothesis.

To test whether or not age has an effect on average salary, we let the null hypothesis be that the average salary is equal for each of the age levels. The appropriate test statistic, which is the mean square of the age factor divided by the mean square for error, is given by

$$F = \frac{MSB}{MSE}.$$

Under the null hypothesis that age has no effect on average salary, the F-test statistic has an F-distribution with 3 numerator degrees of freedom and 12 denominator degrees of freedom. At $\alpha = 0.05$, the null hypothesis will be rejected if the calculated value of the test statistic is greater than or equal to 3.4903.

From Figure 12.14, the calculated value of the test statistic is approximately 13.4435. Figure 12.17 shows the rejection region. Since 13.4435 is larger than 3.4903, we reject the null hypothesis that age has no effect on average salary. There is persuasive evidence at $\alpha = 0.05$ that there is a significant difference between the mean salaries for at least two age levels.

F-Distribution
$df_{num} = 3$, $df_{den} = 12$

Fail to Reject H_0   **Reject H_0**

$\alpha = 0.05$

0   3.4903 F

Figure 12.17

When using a two-way ANOVA, it is important to remember that the assumptions of normality, equal variances, and random sampling must be satisfied in order for the test to produce meaningful results.

12.6 Exercises

Basic Concepts

1. What is the difference between a randomized block design and a factorial design?

2. What is a complete factorial experiment?

3. What is a profile plot? What kind of information does this plot give us?

4. Why is it so important to determine if there is interaction between the two variables of interest in a factorial design?

5. Is it possible to perform a two-way analysis of variance if interaction exists between the two variables of interest? Explain why or why not.

6. Identify the four components that make up the total sum of squares in a complete factorial model. Also, give the acronym associated with each component.

7. Give the degrees of freedom associated with each component of the total sum of squares.

8. What is the test statistic for a test of interaction between factors? What are the degrees of freedom associated with this test statistic?

9. If there is enough evidence to reject the null hypothesis in a test for interaction, may we proceed with the main effects tests? Explain.

10. What is the test statistic for the main effects test for Factor A? What are the degrees of freedom associated with this test statistic?

11. What is the test statistic for the main effects test for Factor B? What are the degrees of freedom associated with this test statistic?

12. What are the rejection rules for the main effects tests? Can P-values be used as rejection criteria?

Exercises

13. The following table contains the results of a survey of daily rental rates of a mid-size car for three major rental car companies at three airport locations on three different days during the year.

Table for Exercise 13 – Daily Rental Rates of Mid-Size Cars ($)			
	New York	**Chicago**	**Miami**
Hertz	93.99	54.99	71.99
	90.99	63.99	87.99
	96.99	57.99	68.99
Avis	58.86	81.99	61.99
	52.10	85.99	70.99
	68.98	71.99	66.99
National	56.00	64.99	66.00
	63.00	67.00	58.99
	52.00	52.99	71.99

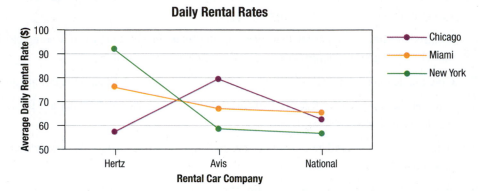

a. Consider the graph of the average daily rental rates for each of the major car rental companies by airport location. Does there appear to be any interaction between the variables airport location and major car rental company?

b. The results of the two-way ANOVA for the study are given in the following table.

ANOVA			
Source of Variation	SS	df	MS
Sample	1011.7730	2	505.8865
Columns	58.7126	2	29.3563
Interaction	2514.3099	4	628.5775
Within	819.1289	18	45.5072
Total	4403.9244	26	

Perform a hypothesis test to determine if there is any interaction between the variables major rental car company and airport location at $\alpha = 0.05$. Does this agree with your observation in part **a.**?

c. If there is no interaction found in part **b.**, is there sufficient evidence to conclude that there is a significant difference among the average daily rental rates for mid-size cars for the three rental car companies at the 0.05 level?

14. A doctor is interested in determining the increase in average heart rate caused by a medication used for treating high blood pressure. The doctor believes that the increase in heart rate will be related to two factors: the age of a person and the weight of a person. To test this theory, the doctor randomly selects two patients in each of the age and weight categories listed in the following table and determines the increase in heart rate (in beats per minute) of each patient 15 minutes after administering the drug. The results of the study are as follows.

Table for Exercise 14 – Increase in Heart Rate (Beats per Minute)			
	25–39 Years	40–54 Years	55–69 Years
100–149 Pounds	2	7	11
	2	6	7
150–199 Pounds	7	11	16
	7	9	12
200–249 Pounds	10	13	18
	8	11	14

a. Consider the following graph of the average increase in heart rate for each of the weight and age categories. Does there appear to be any interaction between the age and weight variables? Explain.

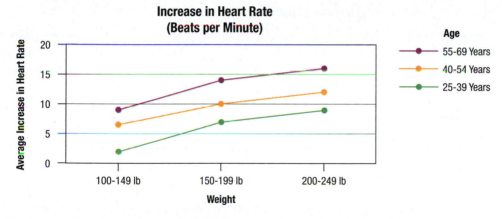

b. The results of the two-way ANOVA for the study are given in the following table.

ANOVA			
Source of Variation	SS	df	MS
Sample	133.0000	2	66.5000
Columns	147.0000	2	73.5000
Interaction	2.0000	4	0.5000
Within	30.5000	9	3.3889
Total	312.5000	17	

Perform a hypothesis test to determine if there is any interaction between the variables age and weight at $\alpha = 0.01$. Does this agree with your observation in part **a.**?

c. Is there sufficient evidence to conclude that there is a significant difference among the average increases in heart rate for the different weight categories? Use $\alpha = 0.01$.

d. Is there sufficient evidence to conclude that there is a significant difference among the average increases in heart rate for the different age groups? Use $\alpha = 0.01$.

15. A supervisor of a manufacturing plant is interested in relating the average number of defects produced per day to two factors: the operator working the machine and the machine itself. The supervisor randomly assigns each operator to use each machine for three days and records the number of defects produced per day. The results of the study are as follows.

Table for Exercise 15 – Number of Defects Produced per Day			
	Operator A	Operator B	Operator C
Machine A	3	7	3
	3	5	2
	3	3	1
Machine B	2	6	2
	2	4	1
	2	2	0
Machine C	1	5	1
	1	3	0
	1	2	1

a. Consider the following graph of the average number of defects produced per day for each of the operators by machine. Does there appear to be any interaction between the variables operator and machine?

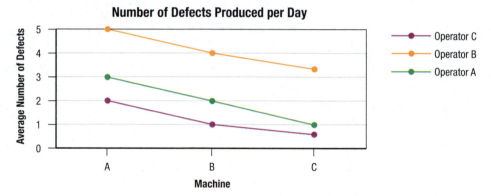

b. The results of the two-way ANOVA for the supervisor's survey of the number of defects produced per day are given in the following table.

ANOVA			
Source of Variation	SS	df	MS
Sample	12.6667	2	6.3333
Columns	40.2222	2	20.1111
Interaction	0.4444	4	0.1111
Within	25.3333	18	1.4074
Total	78.6667	26	

Perform a hypothesis test to determine if there is any interaction between the machine and operator variables. Use $\alpha = 0.10$. Does this agree with your observation in part **a.**?

c. Is there sufficient evidence to conclude that there is a significant difference among the average number of defects produced per day for the different machines? Use $\alpha = 0.10$.

d. Is there sufficient evidence to conclude that there is a significant difference among the average number of defects produced per day for the different operators? Use $\alpha = 0.10$.

16. A dairy farmer thinks that the average weight gain of his cows depends on two factors: the type of grain that they are fed and the type of grass that they are fed. The dairy farmer has four different types of grain from which to choose and three different types of grass from which to choose. He would like to determine if there is a particular combination of grain and grass that would lead to the greatest weight gain on average for his cows. He randomly selects three one-year-old cows and assigns them to each of the possible combinations of grain and grass. After one year he records the weight gain for each cow (in pounds) with the following results.

Table for Exercise 16 – Cow Weight Gain (Pounds)			
	Grass A	**Grass B**	**Grass C**
Grain A	175	225	250
	160	215	240
	185	230	260
Grain B	190	245	275
	185	240	260
	195	255	285
Grain C	210	255	300
	200	245	310
	220	265	295
Grain D	225	275	350
	235	270	360
	220	280	345

a. Consider the following graph of the average weight gain of the cows for each of the possible combinations of grass and grain. Does there appear to be any interaction between the grass and grain variables?

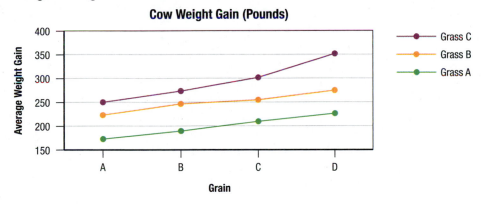

b. The results of the two-way ANOVA for the farmer's study are given in the following table.

ANOVA			
Source of Variation	SS	df	MS
Sample	23097.2222	3	7699.0741
Columns	53272.2222	2	26636.1111
Interaction	3127.7778	6	521.2963
Within	1916.6667	24	79.8611
Total	81413.8889	35	

Perform a hypothesis test to determine if there is any interaction between the variables grass and grain at $\alpha = 0.05$. Does this agree with your observation in part **a.**?

c. If there is no interaction found in part **b.**, is there sufficient evidence to conclude that there is a significant difference in the average weight gains among the cows for the four different types of grain? Use $\alpha = 0.05$.

    d. Is there sufficient evidence to conclude that there is a significant difference in the average weight gains among the cows for the three different types of grass? Use $\alpha = 0.05$.

17. The partially completed analysis of variance table given below is taken from the article, "Power and Status, Exchange, Attribution, and Expectation States (Small Group Research)." The experimenters investigated the effects of power and knowledge on one's emotional reaction in a study involving 52 students selected from a large private university. Each of the factors was run at two levels, with 13 subjects at each of the four different factor combinations.

ANOVA				
Source of Variation	*SS*	*df*	*MS*	*F*
Power	1.2700	1		
Knowledge	0.2500	1		
Interaction		1		
Error	4.1400	48		
Total	5.6700	51		

    a. Complete the ANOVA table.

    b. Can we conclude, with $\alpha = 0.10$, that there is interaction between power and knowledge?

    c. With $\alpha = 0.05$, can we conclude that there is a significant difference in the two levels of power?

    d. With $\alpha = 0.05$, can we conclude that there is a significant difference in the two levels of knowledge?

T   Discovering Technology

Using the TI-84 Plus Calculator

One-Way ANOVA

Use the information from Example 12.1 for this exercise.

1. Press **STAT**, select **EDIT**, **1:Edit**, and press **ENTER**. Then input the four sales for the three sales reps into lists L1, L2, and L3.

2. Press **STAT** and select **TESTS**, then choose option **F: ANOVA** and press **ENTER**.

3. Enter **ANOVA(L1, L2, L3)**. (Note that lists can be entered by pressing **2ND** followed by the number of the list. For example, **2ND** and **1** gives L1.) Press **ENTER**.

4. Observe the output screen for the one-way analysis of variance. Compare the results to those which we found in Example 12.1.

Figure 12.18

Figure 12.19

Using Excel

One-Way ANOVA

Use the information from Example 12.1 for this exercise.

1. Enter the sales for each of the three sales reps into columns A, B, and C. Cells A1, B1, and C1 should contain the column titles **Salesperson 1**, **Salesperson 2**, and **Salesperson 3**, respectively.

2. Under the **Data** tab, choose **Data Analysis**, and **Anova: Single Factor**.

3. Enter the sales data for the Input Range (**$A$1:$C$5**).

4. Be sure that the radio button next to **Columns** is selected, and that Alpha is **0.05**. Check the box **Labels in First Row** since we are also including the column titles in the input range. Click **OK**.

5. Observe the output screen for the one-way analysis of variance. Compare these results to the ones presented in Example 12.1.

▲	A	B	C	D	E	F	G
1	Anova: Single Factor						
2							
3	SUMMARY						
4	*Groups*	*Count*	*Sum*	*Average*	*Variance*		
5	Salesperson 1	4	49	12.25	2.916667		
6	Salesperson 2	4	56	14	4.666667		
7	Salesperson 3	4	51	12.75	2.916667		
8							
9							
10	ANOVA						
11	*Source of Variation*	*SS*	*df*	*MS*	*F*	*P-value*	*F crit*
12	Between Groups	6.5	2	3.25	0.928571	0.429904	4.256495
13	Within Groups	31.5	9	3.5			
14							
15	Total	38	11				

Figure 12.20

Two-Way ANOVA: Randomized Block Design

The Anova: Two-Factor Without Replication data analysis tool will be used for this exercise. This tool is useful when data can be classified along two different dimensions (i.e. training method and experience) and there is only a single observation for each classification pair. Use the information from Table 12.4 for this exercise.

1. Enter the column labels **Years of Experience**, **In-Home Training**, **On-Site Training**, and **Off-Site Training** in cells A1 through D1, respectively. Enter the rest of the data from Table 12.4 in the corresponding columns.

▲	A	B	C	D
1	Years of Experience	In-Home Training	On-Site Training	Off-Site Training
2	1	91	72	79
3	2	95	75	77
4	3	63	63	84
5	4	84	89	77
6	5	72	92	73
7	6	85	94	68
8	7	72	89	76
9	8	79	92	92
10	9	87	71	91
11	10	76	70	68

Figure 12.21

2. Under the **Data** tab, choose **Data Analysis**, and **Anova: Two-Factor Without Replication**.

3. Enter the data as the Input Range (**$A$1:$D$11**).

4. Make sure that **0.05** is entered for Alpha and check the box next to **Labels** since we are including the column titles in the input range. Press **OK**.

5. Observe the output screen for the two-way analysis of variance. Notice that the ANOVA

table is given following the summary statistics. "Rows" corresponds to the blocks (years of experience) and "Columns" corresponds to the treatments (training method). Compare these results to those that we discussed in Section 12.5.

	A	B	C	D	E	F	G
1	Anova: Two-Factor Without Replication						
2							
3	SUMMARY	Count	Sum	Average	Variance		
4	1	3	242	80.66667	92.33333		
5	2	3	247	82.33333	121.3333		
6	3	3	210	70	147		
7	4	3	250	83.33333	36.33333		
8	5	3	237	79	127		
9	6	3	247	82.33333	174.3333		
10	7	3	237	79	79		
11	8	3	263	87.66667	56.33333		
12	9	3	249	83	112		
13	10	3	214	71.33333	17.33333		
14							
15	In-Home Training	10	804	80.4	96.48889		
16	On-Site Training	10	807	80.7	133.3444		
17	Off-Site Training	10	785	78.5	70.05556		
18							
19							
20	ANOVA						
21	Source of Variation	SS	df	MS	F	P-value	F crit
22	Rows	801.4667	9	89.05185	0.844746	0.586668	2.456281
23	Columns	28.46667	2	14.23333	0.135017	0.874577	3.554557
24	Error	1897.533	18	105.4185			
25							
26	Total	2727.467	29				

Figure 12.22

Two-Way ANOVA: Factorial Design

The Anova: Two-Factor With Replication data analysis tool will be used for this exercise. This tool is useful when data can be classified along two different dimensions (i.e. age and experience) and there is more than one observation for each combination of classifications. Use the information from Table 12.6 for this exercise.

1. Enter the age ranges, **25–34**, **35–44**, **45–54**, and **55–64** in cells B1 through E1, respectively.

2. Enter the years of experience, **0–4**, **5–9**, and **10–14** in Column A, cells A2, A4, and A6, respectively.

3. Input the sample data as given in Table 12.6.

	A	B	C	D	E
1		25 – 34	35 – 44	45 – 54	55 – 64
2	0 – 4	22	25	34	37
3		27	35	36	43
4	5 – 9	34	35	42	49
5		36	45	48	51
6	10 – 14	39	40	53	51
7		41	50	57	59

Figure 12.23

4. Under the **Data** tab, choose **Data Analysis**, and **Anova: Two-Factor With Replication**.

5. Enter the data as the Input Range ($A$1:$E$7).

6. Enter **2** for Rows per sample and make sure that **0.05** is entered for Alpha. Press **OK**.

7. Observe the output screen for the two-way analysis of variance. Notice that the ANOVA table is given following the summary statistics that are presented for each group. "Sample" corresponds to years of experience (Factor A) and "Columns" corresponds to age (Factor B). Compare these results to those that we discussed in Section 12.6.

	A	B	C	D	E	F	G
1	Anova: Two-Factor With Replication						
2							
3	SUMMARY	25 – 34	35 – 44	45 – 54	55 – 64	Total	
4	0 – 4						
5	Count	2	2	2	2	8	
6	Sum	49	60	70	80	259	
7	Average	24.5	30	35	40	32.375	
8	Variance	12.5	50	2	18	49.69643	
9							
10	5 – 9						
11	Count	2	2	2	2	8	
12	Sum	70	80	90	100	340	
13	Average	35	40	45	50	42.5	
14	Variance	2	50	18	2	46	
15							
16	10 – 14						
17	Count	2	2	2	2	8	
18	Sum	80	90	110	110	390	
19	Average	40	45	55	55	48.75	
20	Variance	2	50	8	32	61.35714	
21							
22	Total						
23	Count	6	6	6	6		
24	Sum	199	230	270	290		
25	Average	33.16667	38.33333	45	48.33333		
26	Variance	53.36667	76.66667	85.6	57.06667		
27							
28							
29	ANOVA						
30	*Source of Variation*	*SS*	*df*	*MS*	*F*	*P-value*	*F crit*
31	Sample	1092.583	2	546.2917	26.59432	3.89E-05	3.885294
32	Columns	828.4583	3	276.1528	13.44354	0.000382	3.490295
33	Interaction	24.41667	6	4.069444	0.198107	0.971006	2.99612
34	Within	246.5	12	20.54167			
35							
36	Total	2191.958	23				

Figure 12.24

Using Minitab

One-Way ANOVA

Use the information from Example 12.1 for this exercise.

1. Enter the sales data for Salesperson 1, Salesperson 2, and Salesperson 3 from Table 12.3 into columns C1, C2, and C3, respectively.

2. Choose **Stat**, **ANOVA**, and **One-Way (Unstacked)**.

3. In the dialog box, enter **C1**, **C2**, and **C3** in the box for Responses (in separate columns).

4. Be sure that the Confidence level is set to **95.0**, and press **OK**.

5. Observe the output screen for the one-way analysis of variance. Compare these to the results that were presented in Example 12.1.

```
One-way ANOVA: Salesperson 1, Salesperson 2, Salesperson 3

Source  DF    SS    MS    F     P
Factor   2   6.50  3.25  0.93  0.430
Error    9  31.50  3.50
Total   11  38.00

S = 1.871   R-Sq = 17.11%   R-Sq(adj) = 0.00%


                                    Individual 95% CIs For Mean Based on
                                    Pooled StDev
Level          N    Mean   StDev   --+---------+---------+---------+-------
Salesperson 1  4  12.250   1.708   (-------------*-------------)
Salesperson 2  4  14.000   2.160                 (-------------*-------------)
Salesperson 3  4  12.750   1.708       (-------------*-------------)
                                    --+---------+---------+---------+-------
                                    10.5      12.0      13.5      15.0

Pooled StDev = 1.871
```

Figure 12.25

Two-Way ANOVA

Use the information from Table 12.6 for this exercise.

1. Enter the salary data from Table 12.6 into Column C1. Along with each salary entry, enter the corresponding age and experience into columns C2 and C3.

↓	C1	C2-T	C3-T
	Salary	Experience	Age
1	22	0-4	25-34
2	27	0-4	25-34
3	34	5-9	25-34
4	36	5-9	25-34
5	39	10-14	25-34
6	41	10-14	25-34
7	25	0-4	35-44
8	35	0-4	35-44
9	35	5-9	35-44
10	45	5-9	35-44
11	40	10-14	35-44
12	50	10-14	35-44
13	34	0-4	45-54
14	36	0-4	45-54
15	42	5-9	45-54
16	48	5-9	45-54
17	53	10-14	45-54
18	57	10-14	45-54
19	37	0-4	55-64
20	43	0-4	55-64
21	49	5-9	55-64
22	51	5-9	55-64
23	51	10-14	55-64
24	59	10-14	55-64

Figure 12.26

2. Choose **Stat**, **ANOVA**, and **Two-Way**.

3. In the dialog box, enter **C1** for Response, **C2** for Row factor, and **C3** for Column Factor. (Note that in Table 12.6 the ages are column titles and the years of experience are row titles.)

4. Be sure that **95.0** is entered for Confidence level and click **OK**.

5. Observe the output screen for the two-way analysis of variance. Compare these results to those discussed in Section 12.6.

```
Two-way ANOVA: Salary versus Experience, Age

Source         DF       SS       MS      F      P
Experience      2  1092.58  546.292  26.59  0.000
Age             3   828.46  276.153  13.44  0.000
Interaction     6    24.42    4.069   0.20  0.971
Error          12   246.50   20.542
Total          23  2191.96

S = 4.532   R-Sq = 88.75%   R-Sq(adj) = 78.45%
```

Figure 12.27

R   Chapter 12 Review

Key Terms and Ideas

- Experimental Units
- Treatment
- Analysis of Variance (ANOVA)
- Sum of Squares for Treatments (SST)
- Sum of Squares for Error (SSE)
- Sample Variance
- Grand Mean
- Total Sum of Squares (TSS)
- Mean Square for Treatments (MST)
- Mean Square for Error (MSE)
- F-Statistic
- F-Test
- F-Distribution
- Numerator Degrees of Freedom
- Denominator Degrees of Freedom
- Multiple Comparison Procedures
- Completely Randomized Design
- Blocks
- Randomized Block Design

- Sum of Squares for Blocks (SSBL)
- Mean Square for Blocks (MSBL)
- Two-Way Analysis of Variance (Two-Way ANOVA)
- Complete Factorial Experiment
- Profile Plot
- Main Effect for Interaction
- Main Effect for Factor A
- Main Effect for Factor B
- Sum of Squares for Factor A (SSA)
- Sum of Squares for Factor B (SSB)
- Sum of Squares for Interaction (SSAB)
- Test for Interaction between Factors
- Mean Square for Interaction (MSAB)
- Test for Main Effects for Factor A
- Mean Square for Factor A (MSA)
- Test for Main Effects for Factor B
- Mean Square for Factor B (MSB)

Key Formulas

	Section
Grand Mean	
$$\bar{\bar{x}} = \frac{\sum_{j=1}^{k}\sum_{i=1}^{n_j} x_{ij}}{n_T}$$	12.2
Sample Variance	
$$s^2 = \frac{\sum_{j=1}^{k}\sum_{i=1}^{n_j}\left(x_{ij} - \bar{\bar{x}}\right)^2}{n_T - 1}$$	12.2
Sum of Squares for Treatments	
$$\text{SST} = \sum_{j=1}^{k} n_j \left(\bar{x}_j - \bar{\bar{x}}\right)^2$$	12.2

Key Formulas (cont.)

	Section

Mean Square for Treatments

$$MST = \frac{SST}{k-1}$$

12.2

Sum of Squares for Error

$$SSE = \sum_{i=1}^{n_1}\left(x_{i1}-\overline{x}_1\right)^2 + \sum_{i=1}^{n_2}\left(x_{i2}-\overline{x}_2\right)^2 + \cdots + \sum_{i=1}^{n_k}\left(x_{ik}-\overline{x}_k\right)^2$$

12.2

Total Sum of Squares

$$TSS = SST + SSE$$

$$TSS = \sum_{j=1}^{k}\sum_{i=1}^{n_j}\left(x_{ij}-\overline{\overline{x}}\right)^2$$

12.2

Mean Square for Error

$$MSE = \frac{SSE}{n_T - k}$$

12.2

F-Statistic

$$F = \frac{MST}{MSE}$$

12.4

F-Statistic for One-Way ANOVA

$$F = \frac{MST}{MSE}$$

$$= \frac{\dfrac{\sum_{i=1}^{k} n_i\left(\overline{x}_i-\overline{\overline{x}}\right)^2}{k-1}}{\dfrac{\sum_{i=1}^{n_1}\left(x_{i1}-\overline{x}_1\right)^2 + \sum_{i=1}^{n_2}\left(x_{i2}-\overline{x}_2\right)^2 + \cdots + \sum_{i=1}^{n_k}\left(x_{ik}-\overline{x}_k\right)^2}{n_T - k}}$$

12.4

Computational Formula for MST

$$MST =$$

$$\frac{\left[\dfrac{\left(\sum_{i=1}^{n_1} x_{i1}\right)^2}{n_1} + \dfrac{\left(\sum_{i=1}^{n_2} x_{i2}\right)^2}{n_2} + \cdots + \dfrac{\left(\sum_{i=1}^{n_k} x_{ik}\right)^2}{n_k}\right] - \dfrac{\left(\sum_{j=1}^{k}\sum_{i=1}^{nj} x_{ij}\right)^2}{n_T}}{k-1}$$

12.4

Key Formulas (cont.)

Section

Computational Formula for MSE

$$MSE = \frac{\displaystyle\sum_{j=1}^{k}\sum_{i=1}^{n_j} x_{ij}^2 - \left[\frac{\left(\sum_{i=1}^{n_1} x_{i1}\right)^2}{n_1} + \frac{\left(\sum_{i=1}^{n_2} x_{i2}\right)^2}{n_2} + \cdots + \frac{\left(\sum_{i=1}^{n_k} x_{ik}\right)^2}{n_k}\right]}{n_T - k}$$

12.4

F-Statistic for Two-Way ANOVA (Randomized Block Design)

$$F = \frac{MST}{MSE} = \frac{\dfrac{SST}{k-1}}{\dfrac{SSE}{(k-1)(b-1)}}$$

12.5

where $SSE = TSS - SST - SSBL$

Test Statistic for Interaction between Factors

$$F = \frac{\dfrac{SSAB}{(a-1)(b-1)}}{\dfrac{SSE}{ab(r-1)}} = \frac{MSAB}{MSE}$$

12.6

Test Statistic for Main Effects for Factor A

$$F = \frac{\dfrac{SSA}{(a-1)}}{\dfrac{SSE}{ab(r-1)}} = \frac{MSA}{MSE}$$

12.6

Test Statistic for Main Effects for Factor B

$$F = \frac{\dfrac{SSB}{(b-1)}}{\dfrac{SSE}{ab(r-1)}} = \frac{MSB}{MSE}$$

12.6

AE   Additional Exercises

1. An experimenter often uses a randomized block design to reduce variation by comparing the treatments in homogeneous groups of experimental units called blocks. In many cases, differences in treatments are more likely to be detected with such a design than if the blocking factor were ignored. In each of the following situations, give an example of how one would run a randomized block design to make the comparison in each case. Be aware that there may be more than one correct answer in each example.

   a. Three different methods of teaching science are to be analyzed by comparing final exam scores from classes taught by the different methods. Assume that the same final exam is given to each class.

   b. Five hypertensive treatments are to be compared based on their ability to reduce systolic blood pressure. It is felt that the performance of the drugs is affected by the weight of the participant in the study.

   c. It is desired to compare three different ethnic groups on their knowledge of American history. Each person in the study will be given a 50-question multiple choice test to determine their overall knowledge of the subject.

2. A pharmacist is interested in studying the rate at which three different sinus headache drugs are absorbed into the bloodstream. She randomly selects 12 people, and then randomly assigns four people to try each drug. She administers the drug to each participant and measures the time it takes for the drug to be absorbed into the patient's bloodstream (in minutes). The results of the study are as follows.

Table for Exercise 2 – Drug Absorption Time (Minutes)		
Drug 1	Drug 2	Drug 3
5	10	6
4	11	7
6	9	5
3	8	5

   a. Can the pharmacist conclude at $\alpha = 0.01$ that there is a significant difference among the average times required for absorption into the bloodstream for the three drugs?

   b. What assumptions did the pharmacist make in performing the test procedure in part a.? Do the data appear to satisfy these assumptions? Explain.

   c. Describe an alternate design that the pharmacist could have used for the above analysis. What are the advantages and disadvantages of this design?

3. An FDA representative is interested in knowing if there is a difference in the average fat contents of three different brands of margarine. The representative randomly selects six samples of each of the brands of margarine and measures the average fat contents per serving. The results of the study are displayed in the following table.

Table for Exercise 3 – Fat Content per Serving (Grams)		
Margarine #1	**Margarine #2**	**Margarine #3**
6	5	9
7	6	8
6	5	7
8	4	8
6	6	9
8	5	7

a. Do the data indicate a difference among average fat contents per serving for the three brands of margarine at $\alpha = 0.01$?

b. What assumptions were made for the test in part **a.**? Do the data appear to satisfy these assumptions? Explain.

c. Why wouldn't a randomized block design be appropriate for this experiment?

4. Psychological reactance may be viewed as the motivational state resulting when someone's freedom is threatened or eliminated. A study relating psychological reactance to one's age was reported in "Psychological Reactance: Effects of Age and Gender" in the *Journal of Social Psychology*. In order to determine the degree of psychological reactance, participants were asked to fill out a questionnaire which was then scored. The higher the score, the more acute the degree of psychological reactance. The means, standard deviations, and group sizes (for different age groups) are given in the following table.

Table for Exercise 4 – Psychological Reactance			
Age Group	**Mean**	**Standard Deviation**	**Group Size**
18–24	3.36	0.60	1011
24–29	3.28	0.65	321
30–40	3.16	0.64	385

Although the summary statistics were given in the article, the actual data values upon which the statistics were based were not listed. This is standard procedure in many scientific journals.

a. Compute the sums of squares and their degrees of freedom for treatments and error based upon the statistics given in the table.

b. Compute MST and MSE.

c. With $\alpha = 0.01$, can we conclude that there is a significant difference among the degrees of psychological reactance for the different age groups?

d. What assumptions are necessary for performing the test in part **c.**? Can they be checked in this instance?

5. Interviews of fans following an Australian Football League game were summarized in the article "On Being a Sore Loser: How Fans React to Their Team's Failure" in the *Australian Journal of Psychology*. The study divided the fans interviewed into losers (those who supported the losing team), winners (those who supported the winning team), and non-partisans (those who were indifferent to the outcome of the game). Each fan was asked several questions, all dealing with the fan's perceptions of the game. The purpose of the study was to see if the groups differed on their responses to any of the questions. One question asked the fans to rate the umpire's performance on a five-point scale from *very bad* (1) to *very good* (5). The mean responses and group sizes associated with this question are given in the following table.

Table for Exercise 5 – Umpire Performance		
Group	Mean	Group Size
Losers	2.8	49
Winners	3.7	35
Non-Partisans	3.5	57

a. Compute the grand mean and the sum of squares for treatments (SST).

b. Compute the F-statistic for testing for equality of the group means. The sum of squares for error (SSE) was given in the article as 39.3.

c. With $\alpha = 0.01$, can we conclude that there is a significant difference among the groups in the perception of the umpire?

d. What assumptions are necessary for performing the test in part **c.**? Can they be checked in this instance?

6. Consider the following data regarding median starting and mid-career salaries for graduates from schools in different regions of the United States.

School	Starting Median Salary ($)	Mid-Career Median Salary ($)
Table for Exercise 6 – Starting and Mid-Career Salaries by Region		
Midwest		
Notre Dame	52,900	107,000
Carleton College	42,800	98,300
Illinois Institute of Technology	52,000	96,000
Denison University	40,600	94,000
University of Chicago	46,900	92,700
Northwestern University	49,900	88,300
Washington University in St. Louis	51,200	87,700
Northeast		
Princeton	56,900	130,000
Harvard	54,100	116,000
Massachusetts Institute of Technology	69,700	115,000
Dartmouth	51,600	114,000
Bucknell University	52,600	108,000
Manhattan College	53,900	107,000
Williams College	51,800	105,000
South		
Duke University	54,400	113,000
Vanderbilt University	51,300	100,000
Washington and Lee University	48,600	99,800
Wake Forest University	46,000	98,800
Rice University	51,100	97,400
Georgetown University	50,300	96,900
College of William and Mary	45,000	96,500
West		
University of Colorado – Boulder	45,900	90,400
University of Washington	46,700	88,400
Gonzaga University	44,200	87,700
Brigham Young University	47,400	86,800
University of Arizona	45,400	81,600
University of Oregon	39,700	79,200
Santa Clara University	52,900	105,000

Source: Payscale.com

a. Using $\alpha = 0.05$, is there a significant difference among average starting salaries for the four different regions?

b. What is the value of F for this test?

c. Using $\alpha = 0.05$, is there a significant difference among average mid-career salaries for the four different regions?

d. State the assumptions made for the two hypothesis tests in parts a. and c.

e. Do you have any concerns about these data? Explain.

7. Consider the following data regarding the median starting and mid-career salaries by type of major.

Table for Exercise 7 – Starting and Mid-Career Salaries by Major						
Major	Starting Median Pay ($)	Mid-Career Median Pay ($)	Major	Starting Median Pay ($)	Mid-Career Median Pay ($)
Engineering			**Math and Science**		
Petroleum Engineering	97,900	155,000	Applied Mathematics	52,600	98,600
Chemical Engineering	64,500	109,000	Computer Science	56,600	97,900
Electrical Engineering	61,300	103,000	Statistics	49,000	93,800
Aerospace Engineering	60,700	102,000	Mathematics	47,000	89,900
Computer Engineering	61,800	101,000	Physics	49,800	101,000
Nuclear Engineering	65,100	97,800	Biochemistry	41,700	84,700
Biomedical Engineering	53,800	97,800	Food Science	43,300	83,700
Mechanical Engineering	58,400	94,500	Geology	45,300	83,300
Industrial Engineering	57,400	93,100	Molecular Biology	40,500	81,200
Civil Engineering	53,100	90,200	Chemistry	42,000	80,900
Environmental Engineering	51,700	88,600	**Other**		
Business			Economics	47,300	94,700
Finance	46,500	87,300	Film Production	41,600	80,700
Supply Chain Management	50,200	84,700	Political Science	39,900	80,100
International Business	41,600	83,700	International Relations	40,500	79,400
Accounting	44,700	75,700	Philosophy	39,800	75,600
Advertising	37,700	74,700	History	37,800	69,000
Marketing	38,200	73,500	Communications	38,000	66,900
Business	41,000	70,500	Journalism	36,100	66,400
Public Relations	35,500	65,700	Spanish	36,400	58,400

Source: Payscale.com

a. State the null and alternative hypotheses to test if there is a significant difference in starting salary among the four types of majors.

b. Use $\alpha = 0.01$ to test for a significant difference among average starting salaries for the different types of majors.

c. State the null and alternative hypotheses to test if there is a significant difference among average mid-career salaries for the four types of majors.

d. Use $\alpha = 0.01$ to test for a significant difference among average mid-career salaries for the different types of majors.

8. Consider the following partially completed ANOVA table for a 3 × 4 factorial experiment with two replications.

ANOVA				
Source of Variation	*SS*	*df*	*MS*	*F*
Factor A	0.800	2		
Factor B	5.300	3		
Interaction	9.600			
Within				
Total	17.000	26		

a. Complete the ANOVA table.

b. At the 0.05 level, is there evidence of significant interaction between A and B? Justify your answer.

c. At the 0.05 level, is there evidence of a significant Factor A effect? Justify your answer.

d. At the 0.05 level, is there evidence of a significant Factor B effect? Justify your answer.

e. Does the result of the test for interaction suggest further investigation? Justify your answer.

9. In an experiment to determine the best method by which to assess college students, a group of students were exposed to one of three types of tests. The three methods were: all multiple choice questions, all free-response questions, and mixed questions (a mixture of multiple choice and free-response questions). The scores were recorded for each test taken. Fifteen students were used in the study and were grouped by class level (freshman, sophomore, junior, senior, and graduate). The following table contains the results of the experiment.

Table for Exercise 9 – Testing Methods			
Class Level	**Multiple Choice**	**Free-Response**	**Mixed**
Freshman	78	84	90
Sophomore	82	90	95
Junior	90	94	98
Senior	88	96	100
Graduate	95	98	99

a. Graphically plot the test scores by class level and testing method. Discuss the graph.

b. Perform an analysis of the data using the class-level blocks. Are blocking effects significant at the 0.05 level of significance? Explain.

c. Is the experiment useful having been analyzed as a completely randomized block design? Explain.

10. A randomized block design yielded the following ANOVA table.

ANOVA				
Source of Variation	*SS*	*df*	*MS*	*F*
Treatment	500.000	5	100	7.502
Block	230.000	3	76.67	5.752
Error	120.000	9	13.33	
Total	850.000	17		

    a. How many blocks are used in the experiment?

    b. How many treatments are used in the experiment?

    c. How many observations are used in the experiment?

    d. What are the null and alternative hypotheses to test if there is a difference among the treatment means?

    e. What test statistic should be used to conduct the test in part **d.**?

    f. What is the rejection region for the test in parts **d.** and **e.**?

    g. Carry out the test and state your conclusion based on a significance level of 0.05.

11. JAS & Associates, a commercial developer, usually gets three cost estimates for many of the jobs for their building projects. Even though one contractor normally works on each potential job, it is in the best interest of the company to get additional estimates and compare them for consistency, no matter who gets the job. To check the consistency of the estimates, several projects are selected and three contractors are asked to submit estimates. The estimates (in thousands of dollars) for the 10 jobs are given in the following table.

Table for Exercise 11 – Contractor Cost Estimates (Thousands of Dollars)			
Job	Contractor A	Contractor B	Contractor C
1	27	26	28
2	20	18	22
3	14	13	17
4	18	21	20
5	23	20	22
6	19	17	19
7	12	14	15
8	10	12	13
9	16	20	19
10	40	42	47

    a. Perform the appropriate analysis on the data given and generate the ANOVA table for the analysis.

    b. Do the data provide sufficient evidence to conclude that there is a difference among the cost estimates supplied by the contractors? Use a significance level of 0.05 to make your decision.

    c. What is the *P*-value for the test performed in part **a.**? Interpret this value.

12. The following table is a 3 × 3 factorial design with three observations for each factor level.

Table for Exercise 12 – Factorial Design Data			
		Factor B	
Factor A	1	2	3
1	30	47	36
	30	42	37
	30	42	38
2	12	27	35
	14	24	31
	15	22	33
3	10	34	24
	13	31	20
	12	31	22

a. Plot the treatment means using Factor A as the x-axis and Factor B as plotting symbols. Do the means appear to be different? Does interaction between factors A and B appear to be present? Justify your answers.

b. Perform the analysis using a software package, generating the ANOVA table.

c. Test for significant interaction using a 0.05 level of significance. Discuss your findings.

d. Test for A and B effects using a 0.05 level of significance. Discuss your findings.

13. Tech SportsPlex (TSP) is conducting a study to determine the effectiveness of three types of marketing/advertising methods: e-coupons, newspaper ads, and price discounts. Three counties (believed to be of equal size and close driving distance to TSP) were selected for the marketing campaign. Each strategy was used for a three-month period. It is known that the sales would be seasonal (i.e. TSP's management expects less activity during the summer months). The revenue data (in thousands of dollars) from the study are given in the following table.

Table for Exercise 13 – TSP Revenues by Marketing Strategy (Thousands of Dollars)			
Quarter	e-Coupons	Newspaper Ads	Price Discounts
1	48	42	37
2	25	18	21
3	20	15	18
4	40	30	24

a. Specify the null and alternative hypotheses to determine if there is a significant difference among average revenues for the three advertising strategies.

b. Generate the ANOVA table to test the hypotheses in **a.**

c. Conduct the test in **a.** using a significance level of 0.05.

d. Was the variation among the observed revenues significantly reduced by blocking? Explain using $\alpha = 0.05$.

Chapter Thirteen

Regression, Inference, and Model Building

13

Introduction

In Chapter 4, we discussed bivariate data and how to look for patterns in data to help us establish a relationship between two variables. In particular, we discussed establishing a linear relationship between two variables. Businesses are interested in developing relationships between two variables for many reasons such as:

- Predicting sales volume based on the amount of advertising expenditures
- Predicting the number of homes sold as a function of interest rate
- Predicting one's salary or net worth as a function of his or her age
- Determining the relationship between gasoline prices and crude oil prices

In this chapter we will develop a more detailed plan for studying the relationship between two or more variables called **regression analysis**. Regression analysis is a statistical technique that describes the relationship between a **dependent variable** (or **response variable**), which is the variable that we want to predict, and one or more **independent variables** (or **predictor variables**), which are the variables that we will use to make predictions. In this chapter we will discuss **simple linear regression**, which is the analysis used to study the relationship between two quantitative variables. Chapter 14 will discuss multiple regression models, in which we use several independent variables to predict the dependent variable.

13.1 Building a Simple Linear Regression Model

Consider the problem of deciding how long to study for an upcoming test. If we knew the exact relationship between time spent studying and the grade received, it could be useful in allocating study time. But the exact relationship between these variables is unknown and different for each course.

For most students the model relating study time to test performance is subjective, relying on past experience and data from other students. Is there any benefit from defining an exact relationship between study time and the grade received on a test? Essentially, knowing an exact relationship between study time and grade received would allow a student to choose his or her grade by allocating study time appropriately. How do you define an exact relationship? One method of defining a precise relationship between two or more variables is with the use of a mathematical model. Suppose, for example, the relationship between test score and study time was given by the following linear equation.

$$\text{Test Score} = 45 + 3.8(\text{Hours of Study Time})$$

If this mathematical model is accurate, then a student would be able to control his or her destiny. If a person studied for 10 hours, according to the model his or her test score would be:

$$\text{Test Score} = 45 + 3.8\,(10) = 83.$$

Figure 13.1

To get a higher test score, then study for 12 hours.

$$\text{Test Score} = 45 + 3.8\,(12) = 90.6$$

Since the relationship is known in advance, students would be able to control their grade by choosing their study time appropriately. In fact, if the exact relationship were known in advance, taking the test would be necessary only as an intellectual exercise, since the grade would already be determined by the time studied and the model. Admittedly, there is no model that can precisely predict a test score solely on the basis of time studied; since there are many other variables that affect test scores. But suppose a model were available which, although imperfect, fairly reliably predicted test scores based on hours studied.

$$\text{Test Score} = 45 + 3.8\,(\text{Hours of Study Time}) + \text{Error}$$

The new model introduces the error term. Now, if someone studies 10 hours, the model would predict

$$\text{Test Score} = 45 + 3.8\,(10) + \text{Error} = 83 + \text{Error}.$$

The predicted test score would still be 83, but there is an unknown random error associated with the prediction. If the error is reasonably small (say, at most 5 points), then the prediction will still be useful for planning purposes. But if the error is too large, then it will be difficult to rely on the model's predictions. If a model admits the possibility of an error, then gauging the expected magnitude of the error is essential in determining the model's usefulness. Estimating the mean and variance of the errors will be an important part of determining model utility. A model with a mean error of zero and small variation in the error terms would be desirable and should yield useful predictions. The model we have been using is simple. If two variables appear to be related in a straight line manner, we can use a **simple linear regression model** to describe their relationship.

The model-building process begins with a desire to find a relationship between two or more variables. Since there are a great number of possible models that can be selected, choosing the type of model to represent the relationship is a complex problem. A straight line is the simplest relationship between two variables. This straight line relationship is modeled by the simple linear regression model given by the following linear equation.

13

> ### Definition
>
> The **simple linear regression model** is given by the linear equation
>
> $$y_i = \beta_0 + \beta_1 x_i + \varepsilon_i,$$
>
> where
>
> β_0 is the y-intercept for the population data,
>
> β_1 is the slope coefficient for the population data,
>
> x_i is the value of the independent (or predictor) variable for observation i,
>
> ε_i is the random error in y for observation i, and
>
> y_i is the value of the dependent (or response) variable for observation i.

β_0 is pronounced *beta sub-zero* (or *beta naught*) and β_1 is pronounced *beta sub-one*. $y_i = \beta_0 + \beta_1 x_i$ is the portion of the model that represents a straight line. The slope of the line, denoted by β_1, is the expected (or average) change in y for each unit change in x. That is, it represents the average amount that y changes for every one unit that x changes. The y-intercept, denoted by β_0, represents the average value of y when x is equal to zero. The last term in the model, ε_i, represents the random error in y for each observation. That is, ε_i represents the vertical distance that the *actual* value of y_i is above or below the *average* value of y_i on the regression line.

Simple Linear Regression Model

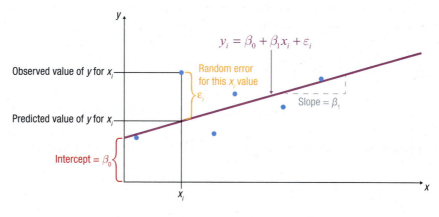

Figure 13.2

The slope and the intercept, β_0 and β_1, respectively, are called the **parameters** of the simple linear regression model. The parameters completely define the equation of the line. Developing a model to fit real world measurements is not trivial. Nature doesn't cooperate by requiring all relationships to be straight lines. Seldom, in fact, do pairs of measurements fall on perfectly straight lines. If a linear relationship exists, the data will have some general tendency to move together, or in opposite directions, as in Figure 13.3. In the following sections we will look at ways of measuring the degree of a linear relationship between two variables, as well as estimating the parameters (slope and intercept) of a line for a specific set of data.

As *x* increases *y* increases.

As *x* increases *y* decreases.

As *x* increases *y* does not change in a predictable way.

Figure 13.3

More often than not, bivariate data for an entire population are not available. Therefore, we must estimate the simple linear regression model using sample data. Using the data collected from the sample, we can estimate the coefficients of the **simple linear regression equation** and use this model for estimation and prediction.

Definition

The **estimated simple linear regression equation** is

$$\hat{y}_i = b_0 + b_1 x_i,$$

where b_0 and b_1 are estimates of their population counterparts. Specifically,

b_0 is an estimate of β_0, and b_1 is an estimate of β_1.

$\hat{y}_i$ is the predicted value of y for a given value of x_i, and is pronounced *y-hat*.

The symbol y_i is reserved for the observed value of y.

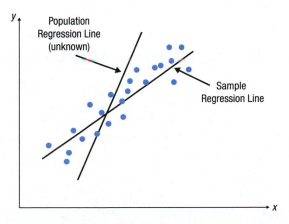

Figure 13.4

13

Essentially the regression model is used as a descriptive tool to summarize the relationship between the two variables. By providing estimates of β_0 and β_1, the simple linear relationship between x and y is specified. After data are collected, the **method of least squares** is the technique used to estimate β_0 and β_1. This method will be discussed further in Section 13.3.

Suppose we want to estimate the relationship between weight and height throughout the world for persons over the age of 18. Assuming the relationship is fundamentally linear, how is the relationship estimated? The number of persons on the earth over age 18 is somewhere in the neighborhood of 5.5 billion. Collecting 5.5 billion heights and weights could be a formidable task. In its current form, the problem is too large. Suppose the problem is reduced by only considering the population of the United States, where there are roughly 250 million persons over age 18. Even though this reduces the size from 5.5 billion to 250 million, it is still too large.

To make progress on this problem requires the use of sampling and inference. It is quite conceivable that a small random sample of bivariate measurements (height and weight) can be drawn from persons over 18 in the United States. Since all 250 million measurements will never be available, the best that we will be able to do is to infer that the relationship constructed from the sample will be representative of the 250 million population members.

If all 250 million bivariate measures were available, then the exact values of β_0 and β_1 could be determined. Since the entire population will never be at our disposal, we are confronted with a familiar sampling problem. Even if a random sample from the population is drawn, there can be no guarantee the sample will be exactly representative of the population. Since there is no guarantee that the sample will be representative, when a linear relationship is estimated from sample data we need to address the accuracy of the sample estimates $\left(b_0 \text{ and } b_1\right)$. Two familiar inferential techniques, confidence intervals and hypothesis testing, will be used to analyze the model's estimated coefficients and predicted values.

13.1   Exercises

Basic Concepts

1. What is regression analysis?

2. Give two examples of why businesses might be interested in studying the relationship between two variables.

3. What is the difference between a dependent and an independent variable?

4. What is a simple linear regression model? Give the equation that describes a simple linear regression model and define all terms in the equation.

5. What is the estimated simple linear regression equation and how is it used?

6. What is $\hat{y}$? How does this differ from y?

7. What is the technique used to estimate the simple linear regression coefficients?

8. What is the relationship between scatterplots and simple linear regression?

9. Why is it often difficult to accurately describe real world situations using a simple linear regression equation?

Exercises

10. Consider the following simple linear regression model. Write the estimated simple linear regression equation that corresponds to this model.

$$y_i = \beta_0 + \beta_1 x_i + \varepsilon_i$$

11. Consider the following estimated simple linear regression equation.

$$\hat{y}_i = b_0 + b_1 x_i$$

   a. What population parameter does b_0 estimate?

   b. What population parameter does b_1 estimate?

   c. Is error incorporated into the estimated model? Explain.

12. Suppose that a company wishes to predict sales volume based on the amount of advertising expenditures. The sales manager thinks that sales volume and advertising expenditures are modeled according to the following linear equation. Both sales volume and advertising expenditures are in thousands of dollars.

$$\text{Estimated Sales Volume} = 49.25 + 0.51(\text{Advertising Expenditures})$$

a. What is the dependent variable in this model? Explain.

b. What is the independent variable in this model? Explain.

c. What is the estimated sales volume for this company when the marketing department spends \$40,000 on advertising?

d. If the company had a target sales volume of \$100,000, how much should the sales manager allocate for advertising in the budget?

e. What is the sales manager forgetting to account for when using this linear equation to determine sales volume? What kinds of problems could this cause for the company?

13. Suppose the following estimated regression equation was determined to predict salary based on years of experience.

$$\text{Estimated Salary} = 25689.10 + 2148.35(\text{Years of Experience})$$

a. What is the dependent variable?

b. What is the independent variable?

c. What is the value that estimates β_0 in this particular equation?

d. What is the value that estimates β_1 in this particular equation?

e. What is the estimated salary for an employee with 15 years of experience?

13.2   Defining a Linear Relationship

In Chapter 4, the correlation coefficient was used to measure the degree of linear relationship between two variables. However, it does not describe the exact linear association between x and y. That is the role of regression analysis. By determining a specific relationship between x and y, we may be able to use x to help predict y.

What does it mean to specify a linear relationship between two variables? Suppose we model the relationship between two variables using $y_i = \beta_0 + \beta_1 x_i + \varepsilon_i$. The estimated regression equation is given by $\hat{y}_i = b_0 + b_1 x_i$.

For example, letting $b_0 = 3$ and $b_1 = 2$ specifies the line $\hat{y} = 3 + 2x$.

Figure 13.5 shows this relationship.

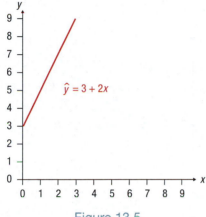

Figure 13.5

Letting $b_0 = 8$ and $b_1 = -2$ specifies the line $\hat{y} = 8 - 2x$.

Figure 13.6 shows this relationship.

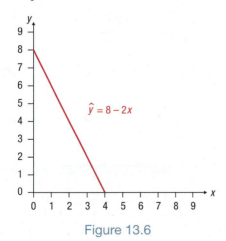

Figure 13.6

Consider the data plotted in Figure 13.7.

Figure 13.7

The data in Figure 13.7 seem to be related. Specifying the relationship between x and y with a linear model means finding a line that best fits the data in some way. The problem is that there are many lines that could be interpreted as fitting the data.

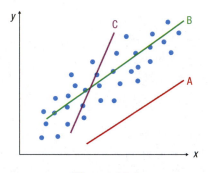

Figure 13.8

Which one of the lines in Figure 13.8 do you think best fits the data? Clearly, Line A doesn't do a very good job. The points don't cluster around the line at all. Although Lines B and C are very different lines, they both seem to go through the data in some sense, but it would be difficult to choose between them. Moreover, neither Line B nor Line C really seems to quite fit the data. To find the best line, we must come up with a method of summarizing how close each line is to the data. The closeness measure can then be used as a criterion to choose between various lines.

How Do We Measure How Close a Line Is to the Data?

One possible method of choosing the best fitting line is to use the line to predict the y-value for each observation. The superior line is the one that does the best job of predicting the observed y-values. Let's look at a small data set.

x	y
2	3
4	2
5	6
8	5
9	8

We plotted these data in Figure 13.9 and then tried to draw a line through the points. There are infinitely many lines that can be fitted to the data. However, there is no straight line that passes through all of the data points. One line that seems to fit the data reasonably well is:

$$y = 1 + 0.7x.$$

x-y Plot

Figure 13.9

Although other lines might do a better job of fitting the data, let's see how well this line predicts the y-values. In order to evaluate how well a given line fits the data we must construct a method of measuring how well any line fits the data. If a reasonable measurement is developed, it can serve as a decision-making criterion which will enable us to look for the *best* line. To develop the appropriate criterion we need to examine the purpose of the regression line

$$\hat{y}_i = b_0 + b_1 x_i.$$

Although the exact method of estimation has not been discussed, once the coefficients of the model $(b_0$ and $b_1)$ are estimated, they become constants. The value of x is selected by the user of the model. Once the coefficients are estimated and the value of x is chosen, the corresponding value of y can be predicted. For instance, if $x = 2$, then using the model $\hat{y}_i = 1 + 0.7x_i$, the predicted y-value is:

$$\hat{y} = 1 + 0.7(2) = 2.4.$$

The purpose of the model is to predict y for some given value of x. According to the model, y should be 2.4 when $x = 2$. But the model is wrong. In the first historical observation, the observed data for $x = 2$ is $y = 3$, not 2.4 as the model predicts. The difference between the observed value of y and the predicted value of y is called the **error, estimated error**, or **residual** (e_i). Remember, the predicted value of y, the dependent variable, is referred to as $\hat{y}$. The symbol y is reserved for the observed value of y. Therefore, the error for each observation is given by

$$\text{Error} = e_i = \text{Observed } y - \text{Predicted } y = y_i - \hat{y}_i.$$

For the first observation, the error is given by

$$e_1 = y_1 - \hat{y}_1 = 3 - 2.4 = 0.6.$$

The errors reflect how far each observation is from the line. Examining the errors suggests how well the line fits the data.

If we take each of the observed values of x in the data and use the regression line to predict the corresponding value of y, we can compare the model's predicted value of y for each x to the actual value of y for each x. These calculations are presented in Table 13.1.

Table 13.1 – Observed versus Predicted Values				
Observed x	Observed y	Predicted y $\hat{y}_i = 1 + 0.7x_i$	Error $y_i - \hat{y}_i$	Squared Error $\left(y_i - \hat{y}_i\right)^2$
2	3	$2.4 = 1 + 0.7(2)$	$3 - 2.4 = 0.6$	0.36
4	2	$3.8 = 1 + 0.7(4)$	$2 - 3.8 = -1.8$	3.24
5	6	$4.5 = 1 + 0.7(5)$	$6 - 4.5 = 1.5$	2.25
8	5	$6.6 = 1 + 0.7(8)$	$5 - 6.6 = -1.6$	2.56
9	8	$7.3 = 1 + 0.7(9)$	$8 - 7.3 = 0.7$	0.49
			$\sum(y_i - \hat{y}_i) = -0.6$	$\sum\left(y_i - \hat{y}_i\right)^2 = 8.90$

Adrien-Marie Legendre 1752 – 1833

In 1806 Legendre was investigating the orbits of comets and published a book on the subject. In the appendix he gave the method of least squares curve fitting. In 1809 Karl Gauss also published his version of the least squares method. Although acknowledging Legendre's work, Gauss claimed priority on the discovery. This greatly hurt Legendre who fought for many years to have his discovery recognized as his contribution to statistics.

One possible criterion that could be used to compare different lines that might fit the data would be to sum all the errors and choose the line with the smallest sum. But as you can see in the previous table some of the larger positive errors were canceled out by equally large negative errors. It is possible to develop a line that has a small sum of errors, but the errors themselves could be quite large. Thus, the sum of the errors is not the criterion we have been looking for, but it is close. By squaring the error terms and then adding the squared errors, there are no negative errors to cancel out the positive ones.

Let's examine the measurement of the **sum of squared errors (SSE)**.

Formula

Sum of Squared Errors (SSE)

The **sum of squared errors (SSE)** is given by

$$\text{SSE} = \sum\left(y_i - \hat{y}_i\right)^2 = \sum\left(y_i - \left(b_0 + b_1 x_i\right)\right)^2.$$

SSE can be used as a criterion for selecting the best fitting line through a set of points. If SSE is zero, then the model fits the data exactly and the observed data must lie in a straight line. If Line A's SSE is larger than Line B's, then Line B fits the data better than Line A. For the model we have been studying, the value of SSE is 8.90. Without studying other possible lines, there is no way to judge whether this is a large or small value for SSE. However, if there were a method that would find the *best* line (the line with the smallest sum of squared errors), then our problem would be solved. This line of best fit is called the **least squares line** since it has the smallest SSE.

13.2   Exercises

Basic Concepts

1. What is the correlation coefficient? Why is the correlation coefficient insufficient when describing an exact linear relationship between x and y?

2. What is the residual of a model?

3. What is the sum of squared errors and what does it measure?

4. Explain why the best line is referred to as the least squares line.

Exercises

5. Plot the following lines.

   a. $y = 2 + 3x$

   b. $y = 4 + 8x$

   c. $y = 9 - 2x$

   d. $y = x$

6. Plot the following lines.

   a. $y = 100 + 50x$

   b. $y = 0.5 + 0.7x$

   c. $y = 20 - 5x$

7. Consider the following estimated regression equation.

   $$\hat{y}_i = 10x_i - 5$$

   a. Complete the following table.

Table for Exercise 7 – Predicted Values	
x	$\hat{y}$
2	
5	
7	
9	
10	

   b. Do these two variables appear to have a positive or negative relationship?

   c. For these two variables, what sign would you expect the correlation coefficient to have? Explain.

8. Consider the following data.

Table A for Exercise 8 – Observed Values	
x	y
0	2
1	4
5	9
6	7
8	8

a. Draw a scatterplot of the data.

b. Draw a line which you believe fits the data.

c. Suppose that $\hat{y}_i = 3 + 0.8x_i$ is a line that fits the data reasonably well. Complete the following table.

Table B for Exercise 8 – Observed and Predicted Values				
Observed x	Observed y	Predicted y	Error	Squared Error
0	2			
1	4			
5	9			
6	7			
8	8			

d. What is the sum of squared errors for these data?

9. Consider the following data regarding home sale prices and square footage.

Table A for Exercise 9 – Housing Prices and Square Footage	
Selling Price (Thousands of Dollars)	Square Footage
199.9	1065
228.0	1254
235.0	1300
285.0	1577
239.0	1600
293.0	1750
285.0	1800
365.0	1870
295.0	1935
290.0	1948
385.0	2254
505.0	2600
425.0	2800
415.0	3000

a. Suppose we want to predict selling price based on square footage. Write the estimated regression equation in terms of selling price and square footage. (Assume the parameters of this model have not been estimated.)

b. Create a scatterplot of the data and draw a line of best fit.

c. Suppose we determine that an equation that fits the data reasonably well is

Estimated Selling Price $= 52.35 + 0.14$ (Square Footage).

Complete the following table.

| Table B for Exercise 9 – Housing Prices and Square Footage | | | | |
Observed Selling Price (Thousands of Dollars)	Observed Square Footage	Predicted Selling Price (Thousands of Dollars)	Error	Squared Error
199.9	1065			
228.0	1254			
235.0	1300			
285.0	1577			
239.0	1600			
293.0	1750			
285.0	1800			
365.0	1870			
295.0	1935			
290.0	1948			
385.0	2254			
505.0	2600			
425.0	2800			
415.0	3000			

d. Compute the sum of squared errors for these data.

13.3 Finding the Least Squares Line

To define a line you must specify its slope and y-intercept. Fortunately, the problem of finding the least squares line was solved by Karl Gauss several hundred years ago. The line that best fits the data is the **least squares line** that is obtained by minimizing the sum of squared errors (SSE). That is, we want to find the values of b_0 and b_1 that minimize

$$\sum \left(y_i - \hat{y}_i \right)^2.$$

Because $\hat{y}_i = b_0 + b_1 x_i$, we want to minimize

$$\sum \left(y_i - \left(b_0 + b_1 x_i \right) \right)^2.$$

In the above equation, we have two unknowns, b_0 and b_1. The **method of least squares** determines the values of b_0 and b_1 that minimize the sum of squared errors around the prediction line. Using calculus, the computational formulas for b_0 and b_1 are given as follows.

Formula

Slope and y-Intercept of the Least Squares Line

The equation for finding the slope is given by

$$b_1 = \frac{SS_{xy}}{SS_{xx}}$$

where

$$SS_{xy} = \sum \left(x_i - \bar{x} \right)\left(y_i - \bar{y} \right) = \sum x_i y_i - \frac{\left(\sum x_i \right)\left(\sum y_i \right)}{n}$$

and

$$SS_{xx} = \sum \left(x_i - \bar{x} \right)^2 = \sum x_i^2 - \frac{\left(\sum x_i \right)^2}{n}.$$

The slope can also be calculated using

$$b_1 = \frac{n \sum x_i y_i - \sum x_i \sum y_i}{n \sum x_i^2 - \left(\sum x_i \right)^2}.$$

The estimate of the intercept is given by

$$b_0 = \bar{y} - b_1 \bar{x} = \frac{1}{n} \left(\sum y_i - b_1 \sum x_i \right).$$

The x_i and y_i referred to in the expressions are the observed data values of x and y, respectively.

You may never have to use these formulas, since many calculators and computer programs compute the coefficients of the least squares line. However, should manual calculation become necessary, the slope coefficient b_1 must be calculated prior to calculating b_0.

The Discovering Technology section at the end of this chapter shows how to estimate simple linear regression equations using a TI-84 Plus calculator, Microsoft Excel, and Minitab. Using statistical software to estimate linear regression equations will give the most accurate results and is highly recommended.

Table 13.2 contains data on the number of items produced during 10 randomly selected weeks in a production factory and the total production costs. It seems obvious that in a production operation there is a relationship between the number of items produced (the independent variable, x) and the total production cost (the dependent variable, y).

Table 13.2 – Weekly Production		
Week	Items Produced	Cost ($)
1	22	3500
2	30	3800
3	36	4500
4	41	4200
5	27	3700
6	45	4600
7	30	3600
8	37	4550
9	32	3990
10	31	3675

Some summary statistics are given below.

$$\sum x_i = 331$$
$$\sum y_i = 40,115$$
$$\sum x_i y_i = 1,350,055$$
$$\sum x_i^2 = 11,369$$

Using the least squares equations and the summary calculations given above, we can estimate the following coefficients:

$$b_1 = \frac{SS_{xy}}{SS_{xx}}$$
$$= \frac{n\sum x_i y_i - \sum x_i \sum y_i}{n\sum x_i^2 - \left(\sum x_i\right)^2}$$
$$= \frac{10(1350055) - 331(40115)}{10(11369) - (331)^2}$$
$$= \frac{222485}{4129} \approx 53.8835$$

and

$$b_0 = \bar{y} - b_1\bar{x}$$

$$= \frac{1}{n}\left(\sum y_i - b_1 \sum x_i\right)$$

$$= \frac{1}{10}\left(40115 - 53.8835(331)\right)$$

$$\approx 2227.9562.$$

Therefore the model to estimate the total production cost is:

$$\text{Estimated Cost} = \$2227.96 + \$53.88(\text{Items Produced})$$

Dependent
Variable

Independent
Variable

Figure 13.10 shows a graph of this model using the estimated coefficients.

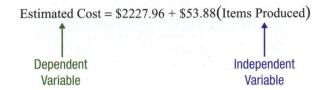

Cost vs. Items Produced Line Fit Plot

Figure 13.10

This model has a smaller SSE for these data than any other line. In Table 13.3 we examine the errors produced by the least squares model. Note that the sum of the errors equals zero.

	Table 13.3 – Errors Produced by the Least Squares Model				
Week	**Items Produced** x	**Cost (\$)** y	**Predicted Cost (\$)** $\hat{y}_i = 2227.96 + 53.88x_i$	**Error** $y_i - \hat{y}_i$	**Squared Error** $\left(y_i - \hat{y}_i\right)^2$
1	22	3500	3413.39	86.61	7500.76
2	30	3800	3844.46	−44.46	1976.79
3	36	4500	4167.76	332.24	110,381.98
4	41	4200	4437.18	−237.18	56,254.21
5	27	3700	3682.81	17.19	295.48
6	45	4600	4652.71	−52.71	2778.74
7	30	3600	3844.46	−244.46	59,761.24
8	37	4550	4221.65	328.35	107,816.56
9	32	3990	3952.23	37.77	1426.71
10	31	3675	3898.34	−223.35	49,882.83
				$\sum\left(y_i - \hat{y}_i\right) \approx 0$	$\sum\left(y_i - \hat{y}_i\right)^2 \approx 398{,}075.30$

As you can see, in the building of the linear model, there is potential for a great deal of calculation. For most problems you will use some type of statistical analysis package or spreadsheet to compute the least squares coefficients as well as related statistics and diagnostic measures. These packages will compute a wealth of summary and diagnostic measures concerning the estimated model.

Using the data and the estimated regression model, if the number of items produced is 0, then the predicted cost of production would be $2227.96, the value of b_0.

$$\text{Estimated Cost} = \$2227.96 + \$53.88(0) = \$2227.96$$

The value of b_0 is always interpreted as the average value of the dependent variable, in this case, the total production cost, when the independent variable is set equal to zero. Since b_1 is the estimated slope of the line, it is interpreted as the average change in the dependent variable (total production cost) for a one unit change in the independent variable (items produced). In our example, the independent variable is expressed in units. Therefore, for every additional unit, the total production cost is expected to increase by $53.88. If this interpretation is correct, the model's predicted cost for 27 items should be $53.88 more than that for 26 items.

If 27 items are produced, then $x = 27$, and

$$\text{Estimated Cost} = \$2227.96 + \$53.88(27) = \$3682.72.$$

If 26 items are produced, then $x = 26$, and

$$\text{Estimated Cost} = \$2227.96 + \$53.88(26) = \$3628.84.$$

According to the model, the difference in cost between producing 27 items and 26 items is exactly equal to the slope.

Table 13.4 – Difference in Production Cost	
Predicted cost for producing 27 items	$3682.72
Predicted cost for producing 26 items	$3628.84
Difference	$53.88

Definition

The intercept coefficient, b_0, is the average value of the dependent variable, y, when the independent variable, x, is equal to zero.

The slope coefficient, b_1, is the average change in the dependent variable, y, for a one unit change in the independent variable, x.

13.3   Exercises

Basic Concepts

1.   What measure should be minimized in order to find the least squares line?

2.   What is the equation for finding the slope of the least squares line?

3.   What is the equation for finding the intercept of the least squares line?

4.   When finding the least squares line manually, which must be calculated first: the slope or the y-intercept?

5.   Interpret the intercept coefficient, b_0.

6.   Interpret the slope coefficient, b_1.

Exercises

7.   Consider the following data.

Table for Exercise 7					
x	1	2	3	4	5
y	1	3	4	4	6

   a.   Plot the data points on a scatterplot.

   b.   Determine the least squares line. Use x as the independent variable.

   c.   Plot the least squares line on the scatterplot.

   d.   Use the model to compute the error for each data point.

8.   Consider the following data.

Table for Exercise 8					
x	−2	−1	0	3	5
y	1	3	5	4	8

   a.   Plot the data points on a scatterplot.

   b.   Determine the least squares line. Use x as the independent variable.

   c.   Plot the least squares line on the scatterplot.

   d.   Use the model to compute the error for each data point.

9.   Comparing the least squares lines in Exercises 7 and 8, which line fits the data better? Explain your answer.

10. Suppose a linear regression analysis produced the following equation relating an individual's salary to the current value of his or her home.

$$\text{Estimated Current Value of Home} = 12331 + 3.14(\text{Annual Salary})$$

    a. Which of the variables in the model is the dependent variable?

    b. Which of the variables in the model is the independent variable?

    c. What would be the predicted current value of home for someone earning a salary of $32,000?

    d. If a person earned $5000 additional income, how much of an increase in home value would be predicted?

    e. In terms of the problem, interpret the estimate of the slope in the model.

    f. In terms of the problem, interpret the estimate of the intercept in the model.

    g. Do you believe annual salary is a causal factor in explaining the price of someone's home? Explain.

11. Suppose a linear regression analysis produced the following equation relating a basketball player's total points scored to the number of minutes played in a season.

$$\text{Estimated Points Scored} = -97.2 + 0.645(\text{Minutes Played})$$

    a. Which of the variables in the model is the dependent variable?

    b. Which of the variables in the model is the independent variable?

    c. What would be the predicted value of total points scored for a basketball player who plays 500 minutes in a season?

    d. If a basketball player played an additional 100 minutes, how much of an increase in total points scored would be predicted?

    e. In the model, which of the coefficients is the slope?

    f. In the model, which of the coefficients is the intercept?

    g. Do you believe the number of minutes played is a causal factor in explaining the total points scored? Explain.

12. Suppose you were studying the educational level of husbands and wives (measured in number of years of education). You have randomly selected 10 couples and have obtained the data in the following table.

Table for Exercise 12 – Education Level										
Husband	12	16	16	18	20	17	23	14	12	16
Wife	14	16	14	16	16	18	18	12	16	20

a. Suppose you wanted to predict the husband's years of education based on the wife's. Use the data to estimate the appropriate model.

b. Use the model in part **b.** to predict the husband's educational level if married to a woman with 16 years of education.

c. Suppose you wanted to predict the years of education for the wife based on the husband's years of education. Use the data to create the appropriate model. Did you get the same model as in part **b.**?

d. Use the model created in part **d.** to predict the wife's educational level if married to a husband with 16 years of education.

e. Do you believe there is a causal relationship between the two variables? If so, which direction is the causality? Does the husband's education cause the wife to have more or less education, or vice versa?

13.4   The Importance of Errors

The usefulness of the estimated model depends on the magnitude of the prediction errors you expect the model to produce. The production model is

$$\text{Cost} = \beta_0 + \beta_1 \left(\text{Items Produced} \right) + \varepsilon_i.$$

Yet we ignored the error component in the previous section when we predicted the production cost for different numbers of items. For instance, when we predicted the price of producing 26 items, we found

$$\text{Estimated Cost} = \$2227.96 + \$53.88(26) = \$3628.84.$$

Since the model is not going to be a perfect predictor, we should incorporate the possibility of error in the model. Thus,

$$\text{Estimated Cost} = \$2227.96 + \$53.88(26) + e_i = \$3628.84 + e_i$$

would have been a more precise statement. It is important to assess the magnitude of the error when the model is used for predictive purposes. If the errors are too large, then it will not be advantageous to use the model for prediction.

How Do We Assess the Magnitude of the Errors?

You may have wondered why we have been trying to predict observations that we already know. For example, an observation in the production data was items produced equals 27 and cost equals $3700. If we already know the cost is $3700, why try to predict it? Using the model to predict the observed outcomes of the dependent variable, y, produces error data that can be used to evaluate the errors the model produces. Knowledge of the errors will provide understanding of the predictive quality of the model.

How Do We Summarize the Errors a Model Produces?

For the production model, the mean error is zero. This is true for all least squares models and hence mean error won't provide any useful information. What about the average variation in the error terms? Large variation in the errors would indicate that a model's prediction is not very reliable. On the other hand, small variation in the errors would indicate the model is capable of producing more trustworthy predictions.

Computing the variation of the error data is not much different from computing the variation of any data set. Recall that the formula for sample variance is

$$s^2 = \frac{\sum \left(x_i - \bar{x} \right)^2}{n-1}.$$

In regression, we are discussing errors. For simple linear regression the degrees of freedom for the errors are $(n-2)$, where the "2" represents the number of parameters estimated in the model (i.e. β_0 and β_1 were estimated). Making this slight adjustment for degrees of freedom produces the definition for the **variance of the error terms**, also known as the **mean square error**.

Formula

Mean Square Error

The variance of the error terms is also known as the **mean square error** and is given by:

$$s_e^2 = \frac{\sum(y_i - \hat{y}_i)^2}{n-2} = \frac{SSE}{n-2}.$$

The numerator is divided by $n - 2$ (the degrees of freedom of the error data) instead of $n - 1$ to account for an additional constraint imposed by least squares estimation. In the production data the mean square error is

$$s_e^2 = \frac{398075.2967}{10-2} \approx 49759.41$$

and the **standard error** (standard deviation of the error terms) is given by

$$s_e = \sqrt{49759.41} \approx 223.0682.$$

	A	B	C	D	E
1	SUMMARY OUTPUT				
2					
3	*Regression Statistics*				
4	Multiple R	0.866441198			
5	R Square	0.75072035			
6	Adjusted R Square	0.719560393			
7	Standard Error	223.0681781			
8	Observations	10			
9					
10	ANOVA				
11		*df*	*SS*	*MS*	*F*
12	Regression	1	1198827.203	1198827	24.09247
13	Residual	8	398075.2967	49759.41	
14	Total	9	1596902.5		
15					
16		*Coefficients*	*Standard Error*	*t Stat*	*P-value*
17	Intercept	2227.955922	370.1487713	6.019082	0.000317
18	Items Produced	53.8835069	10.97779658	4.908408	0.001181

Figure 13.11

Figure 13.11 shows the summary output from Microsoft Excel for the production data. (See the Discovering Technology section at the end of this chapter for information on how to generate a summary output using Excel.) Notice that the mean square error is given in the ANOVA table in the column labeled *MS* and the row labeled *Residual*. The standard error of the model is given in the *Regression Statistics* table and is labeled *Standard Error*.

Should you be satisfied with a model whose standard error is \$223.07? It would certainly be more desirable if the model's standard deviation of error was only \$100. If you know a great deal about the production costs (you have an intuitive model), then perhaps the current model would not be of value. However, if you are uninformed, the current model provides a basic understanding of the relationship between items produced and production cost.

What constitutes small and large variation of the errors is dependent on the nature of the problem. Using a model to predict the diameter of a valve to be used in a heart-lung machine is certainly going to produce a different idea of *small* than estimating production costs. It would be ideal to develop a measure that would summarize the degree of fit on some standardized scale. Such a measure is introduced in the following section.

13.4 Exercises

Basic Concepts

1. Why is the magnitude of the prediction errors important when estimating a regression model?

2. What is the mean error for a least squares model?

3. Describe what the magnitude of the variation in the error terms tells us about the reliability of the regression model.

4. What is mean square error?

5. How many degrees of freedom are associated with the error term in a simple linear regression model?

6. What is the square root of the mean square error known as?

7. Describe where the summary statistics for the standard error and mean square error are found in a standard regression summary output in Microsoft Excel.

8. Is there a universal rule on how large is *large* with regard to standard error in a model?

9. What is estimated by the mean square error and what is estimated by the standard error?

Exercises

10. Consider the following summary output.

	A	B	C	D	E
1	SUMMARY OUTPUT				
2					
3	*Regression Statistics*				
4	Multiple R	0.911653228			
5	R Square	0.831111609			
6	Adjusted R Square	0.79733393			
7	Standard Error	0.253142413			
8	Observations	7			
9					
10	ANOVA				
11		*df*	*SS*	*MS*	*F*
12	Regression	1	1.576737452	1.576737	24.60535
13	Residual	5	0.320405405	0.064081	
14	Total	6	1.897142857		
15					
16		*Coefficients*	*Standard Error*	*t Stat*	*P-value*
17	Intercept	4.021621622	0.181401491	22.16973	3.47E-06
18	X Variable 1	-0.22297297	0.044950802	-4.96038	0.004247

   a. What is the mean square error for these data?

   b. What is the standard error of the model?

11. Consider the following data.

Table A for Exercise 11 – Observed Values	
x	*y*
15	110
18	135
25	150
24	149
26	158
40	169

   a. Suppose that, using statistical software, we determine that $b_0 = 93.2922$ and $b_1 = 2.1030$. Complete the following table.

Table B for Exercise 11 – Observed versus Predicted Values				
Observed *x*	Observed *y*	Predicted *y*	Error	Squared Error
15	110			
18	135			
25	150			
24	149			
26	158			
40	169			

   b. Compute the sum of squared errors.

   c. Compute the mean square error.

   d. Compute the standard error of the model.

   e. Do you believe these estimates of b_0 and b_1 provide a reliable estimated regression equation for these data? Explain.

12. Consider the following data regarding students' college GPAs and high school GPAs.

Table A for Exercise 12 – GPAs	
College GPA	**High School GPA**
2.80	3.42
3.54	3.56
2.88	3.13
2.15	3.27
2.22	3.38
3.31	4.13
2.13	3.95
2.39	3.81
3.01	4.33
2.68	2.85

a. Suppose we want to predict college GPA based on high school GPA. Write the estimated regression equation in terms of college GPA and high school GPA. (Assume the parameters of the model have not been estimated.)

b. Suppose we determine, using statistical software, that the estimated regression equation is

$$\text{Estimated College GPA} = 1.88 + 0.2319(\text{High School GPA}).$$

Complete the following table.

Table B for Exercise 12 – GPAs				
Observed College GPA	**Observed High School GPA**	**Predicted College GPA**	**Error**	**Squared Error**
2.80	3.42			
3.54	3.56			
2.88	3.13			
2.15	3.27			
2.22	3.38			
3.31	4.13			
2.13	3.95			
2.39	3.81			
3.01	4.33			
2.68	2.85			

c. Compute the sum of squared errors for the model.

d. Compute the standard error of the model.

13.5   Evaluating the Fit of a Model

The goal in constructing most linear models is to use the independent variable, x, to explain or predict the dependent variable, y. The question we want to consider is, how much of the variation in y can be explained with the model? Before determining how much variation the model explains, it will be necessary to evaluate how much variability exists in the y-variable. This quantity is called the **total sum of squares (TSS)** and represents the total variation in the dependent variable, y.

Formula

Total Sum of Squares (TSS)

The total variation in y is given by the **total sum of squares (TSS)**.

$$\text{TSS} = \sum \left(y_i - \bar{y} \right)^2$$

If you think TSS looks a great deal like the numerator of the formula for the sample variance, you are right. TSS is the sum of the squared deviations about the mean of the dependent variable, y. If TSS were divided by $n - 1$ it would be the sample variance of y.

What is an Error?

An error $\left(y_i - \hat{y}_i \right)$ represents the model's inability to predict the variation in the dependent variable, y. If y didn't vary, for example if all y's were 6, its value would be easy to predict and the model's errors would all be zero. Adding all of the squared errors accumulates the total of all *unexplained* variation.

$$\text{SSE} = \sum \left(y_i - \hat{y}_i \right)^2$$

The variation in y can be divided into two categories, unexplained and explained. Total variation must equal unexplained variation plus explained variation.

$$\text{TSS} = \text{Unexplained Variation} + \text{Explained Variation}$$

or

$$\text{TSS} = \text{SSE} + \text{Explained Variation}$$

Denoting explained variation as **SSR (sum of squares of regression)** produces

$$\text{TSS} = \text{SSE} + \text{SSR} \quad (\text{TSS is the total unexplained and explained variation in } y).$$

Solving this equation for SSR results in

$$\text{SSR} = \text{TSS} - \text{SSE} \quad (\text{the explained variation, SSR, is equal to the total variation minus the unexplained variation}).$$

Measures of Variation

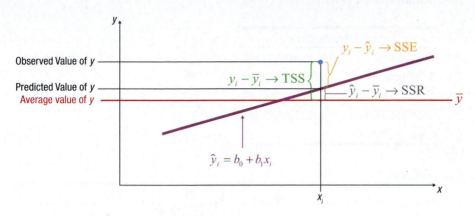

Figure 13.12

Interpreting SSR

It would be delightful if the model would explain all of the variability in y, but unless all of the observed data points fall in a straight line, this will not happen. In virtually all models, there will be errors, i.e. unexplained variation. The difference between the total variation in y and the unexplained variation must be the variation that is explained by the regression model. That's why explained variation is called the sum of squares of regression, SSR.

In the production example,

$$\text{TSS} = 1596902.5 \text{ and } \text{SSE} = 398075.2967.$$

Therefore,

$$\text{SSR} = \text{TSS} - \text{SSE} = 1596902.5 - 398075.2967 = 1198827.2033.$$

These values are found in the ANOVA table of the summary output in the *SS* (sum of squares) column.

◢	A	B	C	D	E
1	SUMMARY OUTPUT				
2					
3	*Regression Statistics*				
4	Multiple R	0.866441198			
5	R Square	0.75072035			
6	Adjusted R Square	0.719560393			
7	Standard Error	223.0681781			
8	Observations	10			
9					
10	ANOVA				
11		*df*	*SS*	*MS*	*F*
12	Regression	1	1198827.203	1198827	24.09247
13	Residual	8	398075.2967	49759.41	
14	Total	9	1596902.5		
15					
16		*Coefficients*	*Standard Error*	*t Stat*	*P-value*
17	Intercept	2227.955922	370.1487713	6.019082	0.000317
18	Items Produced	53.8835069	10.97779658	4.908408	0.001181

Figure 13.13

Of the roughly 1.6 million units of total variation in y, the model explains about 1.2 million. The proportion of the variation explained by the model is called the **coefficient of determination** and is denoted as R^2.

Formula

Coefficient of Determination

The **coefficient of determination**, R^2, is given by

$$R^2 = \frac{\text{SSR}}{\text{TSS}} = 1 - \frac{\text{SSE}}{\text{TSS}}.$$

The coefficient of determination is a value between 0 and 1, inclusive. That is, $0 \le R^2 \le 1$.

The R^2 measurement summarizes the degree of fit on a standardized scale. The largest value R^2 can attain is 1, which will occur when the model explains all of the variation in y and consequently SSR = TSS. The smallest value of R^2 is 0, which occurs when the model does not explain any of the variation in y and, consequently, SSR = 0. Thus, the R^2 value is the proportion of the variation in y explained by the model.

For the production data,

$$R^2 = \frac{1198827.2033}{1596902.5} \approx 0.7507.$$

In other words, the estimated model explains about 75 percent of the variation in costs. That's pretty good. One of the interesting features of the R^2 statistic is the ability to compare the fits of two models. If one model explains 75 percent of the data and another explains 82 percent, then the second model is preferred, all other things being equal.

Example 13.1

The SAT Reasoning Test has been used for years as a predictor of academic success. If SAT scores are predictors of academic success, they should be positively related to the grade point average upon graduation. 27 graduates of a state college were sampled and their grade point averages (GPA) upon graduation and SAT scores reported upon admission are recorded. The data are given in Table 13.5.

Table 13.5 – SAT Scores and Graduating GPA					
Student	SAT Critical Reading	SAT Math	SAT Writing	SAT Total	Graduating GPA
1	440	550	495	1485	2.105
2	390	480	435	1305	2.484
3	410	360	385	1155	2.537
4	390	350	370	1110	2.969
5	490	590	540	1620	3.619
6	400	550	475	1425	2.303
7	450	430	440	1320	2.602
8	420	350	385	1155	2.195
9	370	390	380	1140	2.112
10	460	600	530	1590	3.482
11	370	400	385	1155	2.367
12	410	530	470	1410	2.082
13	470	570	520	1560	2.346
14	490	610	550	1650	3.848
15	540	630	585	1755	2.446
16	560	620	590	1770	2.820
17	440	470	455	1365	2.556
18	440	530	485	1455	3.357
19	580	670	625	1875	3.269
20	360	420	390	1170	2.964
21	440	460	450	1350	2.642
22	290	410	350	1050	2.297
23	440	500	470	1410	2.388
24	510	570	540	1620	2.850
25	320	540	430	1290	2.742
26	470	580	525	1575	2.347
27	550	550	550	1650	3.025

13

Figure 13.14

The scatterplot in Figure 13.14 suggests that as SAT scores increase the GPA tends to increase, although there is a substantial amount of variability in the relationship. The upward sloping pattern of the data suggests a linear model could be constructed. However, a great deal of variation in the model's errors should be expected. What percent of the variation in final grade point average can be explained by the model relating total SAT score to graduating GPA?

Solution

Using the least squares method, the estimated model is given by

$$\text{Estimated Graduating GPA} = 1.355 + 0.00094(\text{Total SAT Score}).$$

◢	A	B	C	D	E
1	SUMMARY OUTPUT				
2					
3	*Regression Statistics*				
4	Multiple R	0.440925279			
5	R Square	0.194415102			
6	Adjusted R Square	0.162191706			
7	Standard Error	0.44340098			
8	Observations	27			
9					
10	ANOVA				
11		*df*	*SS*	*MS*	*F*
12	Regression	1	1.186183794	1.186184	6.0333523
13	Residual	25	4.915110725	0.196604	
14	Total	26	6.101294519		
15					
16		*Coefficients*	*Standard Error*	*t Stat*	*P-value*
17	Intercept	1.355045572	0.551989845	2.454838	0.0213991
18	SAT Total	0.000941501	0.000383302	2.456288	0.0213301

Figure 13.15

When we studied the production model, a list of predicted values and errors for each observed value was given. Instead of providing a list of the errors for each of the observed values, let's summarize the errors from the model. In particular,

$$s_e^2 \approx 0.1966 \quad \text{and} \quad s_e \approx 0.4434.$$

Figure 13.16

One of the differences in the production model and the SAT/GPA model is the manner in which the data seem to fit the model. In the production model, the data seemed to fit closely around the line, while in the SAT/GPA model the data are loosely clustered about the line. While *tight* and *loose* are interesting portrayals of the relative fit of the data, it would be desirable to have a numerical measure to describe fit. R^2 is such a measure.

$$R^2 = \frac{\text{SSR}}{\text{TSS}} \approx \frac{1.1862}{6.1013} \approx 0.1944$$

Thus, approximately 19% of the variation in graduating GPA is explained by this model.

Because the measure is a unit-free measure, it can be used to compare the fits of models. The SAT/GPA model only explains slightly more than 19 percent of the variation in the dependent variable. Compared to the production model, which had an R^2 of approximately 0.7507, this model seems dramatically inferior. Using R^2 as a criterion, the production model seems to have a substantially better fit (0.7507 versus 0.1944) than the SAT/GPA model. The real question is whether you can predict more accurately with the model than other available alternatives. If so, models with relatively low coefficients of determination (such as the SAT/GPA model) are useful. For example, if you could develop a model to predict stock prices, minute-by-minute, achieving an R^2 value of only 0.20, you could be a very wealthy person.

R^2 can also be found using the following computational formula.

Formula

Coefficient of Determination

The **coefficient of determination**, R^2, can be calculated using the equation

$$R^2 = \left(\frac{n \sum x_i y_i - \sum x_i \sum y_i}{\sqrt{\left(n \sum x_i^2 - \left(\sum x_i\right)^2\right)\left(n \sum y_i^2 - \left(\sum y_i\right)^2\right)}} \right)^2 .$$

Normally you will not have to use this formula since calculators and computer programs can calculate the coefficient of determination. Recall the computational formula for the correlation coefficient, discussed in Section 4.8, that measures the degree of linear relationship between two variables.

$$r = \frac{n\sum x_i y_i - \left(\sum x_i\right)\left(\sum y_i\right)}{\sqrt{n\sum x_i^2 - \left(\sum x_i\right)^2}\sqrt{n\sum y_i^2 - \left(\sum y_i\right)^2}}$$

The coefficient of determination is the square of the correlation coefficient. The correlation coefficient can be found by either using the formula given previously or by taking the square root of the coefficient of determination, and adding the sign corresponding to the slope coefficient. Remember that the correlation coefficient takes values between −1 and 1, where negative values indicate a downward sloping relationship and positive values indicate an upward sloping relationship. The coefficient of determination takes values between 0 and 1 where values close to 0 indicate a weak linear relationship and values close to 1 indicate a strong linear relationship.

13.5   Exercises

Basic Concepts

1. What is the total sum of squares?

2. How are the total sum of squares and the sample variance related?

3. Define error in terms of a regression model.

4. What part of the simple linear regression model captures the unexplained variation?

5. Describe the total sum of squares in terms of explained and unexplained variation.

6. What is the sum of squares of regression?

7. Express SSR in terms of the total sum of squares and the sum of squared errors. Interpret this in terms of model variation.

8. Why will there be errors in virtually all regression models?

9. What is the coefficient of determination? What kinds of values can the coefficient of determination take?

10. Suppose that regression analysis is performed and the resulting model has an R^2 value of 0.856. Interpret this value.

11. How is the coefficient of determination related to the correlation coefficient?

Exercises

12. A direct mail marketing company has been experimenting with the effect of price on sales. Five different direct mail prices have been sent to different sets of customers. They have carefully tracked the customers from each group and have recorded the proportion from each price category that purchased the product. The results are given in the following table.

Table for Exercise 12 – Direct Mail	
Proportion That Purchased Product	**Price of Product ($)**
0.032	29.95
0.028	34.95
0.026	39.95
0.015	44.95
0.009	49.95

a. What level of measurement do the two variables in the table possess?

b. Specify the model that the marketing manager would be interested in estimating.

c. Which of the variables is the dependent variable in the model?

d. Which of the variables is the independent variable in the model?

e. Draw a scatterplot of the data.

f. Use the data in the table to estimate the model.

g. Predict the proportion that will buy the product if the price is $35.00.

h. Compute the mean error for the model you estimated in part **f**.

i. Determine the mean square error.

j. What is the coefficient of determination? Interpret this value in terms of the problem.

13.   An economist is studying the relationship between income and savings. He has randomly selected seven subjects and obtained income and savings data from them. He wishes to use a simple linear regression model to predict savings based on annual income.

Table for Exercise 13 – Income and Savings	
Income (Thousands of Dollars)	Savings (Thousands of Dollars)
28	0.2
25	0
34	0.8
43	1.2
48	3.1
39	2.1
74	8.3

a.   What level of measurement do the two variables in the table possess?

b.   Which of the variables is the dependent variable in the model?

c.   Which of the variables is the independent variable in the model?

d.   Draw a scatterplot of the data. Does the scatterplot suggest that a linear model is appropriate? Explain.

e.   Use the data to estimate the appropriate model.

f.   Predict the savings for someone who earns fifty thousand dollars annually.

g.   Interpret the meaning of the slope coefficient in the problem.

h.   What fraction of the variation in savings is explained by income?

14. The Road Warrior Trucking Company has kept careful records on ten hauls. The traffic manager has recorded the haul weight of each truck and its miles per gallon during ten runs with the intent of building a regression model. He wants to predict the miles per gallon for a haul based on the haul weight. The haul weights and miles per gallon information is given in the following table. Haul weights are given in thousands of pounds.

Table for Exercise 14 – Trucking	
Miles per Gallon	Haul Weight (Thousands of Pounds)
4.6	36
4.8	33
5.1	31
4.0	42
4.7	33
5.2	30
4.5	37
4.6	37
4.2	40
4.5	36

a. What is the dependent variable in the model?

b. What is the independent variable in the model?

c. Construct a scatterplot of the data. Based on the scatterplot, does a linear model seem appropriate?

d. Write the model in terms of miles per gallon and haul weight. (Assume the parameters of the model have not been estimated.)

e. Use the data provided and estimate the coefficients of the linear model.

f. Interpret the coefficient of the independent variable.

g. Use the model to predict the miles per gallon for a truck hauling 38,000 pounds.

h. Do you believe there is a causal relationship between haul weight and the miles per gallon? If so, which direction is the causality? Do greater haul weights cause reduced mileage, or vice versa? Does the regression analysis prove the causality?

15. An agricultural research station is trying to determine the relationship between the yield of sunflower seeds and the amount of fertilizer applied. To determine the relationship, three different fields were planted. In each field four different plots were defined. In each plot a different amount of fertilizer was used. The plot assignments for the fertilizer application were randomly selected in each field.

Table for Exercise 15 – Agricultural Research	
Pounds of Fertilizer (per Acre)	Pounds of Sunflower Seeds (per Acre)
200	420
200	445
200	405
400	580
400	540
400	550
600	580
600	600
600	610
800	630
800	620
800	626

a. Are the data developed through a controlled experiment or are the data observational?

b. Draw a scatterplot of the data.

c. If a linear model is developed, which of the variables will be the dependent variable? Why?

d. Use the method of least squares to estimate the appropriate model.

e. Interpret the meaning of the slope coefficient in the model.

f. What fraction of the variation in pounds of sunflower seeds per acre can be explained by the amount of fertilizer used?

g. Predict the sunflower seed yield per acre if 500 pounds of fertilizer are applied.

16. Since 2009, the average term for a new-car loan was nearly 64 months. This leaves the buyer vulnerable to owing more on the car than it is worth. When applying for an automobile loan, it is oftentimes recommended to sign up for the shortest term you can afford. It is believed that along with one's credit rating, the length of the loan will help the buyer get a favorable interest rate. The following table contains interest rates and lengths of loans for 20 randomly selected auto purchases. Using the data in the table, answer the following questions.

Table for Exercise 16 – Lengths of Loans and Interest Rates	
Months Financed	**Interest Rate (%)**
12	4.00
24	4.40
36	5.24
12	3.43
24	4.40
36	5.79
36	5.98
48	6.58
36	5.31
36	5.91
48	6.51
48	6.68
60	7.13
60	7.48
72	8.31
60	7.85
72	8.07
72	8.48
48	6.12
72	8.07

a. Using statistical software, estimate the coefficients of the least squares regression equation.

b. Interpret the meaning of the slope and the intercept in part **a.**

c. Predict the interest rate for a person interested in a four-year auto loan.

d. Should you use the model to predict interest rates for an eight-year loan? Justify your answer.

e. Determine the coefficient of determination and explain its meaning in terms of the problem.

f. Calculate the correlation coefficient for this model. What does it mean?

g. What interest rate would one expect to get if they were planning to apply for a five-year auto loan?

13.6   Fitting a Linear Time Trend

In Chapter 4 we discussed the notion that the mean is not a reasonable descriptor for nonstationary time series data. Recall that nonstationary time series do not meander around some central value. Instead, the data tend to get larger or smaller over time. How can you describe time series data that possess a trend? For some time series, a **linear time trend** is a useful model. A linear time trend is nothing but a line that is used to model the changes in some phenomenon measured over time. In a linear time trend model, the independent variable is always a time index. The following example will illustrate the estimation of a linear time trend model.

Example 13.2

Many analysts believe that college tuition prices may soon be in the same situation as housing prices were when the housing bubble burst (causing home prices to drop significantly). Table 13.6 contains data for the Tuition Consumer Price Index (TCPI) from 1978 to 2009. Use a linear time trend to model the data.

Table 13.6 – Tuition Consumer Price Index			
Year	**TCPI**	**Year**	**TCPI**
1978	60.89	1994	250.80
1979	65.66	1995	249.17
1980	71.80	1996	264.15
1981	80.58	1997	278.42
1982	91.33	1998	307.51
1983	100.73	1999	319.63
1984	110.94	2000	307.80
1985	112.61	2001	324.73
1986	130.63	2002	348.54
1987	140.41	2003	371.42
1988	125.98	2004	343.05
1989	137.86	2005	476.08
1990	175.93	2006	507.90
1991	193.73	2007	456.30
1992	181.14	2008	531.55
1993	234.48	2009	607.60

Source: Bureau of Labor Statistics

Solution

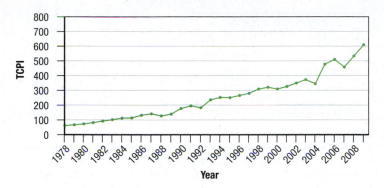

Figure 13.17

A graph of the data reveals an upward trend in the tuition consumer price index. The data appear to be a nonstationary time series with an upward trend. To describe the data, we will model the trend by fitting a line through the data with the notion of capturing how fast (on average) the series is changing over time. Estimating the slope of the line will provide the average rate of change per year in the TCPI. The line is fitted using least squares estimates in exactly the same way as other regression models have been constructed. The independent variable in a linear trend model is always time. In this case, the dependent variable is TCPI.

The estimated least squares equation is

$$\text{Estimated TCPI} = -30609.4994 + 15.4794\,(\text{Year}).$$

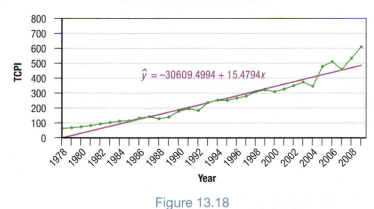

Figure 13.18

The computer output for the problem is given in Figure 13.19.

13

	A	B	C	D	E
1	SUMMARY OUTPUT				
2					
3	*Regression Statistics*				
4	Multiple R	0.964815847			
5	R Square	0.930869619			
6	Adjusted R Square	0.928565273			
7	Standard Error	40.2259548			
8	Observations	32			
9					
10	ANOVA				
11		*df*	*SS*	*MS*	*F*
12	Regression	1	653662.9701	653663	403.9626
13	Residual	30	48543.82319	1618.127	
14	Total	31	702206.7933		
15					
16		*Coefficients*	*Standard Error*	*t Stat*	*P-value*
17	Intercept	-30609.4994	1535.341809	-19.9366	7.38E-19
18	Year	15.47942265	0.770165709	20.09882	5.88E-19

Figure 13.19

The estimate of the slope, 15.4794, tells us that on average the TCPI is increasing at a rate of 15.4794 per year. Given how well the line fits the data $\left(R^2 \approx 0.9309\right)$, the trend line is a good descriptor of the data.

The trend line can also be used for short-term prediction. Suppose you wanted to estimate the TCPI for 2010. If the data are not available, the trend model can be used.

$$\text{Estimated TCPI} = -30609.4994 + 15.4794(2010) = 504.0946$$
$$\left(\text{Prediction of the TCPI for 2010}\right)$$

One of the problems with this prediction, as with all predictions in this chapter, is that the accuracy of the prediction is unknown. It might be very close to the true value or it could be very inaccurate. If there were some knowledge about the accuracy of the prediction it would be more useful. In later sections, we will return to this topic and study inferential methods.

13.6   Exercises

Basic Concepts

1. Why is the mean not a reasonable descriptor for nonstationary time series data?

2. What is a linear time trend?

3. What is the independent variable in a linear trend model?

4. Is there a difference between the way the best fit line is determined for time series data and the way it is determined for other types of data?

5. Identify a problem with predictions that are made using a time trend model.

Exercises

6. Consider the following table containing the Consumer Price Index (CPI) for all urban consumers in the United States from 1990 to 2010. The index is based on 1982–84 prices.

Table for Exercise 6 – Consumer Price Index	
Year	Consumer Price Index (CPI)
1990	130.7
1991	136.2
1992	140.3
1993	144.5
1994	148.2
1995	152.4
1996	156.9
1997	160.5
1998	163.0
1999	166.6
2000	172.2
2001	177.1
2002	179.9
2003	184.0
2004	188.9
2005	195.3
2006	201.6
2007	207.34
2008	215.30
2009	214.54
2010	218.06

Source: Bureau of Labor Statistics

a. Looking at the data in the table, do you believe the trend line will slope upward or downward?

b. Suppose we are interested in constructing a linear trend model for these data. Identify the independent and dependent variables for this model.

c. Write the general equation for the time trend model in terms of year and CPI.

d. Using statistical software, the following least squares model was determined.

$$\text{Estimated CPI} = -8647.4245 + 4.4107\,(\text{Year})$$

Use this model to predict the price level in 2015.

e. Can we determine the accuracy of this prediction? Explain.

7.   Consider the following monthly sales data for an up-and-coming technology company.

Table for Exercise 7 – Sales Data	
Month	**Sales (Thousands of Dollars)**
1	321
2	542
3	540
4	581
5	641
6	700
7	698
8	710
9	799
10	821
11	833
12	850

a.   Identify the independent and dependent variables for the linear time trend model.

b.   Using statistical software, the following summary output was produced.

	A	B	C	D	E
1	SUMMARY OUTPUT				
2					
3	*Regression Statistics*				
4	Multiple R	0.949341195			
5	R Square	0.901248704			
6	Adjusted R Square	0.891373575			
7	Standard Error	51.20789475			
8	Observations	12			
9					
10	ANOVA				
11		*df*	*SS*	*MS*	*F*
12	Regression	1	239318.1818	239318.1818	91.26449427
13	Residual	10	26222.48485	2622.248485	
14	Total	11	265540.6667		
15					
16		*Coefficients*	*Standard Error*	*t Stat*	*P-value*
17	Intercept	403.7575758	31.51628057	12.81107949	1.57569E-07
18	Month	40.90909091	4.282219283	9.553245222	2.41268E-06

Write the estimated regression equation.

c.   What is the mean square error for this model? The standard error?

d.   Using this model, predict the company's sales for the 13th month.

e.   What percent of the variation in sales is explained by the linear time trend model? Does this model seem to accurately fit the data?

8. Consider the following table containing unemployment rates for North Carolina and South Carolina in 2000 through 2010.

Table for Exercise 8 – Unemployment Rates 2000–2010		
	Unemployment Rate (%)	
Year	North Carolina	South Carolina
2000	3.7	3.6
2001	5.6	5.2
2002	6.6	6.0
2003	6.5	6.7
2004	5.5	6.8
2005	5.3	6.8
2006	4.8	6.4
2007	4.7	5.6
2008	6.2	6.8
2009	10.8	11.3
2010	10.6	11.2

Source: Bureau of Labor Statistics

a. Using statistical software, estimate the following linear time trend model:

$$\text{N.C. Unemployment Rate} = \beta_0 + \beta_1(\text{Year}) + \varepsilon_i.$$

Write the estimated regression equation using the least squares estimates for β_0 and β_1.

b. Using statistical software, estimate the following linear time trend model:

$$\text{S.C. Unemployment Rate} = \beta_0 + \beta_1(\text{Year}) + \varepsilon_i.$$

Write the estimated regression equation using the least squares estimates for β_0 and β_1.

c. Use the equations in parts **a.** and **b.** to estimate the unemployment rates for North and South Carolina in the year 2013.

d. What is the coefficient of determination for the regression model in part **a.**?

e. What is the coefficient of determination for the regression model in part **b.**?

f. Do you think that these regression models are reliable in predicting future unemployment rates? Of the two models, which seems to fit the data better?

13

13.7 Assumptions of the Simple Linear Model

Gauss
1777 – 1855

In statistics Karl Friedrich Gauss is best known for invoking the normal error distribution in conjunction with solving the least squares regression problem. Although modern statistical historians find much to criticize in Gauss' work, it was his publication of the normal error curve in relation to the regression problem that led Pierre Simon LaPlace to see the connection between the central limit theorem and linear estimation. For several generations of statistics, linear estimation based on the normal error distribution was a fundamental tool of data analysis.

A number of assumptions are necessary to make inferences about the linear model. An error term was incorporated in the model because virtually no real set of bivariate data is exactly linear. Incorporating the error term in the population regression line produces the simple linear regression model:

$$y_i = \beta_0 + \beta_1 x_i + \varepsilon_i.$$

The error term, ε_i, represents the variation in y not accounted for by the linear regression model. In order to perform inference on the model, some assumptions about the nature of the error term are required.

Assumptions

Assumptions about the Error Term in the Linear Model

1. The error term, ε_i, is presumed to be a normally distributed random variable.
2. The mean, or expected value, of the error term is zero $\left(E(\varepsilon_i) = 0 \right)$.
3. The variance of the error term is constant, denoted by σ_ε^2.
4. The errors are assumed to be independent of each other.

The model's parameters are β_0, β_1, and σ_ε^2. The estimation of these quantities was discussed earlier. In particular,

$$b_1 = \frac{n \sum x_i y_i - \sum x_i \sum y_i}{n \sum x_i^2 - \left(\sum x_i \right)^2},$$

$$b_0 = \bar{y} - b_1 \bar{x} = \frac{1}{n} \left(\sum y_i - b_1 \sum x_i \right),$$

$$s_e^2 = \frac{\sum \left(y_i - \hat{y}_i \right)^2}{n-2} = \frac{\text{SSE}}{n-2}.$$

In addition to the formal assumptions previously stated, a linear model should only be used to fit data that appear to be reasonably linear. Because of the wide availability of computer programs that calculate least squares estimates, you will not need to manually calculate estimates very often.

13.7   Exercises

Basic Concepts

1. Why is there an error term incorporated in the simple linear model?

2. What does the error term represent?

3. List the four assumptions about the error term in the simple linear model.

4. List the parameters of the simple linear regression model, and identify their estimates.

13.8 Inference: How Good is the Estimate of β_1?

Since β_1 specifies the rate of change between x and y, in most linear models the parameter of interest is β_1. Two inferential techniques are useful in evaluating the estimate of β_1. Confidence intervals, similar in structure to those used for means and proportions, will be developed. In addition, a hypothesis testing procedure will be presented to test whether β_1 is equal to some particular value.

The Confidence Interval for β_1

Developing a confidence interval for β_1 requires thinking about the estimate (b_1) as a random variable. Each random sample from the population will produce different data and hence different estimates of b_0 and b_1. The confidence interval will serve two purposes: to place bounds on the location of β_1 and to provide information about the quality of the point estimate, b_1. The form of the confidence interval is familiar.

$$
\begin{pmatrix} \text{Sample} \\ \text{estimate of} \\ \text{parameter} \end{pmatrix} \pm \begin{pmatrix} \text{A certain number of standard} \\ \text{deviation units depending on} \\ \text{the desired confidence} \end{pmatrix} \cdot \begin{pmatrix} \text{The standard} \\ \text{deviation of the} \\ \text{sample estimate} \end{pmatrix}
$$

The sample estimate of β_1 is b_1. The variance of b_1 is given by

$$
\sigma_{b_1}^2 = \frac{\sigma_\varepsilon^2}{\sum (x_i - \bar{x})^2}
$$

but like all population measurements, $\sigma_{b_1}^2$ usually has to be estimated from the data. Notice that the denominator of the expression above is equal to the variance of x, multiplied by the sample size, n. This indicates that the variance of b_1 is reduced if the variance of the error terms decreases, the sample size increases, or the variance of x increases.

The sample estimate of the variance of b_1 is given by

$$
s_{b_1}^2 = \frac{s_e^2}{\sum (x_i - \bar{x})^2} .
$$

The only difference in the computation of $\sigma_{b_1}^2$ and $s_{b_1}^2$ is the replacement of the population variance of the error terms, σ_ε^2, with the corresponding sample statistic, s_e^2. The standard deviation (standard error) of the sample estimate b_1 is

$$
s_{b_1} = \sqrt{\frac{s_e^2}{\sum (x_i - \bar{x})^2}} .
$$

Formula

$100(1 - \alpha)\%$ Confidence Interval for β_1

The $100(1 - \alpha)\%$ confidence interval for β_1 is given by

$$b_1 \pm t_{\alpha/2, df} s_{b_1},$$

where $t_{\alpha/2, df}$ is the critical value for a t-distribution with $n - 2$ degrees of freedom.

The expression $b_1 \pm t_{\alpha/2, df} s_{b_1}$ creates the following interval.

$$b_1 - t_{\alpha/2, df} s_{b_1} \qquad\qquad b_1 \qquad\qquad b_1 + t_{\alpha/2, df} s_{b_1}$$

The expression $t_{\alpha/2, df}$ relates the width of the interval to the amount of confidence required. Recall that the t-distribution is very similar to the z-distribution. To use the t-distribution requires that its degrees of freedom be specified. In this case the degrees of freedom are

$$df = n - 2.$$

If a 95% confidence interval is desired, then $1 - \alpha = 0.95$, which implies $\alpha = 0.05$ and $\alpha/2 = 0.05/2 = 0.025$. Suppose the sample size is $n = 20$. Using the tables for the t-distribution in Appendix A, $t_{\alpha/2, df}$ would be

$$t_{\alpha/2, df} = t_{0.025, 18} = 2.101.$$

df	$t_{0.100}$	$t_{0.050}$	$t_{0.025}$	$t_{0.010}$
1	3.078	6.314	12.706	31.821
2	1.886	2.920	4.303	6.965
⋮				
17	1.333	1.740	2.110	2.567
18	1.330	1.734	2.101	2.552
⋮				

Hence, to be 95% confident in capturing the true value of β_1 in the confidence interval will require placing the interval endpoints 2.101 standard deviation units from the point estimate, b_1.

13

Example 13.3

Table 13.7 – Weekly Production		
Week	Items Produced	Cost ($)
1	22	3500
2	30	3800
3	36	4500
4	41	4200
5	27	3700
6	45	4600
7	30	3600
8	37	4550
9	32	3990
10	31	3675

In Section 13.3, a model relating the number of items produced to total cost was constructed.

$$\text{Cost} = \beta_0 + \beta_1 \left(\text{Items Produced} \right) + \varepsilon_i$$

If the relationship is to be applicable for the entire production process, then a substantial amount of data will be required, more than we could hope to collect. If the data given in Table 13.7 are considered a random sample of weekly production, then a relationship can be constructed from the sample data. Specifically, the estimated least squares regression line relating items produced to total cost is

$$\text{Estimated Cost} = \$2227.96 + \$53.88 \left(\text{Items Produced} \right),$$

where

$b_0 = \$2227.96$ (the sample estimate of β_0, the y-intercept), and

$b_1 = \$53.88$ (the sample estimate of β_1, the slope).

Note: Both estimates were determined using Microsoft Excel and rounded to the nearest hundredth.

To draw conclusions about the relationship between items produced and total cost for the entire population, statistical inference must be applied. We will begin by estimating a 95% confidence interval for β_1.

Solution

Three pieces of information are required to calculate a confidence interval for β_1.

1. A sample estimate of β_1

2. A t-value corresponding to the level of confidence and the degrees of freedom associated with the data used to estimate the model

3. A sample estimate of the standard deviation of b_1, s_{b_1}

Estimating the least squares line from the sample data produces $b_1 = 53.88$.

Since 95% confidence is required, $\alpha = 0.05$, and $\alpha/2 = 0.05/2 = 0.025$. Degrees of freedom will be

$$df = n - 2 = 10 - 2 = 8.$$

Thus, $t_{0.025,8} = 2.306$.

The remaining piece of missing information is the standard deviation of b_1,

$$s_{b_1} = \sqrt{\frac{s_e^2}{\sum (x_i - \bar{x})^2}}.$$

Table 13.8 – Squared Deviation	
Items Produced	**Squared Deviation** $(x_i - \bar{x})^2$
22	$(22 - 33.1)^2 = 123.21$
30	$(30 - 33.1)^2 = 9.61$
36	$(36 - 33.1)^2 = 8.41$
41	$(41 - 33.1)^2 = 62.41$
27	$(27 - 33.1)^2 = 37.21$
45	$(45 - 33.1)^2 = 141.61$
30	$(30 - 33.1)^2 = 9.61$
37	$(37 - 33.1)^2 = 15.21$
32	$(32 - 33.1)^2 = 1.21$
31	$(31 - 33.1)^2 = 4.41$
Total	$\sum (x_i - \bar{x})^2 = 412.90$

Determining s_{b_1} will require the knowledge of two quantities: the mean square error, s_e^2, and the sum of the squared deviations of the independent variable, $\sum (x_i - \bar{x})^2$.

The mean square error was computed in Section 13.4 as follows.

$$s_e^2 = \frac{\sum (y_i - \hat{y}_i)^2}{n - 2} = \frac{398075.2967}{10 - 2} \approx 49759.41$$

The sum of the squared deviations of the independent variable (items produced), $\sum (x_i - \bar{x})^2$, is given in Table 13.8. Thus, the standard deviation of b_1 is calculated as follows.

$$s_{b_1} = \sqrt{\frac{s_e^2}{\sum (x_i - \bar{x})^2}} = \sqrt{\frac{49759.41}{412.90}} \approx 10.9778$$

13

	A	B	C	D	E	F	G
1	SUMMARY OUTPUT						
2							
3	*Regression Statistics*						
4	Multiple R	0.866441198					
5	R Square	0.75072035					
6	Adjusted R Square	0.719560393					
7	Standard Error	223.0681781					
8	Observations	10					
9							
10	ANOVA						
11		*df*	*SS*	*MS*	*F*	*Significance F*	
12	Regression	1	1198827.203	1198827	24.09247121	0.00118116	
13	Residual	8	398075.2967	49759.41			
14	Total	9	1596902.5				
15							
16		*Coefficients*	*Standard Error*	*t Stat*	*P-value*	*Lower 95%*	*Upper 95%*
17	Intercept	2227.955922	370.1487713	6.019082	0.000316596	1374.391324	3081.520519
18	Items Produced	53.8835069	10.97779658	4.908408	0.00118116	28.5686626	79.1983512

Figure 13.20

The manual calculation of s_{b_1} is tedious. Virtually every statistical analysis program that performs regression analysis calculates s_{b_1}. The summary output from Microsoft Excel is given in Figure 13.20. Most software packages will automatically include a confidence interval for β_1 or it will include the pieces required to compute a confidence interval. Microsoft Excel automatically displays the 95% confidence interval for β_1, and is capable of displaying an interval for any level of confidence you choose.

95% Confidence Interval for β_1

$$b_1 \pm t_{\alpha/2, df}\, s_{b_1}$$
$$53.88 \pm 2.306(10.9778)$$
$$53.88 \pm 25.3148$$
$$28.57 \text{ to } 79.19$$

28.57	53.88	79.19

Putting the pieces together results in an interval which spans from \$28.57 to \$79.19. We are 95% confident that this interval contains the true value of β_1. There are two possible interpretations of this interval.

1. We are 95% confident that the true increase in total cost from producing one additional item is between \$28.57 and \$79.19.

2. We are 95% confident the maximum error of the point estimate $\left(b_1 = 53.88\right)$ in estimating the unknown β_1 (the true increase in total cost from producing one additional item) is at most \$25.31.

Notes: The "confidence" we are discussing is in the procedure, not in the specific interval 28.57 to 79.19. Since the hand calculation of the confidence interval for β_1 used values that were rounded, the interval varies slightly from what is reported by Microsoft Excel. Using unrounded values in Excel, the confidence interval for β_1 is given as 28.57 to 79.20, rounded to two decimal places (see Figure 13.20).

Example 13.4

Construct a 99% confidence interval for β_1 using the model described in Example 13.3.

Solution

The difference between a 95% confidence interval and a 99% confidence interval is expressed in the value of $t_{\alpha/2, df}$. For 99% confidence, $1 - \alpha = 0.99$, which implies that $\alpha = 0.01$. Hence, $\alpha/2 = 0.01/2 = 0.005$.

The degrees of freedom will remain $df = 10 - 2 = 8$.

The appropriate t-value is $t_{\alpha/2, df} = t_{0.005, 8} = 3.355$.

99% Confidence Interval for β_1

$$b_1 \pm t_{\alpha/2, df} s_{b_1}$$
$$53.88 \pm 3.355(10.9778)$$
$$53.88 \pm 36.8305$$
$$17.05 \text{ to } 90.71$$

| 17.05 | 53.88 | 90.71 |

Note: The confidence interval in this example was calculated using rounded values from the summary output. Microsoft Excel calculates the confidence interval using unrounded values as 17.05 to 90.72.

Requiring more confidence results in a wider interval. The trade-off is less precision (larger interval) for more confidence.

It is important to remember that the assumptions of the linear model given in Section 13.7 must hold in order to retain the specified degree of confidence in the interval.

Example 13.5

In Example 13.1 we examined a model relating SAT score to graduating GPA. Assuming the goal is to estimate a model that could be applied to all students that have graduated from college during the last five years, the population under consideration would be quite large, too large to obtain all the measurements. If we assume the data are from a sample, then we can estimate a model using the sample data and make inferences about the population. The population regression model to be estimated is

$$\text{Estimated Graduating GPA} = \beta_0 + \beta_1 (\text{Total SAT Score}) + \varepsilon_i.$$

We will begin by making an inference concerning β_1. Construct a 95% confidence interval for β_1.

Solution

	A	B	C	D	E	F	G
1	SUMMARY OUTPUT						
2							
3	*Regression Statistics*						
4	Multiple R	0.440925279					
5	R Square	0.194415102					
6	Adjusted R Square	0.162191706					
7	Standard Error	0.44340098					
8	Observations	27					
9							
10	ANOVA						
11		*df*	*SS*	*MS*	*F*	*Significance F*	
12	Regression	1	1.186183794	1.186183794	6.033352349	0.021330078	
13	Residual	25	4.915110725	0.196604429			
14	Total	26	6.101294519				
15							
16		*Coefficients*	*Standard Error*	*t Stat*	*P-value*	*Lower 95%*	*Upper 95%*
17	Intercept	1.355045572	0.551989845	2.454837865	0.021399134	0.218201206	2.491889938
18	SAT Score	0.000941501	0.000383302	2.456288328	0.021330078	0.000152075	0.001730927

Figure 13.21

Using a statistical analysis package to estimate the linear regression model, the output resembles that given in Figure 13.21.

Using the computer output will make the job of calculating a confidence interval easy. Two of the three pieces of information required to calculate a confidence interval are given in the output.

- The sample estimate of β_1 is given in the *Coefficients* column associated with the variable SAT Score. That is, b_1 is estimated to be 0.00094.

- The standard deviation of b_1 is given in the *Standard Error* column associated with the variable SAT Score. That is, s_{b_1} is estimated to be 0.00038.

The missing piece of information is the value of $t_{\alpha/2, df}$. Determining $t_{\alpha/2, df}$ requires the calculation of $\alpha/2$ and the degrees of freedom. Since the level of confidence is specified to be 95%, $\alpha = 0.05$ and $\alpha/2 = 0.05/2 = 0.025$. The degrees of freedom are

$$df = n - 2 = 27 - 2 = 25.$$

Consequently,

$$t_{\alpha/2, df} = t_{0.025, 25} = 2.060.$$

Now, all the pieces are assembled.

95% Confidence Interval for β_1

$$b_1 \pm t_{\alpha/2, df} s_{b_1}$$

$$0.00094 \pm 2.060(0.00038)$$

$$0.00094 \pm 0.00078$$

$$0.00016 \text{ to } 0.00172$$

| 0.00016 | 0.00094 | 0.00172 |

We are 95% confident that the interval contains the true value of β_1. There are two possible interpretations of this interval.

1. We are 95% confident that the true increase in GPA for a one point increase in SAT score is between 0.00016 and 0.00172.

2. We are 95% confident the maximum error of the point estimate $(b_1 = 0.00094)$ in estimating the unknown β_1 (the true increase in GPA for a one point increase in SAT score) is at most 0.00078.

Note: The confidence interval in this example was calculated using rounded values from the summary output. Microsoft Excel calculates the confidence interval using unrounded values as 0.00015 to 0.00173 (see Figure 13.21).

13.8 Exercises

Basic Concepts

1. Give an example of a practical application of the confidence interval for β_1.

2. Identify two purposes that confidence intervals for the estimated regression coefficients serve.

3. What is the sampling distribution for b_1? Give the mean and standard deviation.

4. What is the expression for determining the $100(1-\alpha)\%$ confidence interval for β_1?

5. Suppose a 95% confidence interval for β_1 is found to be $(15.11,\ 20.11)$. Give two interpretations of this interval.

Exercises

6. Consider the summary output for the monthly sales data given in Exercise 7 in Section 13.6.

	A	B	C	D	E
1	SUMMARY OUTPUT				
2					
3	*Regression Statistics*				
4	Multiple R	0.949341195			
5	R Square	0.901248704			
6	Adjusted R Square	0.891373575			
7	Standard Error	51.20789475			
8	Observations	12			
9					
10	ANOVA				
11		*df*	*SS*	*MS*	*F*
12	Regression	1	239318.1818	239318.1818	91.26449427
13	Residual	10	26222.48485	2622.248485	
14	Total	11	265540.6667		
15					
16		*Coefficients*	*Standard Error*	*t Stat*	*P-value*
17	Intercept	403.7575758	31.51628057	12.81107949	1.57569E-07
18	Month	40.90909091	4.282219283	9.553245222	2.41268E-06

a. Compute a 90% confidence interval for β_1.

b. Interpret this interval.

7. Consider the data in the following table regarding the age of a particular model of car and the asking price for that car.

Table for Exercise 7 – Car Data	
Age (Years)	**Asking Price ($)**
1	11,875
1	10,995
2	9995
2	8500
3	8995
4	6995
5	4450
5	5500
6	4400
6	4800

a. Using statistical software, determine the sample estimate of β_1.

b. What is the standard error of b_1?

c. Find a 99% confidence interval for β_1.

d. Interpret the confidence interval found in part **c.**

8. An economist is studying the relationship between income and IRA contributions. He has randomly selected eight subjects and obtained annual income and IRA contribution data from them. He wishes to predict the amount of money contributed to an IRA based on annual income.

Table for Exercise 8 – Income and IRA Contributions	
Annual Income (Thousands of Dollars)	IRA Contribution (Thousands of Dollars)
28	0.3
25	0
34	1.0
43	1.3
48	3.3
39	2.2
74	8.5

a. Draw a scatterplot of the data. Describe the relationship that you observe between income and IRA contribution.

b. Estimate the parameters of the following model using statistical software.

$$\text{IRA Contribution} = \beta_0 + \beta_1 (\text{Income}) + \varepsilon_i$$

c. Calculate and interpret a 95% confidence interval for β_1.

d. What assumptions are being made in the construction of the confidence interval for β_1?

13.9 Testing a Hypothesis Concerning β_1

In constructing a regression model we must ask the question: *Does a linear relationship exist between y and x?* In answering this question, remember the linear model connects x to y through the slope parameter, β_1:

$$y_i = \beta_0 + \beta_1 x_i + \varepsilon_i.$$

If $\beta_1 = 0$, then there is no linear relationship between x and y since the term $\beta_1 x_i = 0$. Regardless of the value of x, the model becomes

$$y_i = \beta_0 + \varepsilon_i.$$

This says that y is equal to a constant, β_0, plus a random error. Most of the time when developing a linear model used for predictive purposes, discovering that $\beta_1 = 0$ is bad news. Essentially this says that x is not a useful predictor of y. Since β_1 is a model parameter and cannot be known unless all the bivariate population measurements are obtained, the sample estimate, b_1, will be used to make an inference concerning β_1.

If the assumptions of the linear model have been met sufficiently, statistical inference methods can be used to aid in answering the question, *is b_1 close enough to 0 to believe that $\beta_1 = 0$?* We will follow the hypothesis testing procedure used in Chapter 10.

Example 13.6

Using the data in Example 13.3, determine if there is overwhelming evidence at the $\alpha = 0.05$ level of a relationship between the number of items produced and the total production cost.

Solution

Step 1: State the hypotheses in plain English.

- **Null Hypothesis:** There is not a linear relationship between the number of items produced and total production cost.

- **Alternative Hypothesis:** There is a linear relationship between the number of items produced and total production cost.

Step 2: Select the appropriate statistical measure.

Since we are interested in determining if $\beta_1 = 0$, the sample estimate of β_1, namely b_1, will be used to evaluate whether the hypothesis $\beta_1 = 0$ is reasonable.

Step 3: Determine whether the hypothesis should be one-sided or two-sided.

The alternative hypothesis should be two-sided since we are interested in discovering any relationship (positive or negative) between items produced and total cost.

Step 4: Specify the hypotheses using the appropriate statistical measure.

$H_0: \beta_1 = 0$ (No linear relationship exists between items produced and total cost.)

$H_a: \beta_1 \neq 0$ (A linear relationship exists between items produced and total cost.)

Step 5: Specify the level of the test.

The level of the test has been given in the problem statement as the 0.05 level.

Step 6: Select the appropriate test statistic.

Formula

Test Statistic for Testing the Hypothesis $\beta_1 \neq 0$

The test statistic for testing the hypothesis $\beta_1 \neq 0$ is given by

$$t = \frac{b_1 - 0}{s_{b_1}} = \frac{b_1}{s_{b_1}}.$$

The test statistic follows a t-distribution with $n - 2$ degrees of freedom.

The test statistic is similar in nature to the other test statistics developed in Chapter 10. It measures how far b_1 is from the hypothesized value of β_1, which is 0. This distance is measured in standard deviation units. If t is close to 0, then b_1 is close to 0 and $H_0: \beta_1 = 0$ is the more reasonable conclusion. However, if t is far from zero, then b_1 is far from its hypothesized value and $H_a: \beta_1 \neq 0$ would seem more reasonable. This criterion is defined by the critical value of the test statistic.

Step 7: Determine the critical value.

The test is two-tailed and the level of the test is specified to be 0.05, which implies

$$\alpha = 0.05 \quad \text{and} \quad \alpha/2 = 0.05/2 = 0.025.$$

The test statistic has a t-distribution with

$$df = n - 2 = 10 - 2 = 8.$$

The critical value corresponds to $t_{0.025,8} = 2.306$.

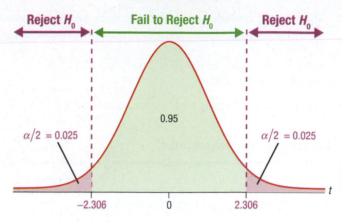

t-Distribution, df = 8

Figure 13.22

Step 8: Compute the test statistic.

Table 13.9 – Regression Results			
Predictor	**Coefficient**	**Standard Deviation of Coefficient**	**t-value**
Intercept	2227.96	370.1488	6.019
Items Produced	53.88	10.9778	4.908

$$t = \frac{b_1 - 0}{s_{b_1}} = \frac{b_1}{s_{b_1}} \approx \frac{53.88}{10.9778} \approx 4.908$$

The estimated value of b_1 is almost five standard deviations above zero. This is very persuasive evidence that $\beta_1 \neq 0$.

Step 9: Make the decision.

Since the value of the test statistic falls into the rejection region, reject the null hypothesis in favor of the alternative.

Step 10: State the conclusion in terms of the original question.

There is overwhelming evidence at the 0.05 level that $\beta_1 \neq 0$ so we reject the null hypothesis in favor of the alternative. This implies that it is reasonable to believe (at the 0.05 level) that there is a linear relationship between the number of items produced and total cost. In fact, there appears to be a positive linear relationship between items produced and production cost. However, our hypothesis test did not address the issue of a *positive* relationship, so we cannot make this conclusion.

Thus far, the focus has been on inference about β_1. What about β_0? Since β_0 is merely a constant term, in most problems its value is not of great concern. However, if a confidence interval or test of hypothesis is needed, the methods used would be virtually identical to those presented for analyzing β_1.

Using *P*-Values to Test a Hypothesis Concerning β_1

The simplest method of testing any hypothesis is the *P*-value method. Since most computer outputs for regression analysis contain *P*-values, there are no calculations to make. Choosing α is all that is required to perform the test.

	A	B	C	D	E	F	G
1	SUMMARY OUTPUT						
2							
3	*Regression Statistics*						
4	Multiple R	0.866441198					
5	R Square	0.75072035					
6	Adjusted R Square	0.719560393					
7	Standard Error	223.0681781					
8	Observations	10					
9							
10	ANOVA						
11		*df*	*SS*	*MS*	*F*	*Significance F*	
12	Regression	1	1198827.203	1198827	24.09247121	0.00118116	
13	Residual	8	398075.2967	49759.41			
14	Total	9	1596902.5				
15							
16		*Coefficients*	*Standard Error*	*t Stat*	*P-value*	*Lower 95%*	*Upper 95%*
17	Intercept	2227.955922	370.1487713	6.019082	0.000316596	1374.391324	3081.520519
18	Items Produced	53.8835069	10.97779658	4.908408	0.00118116	28.5686626	79.1983512

Figure 13.23

The *P*-value measures the probability that the test statistic is as large as it is (in magnitude) under the assumption that the null hypothesis is true. Specifically, a *P*-value is the probability of observing a test statistic as large or larger (in absolute value) than what has been observed, given that the null hypothesis is true. In Example 13.6, the value of the test statistic was 4.908. The probability of observing a test statistic this large (in absolute value) or larger, given that the true value of the slope is zero, is very small. Fortunately, virtually all statistical analysis programs that perform regression analysis calculate *P*-values for the two-tailed test of hypothesis

$$H_0: \beta_1 = 0$$
$$H_a: \beta_1 \neq 0.$$

Figure 13.23 shows the *P*-value of b_1 to be approximately 0.0012. A *P*-value of 0.0012 is persuasive evidence that $\beta_1 \neq 0$. The null hypothesis is rejected if the *P*-value is less than or equal to α. In Example 13.6, the level of the test was set at $\alpha = 0.05$. Since the *P*-value = 0.0012 $\leq$ 0.05, the null hypothesis is rejected in favor of the alternative hypothesis $(H_a: \beta_1 \neq 0)$.

If the *P*-value is used, how does the procedure for testing a hypothesis change? In Step 7 all that is necessary is to specify α. It serves as a critical value. In Step 9, the *P*-value is compared to α. Everything else remains the same, provided the *P*-value has been calculated for you.

If a data analyst feels that the assumptions of the simple linear model have been met and decides to make an inference about the model, the *P*-value of b_1 will be one of the first pieces of the computer output that will be examined.

13

13.9   Exercises

Basic Concepts

1. If there is no linear relationship between two variables, what is the value of β_1? Explain.

2. What is the test statistic for testing the hypothesis that $\beta_1 \neq 0$? Describe how this test statistic is similar to other test statistics used in hypothesis testing.

3. What are the degrees of freedom associated with the simple linear regression model?

4. Can we make inferences about β_0? Explain why we are more interested in inferences about β_1.

5. Describe why the P-value corresponding to b_1, which is displayed by many regression summary outputs, is one of the first values examined by data analysts.

Exercises

6. Consider the following summary output, which was generated from a sample of 8 employees relating age to annual salary.

	A	B	C	D	E
1	SUMMARY OUTPUT				
2					
3	*Regression Statistics*				
4	Multiple R	0.732431223			
5	R Square	0.536455496			
6	Adjusted R Square	0.459198079			
7	Standard Error	15.60374155			
8	Observations	8			
9					
10	ANOVA				
11		*df*	*SS*	*MS*	*F*
12	Regression	1	1690.639497	1690.639	6.943741
13	Residual	6	1460.860503	243.4768	
14	Total	7	3151.5		
15					
16		*Coefficients*	*Standard Error*	*t Stat*	*P-value*
17	Intercept	-2.132440745	20.99597109	-0.10156	0.922412
18	Age	1.564320608	0.593648001	2.635098	0.038794

a. What is the estimated regression equation?

b. Is there evidence of a linear relationship between age and salary at the 0.05 level?

c. Does the decision in part b. change at the 0.01 level? Explain.

d. What percentage of the variation in annual salary is explained by the model?

7. The college placement office is developing a model to relate grade point average (GPA) to starting salary for liberal arts majors. Ten recent graduates have been randomly selected, and their graduating GPAs and starting salaries were recorded.

Table for Exercise 7 – GPA and Starting Salary	
GPA	Starting Salary (Thousands of Dollars)
2.2	35.1
3.5	45.2
2.1	36.3
2.8	39.3
3.2	41.4
2.5	37.6
2.4	34.8
2.9	25.7
3.1	40.1
3.7	39.5

a. Plot the data. Describe the relationship you observe between GPA and starting salary.

b. Using statistical software, estimate the parameters of the model

$$\text{Starting Salary} = \beta_0 + \beta_1 (\text{GPA}) + \varepsilon_i.$$

c. Is there evidence of a linear relationship between GPA and starting salary? Test at the 0.05 level.

d. Predict the starting salary for a student with a GPA of 2.5.

e. Interpret the coefficient of GPA in the model.

f. What fraction of the variation in starting salaries is explained by GPA?

g. To perform statistical inference on the model, what assumptions are being made?

8. An experienced census official feels that she can accurately estimate the number of inhabitants of a city block by simply noting the size of the block and the types of buildings (single family homes, apartments, businesses, etc.) that are found on the block. This procedure, if accurate, would be much quicker and cheaper than visiting each residence and taking a survey of its inhabitants. The data below are estimates of block populations provided by the official for 10 blocks in a large city. Also given are the actual numbers of inhabitants for the same 10 blocks. These were found at a later point in time by conventional methods.

Table for Exercise 8 – Inhabitants of City Blocks									
Estimate	115	234	215	97	78	134	78	129	170	67
Actual	100	225	190	99	92	125	75	130	155	82

a. Draw a scatterplot of points of the actual number against the estimated number of inhabitants. Does the relationship appear to be linear?

b. Estimate the slope and intercept of the following regression equation using statistical software.

$$\text{Actual Inhabitants} = \beta_0 + \beta_1 \left(\text{Estimated Inhabitants}\right) + \varepsilon_i$$

c. Is there evidence of a linear relationship between the actual number and the estimated number? Test at the $\alpha = 0.01$ level.

d. Interpret the regression coefficient for the estimated number of inhabitants.

e. Construct a 95% confidence interval for the slope of the regression equation. Interpret the interval.

f. Compute and interpret the R^2 value.

g. Predict the actual number of inhabitants on a block when the estimated number is 150. Round your answer to the nearest whole number.

9. A statistics professor would like to build a model relating student scores on the first test to the scores on the second test. The test scores from a random sample of 21 students who have previously taken the course are given in the table.

Student	First Test Grade	Second Test Grade	Student	First Test Grade	Second Test Grade
1	69	73	12	54	67
2	66	56	13	57	65
3	69	65	14	85	67
4	75	51	15	75	67
5	57	59	16	79	77
6	75	76	17	44	51
7	75	76	18	82	84
8	82	76	19	57	81
9	91	82	20	75	90
10	66	73	21	69	73
11	88	67			

Table for Exercise 9 – Test Scores

a. Using statistical software, estimate the parameters of the model

$$\text{Second Test Grade} = \beta_0 + \beta_1 (\text{First Test Grade}) + \varepsilon_i.$$

b. What fraction of the variation in the grades on the second test is explained by the grades on the first test?

c. Is there a linear relationship between the first test grades and the second test grades? Test at the 0.05 level.

d. Suppose you're enrolled in the professor's course this semester. If you scored a 75 on the first test, use the model to predict your second test score. Round your answer to the nearest whole number.

13.10 Inference Concerning the Model's Prediction

Many regression models are developed for predictive purposes. For example, if you built the model relating the number of items produced to total cost, it was probably because you want to use it to predict total cost. While it is important to evaluate b_1, the estimate of the slope, the real concern of the model builder is the accuracy of the model's predictions. In the case of the production model, how accurate are the costs the model predicts? If the assumptions of the linear model (detailed in Section 13.7) have been met, then it is possible to make inferences as to the quality of a model's predictions.

The Regression Line as the Mean Value of *y* Given *x*

Examining the production data in Table 13.7 reveals two weeks in which 30 items were produced. For a given value of items produced (say 30) the costs of producing 30 items were $3800 and $3600. For anyone who has observed a production process, price variation is not unexpected. If you use the model

$$\text{Estimated Cost} = \$2227.96 + \$53.88\left(\text{Items Produced}\right)$$

for predictive purposes, then the predicted cost of producing 30 items will be

$$\text{Estimated Cost} = \$2227.96 + \$53.88(30) = \$3844.36.$$

Using this model, producing 30 items will have a predicted cost of $3844.36. Since the actual cost of producing 30 items varies, how do you interpret the predicted cost of $3844.36, which happens to be above the two observed values? The model's predicted value when 30 items are produced is considered to be the average cost of producing 30 items. In other words, it is the mean value of y (cost) when x (items produced) equals 30. But wait a minute! The costs of producing 30 items were $3800 and $3600, and the average of these numbers is not $3844.36. What kind of average is this? What we are essentially saying is that if we are willing to acknowledge that the relationship between x and y is linear, then all data (not just the data for $x = 30$) will be useful in establishing the estimated relationship. Once the relationship is established it will be used to estimate the mean value of y for any given x. In fact, if items produced = 35, the predicted cost would be

$$\text{Estimated Cost} = \$2227.96 + \$53.88(35) = \$4113.76,$$

which would be interpreted to be the average cost of producing 35 items.

The entire regression line can be considered a collection of means of y for different values of x. Unfortunately, the true linear relationship is unknown since all the data in the population will not be available. Since only the estimated model is available, only estimated values of the mean value of y for some given x will be available. How good are these estimates? How good is the estimate of a mean value of $y = \$4113.76$ when $x = 35$? To answer this question, we will once again rely on the notion of a confidence interval.

Confidence Intervals for the Mean Value of *y* Given *x*

Formula

100(1 − α)% Confidence Interval for the Mean Value of *y* Given *x*

If x_p is a value of *x* for which we wish to know the mean value of *y*, then the $100(1-\alpha)\%$ confidence interval is given by

$$\hat{y}_p \pm t_{\alpha/2,df}\, s_e \sqrt{\frac{1}{n} + \frac{\left(x_p - \bar{x}\right)^2}{\sum\left(x_i - \bar{x}\right)^2}}, \quad \text{where}$$

$\hat{y}_p$ is the predicted value of *y* when $x = x_p$, i.e. $\hat{y}_p = b_0 + b_1 x_p$,

$t_{\alpha/2,df}$ is the *t*-value associated with $100(1-\alpha)\%$ confidence (the same *t* used in constructing confidence intervals for β_1),

s_e is the standard deviation of the error terms, and

$\dfrac{\left(x_p - \bar{x}\right)^2}{\sum\left(x_i - \bar{x}\right)^2}$ measures how far x_p is from $\bar{x}$ in relation to the total variation of *x*. The further x_p is from $\bar{x}$, the larger this ratio will be, resulting in a wider confidence interval.

Example 13.7

For the production model, calculate the 95% confidence interval for the average cost of producing 35 items.

Solution

Four pieces of information are required to calculate the 95% confidence interval for the mean value of *y* given $x = x_p$.

1. Use the estimated regression line to calculate $\hat{y}_p$ for the value given for $x = x_p$.

2. Find a *t*-value corresponding to the level of confidence and the degrees of freedom associated with the data used to estimate the model.

3. Determine the standard deviation of the error terms.

4. Compute the term $\dfrac{\left(x_p - \bar{x}\right)^2}{\sum\left(x_i - \bar{x}\right)^2}$.

In Example 13.3, Estimated Cost $= \$2227.96 + \$53.88\,(\text{Items Produced})$, $x_p = 35$, and the predicted value of *y* for this given *x* is

$$\hat{y}_p = b_0 + b_1 x_p = \$2227.96 + \$53.88\,(35) = \$4113.76.$$

For a 95% confidence interval, $\alpha = 0.05$, and the degrees of freedom are $n - 2 = 10 - 2 = 8$. Thus, $t_{\alpha/2,df} = t_{0.025,8} = 2.306$.

The standard deviation of the error terms is found in the output in Figure 13.20. In particular, $s_e \approx 223.0682$. Note that s_e can also be computed by taking the square root of the mean square error, $s_e^2 \approx 49759.41$, which was computed in Section 13.8.

The last piece of information is the quotient

$$\frac{\left(x_p - \bar{x}\right)^2}{\sum \left(x_i - \bar{x}\right)^2} = \frac{\left(35 - 33.1\right)^2}{412.90} \approx 0.0087.$$

Table 13.10 contains the calculation for the sum of the squared deviations, $\sum \left(x_i - \bar{x}\right)^2$.

Table 13.10 – Squared Deviation	
Items Produced	Squared Deviation $\left(x_i - \bar{x}\right)^2$
22	$\left(22 - 33.1\right)^2 = 123.21$
30	$\left(30 - 33.1\right)^2 = 9.61$
36	$\left(36 - 33.1\right)^2 = 8.41$
41	$\left(41 - 33.1\right)^2 = 62.41$
27	$\left(27 - 33.1\right)^2 = 37.21$
45	$\left(45 - 33.1\right)^2 = 141.61$
30	$\left(30 - 33.1\right)^2 = 9.61$
37	$\left(37 - 33.1\right)^2 = 15.21$
32	$\left(32 - 33.1\right)^2 = 1.21$
31	$\left(31 - 33.1\right)^2 = 4.41$
Total	$\sum \left(x_i - \bar{x}\right)^2 = 412.90$

Assembling the pieces for the 95% confidence interval we have the following.

95% Confidence Interval for the Mean Value of _y_ Given _x_ = 35

$$\hat{y}_p \pm t_{\alpha/2, df} \, s_e \sqrt{\frac{1}{n} + \frac{\left(x_p - \bar{x}\right)^2}{\sum \left(x_i - \bar{x}\right)^2}}$$

$$4113.76 \pm 2.306 \left(223.0682 \sqrt{\frac{1}{10} + 0.0087} \right)$$

$$4113.76 \pm 169.5945$$

$$3944.17 \text{ to } 4283.35$$

3944.17   4113.76   4283.35

The confidence interval can be interpreted in two ways.

1. We are 95% confident that the average cost of producing 35 items is between $3944.17 and $4283.35. (Note that like all confidence intervals, the confidence is in the method not in a particular interval.)

2. We are 95% confident that the maximum error of estimation for the average cost of producing 35 items is $169.59.

Figure 13.24 shows the fitted line plot of the production data along with the 95% confidence interval. Notice that as you approach the edges of the data, the confidence interval gets wider. This illustrates why it is important to only make predictions within the scope of the sample data.

Figure 13.24

Many statistical analysis packages compute a confidence interval for the mean value of y for a given value of x. Figure 13.25 is an example of a Minitab output. The output uses the estimated production model when items produced = 35. The value labeled "Fit" is the predicted value of y when $x = 35$. (Note that values differ slightly to the ones previously presented due to rounding.) The value labeled "SE Fit" is the standard error of the fitted value when $x = 35$. The "95% CI" is the 95% confidence interval for the mean value of y when $x = 35$. The "95% PI" will be discussed next.

```
Predicted Values for New Observations

New Obs    Fit  SE Fit       95% CI             95% PI
      1  4113.9    73.6  (3944.3, 4283.5)  (3572.2, 4655.5)


Values of Predictors for New Observations

             Items
New Obs   Produced
      1       35.0
```

Figure 13.25

Confidence Intervals for the Predicted Value of *y* Given *x*

We previously discussed confidence intervals for the mean value of *y* given *x*, rather than individual outcomes. Suppose you wish to produce 35 items, and you wish to predict the total cost of production and compute a 95% confidence interval for the total cost. Using the model,

$$\text{Estimated Cost} = \$2227.96 + \$53.88(35) = \$4113.76$$

for producing 35 items. The actual cost of producing 35 items in a particular week may be above or below average, the model has no way of knowing. Consequently, the actual cost will likely be different from the average value, and that difference will be an error in the model's prediction. That error needs to be accounted for in the confidence interval for the predicted value of *y* given *x*.

The expression for the confidence interval for a predicted value is very similar to the confidence interval for the mean value of *y* given *x*. However, instead of calling this interval a confidence interval (which it is), let's call it a **prediction interval** to distinguish the interval from the confidence interval for the mean value of *y* given *x*.

Formula

100(1 – α)% Prediction Interval for the Value of *y* Given *x*

A 100(1 – α)% confidence interval for the predicted value of *y* given *x*, also known as a **prediction interval**, is given by

$$\hat{y}_p \pm t_{\alpha/2,df}\, s_e \sqrt{1 + \frac{1}{n} + \frac{\left(x_p - \bar{x}\right)^2}{\sum\left(x_i - \bar{x}\right)^2}}.$$

Since the only difference between the prediction interval and the confidence interval for the mean value of *y* given *x* is the "1" inside the square root, the same information is required to compute the interval. However, since many statistical analysis packages will compute the prediction interval, you may not have to perform the computations very often.

Example 13.8

Suppose you, as the production manager, wish to produce 35 items this week. Compute the 95% prediction interval for the total cost of production.

Solution

The predicted cost of producing 35 items is given by

$$\text{Estimated Cost} = \$2227.96 + \$53.88(35) = \$4113.76,$$

$$t_{\alpha/2,df} = t_{0.025,8} = 2.306, \; s_e = 223.0682, \text{ and}$$

$$\frac{\left(x_p - \bar{x}\right)^2}{\sum\left(x - \bar{x}\right)^2} = \frac{(35 - 33.1)^2}{412.90} \approx 0.0087.$$

Computing the interval we have the following.

95% Prediction Interval for y Given $x = 35$

$$\hat{y}_p \pm t_{\alpha/2,df}\, s_e \sqrt{1+\frac{1}{n}+\frac{\left(x_p-\bar{x}\right)^2}{\sum\left(x_i-\bar{x}\right)^2}}$$

$$4113.76 \pm 2.306\left(223.0682\sqrt{1+\frac{1}{10}+0.0087}\right)$$

$$4113.76 \pm 541.6316$$

$$3572.13 \text{ to } 4655.39$$

We are 95% confident that the actual cost of production will be within $541.63 of the average price for producing 35 items ($4113.76). Specifically, we are 95% confident that the production cost will be between $3572.13 and $4655.39.

Figure 13.26 shows a fitted line plot of the production data along with the 95% prediction interval. Notice the difference between the plots of the 95% confidence interval (in Figure 13.24) and the 95% prediction interval.

Figure 13.26

The graphs below compare the confidence intervals for the mean value of y given $x = 35$ and the predicted value of y given $x = 35$.

Confidence Interval for the <u>Mean Value</u> of y

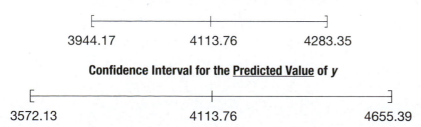

Confidence Interval for the <u>Predicted Value</u> of y

13

The prediction interval is more than three times wider than the confidence interval for the average value. A high price has been paid in order to account for individual variability.

Using the model for prediction outside the range of x values used to create the model can be dangerous. The nature of the relationship may not be linear outside of the range of x used to define the model. In the production example, the range of x-values spans from 22 to 45 items produced. Using the model to predict the cost of producing 100 items would no doubt have a sizable error. Inferential methods are not valid outside the range of x. Figure 13.27 illustrates the uncertainty associated with the regression line outside the range of the data.

Figure 13.27

13.10 Exercises

Basic Concepts

1. What is the main concern for the model builder when performing regression analysis?

2. Distinguish between the mean value of y given x and the predicted value of y given x.

3. Distinguish between a confidence interval and a prediction interval. Which interval is wider? Explain why.

4. Is there a particular range of x-values for which using the regression model for prediction is appropriate? Explain your answer.

Exercises

5. In Nevada, many forms of gambling are legal and very profitable. Sports betting amounts to billions of dollars annually. In football, a customer will bet on one of the teams to win the contest. However, in an attempt to even the game (from a betting point of view) one of the teams is selected as the favorite. The favorite's score in the game is reduced by an amount called the line. For example, if the Cowboys are favored over the Falcons by four points, then four points are subtracted from the Cowboys' score to determine the outcome of the game for betting purposes. Thus, if the Cowboys defeat the Falcons 32 to 30, in so far as settling any bets, the Cowboys score would be reduced by the spread and the Cowboys would be the loser $32 - 4 = 28$ to 30. Where does the betting line come from? The line is created by a betting market. If too many people are betting on the Cowboys before the game starts, the bookmaker will try to make the game more attractive to potential Falcon bettors by increasing the spread say from four points to five points. On the other hand, if too many people are betting on the Falcons, the spread will diminish from four to perhaps three points. How accurate is the betting spread at predicting the actual spread, which is the actual difference in points between the favorite and the underdog? In the example of the Cowboys and the Falcons, the actual spread was +2 (32 − 30). To examine this question, we want to build the following model:

$$\text{Actual Point Spread} = \beta_0 + \beta_1 (\text{Betting Spread}) + \varepsilon_i.$$

If the betting spread is a good predictor of the actual spread, it should be able to account for a substantial portion of the variation in the actual spreads. The following table contains betting and actual spreads from 15 randomly selected football games.

Table for Exercise 5 – Betting vs. Actual Spreads														
Betting	4	1	3	2	1	2	5	5	3	4	2	3	5	7	6
Actual	12	−2	6	7	3	1	14	3	−7	5	14	9	2	21	8

a. Draw a scatterplot of the data. Describe the relationship you observe between actual point spread and the betting spread.

b. Estimate the parameters of the model using statistical software.

c. Is there evidence at the 0.05 level of a linear relationship between the betting spread and the actual spread?

d. What fraction of the variation in the actual point spread is explained by the betting spread?

e. Interpret the coefficient of the betting spread in the model (β_1).

f. Construct and interpret a 95% confidence interval for β_1.

g. If the betting spread is five, what is the predicted actual spread?

h. Construct and interpret a 95% prediction interval for a betting spread of five.

i. Construct a 95% confidence interval for the average value of the actual spread when the betting spread is five.

6. Net income is the level of actual profit that a company reports for the year. Net sales is the total sales less adjustment for returns. What is the relationship between net income and net sales for large corporations? Suppose a random sample of 27 large corporations has been selected, and the net income and net sales have been recorded. A regression analysis has been performed to estimate the model, and the output is given.

$$\text{Net Income} = \beta_0 + \beta_1 (\text{Net Sales}) + \varepsilon_i$$

```
Regression Analysis: Income versus Sales

The regression equation is
Income = 84 + 18.4 Sales


Predictor     Coef  SE Coef      T      P
Constant      83.6    118.1   0.71  0.486
Sales       18.434    4.446   4.15  0.000


S = 372.478   R-Sq = 40.7%   R-Sq(adj) = 38.4%


Analysis of Variance

Source          DF      SS       MS      F      P
Regression       1 2384660  2384660  17.19  0.000
Residual Error  25 3468497   138740
Total           26 5853157


Predicted Values for New Observations

New Obs     Fit  SE Fit       95% CI              95% PI
      1  1005.3   147.1  (702.4, 1308.2)  (180.5, 1830.0)


Values of Predictors for New Observations

New Obs  Sales
      1   50.0
```

a. Find and interpret the standard deviation of the error terms in the output.

b. Interpret the slope coefficient. (The data used to estimate the model was in millions of dollars.)

c. What fraction of the variation in net income is explained by net sales?

d. Is there evidence of a linear relationship between net income and net sales? Test at the 0.05 level.

e. Construct and interpret a 95% confidence interval for β_1, the slope of the line.

f. The output also contains a predicted value for net income when sales are $50,000,000. Find the predicted value of net income when sales are $50,000,000. (Note that in the original data all observations were measured in millions of dollars. Thus a predicted value of 10,000,000 would be displayed in the output as 10.)

g. Find and interpret the 95% confidence interval for the average value of net income given that sales are $50,000,000.

h. Suppose your firm generated $50,000,000 in sales. What would be the 95% prediction interval for your firm's net income?

i. Use the model to predict net income for a company with $60,000,000 in sales. (Note that you must compute this manually.)

7. The personnel director of a large hospital is interested in determining the relationship (if any) between an employee's age and the number of sick days the employee takes per year. The director randomly selects eight employees and records their age and the number of sick days which they took in the previous year.

Table for Exercise 7 – Sick Days and Age								
Employee	1	2	3	4	5	6	7	8
Age	30	50	40	55	30	28	60	25
Sick Days	7	4	3	2	9	10	0	8

A regression analysis has been performed to estimate the model and the output is given.

$$\text{Sick Days} = \beta_0 + \beta_1 (\text{Age}) + \varepsilon_i$$

```
Regression Analysis: Sick Days versus Age

The regression equation is
Sick Days = 15.2 - 0.247 Age


Predictor      Coef   SE Coef      T      P
Constant     15.186     1.713   8.86  0.000
Age         -0.24681   0.04105  -6.01  0.001


S = 1.47652   R-Sq = 85.8%   R-Sq(adj) = 83.4%


Analysis of Variance

Source          DF      SS       MS       F      P
Regression       1  78.794   78.794   36.14  0.001
Residual Error   6  13.081    2.180
Total            7  91.875


Predicted Values for New Observations

New Obs    Fit   SE Fit        95% CI              95% PI
      1  6.547    0.557   (5.184, 7.911)    (2.686, 10.409)


Values of Predictors for New Observations

New Obs    Age
      1   35.0
```

a. Draw a scatterplot of the data. Describe the relationship you observe between the number of sick days and age.

b. Find and interpret the standard deviation of the error terms in the output.

c. Interpret the slope coefficient.

d. What fraction of the variation in the number of sick days an employee takes per year is explained by age?

e. Is there evidence of a linear relationship between the number of sick days an employee takes per year and age? Test at the 0.05 level.

f. Construct and interpret a 95% confidence interval for β_1, the slope of the line.

g. Find the predicted value of the number of sick days an employee will take per year if the employee is 35 years old.

h. Find and interpret the 95% confidence interval for the average number of sick days an employee will take per year, given the employee is 35.

i. Suppose a new employee is 35. Find a 95% prediction interval for the number of sick days this employee will take this year.

j. Use the model to predict the number of sick days per year for an employee who is 45 years old. Round to the nearest whole number.

8. A manufacturing company that produces laminate for countertops is interested in studying the relationship between the number of hours of training that an employee receives and the number of defects per countertop produced. Ten employees are randomly selected. The number of hours of training each employee has received is recorded and the number of defects on the most recent countertop produced is determined. The results are as follows.

| Table for Exercise 8 – Training Hours and Countertop Defects ||
Hours of Training	Defects per Countertop
1	1
4	4
7	0
3	3
2	5
2	4
5	3
5	2
1	5
6	1

A regression analysis has been performed to estimate the model, and the following output is produced.

$$\text{Defects per Countertop} = \beta_0 + \beta_1 \left(\text{Hours of Training}\right) + \varepsilon_i$$

```
Regression Analysis: Defects per Countertop versus Hours of Training

The regression equation is
Defects per Countertop = 4.65 - 0.515 Hours of Training


Predictor               Coef  SE Coef      T      P
Constant              4.6535   0.9426   4.94  0.001
Hours of Training    -0.5149   0.2286  -2.25  0.054


S = 1.45306    R-Sq = 38.8%    R-Sq(adj) = 31.2%


Analysis of Variance

Source            DF      SS      MS      F      P
Regression         1  10.709  10.709   5.07  0.054
Residual Error     8  16.891   2.111
Total              9  27.600


Predicted Values for New Observations

New Obs    Fit  SE Fit      95% CI            95% PI
     1   2.594   0.469  (1.514, 3.674)  (-0.927, 6.115)


Values of Predictors for New Observations

           Hours of
New Obs    Training
      1        4.00
```

a. Draw a scatterplot of the data. Describe the relationship you observe between the number of defects per countertop and hours of training. Are there any unusual observations?

b. Find and interpret the standard deviation of the error terms in the output.

c. Interpret the slope coefficient.

d. What fraction of the variation in the number of defects per countertop is explained by the hours of training? What other factors might affect the number of defects?

e. Is there evidence of a linear relationship between the number of hours of training and the number of defects per countertop? Test at the 0.05 level and the 0.10 level.

f. Construct and interpret a 95% confidence interval for β_1, the slope coefficient.

g. Find the predicted value of the number of defects per countertop for an employee who has had 4 hours of training.

h. Find and interpret the 95% confidence interval for the average number of defects per countertop for employees who have had 4 hours of training.

i. Suppose a new employee has had 4 hours of training. What would be the 95% prediction interval for the number of defects per countertop?

j. Use the model to predict the number of defects per countertop for an employee who has had 7 hours of training. Round your answer to the nearest whole number.

T # Discovering Technology

Using the TI-84 Plus Calculator

Regression Analysis

For this exercise, use the production data from Example 13.3.

1. Before performing regression analysis using the TI-84 Plus calculator, we need to turn on diagnostics so that the coefficient of determination is displayed in the output. Press **2ND** and **0** to access the catalog.

2. Scroll down until you reach the option that reads **DiagnosticOn**. With DiagnosticOn selected, press **ENTER**. Press **ENTER** again. You are now ready to perform regression analysis.

3. Press **STAT**, select **EDIT**, **1:Edit**, and **ENTER**. Then input the *x* and *y* data into lists **L1** and **L2**.

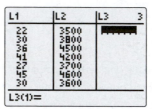

Figure 13.28

4. Press **STAT** and select **CALC**, then choose option **4: LinReg(ax+b)** and press **ENTER**.

5. Press **2ND** and **1** to select L1 as the independent variable. Press the **,** button and then **2ND** and **2** to select L2 as the dependent variable. Press **ENTER**.

6. Observe the output screen for the regression analysis. The estimate for β_1 is given as 53.88 and the estimate for β_0 is given as 2227.96 (rounded to two decimal places). The coefficient of determination is also given as 0.7507. Compare these results to those we calculated for the production data.

Figure 13.29

Using Excel

Regression Analysis

For this exercise, use the production data from Example 13.3.

1. Enter the *x* (items produced) and *y* (cost) data into columns A and B.

	A	B
1	Items Produced (*x*)	Cost (*y*)
2	22	3500
3	30	3800
4	36	4500
5	41	4200
6	27	3700
7	45	4600
8	30	3600
9	37	4550
10	32	3990
11	31	3675

Figure 13.30

2. Under the **Data** tab, choose **Data Analysis**. In the Data Analysis dialog box, choose **Regression** from the list of Analysis Tools.

3. Select the data in Column B (**$B$2:$B$11**) for Input Y Range and the data in Column A (**$A$2:$A$11**) for Input X Range.

4. Excel automatically produces a 95% confidence interval for the estimate of β_1. If you require another level of confidence, check the box next to **Confidence Level** and enter the desired confidence level in the box. For this exercise, enter **99** in the Confidence Level input box.

5. Select the radio button next to **Output Range** and enter **$D$1**.

Figure 13.31

6. Press **OK**. Observe the summary output. The *Regression Statistics* table contains the coefficient of determination, standard error of the model, and the number of observations. The ANOVA table contains the degrees of freedom, sums of squares, and mean squares values for regression and error (residual). Finally, the least squares estimates of the coefficients are given along with their standard errors, *t*-statistics, and *P*-values. Notice that the 95% and 99% confidence intervals for β_1 are given in the *Lower 95%, Upper 95%, Lower 99.0%,* and *Upper 99.0%* columns of the output.

D	E	F	G	H	I	J	K	L
SUMMARY OUTPUT								
Regression Statistics								
Multiple R	0.866441198							
R Square	0.75072035							
Adjusted R Square	0.719560393							
Standard Error	223.0681781							
Observations	10							
ANOVA								
	df	*SS*	*MS*	*F*	*Significance F*			
Regression	1	1198827.203	1198827	24.09247	0.00118116			
Residual	8	398075.2967	49759.41					
Total	9	1596902.5						
	Coefficients	*Standard Error*	*t Stat*	*P-value*	*Lower 95%*	*Upper 95%*	*Lower 99.0%*	*Upper 99.0%*
Intercept	2227.955922	370.1487713	6.019082	0.000317	1374.391325	3081.520518	985.9634238	3469.948419
Items Produced	53.8835069	10.97779658	4.908408	0.001181	28.56866263	79.19835118	17.04874735	90.71826645

Figure 13.32

Using Minitab

Regression Analysis

For this exercise, use the production data from Example 13.3.

1. Enter the *x* (items produced) and *y* (cost) data into columns C1 and C2.

↓	C1	C2
	Items Produced	**Cost**
1	22	3500
2	30	3800
3	36	4500
4	41	4200
5	27	3700
6	45	4600
7	30	3600
8	37	4550
9	32	3990
10	31	3675

Figure 13.33

2. Choose **Stat, Regression**, and **Fitted Line Plot**. Enter **C2** for Response (Y) and enter **C1** for Predictor (X). Press **OK**.

3. Observe the least squares coefficients, R^2 value, and standard error, as well as the regression plot. Notice that the coefficient of determination is given as a percentage.

Figure 13.34

4. Choose **Stat**, **Regression**, and **Regression** for additional output. Enter **C1** for Response and **C2** for Predictors and press **OK**.

5. Observe the regression analysis output. The regression equation is given at the top. Then the estimated coefficients are given along with their standard errors, *t*-statistics, and *P*-values. Then, the standard error for the regression model and the coefficient of determination are given. Finally, the ANOVA table is given, which lists the degrees of freedom, sums of squares, and mean squares values for regression and error.

```
Regression Analysis: Cost versus Items Produced

The regression equation is
Cost = 2228 + 53.9 Items Produced


Predictor          Coef   SE Coef     T      P
Constant         2228.0     370.1   6.02  0.000
Items Produced    53.88     10.98   4.91  0.001


S = 223.068    R-Sq = 75.1%    R-Sq(adj) = 72.0%


Analysis of Variance

Source           DF       SS        MS      F      P
Regression        1  1198827   1198827  24.09  0.001
Residual Error    8   398075     49759
Total             9  1596903
```

Figure 13.35

Confidence and Prediction Intervals

Minitab will compute confidence intervals for the mean value of *y* given *x* and prediction intervals for the individual value of *y* given *x*. Use the production data from Example 13.3 for this exercise.

1. Enter the *x* (items produced) and *y* (cost) data into columns C1 and C2.

2. Choose **Stat**, **Regression**, and **Regression**.

3. Enter **C2** for Response (Y) and enter **C1** for Predictor (X).

4. In the Regression dialog box, choose **Options**.

5. In the Regression – Options dialog box, enter **35** in the Prediction intervals for new observations box.

6. Set the Confidence level at **99**.

7. Check the boxes next to Fits and SEs of fits. Press **OK**.

8. Press **OK** again. Observe the output to find the 99% confidence interval (CI) and 99% prediction interval (PI).

```
Regression Analysis: Cost (y) versus Items Produced (x)

The regression equation is
Cost (y) = 2228 + 53.9 Items Produced (x)


Predictor             Coef  SE Coef     T      P
Constant            2228.0    370.1  6.02  0.000
Items Produced (x)   53.88    10.98  4.91  0.001


S = 223.068   R-Sq = 75.1%   R-Sq(adj) = 72.0%


Analysis of Variance

Source          DF       SS       MS      F      P
Regression       1  1198827  1198827  24.09  0.001
Residual Error   8   398075    49759
Total            9  1596903


Predicted Values for New Observations

New Obs    Fit  SE Fit        99% CI               99% PI
      1  4113.9    73.6  (3867.1, 4360.7)   (3325.8, 4902.0)


Values of Predictors for New Observations

               Items
            Produced
New Obs        (x)
      1        35.0
```

Figure 13.36

R   Chapter 13 Review

Key Terms and Ideas

- Regression Analysis
- Dependent Variable (Response Variable)
- Independent Variable (Predictor Variable)
- Simple Linear Regression
- Simple Linear Regression Model
- Parameters
- Error Term
- Estimated Simple Linear Regression Equation
- Method of Least Squares
- Correlation Coefficient
- Residual
- Sum of Squared Errors
- Least Squares Line

- Variance of the Error Terms
- Mean Square Error
- Standard Error
- Total Sum of Squares
- Sum of Squares of Regression
- Coefficient of Determination
- Linear Time Trend
- Assumptions about the Error Term
- Confidence Interval for β_1
- Testing a Hypothesis Concerning β_1
- Confidence Interval for the Mean Value of y Given x
- Prediction Interval for the Value of y Given x

Key Formulas

	Section
Simple Linear Regression Model $$y_i = \beta_0 + \beta_1 x_i + \varepsilon_i$$	13.1
Estimated Simple Linear Regression Equation $$\hat{y}_i = b_0 + b_1 x_i$$	13.1
Sum of Squared Errors $$\text{SSE} = \sum \left(y_i - \hat{y}_i \right)^2 = \sum \left(y_i - \left(b_0 + b_1 x_i \right) \right)^2$$	13.2

Key Formulas (cont.)

Section

Slope of the Least Squares Line

$$b_1 = \frac{SS_{xy}}{SS_{xx}}$$

where

$$SS_{xy} = \sum (x_i - \bar{x})(y_i - \bar{y}) = \sum x_i y_i - \frac{\left(\sum x_i\right)\left(\sum y_i\right)}{n}$$

and

13.3

$$SS_{xx} = \sum (x_i - \bar{x})^2 = \sum x_i^2 - \frac{\left(\sum x_i\right)^2}{n}$$

or

$$b_1 = \frac{n\sum x_i y_i - \sum x_i \sum y_i}{n\sum x_i^2 - \left(\sum x_i\right)^2}.$$

y-Intercept of the Least Squares Line

13.3

$$b_0 = \bar{y} - b_1 \bar{x} = \frac{1}{n}\left(\sum y_i - b_1 \sum x_i\right)$$

Mean Square Error

$$s_e^2 = \frac{\sum (y_i - \hat{y}_i)^2}{n-2} = \frac{SSE}{n-2}$$

13.4

Standard Error

$$s_e = \sqrt{\frac{\sum (y_i - \hat{y}_i)^2}{n-2}} = \sqrt{\frac{SSE}{n-2}}$$

13.4

Total Sum of Squares

$$TSS = \sum (y_i - \bar{y})^2$$

13.5

$$TSS = SSE + SSR$$

Sum of Squares of Regression

$$SSR = TSS - SSE$$

13.5

Key Formulas (cont.)

	Section
Coefficient of Determination	

$$R^2 = \frac{\text{SSR}}{\text{TSS}} = 1 - \frac{\text{SSE}}{\text{TSS}}$$

or

$$R^2 = \left(\frac{n\sum x_i y_i - \sum x_i \sum y_i}{\sqrt{\left(n\sum x_i^2 - \left(\sum x_i\right)^2\right)\left(n\sum y_i^2 - \left(\sum y_i\right)^2\right)}} \right)^2$$

13.5

Correlation Coefficient

$$r = \frac{n\sum x_i y_i - \sum x_i \sum y_i}{\sqrt{\left(n\sum x_i^2 - \left(\sum x_i\right)^2\right)\left(n\sum y_i^2 - \left(\sum y_i\right)^2\right)}}$$

13.5

Sample Estimate of the Variance of b_1

$$s_{b_1}^2 = \frac{s_e^2}{\sum \left(x_i - \bar{x}\right)^2}$$

13.8

Sample Estimate of the Standard Deviation (Standard Error) of b_1

$$s_{b_1} = \sqrt{\frac{s_e^2}{\sum \left(x_i - \bar{x}\right)^2}}$$

13.8

$100\left(1-\alpha\right)\%$ Confidence Interval for β_1

$$b_1 \pm t_{\alpha/2,df}\, s_{b_1}$$

13.8

Test Statistic for Testing the Hypothesis $\beta_1 \neq 0$

$$t = \frac{b_1 - 0}{s_{b_1}} = \frac{b_1}{s_{b_1}}$$

13.9

Key Formulas (cont.)

Section

$100(1-\alpha)\%$ Confidence Interval for the Mean Value of y Given x

$$\hat{y}_p \pm t_{\alpha/2,df} s_e \sqrt{\frac{1}{n} + \frac{\left(x_p - \bar{x}\right)^2}{\sum\left(x_i - \bar{x}\right)^2}}$$

13.10

$100(1-\alpha)\%$ Prediction Interval for the Value of y Given x

$$\hat{y}_p \pm t_{\alpha/2,df} s_e \sqrt{1 + \frac{1}{n} + \frac{\left(x_p - \bar{x}\right)^2}{\sum\left(x_i - \bar{x}\right)^2}}$$

13.10

AE   Additional Exercises

1. A pharmacist is interested in studying the relationship between the amount of a particular drug in the bloodstream (in mg) and reaction time (in seconds) of subjects taking the drug. Ten subjects are randomly selected and administered various doses of the drug. The reaction times (in seconds) are measured 15 minutes after the drug is administered with the following results.

Table for Exercise 1 – Reaction Times			
Amount of Drug (mg)	Reaction Time (Seconds)	Amount of Drug (mg)	Reaction Time (Seconds)
1	0.5	6	0.8
2	0.7	7	0.9
3	0.6	8	0.6
4	0.7	9	0.9
5	0.8	10	1.0

A regression analysis has been performed to estimate the model, and the following output was produced.

$$\text{Reaction Time} = \beta_0 + \beta_1\left(\text{Amount of Drug}\right) + \varepsilon_i$$

```
Regression Analysis: Reaction Time (Seconds) versus Amount of Drug (mg)

The regression equation is
Reaction Time (Seconds) = 0.533 + 0.0394 Amount of Drug (mg)


Predictor               Coef   SE Coef    T      P
Constant             0.53333   0.07521  7.09  0.000
Amount of Drug (mg)  0.03939   0.01212  3.25  0.012


S = 0.110096   R-Sq = 56.9%   R-Sq(adj) = 51.5%


Analysis of Variance

Source         DF      SS       MS       F      P
Regression      1  0.12803  0.12803  10.56  0.012
Residual Error  8  0.09697  0.01212
Total           9  0.22500


Predicted Values for New Observations

New Obs    Fit   SE Fit      95% CI              95% PI
      1  0.6909  0.0393  (0.6003, 0.7815)  (0.4214, 0.9605)


Values of Predictors for New Observations

          Amount of
New Obs   Drug (mg)
      1       4.00
```

a. Draw a scatterplot of the data. Describe the relationship you observe between the reaction time and the amount of drug in the bloodstream. Are there any unusual observations?

b. Find and interpret the standard deviation of the error terms in the output.

c. Interpret the slope coefficient.

d. What fraction of the variation in reaction time is explained by the amount of drug in the bloodstream? What other factors might affect reaction time?

13

e. Is there evidence of a linear relationship between the amount of drug in the bloodstream and reaction time? Test at the 0.05 level and the 0.01 level.

f. Construct and interpret a 95% confidence interval for β_1, the slope of the line.

g. Find the predicted value of the reaction time of an individual who has 4 mg of the drug in the bloodstream.

h. Find and interpret a 95% confidence interval for the average reaction time of all individuals who have 4 mg of the drug in their bloodstreams.

i. Suppose a particular individual has 4 mg of the drug in the bloodstream. What would be the 95% prediction interval for the reaction time?

2. A sample of 11 lonely hearts advertisements, all placed by males, was selected from the local newspaper. In each of the selected ads, the males gave their heights, along with other physical characteristics and preferences. Some of the males obviously felt that being taller than average might result in more responses to the ad. Suppose that y, the number of responses to the ad over the next 30 days, was determined for each male. The following table contains the data.

Table for Exercise 2 – Height and Response										
Height (Inches)	70	62	67	75	78	69	70	64	66	69	75
y	14	7	10	18	17	12	15	9	12	14	17

a. Draw a scatterplot of the data. Does the relationship appear to be linear?

b. Estimate the slope and intercept of the regression equation using statistical software.

c. Is there evidence of a linear relationship between the number of responses and height? Test at the $\alpha = 0.01$ level.

d. Interpret the regression coefficient corresponding to height.

e. Construct a 95% confidence interval for the slope.

f. Compute R^2 and interpret this value.

g. Estimate the number of responses for a male 6 feet tall. Round your answer to the nearest whole number.

h. Construct and interpret a 95% prediction interval for the number of responses for a male who is 6 feet tall.

i. Construct and interpret a 95% confidence interval for the average number of responses for a male who is 6 feet tall.

3. It is believed that when one is in the process of buying a home, the interest rate that is given on the loan is a function of his or her credit score. The Fair Isaac Corporation (FICO) is a major producer of credit scores. They have collected data from major lenders about buyers' history of borrowing and paying back credit. The following table contains 20 randomly selected loan applicants along with their FICO scores and the interest rate that they were given when financing their homes. With the data given, answer the following questions.

Table for Exercise 3 – Credit Scores and Interest Rates		
Observation	FICO Score	Interest Rate (%)
1	756	6.32
2	679	7.85
3	527	10.20
4	839	5.52
5	677	7.30
6	686	7.37
7	512	9.67
8	590	8.40
9	765	5.82
10	502	10.01
11	819	5.86
12	630	8.51
13	704	6.83
14	679	7.72
15	663	7.68
16	542	9.53
17	575	6.86
18	508	9.65
19	689	7.75
20	750	6.89

a. Draw a scatterplot of the data. Does there appear to be a linear relationship between FICO score and interest rate?

b. Estimate the simple linear regression equation using statistical software.

c. What is the estimate of the mean square error? Interpret this value.

d. Test at the 5% level if a linear relationship exists between FICO scores and interest rates.

e. Interpret the regression coefficient corresponding to FICO score.

f. Construct a 95% confidence interval for the slope. Interpret the interval.

g. Compute the coefficient of determination. Interpret this value.

h. Calculate the correlation coefficient. Interpret this value.

i. What is the average interest rate for a credit score of 725?

j. Construct a 90% confidence interval for the average interest rate for people who have FICO scores of 725. Interpret this interval.

k. Construct a 90% prediction interval for the interest rate for a person with a FICO score of 725. Interpret this interval.

4. It appears that many cellular phone service providers are making huge profits from customers using their messaging services such as text and multimedia messaging services (MMS). To that end, the cellular phone companies are using their marketing campaigns to target kids rather than adults. The belief is that kids tend to utilize their messaging services much more than adults. In fact, it is the belief that the younger one is, the more texts and MMS sent via his or her cell phone. Using the data given which reports the number of monthly messages sent by age, formulate a simple linear regression model to answer the following questions.

Table for Exercise 4 – Age and Message Use			
Age	Number of Messages	Age	Number of Messages
78	7	37	1541
36	1607	69	6
11	3037	69	25
69	26	55	517
56	491	39	1439
74	0	20	2505
22	2373	14	2845
74	5	10	3048
10	3059	80	0
26	2155	59	295
18	2619	40	1374
68	17	67	35
10	3067		

a. Draw a scatterplot of the data. Does there appear to be a linear relationship between age and the number of messages that one sends?

b. What is the estimated simple linear regression equation?

c. What is the estimate of the coefficient of determination? Interpret this value.

d. Test at the 5% level if a linear relationship exists between age and the number of messages sent via a cellular phone.

e. Interpret the regression coefficient corresponding to age.

f. Construct a 95% confidence interval for the slope. Interpret this interval.

g. Calculate the correlation coefficient. Interpret this value.

h. What is the average number of messages sent by a 15-year-old? Round your answer to the nearest whole number.

i. Construct a 95% confidence interval for the average number of messages sent by a 15-year-old. Interpret this interval.

j. Suppose Jacob's parents are contemplating giving him a cell phone but with a limited messaging plan at 500 per month. Eager to get the cell phone, Jacob, at 15 years old, promises that he won't send more than 500 messages per month and he'll also limit the number of friends that will have his phone number. In spite of Jacob's honesty and loyalty, should his parents believe that he won't send more than 500 messages per month? Explain your answer.

5. For the last 10 years, the Virginia Department of Mines, Minerals, and Energy (VDMME) has been promoting Energy Star, a resource for energy-efficient products and solutions. VDMME wants all energy consumers to take responsibility and exercise leadership by practicing conservation and efficiency on a daily basis. The average annual energy usage for a 1800 square foot home is 18,000 kilowatt hours. VDMME believes that this number can be significantly reduced if consumers started using Energy Star appliances. Answer the following questions based on data of 25 randomly selected homes with Energy Star appliances built within the last five years.

Table for Exercise 5 – Home Size and Energy Usage			
Home Size (Square Feet)	Annual Energy Usage (kWh)	Home Size (Square Feet)	Annual Energy Usage (kWh)
2895	15,200	2180	13,227
3650	17,333	4492	19,492
2927	15,050	6450	25,353
6289	24,763	1583	11,075
7252	27,098	4170	18,557
4147	18,291	4189	18,636
6505	25,028	3920	18,210
1413	11,099	6833	26,075
2279	13,110	4469	19,232
3251	15,844	6141	24,225
2992	14,904	5084	21,530
6912	26,329	6746	26,333
2503	13,765		

a. Draw a scatterplot of the data. Does there appear to be a linear relationship between home size and the amount of annual kWh used?

b. What is the estimated simple linear regression equation?

c. What is the estimate of the coefficient of determination? Interpret this value.

d. Test at the 5% level if a linear relationship exists between home size and the annual amount of kWh used.

e. Interpret the regression coefficient corresponding to home size.

f. Construct a 99% confidence interval for the slope. Interpret the interval.

g. Calculate the correlation coefficient. Interpret this value.

h. Suppose the James family constructed a 3200 square foot home using all Energy Star appliances. How many kilowatt hours should they expect to use in their first year in the home?

i. Construct a 95% confidence interval for the average number of kWh that will be used by the James family. Interpret this interval.

6. With grade inflation being a major problem in many U.S. high schools, college admissions offices are beginning to look at other performance measures when evaluating student applications. It is believed that many students with high grade point averages in high school will not necessarily score high on the SAT. Using the data of 30 randomly selected students that took the SAT, answer the following questions to determine if there is a linear relationship between high school GPA and SAT score.

Table for Exercise 6 – High School GPA and SAT Score			
High School GPA	**SAT Score**	**High School GPA**	**SAT Score**
3.21	1448	4.95	1960
2.23	1435	4.69	1717
2.89	1411	2.49	1365
1.84	1291	2.45	1561
3.34	1462	2.57	1474
2.42	1357	1.28	1328
2.75	1396	1.94	1302
2.35	1549	4.75	1622
4.80	1829	1.91	1499
1.98	1508	4.25	1566
2.92	1514	1.15	1413
4.18	1658	2.17	1428
4.50	1694	4.73	1720
4.42	1686	4.39	1783
4.78	1840	2.92	1614

a. Draw a scatterplot of the data. Does there appear to be a linear relationship between high school GPA and SAT score?

b. What is the estimated simple linear regression equation?

c. What is the coefficient of determination? Interpret this value.

d. Test at the 5% level if a linear relationship exists between high school GPA and SAT score.

e. Interpret the regression coefficient corresponding to high school GPA.

f. Construct a 95% confidence interval for the slope. Interpret the interval.

g. Calculate the correlation coefficient. Interpret this value.

h. What SAT score would you expect for students with a GPA of 3.5? Round your answer to the nearest whole number.

P   Discovery Project

Data sets to use for this Discovery Project are available for download from the following website. The data are available in Excel and Minitab format.

http://www.hawkeslearning.com/Support/Downloads.htm

Automobile Prices

The data set titled AUTOMOBILE contains information that is intended to be used to predict the price of a vehicle (dependent variable) as a function of a number of variables such as the make of the vehicle, fuel-type, wheel-base, city-mpg, etc. Using the data provided, answer the following questions.

a. If one were to fit a simple linear regression (SLR) model to predict the price of a vehicle, what is the most important variable in the data set? Justify your answer.

b. Fit a SLR model using the variable that you selected in **a.**

c. Test that the slope of the SLR model in **b.** is significantly different from zero at the 5% level.

d. Discuss the coefficient of determination for the model in **b.**

e. Interpret the estimate of the slope and intercept in the model used in **b.**

Housing Values

The data set titled HOUSING contains information about the housing values in Boston, MA, in the late 1970s. The variables in the data are used to predict the median value of owner-occupied homes. Using the data provided, answer the following questions.

a. What is the single most important variable in the data set that can be used to predict the median value of owner-occupied homes? Justify your answer.

b. Fit a SLR model using the variable that you selected in **a.** Discuss the quality of fit of the SLR model.

c. Discuss the coefficient of determination for the model in **b.**

d. Test that the slope of the SLR model in **b.** is significantly different from zero at the 5% level.

e. Interpret the estimate of the slope and intercept in the model used in **b.**

Housing Starts

The data set titled HOUSING2 contains data that can be used to predict the number of housing starts in the United States as a function of unemployment rate, 30-year mortgage interest rate, and national consumer price index (CPI). Using the data provided, answer the following questions.

a. Fit a SLR model using each of the independent variables in the model.

b. Which variable best explains the amount of variability in housing starts? Justify your answer.

c. Test that the slopes of each of the SLR models in **a.** are significantly different from zero at the 5% level.

d. Calculate a 95% confidence interval for the slope of your "best" model selected in part **b.** Interpret this interval.

Chapter Fourteen
Multiple Regression

14

14.1   Multiple Regression

Previously, we constructed linear regression models in which one independent variable, x, is related to one dependent variable, y. For example, in the production model from the previous chapter, only one independent variable (items produced) is used to explain total cost. But there are many other variables, the number of labor hours required or energy costs, for example, which may also be useful in explaining total cost. If a model is to accurately represent real world phenomena, then the model must potentially be able to accommodate multiple independent variables. **Multiple regression** is an extension of simple regression techniques, allowing more than one independent variable in the regression equation.

> ### Definition
>
> The **multiple regression model** is given by
>
> $$y_i = \beta_0 + \beta_1 x_{1i} + \beta_2 x_{2i} + \cdots + \beta_k x_{ki} + \varepsilon_i$$
>
> where
>
> β_0, β_1, β_2, ..., β_k are the model's parameters. (They are unknown constants that will require estimation.)
>
> x_{1i}, x_{2i}, ..., x_{ki} are independent variables which are measured without error.
>
> ε_i is a random error which is normally distributed with a mean of zero and a standard deviation σ_ε. (The errors are independent of each other.)

The multiple regression model contains additional parameters that must be estimated. Recall from Chapter 13 that the **method of least squares** is used to find the estimated regression line that minimizes the sum of squared errors, $\left(\text{SSE} = \sum\left(y_i - \hat{y}_i\right)^2\right)$. Even though we are now discussing multiple regression, least squares will remain the method of estimation. Because of the complexity of the calculations, formulas for the least squares coefficients will not be presented. Instead, estimates of the model's parameters will be obtained from one of the many statistical analysis programs that perform multiple regression analysis. Although the format of the output varies among the programs, the program outputs mostly contain the same fundamental information. The Discovering Technology section at the end of this chapter illustrates how to perform and interpret multiple regression equations using Microsoft Excel and Minitab. Throughout the examples in this chapter, we will be using outputs from these statistical programs to construct and interpret multiple regression models.

Just as with the simple linear regression model, it is not often that an entire population is available. Therefore, we must estimate the parameters of the multiple regression model using sample data. Using data from the sample, we can estimate the coefficients of the **estimated multiple regression equation**.

Definition

The **estimated multiple regression equation** is

$$\hat{y}_i = b_0 + b_1 x_{1i} + b_2 x_{2i} + \cdots + b_k x_{ki}$$

where b_0, b_1, b_2, ..., b_k are estimates of their population counterparts. Specifically, b_0 is an estimate of β_0, b_1 is an estimate of β_1, b_2 is an estimate of β_2, etc.

$\hat{y}_i$ is the predicted value of y for given values of x_1, x_2, ..., x_k, and is pronounced y-hat. The symbol y_i is reserved for the observed value of y.

Model building is a process. The greatest challenge in building a multiple regression model is in determining the appropriate independent variables needed to explain the dependent variable. In practical applications, this usually requires a great deal of experimentation with the model. Analysts examine the effects of adding and removing independent variables in order to determine which model is the best predictor of the dependent variable.

A pizza delivery manager is analyzing the delivery routes in her system. She is interested in predicting the amount of time required for the driver to deliver the pizzas on a specific route. The driver has to make several stops on the route because the manager doesn't want to send several vehicles/drivers to cover the same area/route.

Table 14.1 – Pizza Delivery Time Data

Observation	Delivery Time (Minutes)	Number of Pizzas	Distance (Miles)
1	16.68	7	5.60
2	11.50	3	2.20
3	12.03	3	3.40
4	14.88	8	0.80
5	13.75	6	1.50
6	18.11	7	3.30
7	8.00	2	1.10
8	17.83	7	2.10
9	79.24	30	14.60
10	21.50	5	6.05
11	40.33	16	6.88
12	21.00	10	2.15
13	13.50	4	2.55
14	19.75	6	4.62
15	24.00	9	4.48
16	29.00	10	7.76
17	15.35	6	2.00
18	19.00	7	1.32
19	9.50	3	0.36
20	35.10	17	7.70
21	17.90	10	1.40
22	52.32	26	8.10
23	18.75	9	4.50
24	19.83	8	6.35
25	10.75	4	1.50

The linear multiple regression model is as follows.

$$\text{Delivery Time} = \beta_0 + \beta_1 \left(\text{Number of Pizzas}\right) + \beta_2 \left(\text{Distance}\right) + \varepsilon_i$$

To have a useful model, the parameters β_0, β_1, and β_2, must be estimated. Estimating these parameters requires the collection of historical data on pizza delivery times. If inferences concerning the model's predicted values or parameters are desired, then random sampling methods must be used during data collection. Assume that the data in Table 14.1 have been collected using random sampling techniques. The summary output from Microsoft Excel is given in Figure 14.1.

	A	B	C	D	E	F	G
1	SUMMARY OUTPUT						
2							
3	*Regression Statistics*						
4	Multiple R	0.981846095					
5	R Square	0.964021754					
6	Adjusted R Square	0.960751004					
7	Standard Error	3.075694294					
8	Observations	25					
9							
10	ANOVA						
11		*df*	*SS*	*MS*	*F*	*Significance F*	
12	Regression	2	5576.424901	2788.212	294.7403	1.30749E-16	
13	Residual	22	208.1176986	9.459895			
14	Total	24	5784.5426				
15							
16		*Coefficients*	*Standard Error*	*t Stat*	*P-value*	*Lower 95%*	*Upper 95%*
17	Intercept	1.792903307	1.048779122	1.709515	0.101425	-0.382131459	3.967938072
18	Number of Pizzas	1.589101026	0.156345691	10.16402	8.97E-10	1.26485991	1.913342142
19	Distance	1.567708058	0.327534513	4.78639	8.85E-05	0.888443055	2.246973061

Figure 14.1

The estimated model parameters are $b_0 = 1.7929$, $b_1 = 1.5891$, and $b_2 = 1.5677$. The estimated multiple regression model is as follows.

$$\text{Estimated Delivery Time} = 1.7929 + 1.5891\left(\text{Number of Pizzas}\right) + 1.5677\left(\text{Distance}\right)$$

There are numerous questions to ask about this model.

- Can the model explain a substantial portion of the variation in delivery times? If not, it will not be very useful.

- Do the signs and magnitudes of the estimated coefficients appear to be reasonable? The answer to this question is a function of the data. For the pizza delivery example, the estimate of the coefficient on number of pizzas is 1.5891. This means that for a fixed distance, we would expect the average delivery time to increase by approximately 1.59 minutes for each additional pizza that is ordered. Does this make sense? Certainly, one would expect the average delivery time to increase as the number of pizzas increases. Therefore, the sign (positive, in this case) and magnitude are reasonable.

- Are the estimates of the coefficients reliable or do the estimates have substantial sampling variation?

- Are both independent variables necessary? Do any of the variables not contribute to the explanation of delivery times? This question will be addressed using the familiar hypothesis testing procedure.

- Are there other independent variables that could be included that would enhance the model's ability to accurately predict delivery times?

- Can the model make a useful prediction of delivery time? How much confidence can be placed in the prediction?

We will address these questions as we continue to discuss multiple regression throughout this chapter.

14.1   Exercises

Basic Concepts

1. Explain why a simple linear regression model might not always suffice when attempting to establish a relationship between variables in a business environment.

2. What is the general multiple regression model?

3. What are the assumptions about the error term in a multiple regression model? Are these different from the assumptions required for the simple linear model?

4. What method is used to find the estimated regression equation? Is this method different from the one used to find the simple linear regression equation?

5. What is the greatest challenge in building a multiple regression model?

6. What are some questions that should be asked once a multiple regression model is estimated? Give at least four.

Exercises

7. Consider the following computer output of a multiple regression analysis relating annual salary to years of education and years of work experience.

	A	B	C	D	E	F	G
1	SUMMARY OUTPUT						
2							
3	*Regression Statistics*						
4	Multiple R	0.566946595					
5	R Square	0.321428441					
6	Adjusted R Square	0.29192533					
7	Standard Error	10909.996					
8	Observations	49					
9							
10	ANOVA						
11		*df*	*SS*	*MS*	*F*	*Significance F*	
12	Regression	2	2593556200	1296778100	10.89473033	0.000133875	
13	Residual	46	5475288584	119028012.7			
14	Total	48	8068844784				
15							
16		*Coefficients*	*Standard Error*	*t Stat*	*P-value*	*Lower 95%*	*Upper 95%*
17	Intercept	11214.19915	5625.172956	1.993574106	0.052147881	-108.6867382	22537.08504
18	Education (Years)	2854.891271	689.6666061	4.139523715	0.000146836	1466.664395	4243.118147
19	Experience (Years)	839.6360369	261.7094444	3.208275646	0.002433357	312.842248	1366.429826

a. Identify the estimated values of the coefficients b_0, b_1, and b_2.

b. Write the estimated multiple regression equation.

c. Can you think of other independent variables that may be useful in predicting annual salary?

8. The manager of a publishing company would like to conduct cost analysis on the most recent books the company has published. He would like to estimate a multiple regression model to relate the cost of printing (per book) to the number of pages in the book and the number of copies printed. A computer output of the multiple regression model for the manager's data is given in the following table.

	A	B	C	D	E	F	G
1	SUMMARY OUTPUT						
2							
3	*Regression Statistics*						
4	Multiple R	0.987606014					
5	R Square	0.975365639					
6	Adjusted R Square	0.972467479					
7	Standard Error	0.445885396					
8	Observations	20					
9							
10	ANOVA						
11		*df*	*SS*	*MS*	*F*	*Significance F*	
12	Regression	2	133.8201656	66.91008281	336.5464936	2.12863E-14	
13	Residual	17	3.379834375	0.198813787			
14	Total	19	137.2				
15							
16		*Coefficients*	*Standard Error*	*t Stat*	*P-value*	*Lower 95%*	*Upper 95%*
17	Intercept	6.134155476	3.993435752	1.536059638	0.142925974	-2.291257484	14.55956844
18	Number of Pages	0.010801	0.004147682	2.604105041	0.018522101	0.002050156	0.019551845
19	Number of Copies	-0.009954478	0.005271436	-1.888380579	0.07616193	-0.021076236	0.00116728

a. Identify the estimated regression coefficients.

b. Write the estimated multiple regression equation.

c. Do the magnitudes and signs of the coefficients seem reasonable? Explain.

d. What other variables do you think could be useful in explaining printing cost per book?

9. A nutritionist wishes to study body weight based on height, age, average calories consumed per day, and the average number of minutes spent exercising per day.

a. Write the multiple regression model the nutritionist is interested in in terms of weight, height, age, calories, and exercise. Assume the coefficients have not yet been estimated.

b. Identify the independent variables in the multiple regression model.

c. Predict the sign of the coefficient for each of the independent variables in the model. Explain your answers.

d. Can you think of any other variables that might be useful for the nutritionist to take into account before performing the regression analysis?

10. Suppose the CEO of an electronics company wants to study the effects of various business practices on annual revenue.

    a. Make a list of independent variables the CEO might be interested in studying.

    b. Suppose the CEO has narrowed his list of factors down, and decided he wants to mainly study the effects of research and development expenditures, advertising expenditures, and the average annual salary paid to employees. Write the multiple regression model in terms of the dependent and independent variables, assuming the coefficients have not yet been estimated.

    c. Make a guess of the sign of the coefficient of research and development expenditures. Explain your prediction.

    d. Why should the CEO be cautious when using this model for revenue estimation and prediction?

14.2 The Coefficient of Determination and Adjusted R^2

Just as for simple linear regression, we will discuss methods that can be used to evaluate the overall effectiveness of multiple regression models. For the pizza delivery model in the previous section, one of the questions to ask is, how do we determine whether the model explains a substantial portion of the variation in the delivery times? The overall effectiveness and usefulness of multiple regression models can be addressed using the coefficient of determination $\left(R^2\right)$ and the adjusted R^2 $\left(R_a^2\right)$ statistics.

Coefficient of Determination (R^2)

Recall our discussion about the coefficient of determination $\left(R^2\right)$ in the previous chapter. In Section 13.5, we defined the R^2 statistic as the statistic that directly measures the degree to which the model explains the dependent variable. In multiple regression, the **coefficient of determination** (sometimes called the **multiple coefficient of determination**), still denoted by R^2, represents the proportion of variation in the response variable, y, that is explained by the set of independent variables, x_1, x_2,..., x_k. R^2 is defined in the same way as for the simple linear regression model.

Formula

Coefficient of Determination

The coefficient of determination, R^2, is given by

$$R^2 = \frac{\text{SSR}}{\text{TSS}} = 1 - \frac{\text{SSE}}{\text{TSS}}$$

where SSR = the sum of squares of regression, TSS = the total sum of squares, and SSE = the sum of squared errors. R^2 represents the proportion of variation in the dependent variable explained by the set of independent variables in a multiple regression model.

Just as for simple linear regression,

$$0 \leq R^2 \leq 1.$$

If all of the slopes are zero $\left(b_i = 0 \text{ for } i = 1,\ 2,...,\ k\right)$ then R^2 is also zero, indicating that there is no relationship between the x_i's and the response variable, y. Similarly, if $y_i = \hat{y}_i$ for all i, the value of the coefficient of determination is one $\left(R^2 = 1\right)$. Note that since we are fitting a multiple regression model, we are not fitting a line but a plane or surface.

The output in Figure 14.1 reveals $R^2 \approx 0.9640$. Thus, in the pizza delivery model, the two independent variables (number of pizzas and distance) can explain approximately 96.4% of the variation in delivery times. Accounting for such a large amount of variation in the dependent variable would seem to demonstrate substantial explanatory power.

It should be noted, however, that a large value of R^2 does not necessarily imply that the fitted model is a useful one. For instance, observations may have been taken at only a few levels of the independent variable(s). Despite a large R^2 in this case, the fitted model may not be useful because most predictions would require extrapolations outside the region of observations. Again, even though R^2 is large, the mean square error (MSE) may still be too large for inferences to be useful when high precision is required. Thus, one should examine more than one statistic when evaluating the adequacy of a regression model.

Adjusted R^2 (R_a^2)

Adding more independent variables to a regression model will always increase the R^2 value. R^2 will never decrease as variables are added because the SSE can never become smaller with the addition of independent variables and the TSS is always the same for a given set of responses. Since R^2 can be made larger by including a large number of independent variables, it is sometimes suggested that a modified measure be used that adjusts for the number of independent variables in the model. The **adjusted coefficient of determination** (denoted by R_a^2) adjusts R^2 by dividing each sum of squares by its associated degrees of freedom. Thus, R_a^2 is given by the following formula.

Formula

Adjusted R^2

The adjusted R^2 statistic takes into account the number of independent variables in the model by dividing each sum of squares by its associated degrees of freedom.

$$R_a^2 = 1 - \left(\frac{n-1}{n-k-1} \right) \frac{\text{SSE}}{\text{TSS}}$$

where n is the number of observations and k is the number of independent variables in the model.

For example, if one were to fit a simple linear regression model to the pizza delivery time using only number of pizzas as the independent variable, we would get $R^2 = 0.9265$ and $R_a^2 = 0.9233$. However, when we add another independent variable, distance, to the model, we have $R^2 = 0.9640$ and $R_a^2 = 0.9607$. In this case, the value of R^2 increased by 0.0375, indicating that adding the variable distance to the model helped explain more variability in delivery times. On the other hand, the value of R_a^2 increased by slightly less (0.0374). As the number of independent variables increases, the difference between the R^2 and adjusted R^2 values also increases. R_a^2 is commonly used as a method of comparison between multiple regression models when one is attempting to find the model that best fits the data. Unlike the R^2 value, the adjusted coefficient of determination may actually become smaller when another independent variable is added to the model. Thus, the adjusted R^2 value is most useful when comparing multiple regression models with different numbers of independent variables.

14.2   Exercises

Basic Concepts

1. What is the purpose of the R^2 and adjusted R^2 statistics?

2. What values can the coefficient of determination take?

3. If a particular regression model explains 68% of variation in the dependent variable, what is the value of R^2?

4. If the coefficient of determination has a value of zero, is it possible for a regression coefficient to have a value other than zero? Explain why.

5. If the coefficient of determination has a value of one, what is the relationship between the sum of squares of regression and the total sum of squares? Explain why.

6. Does a large value of R^2 always indicate that the fitted model is useful? Explain.

7. Explain the difference between R^2 and adjusted R^2.

8. Explain why the adjusted R^2 statistic is sometimes a better measure to use to evaluate the fit of a regression model.

9. Will there ever be a situation in which the adjusted R^2 statistic is greater than R^2 statistic? Explain your answer.

Exercises

10. Consider the following ANOVA table for a multiple regression model relating housing prices (in thousands of dollars) to the number of bedrooms in the house and the size of the lot on which the house was built (in square feet). There were 88 total observations.

$$\text{Estimated Price} = 63.26 + 57.31(\text{Bedrooms}) + 0.0029(\text{Lot Size})$$

10	ANOVA					
11		df	SS	MS	F	Significance F
12	Regression	2	309148.8902	154574.4451	21.58486376	2.62677E-08
13	Residual	85	608705.618	7161.242565		
14	Total	87	917854.5083			

a. Identify the values of SSR, SSE, and TSS from the table.

b. What is the coefficient of determination for this model? Interpret this value in terms of the problem.

c. What is R_a^2 ? Interpret this value.

d. Compare the R^2 and R_a^2 values. Which value should be used to evaluate the fit of the multiple regression model? Explain why.

11. Suppose an additional variable, Square Feet, was added to the housing price model from Exercise 10. The summary output is given below.

	A	B	C	D	E	F	G
1	SUMMARY OUTPUT						
2							
3	*Regression Statistics*						
4	Multiple R	0.819976968					
5	R Square	0.672362228					
6	Adjusted R Square	0.660660879					
7	Standard Error	59.83347988					
8	Observations	88					
9							
10	ANOVA						
11		*df*	*SS*	*MS*	*F*	*Significance F*	
12	Regression	3	617130.7018	205710.2339	57.46023188	2.69597E-20	
13	Residual	84	300723.8065	3580.045315			
14	Total	87	917854.5083				
15							
16		*Coefficients*	*Standard Error*	*t Stat*	*P-value*	*Lower 95%*	*Upper 95%*
17	Intercept	-21.7703086	29.47504196	-0.738601446	0.462207782	-80.38466199	36.84404478
18	Bedrooms	13.85252186	9.010145446	1.537435988	0.127945059	-4.065140472	31.7701842
19	Lot Size	0.002067707	0.000642126	3.220095719	0.001822929	0.000790769	0.003344644
20	Square Feet	0.122778185	0.013237407	9.275092996	1.65802E-14	0.096454149	0.149102222

a. What is R_a^2 for this model?

b. How does the adjusted R^2 value for this model compare to the adjusted R^2 value for the model in Exercise 10?

c. Do you think adding the additional independent variable, Square Feet, improved the model? Explain your answer.

12. The owner of a new pizzeria in town wants to study the relationship between weekly revenues and advertising expenditures. Both measures were recorded in thousands of dollars. The computer output for the simple linear regression model is given below.

	A	B	C	D	E	F	G
1	SUMMARY OUTPUT						
2							
3	*Regression Statistics*						
4	Multiple R	0.858179902					
5	R Square	0.736472743					
6	Adjusted R Square	0.692551534					
7	Standard Error	1.058296197					
8	Observations	8					
9							
10	ANOVA						
11		*df*	*SS*	*MS*	*F*	*Significance F*	
12	Regression	1	18.78005496	18.78005496	16.76804334	0.006394067	
13	Residual	6	6.719945042	1.11999084			
14	Total	7	25.5				
15							
16		*Coefficients*	*Standard Error*	*t Stat*	*P-value*	*Lower 95%*	*Upper 95%*
17	Intercept	74.69887795	7.104358625	10.51451396	4.34789E-05	57.31513863	92.08261726
18	Advertising Expenditures	1.854820243	0.452960815	4.094880138	0.006394067	0.746465058	2.963175428

a. Write the estimated regression equation.

b. What is the coefficient of determination for this model? Interpret this value.

c. What is the value of the adjusted R^2 statistic? Is this statistic useful for the pizzeria owner as he studies this model? Explain.

d. Do you believe this model is useful in explaining revenues based on advertising expenditures? Explain your answer.

e. How could the restaurant owner improve this model? Are there other independent variables that he should consider including?

13. The owner of the pizzeria discussed in Exercise 12 wishes to build on the model relating revenues to advertising expenditures by breaking the advertising expenditures into three categories: television advertising, newspaper advertising, and direct mail advertising.

   a. Write the new regression model in terms of television, newspaper, and mail expenditures. Assume the coefficients have not yet been estimated.

   b. Consider the following summary output for the new model. Write the estimated multiple regression equation.

	A	B	C	D	E	F	G
1	SUMMARY OUTPUT						
2							
3	*Regression Statistics*						
4	Multiple R	0.967040091					
5	R Square	0.935166537					
6	Adjusted R Square	0.88654144					
7	Standard Error	0.64289449					
8	Observations	8					
9							
10	ANOVA						
11		*df*	*SS*	*MS*	*F*	*Significance F*	
12	Regression	3	23.8467467	7.948915566	19.23217829	0.007708883	
13	Residual	4	1.653253302	0.413313326			
14	Total	7	25.5				
15							
16		*Coefficients*	*Standard Error*	*t Stat*	*P-value*	*Lower 95%*	*Upper 95%*
17	Intercept	73.93199827	4.523870838	16.34264127	8.20538E-05	61.37171922	86.49227731
18	Television	2.383047934	0.318133378	7.490719616	0.001698799	1.499768074	3.266327793
19	Newspaper	1.454439994	0.355820285	4.087569076	0.015004989	0.466524505	2.442355483
20	Mail	1.815990841	0.276487962	6.568064755	0.002780349	1.048337191	2.58364449

   c. Interpret the coefficient for television advertising expenditures. Remember that revenues and expenditures are in thousands of dollars.

   d. What is the adjusted coefficient of determination? Interpret this value.

   e. How does the coefficient of determination of this model compare to the coefficient of determination for the simple linear regression model in Exercise 12? Does this appear to be a more useful model? Explain.

   f. What is the value of the R^2 statistic for this model? Should we use the R^2 value or the adjusted R^2 value when evaluating the usefulness of this model? Explain why.

14.3 Interpreting the Coefficients of the Multiple Regression Model

In interpreting the coefficients of the model, we ask the question, *do the signs and magnitudes of the estimated coefficients appear to be reasonable?*

In the simple linear regression model, the estimated coefficient, b_1, is the slope of the line. It is interpreted to be the average change in the dependent variable associated with a one-unit change in the independent variable. This interpretation remains basically valid for the multiple regression model as well. For the pizza delivery model, the coefficient b_1 is the estimated change in delivery time for a one-unit increase in the number of pizzas, given that distance traveled is constant. Is it reasonable to believe that each additional pizza would add approximately 1.6 minutes to the delivery time? While the delivery time varies, 1.6 minutes seems sensible.

The coefficient b_2 is the estimated change in delivery time for a one unit increase in distance (measured in miles), given a specific number of pizzas (i.e. the number of pizzas to be delivered is constant). That is, for each additional mile added to the delivery, we should expect the average delivery time to increase by approximately 1.57 minutes. All other conditions being equal, do we find that the signs of the coefficients are reasonable? If we add more pizzas and distance traveled to the delivery route, it seems reasonable to expect the delivery time to increase. Thus, the positive signs on the coefficients seem to make sense. Were the signs of the coefficients unexpected, a reasonable question would be, *is the estimate accurate?* Are there other factors that have not been considered that could reasonably change the signs of the coefficients? We will consider this question later in Section 14.8.

14.3 Exercises

Basic Concepts

1. In the simple linear regression model, what is the interpretation of b_1? Does this interpretation change in the multiple regression model?

2. When interpreting the coefficient of an independent variable in a multiple regression model, what assumption are we making regarding the other independent variables?

3. What two aspects of the model coefficients are usually analyzed first when studying a multiple regression model?

Exercises

4. Consider the following estimated multiple regression equation relating the number of study hours and GPA to a student's ACT score.

$$\text{Estimated ACT Score} = 8.35 + 1.53(\text{Study Hours}) + 0.30(\text{GPA})$$

   a. Identify the values of b_0, b_1, and b_2.
   b. Interpret the value of b_0 in terms of the problem.
   c. Interpret the value of b_1 in terms of the problem.
   d. Interpret the value of b_2 in terms of the problem.

5. Consider the following estimated regression model relating annual salary to years of education and work experience, which was presented in Section 14.1, Exercise 7.

$$\text{Estimated Salary} = 11214.20 + 2854.89(\text{Education}) + 839.64(\text{Experience})$$

a. Consider the coefficient for the education variable. Do the sign and magnitude of the coefficient seem to make sense? Explain.

b. Consider the coefficient for the experience variable. Do the sign and magnitude of the coefficient seem to make sense? Explain.

c. Interpret the regression coefficient for years of experience.

d. Suppose an employee with 8 years of education (note that education years are the number of years after 8th grade) has been with the company for 5 years. According to this model, what is his estimated annual salary?

e. How would you expect his salary to change if he stays at the company for another year?

f. Suppose two employees at the company have been working there for five years. One has a bachelor's degree (8 years of education) and one has a master's degree (10 years of education). Which employee would you expect to earn a higher salary? How much more money would you expect him to make?

6. Suppose the owner of a car dealership wishes to study how certain factors affect the number of new cars sold. Specifically, he wishes to construct a multiple regression model relating car price, average income per capita in the surrounding area, and the interest rate to the quantity of new cars sold. After compiling historical data, he obtains the following summary output for the multiple regression model. In the original data, car price and average income per capita were in thousands of dollars, and interest rates were reported as percentages.

	A	B	C	D	E	F	G
1	SUMMARY OUTPUT						
2							
3	*Regression Statistics*						
4	Multiple R	0.581495041					
5	R Square	0.338136482					
6	Adjusted R Square	0.305043307					
7	Standard Error	277.5372802					
8	Observations	64					
9							
10	ANOVA						
11		*df*	*SS*	*MS*	*F*	*Significance F*	
12	Regression	3	2361116.922	787038.9742	10.21771025	1.57438E-05	
13	Residual	60	4621616.515	77026.94192			
14	Total	63	6982733.438				
15							
16		*Coefficients*	*Standard Error*	*t Stat*	*P-value*	*Lower 95%*	*Upper 95%*
17	Intercept	-308.6097834	971.3332832	-0.317717707	0.751802188	-2251.565634	1634.346068
18	Price	-18.29845156	6.196928308	-2.952826085	0.004489897	-30.69415375	-5.902749358
19	Income	458.1641011	137.8510094	3.323618036	0.001518198	182.4210273	733.907175
20	Interest Rate	-26.29956832	11.93107528	-2.204291541	0.031350669	-50.16527221	-2.43386442

a. Write the estimated multiple regression equation.

b. Consider the coefficient of the price variable. Do you think the magnitude and sign of this coefficient make sense? Explain your answer.

c. Consider the coefficient of the income variable. Do you think the magnitude and sign of this coefficient make sense? Explain your answer.

d. Consider the coefficient of the interest rate variable. Do you think the magnitude and sign of this coefficient make sense? Explain your answer.

e. How many additional cars would the dealership owner expect to sell if per capita income for the area increased by $1000?

f. How would the dealership owner expect the quantity of cars sold to change if the interest rate increased by 2 percentage points?

14.4   The *F*-Distribution

In Chapter 12 the *F*-distribution was used in conjunction with the *F*-test to determine if there was a significant difference among population means when performing an analysis of variance. We will also use the *F*-statistic to determine if an overall regression model is significant. In general, the *F*-statistic is a ratio of explained variation to unexplained variation. Let's review the characteristics of the *F*-distribution.

The *F*-distribution is not symmetrical, rather, it is skewed to the right. Like the *t*-distribution, its parameters are degrees of freedom. *F*-distributions are associated with test statistics that are quotients. What distinguishes the *F*-distribution is that it has a pair of values for its degrees of freedom. The number of degrees of freedom associated with the numerator is df_{num} and the number of degrees of freedom associated with the denominator is df_{den}.

The shape of the *F*-distribution is a function of its degrees of freedom. In Figure 14.2, the *F*-distributions for $df_{num} = 6$, $df_{den} = 40$ and $df_{num} = 10$, $df_{den} = 4$ are given.

F-Distributions

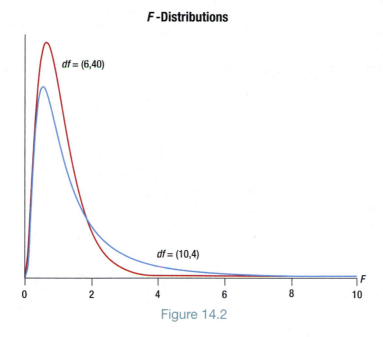

Figure 14.2

Like the *t*-distribution, tables have been compiled (see Appendix A) that show critical values of the *F*-distribution at common levels of significance. The notation F_{α} will indicate the value of *F* such that α of the area under the curve lies to the right of this value.

Example 14.1

Find a point on the F-distribution with $df_{\text{num}} = 4$ and $df_{\text{den}} = 20$ such that 0.05 of the area lies to the right of this value. (See Figure 14.3.)

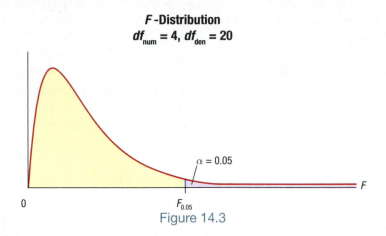

F-Distribution
$df_{\text{num}} = 4$, $df_{\text{den}} = 20$

$\alpha = 0.05$

0          $F_{0.05}$ F

Figure 14.3

Solution

Use the F-table corresponding to $\alpha = 0.05$. The leftmost column of the table corresponds to the degrees of freedom in the denominator, while the top row of the table corresponds to the degrees of freedom in the numerator. Looking up the critical value in the 0.05 table corresponding to $df_{\text{num}} = 4$ and $df_{\text{den}} = 20$ produces an F-value of 2.8661.

Numerator Degrees of Freedom

$\alpha = 0.05$	1	2	3	4	5	...
1	161.4476	199.5000	215.7073	224.5832	230.1619	
2	18.5128	19.0000	19.1643	19.2468	19.2964	
3	10.1280	9.5521	9.2766	9.1172	9.0135	
⋮						
20	4.3512	3.4928	3.0984	2.8661	2.7109	
21	4.3248	3.4668	3.0725	2.8401	2.6848	
⋮						

Denominator Degrees of Freedom (left side label)

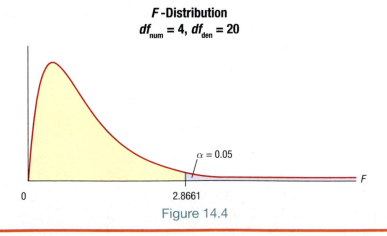

F-Distribution
$df_{\text{num}} = 4$, $df_{\text{den}} = 20$

$\alpha = 0.05$

0   2.8661 F

Figure 14.4

Example 14.2

Find a point on the *F*-distribution with 4 numerator degrees of freedom and 10 denominator degrees of freedom such that 0.01 of the area lies to the right.

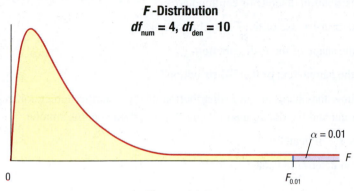

F-Distribution
$df_{num} = 4$, $df_{den} = 10$

Figure 14.5

Solution

Using the *F*-table for $\alpha = 0.01$ corresponding to $df_{num} = 4$ and $df_{den} = 10$ results in an *F*-value of 5.9943.

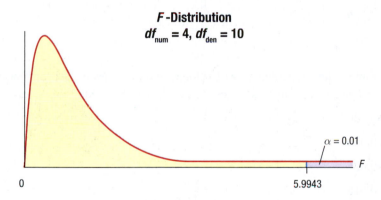

F-Distribution
$df_{num} = 4$, $df_{den} = 10$

Figure 14.6

14.4   Exercises

Basic Concepts

1. Is the F-distribution discrete or continuous?

2. What is the primary use of the F-distribution?

3. Describe the shape of the F-distribution.

4. What are the parameters of the F-distribution?

5. Describe how the shape of the F-distribution changes as the numerator degrees of freedom become larger and the denominator degrees of freedom become smaller.

6. What does F_α indicate?

7. How do you determine values of F_α?

Exercises

8. Find a point on the F-distribution with 7 numerator degrees of freedom and 22 denominator degrees of freedom such that the following area lies to the right of this value.

    a. $\alpha = 0.100$

    b. $\alpha = 0.050$

    c. $\alpha = 0.025$

    d. $\alpha = 0.010$

9. Find a point on the F-distribution with 30 numerator degrees of freedom and 8 denominator degrees of freedom such that the following area lies to the right of this value.

    a. $\alpha = 0.100$

    b. $\alpha = 0.050$

    c. $\alpha = 0.025$

    d. $\alpha = 0.010$

10. Find $F_{0.025}$ for an F-distribution with the following parameters.

    a. 1 numerator degree of freedom, 25 denominator degrees of freedom

    b. 6 numerator degrees of freedom, 11 denominator degrees of freedom

    c. 8 numerator degrees of freedom, 40 denominator degrees of freedom

    d. 3 numerator degrees of freedom, 18 denominator degrees of freedom

11. Find $F_{0.010}$ for an F-distribution with the following parameters.

    a. 15 numerator degrees of freedom, 19 denominator degrees of freedom

    b. 10 numerator degrees of freedom, 29 denominator degrees of freedom

    c. 60 numerator degrees of freedom, 24 denominator degrees of freedom

    d. 12 numerator degrees of freedom, 21 denominator degrees of freedom

14.5   Inference Concerning the Multiple Regression Model and its Coefficients

When fitting a multiple regression model, we must ask the question, *is the overall model worthwhile?*

If the model has no redeeming merit, then the set of independent variables will not be related to the dependent variable. Consider the multiple regression model

$$y_i = \beta_0 + \beta_1 x_{1i} + \beta_2 x_{2i} + \cdots + \beta_k x_{ki} + \varepsilon_i.$$

Suppose all of the $\beta_1 = \beta_2 = \cdots = \beta_k = 0$, then the regression equation becomes

$$y_i = \beta_0 + 0 x_{1i} + 0 x_{2i} + \cdots + 0 x_{ki} + \varepsilon_i.$$

Since multiplying any number by 0 results in 0, all of the independent variables disappear from the model, and we are left with

$$y_i = \beta_0 + \varepsilon_i,$$

which says that y is a constant, β_0, plus a random error, ε_i. So, if $\beta_1 = \beta_2 = \cdots = \beta_k = 0$, then the model is not useful. We will develop a methodology for testing the hypotheses

$$H_0: \beta_1 = \beta_2 = \cdots = \beta_k = 0$$
$$H_a: \text{At least one } \beta_i \neq 0.$$

Steps in the Test of Hypothesis

Step 1: State the hypotheses in plain English.

- **Null Hypothesis:** The overall model is not useful in explaining variation in the dependent variable.

- **Alternative Hypothesis:** The overall model is useful in explaining variation in the dependent variable. That is, at least one of the coefficients is different from zero, and thus, the model can be useful in explaining the variation in y.

Step 2: Select the appropriate statistical measure.

If some of the model's independent variables are useful predictors of y, then the coefficients, β_i, of these variables will have nonzero values.

Step 3: Determine whether the hypothesis should be one-sided or two-sided.

This determination will be postponed until the test statistic is presented in Step 6.

Step 4: Specify the hypotheses using the appropriate statistical measure.

$H_0: \beta_1 = \beta_2 = \cdots = \beta_k = 0$ (Implies the overall model is not useful. Specifically, that the independent variables (as a group) are not useful predictors of the dependent variable.)

H_a: At least one $\beta_i \neq 0$.   (Implies at least one of the independent variables is a useful predictor of the dependent variable.)

Step 5: Specify the level of the test.

The level of the test is specified in the problem or arbitrarily chosen by the analyst.

Step 6: Select the appropriate test statistic.

In all of the previous hypothesis tests, the test statistic involved the sample estimate of the measure specified in Step 2. This test statistic will be different. If none of the independent variables are useful predictors of y, then the model will not explain much (if any) of the variation in the dependent variable. One way of testing whether the overall model is useful is to examine whether the model explains a sufficient portion of the variation in the dependent variable. We have already studied R^2, which measures the fraction of variation explained by the model. However, the sampling distributions of R^2 and R_a^2 for the multiple regression model are too complex. Instead, we will use the **F-statistic**.

Formula

F-Statistic

The **F-statistic** is the test statistic used to test the hypothesis H_a: At least one $\beta_i \neq 0$.

$$F = \dfrac{\dfrac{\text{Sum of Squares of Regression}}{k}}{\dfrac{\text{Sum of Squared Errors}}{n-(k+1)}}$$

$$= \dfrac{\dfrac{SSR}{k}}{\dfrac{SSE}{n-(k+1)}} = \dfrac{\text{Mean Square Regression}}{\text{Mean Square Error}}$$

The F-statistic is a ratio which compares the variation explained by the model (SSR) to the unexplained variation (SSE). The **sum of squares of regression** (SSR) and **the sum of squared errors** (SSE) were discussed in Section 13.5. The symbol k represents the number of independent variables in the model and n represents the number of observations.

A small value of F means that **mean square regression** (MSR) is small in relation to **mean square error** (MSE). This would mean that the SSR, which is the amount of variation explained by the model, is small in relation to SSE (which represents the variation not explained by the model). A small value of F would indicate that the model is of little value in explaining the variation in the dependent variable, y. On the other hand, a large F-value indicates that the variation explained by the model is large in relation to the unexplained variation. The fundamental question is how large must the F-value be in order to believe the model has some explanatory power. This question is complicated by sampling variation. Even if there is no relationship between the independent variables and the dependent variable, sampling variation will produce a model that explains some portion of the variation in y.

Calculating Degrees of Freedom in Multiple Regression Models

The F-statistic has an F-distribution with k numerator degrees of freedom and $n-(k+1)$ denominator degrees of freedom, where k is the number of independent variables in the model. We will reject the null hypothesis if the F-statistic is larger than the F-value corresponding to a one-tailed test for some prescribed significance level, α.

Step 7: Determine the critical value.

In Example 14.2, we observed that for $df_{num} = 4$, $df_{den} = 10$, and $\alpha = 0.01$, the F-value was 5.9943. If the null hypothesis is true $(\beta_1 = \beta_2 = \cdots = \beta_k = 0)$, and there is no relationship between the independent variables and the dependent variable, sampling variation would cause F to be as large or larger than 5.9943 only 1% of the time. If an F-value larger than 5.9943 is observed, then there is a choice to make. Is the null hypothesis really true, and we have observed a rare phenomenon (occurs less than 1% of the time due to normal sampling variation)? Or is the null hypothesis false? Suppose the following decision rule is used.

The decision rule defines "rareness" for the F-statistic assuming the null hypothesis is true. Essentially what the rule says is that F-values as large or larger than 5.9943 would be too rare for us to believe that the null is reasonable. When such a value occurs, the null hypothesis will be rejected in favor of the alternative.

Step 8: Compute the test statistic.

Fortunately, the F-statistic is computed by virtually every statistical analysis program. It is found in the analysis of variance (ANOVA) table under the heading F in Microsoft Excel, for example.

Step 9: Make the decision.

If the value of the F-statistic falls in the rejection region, then reject the null hypothesis. If not, then we will fail to reject the null hypothesis in favor of the alternative.

Step 10: State the conclusion in terms of the original question.

Anyone trying to build a model for predictive purposes hopes the null hypothesis $(H_0: \beta_1 = \beta_2 = \cdots = \beta_k = 0)$ is rejected, since rejecting the null implies the model can explain some of the variation in the dependent variable.

Example 14.3

For the pizza delivery model from Section 14.1, determine if there is sufficient evidence at the 0.05 level that the overall model is useful in explaining the variation in delivery times.

Solution

Step 1: State the hypotheses in plain English.

- **Null Hypothesis:** The overall model is not useful in explaining the variation in delivery times.

- **Alternative Hypothesis:** The overall model is useful in explaining the variation in delivery times.

Step 2: Select the appropriate statistical measure.

If some of the model's independent variables are useful predictors of y, then the coefficients (β_i) of these variables will have nonzero values.

Step 3: Determine whether the hypothesis should be one-sided or two-sided.

The F-test is always a one-sided test, thus the hypothesis is one-sided.

Step 4: Specify the hypotheses using the appropriate statistical measure.

$H_0: \beta_1 = \beta_2 = 0$ (Implies the overall model is not useful. Specifically, that the independent variables (as a group) are not useful predictors of delivery time.)

$H_a:$ At least one $\beta_i \neq 0$.   (Implies at least one of the variables (number of pizzas or distance) is a useful predictor of delivery time.)

Step 5 Specify the level of the test.

The level of the test is given as $\alpha = 0.05$.

Step 6: Select the appropriate test statistic.

We will use the F-statistic as the test statistic for this hypothesis.

Step 7: Determine the critical value.

Since there are two independent variables (number of pizzas and distance) in the model, the degrees of freedom are $df_{num} = 2$ and $df_{den} = 25 - (2+1) = 22$. The critical value for F at the 0.05 level with 2 numerator degrees of freedom and 22 denominator degrees of freedom is 3.4434.

An F-value larger than 3.4434 will indicate a value of F that is too rare to have occurred by chance if the null were true, and thus we will reject the null hypothesis.

Step 8: Compute the test statistic.

The ANOVA table for the pizza delivery model is given in Figure 14.7. The value of F is given to be approximately 294.7403.

◢	A	B	C	D	E	F
1	SUMMARY OUTPUT					
2						
3	*Regression Statistics*					
4	Multiple R	0.981846095				
5	R Square	0.964021754				
6	Adjusted R Square	0.960751004				
7	Standard Error	3.075694294				
8	Observations	25				
9						
10	ANOVA					
11		*df*	*SS*	*MS*	*F*	*Significance F*
12	Regression	2	5576.424901	2788.212	294.7403	1.30749E-16
13	Residual	22	208.1176986	9.459895		
14	Total	24	5784.5426			

Figure 14.7

Step 9: Make the decision.

Since the value of F, 294.7403, falls in the rejection region, the null hypothesis will be rejected in favor of the alternative.

Step 10: State the conclusion in terms of the original question.

The null hypothesis is rejected in favor of the alternative, which states that at least one of the independent variables is useful in explaining the variation in delivery times.

Testing Hypotheses Concerning Individual β_i

In Section 13.9, we discussed the implications of testing whether the slope coefficient $\beta_1 = 0$. In the case of the simple linear model, if $\beta_1 = 0$, then the independent variable is not related to the dependent variable. However, in multiple regression there is more than one independent variable. In multiple regression models, deciding whether a particular independent variable is a useful predictor of the dependent variable is an important part of the model-building process. For example, is the distance variable useful in predicting delivery time? The rationale for testing whether an independent variable is useful in predicting the dependent variable is the same as that used in testing whether $\beta_1 = 0$ for the simple linear model.

Formula

Test Statistic for Testing the Hypothesis $\beta_i \neq 0$

In a test of hypothesis about an individual coefficient, β_i, the test statistic is computed as

$$t = \frac{b_i - 0}{s_{b_i}} = \frac{b_i}{s_{b_i}},$$

where b_i is the estimated coefficient and s_{b_i} is the standard deviation (standard error) of the estimated coefficient. The test statistic follows a t-distribution with $n - (k+1)$ degrees of freedom.

The technique will be illustrated with an example.

Example 14.4

For the pizza delivery model,

$$\text{Delivery Time} = \beta_0 + \beta_1 (\text{Number of Pizzas}) + \beta_2 (\text{Distance}) + \varepsilon_i.$$

Is the distance variable a useful predictor of delivery time? Use $\alpha = 0.05$ as the level of significance.

Solution

Step 1: State the hypotheses in plain English.

- **Null Hypothesis:** Distance is not a useful predictor of delivery time.
- **Alternative Hypothesis:** Distance is a useful predictor of delivery time.

Step 2: Select the appropriate statistical measure.

For the pizza delivery model, if distance is **not** a useful predictor of delivery time, then its coefficient in the model (β_2) will equal 0. The sample estimate of β_2, namely b_2, will be used to evaluate the reasonableness of the null hypothesis.

Step 3: Determine whether the hypothesis should be one-sided or two-sided.

The hypothesis should be two-sided since the relationship between distance and delivery time can be positive or negative. Thus, we will conduct a two-tailed test.

Step 4: Specify the hypotheses using the appropriate statistical measure.

$$H_0: \beta_2 = 0 \qquad \text{(Implies distance is not a useful predictor of delivery time.)}$$
$$H_a: \beta_2 \neq 0. \qquad \text{(Implies distance is a useful predictor of delivery time.)}$$

Step 5: Specify the level of the test.

The level of the test is specified in the problem to be $\alpha = 0.05$.

Step 6: Select the appropriate test statistic.

The test statistic is $t = \dfrac{b_2 - 0}{s_{b_2}} = \dfrac{b_2}{s_{b_2}}$. The test statistic is identical in form to the test statistic used in Section 13.9 for the simple linear model. If the value of the test statistic is near zero, then there is evidence that the distance variable is not a significant predictor of delivery time.

Step 7: Determine the critical value.

Since the test is two-tailed and the level of significance is specified to be $\alpha = 0.05$, $\alpha/2 = 0.05/2 = 0.025$. The test statistic has a t-distribution with $df = 25 - (2+1) = 22$. The critical values correspond to $t_{0.025, 22} = \pm 2.074$.

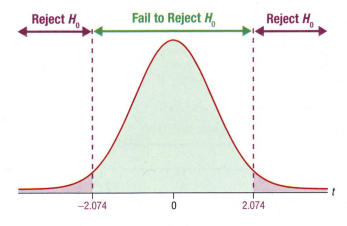

Figure 14.8

Step 8: Compute the test statistic.

The summary output for the pizza delivery model is given in Figure 14.9.

	A	B	C	D	E
1	SUMMARY OUTPUT				
2					
3	*Regression Statistics*				
4	Multiple R	0.981846095			
5	R Square	0.964021754			
6	Adjusted R Square	0.960751004			
7	Standard Error	3.075694294			
8	Observations	25			
9					
10	ANOVA				
11		*df*	*SS*	*MS*	*F*
12	Regression	2	5576.424901	2788.212451	294.7403048
13	Residual	22	208.1176986	9.45989539	
14	Total	24	5784.5426		
15					
16		*Coefficients*	*Standard Error*	*t Stat*	*P-value*
17	Intercept	1.792903307	1.048779122	1.709514681	0.101425505
18	Number of Pizzas	1.589101026	0.156345691	10.16402191	8.97113E-10
19	Distance	1.567708058	0.327534513	4.786390423	8.84845E-05

Figure 14.9

Using the values of b_2 and the standard error of b_2 from summary output,

$$t = \frac{b_2}{s_{b_2}} \approx \frac{1.56771}{0.32753} \approx 4.786.$$

Note: The t-statistic is also given in the t *Stat* column in the Excel summary output.

The estimated value of b_2 is approximately 4.786 standard deviations from zero. Is this persuasive evidence that $\beta_2 \neq 0$?

Step 9: Make the decision.

Since the value of the test statistic falls into the rejection region, we reject the null hypothesis, $H_0: \beta_2 = 0$, at the 0.05 level.

Step 10: State the conclusion in terms of the original question.

Since we rejected the null hypothesis, $H_0: \beta_2 = 0$, the distance variable is a significant predictor of delivery time, given that the other variable currently in the model (number of pizzas) is constant.

We can apply the exact same t-test to the other variable in the model. The estimated coefficient of the number of pizzas variable, b_1, is also significant $(t = 10.164)$. This suggests that the two-variable model is a reasonable model to use to predict delivery times.

Confidence Intervals for Individual Coefficients

Confidence intervals for individual β_i in a multiple regression model are almost identical to the confidence interval for β_1 in the simple regression model introduced in Section 13.8.

Formula

$100(1 - \alpha)\%$ Confidence Interval for an Individual Coefficient, β_i

The $100(1-\alpha)\%$ confidence interval for each coefficient, β_i, in a multiple regression model is given by

$$b_i \pm t_{\alpha/2, df} s_{b_i}.$$

The computation of the confidence interval requires three pieces of information:

1. An estimate of the coefficient β_i, namely b_i.

2. The standard deviation of the estimate, s_{b_i}. Both b_i and s_{b_i} are reported in the summary output.

3. The value of $t_{\alpha/2, df}$, which is the number of standard deviations the endpoint of the interval is from the point estimate. (The degrees of freedom for this t is $n - (k+1)$, which is the same as the degrees of freedom associated with the error terms (residuals) in the ANOVA table.)

Example 14.5

Compute and interpret the 95% confidence interval for β_1 in the pizza delivery model.

Solution

The β_1 model coefficient relates to the number of pizzas variable in the pizza delivery model. The output in Figure 14.10 provides the information necessary to develop a confidence interval for β_1.

	A	B	C	D	E	F	G
1	SUMMARY OUTPUT						
2							
3	*Regression Statistics*						
4	Multiple R	0.981846095					
5	R Square	0.964021754					
6	Adjusted R Square	0.960751004					
7	Standard Error	3.075694294					
8	Observations	25					
9							
10	ANOVA						
11		*df*	*SS*	*MS*	*F*	*Significance F*	
12	Regression	2	5576.424901	2788.212451	294.7403048	1.30749E-16	
13	Residual	22	208.1176986	9.45989539			
14	Total	24	5784.5426				
15							
16		*Coefficients*	*Standard Error*	*t Stat*	*P-value*	*Lower 95%*	*Upper 95%*
17	Intercept	1.792903307	1.048779122	1.709514681	0.101425505	-0.382131469	3.967938082
18	Number of Pizzas	1.589101026	0.156345691	10.16402191	8.97113E-10	1.264859909	1.913342144
19	Distance	1.567708058	0.327534513	4.786390423	8.84845E-05	0.888443052	2.246973064

Figure 14.10

b_1 is given in the output to be 1.5891.

s_{b_1} is given in the output to be 0.1563.

In order to compute the *t*-value, the degrees of freedom must be determined.

$$df = n - (k+1) = 25 - (2+1) = 22$$

For a 95% confidence interval, $t_{\alpha/2, df}$ will be $t_{0.025, 22} = 2.074$.   The resulting confidence interval will be

$$1.5891 \pm 2.074(0.1563)$$
$$1.5891 \pm 0.3242$$
$$1.2649 \text{ to } 1.9133.$$

We are 95% confident that the true value of β_1, the increase in the delivery time for each additional pizza (given that distance is held constant), will be between 1.2649 and 1.9133 minutes.

Notice that Excel automatically generates confidence intervals for each individual coefficient. The endpoints for the 95% confidence intervals are given in the *Lower 95%* and *Upper 95%* columns of the output. Compare the upper and lower limits given by Microsoft Excel to the ones just calculated by hand. Excel has the ability to calculate confidence intervals for individual coefficients for any level of significance.

14.5   Exercises

Basic Concepts

1. If the overall multiple regression model is not useful, what does this tell us about the coefficients of the independent variables?

2. What is the hypothesis being tested when we test to determine if the overall multiple regression model is useful?

3. When testing the overall model, describe the null and alternative hypotheses in plain English.

4. Why is the R^2 value not used in the test statistic for a hypothesis test to determine if a multiple regression model is significant?

5. What is the test statistic used in a hypothesis test to determine if an overall model is significant? What is the distribution of this test statistic?

6. Give two equivalent formulas for the test statistic in a hypothesis test about an overall regression model.

7. Explain the significance of the ratio of the mean square regression to the mean square error.

8. True or false: Even if there is no relationship between any of the independent variables and the dependent variable, sampling variation will explain some portion of the variation in the dependent variable.

9. How are the degrees of freedom calculated for a multiple regression model?

10. When testing the overall model for significance, do you perform a one or two-tailed test?

11. What is the rejection rule in tests of hypothesis for model significance?

12. What is the expression for a confidence interval for an individual coefficient, β_i?

13. Outline the three pieces of information needed to compute a confidence interval for an individual coefficient.

14. What is the test statistic used to test a hypothesis about an individual coefficient in a multiple regression model? How many degrees of freedom are associated with this test statistic?

15. If we fail to reject the null hypothesis in a hypothesis test about an individual coefficient, should this variable remain in the regression model? Explain.

14

Exercises

16. An article appearing in the *Journal of Wildlife Management* summarized the percent fat of 75 arctic foxes. According to the authors, "Storage of fat to provide energy during regular periods of food shortage, or to insulate against low ambient temperatures, is essential for survival in severe arctic homeotherms." Computing percent fat was a laborious process. In one analysis, the author regressed y = percent fat on rump fat thickness (RFT), which was measured in millimeters and was much easier to determine than percent fat. It was noted that a plot of percent fat versus RFT was indeed linear, and the resulting regression equation was $y = 7.40 + 1.36(\text{RFT})$. R^2 and s_e^2 were determined to be 0.88 and 3.70, respectively.

10	ANOVA				
11		df	SS	MS	F
12	Regression	k	SSR	SSR/k	MSR/MSE
13	Residual	$n-(k+1)$	SSE	SSE/$(n-(k+1))$	
14	Total	$n-1$	SST		

Use the table shown to help answer the following questions.

a. Compute the sum of squares of regression and the sum of squared errors. Note that R^2 is SSR/TSS and s_e^2 is the same as MSE.

b. Give the degrees of freedom for the regression and for the error (residual).

c. Compute MSR and MSE.

d. Compute the F ratio for testing the significance of the regression line. With $\alpha = 0.05$, can we conclude that the relationship between percent fat and RFT is significant?

e. Compute a point estimate for percent fat if RFT = 10 millimeters.

f. For a 2-millimeter increase in RFT, what would be the expected change in percent fat?

17. Consider the model from Exercise 7 in Section 14.1 relating annual salary to years of work experience and years of education.

	A	B	C	D	E	F	G
1	SUMMARY OUTPUT						
2							
3	*Regression Statistics*						
4	Multiple R	0.566946595					
5	R Square	0.321428441					
6	Adjusted R Square	0.29192533					
7	Standard Error	10909.996					
8	Observations	49					
9							
10	ANOVA						
11		*df*	*SS*	*MS*	*F*	*Significance F*	
12	Regression	2	2593556200	1296778100	10.89473033	0.000133875	
13	Residual	46	5475288584	119028012.7			
14	Total	48	8068844784				
15							
16		*Coefficients*	*Standard Error*	*t Stat*	*P-value*	*Lower 95%*	*Upper 95%*
17	Intercept	11214.19915	5625.172956	1.993574106	0.052147881	-108.6867382	22537.08504
18	Education (Years)	2854.891271	689.6666061	4.139523715	0.000146836	1466.664395	4243.118147
19	Experience (Years)	839.6360369	261.7094444	3.208275646	0.002433357	312.842248	1366.429826

a. Formulate the hypotheses for testing the multiple regression model for overall significance.

b. Find the value of the test statistic for a hypothesis test about the overall model.

c. Is there evidence at the 5% level of significance that the overall model is useful in predicting annual salary?

d. Consider the coefficient for years of education. Find a 95% confidence interval for the value of β_1. Interpret this interval.

e. Formulate the hypotheses for testing the significance of the coefficient β_1.

f. Is there sufficient evidence at the 0.05 level that years of education is useful in predicting annual salary?

14

18. Consider the printing cost model discussed in Exercise 8 of Section 14.1.

	A	B	C	D	E	F	G
1	SUMMARY OUTPUT						
2							
3	*Regression Statistics*						
4	Multiple R	0.987606014					
5	R Square	0.975365639					
6	Adjusted R Square	0.972467479					
7	Standard Error	0.445885396					
8	Observations	20					
9							
10	ANOVA						
11		*df*	*SS*	*MS*	*F*	*Significance F*	
12	Regression	2	133.8201656	66.91008281	336.5464936	2.12863E-14	
13	Residual	17	3.379834375	0.198813787			
14	Total	19	137.2				
15							
16		*Coefficients*	*Standard Error*	*t Stat*	*P-value*	*Lower 95%*	*Upper 95%*
17	Intercept	6.134155476	3.993435752	1.536059638	0.142925974	-2.291257484	14.55956844
18	Number of Pages	0.010801	0.004147682	2.604105041	0.018522101	0.002050156	0.019551845
19	Number of Copies	-0.009954478	0.005271436	-1.888380579	0.07616193	-0.021076236	0.00116728

a. What percentage of the variation in printing price is explained by the two independent variables number of pages and number of copies?

b. Is the overall model significant at the 1% level?

c. Consider the estimated regression coefficient for the number of pages. Construct a 99% confidence interval for β_1. Interpret this interval.

d. Is the number of pages variable useful in predicting printing cost at the 5% level? Would the decision change at the 1% level?

e. Construct a 95% confidence interval for β_2. Interpret this interval.

f. Is the number of copies useful in explaining the variation in printing cost at the 5% level of significance? Do you think the publisher should consider removing this variable from the model? Explain your answer.

19. The following table contains data from selected cities regarding rental rates of two-bedroom apartments, city populations, and median incomes. Monthly rent is given in dollars, population is given in thousands of people, and median income is given in thousands of dollars. Suppose we wish to build a multiple regression model to predict the cost of rent based on population and median income.

City	Monthly Rent ($)	2010 Population (Thousands)	2010 Median Income (Thousands of Dollars)
Denver, CO	868	600.158	45.438
Birmingham, AL	711	212.237	31.704
San Diego, CA	1414	1307.402	61.962
Gainesville, FL	741	124.354	28.653
Winston-Salem, NC	707	229.617	41.979
Memphis, TN	819	646.889	36.535
Austin, TX	966	790.390	50.236
Seattle, WA	1219	608.660	58.990
Richmond, VA	735	204.214	37.735
Charleston, SC	812	120.083	47.799
College Park, MD	1407	30.413	66.900
Savannah, GA	789	136.286	33.778
Minneapolis, MN	988	382.578	45.625
Detroit, MI	805	713.777	29.447
Baton Rouge, LA	827	229.493	35.436

Table for Exercise 19 – Monthly Rent, Population, and Median Income in Selected Cities

Source: U.S. Census Bureau

a. Write the multiple regression model in terms of rent, population, and income. Assume the regression coefficients have not yet been estimated.

b. Predict the signs of the coefficients β_1 and β_2. Explain your answers.

c. Using statistical software, estimate the multiple regression equation. Identify the values of b_0, b_1, and b_2 and write the estimated multiple regression equation. Interpret the estimated coefficients.

d. At the 1% level of significance, is the overall model useful in predicting monthly rent? Identify the test statistic for this test.

e. Find a 95% confidence interval for β_2. Interpret this interval.

f. Determine if each independent variable is related to the dependent variable at the 0.05 level of significance.

g. Should we consider removing any independent variables from this regression model? If yes, identify the variable(s) that should be removed and explain why.

14

20. Using the information from Exercise 19, estimate the simple linear regression equation relating monthly rent to median income only.

    a. Write the estimated simple regression equation.

    b. Is the simple linear regression model significant at $\alpha = 0.01$?

    c. Is median income related to the monthly rental rate at $\alpha = 0.01$? Identify the test statistic used in this hypothesis test.

    d. What percent of the variation in monthly rent is explained by median income? Compare this to the percent of variation in monthly rent explained by both population and median income in Exercise 19.

    e. Which model do you think is a better model to use to predict monthly rental rates? Explain your answer.

14.6 Inference Concerning the Model's Prediction

Many regression models are developed solely to predict the dependent variable. To use the multiple regression model for prediction, insert the values of the independent variables in the model and calculate the predicted value. Recall the estimated multiple regression equation for our pizza delivery model:

$$\text{Delivery Time} = 1.7929 + 1.5891(\text{Number of Pizzas}) + 1.5677(\text{Distance}).$$

For 5 pizzas being delivered 6 miles away, the model would predict the delivery time to be

$$\text{Estimated Delivery Time} = 1.7929 + 1.5891(5) + 1.5677(6) = 19.1446 \text{ minutes}.$$

This is the point estimate. How good is this estimate? The answer to this question depends on what you are trying to predict. Are you trying to predict the average delivery time for delivering 5 pizzas 6 miles away, or are you trying to predict the delivery time for a particular order?

Confidence Interval for the Mean Value of y Given x

In Section 13.10, we discussed a confidence interval for the mean value of y given $x = x_p$ for the simple regression model. In our multiple regression model, the point estimate, 19.1446 minutes, is the average value of y given $x_1 = 5$ and $x_2 = 6$. In other words, the delivery time 19.1446 minutes is the estimated average delivery time for *all* deliveries of 5 pizzas to locations 6 miles away. Since we do not have all the deliveries in the sample, the predicted average delivery time of 19.1446 is only an estimate of the true mean value. How good is the estimate? For multiple regression, the expression for the confidence interval of the mean value of y given x is beyond the scope of this text.

Fortunately, statistical analysis programs such as Minitab will compute a confidence interval for the mean value of y given x for the multiple regression model. (See Figure 14.11.)

```
Regression Analysis: Delivery Time versus Number of Pizzas, Distance

The regression equation is
Delivery Time = 1.79 + 1.59 Number of Pizzas + 1.57 Distance


Predictor              Coef   SE Coef      T      P
Constant              1.793     1.049   1.71  0.101
Number of Pizzas     1.5891    0.1563  10.16  0.000
Distance (Miles)     1.5677    0.3275   4.79  0.000


S = 3.07569   R-Sq = 96.4%   R-Sq(adj) = 96.1%


Analysis of Variance

Source             DF       SS      MS       F      P
Regression          2   5576.4  2788.2  294.74  0.000
Residual Error     22    208.1     9.5
Total              24   5784.5

Predicted Values for New Observations

New Obs     Fit  SE Fit         95% CI              95% PI
      1  19.145   1.328  (16.391, 21.898)   (12.197, 26.092)


Values of Predictors for New Observations

          Number
              of  Distance
New Obs   Pizzas    (Miles)
      1     5.00       6.00
```

Figure 14.11

According to the Minitab output given in Figure 14.11, the 95% confidence interval for the mean delivery time for 5 pizzas being delivered 6 miles away is 16.391 minutes to 21.898 minutes.

16.391   19.1446   21.898

Confidence Interval for the Predicted Value of *y* Given *x*

A caller asks to speak to the manager of the pizza restaurant. The caller wants to know how long it would take to deliver five pizzas to his specific location, which is six miles away. As the manager, you would like to guarantee how long it will take to make this delivery of 5 pizzas to this customer. You are not especially interested in the *average* delivery time for such a delivery. Instead, it would be preferable to create a confidence interval for the time it is going to take for this particular order to be delivered. Once again, for multiple regression, the expression for the prediction interval is beyond the scope of this text. Fortunately, statistical analysis programs such as Minitab will also produce a prediction interval. Using the output shown in Figure 14.11, the 95% prediction interval for the delivery of 5 pizzas to a location 6 miles away is 12.197 minutes to 26.092 minutes. As we observed in Section 13.10, to account for individual variation, the prediction interval for *y* given *x* is substantially wider than the confidence interval for the mean value of *y* given *x*.

Confidence Interval for Average Delivery Time

16.391   19.1446   21.898

Prediction Interval for Individual Delivery Time

12.197   19.1446   26.092

Can the model make a useful prediction of the delivery time? Although the model has an R^2 of 0.9640, the prediction interval is fairly wide. This indicates that not a great deal of confidence can be placed in the estimated value of 19.1446 minutes as a delivery time for 5 pizzas traveling 6 miles.

14.6   Exercises

Basic Concepts

1. What is a point estimate for a multiple regression model?

2. Explain how a point estimate is interpreted as an "average" value.

3. Distinguish between a confidence interval and a prediction interval for a multiple regression model.

4. What is the price that is paid when making predictions regarding individual values?

5. Suppose an estimated multiple regression model, $\hat{y}_i = b_0 + b_1 x_{1i} + b_2 x_{2i}$, produces a 95% confidence interval of (3.292, 7.072) and a 95% prediction interval of (0.364, 10.000) when $x_1 = 6$ and $x_2 = 6$. Interpret both of these intervals.

Exercises

6. In Section 13.5, the SAT scores and graduating GPAs of 27 students were given. In Chapter 13, we estimated a model using only total SAT score to predict graduating GPA. However, with multiple regression, the SAT critical reading and SAT math scores can be treated as separate variables in the model. Computer output of the model

$$GPA = \beta_0 + \beta_1(\text{SAT Reading}) + \beta_2(\text{SAT Math}) + \varepsilon_i$$

is given.

```
Regression Analysis: Graduating GPA versus SAT Critical Rea, SAT Math

The regression equation is
Graduating GPA = 1.35 + 0.00151 SAT Critical Reading + 0.00134 SAT Math


Predictor                   Coef    SE Coef     T      P
Constant                  1.3480     0.5760   2.34  0.028
SAT Critical Reading    0.001510   0.001757   0.86  0.399
SAT Math                0.001341   0.001337   1.00  0.326


S = 0.452511    R-Sq = 19.5%    R-Sq(adj) = 12.7%


Analysis of Variance

Source            DF     SS      MS      F      P
Regression         2  1.1869  0.5934   2.90  0.075
Residual Error    24  4.9144  0.2048
Total             26  6.1013


Predicted Values for New Observations

New Obs    Fit  SE Fit      95% CI            95% PI
      1  2.7736  0.1416  (2.4814, 3.0659)  (1.7950, 3.7522)


Values of Predictors for New Observations

          SAT Critical   SAT
New Obs         Reading  Math
      1             500   500
```

a. Find the standard deviation of the error terms in the output.

b. Interpret the coefficient of SAT Critical Reading. What would it mean if the coefficient was negative?

c. Determine if the overall model is useful in explaining GPA. Test at the 0.05 level.

d. What fraction of the variation in GPA is explained by the model?

e. Determine if the SAT Critical Reading variable is a useful predictor of GPA. Test at the 0.05 level.

f. The output includes a predicted GPA for someone scoring 500 on both the SAT Critical Reading and SAT Math portions. Find the predicted value in the output.

g. What is the average GPA for an individual who scored 500 on both the SAT Critical Reading and SAT Math sections? Find the 95% confidence interval for this average. Interpret this interval.

h. Suppose your nephew scored 500 on both the critical reading and math sections. What would be the model's prediction for his graduating GPA? Find the 95% prediction interval for your nephew in the output. Interpret this interval.

i. Why is the prediction interval so much wider than the confidence interval in part **g.**?

j. Summarize the strengths and weaknesses of the estimated model.

7. How tall will your child be? A researcher has collected a random sample of heights of parents and their female children (all heights are in inches). The heights of the mother, father, and daughter are recorded in the following table.

Table for Exercise 7 – Heights of Parents and Daughters (Inches)												
Mother	64	66	62	70	70	58	66	66	64	67	65	66	68
Father	73	70	72	72	72	63	75	75	72	69	77	70	74
Daughter	65	65	61	69	67	59	69	70	68	70	70	65	70

a. Create two scatterplots using the mother with the daughter and the father with the daughter. Does there appear to be a linear relationship in either of the plots?

b. Using statistical software, estimate the parameters of the following regression model.

$$\text{Daughter Height} = \beta_0 + \beta_1 (\text{Mother Height}) + \beta_2 (\text{Father Height}) + \varepsilon_i$$

c. Is the overall model useful in explaining the variation in daughter height? Test at the 0.05 level.

d. Is the father's height useful in explaining the daughter's height? Test at the 0.05 level.

e. Is the mother's height useful in explaining the daughter's height? Test at the 0.01 level.

f. Interpret each of the regression coefficients.

g. Construct and interpret 95% confidence intervals for β_1 and β_2. Interpret these intervals.

h. Predict the height of a daughter whose father is six feet two inches tall and whose mother is five feet four inches tall.

i. Find a 95% prediction interval for the height of a daughter whose father is six feet two inches tall and whose mother is five feet four inches tall. Interpret this interval.

j. Find a 95% confidence interval for the average height of a daughter whose father is six feet two inches tall and whose mother is five feet four inches tall.

8. On Sunday, January 2, 2011, 16 games were played in the National Football League. The number of rushing yards, passing yards, first downs, and points for the 32 teams participating in these games is given in the table.

Team	Rushing Yards	Passing Yards	First Downs	Points
Table for Exercise 8 – Team Data: January 2, 2011				
Miami	44	240	16	7
New England	181	321	24	38
Buffalo	37	130	6	7
New York	276	119	17	38
Cincinnati	90	305	20	7
Baltimore	98	125	10	13
Pittsburgh	100	325	24	41
Cleveland	43	210	17	9
Oakland	209	160	21	31
Kansas City	115	142	17	10
Minnesota	74	145	16	13
Detroit	107	258	22	20
Carolina	137	182	12	10
Atlanta	99	256	24	31
Tampa Bay	84	255	18	23
New Orleans	106	212	20	13
Jacksonville	198	140	23	17
Houston	244	253	22	34
Dallas	159	127	14	14
Philadelphia	121	162	14	13
New York	82	243	14	17
Washington	67	336	20	14
San Diego	164	313	20	33
Denver	146	205	18	28
Arizona	78	242	19	7
San Francisco	100	276	16	38
Chicago	110	168	13	3
Green Bay	60	229	14	10
Tennessee	51	300	17	20
Indianapolis	101	264	24	23
Saint Louis	47	155	10	6
Seattle	141	192	19	16

Source: National Football League

a. In order to predict a team's points from rushing yards, passing yards, and first downs, a multiple regression analysis is performed on the data with points as the dependent variable. The associated regression output is given. Write the estimated regression equation for predicting points based on the three predictor variables.

	A	B	C	D	E	F
1	SUMMARY OUTPUT					
2						
3	*Regression Statistics*					
4	Multiple R	0.774540403				
5	R Square	0.599912836				
6	Adjusted R Square	0.557046354				
7	Standard Error	7.508433133				
8	Observations	32				
9						
10	ANOVA					
11		*df*	*SS*	*MS*	*F*	*Significance F*
12	Regression	3	2366.956093	788.9853643	13.99492	9.23535E-06
13	Residual	28	1578.543907	56.37656811		
14	Total	31	3945.5			
15						
16		*Coefficients*	*Standard Error*	*t Stat*	*P-value*	
17	Intercept	-15.6150327	5.982932568	-2.609929582	0.014378	
18	Rushing Yards	0.127133936	0.029615758	4.292780033	0.000191	
19	Passing Yards	0.081411959	0.029307161	2.777886231	0.009655	
20	First Downs	0.121492053	0.460140758	0.264032366	0.793689	

b. Find the standard deviation of the error terms in the output.

c. Determine if the overall model is useful in predicting points scored. Use $\alpha = 0.05$.

d. What fraction of the total variation in points is explained by the model?

e. Is the rushing yards variable useful in predicting points scored at the 0.01 level?

f. Is the passing yards variable useful in predicting points scored at the 0.01 level?

g. Is the first downs variable useful in predicting points scored at the 0.01 level?

h. The coefficient of rushing yards in the regression equation is 0.1271. Interpret this value.

i. Should any variables be removed from this model? Explain.

9. In the previous exercise, total points was predicted based on rushing yards, passing yards, and first downs. It is noted from the summary output that both rushing yards and passing yards have P-values of less than 0.01. However, first downs does not appear to be significant as an independent variable. Perhaps a simpler model would be better.

a. Using the data from the previous exercise, estimate the regression equation

$$\text{Points} = \beta_0 + \beta_1 \left(\text{Rushing Yards} \right) + \beta_2 \left(\text{Passing Yards} \right) + \varepsilon_i.$$

b. Is the overall model significant in predicting total points? Test at $\alpha = 0.01$.

c. What percentage of the variation in total points is explained by rushing yards and passing yards? Compare this to the percentage of the variation in total points that was explained by the three independent variables rushing yards, passing yards, and first downs.

d. Which model do you think would be better to use for estimation and prediction of total points; the model from Exercise 8 or the model in this exercise? Explain your answer.

e. Suppose that in preparation for the upcoming game against Miami, the coach of Buffalo wishes to predict the points that will be scored. He has studied Miami's defense in previous games, and predicts that the Buffalo offense will have approximately 102 rushing yards and 63 passing yards. How many points, according to the model, should Buffalo score in the next game?

f. Construct at 95% confidence interval for the average number of points that will be scored in the game against Miami. Interpret this interval.

g. Construct a 95% prediction interval for the number of points that will be scored in the game against Miami. Interpret this interval.

14.7 Models with Qualitative Independent Variables

Throughout Chapter 13 and this chapter, we have discussed quantitative variables in the regression models. Quantitative variables take on values on a well-defined scale, such as number of pizzas, miles to destination, income, age, and temperature. Many variables of interest in business and economics, however, are not quantitative, but qualitative. Examples of qualitative variables are gender (male or female), firm size (small, medium, or large), and type of investment (stock or mutual fund).

In order to use qualitative variables in regression analysis, we need to identify the classes of the qualitative variable quantitatively. To do this, we will use **indicator** (or **dummy**) **variables** that take on values of 0 and 1. If we have a qualitative variable with c classes, that variable will be represented by $c - 1$ indicator variables in the regression model, with each indicator variable taking on a value of 0 or 1.

Suppose we added a variable to our pizza model that asked the customer if they lived in town or out of town. The variable, let's call it town (in or out), has $c = 2$ classes. Thus, the variable town will be represented by $c - 1 = 1$ indicator variable in the model. Town could be modeled as follows.

$$x_3 = \begin{cases} 1 & \text{if in town} \\ 0 & \text{otherwise} \end{cases}$$

The regression model would then be

$$y_i = \beta_0 + \beta_1 x_{1i} + \beta_2 x_{2i} + \beta_3 x_{3i} + \varepsilon_i.$$

To understand the meaning of the regression coefficients in this model, consider the difference in the model for out-of-town $(x_3 = 0)$ and in-town $(x_3 = 1)$ deliveries. For an out-of-town delivery, $x_3 = 0$, and we have

$$\begin{aligned} y_i &= \beta_0 + \beta_1 x_{1i} + \beta_2 x_{2i} + \beta_3 (0) + \varepsilon_i \\ &= \beta_0 + \beta_1 x_{1i} + \beta_2 x_{2i} + \varepsilon_i. \end{aligned} \qquad \text{Out of Town}$$

Similarly, for an in-town delivery, $x_3 = 1$, and the model becomes

$$\begin{aligned} y_i &= \beta_0 + \beta_1 x_{1i} + \beta_2 x_{2i} + \beta_3 (1) + \varepsilon_i \\ &= (\beta_0 + \beta_3) + \beta_1 x_{1i} + \beta_2 x_{2i} + \varepsilon_i. \end{aligned} \qquad \text{In Town}$$

Notice the difference in the two models. The out-of-town model has an intercept of β_0 and the in-town model has an intercept of $(\beta_0 + \beta_3)$. After examining the two models, the difference is clear. The coefficient β_3 indicates how much more (or less) the delivery time will be for deliveries out of town versus deliveries in town, for any given number of pizzas (x_1) and distance (x_2). Thus, β_3 measures the differential effect of the in-/out-of-town delivery. In general, β_3 shows how much higher (or lower) the line of best fit is for the class coded 1 than the class coded 0, for any given level of x_1 and x_2.

Example 14.6

Table 14.2 contains the pizza delivery data set, but with the added column of the variable town, indicating whether the pizza is to be delivered in town or out of town.

	Table 14.2 – Pizza Delivery Time Data				
Observation	Delivery Time (Minutes)	Number of Pizzas	Distance (Miles)	In Town Delivery?	Town (= 1 if in town, = 0 otherwise)
1	16.68	7	5.60	Yes	1
2	11.50	3	2.20	No	0
3	12.03	3	3.40	Yes	1
4	14.88	8	0.80	No	0
5	13.75	6	1.50	No	0
6	18.11	7	3.30	Yes	1
7	8.00	2	1.10	Yes	1
8	17.83	7	2.10	Yes	1
9	79.24	30	14.60	No	0
10	21.50	5	6.05	No	0
11	40.33	16	6.88	Yes	1
12	21.00	10	2.15	No	0
13	13.50	4	2.55	No	0
14	19.75	6	4.62	No	0
15	24.00	9	4.48	No	0
16	29.00	10	7.76	No	0
17	15.35	6	2.00	No	0
18	19.00	7	1.32	Yes	1
19	9.50	3	0.36	No	0
20	35.10	17	7.70	Yes	1
21	17.90	10	1.40	Yes	1
22	52.32	26	8.10	No	0
23	18.75	9	4.50	Yes	1
24	19.83	8	6.35	No	0
25	10.75	4	1.50	Yes	1

You may recall that we are interested in modeling the delivery time (in minutes) as a function of the following variables.

$$x_1 = \text{Number of Pizzas}$$
$$x_2 = \text{Distance (miles)}$$
$$x_3 = \begin{cases} 1 & \text{if in town} \\ 0 & \text{otherwise} \end{cases}$$

The multiple regression model is given by

$$y_i = \beta_0 + \beta_1 x_{1i} + \beta_2 x_{2i} + \beta_3 x_{3i} + \varepsilon_i.$$

The fitted regression model is as follows.

Estimated Delivery Time $= 2.4589 + 1.5936(\text{Number of Pizzas}) + 1.5294(\text{Distance}) - 1.2493(\text{Town})$

	A	B	C	D	E	F	G
1	SUMMARY OUTPUT						
2							
3	*Regression Statistics*						
4	Multiple R	0.982670531					
5	R Square	0.965641372					
6	Adjusted R Square	0.960732997					
7	Standard Error	3.076399777					
8	Observations	25					
9							
10	ANOVA						
11		*df*	*SS*	*MS*	*F*	*Significance F*	
12	Regression	3	5585.793653	1861.931	196.733397	1.58321E-15	
13	Residual	21	198.7489474	9.464236			
14	Total	24	5784.5426				
15							
16		*Coefficients*	*Standard Error*	*t Stat*	*P-value*	*Lower 95%*	*Upper 95%*
17	Intercept	2.458883847	1.244384484	1.975984	0.061441395	-0.128955355	5.046723048
18	Number of Pizzas	1.593644691	0.156448219	10.1864	1.4007E-09	1.268292809	1.918996573
19	Distance	1.529396057	0.329864891	4.636432	0.000142011	0.843404464	2.21538765
20	Town	-1.249333348	1.255683682	-0.99494	0.33109102	-3.860670518	1.362003822

Figure 14.12

Now that we are fitting the model with town as the indicator variable, we are interested in the effect that delivering pizzas in town or out of town has on the average delivery time. The estimate of β_3 is −1.2493. This means that the difference in average delivery time is approximately 1.25 minutes. Specifically, the average delivery time in town is about 1.25 minutes less than out of town.

Another important question would be, *is the slope associated with the town variable significant?* The formal test can be carried out just as we did in Section 14.5. That is, our hypotheses are the following.

$$H_0: \beta_3 = 0 \quad \text{(Implies town is not a useful predictor of delivery time.)}$$
$$H_a: \beta_3 \neq 0 \quad \text{(Implies town is a useful predictor of delivery time.)}$$

According to the Excel output, we have a *t*-statistic of −0.995 with a *P*-value of 0.3311. Given the large *P*-value associated with this test, we would fail to reject the null hypothesis and conclude that the town variable is not a useful predictor of delivery time.

Example 14.7

A real estate investor wants to study the relationship between annual return on his commercial retail shops (measured in thousands of dollars) as it relates to their location and the number of homes near the shops. Specifically, the investor has collected data on the annual return of the shops, the number of households within 15 miles of the shops (measured in thousands), and the location of the shops (whether the shops are in a suburban area, near a shopping mall, or downtown). The annual return data are given in the following table.

		Table 14.3 – Shop Data	
Shop	Location	Annual Return (Thousands of Dollars)	Number of Households (Thousands)
1	Mall	185.69	163
2	Suburban	203.00	215
3	Mall	245.81	232
4	Mall	137.07	108
5	Suburban	207.36	220
6	Suburban	146.12	150
7	Downtown	111.21	102
8	Suburban	188.19	198
9	Downtown	152.23	149
10	Suburban	182.23	192
11	Mall	198.88	179
12	Mall	139.65	111
13	Mall	156.22	130
14	Downtown	224.17	232
15	Downtown	195.62	199
16	Suburban	210.38	224
17	Downtown	209.16	215
18	Mall	260.82	250
19	Mall	200.16	180
20	Suburban	127.66	129
21	Downtown	170.59	170
22	Mall	219.93	203
23	Downtown	166.61	166
24	Downtown	225.71	234
25	Downtown	143.25	139
26	Downtown	239.55	250
27	Downtown	219.67	227
28	Suburban	224.84	241
29	Mall	232.32	217
30	Downtown	163.95	163

Solution

Given the data, we can define our variables as follows.

$$x_1 = \text{Number of Households}$$

Since location is a qualitative variable and there are three locations, we will use $c - 1 = 3 - 1 = 2$ variables to represent the locations. The variables representing locations are

$$\text{Mall:} \quad x_2 = \begin{cases} 1 & \text{if mall} \\ 0 & \text{otherwise} \end{cases}$$

$$\text{Downtown:} \quad x_3 = \begin{cases} 1 & \text{if downtown} \\ 0 & \text{otherwise.} \end{cases}$$

The suburban location is known as the **base level variable** since for suburban locations $x_2 = 0$ and $x_3 = 0$. The coefficients estimated for the x_2 and x_3 indicator variables will represent a difference between the mall and downtown locations and the suburban location.

The multiple regression model for annual return is given by

$$y_i = \beta_0 + \beta_1 x_{1i} + \beta_2 x_{2i} + \beta_3 x_{3i} + \varepsilon_i.$$

The fitted regression model is

$$\text{Estimated Annual Return} = 15.5071 + 0.8704(\text{Number of Households}) + 27.8186(\text{Mall})$$
$$+ 6.7185(\text{Downtown}).$$

Let's discuss the model and its implications. First of all, we want to know if the location and the number of households in the area are useful predictors of annual return of the shops. You may recall that the test of the model's significance is conducted using the F-test that tests the following hypothesis.

$$H_0: \beta_1 = \beta_2 = \beta_3 = 0$$
$$H_a: \text{At least one } \beta_i \neq 0.$$

◢	A	B	C	D	E	F	G
1	SUMMARY OUTPUT						
2							
3	*Regression Statistics*						
4	Multiple R	0.999974838					
5	R Square	0.999949677					
6	Adjusted R Square	0.999943871					
7	Standard Error	0.288707033					
8	Observations	30					
9							
10	ANOVA						
11		*df*	*SS*	*MS*	*F*	*Significance F*	
12	Regression	3	43062.89567	14354.29856	172213.5214	5.5533E-56	
13	Residual	26	2.167145526	0.083351751			
14	Total	29	43065.06282				
15							
16		*Coefficients*	*Standard Error*	*t Stat*	*P-value*	*Lower 95%*	*Upper 95%*
17	Intercept	15.50707387	0.261037514	59.40553771	2.6636E-29	14.97050357	16.04364416
18	Number of Households	0.870441943	0.001225001	710.5645401	2.77306E-57	0.867923919	0.872959968
19	Mall	27.81856958	0.138873816	200.3154397	5.45545E-43	27.53311037	28.1040288
20	Downtown	6.718542409	0.13223228	50.80864057	1.49967E-27	6.446735064	6.990349754

Figure 14.13

Examining the Excel output in Figure 14.13, we see that the F-statistic for the test is 172,213.5214 with a P-value near zero. This is an indication that we reject H_0 and conclude that at least one of the slopes in the model is significantly different from zero, and the model can be useful.

Having concluded that the model is significant, we focus our attention on the individual regression coefficients. That is, we can test the following hypothesis.

$$H_0: \beta_i = 0$$
$$H_a: \beta_i \neq 0$$

If we examine the *P*-values in the Excel output for each of the coefficients, it is noted that each of them is near zero, indicating that each coefficient is different from zero, and each independent variable is useful in predicting annual return.

Now we can examine the average annual return of the investments as a function of the location of the shops (i.e. whether they are near a mall, downtown, or in a suburban area). The estimated models can be written as follows.

For the mall location $(x_2 = 1, \ x_3 = 0)$:

$$\text{Estimated Annual Return} = (b_0 + b_2) + b_1 (\text{Number of Households})$$
$$= (15.5071 + 27.8186) + 0.8704 (\text{Number of Households}).$$

For the downtown location $(x_2 = 0, \ x_3 = 1)$:

$$\text{Estimated Annual Return} = (b_0 + b_3) + b_1 (\text{Number of Households})$$
$$= (15.5071 + 6.7185) + 0.8704 (\text{Number of Households}).$$

For the suburban location $(x_2 = 0, \ x_3 = 0)$:

$$\text{Estimated Annual Return} = b_0 + b_1 (\text{Number of Households})$$
$$= 15.5071 + 0.8704 (\text{Number of Households}).$$

Notice that the best fit line for each location has a different intercept value, but the same slope. Plotting the three lines illustrates this difference. (See Figure 14.14.)

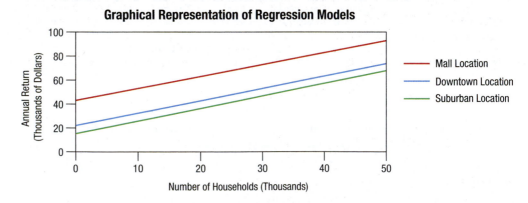

Figure 14.14

So, modeling with the qualitative variables allows us to compare the average difference in annual return of the shops by location. That is, in order to compare the mall and suburban locations, the difference (as can be seen in the model) is

$$(\beta_0 + \beta_1 x_1 + \beta_2) - (\beta_0 + \beta_1 x_1) = \beta_2.$$

Therefore, the estimate of β_2 (which is 27.8186) represents the difference between the average annual return for shops in the mall locations having x_1 households in the area and the average annual return for shops in the suburban locations having x_1 households in the area. Thus, we can say that for any number of households in a given area, the average annual return in a mall location will be approximately $27,818.60 greater than the average annual return in a suburban location.

Similarly, to compare downtown and suburban returns, we get

$$\left(\beta_0 + \beta_1 x_1 + \beta_3\right) - \left(\beta_0 + \beta_1 x_1\right) = \beta_3.$$

Thus, the estimate of β_3 (which is 6.7185) represents the difference between the average annual return for shops in the downtown locations having x_1 households in the area and the average annual return for shops in the suburban locations having x_1 households in the area. So, we can say that for any number of households in a given area, the average annual return in a downtown location will be $6718.50 greater than the average annual return in a suburban location.

Lastly, to compare mall and downtown locations, we look at

$$\left(\beta_0 + \beta_1 x_1 + \beta_2\right) - \left(\beta_0 + \beta_1 x_1 + \beta_3\right) = \beta_2 - \beta_3.$$

The estimated difference between β_2 and β_3 (which is 21.1001) represents the difference between the average annual return for shops in the mall locations having x_1 households in the area and the average annual return for shops in the downtown locations having x_1 households in the area. In terms of the problem, we can say that for any number of households in a given area, the average annual return in a mall area will be $21,100.10 greater than the average annual return in a downtown location.

There are three potential issues to keep in mind when using these regression results.

1. These results are only meaningful within the relevant range of the data that was used to estimate the regression equation. For example, using the model in Example 14.7 to predict annual return for a shop with 1000 households or 1,000,000 households in the surrounding area would likely yield an unreliable point estimate.

2. The regression lines for the three locations in Example 14.7 are assumed to have the same slope, but in reality they could have very different slopes. Using a regression model with **interaction terms** allows the slopes for the regression lines to differ.

3. The regression lines estimated in this example are all linear. This implies that annual return increases by the same amount for each additional thousand households within 15 miles of the shops. This assumption is sometimes unrealistic. This issue can be addressed using **polynomial (or nonlinear) regression models**.

Regression models with interaction terms and polynomial regression models are beyond the scope of this text and are not discussed in detail. Multiple regression is a complex topic that involves many methods of estimation. We only present the basics in this text.

14.7   Exercises

Basic Concepts

1.   Explain why qualitative variables have not been used in the regression models we have discussed in previous sections.

2.   Give three examples of qualitative independent variables that may be of interest to someone performing regression analysis to predict annual salary.

3.   Explain how qualitative variables are transformed into quantitative variables in order to estimate a regression model.

4.   If a qualitative variable has c classes, how many indicator (dummy) variables will there be in the model? Explain why this is the case.

5.   When an indicator (dummy) variable is equal to one, does this represent a difference in the slope or the intercept of the model? Explain.

6.   What is a base level variable? Interpret the value of an estimated coefficient for an indicator variable in terms of the base level variable.

7.   Identify three potential issues to keep in mind when constructing regression models involving indicator variables. Also suggest how these issues can be addressed.

Exercises

8.   Consider the following estimated multiple regression model relating GPA to the number of classes attended and the final exam score in a particular class, and if the student is a freshman ($= 1$ if freshman, $= 0$ otherwise).

$$GPA = -0.8777 + 0.0672(\text{Attendance})$$
$$+ 0.0678(\text{Exam Score}) - 0.1436(\text{Freshman})$$

a.   Are the signs of the estimated coefficients what you would expect for these three independent variables? Explain.

b.   Interpret the coefficient for the attendance variable.

c.   Interpret the coefficient for the exam score variable.

d.   Interpret the coefficient for the freshman variable.

e.   Suppose two students, one a freshman and one a senior, attended the same number of classes and both got a score of 88 on the final exam. What would be the expected difference in GPA for the two students?

9. Consider the following computer output for the multiple regression model discussed in the previous exercise.

	A	B	C	D	E	F
1	SUMMARY OUTPUT					
2						
3	*Regression Statistics*					
4	Multiple R	0.714589997				
5	R Square	0.510638864				
6	Adjusted R Square	0.508467143				
7	Standard Error	0.516416069				
8	Observations	680				
9						
10	ANOVA					
11		*df*	*SS*	*MS*	*F*	*Significance F*
12	Regression	3	188.1180981	62.70603271	235.1309671	1.8485E-104
13	Residual	676	180.2794359	0.266685556		
14	Total	679	368.397534			
15						
16		*Coefficients*	*Standard Error*	*t Stat*	*P-value*	*Lower 95%*
17	Intercept	-0.877712645	0.138557037	-6.334666683	4.34037E-10	-1.149766538
18	Attendance	0.067163994	0.003669275	18.3044333	4.30384E-61	0.059959449
19	Exam Score	0.067820136	0.004265782	15.89864106	1.37161E-48	0.059444361
20	Freshman	-0.143623671	0.047077779	-3.050774156	0.002371853	-0.236059922

a. Test the usefulness of the overall model in predicting GPA using a 5% significance level.

b. What percentage of the variation in GPA is explained by the three independent variables?

c. Is the qualitative independent variable, freshman, useful in predicting GPA? Use $\alpha = 0.05$.

d. Do you believe this is a good model to use to predict GPA? Why or why not?

e. Can you think of other variables that could be added to the model? Name one quantitative variable and one qualitative variable that might be useful.

14

10. A personnel director is interested in studying the effects which age, experience, and gender have on salary. Eight employees are randomly selected and each employee's salary, age, experience, and gender (= 0 if male, = 1 if female) are recorded.

Table for Exercise 10 – Employee Data

Salary ($)	Age	Experience (Years)	Gender
27,000	25	2	1
50,000	55	20	0
48,000	27	5	0
35,000	30	7	1
29,000	22	3	1
58,000	33	8	1
23,000	19	1	0
43,000	45	15	1

a. Create three scatterplots using salary with age, salary with experience, and salary with gender. Does each of the plots have a linear relationship?

b. Using statistical software, estimate the parameters of the following regression model:

$$\text{Salary} = \beta_0 + \beta_1(\text{Age}) + \beta_2(\text{Experience}) + \beta_3(\text{Gender}) + \varepsilon_i.$$

c. Is the overall model useful in explaining salary? Test at the 0.05 level.

d. Is age useful in explaining salary? Test at the 0.05 level.

e. Is experience useful in explaining salary? Test at the 0.01 level.

f. Is gender useful in explaining salary? Test at the 0.10 level.

g. Interpret each of the regression coefficients.

h. Construct and interpret 95% confidence intervals for each of the regression coefficients. Do you think that the company discriminates in the salary paid based on gender?

i. Predict the salary of a female employee who is 35 years old with 10 years of experience.

j. Construct and interpret a 95% prediction interval for a female employee who is 35 years old with 10 years of experience. How useful is this interval?

k. Construct and interpret a 95% confidence interval for the average salary of a female employee who is 35 years old with 10 years of experience. How useful is this interval?

11. Consider the following crime data from select college campuses. The table contains the number of crimes committed, the number of campus police employed on campus, the total enrollment of the college, and whether or not the college is private.

Table for Exercise 11 – Campus Crime Data			
Number of Crimes	**Number of Police**	**Total Enrollment**	**Private School**
64	12	1131	Yes
138	21	12,954	No
141	32	16,009	No
84	22	1682	Yes
86	35	2888	Yes
141	45	17,407	No
135	42	3028	Yes
174	50	4306	Yes
201	75	34,511	No
203	84	37,240	No
125	36	2918	Yes
234	109	39,414	No
143	45	4000	Yes
148	50	20,950	No
152	48	4277	Yes
158	52	26,519	No
174	69	27,687	No
84	26	2810	Yes
173	58	27,619	No
193	56	4563	Yes

a. Create an indicator (dummy) variable for whether or not the college is private. Let Private $= 1$ if the school is private and Private $= 0$ if the school is public.

b. Suppose education officials wish to predict the number of crimes on college campuses based on the number of police employed and total enrollment. They would also like to know whether there are fewer crimes committed on private campuses than public ones. Use statistical software to estimate the following regression model.

$$\text{Crimes} = \beta_0 + \beta_1 \left(\text{Police}\right) + \beta_2 \left(\text{Enrollment}\right) + \beta_3 \left(\text{Private}\right) + \varepsilon_i$$

Write the estimated multiple regression equation.

c. Is the overall model useful in predicting the number of crimes? Use $\alpha = 0.05$.

d. Are the signs of the coefficients of the independent variables what you would expect for these data? Explain.

e. Is there evidence to support the officials' belief that there are fewer crimes committed at private schools than at public schools? Test using $\alpha = 0.05$. Would this decision change if $\alpha = 0.01$?

12. Consider the following sales data regarding weekly sales, the number of sales reps, and whether or not the sales were made in the first, second, third, or fourth quarter of the year. For each column containing an indicator variable, the variable is equal to 1 if that particular week was in that particular quarter, and equal to zero otherwise. For example, if the weekly data were recorded in January, the 1st quarter indicator variable would be equal to 1 and the indicator variables for the 2nd, 3rd, and 4th quarters would be equal to zero. The first quarter comprises January through March, the second quarter April through June, the third quarter July through September, and the fourth quarter October through December.

Table for Exercise 12 –Weekly Sales by Quarter					
Weekly Sales ($)	Number of Sales Reps	1st Quarter	2nd Quarter	3rd Quarter	4th Quarter
4272.90	3	1	0	0	0
5069.70	9	1	0	0	0
6067.70	11	1	0	0	0
6680.55	17	1	0	0	0
9725.05	20	1	0	0	0
4107.10	3	0	1	0	0
7520.25	9	0	1	0	0
12,135.00	11	0	1	0	0
13,016.55	17	0	1	0	0
13,673.90	20	0	1	0	0
3272.05	3	0	0	1	0
5074.40	9	0	0	1	0
7505.45	11	0	0	1	0
8272.75	17	0	0	1	0
10,020.40	20	0	0	1	0
4925.75	3	0	0	0	1
10,018.10	9	0	0	0	1
12,505.85	11	0	0	0	1
15,329.05	17	0	0	0	1
19,477.20	20	0	0	0	1

a. How many indicator variables should be included in the multiple regression model relating weekly sales to the number of sales reps and the quarter of the year? Explain why.

b. What sign would you expect the coefficient for the sales reps variable to have? Explain your reasoning.

c. Using statistical software, estimate the following multiple regression model.

$$\text{Sales} = \beta_0 + \beta_1 (\text{Reps}) + \beta_2 (\text{Quarter 1}) + \beta_3 (\text{Quarter 2}) + \beta_4 (\text{Quarter 3}) + \varepsilon_i$$

Write the estimated multiple regression equation.

d. Interpret the coefficient of the indicator variable representing the first quarter.

e. Is there sufficient evidence that sales in the second quarter tend to be different from the sales in the fourth quarter? Use $\alpha = 0.05$.

f. What concerns should we have when predicting weekly sales using this model?

14.8   Pitfalls and Concerns in Regression Modeling

When fitting multiple regression models, there are some potential problems that one should be aware of and that one must consider. One of the problems that must be considered is that of **multicollinearity**. Multicollinearity is when some of the independent variables $(x_1, x_2,..., x_k)$ are similar to each other (i.e. highly correlated). When this is the case, the individual regression coefficients are not properly estimated because they are so similar to each other. That is, since you have two or more variables that are related, perhaps these variables are measuring the same information, and thus you only need one of these variables in the model to explain the dependent variable. Thus, omitting highly correlated variables will help minimize the risk of multicollinearity. Multicollinearity can be detected by studying a correlation matrix for the variables in question, which is easily generated using technology such as Microsoft Excel or Minitab. If the correlation coefficient between two of the independent variables is close to 1 or -1, it is likely that multicollinearity could be obscuring the regression results.

Another concern in multiple regression is **parameter estimability**. Parameter estimability is the inability to estimate the parameters of the regression model because the data are concentrated in one area. To ensure that the parameters are estimable, the data must include at least one more level of the independent variable than the highest order of the independent variable included in the model. This means that the sample size, n, must exceed $k + 1$ to ensure that the degrees of freedom are not equal to zero.

Another concern or pitfall of multiple regression involves **variable selection**. The problem of variable selection arises when you have a large number of independent variables and need to decide which ones to include in the model. As we have seen in earlier sections, the R^2 value increases as more variables are added to the model. However, the quality of the model may degrade as more variables are added to the model because information is being wasted when estimating unnecessary parameters. Therefore, it would be best to include only important variables in the model or variables that are clearly necessary. A solution to avoid over-fitting a model is to utilize a model-building procedure such as **stepwise regression**. Stepwise regression involves selecting independent variables using an automated procedure, which is beyond the scope of this text.

Some other issues with fitting multiple regression models are **extrapolation** (which was discussed in Chapter 13) and **correlated errors**. Extrapolation can be a concern when the regression model is used to predict values outside the range of the data used to estimate the model. Be sure to only use the model within an appropriate range of x-values. The problem with correlated errors arises when measurements of the dependent variable are correlated. That is, since the observations (the responses) of the regression model are assumed to be independent, it is problematic if there is a relationship between responses. One will often see this type of dependency with time series data. Current measurements are often dependent on measurements in the previous time period.

14.8   Exercises

Basic Concepts

1. Define multicollinearity. How can you detect if multicollinearity exists in a regression model?

2. Why is multicollinearity a concern when performing regression analysis?

3. How can you attempt to correct the problem of multicollinearity?

4. What is parameter estimability?

5. How can you alleviate concerns about parameter estimability?

6. Why is variable selection difficult when building a multiple regression model?

7. Does the R^2 value always increase as additional variables are added? Does this mean that adding additional variables always produces a more useful model? Explain.

8. What is extrapolation? Why is this a concern?

9. With what type of data do you often encounter issues with correlated errors?

Exercises

10. In Exercise 8 of Section 14.6, we modeled the relationship between total points and rushing yards, passing yards, and first downs.

   a. Using the correlation matrix below, discuss whether collinearity might play a role in estimating total points using rushing yards, passing yards, and first downs in the model.

Table for Exercise 10 – Correlation Matrix			
	Rushing Yards	**Passing Yards**	**First Downs**
Rushing Yards	1.0000	−0.1943	0.3789
Passing Yards		1.0000	0.5744
First Downs			1.0000

   b. How would you determine if there is a relationship (and if so, the strength of such relationship) between the independent variables in the model?

   c. Given the multiple regression model that was fit in Exercise 8, what would the total points be if a team had 30 rushing yards, 100 passing yards, and 5 first downs?

   d. Should you have any concerns about the estimate in part **c.**? Explain your answer.

11. In Exercise 10 of Section 14.7, salary was modeled as a function of age, experience, and gender.

   a. Discuss how collinearity might play a role in estimating salary using age, experience, and gender in the model.

   b. How would you determine if there is a relationship (and if so, the strength of such relationship) between the independent variables in the model?

   c. Given the multiple regression model that was fit in the problem, what would the expected salary be for a 60-year-old male employee with 25 years of experience?

   d. Should you have any concerns about the estimate in part **c.**? Explain your answer.

12.   In Exercise 11 of Section 14.7, we attempted to predict the number of crimes on a college/university campus based on the number of police, the enrollment at the university, and if it was a private institution.

   a.   Examine the correlation matrix below and discuss whether collinearity might play a role in estimating the number of crimes based on the number of police, enrollment, and if the institution is private.

Table for Exercise 12 – Correlation Matrix			
	Police	**Enrollment**	**Private**
Police	1.0000	0.8042	−0.4942
Enrollment		1.0000	−0.8795
Private			1.0000

   b.   How would you determine if there is a relationship (and if so, the strength of such relationship) between the independent variables in the model?

   c.   Given the multiple regression model that was fit in the problem, what would the expected number of crimes be for a private university with a police force of 100 officers and an enrollment of 50,000?

   d.   Should you have any concerns about the estimate in part **c.**? Explain your answer.

13.   Suppose you fit a multiple regression model of the form

$$y_i = \beta_0 + \beta_1 x_{1i} + \beta_2 x_{2i} + \beta_3 x_{3i} + \beta_4 x_{4i} + \varepsilon_i.$$

The correlation matrix for the pairs of independent variables is given in the following table. Discuss if you detect multicollinearity between any of the variables.

Table for Exercise 13 – Correlation Matrix			
	x_1	x_2	x_3	x_4
x_1	1.00	0.18	0.86	0.45
x_2		1.00	0.35	0.22
x_3			1.00	0.50
x_4				1.00

T   Discovering Technology

Using Excel

Regression

Use the data in Table 14.1 to perform multiple regression analysis.

1. Enter the labels **Delivery Time, Number of Pizzas,** and **Distance** in cells A1, B1, and C1, respectively.

2. Enter the y (Delivery Time), x_1 (Number of Pizzas), and x_2 (Distance) data into columns A, B, and C, respectively.

	A	B	C
1	Delivery Time	Number of Pizzas	Distance
2	16.68	7	5.6
3	11.5	3	2.2
4	12.03	3	3.4
5	14.88	8	0.8
6	13.75	6	1.5
7	18.11	7	3.3
8	8	2	1.1
9	17.83	7	2.1
10	79.24	30	14.6
11	21.5	5	6.05
12	40.33	16	6.88
13	21	10	2.15
14	13.5	4	2.55
15	19.75	6	4.62
16	24	9	4.48
17	29	10	7.76
18	15.35	6	2
19	19	7	1.32
20	9.5	3	0.36
21	35.1	17	7.7
22	17.9	10	1.4
23	52.32	26	8.1
24	18.75	9	4.5
25	19.83	8	6.35
26	10.75	4	1.5

Figure 14.15

3. Under the **Data** tab, choose **Data Analysis**, and **Regression**.

4. In the dialog box, select the delivery time data in Column A for the Input Y Range (**$A$1:$A$26**).

5. For the Input X Range, select the data in columns B and C for number of pizzas and distance (**$B$1:$C$26**). Check the box next to **Labels** since the column titles are included in the selection. If you would like Excel to compute confidence intervals for individual coefficients at a level other than 95%, you can check the box next to **Confidence level** and enter the desired level of confidence as a percentage (i.e. for a 99% confidence interval you would enter 99). Click **OK**.

6. Observe the summary output for the regression analysis. In the *Regression Statistics* table you will find Multiple R, which is the positive square root of the coefficient of determination, the coefficient of determination, R^2, the adjusted R^2 value, the standard error for the model, and the total number of observations. The ANOVA table gives the degrees of freedom for regression and error (residual), along with SSR, SSE, TSS, MSR, MSE, and the *F*-statistic. The *Significance F* column contains the *P*-value corresponding to the *F*-statistic. This is the *P*-value we are interested in when testing the overall model for significance. Finally, the estimated values of the coefficients are reported along with the standard error for each coefficient, the *t*-statistic, *P*-value, and 95% confidence interval for the individual coefficient.

	A	B	C	D	E	F	G
1	SUMMARY OUTPUT						
2							
3	*Regression Statistics*						
4	Multiple R	0.981846095					
5	R Square	0.964021754					
6	Adjusted R Square	0.960751004					
7	Standard Error	3.075694294					
8	Observations	25					
9							
10	ANOVA						
11		*df*	*SS*	*MS*	*F*	*Significance F*	
12	Regression	2	5576.424901	2788.212451	294.7403048	1.30749E-16	
13	Residual	22	208.1176986	9.45989539			
14	Total	24	5784.5426				
15							
16		*Coefficients*	*Standard Error*	*t Stat*	*P-value*	*Lower 95%*	*Upper 95%*
17	Intercept	1.792903307	1.048779122	1.709514681	0.101425497	-0.382131459	3.967938072
18	Number of Pizzas	1.589101026	0.156345691	10.16402191	8.97113E-10	1.26485991	1.913342142
19	Distance	1.567708058	0.327534513	4.786390423	8.84845E-05	0.888443055	2.246973061

Figure 14.16

Using Minitab

Regression

Use the data in Table 14.1 to perform multiple regression analysis.

1. Enter the labels **Delivery Time**, **Number of Pizzas**, and **Distance** in columns C1, C2, and C3, respectively.

2. Enter the y (Delivery Time), x_1 (Number of Pizzas), and x_2 (Distance) data into columns C1, C2, and C3, respectively.

3. Choose **Stat**, **Regression**, and **Regression**.

4. Enter **C1** (Delivery Time) in the Response box, and **C2** and **C3** (Number of Pizzas and Distance) in the Predictors box. Press **OK**.

5. Observe the output for the regression analysis. The estimated regression equation is written first. Then, the estimated values of the coefficients are reported along with their corresponding standard errors, t-statistics, and P-values. Then the standard error of the model is given, S, along with the coefficient of determination, R-Sq, and the adjusted coefficient of determination, R-Sq(adj). Finally, the ANOVA table is given with the degrees of freedom for regression and error, along with SSR, SSE, TSS, MSR, MSE, and the F-statistic. The P column contains the P-value corresponding to the F-statistic. This is the P-value we are interested in when testing the overall model for significance.

Regression Analysis: Delivery Time versus Number of Pizzas, Distance

```
The regression equation is
Delivery Time = 1.79 + 1.59 Number of Pizzas + 1.57 Distance


Predictor              Coef  SE Coef      T      P
Constant              1.793    1.049   1.71  0.101
Number of Pizzas     1.5891   0.1563  10.16  0.000
Distance             1.5677   0.3275   4.79  0.000


S = 3.07569   R-Sq = 96.4%   R-Sq(adj) = 96.1%


Analysis of Variance

Source            DF      SS      MS      F      P
Regression         2  5576.4  2788.2  294.74  0.000
Residual Error    22   208.1     9.5
Total             24  5784.5
```

Figure 14.17

Confidence and Prediction Intervals

Minitab will compute confidence intervals for the mean value of y given x and prediction intervals for individual values of y given x. Use the data in Table 14.1 for this exercise.

1. Enter the labels **Delivery Time**, **Number of Pizzas**, and **Distance** in columns C1, C2, and C3, respectively.

2. Enter the y (Delivery Time), x_1 (Number of Pizzas), and x_2 (Distance) data into columns C1, C2, and C3, respectively.

3. Choose **Stat**, **Regression**, and **Regression**.

4. Enter **C1** (Delivery Time) in the Response box, and **C2** and **C3** (Number of Pizzas and Distance) in the Predictors box.

5. In the Regression dialog box, choose **Options**.

6. In the Regression – Options dialog box, enter **5, 6** in the Prediction intervals for new observations box. This tells Minitab to compute a point estimate and confidence and prediction intervals for delivering 5 pizzas to a location 6 miles away.

7. Set the Confidence level to **99**.

8. Check the boxes next to **Fits** and **SEs of fits**. Press **OK**.

9. Press **OK** again. Observe the output to find the 99% confidence interval and 99% prediction interval. Under Predicted Values for New Observations, The Fit column contains the predicted value of y for delivering 5 pizzas to a location 6 miles away. The confidence interval is then given in the 95% CI column, and the prediction interval is given in the 95% PI column.

```
Regression Analysis: Delivery Time versus Number of Pizzas, Distance

The regression equation is
Delivery Time = 1.79 + 1.59 Number of Pizzas + 1.57 Distance


Predictor            Coef  SE Coef      T      P
Constant            1.793    1.049   1.71  0.101
Number of Pizzas  1.5891   0.1563  10.16  0.000
Distance          1.5677   0.3275   4.79  0.000


S = 3.07569   R-Sq = 96.4%   R-Sq(adj) = 96.1%


Analysis of Variance

Source          DF      SS      MS       F      P
Regression       2  5576.4  2788.2  294.74  0.000
Residual Error  22   208.1     9.5
Total           24  5784.5


Predicted Values for New Observations

New Obs    Fit  SE Fit        99% CI              99% PI
      1 19.145   1.328  (15.402, 22.887)  (9.702, 28.588)


Values of Predictors for New Observations

          Number
            of
New Obs  Pizzas  Distance
      1    5.00      6.00
```

Figure 14.18

R | Chapter 14 Review

Key Terms and Ideas

- Multiple Regression
- Multiple Regression Model
- Method of Least Squares
- Estimated Multiple Regression Equation
- Coefficient of Determination (Multiple Coefficient of Determination)
- Adjusted R^2
- F-Distribution
- Numerator Degrees of Freedom
- Denominator Degrees of Freedom
- F-Statistic
- Sum of Squares of Regression
- Sum of Squared Errors
- Total Sum of Squares
- Mean Square Regression
- Mean Square Error
- Calculating Degrees of Freedom in Multiple Regression Models
- Hypothesis Tests Concerning Individual Coefficients
- Test Statistic for Testing the Hypothesis $\beta_i \neq 0$
- Confidence Intervals for Individual Coefficients
- Confidence Interval for the Mean Value of y Given x
- Confidence Interval for the Predicted Value of y Given x
- Indicator (Dummy) Variable
- Base Level Variable
- Interaction Terms
- Polynomial (Nonlinear) Regression Models
- Multicollinearity
- Parameter Estimability
- Variable Selection
- Stepwise Regression
- Extrapolation
- Correlated Errors

Key Formulas

	Section

Multiple Regression Model

$$y_i = \beta_0 + \beta_1 x_{1i} + \beta_2 x_{2i} + \ldots + \beta_k x_{ki} + \varepsilon_i$$

where β_0, β_1, β_2, ..., β_k are the model's parameters, x_{1i}, x_{2i}, ..., x_{ki} are the independent variables, and ε_i is a random error.

14.1

Estimated Multiple Regression Equation

$$\hat{y}_i = b_0 + b_1 x_{1i} + b_2 x_{2i} + \cdots + b_k x_{ki}$$

where b_0, b_1, b_2, ..., b_k are estimates of their population counterparts.

14.1

Key Formulas (cont.)

	Section

Sum of Squared Errors

$$SSE = \sum \left(y_i - \hat{y}_i \right)^2$$

14.1

Coefficient of Determination

$$R^2 = \frac{SSR}{TSS} = 1 - \frac{SSE}{TSS}$$

14.2

Adjusted R^2

$$R_a^2 = 1 - \left(\frac{n-1}{n-k-1} \right) \frac{SSE}{TSS}$$

14.2

where n is the number of observations and k is the number of independent variables in the model.

F-Statistic

$$F = \frac{\dfrac{\text{Sum of Squares of Regression}}{k}}{\dfrac{\text{Sum of Squared Errors}}{n-(k+1)}}$$

$$= \frac{\dfrac{SSR}{k}}{\dfrac{SSE}{n-(k+1)}} = \frac{\text{Mean Square Regression}}{\text{Mean Square Error}}$$

14.5

Test Statistic for Testing the Hypothesis $\beta_i \neq 0$

$$t = \frac{b_i - 0}{s_{b_i}} = \frac{b_i}{s_{b_i}}$$

14.5

where b_i is the estimated coefficient and s_{b_i} is the standard deviation (standard error) of the estimated coefficient.

$100(1-\alpha)\%$ Confidence Interval for an Individual Coefficient, β_i

$$b_i \pm t_{\alpha/2, df} s_{b_i}$$

14.5

AE **Additional Exercises**

1. Drew is undecided about whether to go back to school and get his master's degree. He is trying to perform a cost-benefit analysis to determine whether the cost of attending the school of his choice will be outweighed by the increase in salary he will receive after he attains his degree. He does research and compiles data on annual salaries in the industry he currently works in (he has been working for 10 years), along with the years of experience for each employee and whether or not the employee has a master's degree. Earning his master's degree will require him to take out approximately $20,000 worth of student loans. He has decided that if the multiple regression model shows, with 95% confidence, that earning a master's degree is significant in predicting annual salary, and the estimated increase in salary is at least $10,000, he will enroll in a degree program.

Table for Exercise 1 — Industry Salaries		
Salary ($)	**Years of Experience**	**Master's Degree**
37,620	22	No
67,080	27	Yes
31,280	15	No
21,500	2	No
75,120	28	Yes
59,820	25	Yes
40,180	15	Yes
81,360	32	Yes
35,080	19	No
36,080	12	Yes
36,680	22	No
29,200	11	Yes
33,040	18	No
30,060	14	No
53,300	21	Yes
22,820	7	No
72,900	31	Yes
55,920	22	Yes
19,280	0	No
26,000	7	No

a. Create an indicator variable, degree, that is equal to 1 if the employee has a master's degree and equal to 0 if the employee does not have a master's degree.

b. Using statistical software, estimate the following multiple regression model.

$$\text{Salary} = \beta_0 + \beta_1 \left(\text{Experience}\right) + \beta_2 \left(\text{Degree}\right) + \varepsilon_i$$

Write the estimated multiple regression equation.

c. According to the model, how much does salary increase on average with each additional year of experience?

d. According to this model, will Drew decide to enroll in a master's program? Explain your answer.

e. Why should Drew be cautious when using this model to make his decision?

2. A chain of sports clubs wishes to use regression analysis to help determine which features should be included in their new location. They believe that median income in the area is a significant factor in determining the number of people who join a neighborhood sports club. The CEO of the chain gathered data from existing sports clubs regarding the number of members each club had, the median income in the area in which they were located, and whether or not the clubs had a pool, racquetball courts, or group fitness classes. If management can determine with 90% confidence that a pool, racquetball courts, or group fitness classes produces significantly more memberships than sports clubs without those features, they will include them in the new location.

Table for Exercise 2 – Sports Club Membership				
Number of Members	Median Income ($)	Pool?	Racquetball Courts?	Fitness Classes?
1258	32,223	No	No	No
1479	34,975	No	No	No
1480	43,187	No	Yes	No
1701	44,337	No	No	No
2014	52,167	No	No	Yes
2271	57,521	No	No	Yes
2615	58,347	No	Yes	No
2632	60,960	Yes	No	No
2737	62,201	Yes	No	Yes
2810	67,993	No	No	Yes
3563	68,770	No	No	Yes
3765	81,289	Yes	Yes	Yes
3792	83,902	No	No	Yes
4069	84,594	Yes	No	Yes
4393	86,855	Yes	Yes	Yes
4787	88,381	Yes	Yes	Yes

a. What sign do you expect the coefficient of median income to have? Explain why.

b. Create three dummy variables, pool, courts, and classes, that are equal to 1 if the observation contains this feature and equal to 0 if the observation does not contain this feature.

c. Use statistical software to estimate the following regression models. In each case, write the estimated regression equation and state whether the coefficient of the independent variable is significant at the 0.10 level.

   i. $\text{Members} = \beta_0 + \beta_1(\text{Pool}) + \varepsilon_i$

   ii. $\text{Members} = \beta_0 + \beta_1(\text{Courts}) + \varepsilon_i$

   iii. $\text{Members} = \beta_0 + \beta_1(\text{Classes}) + \varepsilon_i$

d. Estimate the following multiple regression model.

   $$\text{Members} = \beta_0 + \beta_1(\text{Income}) + \beta_2(\text{Pool}) + \beta_3(\text{Courts}) + \beta_4(\text{Classes}) + \varepsilon_i$$

   Write the estimated regression equation.

e. Are any of the coefficients of the indicator variables significant at the 0.10 level?

f. Explain why it is important to include the income variable in the regression model.

g. After studying these regression results, how would you suggest the management of the sports club chain go about building their new location? Should they use any of the regression models you have estimated? Explain why or why not.

3. The amount of a certain additive injected into a chemical process has a direct effect on the yield. The following table contains data on the amount of additive and yield.

Table for Exercise 3 – Amount of Additive and Yield										
Additive	12.0	6.7	5.6	13.2	8.9	7.8	12.9	16.4	4.5	9.6	5.8
Yield	96	50	42	82	76	70	89	94	15	75	32

a. Assuming that yield is the dependent variable, plot yield against additive. Does the relationship appear to be linear?

b. Using statistical software, estimate the simple linear regression model. Identify R^2 and s_e^2.

c. In instances such as this where linearity does not hold, polynomial regression can be used to provide a better fit to the data. Polynomial regression is a special case of multiple regression where new predictor variables are formed by raising other predictor variables to integral powers. In this exercise, a new predictor will be formed by squaring the values of additive (Add_sq). Yield will then be fitted to the predictors Additive and Add_sq. The prediction equation based upon the polynomial regression is Estimated Yield $= -67.53 + 23.04(\text{Additive}) - 0.82(\text{Add\_sq})$. R^2 and s_e^2 are 0.95 and 47.53, respectively. Predict the yield when Additive $= 16$. Make this prediction using both the linear and polynomial fits. Compare your results.

d. Compare the linear and polynomial fits to the data by the values for R^2 and s_e^2.

e. Which model do you believe is best to use for estimation and prediction? Explain your answer.

4. Suppose that an association of real estate professionals has reported home sales for 2011 in a data set titled HOME SALES. The table contains the current sales by region and the inventory for existing-home sales (single-family and condos/co-ops). An excerpt of the full table is given below. The full table can be downloaded from http://www.hawkeslearning.com/Support/Downloads.htm.

Table for Exercise 4 – Home Sales							
Sale Price	Region	Home Type	Inventory	Sale Price	Region	Home Type	Inventory
$237,000	NE	Condo/Co-op	185,000	$239,600	NE	Single-Family	550,000
$225,400	NE	Condo/Co-op	188,000	$242,400	NE	Single-Family	550,000
$235,200	NE	Condo/Co-op	205,000	$244,600	NE	Single-Family	560,000
$144,900	MW	Condo/Co-op	80,000	$138,800	MW	Single-Family	850,000
$145,000	MW	Condo/Co-op	79,000	$138,900	MW	Single-Family	840,000
$139,200	MW	Condo/Co-op	82,000	$138,600	MW	Single-Family	900,000
$110,400	S	Condo/Co-op	194,000	$153,400	S	Single-Family	1,520,000
$108,100	S	Condo/Co-op	176,000	$153,100	S	Single-Family	1,520,000
$112,100	S	Condo/Co-op	200,000	$150,800	S	Single-Family	1,570,000
$154,600	W	Condo/Co-op	90,000	$223,100	W	Single-Family	940,000
$152,900	W	Condo/Co-op	91,000	$216,300	W	Single-Family	940,000
$146,800	W	Condo/Co-op	74,000	$216,900	W	Single-Family	1,050,000

a. Suggest a regression model that would allow you to predict sale price as a function of inventory, region, and whether you have a condo/co-op or a single-family home.

b. Estimate the model that you suggested in part **a.**

c. What is the estimated equation for predicting sale price by region?

d. What is the estimated equation for predicting sale price by type of home?

e. Is your model estimated in part **b.** statistically useful for predicting sale price at a 1% significance level? Explain your answer.

5. Given the digital revolution in the United States and the trend towards textbooks and lecture materials being made available through electronic means, one wonders if having these technologies will help improve students' grades. To help answer this question, a survey was taken on a university campus inquiring if students used a tablet for their classes, which tablet was used, and their grade point average, household income, and the highest level of education attained by their parents. Using the data from the survey presented in the following table, answer the following questions. (The data presented here are also available for download from http://www.hawkeslearning.com/Support/Downloads.htm in the spreadsheet titled TABLET.)

Key for Student Survey		
Tablet:	**Household Income:**	**Parent's Highest Level of Education:**
1 = Motorola Xoom	1 = < $30,000	1 = Some High School
2 = Samsung Galaxy	2 = $30,000–$49,999	2 = High School Diploma
3 = Apple iPad	3 = $50,000–$74,999	3 = Some College
4 = No Tablet	4 = > $75,000	4 = College Graduate

Table for Exercise 5 – Student Data							
GPA	**Tablet**	**Income**	**Education**	**GPA**	**Tablet**	**Income**	**Education**
3.9622	4	3	1	3.0530	2	3	4
2.9555	1	2	3	3.8034	2	3	3
3.2058	4	2	2	2.3986	2	2	4
3.6487	1	4	2	3.1191	1	3	2
3.4459	1	4	4	2.9556	1	3	3
3.5222	4	1	1	3.3100	2	3	1
3.9964	2	1	3	3.5477	1	3	2
2.4374	1	3	4	3.7710	3	3	1
2.9262	4	3	3	3.2706	1	3	1
3.8684	2	2	3	3.8039	2	1	3
3.2102	1	3	3	3.5149	1	1	2
3.4394	2	3	4	2.0290	2	3	1
3.8169	4	1	3	3.2985	3	1	4
3.6107	2	2	1	2.2877	2	2	2
2.7475	3	2	3	2.4569	3	2	1

a. Suggest a multiple regression model to predict GPA from tablet use and which type of tablet is being used.

b. What is the estimated regression equation for the model proposed in part **a.**?

c. Is the model useful in predicting GPA from whether a student uses a tablet or not and which type of tablet is being used at the 0.05 level? Justify your answer.

d. Discuss whether there is a difference between tablet use and household income regarding how the variables affect one's GPA.

e. Discuss whether there is a difference in the extent to which each type of tablet affects one's GPA.

6. The following table contains a list of high-dividend exchange-traded funds (ETFs). Exchange-traded funds are investment funds traded on stock exchanges, much like stocks. ETFs are traditionally index funds, but ETFs can hold assets such as stocks, commodities, or bonds, and trade at approximately the same price as the net asset value of their underlying assets over the course of the trading day. ETFs may be attractive as investments because of their low costs, tax efficiency, and stock-like features. (The data presented here are also available for download from http://www.hawkeslearning.com/Support/Downloads.htm in the spreadsheet titled ETF.)

Table for Exercise 6 – Exchange-Traded Funds						
ETF	**Share Price ($)**	**Dividend Per Share ($)**	**Dividend Yield (%)**	**ETF**	**Share Price ($)**	**Dividend Per Share ($)**	**Dividend Yield (%)**
1	4.32	0.28	6.49	26	23.82	0.78	3.28
2	15.38	0.92	6.02	27	19.58	0.64	3.26
3	25.25	1.43	5.66	28	136.94	4.41	3.22
4	21.28	1.13	5.31	29	83.57	2.65	3.17
5	698.75	36.88	5.28	30	6.44	0.20	3.12
6	120.55	6.22	5.16	31	13.50	0.42	3.07
7	23.00	1.09	4.74	32	47.47	1.45	3.06
8	13.34	0.62	4.66	33	30.28	0.92	3.03
9	24.82	1.15	4.63	34	14.56	0.43	2.94
10	22.96	1.04	4.53	35	36.06	1.06	2.93
11	78.51	3.21	4.09	36	14.26	0.41	2.90
12	24.22	0.99	4.08	37	369.35	10.56	2.86
13	8.70	0.35	4.02	38	7.07	0.20	2.83
14	15.08	0.58	3.86	39	20.12	0.57	2.82
15	23.44	0.90	3.84	40	1.00	0.03	2.81
16	22.81	0.87	3.80	41	104.15	2.91	2.79
17	22.80	0.84	3.70	42	32.4	0.89	2.75
18	113.84	4.21	3.69	43	47.58	1.30	2.73
19	13.62	0.49	3.60	44	18.43	0.49	2.64
20	22.86	0.82	3.58	45	27.78	0.73	2.63
21	15.58	0.56	3.57	46	30.10	0.76	2.52
22	121.33	4.16	3.43	47	26.55	0.66	2.49
23	15.27	0.51	3.34	48	19.53	0.46	2.35
24	8.70	0.29	3.30	49	27.43	0.62	2.26
25	45.40	1.50	3.30	50	30.33	0.62	2.04

a. Using the data in the table, can dividend yield be predicted by share price and dividend per share? Is it a useful model? Justify your answers.

b. Which variable explains the greatest amount of variability in dividend yield? Explain your answer.

c. Can you detect any multicollinearity in the model containing share price and dividend per share? Explain your answer.

7. The Supplemental Nutrition Assistance Program (SNAP) provides monthly benefits that help eligible low-income households buy the food they need for good health. For most households, SNAP funds account for only a portion of their food budgets, so they must also use their own funds to buy enough food to last throughout the month. Eligible households can receive food assistance through regular SNAP or through the Louisiana Combined Application Project (LaCAP). Using the data in the table, answer the following questions to help predict monthly benefits to eligible households. (The data presented here are also available for download from http://www.hawkeslearning.com/Support/Downloads.htm in the spreadsheet titled SNAP.)

| \multicolumn{6}{c}{Table for Exercise 7 – SNAP Benefits} |
Monthly Benefit ($)	Family Size	Gross Monthly Income ($)	Monthly Benefit ($)	Family Size	Gross Monthly Income ($)
603.41	5	3753	556.42	1	3098
560.69	3	3778	569.05	8	3707
623.24	6	3609	365.80	8	2071
416.12	5	2262	489.08	5	3166
323.90	1	1966	495.86	4	3126
418.78	4	2736	642.77	4	3933
506.46	2	3274	364.81	8	1925
552.53	2	3480	619.30	6	3736
586.46	7	3741	238.71	1	1453
637.18	8	3684	378.94	4	2538
244.49	2	1476	302.58	1	1798
507.19	5	2835	231.74	8	1189
512.56	5	2873	428.67	6	2247
312.89	4	1618	286.99	5	1460
329.05	4	1565	268.81	1	1567
243.49	6	1582	329.81	6	1622
560.37	8	3380	627.25	3	3828
599.90	3	3922	421.52	6	2782
657.09	5	3845	656.38	2	3978
394.82	5	2233	400.64	3	2493

a. Suggest a regression model that will assist SNAP administrators in providing a monthly benefit to eligible households.

b. Fit the model that you suggested in part a. Is this model useful in predicting monthly benefits? Justify your answer.

c. Are all independent variables in the model helpful in explaining the variation in monthly benefits? Explain your answer.

d. Give a 95% confidence interval for average monthly benefits for a four-member household with a gross monthly income of $2500. Interpret this interval.

e. Provide a 99% prediction interval for a four-member household with a gross monthly income of $2500. Interpret this interval.

f. What is the difference between the intervals found in parts d. and e.?

P   Discovery Project

Automobile Prices

Data sets to use for this Discovery Project are available for download from the following website. The data are available in Excel and Minitab format.

http://www.hawkeslearning.com/Support/Downloads.htm

The data set titled AUTOMOBILE contains information that is intended to be used to predict the price of a vehicle (dependent variable) as a function of a number of variables such as the make of the vehicle, fuel-type, wheel-base, city-mpg, etc. Using the data provided, answer the following questions.

a. List the quantitative and qualitative variables in the data.

b. Fit a full (using all independent variables) multiple regression model to the data. What is the coefficient of determination for this model?

c. Are there any variables that are insignificant at the 5% level of significance? If so, remove these variables from the model, and re-run the multiple regression analysis.

d. What is the coefficient of determination for the model in part **c.**? Did you expect it to be larger or smaller than the coefficient of determination in part **b.**? Explain.

e. Having two models in parts **b.** and **c.**, is the coefficient of determination the best statistic to use when comparing the quality of fit? Explain.

f. Develop the "best" and the "simplest" multiple linear regression model to predict the price of a vehicle. Justify the chosen model.

g. Give a general interpretation of the estimate of the slopes in the multiple linear regression model that you've chosen. That is, select any one slope and give its interpretation.

h. Test that all of the slopes of your "best" model are significantly different from zero at the 5% level. What is your conclusion to this test?

i. Which variable plays the biggest role in your model for predicting the price of a vehicle? Justify your answer.

Housing Values

The data set titled HOUSING contains information about the housing values in Boston, MA, in the late 1970s. The variables in the data are used to predict the median value of owner-occupied homes. Using the data provided, answer the following questions.

a. List the quantitative and qualitative variables in the data.

b. Fit a full (using all independent variables) multiple regression model to the data. What is the coefficient of determination for this model?

c. Are there any variables that are insignificant at the 5% level of significance? If so, remove these variables from the model, and re-run the multiple regression analysis.

d. What is the coefficient of determination for the model in part **c.**? Did you expect it to be larger or smaller than the coefficient of determination in part **b.**? Explain.

e. Having two models in parts **b.** and **c.**, is the coefficient of determination the best statistic to use when comparing the quality of fit? Explain.

f. Develop the "best" and the "simplest" multiple linear regression model to predict the median value of a home. Justify the chosen model.

g. Give a general interpretation of the estimate of the slopes in the multiple linear regression model that you've chosen. That is, select any one slope and give its interpretation.

h. Test that all of the slopes of your "best" model are significantly different from zero at the 5% level. What is your conclusion to this test?

i. Which variable plays the biggest role in your model for predicting the median value of a home? Justify your answer.

Chapter Fifteen

Looking for Relationships in Qualitative Data

15

| D | **Discovering the Real World** |

Opinion Polls

In February 2009, shortly after Barack Obama was sworn in as the 44th president of the United States, CNN conducted a survey of 1008 adults (at least 18 years old) asking if they approved or disapproved of the way Obama was handling his job as president. The same survey was given 14 months later in April 2010. The table below summarizes the results of both nationwide polls.

Table 15.1 – CNN/Opinion Research Corporation Poll					
	Approve Strongly	**Approve Moderately**	**Disapprove Moderately**	**Disapprove Strongly**	**Unsure**
February 2009	40%	27%	10%	19%	4%
April 2010	29%	22%	13%	34%	2%

Source: CNN/Opinion Research Corporation

For this survey, the variable of interest is a qualitative variable, opinion. Based on the survey, the ordinal variable (opinion), can take on five categorical values: Approve Strongly, Approve Moderately, Disapprove Moderately, Disapprove Strongly, or Unsure.

In evaluating the survey presented in Table 15.1, we might be interested in knowing if the American opinion about how Barack Obama is handling his job as president had changed significantly from February 2009 to April 2010. There could be a number of reasons for this significant change but most analysts would point to the condition of the economy. President Obama was sworn into office during an economic crisis not seen since the 1930s. Although President Obama "inherited" the state of the economy, perhaps this poll suggests that the American people expected to have seen some significant change in the 14 months between surveys. Given these assertions, there are some important things to remember regarding the survey and its results:

- We do not know the true percentages for all Americans in February 2009 (we only have sample estimates).

- We do not know the true percentages for all Americans in April 2010 (we only have sample estimates).

Any conclusion we wish to draw concerning the relationship between the two opinion polls is complicated by the fact that we have sample estimates rather than population parameters. It could be that the difference in percentages is simply due to sampling variation. Population percentages for Americans could be unchanged and the sole cause of the observed differences could be that different subjects were chosen for the two samples. Thus, an important question is whether or not the difference in percentages is large enough such that it is unlikely to be due to sampling variation alone. This question will be analyzed with the **chi-square test for goodness of fit**, which is discussed in Section 15.2.

A slightly different but similar area of interest is the relationship between two types of qualitative data. For example, a brokerage firm may be interested in determining the relationship between the age of investors and their risk tolerance. A political candidate may be interested in whether or not there is a relationship between the gender of a particular voter and a voter's opinion about abortion. Here, the political candidate is interested in evaluating the relationship between two qualitative variables, gender and opinion on abortion.

In each of the above examples we are interested in determining if a relationship exists between two qualitative variables. The **chi-square test for association** between two qualitative variables, introduced in Section 15.3, will provide the methodology to test the existence of these types of qualitative relationships.

Introduction

In Chapters 5 and 13 we discussed techniques for describing relationships between quantitative variables. However, up to this point we have not discussed any means of effectively summarizing relationships between qualitative variables. This chapter focuses on two methods of summarizing these qualitative relationships. The first method compares the actual proportion of observations appearing in a particular category with the proportion of observations which are expected to appear in that category. The second method determines whether or not two categorical variables are related. Both inferential methods rely on the **chi-square distribution**, which was briefly discussed in Chapter 10.

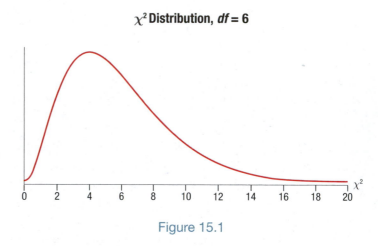

χ^2 Distribution, $df = 6$

Figure 15.1

<table>
<tr><td>15.1</td><td>

The Chi-Square Distribution

</td></tr>
</table>

In Section 10.8 we briefly discussed the **chi-square distribution**, in the context of developing a hypothesis testing procedure for testing claims about a population variance. But the chi-square distribution has much broader applications. Before discussing these applications, let's review a few of the basics about the chi-square distribution.

Formula

Chi-Square Statistic

If n observations are randomly selected from a normal population with variance σ^2, and s^2 is computed for the sample, then the **chi-square statistic**

$$\chi^2 = \frac{(n-1)s^2}{\sigma^2}$$

has a chi-square distribution with $n-1$ degrees of freedom.

χ^2 Distributions

Figure 15.2

A chi-square distribution is a continuous distribution. Unlike the normal distribution and the t-distribution, the chi-square distribution is not symmetric. In fact, for small values of n, it is very skewed, as you can see from Figure 15.2. Another property of the chi-square distribution which is different from the normal and t-distributions is that it only takes on values which are nonnegative. This seems reasonable given that the chi-square distribution is used to represent a ratio which is composed of variances $\left(s^2 \text{ and } \sigma^2\right)$ which must be nonnegative. In fact, σ^2 must be greater than zero for any population in which variation exists.

There are infinitely many chi-square distributions. Each chi-square distribution is uniquely defined by its degrees of freedom. Figure 15.2 shows chi-square distributions for various sample sizes. It is interesting to notice that as n becomes large, the chi-square distribution becomes more and more symmetrical, almost normal-looking.

When analyzing the chi-square distribution, we are usually interested in determining critical values rather than particular probabilities. The concept of a critical value was first introduced in the context of constructing a confidence interval and has been used in all of the chapters which have dealt with inference. The critical value for the chi-square distribution is denoted by χ_α^2. It is a value such that the probability that the chi-square random variable is greater than or equal to that value is equal to α. In hypothesis testing, α, our tolerance for making a Type I error, is specified by the researcher. The chi-square tables contain critical values for various levels of α and many possible degrees of freedom. Tables containing critical values for the chi-square distribution are found in Appendix A.

Determining the Chi-Square Critical Values

Suppose a significance level of $\alpha = 0.05$ has been specified and our sample size is 20. The chi-square distribution has one parameter, degrees of freedom, which is equal to $n - 1$. If the null hypothesis is rejected for large values of the test statistic, we look in the table under the column labeled $\chi_{0.050}^2$ and find the critical value corresponding to $20 - 1$, or 19 degrees of freedom. The corresponding critical value is 30.144, as shown in Figure 15.3.

df	...	$\chi_{0.050}^2$	$\chi_{0.025}^2$	$\chi_{0.010}^2$
1		3.841	5.024	6.635
2		5.991	7.378	9.210
3		7.815	9.348	11.345
⋮				
19		30.144	32.852	36.191
20		31.410	34.170	37.566
⋮				

15

χ^2 **Distribution, df = 19**

$\alpha = 0.05$

0   30.144

Figure 15.3

Suppose a significance level of $\alpha = 0.01$ has been specified and our sample size is 14. If we reject the null hypothesis for large values of the test statistic, we look in the table under the column labeled $\chi^2_{0.010}$ and find the critical value corresponding to $14 - 1$, or 13 degrees of freedom. The corresponding critical value is 27.688, as shown in Figure 15.4.

df	...	$\chi^2_{0.050}$	$\chi^2_{0.025}$	$\chi^2_{0.010}$
1		3.841	5.024	6.635
2		5.991	7.378	9.210
3		7.815	9.348	11.345
⋮				
13		22.362	24.736	27.688
14		23.685	26.119	29.141
⋮				

χ^2 **Distribution, df = 13**

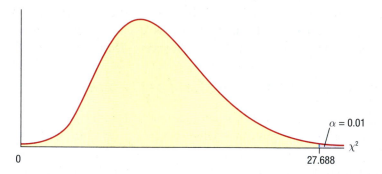

$\alpha = 0.01$

0   27.688

Figure 15.4

The analysis in the remainder of the chapter deals with comparing the actual number of observations falling into a particular category with the number of observations that is expected to fall in that category, based on our hypothesis. In certain circumstances, this type of formulation can be evaluated with a chi-square distribution.

15.1   Exercises

Basic Concepts

1. Describe the shape of the chi-square distribution.

2. What is the sampling distribution of the sample variance?

3. What are the degrees of freedom associated with the chi-square distribution?

4. Can a chi-square statistic ever be negative? Explain why or why not.

5. Describe how the chi-square distribution changes in shape as n becomes large.

6. Explain the meaning of χ_α^2.

7. Explain the procedure for determining chi-square critical values.

Exercises

8. Find the chi-square critical value for each of the following.

   a. $\alpha = 0.01$, $df = 14$

   b. $\alpha = 0.01$, $df = 26$

   c. $\alpha = 0.05$, $df = 4$

   d. $\alpha = 0.05$, $df = 9$

   e. $\alpha = 0.005$, $df = 12$

9. Find the chi-square critical value for each of the following.

   a. $\alpha = 0.005$, $df = 21$

   b. $\alpha = 0.025$, $df = 16$

   c. $\alpha = 0.025$, $df = 1$

   d. $\alpha = 0.10$, $df = 90$

   e. $\alpha = 0.10$, $df = 17$

10. Find the chi-square critical value for each of the following.

    a. $\alpha = 0.01$, $df = 10$

    b. $\alpha = 0.01$, $df = 21$

    c. $\alpha = 0.05$, $df = 6$

    d. $\alpha = 0.05$, $df = 11$

    e. $\alpha = 0.005$, $df = 29$

11. Find the chi-square critical value for each of the following.

    a. $\alpha = 0.005$, $df = 40$

    b. $\alpha = 0.025$, $df = 15$

    c. $\alpha = 0.025$, $df = 2$

    d. $\alpha = 0.10$, $df = 24$

    e. $\alpha = 0.10$, $df = 50$

12. Suppose that a marketing manager is studying sales data for products that are not available in stores and only sold on television. She collects the following weekly sales data for 10 products not sold in stores. Assume the population standard deviation for these data is $5000.

Table for Exercise 12 – Weekly Sales Figures	
Product	Weekly Sales ($)
1	26,259
2	18,514
3	21,579
4	18,739
5	27,821
6	22,511
7	29,753
8	20,235
9	16,258
10	15,990

a. Compute the sample standard deviation for these data. Round your answer to the nearest dollar.

b. Compute the value of χ^2. Round your answer to three decimal places.

c. How many degrees of freedom are associated with this chi-square distribution?

d. What is the value of $\chi^2_{0.05}$ for these data?

13. Michael is studying 30-year fixed mortgage rates in Myrtle Beach, SC. He got quotes from 8 lenders, and the APR rates that were quoted to him are given in the following table.

Table for Exercise 13 – 30-Year Fixed Mortgage Rates	
Lender	APR (%)
EverBank	3.918
AimLoan	3.925
Great Western	4.062
Greenlight	4.353
Flagstar	4.350
AuroraBank	4.040
Quicken	4.458
Roundpoint	4.125

a. Calculate the variance of the sample. Round your answer to six decimal places.

b. Assuming the population standard deviation for the rates is 0.1%, calculate the value of χ^2.

c. Determine the value of $\chi^2_{0.025}$ for these data.

15.2   The Chi-Square Test for Goodness of Fit

An owner of a grocery store chain suspects that there has been a change in the shopping pattern of his customers. He has always believed that his stores are equally busy regardless of the day of the week, and he has geared his weekly advertised specials accordingly.

If the shopping pattern has changed and some days are in fact busier shopping days than others, he estimates that this could be costing him thousands of dollars in lost sales due to improperly scheduled advertisements. He decides to use statistical inference to test his theory. He randomly samples 105 customers from his customer database and asks them which day of the week they go grocery shopping. The results of the survey are listed in Table 15.2. Based on the data, what criteria can the grocer use to decide if some days of the week are preferred for grocery shopping over others?

Table 15.2 – Customer Shopping Data						
Monday	Tuesday	Wednesday	Thursday	Friday	Saturday	Sunday
10	15	14	16	11	20	19

In 1900, Karl Pearson developed the **chi-square test for goodness of fit**. The test statistic for the chi-square test for goodness of fit is a summary measure which compares the actual percentage of observations which fall into a particular category with the expected percentage for that category.

In order to use the chi-square test for goodness of fit for inference, our experiment must satisfy some basic conditions. First, we must be able to assume that the underlying distribution of the number of shoppers shopping on the various days of the week has a **multinomial probability distribution**. The multinomial probability distribution is simply an extension of the familiar binomial probability distribution. The properties of the multinomial experiment are listed in the following box.

Properties

Multinomial Experiment

1. The experiment consists of n independent, identical trials.

2. There are k possible outcomes for each trial.

3. The probabilities of the k outcomes, $p_1, p_2, ..., p_k$, are constant from trial to trial.

4. The random variables of interest are the counts in each of the k possible outcomes, $n_1, n_2, ..., n_k$.

The grocer's survey of the number of shoppers shopping on the various days of the week seems to satisfy the conditions of a multinomial experiment. Since the customers were randomly selected and each of them was asked the same question, there were 105 independent, identical trials. The true proportions of shoppers shopping on each day of the week are unknown and constant from trial to trial.

The proportion of shoppers shopping on each day of the week form a probability distribution whose probabilities sum to one.

15

Using the procedure outlined in Section 10.3 for testing a hypothesis, we must first define the hypotheses to be tested in plain English. Our grocer might formulate his hypotheses as follows.

- **Null Hypothesis:** The proportion of shoppers does not vary by day of the week

- **Alternative Hypothesis:** The proportion of shoppers varies by day of the week.

Next, select the appropriate statistical measure. In this situation, the proportion is the obvious measure of choice. However, there are several proportions of interest, namely, the proportion of shoppers shopping on each day of the week. Therefore, let

$$p_1 = \text{proportion of shoppers shopping on Monday,}$$
$$p_2 = \text{proportion of shoppers shopping on Tuesday,}$$
$$\vdots$$
$$p_7 = \text{proportion of shoppers shopping on Sunday.}$$

Step 3 directs us to decide if the test will be one-tailed or two-tailed. As we will see in Step 6, this will be a one-tailed test.

Note: This type of chi-square test is also referred to as the goodness of fit test for homogeneity. It is used to determine whether frequency counts are distributed identically across different populations.

For Step 4, we specify the hypotheses as follows.

$$H_0: p_1 = p_2 = p_3 = p_4 = p_5 = p_6 = p_7 = \frac{1}{7}$$

(The proportion of shoppers does not vary by day of the week.)

$$H_a: \text{More than one proportion does not equal } \frac{1}{7}.$$

(The proportion of shoppers varies by day of the week.)

For Step 5, our grocer must choose a level of the test. Let's assume that he is willing to live with a 5% chance of making a Type I error. Namely, concluding that the proportion of shoppers varies by day of the week, when in fact it doesn't. Therefore, he chooses a level of $\alpha = 0.05$.

To determine the appropriate test statistic required for Step 6, let's consider the statement of the null hypothesis. If the null hypothesis is true, then

$$H_0: p_1 = p_2 = p_3 = p_4 = p_5 = p_6 = p_7 = \frac{1}{7},$$

and 15 $\left(\frac{1}{7} \text{ of } 105 \right)$ of the 105 customers would be expected to shop on each day of the week. If we let $n_1, n_2, ..., n_7$ represent the number of customers who shop on the seven days of the week, respectively, then this expectation would be formally stated as follows.

$$E(n_1) = n \cdot p_1 = 105 \cdot \frac{1}{7} = 15$$

Similarly, $E(n_2) = E(n_3) = \cdots = E(n_7) = 15.$

To test the hypothesis we need to measure how close reality is (the data given by the shoppers) to what is stated in the null hypothesis. This measure will become a test statistic. The test statistic will be developed as part of a criterion which will be used to make a decision whether or not to reject the null hypothesis. Intuitively, the test statistic should compare the expected number of shoppers for each day of the week given that our null hypothesis is true, $(E(n_i))$, to the actual number of shoppers reported in the grocer's survey (n_i).

A reasonable test statistic would consist of the differences between the number of shoppers the grocer expected to see each day and the number of shoppers the grocer actually observed each day. In attempting to summarize this information, it is tempting to simply add these differences. However, negative and positive differences will cancel each other out. Further, we have not yet adjusted these differences for any potential differences in scale among the days of the week. (**Note:** In our problem the null hypothesis implies there is no difference in scale. However, the methodology is designed to handle hypothesized differences in scale among the categories.) To compensate for these problems, square the differences so that positive and negative differences will not cancel each other out, and divide each squared difference by the appropriate $E(n_i)$ to adjust for any differences in scale. This will give us the following test statistic.

$$\chi^2 = \frac{\left[n_1 - E(n_1)\right]^2}{E(n_1)} + \frac{\left[n_2 - E(n_2)\right]^2}{E(n_2)} + \cdots + \frac{\left[n_7 - E(n_7)\right]^2}{E(n_7)}$$

If the null hypothesis is true and the data satisfy the conditions of a multinomial distribution, this test statistic has an approximate chi-square distribution with $7 - 1 = 6$ degrees of freedom.

If the differences between what is observed and what is expected are large enough, the chi-square test statistic quantity will become large, causing us to doubt the reasonableness of the null hypothesis. Thus, for **Step 7** we will reject the null hypothesis for large values of the test statistic (we have a one-tailed test). How large is *large*?

How large must the value of χ^2 be in order to conclude the null hypothesis is not reasonable? We must have a criterion. The critical value for a chi-square distribution with a level of significance of 0.05, $\chi^2_{0.050}$, with 6 degrees of freedom is 12.592. Thus, if the null hypothesis is true, a test statistic as large or larger than 12.592 will occur only 5% of the time due to sampling variation. Hence, if we observe a value of the test statistic that is greater than or equal to 12.592 we will reject the null hypothesis. This rejection region is drawn in Figure 15.5.

Karl Pearson
1857 – 1936
Pearson was educated at home until he was nine, and then he was sent to University College where he eventually earned a degree in mathematics. After studying in Germany he returned to University College as a teacher and lecturer. Karl Pearson is regarded as one of the founders of modern statistics. In addition to developing the correlation coefficient, he developed the chi-square test (which included the development of the chi-square sampling distribution) as a means for assessing relationships for categorical data.

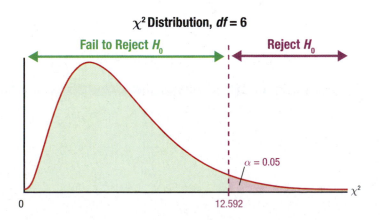

Figure 15.5

For Step 8, the resulting value of the test statistic is as follows.

$$\chi^2 = \frac{[10-15]^2}{15} + \frac{[15-15]^2}{15} + \frac{[14-15]^2}{15} + \frac{[16-15]^2}{15}$$
$$+ \frac{[11-15]^2}{15} + \frac{[20-15]^2}{15} + \frac{[19-15]^2}{15}$$
$$= 5.6$$

Notice that the value of the test statistic does not fall in the rejection region since 5.6 is less than 12.592. Thus for Step 9, we conclude that the data do not provide enough evidence to reject the null hypothesis.

Figure 15.6

Note: The P-value of the χ^2 test statistic in this case is approximately 0.4695.

Our conclusion for Step 10 is that the grocer's suspicion that some shopping days are busier than others was not substantiated. Based on the evidence gathered in his study he should not change his advertising practice.

The basic steps in performing a chi-square test for goodness of fit are outlined in the following box. It is important to note that in order to use this test we must assume that the sample size, n, is large enough that there will be at least five expected observations per category and that the underlying distribution is a multinomial probability distribution.

Procedure

Chi-Square Test for Goodness of Fit

Hypotheses:

$$H_0: p_1 = p_{1,0}, \; p_2 = p_{2,0}, \; \ldots, \; p_k = p_{k,0}$$
$$H_a: \text{Any possible difference.}$$

Test Statistic:

$$\chi^2 = \sum_{i=1}^{k} \frac{\left[n_i - E(n_i)\right]^2}{E(n_i)}$$

where n_i is the actual number of observations observed in each category, and $E(n_i)$ is the expected number of observations for each category given that the null hypothesis is true.

Assuming that the null hypothesis is true, and n is large enough so that $E(n_i)$ is at least 5 for each category, the test statistic has a chi-square distribution with $k-1$ degrees of freedom.

Rejection Region:

Reject H_0 if $\chi^2 \geq \chi_\alpha^2$ with $k-1$ degrees of freedom.

Example 15.1

Let's return to the survey about the approval rating of Barack Obama as president in February 2009 and April 2010. A cursory look at Table 15.3 appears to suggest that the American public's perception of his job performance decreased over time. This difference may simply be due to random variation in the samples or it could represent a real shift in the public sentiment. We will rely on statistical inference methods to determine if the differences are too large to be attributed to sampling variation alone. The chi-square test for goodness of fit can be used to answer the question of whether or not there had been any change in the American sentiment towards Barack Obama.

Table 15.3 – Obama Approval Ratings		
Opinion	February 2009	April 2010
Approve Strongly	40%	29%
Approve Moderately	27%	22%
Disapprove Moderately	10%	13%
Disapprove Strongly	19%	34%
Unsure	4%	2%

The question: *Do you approve or disapprove of the way Barack Obama is handling his job as president?* If approve: *Do you approve strongly, or only moderately?* If disapprove: *Do you disapprove strongly, or only moderately?*

There are 1008 respondents in the sample. Is there sufficient evidence to conclude at $\alpha = 0.10$ that American sentiment towards Barack Obama changed from February 2009 to April 2010?

Solution

Step 1: State the hypotheses in plain English.

- **Null Hypothesis:** American sentiment did not change from February 2009 to April 2010.

- **Alternative Hypothesis:** American sentiment did change from February 2009 to April 2010.

Step 2: Select the appropriate statistical measure.

The variable to be analyzed is opinion, which has five possible values: Approve Strongly, Approve Moderately, Disapprove Moderately, Disapprove Strongly, and Unsure. Further, our interest is in the proportion (or percent) of Americans that falls into each of the opinion categories:

$$p_1 = \text{proportion of Americans who Approve Strongly.}$$
$$p_2 = \text{proportion of Americans who Approve Moderately.}$$
$$p_3 = \text{proportion of Americans who Disapprove Moderately.}$$
$$p_4 = \text{proportion of Americans who Disapprove Strongly.}$$
$$p_5 = \text{proportion of Americans who are Unsure.}$$

Step 3: Determine whether the hypothesis should be one-sided or two-sided.

The chi-square test for goodness of fit is always a one-tailed test because of the way the test statistic is constructed. We will reject the null hypothesis in favor of the alternative hypothesis for large values of the test statistic.

Step 4: Specify the hypotheses using the appropriate statistical measure.

The resulting hypotheses would be:

$$H_0: p_1 = 0.40, \ p_2 = 0.27, \ p_3 = 0.10, \ p_4 = 0.19, \ p_5 = 0.04$$

(The American sentiment did not change from February 2009 to April 2010.)

$$H_a: \text{Any possible difference.}$$

Step 5: Specify the level of the test.

The level of the test is specified in the statement of the problem as $\alpha = 0.10$.

Step 6: Select the appropriate test statistic.

The number of Americans falling into a particular sentiment category satisfies the properties of a multinomial probability distribution, and the expected number of observations in each category is at least 5. Thus, we can use the chi-square test for goodness of fit. The chi-square test statistic is given by

$$\chi^2 = \frac{\left[n_1 - E(n_1)\right]^2}{E(n_1)} + \frac{\left[n_2 - E(n_2)\right]^2}{E(n_2)} + \frac{\left[n_3 - E(n_3)\right]^2}{E(n_3)}$$
$$+ \frac{\left[n_4 - E(n_4)\right]^2}{E(n_4)} + \frac{\left[n_5 - E(n_5)\right]^2}{E(n_5)}.$$

This test statistic has an approximate chi-square distribution with $5 - 1 = 4$ degrees of freedom, assuming the null hypothesis is true.

Step 7: Determine the critical value.

Since the level of the test is $\alpha = 0.10$ and we reject the null hypothesis for large values of the test statistic, the chi-square critical value is $\chi^2_{0.100}$ with 4 degrees of freedom, or 7.779.

Figure 15.7

Step 8: Compute the test statistic.

To compute the test statistic, the expected number of observations in each category must be computed. Using the proportions from the February 2009 data, we would expect the number of respondents in each category in April 2010 to be as follows.

$$E(n_1) = 1008(0.40) \approx 403$$
$$E(n_2) = 1008(0.27) \approx 272$$
$$E(n_3) = 1008(0.10) \approx 101$$
$$E(n_4) = 1008(0.19) \approx 192$$
$$E(n_5) = 1008(0.04) \approx 40$$

The actual number of respondents in each category in the April 2010 sample is as follows.

$$n_1 = 1008(0.29) \approx 292$$
$$n_2 = 1008(0.22) \approx 222$$
$$n_3 = 1008(0.13) \approx 131$$
$$n_4 = 1008(0.34) \approx 343$$
$$n_5 = 1008(0.02) \approx 20$$

Our test statistic is calculated as

$$\chi^2 = \frac{[292-403]^2}{403} + \frac{[222-272]^2}{272} + \frac{[131-101]^2}{101} + \frac{[343-192]^2}{192} + \frac{[20-40]^2}{40}$$
$$\approx 177.430.$$

Step 9: Make the decision.

Figure 15.8

If the null hypothesis is true, the test statistic will be greater than or equal to the critical value of 7.779 only 10% of the time. Since $\chi^2 \approx 177.430$ is larger than 7.779, we will reject H_0. Large values of the test statistic indicate that the proportion of Americans falling into a category changed from February 2009 to April 2010, and the change is too great to be due to ordinary sampling variation.

Step 10: State the conclusion in terms of the original question.

At the 10% level of significance, there is sufficient evidence to conclude that American sentiment towards Barack Obama changed from February 2009 to April 2010. The difference in sentiment from February 2009 to April 2010 is much too great to be attributed to ordinary sampling variation alone.

15.2 Exercises

Basic Concepts

1. Describe what the test statistic for the chi-square test for goodness of fit measures.

2. What is a multinomial probability distribution? What more familiar probability distribution discussed previously in the text is a multinomial probability distribution related to?

3. List the four requirements for a multinomial experiment.

4. What are the null and alternative hypotheses for a chi-square test for goodness of fit?

5. What is the test statistic for a chi-square test for goodness of fit?

6. How many degrees of freedom does the test statistic for the chi-square test for goodness of fit have?

7. What assumptions are necessary for a chi-square test for goodness of fit?

8. How are the expected values determined in a chi-square test for goodness of fit?

Exercises

9. A telephone company claims that the service calls they receive are equally distributed among the five working days of the week. A survey of 85 randomly selected service calls produced the following results.

Table for Exercise 9 – Service Calls				
	Monday	Tuesday	Wednesday	Thursday	Friday
Number of Calls	15	20	25	15	10

a. Is the company's claim refuted by the data at $\alpha = 0.05$?

b. What assumptions were made in the test for part **a.**?

10. Suppose a consumer affairs representative for Mars Incorporated claims that M&M's plain chocolate candies are mixed such that each large production batch has "precisely" the following ratios of colored candies: 30% brown, 20% yellow, 20% red, 10% orange, 10% green, and 10% blue. To test this claim, a professor distributed small sample bags of M&M's to students and had them count the number of candies of each color. The counts of the students were then pooled with the following results.

Table for Exercise 10 – Candy Colors						
	Brown	Yellow	Red	Orange	Green	Blue	Total
Number of Candies	84	79	75	49	36	47	370

a. If the representative's claim is true, what would be the expected number of candies in each of the color categories for 370 candies?

b. Is the representative's claim refuted by the data at $\alpha = 0.01$?

c. What assumptions were made in performing the test for part **b.**?

11. A highway department executive claims that the number of fatal accidents which occur in her state does not vary from month to month. A survey of 170 fatal accidents produced the following results.

Table for Exercise 11 – Accidents											
	Jan.	Feb.	Mar.	Apr.	May	Jun.	July	Aug.	Sept.	Oct.	Nov.	Dec.
Accidents	18	16	7	5	8	12	15	18	15	11	20	25

a. Is the executive's claim refuted by the data at $\alpha = 0.01$?

b. What assumptions were made in the test for part **a.**?

12. The market research firm Nielson recently published market share figures for the operating systems in smartphones. The report stated the following results.

Table A for Exercise 12 – Market Share for Smartphone Operating Systems	
Operating System	**Market Share (%)**
Android OS	29
iPhone OS	27
Blackberry OS	27
Microsoft Windows Mobile	10
HP Palm/WebOS	4
Symbian OS	2
Other	1

Source: Nielson.com

Suppose that a marketing manager for a telecommunications company that uses one of the above operating systems doubts the Nielson findings. He collects his own data by surveying 400 people at a local mall. His findings are given in the following table.

Table B for Exercise 12 – Survey Results	
Operating System	**Number of People**
Android OS	125
iPhone OS	115
Blackberry OS	99
Microsoft Windows Mobile	56
HP Palm/WebOS	5
Symbian OS	0
Other	0

a. Compute the expected number of observations for each category for the survey conducted by the telecommunications marketing manager.

b. State the null and alternative hypotheses for the chi-square test for goodness of fit.

c. Using $\alpha = 0.05$, perform a goodness of fit test to determine if the survey conducted by the marketing manager is evidence that the market shares reported by Nielson have changed.

d. What assumptions were made in the test for part **c.**?

e. Do you have any concerns about the way in which the marketing manager's survey was conducted? Explain.

13. A psychologist conducted an attitude survey of 200 randomly selected individuals several years ago. The individuals were asked to pick the one category which most accurately described their attitudes. The results of the survey were as follows.

Table A for Exercise 13 – 1st Attitude Survey	
Attitude	**Percent of Respondents**
Optimistic	15%
Slightly Optimistic	30%
Slightly Pessimistic	30%
Pessimistic	25%

The psychologist believes that these attitudes have changed over time. To test this theory, he randomly selects 200 individuals and asks them the same questions. The results of the second survey are as follows.

Table B for Exercise 13 – 2nd Attitude Survey	
Attitude	**Percent of Respondents**
Optimistic	20%
Slightly Optimistic	40%
Slightly Pessimistic	30%
Pessimistic	10%

a. Can the psychologist conclude that the attitudes have changed over time at $\alpha = 0.01$?

b. What assumptions were made in the test for part **a.**?

15.3 The Chi-Square Test for Association between Two Qualitative Variables

Sometimes, our interest extends beyond one variable to summarizing the relationship between two qualitative variables. For example, a radio executive might be interested in knowing if the gender of an individual affects that person's preference for music. Here the qualitative variables of interest are gender, which can take on the values Male and Female, and preference for music which can take on the values Classical, Jazz, Easy Listening, Rock, Rap, and Country. Other examples of relationships between two qualitative variables which might be of interest are age and political preference, education level and job performance, income level and occupation, etc.

When we are interested in this type of relationship, we often make use of a **contingency table**. A contingency table organizes data on two characteristics simultaneously. Each cell in a contingency table contains either a count or a proportion which represents the number of observations falling into that combination of categories. It is important to note that contingency tables are composed of two qualitative variables and each variable satisfies the properties of a multinomial distribution.

Table 15.4 shows the general form of a contingency table.

Table 15.4 – General Form of a Contingency Table			
	Factor A		
	Level A1	**Level A2**	**Total**
Factor B **Level B1**	n_1	n_2	$n_1 + n_2$
Level B2	n_3	n_4	$n_3 + n_4$
Total	$n_1 + n_3$	$n_2 + n_4$	$n = n_1 + n_2 + n_3 + n_4$

For example, the editor of a regional newspaper is concerned about dwindling sales. Due to her concerns, she's interested in the primary method by which people receive news. She surveyed 990 residents (in three different age groups) in the region and asked them how they get their news (newspaper, radio, television, or the Internet). For those who responded, the contingency table shown in Table 15.5 resulted.

Table 15.5 – Method of Receiving News					
Age Group	**Newspaper**	**Radio**	**Television**	**Internet**	**Total**
18–24	30	20	60	200	310
25–40	100	75	75	150	400
Over 40	125	50	75	30	280
Total	255	145	210	380	990

One of the interesting features of this table is that the age variable, whose level of measurement would ordinarily be at the ratio level, has been divided into three categories. Age is now a qualitative variable measured at the ordinal level. What is the advantage of reducing the level of measurement of a variable? To use the test we are about to develop, both variables must be qualitative (i.e. nominal or ordinal levels of measurement). The other variable in the table is the method by which the person receives his or her news, which is also qualitative.

Notice that the column totals and the row totals both sum to the number of readers who responded to the survey, 990. This means that each respondent was allowed to choose only one category. This is necessary if the table is to satisfy the conditions of a multinomial experiment. If the survey is designed such that respondents are allowed to choose more than one category, the following analysis will not apply.

Suppose we are interested in knowing if there is some relationship between age group and method of receiving the news. Another way of thinking about the same problem is to ask if the primary method of receiving the news is dependent on the age group. Rephrasing the question in this manner is important because a formal definition of dependence was developed in Chapter 5. We will use this definition in the development of the hypothesis test. Further, stating that two variables are dependent implies that they are related.

Recall the multiplication rule for independent events from Section 5.6. It states that if two events A and B are independent, then $P(A \cap B) = P(A)P(B)$. If we consider each cell in our contingency table to be the intersection of two events, age group and primary method of receiving news, we can use this multiplication rule to help determine whether or not these two events are dependent. To see this, it is helpful to express our contingency table in terms of relative frequencies or proportions as shown in Table 15.6.

Table 15.6 – Method of Receiving News				
Age Group	**Newspaper**	**Radio**	**Television**	**Internet**	**Total**
18–24	$p_{1N} = \dfrac{30}{990}$	$p_{1R} = \dfrac{20}{990}$	$p_{1T} = \dfrac{60}{990}$	$p_{1I} = \dfrac{200}{990}$	$p_1 = \dfrac{310}{990}$
25–40	$p_{2N} = \dfrac{100}{990}$	$p_{2R} = \dfrac{75}{990}$	$p_{2T} = \dfrac{75}{990}$	$p_{2I} = \dfrac{150}{990}$	$p_2 = \dfrac{400}{990}$
Over 40	$p_{3N} = \dfrac{125}{990}$	$p_{3R} = \dfrac{50}{990}$	$p_{3T} = \dfrac{75}{990}$	$p_{3I} = \dfrac{30}{990}$	$p_3 = \dfrac{280}{990}$
Total	$p_N = \dfrac{255}{990}$	$p_R = \dfrac{145}{990}$	$p_T = \dfrac{210}{990}$	$p_I = \dfrac{380}{990}$	1

The relative frequencies in Table 15.6 were determined in the following manner.

$$p_{1N} = P(18-24 \cap \text{Newspaper}) = \frac{30}{990} = \text{ the proportion of residents ages 18–24 that used the newspaper as their primary news source.}$$

$$p_N = P(\text{Newspaper}) = \frac{255}{990} = \text{ the proportion of residents that used the newspaper as their primary news source.}$$

$$p_1 = P(18-24) = \frac{310}{990} = \text{ the proportion of residents ages 18–24 in the sample.}$$

If age group and method of receiving the news are independent, then

$$p_{1N} = p_1 \cdot p_N$$
$$p_{1R} = p_1 \cdot p_R$$
$$\vdots$$
$$p_{3I} = p_3 \cdot p_I.$$

Given that we are interested in determining whether or not the primary method of receiving the news is dependent on the age group, and we have developed some relationships that should hold true if the events are independent, it seems reasonable to formulate our hypotheses as follows.

- **Null Hypothesis:** Age group and method of receiving the news are independent.

- **Alternative Hypothesis:** Age group and method of receiving the news are dependent.

Assuming that the null hypothesis is true (age group and method of receiving the news are independent), then we would expect the number of respondents who received their news by newspapers and were in the 18–24 age group, $E(n_{1N})$, to be equal to the total number of people surveyed (n) times the true probability that a person is between 18–24 years old (p_1) times the true probability that a person receives his or her news via newspaper (p_N). Since the true probabilities of the age group 18–24 and that people receive their news from newspapers are unknown, the results of the survey can be used to estimate these probabilities. This can be written symbolically as

$$E(n_{1N}) = (n)(p_1)(p_N) = (990)\left(\frac{310}{990}\right)\left(\frac{255}{990}\right) \approx 79.85.$$

This is equivalent to multiplying the row total by the column total and dividing by the total number of observations. We will use the following formula to calculate the expected value for each cell (see Table 15.7).

$$E(n_{ij}) = \frac{(\text{row } i \text{ total})(\text{column } j \text{ total})}{\text{total number of observations}}$$

Table 15.7 – Expected Value Calculation

Age Group	Newspaper	Radio	Television	Internet	Total
18–24	$\frac{310 \cdot 255}{990} \approx 79.85$				310
25–40					400
Over 40					280
Total	255	145	210	380	990

Table 15.8 displays the expected values for each cell and Table 15.9 displays the observed values alongside the expected values.

Table 15.8 – Expected Values

Age Group	Newspaper	Radio	Television	Internet	Total
18–24	$\frac{310 \cdot 255}{990} \approx 79.85$	$\frac{310 \cdot 145}{990} \approx 45.40$	$\frac{310 \cdot 210}{990} \approx 65.76$	$\frac{310 \cdot 380}{990} \approx 118.99$	310
25–40	$\frac{400 \cdot 255}{990} \approx 103.03$	$\frac{400 \cdot 145}{990} \approx 58.59$	$\frac{400 \cdot 210}{990} \approx 84.85$	$\frac{400 \cdot 380}{990} \approx 153.54$	400
Over 40	$\frac{280 \cdot 255}{990} \approx 72.12$	$\frac{280 \cdot 145}{990} \approx 41.01$	$\frac{280 \cdot 210}{990} \approx 59.39$	$\frac{280 \cdot 380}{990} \approx 107.47$	280
Total	255	145	210	380	990

	Newspaper		Radio		Television		Internet	
Age Group	**Observed**	**Expected**	**Observed**	**Expected**	**Observed**	**Expected**	**Observed**	**Expected**
18–24	30	79.85	20	45.40	60	65.76	200	118.99
25–40	100	103.03	75	58.59	75	84.85	150	153.54
Over 40	125	72.12	50	41.01	75	59.39	30	107.47

Table 15.9 – Observed versus Expected Values

Using the same reasoning employed in the chi-square test for goodness of fit, we can construct a chi-square statistic using the differences between the observed and expected values. Notationally, this statistic can be written as follows.

$$\chi^2 = \sum \frac{(\text{Observed} - \text{Expected})^2}{\text{Expected}}$$

Given Table 15.9, the chi-square statistic is calculated as follows.

$$\chi^2 = \frac{\left[n_{1N} - E(n_{1N})\right]^2}{E(n_{1N})} + \frac{\left[n_{2N} - E(n_{2N})\right]^2}{E(n_{2N})} + \cdots + \frac{\left[n_{3I} - E(n_{3I})\right]^2}{E(n_{3I})}$$

Big differences between the observed and expected values will cause the χ^2 statistic to become large. If the statistic is too large to be due to ordinary sampling variation alone, this will cast doubt on the null hypothesis and cause us to reject the belief that age group and the method of receiving the news are independent. If the assumption of independence is true and the expected number of observations in each cell is at least five, then the sampling distribution of the test statistic has an approximate chi-square distribution with $(r-1)(c-1)$ degrees of freedom where r is the number of rows and c is the number of columns.

If we choose $\alpha = 0.01$, we will reject the null hypothesis if the chi-square test statistic is larger than or equal to the chi-square critical value of 16.812 which is $\chi^2_{0.01}$ with $(3-1)(4-1) = 6$ degrees of freedom.

χ^2 **Distribution, df = 6**

$\alpha = 0.01$

0   16.812 χ^2

Figure 15.9

For our data, the chi-square test statistic is given by

$$\chi^2 = \frac{[30 - 79.85]^2}{79.85} + \frac{[100 - 103.03]^2}{103.03} + \cdots + \frac{[30 - 107.47]^2}{107.47} \approx 207.590$$

15

Figure 15.10

Since the value of the test statistic falls in the rejection region, we will reject the null hypothesis that age group and preferred method of getting the news are independent. There is sufficient evidence at the 0.01 level to conclude that age group and preferred method of receiving the news are dependent. In other words, there is a relationship between the two variables. A summary of the chi-square test for association between two qualitative variables is given in the following box. It is important to note that n should be large enough such that the expected cell count (number of observations) in each cell will be equal to five or more.

Procedure

Chi-Square Test for Association between Two Qualitative Variables

Hypotheses:

H_0: The two qualitative variables are independent (not related).

H_a: The two qualitative variables are dependent (related).

Test Statistic:

$$\chi^2 = \sum_{i=1}^{r}\sum_{j=1}^{c}\frac{\left[n_{ij} - E\left(n_{ij}\right)\right]^2}{E\left(n_{ij}\right)}$$

where r = the number of rows and c = the number of columns.

$$E\left(n_{ij}\right) = np_i p_j = \frac{\left(\text{row } i \text{ total}\right)\left(\text{column } j \text{ total}\right)}{n}$$

where n is the total number of observations and p_i and p_j are estimates of the true population proportions calculated as follows.

$$p_i = \frac{\text{the number of observations in row } i}{n}$$

$$p_j = \frac{\text{the number of observations in column } j}{n}$$

If the null hypothesis is true and n is large enough so that the expected number of observations in each cell is at least 5, then the test statistic has an approximate chi-square distribution with $(r-1)(c-1)$ degrees of freedom.

Rejection Region:

Reject H_0 if $\chi^2 \geq \chi_\alpha^2$ with $(r-1)(c-1)$ degrees of freedom.

Example 15.2

Many grocery stores have self-checkout scanners so that customers can check out faster. Some customers are concerned that the scanners will be wrong and overcharge them. However, the manufacturer of the scanners indicates that the scanners will err in favor of the customer most of the time. In a random sample of customers checking themselves out of the grocery store as well as those that allow the employees to check them out, we have the following contingency table.

Table 15.10 – Self Checkout versus Employee Checkout at the Grocery Store			
Price Charged	Self Checkout	Employee Checkout	Total
Undercharge	20	10	30
Overcharge	15	30	45
Correct Price	200	225	425
Total	235	265	500

Based on the data, can we conclude that the price charged using scanners is independent of whether a customer checks themselves out or if a store employee checks them out?

Solution

Step 1: State the hypotheses in plain English.

- **Null Hypothesis:** The price charged at checkout and method of checkout are independent (not related).

- **Alternative Hypothesis:** The price charged at checkout and method of checkout are dependent (related).

Step 2: Select the appropriate statistical measure.

For this example, we are interested in the proportion of data in each of the Price Charged and Checkout Method categories.

Step 3: Determine whether the hypothesis should be one-sided or two-sided.

The chi-square test for association between two qualitative variables is always a one-tailed test because of the way we construct the test statistic. We will reject the null hypothesis that price charged at checkout is independent of checkout method for large values of the test statistic.

Step 4: Specify the hypotheses using the appropriate statistical measure.

The resulting hypotheses would be the following.

H_0: Price charged at checkout and method of checkout are independent (not related).

H_a: Price charged at checkout and method of checkout are dependent (related).

Step 5: Specify the level of the test.

We must specify the level of the test ourselves since we are not given one in the problem. Let's choose $\alpha = 0.10$.

Step 6: Select the appropriate test statistic.

Since each piece of data cannot belong to more than one category and the expected cell count for each category is at least 5, we can use the chi-square test for association between two qualitative variables. The formula for the test statistic is given as follows.

$$\chi^2 = \frac{\left[n_{US} - E\left(n_{US}\right)\right]^2}{E\left(n_{US}\right)} + \frac{\left[n_{OS} - E\left(n_{OS}\right)\right]^2}{E\left(n_{OS}\right)} + \frac{\left[n_{CS} - E\left(n_{CS}\right)\right]^2}{E\left(n_{CS}\right)}$$
$$+ \frac{\left[n_{UE} - E\left(n_{UE}\right)\right]^2}{E\left(n_{UE}\right)} + \frac{\left[n_{OE} - E\left(n_{OE}\right)\right]^2}{E\left(n_{OE}\right)} + \frac{\left[n_{CE} - E\left(n_{CE}\right)\right]^2}{E\left(n_{CE}\right)}$$

This test statistic has an approximate chi-square distribution with $(3-1)(2-1)=2$ degrees of freedom, assuming the null hypothesis is true.

Step 7: Determine the critical value.

Since the level of the test is $\alpha = 0.10$ and we will reject the null hypothesis for large values of the test statistic, the chi-square critical value is $\chi^2_{0.100}$ with $(3-1)(2-1)=2$ degrees of freedom, or 4.605. Values of the test statistic greater than or equal to 4.605 would indicate that the dependence of the two classifications is not likely due to ordinary sampling variation alone.

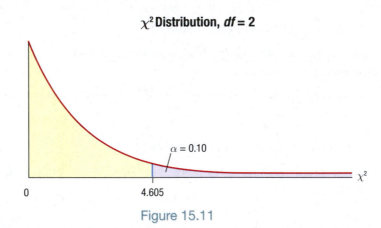

χ^2 **Distribution, df = 2**

$\alpha = 0.10$

0   4.605 χ^2

Figure 15.11

Step 8: Compute the test statistic.

The expected values are calculated in Table 15.11, and Table 15.12 gives the actual data values in each category versus the expected values in each category.

Table 15.11 – Self Checkout versus Employee Checkout at the Grocery Store: Expected Values			
Price Charged	Self Checkout	Employee Checkout	Total
Undercharge	$\dfrac{30 \cdot 235}{500} = 14.10$	$\dfrac{30 \cdot 265}{500} = 15.90$	30
Overcharge	$\dfrac{45 \cdot 235}{500} = 21.15$	$\dfrac{45 \cdot 265}{500} = 23.85$	45
Correct Price	$\dfrac{425 \cdot 235}{500} = 199.75$	$\dfrac{425 \cdot 265}{500} = 225.25$	425
Total	235	265	500

Table 15.12 – Self Checkout versus Employee Checkout at the Grocery Store: Observed and Expected Values				
Price Charged	Self Checkout		Employee Checkout	
	Observed	Expected	Observed	Expected
Undercharge	20	14.10	10	15.90
Overcharge	15	21.15	30	23.85
Correct Price	200	199.75	225	225.25

Based on these results, the test statistic is calculated as follows.

$$\chi^2 = \frac{\left[20-14.10\right]^2}{14.10} + \frac{\left[15-21.15\right]^2}{21.15} + \frac{\left[200-199.75\right]^2}{199.75}$$
$$+ \frac{\left[10-15.90\right]^2}{15.90} + \frac{\left[30-23.85\right]^2}{23.85} + \frac{\left[225-225.25\right]^2}{225.25}$$
$$\approx 8.033$$

Step 9: Make the decision.

Since the χ^2 test statistic is greater than the critical value and falls in the rejection region, the null hypothesis should be rejected.

Figure 15.12

Step 10: State the conclusion in terms of the original question.

There is significant evidence at the 0.10 level to conclude that the price charged at checkout and checkout method are dependent. The difference is much too great to be attributed to ordinary sampling variation alone.

15

15.3 Exercises

Basic Concepts

1. Give two examples of relationships between qualitative variables that would be of interest to a manager in a business setting.

2. Explain the difference between the chi-square test for goodness of fit and the chi-square test for association.

3. What is a contingency table?

4. Describe the information that each cell in a contingency table gives.

5. What properties must the two categories of the contingency table possess?

6. What level(s) of measurement may the categories of a contingency table have?

7. Consider the variable income. Describe how this variable could be transformed to be included in a contingency table. Is information lost during the transformation?

8. Explain why a test for association is not valid if single data points are allowed to belong to more than one category.

9. Restate the multiplication rule for independent events. Explain how this rule pertains to the chi-square test for association.

10. State the null and alternative hypotheses for a chi-square test for association between two qualitative variables.

11. What is the test statistic for the chi-square test for association?

12. How many degrees of freedom are associated with the test statistic given in Exercise 11?

Exercises

13. A political analyst is interested in studying the relationship between age and political affiliation. The analyst randomly selects 200 people and determines their age and political affiliation. The number of responses in each of the categories is as follows.

Table for Exercise 13 – Age and Political Affiliation

Age	Political Affiliation		
	Democrat	Republican	Independent
18–34	50	10	15
35–51	15	25	15
52–68	25	35	10

a. Can the analyst conclude that age and political affiliation are dependent at $\alpha = 0.05$?

b. What assumptions were made in the test for part **a.**?

14. A women's rights activist believes that salary and gender are dependent. To test this theory, she randomly selects 400 individuals and determines their salary and gender. The number of responses in each of the categories is as follows.

Table for Exercise 14 – Salary and Gender		
	Gender	
Salary	Male	Female
Below $25,000	25	100
$25,000–$50,000	75	70
$50,000–$75,000	75	20
Above $75,000	25	10

a. Is the activist's claim substantiated by the data at $\alpha = 0.01$?

b. What assumptions did you make in performing the test for part a.?

c. Can you think of any additional factors that could contribute to the relationship between salary and gender?

15. A sociologist is interested in studying the relationship between education and crime. She randomly selects 150 people and asks their education level and whether or not they have ever been convicted of a felony. The following table displays the number of respondents in each category.

Table for Exercise 15 – Education and Crime		
Have you ever been convicted of a felony?		
	Response	
Education Level	Yes	No
Less Than 9 Years	2	35
9 Years to 12 Years	4	31
12 Years to 16 Years	1	31
16+ Years	4	42

a. Can the sociologist conclude that education level and crime are dependent at $\alpha = 0.10$?

b. What assumptions were made in the test for part a.?

16. A psychologist is preparing his thesis on child abuse. He thinks that there may be a relationship between various types of child abuse and the gender of the child. To study this, he randomly selects the records of 197 abused children and determines both the gender of the child and the documented type of child abuse. The results of the study are as follows.

Table for Exercise 16 – Child Abuse		
	Gender	
Type of Abuse	Male	Female
Neglect	50	50
Physical	20	30
Sexual	10	19
Emotional	10	8

a. Can the psychologist conclude that the type of child abuse and gender of a child are dependent at $\alpha = 0.05$?

b. What assumptions were made in the test for part a.?

15

17. The National Fire Protection Association is interested in studying the relationship between the causes of fires and the region of the country in which the fires occur. They randomly select 500 fires and determine the region of the country in which the fire occurred and cause of the fire with the following results.

Table for Exercise 17 – Fires				
	Region			
Cause of Fire	North	South	East	West
Smoking	37	38	40	35
Heating Equipment	25	20	18	19
Arson	17	15	16	15
Electrical	12	13	12	13
Children at Play	10	11	12	11
Other	27	28	29	27

a. Can the association conclude that the cause of the fire and the region of the fire are dependent at $\alpha = 0.01$?

b. What assumptions were made in the test for part a.?

T   Discovering Technology

Using the TI-84 Plus Calculator

Chi-Square Test for Association

Use the information from Example 15.2 for this exercise.

1. Access the **MATRIX** menu by pressing 2ND and $[x^{-1}]$.

2. Choose **EDIT**, and **1: [A]**. Press ENTER.

3. Enter 3×2 as the matrix dimensions, and enter the data from Table 15.10 into this matrix. The total row and column are not needed.

Figure 15.13

4. Press STAT, TESTS, and **C: χ^2 – Test**.

5. Press ENTER. Choose **Calculate**. Press ENTER Observe the chi-square results. Compare these to the results which we determined in Example 15.2. Notice that the test statistic, *P*-value for the test, and the degrees of freedom are given in the calculator output. Since the *P*-value is less than the level of significance, we reject the null hypothesis and conclude that price charged at checkout and checkout method are dependent.

Figure 15.14

Using Excel

Chi-Square Test for Association

Use the information from Example 15.2 for this exercise.

1. Enter the data from Table 15.10 into a new worksheet in Excel.

	A	B	C	D
			Employee	
1	Price Charged	Self Checkout	Checkout	Total
2	Undercharge	20	10	30
3	Overcharge	15	30	45
4	Correct Price	200	225	425
5	Total	235	265	500

Figure 15.15

2. The statements of the hypotheses are as follows.

H_0:   Price charged at checkout and method of checkout are independent (not related).

H_a: Price charged at checkout and method of checkout are dependent (related).

3. Specify the critical value. The level of the test in Example 15.2 is $\alpha = 0.10$ with $(3-1)(2-1) = 2$ degrees of freedom. In Table G of Appendix A we find that the chi-square critical value of $\chi^2_{0.100}$ with 2 degrees of freedom is 4.605. Values of the test statistic greater than or equal to 4.605 would indicate that the apparent dependence of the two classifications is not likely due to ordinary sampling variation alone.

4. Compute the expected values. In cells F1 and G1 enter the labels **Expected Self** and **Expected Employee**. Compute the expected value of undercharges using self checkout in cell F2.

$$=D5*(D2/D5)*(B5/D5)$$

Compute the other expected values by substituting the appropriate values in the formula.

	A	B	C	D	E	F	G
			Employee			Expected	Expected
1	Price Charged	Self Checkout	Checkout	Total		Self	Employee
2	Undercharge	20	10	30		14.1	15.9
3	Overcharge	15	30	45		21.15	23.85
4	Correct Price	200	225	425		199.75	225.25
5	Total	235	265	500			

Figure 15.16

5. Compute the test statistic. In cell A8, enter the formula for $\dfrac{\left[n_{US} - E\left(n_{US}\right)\right]^2}{E\left(n_{US}\right)}$.

$$=((B2-F2)^2)/F2$$

Copy this formula to cells A9, A10, B8, B9, and B10, substituting the appropriate values to compute the remaining $\dfrac{\left[n_{ij} - E\left(n_{ij}\right)\right]^2}{E\left(n_{ij}\right)}$ components of the test statistic.

6. Sum cells A8 through B10 to compute the test statistic by entering the formula **=sum(A8:B10)** into cell A12. The result is the chi-square test statistic. Compare this to that which we found in Example 15.2.

	A	B	C	D	E	F	G
			Employee			Expected	Expected
1	Price Charged	Self Checkout	Checkout	Total		Self	Employee
2	Undercharge	20	10	30		14.1	15.9
3	Overcharge	15	30	45		21.15	23.85
4	Correct Price	200	225	425		199.75	225.25
5	Total	235	265	500			
6							
7							
8	2.468794326	2.189308176					
9	1.788297872	1.585849057					
10	0.000312891	0.000277469					
11							
12	8.032839792						

Figure 15.17

7. Make a decision. The chi-square test statistic is approximately 8.033. Since the value of the test statistic is greater than the critical value and falls in the rejection region, the null hypothesis should be rejected.

Using Minitab

Chi-Square Test for Goodness of Fit

Use the information from Example 15.1 for this exercise.

1. Enter the opinion categories in Column C1, the expected number of respondents for each category based on the 2009 data in Column C2, and the actual number of respondents for each category in 2010 in Column C3 as shown.

↓	C1-T	C2	C3
	Opinion	2009	2010
1	Approve Strongly	403	292
2	Approve Moderately	272	222
3	Disapprove Moderately	101	131
4	Disapprove Strongly	192	343
5	Unsure	40	20

Figure 15.18

2. Choose **Stat**, **Tables**, and **Chi-Square Goodness-of-Fit Test (One Variable)**.

3. Enter **C3** for the Observed counts, be sure that the radio button next to **Proportions specified by historical counts** is selected, and enter **C2** as the Input column.

4. Press **OK**. Minitab performs the chi-square test for goodness of fit and also gives two corresponding charts. One chart shows the contribution to the chi-square value by category and one shows the observed and expected values for each category. The output in the session window shows each category's observed value, historical count, test proportion, and expected value. The value of the chi-square test statistic is given below these statistics along with the corresponding P-value. Compare these results to those which we found in Example 15.1.

```
Chi-Square Goodness-of-Fit Test for Observed Counts in Variable: 2010

                        Historical       Test                   Contribution
Category   Observed        Counts  Proportion   Expected          to Chi-Sq
1               292           403    0.399802        403             30.573
2               222           272    0.269841        272              9.191
3               131           101    0.100198        101              8.911
4               343           192    0.190476        192            118.755
5                20            40    0.039683         40             10.000


   N  DF   Chi-Sq  P-Value
1008   4  177.430    0.000
```

Figure 15.19

Chi-Square Test for Association

Use the information from Example 15.2 for this exercise.

1. Enter the data from Table 15.10 into a new Minitab worksheet. The "Total" row and "Total" column are not needed.

↓	C1-T	C2	C3
	Price Charged	Self Checkout	Employee Checkout
1	Undercharge	20	10
2	Overcharge	15	30
3	Correct Price	200	225

Figure 15.20

2. The statements of the hypotheses are as follows.

H_0:   Price charged at checkout and method of checkout are independent (not related).

H_a: Price charged at checkout and method of checkout are dependent (related).

3. Specify the critical value of the test. The level of the test given in Example 15.2 is $\alpha = 0.10$ with $(3-1)(2-1) = 2$ degrees of freedom. In Table G of Appendix A we find that the chi-square critical value of $\chi^2_{0.100}$ with 2 degrees of freedom is 4.605. Values of the test statistic greater than or equal to 4.605 would indicate that the apparent dependence of the two classifications is not likely due to ordinary sampling variation alone.

4. Compute the test statistic. Choose **Stat, Tables,** and **Chi-Square Test (Two-Way Table in Worksheet)**. Choose **C2** and **C3** as the Columns containing the table. Press **OK**.

5. Observe the Minitab output. The test statistic is computed as 8.033. Compare this to the test statistic which we computed in Example 15.2.

```
┌─────────────────────────────────────────────────────────────────┐
│ Chi-Square Test: Self Checkout, Employee Checkout                 │
│                                                                   │
│ Expected counts are printed below observed counts                 │
│ Chi-Square contributions are printed below expected counts        │
│                                                                   │
│             Self   Employee                                       │
│          Checkout  Checkout   Total                               │
│     1         20        10      30                                │
│            14.10     15.90                                         │
│            2.469     2.189                                         │
│                                                                   │
│     2         15        30      45                                │
│            21.15     23.85                                         │
│            1.788     1.586                                         │
│                                                                   │
│     3        200       225     425                                │
│           199.75    225.25                                        │
│            0.000     0.000                                        │
│                                                                   │
│ Total        235       265     500                                │
│                                                                   │
│ Chi-Sq = 8.033, DF = 2, P-Value = 0.018                           │
└─────────────────────────────────────────────────────────────────┘
```

Figure 15.21

6. **Make a decision.** Since the test statistic falls into the rejection region, the null hypothesis should be rejected. Also notice that Minitab reports the P-value for the chi-square test statistic with 2 degrees of freedom as 0.018. Since the P-value is less than the level of significance, $\alpha = 0.10$, there is sufficient evidence to reject the null hypothesis.

R	**Chapter 15 Review**

Key Terms and Ideas

- Chi-Square Test for Goodness of Fit
- Chi-Square Test for Association between Two Qualitative Variables
- Chi-Square Distribution
- Chi-Square Critical Values
- Multinomial Probability Distribution
- Multinomial Experiment
- Contingency Table
- Multiplication Rule for Independent Events

Key Formulas

Section

Chi-Square Statistic

$$\chi^2 = \frac{(n-1)s^2}{\sigma^2}$$

15.1

Test Statistic for the Chi-Square Test for Goodness of Fit

$$\chi^2 = \sum_{i=1}^{k} \frac{\left[n_i - E(n_i) \right]^2}{E(n_i)}$$

15.2

where n_i is the actual number of observations for each category, and $E(n_i)$ is the expected number of observations for each category.

Multiplication Rule for Independent Events

$$P(A \cap B) = P(A)P(B)$$

15.3

Test Statistic for the Chi-Square Test for Association between Two Qualitative Variables

$$\chi^2 = \sum_{i=1}^{r} \sum_{j=1}^{c} \frac{\left[n_{ij} - E(n_{ij}) \right]^2}{E(n_{ij})}$$

15.3

where r = the number of rows, c = the number of columns, and

$$E(n_{ij}) = np_i p_j = \frac{(\text{row } i \text{ total})(\text{column } j \text{ total})}{n}$$ where n is the total number of

observations and p_i and p_j are estimates of the true population proportions.

AE   Additional Exercises

1. A sales manager for an insurance company believes that customers have the following preferences for life insurance products: 50% prefer Whole Life, 25% prefer Universal Life, and 25% prefer Life Annuities. A survey of 250 customers produced the following results.

Table for Exercise 1 – Insurance Preferences	
Product	**Number**
Whole Life	60
Universal Life	100
Life Annuities	90

   a. Is the sales manager's claim refuted by the data at $\alpha = 0.05$?

   b. What assumptions were made in the test for part **a.**?

2. Consider the following proportions of wives having affairs and husbands having affairs reported in *What the Odds Are* by Les Kranz.

Table for Exercise 2 – Affairs		
	Wives	**Husbands**
Have an Affair Before 2 Years of Marriage	0.13	0.14
Have an Affair Between 2 and 10 Years of Marriage	0.20	0.25
Have an Affair After 10 Years of Marriage	0.20	0.33
Have Not Had an Affair	0.47	0.28

   a. If the results were based on a survey of 200 wives and 200 husbands, do the data suggest that the proportion of wives in the various categories who have had affairs differs significantly from the proportion of husbands in the various categories who have had affairs at $\alpha = 0.10$?

   b. What assumptions were made in the test for part **a.**?

3. Do you think there are people somewhat like ourselves living on other planets in the universe? The responses to this question, which was asked in several different calendar years, were summarized in the *Gallup Poll Monthly*. For the year 1978, 51% answered "Yes," 33% "No," and the remainder had no opinion. Suppose that a sample of 100 people is chosen in 2011 in order to determine if opinions have changed concerning extraterrestrial life. Assume that the same question is asked and that 42 answer "Yes," 30 "No," and that the rest have no opinion. Assuming that the percentages given above accurately represent the attitudes of the people in 1978, can we conclude with $\alpha = 0.05$ that people's attitudes of toward extraterrestrial life have changed since 1978?

4. A traffic engineer feels that on a certain four-lane highway, the probability of being in the innermost lane is twice as great as any of the other lanes. Assume the other lanes have equal probabilities. A random sample of 200 motorists is chosen and the lanes in which they are traveling in are noted. The results (Lane 1 is the innermost lane) are given in the following table.

Table for Exercise 4 – Traffic Lanes				
Lane	1	2	3	4
Frequency	55	45	62	38

    a. Find the probabilities implied by the engineer's claim that a randomly chosen motorist will be in each of the four lanes.

    b. With $\alpha = 0.05$, can we refute the claim of the traffic engineer?

5. According to the Statistical Abstract of the United States, 22.6% of those 18 and over in the U.S. in 1992 were never married, 61.1% were married, 7.5% were widowed, and 8.8% were divorced. Suppose the following table summarizes the marital status of 90 randomly chosen adults in 2011.

Table for Exercise 5 – Marital Status				
	Never Married	Married	Widowed	Divorced
Frequency	27	42	10	11

With $\alpha = 0.01$, can we conclude that in 2011 the U.S. is different than in 1992, with respect to marital status?

6. The National Restaurant Association is interested in determining if there is a relationship between the type of pizza pie Americans prefer and the region of the country in which they live. The association randomly selects 285 Americans and records the category of pizza pie which best describes their preference and the region of the country in which they live with the following results.

Table for Exercise 6 – Pizza Preference				
	Region			
Type of Pizza Pie Preferred	North	South	East	West
Thin Crust	40	30	35	45
Thick Crust	17	15	21	22
Pan Pizza	15	15	15	15

    a. Can the association conclude that the type of pizza pie Americans prefer and the region of the country in which they live are dependent at $\alpha = 0.10$?

    b. What assumptions were made in the test for part **a.**?

7. Do you think marriages between homosexuals should or should not be recognized by law as valid with the same rights as traditional marriages? The responses to this question and the region of the country where the respondent lived were summarized in the *Gallup Poll Monthly*. The following contingency table summarizes the 799 responses.

Table for Exercise 7 – Poll Results			
		Response	
Region	Should	Should Not	No Opinion
East	78	136	16
Midwest	63	183	16
South	74	23	16
West	50	138	6

With $\alpha = 0.01$, can we conclude that attitude about homosexual marriages is dependent on region?

8. Consider the following data regarding the percentage of mobile application usage and smartphone operating system. Suppose the data were based off of 100 responses.

Table for Exercise 8 – Mobile App Usage (%)			
Operating System	Multiple Times a Day	Once a Day	Several Times a Week
Apple iOS	68	11	16
Android OS	60	12	21
Palm OS	48	14	25
Blackberry OS	45	13	29
Microsoft Windows Mobile	29	21	43
Other	47	13	21

a. Is there evidence that the frequency of mobile app usage is dependent on the smartphone operating system? Test at $\alpha = 0.05$.

b. What assumptions are made for the test in part **a.**? Are the assumptions reasonable for this problem? Why or why not?

Chapter Sixteen
Nonparametrics

16

Introduction

One of the characteristics of **parametric statistics** is that hypothesis testing procedures rely on assumptions concerning the distribution of the population from which the sample is drawn. For example, when using test statistics which rely on the *t*-distribution or the *F*-distribution, we assume that the distribution of the underlying population from which the data are drawn is approximately normal. Figures 16.1 and 16.2 show examples of normal and non-normal distributions, respectively. Inferences made from parametric statistics are valid only to the extent that the assumptions made about the underlying distribution of the population are correct.

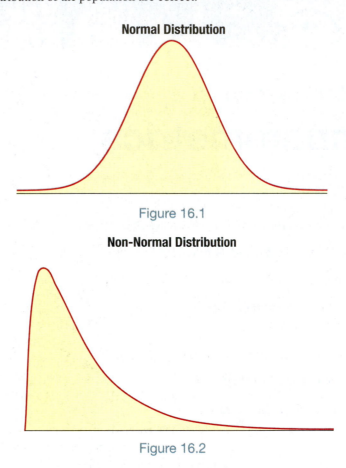

Normal Distribution

Figure 16.1

Non-Normal Distribution

Figure 16.2

In this chapter the focus is on **nonparametric statistical methods**. There are three characteristics of nonparametric statistical methods:

1. They do not involve the estimation of specific population parameters.

2. These methods are tests of hypotheses using data measured on an ordinal scale (in most instances).

3. Assumptions regarding the underlying distribution of the data are not required.

The fact that a distributional assumption about the data is not necessary to make valid inferences is one of the significant advantages of nonparametric techniques.

However, there are some disadvantages of nonparametric statistics. Because nonparametric tests tend to waste information, they are not as powerful as their parametric counterparts. Some nonparametric tests reduce exact ratio level measurements to ordinal level measurements. For example, taking an income measurement of $48,200 (a ratio measurement) and converting it into an ordinal variable consisting of three categories (low, medium, high) results in a significant loss of information.

Several nonparametric techniques are discussed in this chapter. The first two, the **sign test** and the **Wilcoxon signed-rank test**, are used to conduct hypothesis tests involving paired data experiments. Paired data experiments are designed to detect the effect of a treatment on some population measurement. In a paired data experiment, a treatment is applied to a population or sample. There is a comparison of some population characteristic of interest, before and after the treatment is applied. Suppose, for example, database entry personnel are given a course in speed-typing. If the course is successful, there should be a shift in the typing speed of the database personnel. A *before* and *after* measurement of typing speed would be desirable to measure the effect of the typing training. Since there are different typing speeds in the population of typists, how can we compare the average typing speed before and after the training? One way is to compare the population medians. If the distribution of typing speeds has shifted because of the treatment (training), then the median will also have shifted. The sign test is used for testing a hypothesis concerning two population medians or testing the hypothesis that a population median is equal to some predetermined value.

Other nonparametric tests introduced in this chapter include:

- The Wilcoxon rank-sum test, used for comparing two independent samples
- The Kruskal-Wallis test, a test similar to the ANOVA F-test
- The rank correlation test, a test similar to the parametric correlation test
- The runs test for randomness, a test to determine if a sequence exhibits randomness

16.1 The Sign Test

Sign tests are used in the comparison of paired data when it is not possible to assume that the differences have an approximately normal distribution. Suppose a company has two developmental gasoline products, say, Gasoline A and Gasoline B. The company wants to know if there is a significant difference in miles per gallon (mpg) between the two types of gasoline. Ten vehicles are randomly selected and driven using each of the gasolines in order to track the mpg. The data are as follows.

Table 16.1 – Miles per Gallon		
Car	MPG for Gasoline A	MPG for Gasoline B
1	35	40
2	38	35
⋮	⋮	⋮
10	41	38

Traditionally, to determine if there was a difference between the mpg between the two gasolines, the company would perform a paired t-test. However, since the company is not comfortable making assumptions about the distribution of the given data (i.e., assuming that the differences follow a normal distribution), the sign test is more appropriate to compare the differences in mpg between the two gasolines.

The methodology for the sign test is based on the assumption that the two populations have the same median. Suppose we were to subtract the paired sample values. If the population medians are identical, the resulting differences would be centered around zero. Further, we would expect the number of positive difference to be roughly equal to the number of negative differences. If the number of positive differences significantly outnumbers the number of negative differences or vice-versa, this will cause us to doubt that the two populations have the same median.

Using the Sign Test on Paired Data

We will present the test procedure for the sign test by example. For the paired difference experimental design discussed in Section 11.3, there is an assumption that the differences have an approximately normal distribution. The sign test makes no such assumption.

Example 16.1

Are the demands of college significantly less than the demands of high school? Students from Tech University were surveyed asking them the amount of time they spent watching TV per day in high school and the amount of time they spent watching TV in college. The data are listed in the following table. Can the researcher conclude at the 1% level of significance that the median TV-watching time (in minutes) is greater in college than in high school?

Table 16.2 – TV-Watching Times (Minutes)				
Student	High School	College	Difference	Sign
1	30	120	−90	−
2	60	60	0	Ignore
3	45	180	−135	−
4	120	240	−120	−
5	60	75	−15	−
6	180	240	−60	−
7	45	60	−15	−
8	90	90	0	Ignore
9	60	90	−30	−
10	90	120	−30	−

Solution

Normally, we would analyze these data with a paired difference test (see Section 11.3). However, if we doubt that the differences follow a normal distribution, we will use the sign test to determine if the populations are different.

The sign test is based on the idea that if two data sets have the same median, the number of differences with positive signs should approximately equal the number of differences with negative signs. We will conclude that the median number of minutes of TV-watching time increases in college if the number of negative signs is larger than the number of positive signs. How large is large? We will answer this question in Step 7.

Step 1: State the hypotheses in plain English.

- **Null Hypothesis:** The median TV-watching time in college is less than or equal to the median TV-watching time in high school.

- **Alternative Hypothesis:** The median TV-watching time is greater in college than in high school.

Step 2: Select the appropriate statistical measure.

The procedure we are using is called the sign test because it relies on counting the number of positive differences and the number of negative differences. Only the sign of the difference matters, the magnitude is ignored. Differences of zero are also ignored.

Step 3: Determine whether the hypothesis should be one-sided or two-sided.

Since the researcher is interested in whether the TV-watching time is longer in college than in high school, this is a one-tailed test. Thus, the alternative hypothesis will be one-sided.

Step 4: Specify the hypotheses using the appropriate statistical measure.

The researcher wants to determine if the median TV-watching time for college students is longer than that of high school students. The statistical measures being used are counts of the number of positive signs and negative signs. If the amount of TV-watching time is greater in college than in high school, then we would expect the negative differences to outnumber the positive differences. If the amounts of TV-watching time in high school and college are identical, then we would expect that any difference would be due to random variation, in which case we would have about the same number of positive and negative differences.

Using these ideas, we can translate our plain English hypotheses into:

H_0: The number of negative signs is less than or equal to the number of positive signs. (The median TV-watching time is the same or less in college as it was in high school.)

H_a: The number of negative signs is greater than the number of positive signs. (The median TV-watching time is greater in college than in high school.)

Step 5: Specify the level of the test.

The level of the test is specified in the problem to be $\alpha = 0.01$.

Step 6: Select the appropriate test statistic.

Our discussion has focused on the signs of the differences. Suppose we let

$$X = \text{the number of times the less frequent sign occurs.}$$

Assuming that the null hypothesis is true, the number of times the less frequent sign is observed in the paired difference experiment has a binomial distribution with

$$p = 0.5 \text{ and}$$

$$n = \text{the number of nonzero differences.}$$

If X is too small, then either H_0 is true and a rare phenomenon has been observed, or H_0 is not true. To make the decision whether X is too small, a decision rule must be developed (see Step 7).

If $n \leq 25$, where $n =$ the number of nonzero differences, the test statistic of the sign test is the variable X described previously.

If $n > 25$, X can be approximated by a normal distribution and the test statistic is given by

$$z = \frac{X + 0.5 - \left(\dfrac{n}{2}\right)}{\dfrac{\sqrt{n}}{2}}.$$

Assuming the null hypothesis is true, z has an approximately standard normal distribution.

Step 7: Determine the critical value.

The role of the critical value in this test is exactly the same as for all of the hypothesis tests discussed earlier. It defines a range of values for the test statistic, the rejection region, such that if the test statistic falls in this region it is unlikely to be due to ordinary sampling variation.

The level of the test defines the rareness criterion that will be used and implicitly determines the size of the rejection region. Should the computed value of the test statistic fall in the rejection region, its value will be presumed to be too rare to have occurred because of ordinary sampling variation, and the null hypothesis will be rejected.

If $n \leq 25$, the critical values are given in Appendix A, Table I for the sign test. Because the test statistic is defined to be the number of times the less frequent sign occurs, we will always reject the null hypothesis for small values of the test statistic. If the alternative hypothesis is *greater than* or *less than*, use the column in Table I labeled α *for a one-tailed test*. If the alternative is *not equal to*, use the column in Table I labeled α *for a two-tailed test*.

There are $n = 8$ nonzero differences in the TV-watching example, $\alpha = 0.01$, and the alternative hypothesis is *greater than*. From Table I in the Appendix, we find that the critical value is 0. The rejection region is drawn in Figure 16.3.

Figure 16.3

If $n > 25$, the critical values are determined in a similar manner to other hypothesis tests whose test statistics have standard normal distributions. However, because the test statistic is defined to be the number of times the less frequent sign occurs, we will always reject the null hypothesis for small values of the test statistic. Thus, the rejection region will always be established in the left tail, with the critical value defined by α for one-tailed tests, and $\alpha/2$ for two-tailed tests.

Note: Because of the manner in which the test statistic is defined, fail to reject the null hypothesis and do not proceed with the test procedure if the computed value of the test statistic does not tend to support the alternative hypothesis.

Step 8: Compute the test statistic.

Since $n \leq 25$, the value of the test statistic is given by

$$X = \text{the number of times the less frequent sign occurs.}$$

The less frequent sign in this case is the positive sign. In fact, the positive sign does not occur at all in these data, meaning that the TV-watching time in college was always greater than in high school. Since there are eight negative signs ($-$) and zero positive signs ($+$), the value of X is

$$X = 0$$

Note that differences of zero are ignored.

Since the number of positive signs is zero, this tends to support the alternative hypothesis that the number of minutes watching TV in college is greater than in high school. It is safe to proceed with the test. Suppose there were not any negative signs, meaning that all of the TV-watching times in college were less than or equal to the times in high school. This evidence would not support the alternative hypothesis, and the test must be halted at this point with a conclusion that the null hypothesis cannot be rejected. This stage of a one-sided sign test will always require thinking to be sure that the sign with the minimum value is the sign that supports the alternative hypothesis.

Step 9: Make the decision.

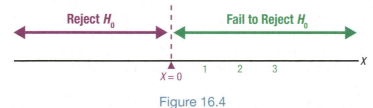

Figure 16.4

As shown in Figure 16.4, the value of the test statistic ($X = 0$) falls in the rejection region. Thus, we reject the null hypothesis at $\alpha = 0.01$ It is unlikely that the small number of positive differences could be attributed to ordinary sampling variation.

Step 10: State the conclusion in terms of the original question.

There is sufficient evidence for the researcher to conclude at $\alpha = 0.01$ that the median TV-watching time for college students is longer than the median TV-watching time when those students were in high school.

The test procedure for the sign test is outlined in the following box.

Nonparametric Rejection Regions

Most of the rejection regions we will construct for nonparametric tests will look different from those we have previously constructed. Because the test statistics are counts or sums of rank data, the test statistic will usually be a discrete value, so the rejection "region" will not be an interval. Most of the test statistics in this chapter will have rejection regions with only a few points in them, sometimes only one point.

Procedure

Sign Test

Hypotheses:

H_0: The number of positive [negative] signs is less than or equal to, greater than or equal to, or equal to the number of negative [positive] signs

H_a: The number of negative [positive] signs is greater than, less than, or not equal to the number of positive [negative] signs (depending on the claim which is being tested).

Note: No sign is given if the difference is zero and the difference is ignored.

Test Statistic:

If $n \leq 25$, then

X = the number of times the less frequent sign occurs.

If $n > 25$, then

$$z = \frac{X + 0.5 - \left(\frac{n}{2}\right)}{\frac{\sqrt{n}}{2}},$$

where X is defined just as when $n \leq 25$.

Critical Value(s):

If the data tend to support the alternative hypothesis and

$n \leq 25$:

Reject the null hypothesis if the test statistic is less than or equal to the critical value in Appendix A, Table I.

$n > 25$:

The critical values are based on the standard normal distribution with the critical values defined such that we reject the null hypothesis for small values of the test statistic.

Assumptions:

The pairs of data are randomly selected.

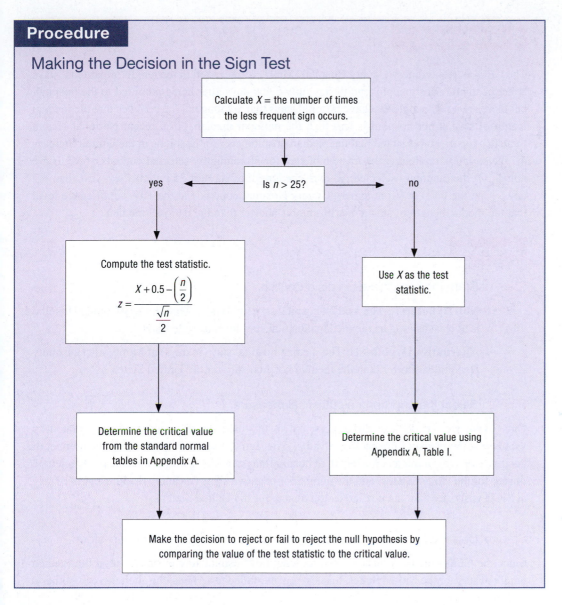

Procedure

Making the Decision in the Sign Test

Calculate X = the number of times the less frequent sign occurs.

Is $n > 25$?

yes

no

Compute the test statistic.

$$z = \frac{X + 0.5 - \left(\frac{n}{2}\right)}{\frac{\sqrt{n}}{2}}$$

Use X as the test statistic.

Determine the critical value from the standard normal tables in Appendix A.

Determine the critical value using Appendix A, Table I.

Make the decision to reject or fail to reject the null hypothesis by comparing the value of the test statistic to the critical value.

Testing a Population Median with the Sign Test

The median is a more useful measure of the center for a population than the mean in certain situations such as the following.

- When there are extreme data values (outliers) which significantly skew the distribution of the data.

- When the mean cannot be used as a measure of center of the data, as with ordinal level data.

The sign test can be used to test claims about the median of a population. The procedure for using the sign test to evaluate claims about a single population median is presented in the following example.

Example 16.2

The CEO of Montgomery County Hospital (MCH) is interested in the length of patients' stays at her hospital. Most people tend to stay only a few days after being admitted to the hospital, but there are a few people that have hospital stays that are longer. She believes that the median length of stay at her hospital is less than the national median. In a recent National Health Statistics report, it was stated that the median hospital stay for hospitals in the United States is 4.7 days. Upon discharge, the lengths of stay for 50 randomly selected patients of MCH are noted. Of the randomly selected patients, the length of stay for 35 patients was less than 4.7 days. Can the CEO of Montgomery County Hospital conclude at the 10% significance level that the median length of stay at her hospital is shorter than the national median?

Solution

Step 1: State the hypotheses in plain English.

- **Null Hypothesis:** The median hospital stay (in days) at Montgomery County Hospital is at least as long as the median hospital stay for the United States.

- **Alternative Hypothesis:** The median hospital stay (in days) at Montgomery County Hospital is shorter than the median hospital stay for the United States.

Step 2: Select the appropriate statistical measure.

Once again we will look at differences, but in this case the differences will be created by subtracting the median from each data value. The sign test will rely on counting the number of positive signs (meaning that the length of hospital stay at MCH is longer than the median length of stay for the United States) and the number of negative signs (meaning that the length of stay at MCH is shorter than the median length of stay for the United States).

Step 3: Determine whether the hypothesis should be one-sided or two-sided.

Since the CEO's claim is that the median length of hospital stay is *shorter* than the median length of stay for the United States, this test is a one-tailed test. Thus, the alternative hypothesis will be one-sided.

Step 4: Specify the hypotheses using the appropriate statistical measure.

H_0: The number of positive signs is greater than or equal to the number of negative signs. (The median length of hospital stay at Montgomery County Hospital is at least as long as the median length of hospital stay for the United States.)

H_a: The number of negative signs outnumbers the number of positive signs. (The median length of hospital stay at Montgomery County Hospital is shorter that the median length of hospital stay in the United States.)

Step 5: Specify the level of the test.

The level of the test is specified in the problem to be $\alpha = 0.10$.

Step 6: Select the appropriate test statistic.

Since $n > 25$, the test statistic is given by

$$z = \frac{X + 0.5 - \left(\dfrac{n}{2}\right)}{\dfrac{\sqrt{n}}{2}}.$$

Assuming the null hypothesis is true, z has an approximately standard normal distribution.

Step 7: Determine the critical value.

More than half of the hospital stays were shorter than the United States median length of stay, which supports the alternative hypothesis. Thus, it is appropriate to proceed with the test procedure.

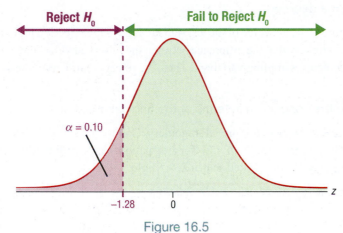

Figure 16.5

Since $n > 25$, the critical value is determined from the standard normal distribution. Because of the way the test statistic is defined, we will always reject the null hypothesis for small values of the test statistic. If the alternative hypothesis is *greater than* or *less than*, use an area of α in the left tail to determine the critical value. If the alternative is *not equal to*, use an area of $\alpha/2$ in the left tail to determine the critical value. In this example, the alternative is *less than* and $\alpha = 0.10$. The rejection region is shown in Figure 16.5. We will reject the null hypothesis if the calculated value of the test statistic is less than or equal to -1.28.

Step 8: Compute the test statistic.

Negative signs will occur if the length of stay for patients in MCH is shorter than the median length of stay for the United States. 35 of the sampled patients had an actual length of stay shorter than the median length of stay for the United States. Thus, there will be 35 negative signs.

Positive signs will occur if the length of stay for patients in MCH is longer than the median length of stay for the United States. 15 of the sampled patients had an actual length of stay longer than the median length of stay for the United States. Thus, there will be 15 positive signs.

Since $X=$ the number of times the less frequent sign occurs, $X = 15$, and the calculated value of the test statistic is

$$z = \frac{15 + 0.5 - \left(\dfrac{50}{2}\right)}{\dfrac{\sqrt{50}}{2}} \approx -2.69.$$

Figure 16.6

Step 9: Make the decision.

As shown in Figure 16.6, the value of the test statistic falls in the rejection region (-2.69 is less than -1.28). It is unlikely that the difference between the observed value and the hypothesized value is due to ordinary sampling variation. Thus, we reject the null hypothesis at $\alpha = 0.10$.

Step 10: State the conclusion in terms of the original question.

There is sufficient evidence for the Montgomery County Hospital CEO to conclude at $\alpha = 0.10$ that the median length of stay for her patients is significantly shorter than the median length of stay for the United States as a whole.

16.1   Exercises

Basic Concepts

1. What are parametric statistics?

2. Identify and explain the main disadvantage of the sign test.

3. Under what conditions are parametric statistical methods not appropriate for data analysis?

4. Identify the three characteristics of nonparametric statistical methods.

5. What are the disadvantages of nonparametric statistics?

6. The sign test and the Wilcoxon signed-rank test are designed to conduct hypothesis tests involving which kind(s) of experiments? What is the corresponding parametric statistical technique used to analyze these types of experiments?

7. What assumptions are made when conducting the sign test?

8. Name the two ways that the sign test can be used to perform hypothesis tests.

9. How do the rejection regions for nonparametric tests differ from those for parametric tests? Explain.

10. What is done with measurements that have a difference of zero in a paired difference experiment? Why is this the case?

11. What are the null and alternative hypotheses associated with the sign test?

12. What is the test statistic for the sign test for small samples? How small is a *small* sample?

13. What is the test statistic for the sign test for large samples? How large is a *large* sample?

14. Identify the critical values and rejection rules for both small and large samples with regard to the sign test.

Exercises

15. Hurricane Hugo swept through the Lowcountry in South Carolina causing billions of dollars of damage. In the past, the median claim for homes damaged by hurricanes for an insurance company in the Lowcountry had been $25,000. The insurance company believes that the median claim will be significantly larger for homes damaged by Hugo than past hurricanes. In order to investigate this theory, the insurance company randomly selects 55 homes and sends adjusters to settle the claims. In the sample of 55 homes, 40 of the homes had a claim in excess of the historical median. Is there overwhelming evidence at $\alpha = 0.10$ that the median claim for home damage from Hurricane Hugo was greater than the historical median?

16. The manufacturer of Brand X floor polish is developing a new polish that they hope will dry faster than the competition's polish. The competition's polish is advertised to have an average (median) drying time of 10 minutes. In a random sample of 1000 polishes with the new polish, 700 of the polishes dried in less than 10 minutes. Based on the data, can the manufacturer conclude that the median drying time for Brand X is faster than the competition's brand at a 0.05 level of significance?

17. NarStor, a computer disk drive manufacturer, claims that the median time until failure for their hard drives is 14,400 hours. You work for a consumer group that has decided to examine this claim. Technicians ran 16 NarStor hard drives continuously for almost three years. Recently the last drive failed. The times to failure (in hours) are given in the following table.

Table for Exercise 17 – Time Until Hard Drive Failure (Hours)							
330	620	1870	2410	4620	6396	7822	8102
8309	12,882	14,419	16,092	18,384	20,916	23,812	25,814

a. Is there overwhelming evidence that the median time until failure is less than the manufacturer claims? Use $\alpha = 0.10$

b. What assumption did you make in performing the test in part a.?

18. A.C. Bone has developed a duck hunting boot which it claims can remain immersed for more than 12 hours without leaking. 15 of the boots are tested and the time until first leakage is measured. Nine of the boots last more than 12 hours without leaking.

    a. Do the data substantiate A.C. Bone's claim at $\alpha = 0.05$?

    b. What assumption did you make in performing the test in part **a.**?

19. Given that most textbooks can now be purchased online, one wonders if students can save money by comparison shopping for textbooks at online retailers and at their local bookstores. To investigate, students at Tech University randomly sampled 25 textbooks on the shelves of their local bookstores. The students then found the "best" available price for the same textbooks via online retailers. The prices for the textbooks are listed in the following table.

Table for Exercise 19 – Textbook Prices		
	Price ($)	
Textbook	Bookstore	Online Retailer
1	70	60
2	38	36
3	88	89
4	165	149
5	80	136
6	103	95
7	42	50
8	98	111
9	89	65
10	97	86
11	140	130
12	40	30
13	175	150
14	85	75
15	100	85
16	68	62
17	67	69
18	140	142
19	49	40
20	149	127
21	126	130
22	92	93
23	144	129
24	98	84
25	40	52

Using the data in the table, and without making any distributional assumptions, is it less expensive for the students to purchase textbooks from the online retailers than the local bookstores? Use $\alpha = 0.01$.

20. The management for a large grocery store chain would like to determine if a new cash register will enable cashiers to process a larger number of items on average than the cash register which they are currently using. Seven cashiers are randomly selected, and the number of grocery items which they can process in three minutes is measured for both the old cash register and the new cash register. The results of the test are as follows.

Table for Exercise 20 – Number of Grocery Items Processed in Three Minutes							
Cashier	1	2	3	4	5	6	7
Old Register	60	70	55	75	62	52	58
New Register	65	71	55	75	65	57	57

Without making any assumptions about the distribution, can management conclude that the new cash register will allow cashiers to process a significantly larger number of items on average than the old cash register at $\alpha = 0.05$.

21. An auto dealer is marketing two different models of a high-end sedan. Since customers are particularly interested in the safety features of the sedans, the dealer would like to determine if there is a difference in the braking distance (the number of feet required to go from 60 mph to 0 mph) of the two sedans. Six drivers are randomly selected and asked to participate in a test to measure the braking distance for both models. Each driver is asked to drive both models and brake once they have reached exactly 60 mph. The distance required to come to a complete halt is then measured in feet. The results of the test are as follows.

Table for Exercise 21 – Braking Distance of High-End Sedans (in Feet)						
Driver	1	2	3	4	5	6
Model A	150	145	160	155	152	153
Model B	152	146	160	157	154	155

Without making assumptions about the distribution of the data, can the auto dealer conclude that there is a significant difference in the braking distance of the two models of high end sedans? Use $\alpha = 0.10$.

22. A nutritionist is interested in determining the decrease in cholesterol level which a person can achieve by following a particular diet which is low in fat and high in fiber. Seven subjects are randomly selected to try the diet for six months, and their cholesterol levels are measured both before and after the diet. The results of the study are as follows.

Table for Exercise 22 – Cholesterol Levels							
Subject	1	2	3	4	5	6	7
Before Diet	155	170	145	200	162	180	160
After Diet	152	168	148	195	162	178	157

Can the nutritionist conclude that there is a significant decrease in average cholesterol level when the diet is used? We don't have any knowledge about the distribution of the data. Use $\alpha = 0.01$.

16.2   The Wilcoxon Signed-Rank Test

A disadvantage of the sign test is that it wastes information. The sign test merely counts the number of positive or negative signs in a paired difference experiment and ignores the magnitude of the differences. The **Wilcoxon signed-rank test** is a nonparametric technique which can also be used to evaluate a paired difference experiment. This test is designed to detect populations whose centers are shifted to the right or the left of each other. As with the sign test, no distributional assumption is required. However, the pairs of data must have been randomly selected, and it must be possible to rank the differences.

An advantage of the Wilcoxon signed-rank test is that it does not ignore the magnitudes of the differences. However, it does not take the magnitude directly into account. Instead, the ranks of the data are analyzed.

Ranking is nothing new. It simply requires putting the data in order from smallest to largest and attaching a rank to each data item. In general, the lowest value is assigned a rank of one and the highest value is assigned a rank of n, where n is the number of nonzero differences. How do we handle ties? If there are two or more values with the same magnitude, these values will each be assigned the same rank, which is equal to the average of the ranks which would have been assigned to these values if they had slightly different consecutive values. The ranking procedure is explained more fully in the following example.

Example 16.3

Rank the following stocks, traded on the New York Stock Exchange, from smallest price to largest price.

Table 16.3 – Stock Prices	
Stock	**Price per Share ($)**
3M Co.	79.35
AT&T, Inc.	28.05
Citigroup, Inc.	28.40
Exxon Mobil Corp.	72.14
Hewlett-Packard Co.	24.34
Merck & Co. Inc.	32.46
Procter and Gamble Co.	62.55
Wal-Mart Stores Inc.	52.03
Walt Disney Co.	32.46

Solution

The lowest priced stock is Hewlett-Packard Co. at $24.34. It is assigned a rank of one. Rankings of two and three go to AT&T, Inc. and Citigroup, Inc., respectively, because they have the next lowest prices. Fourth place is a tie between Merck & Co. Inc. and Walt Disney Co. To assign their ranks, compute the average of the ranks of the positions for which they are tied. The rank for each is determined as the average of four and five. Thus,

$$\text{Rank} = \frac{4+5}{2} = 4.5.$$

Table 16.4 – Ranked Stock Prices			
Stock	Price per Share ($)	Rank	
3M Co.	79.35	9	
AT&T, Inc.	28.05	2	
Citigroup, Inc.	28.40	3	
Exxon Mobil Corp.	72.14	8	
Hewlett-Packard Co.	24.34	1	
Merck & Co. Inc.	32.46	4.5	Tied for 4th and 5th
Procter and Gamble Co.	62.55	7	
Wal-Mart Stores Inc.	52.03	6	
Walt Disney Co.	32.46	4.5	Tied for 4th and 5th

There are no other ties. The ranking resumes with Wal-Mart Stores Inc. being ranked sixth.
The highest ranked stock is 3M Co., which has a rank of nine.

The test procedure for the Wilcoxon signed-rank test will be developed in the following example.

Example 16.4

It is believed by many economists that the United States entered into a recession in December
of 2007. In tough economic times, most families tend to cut back on dining outside the home
in an effort to save money. Thus, one indication of a weak economy is the reduction in the
number of times a family dines out. Randomly sampling nine customers at the local grocer,
the patrons were asked the number of times that they dined out between January 2006 and
December 2006 (prior to the recession) and from January 2008 to December 2008 (12 months
after entering the recession). The results of the survey are shown in Table 16.5. Based on
the data, can it be concluded that people tend to dine out less during tough economic times
(i.e., during a recession) at $\alpha = 0.05$?

Table 16.5 – Number of Times Dining Out						
Customer	2006	2008	Difference 2006–2008	Absolute Value of Difference	Rank of Absolute Value of Difference	Signed Rank
1	52	47	5	5	7	7
2	68	68	0	0	Ignore	Ignore
3	72	66	6	6	8	8
4	64	67	−3	3	4	−4
5	70	66	4	4	5.5	5.5
6	34	36	−2	2	2.5	−2.5
7	64	60	4	4	5.5	5.5
8	66	64	2	2	2.5	2.5
9	34	33	1	1	1	1
					Total +	29.5
					Total −	6.5

Solution

Step 1: State the hypotheses in plain English.

- **Null Hypothesis:** The number of times that people dined out in 2006 (pre-recession) is greater than or equal to the number of times that people dined out in 2008 (recession).

- **Alternative Hypothesis:** The number of times that people dined out in 2008 is less than the number of times that people dined out in 2006.

Step 2: Select the appropriate statistical measure.

The Wilcoxon signed-rank test compares the probability distributions for 2006 and 2008 (i.e., before recession and during recession). How this is accomplished will be discussed in Step 6.

Step 3: Determine whether the hypothesis should be one-sided or two-sided.

Since we are interested in knowing if the number of times people dined out has decreased, this test is a one-tailed test. Thus, the hypothesis will be one-sided.

Step 4: Specify the hypotheses using the appropriate statistical measure.

H_0: The distribution of those dining out in 2008 after the U.S. economy entered into a recession is the same or shifted to the right of the distribution of those dining out in 2006.

H_a: The distribution of those dining out in 2008 after the U.S. economy entered into a recession is shifted to the left of the distribution of those dining out in 2006.

Step 5: Specify the level of the test.

The level of the test is specified in the problem to be $\alpha = 0.05$.

Step 6: Select the appropriate test statistic.

If we let $n =$ the number of nonzero differences, the test statistic for the Wilcoxon signed-rank test is determined in the following manner.

1. Compute the difference for each of the pairs.

2. Rank the absolute values of the differences from lowest to highest (ignoring zero differences).

3. Calculate $T_+ =$ the sum of the ranks associated with positive differences.

4. Calculate $T_- =$ the sum of the ranks associated with negative differences.

5. Determine the test statistic based on the following criteria.

The test statistic will vary depending on the sample size, n, and the alternative hypothesis.

Note: In the following discussion, Population X will be the group in 2008 (during recession) and Population Y will be the group in 2006 (before recession).

If $n \leq 25$:

- If the alternative hypothesis is that Population X is shifted to the right of Population Y, use

   $T_+ =$ the sum of the ranks associated with positive differences, where differences are defined as observations in Population Y minus observations in Population X.

The test statistic is defined in this manner because we expect the number of negative differences to outnumber the number of positive differences, and thus expect T_+ to be the rank sum with the smallest value.

Note: This is analogous to the *greater than* alternative hypothesis.

- If the alternative hypothesis is that Population X is shifted to the left of Population Y, use

  T_- = the sum of ranks associated with negative differences, where differences are
  defined as observations in Population Y minus observations in Population X.

The test statistic is defined in this manner because we expect the number of positive differences to outnumber the number of negative differences, and thus expect T_- to be the rank sum with the smallest value.

Note: This is analogous to the *less than* alternative hypothesis.

- If the alternative hypothesis is that Population X is shifted to the left or to the right of Population Y, use

$$T = \min(T_+, T_-) = \text{the rank sum with the smallest value.}$$

Note: This is analogous to the *not equal to* alternative hypothesis.

It is important to note that if the smaller sum of the ranks is not as expected, the test should be halted and the decision should be to fail to reject the null hypothesis since there is no evidence in favor of the alternative hypothesis.

If $n > 25$, the test statistic can be approximated by a standard normal distribution and is specified in the procedure for the signed-rank test following this example.

For the example at hand, $n \leq 25$ and the alternative hypothesis is that the probability distribution for dining out in 2008 (recession) is shifted to the left of dining out in 2006 (pre-recession). This implies that dining out in 2006 minus dining out in 2008 should generally be positive. If the rank sum with the smallest value is T_+, fail to reject the null hypothesis since there is no evidence that the alternative hypothesis is more reasonable. Otherwise, the test statistic is given by

$$T = T_- = \text{the sum of the ranks associated with negative differences.}$$

Step 7: Determine the critical value.

If $n \leq 25$ the critical values are determined from Appendix A, Table J for the Wilcoxon signed-rank test. The critical value is specified by the level of α, n = number of nonzero differences in the test, and the alternative hypothesis. If the alternative hypothesis is *greater than* or *less than*, use the column in Table J labeled α *for a one-tailed test*. If the alternative hypothesis is *not equal to*, use the column in Table J labeled α *for a two-tailed test*. Because of the way the test statistic is defined, the null hypothesis is always rejected if the test statistic is calculated from the data, T, is less than or equal to the critical value in Table J, T_c.

Figure 16.7

For the dining out example, $n = 8$ (recall there was one difference which was zero), $\alpha = 0.05$, and the test is one-tailed. Based on these specifications, the critical value determined from Table J is $T_c = 6$. The null hypothesis will be rejected if $T \leq 6$. This rejection region is displayed in Figure 16.7.

If $n > 25$, the critical values are determined in a similar manner to test statistics which have an approximate standard normal distribution under the null hypothesis. However, because of the way the test statistic is defined, we will always reject the null hypothesis for small values of the test statistic. Thus, the rejection region will always be established in the left tail, with the critical value defined by α for one-tailed tests, and $\alpha/2$ for two-tailed tests.

Step 8: Compute the test statistic.

In order to calculate the test statistic, the signed ranks must be determined. To determine the signed ranks, first compute the difference for each pair of data values. Next, find the absolute value of each difference. Rank the absolute values of the differences using the ranking technique illustrated in Example 16.3 (ignore differences of zero). Once the absolute values of the differences have been ranked, reassign the ranks the sign which each associated difference had before the absolute value was computed. Finally, add all of the ranks with positive signs to determine T_+, and add all of the ranks with negative signs to determine T_-. This procedure is illustrated in Table 16.5.

Using the results from the table, the test statistic is given by

$$T = T_- = \text{the sum of the ranks associated with negative differences} = 6.5.$$

Step 9: Make the decision.

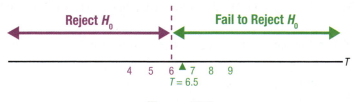

Figure 16.8

As shown in Figure 16.8, the value of the test statistic does not fall in the rejection region (that is, $T_- = 6.5$ is greater than $T_c = 6$). Thus, we fail to reject the null hypothesis at $\alpha = 0.05$. It is likely that the difference in the distributions of the two groups is due to ordinary sampling variation.

Step 10: State the conclusion in terms of the original question.

There is insufficient evidence to conclude at $\alpha = 0.05$ that the number of times people dined out was significantly higher before the recession than during the recession.

Procedure

Wilcoxon Signed-Rank Test

Hypotheses:

H_0: The distributions of the two populations of interest are the same.

H_a: > One-Tailed: The distribution of Population X is shifted to the *right* of the distribution of Population Y (Difference = Y – X).

   < One-Tailed: The distribution of Population X is shifted to the *left* of the distribution of Population Y (Difference = Y – X).

   ≠ Two-Tailed: The distribution of Population X is shifted to the left or to the right of the distribution of Population Y.

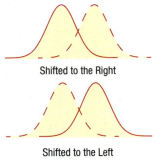

Shifted to the Right

Shifted to the Left

Test Statistic:

If $n \le 25$:

If H_a is > One-Tailed then

$T = T_+$ = the sum of the ranks associated with the positive differences.

If H_a is < One-Tailed then

$T = T_-$ = the sum of the ranks associated with the negative differences.

If H_a is ≠ Two-Tailed, then $T = \mathrm{Min}\left(T_+, T_-\right)$.

If $n > 25$ then $T = \mathrm{Min}\left(T_+, T_-\right)$, and the test statistic is given by

$$z = \frac{T - \dfrac{n(n+1)}{4}}{\sqrt{\dfrac{n(n+1)(2n+1)}{24}}}.$$

Critical Value(s):

If $n \le 25$, reject H_0 if $T \le T_c$, where T_c is the critical value found in Appendix A, Table J.

If $n > 25$:

One-Tailed Test: reject H_0 if $z \le -z_\alpha$.

Two-Tailed Test: reject H_0 if $z \le -z_{\alpha/2}$.

Assumptions:

Pairs of data have been randomly selected and are such that the absolute values of their differences can be ranked.

16

16.2   Exercises

Basic Concepts

1. What assumptions are required for the Wilcoxon signed-rank test?

2. The Wilcoxon signed-rank test is primarily used to perform hypothesis tests about what type of experiment?

3. What are the advantages and disadvantages of the Wilcoxon signed-rank test?

4. Describe the procedure for assigning ranks to data in order to perform a Wilcoxon signed-rank test. What is to be done when two values are the same?

5. Describe how to calculate the rank sums for a paired difference experiment in order to perform a Wilcoxon signed-rank test.

6. If the sample size is less than or equal to 25, identify the three possible test statistics used for the Wilcoxon signed-rank test. How do you choose which statistic to use?

7. What are the null and alternative hypotheses associated with the Wilcoxon signed-rank test?

8. Explain why the population distributions are important when performing a Wilcoxon signed-rank test.

9. What is the test statistic for the Wilcoxon signed-rank test if the sample size is large? How large is *large* with regard to sample size?

10. Identify the critical values and rejection regions for both large and small samples with regard to the Wilcoxon signed-rank test.

Exercises

11. Rank the following emerging markets mutual funds from lowest to highest price using the methodology presented for the signed-rank test.

Table for Exercise 11 – Emerging Markets Mutual Funds	
Mutual Fund	**Price ($)**
American Funds	24.40
Columbia Management	9.41
Morgan Stanley	23.74
Fidelity Investments	24.40
John Hancock	9.41
DWS Investments	15.57
UBS	12.15
Prudential Investments	9.23
Value Line Funds	32.82
The Vanguard Group	34.72

12. Rank the following consumer price indexes (CPI) for selected groups of goods and services in September 2011 using the methodology presented for the signed-rank test. The data in the table represent the unadjusted percent change in price level from September 2010 to September 2011.

Table for Exercise 12 – Percent Change in CPI

Expenditure Category	CPI (% Change 9/10 to 9/11)
Food	4.7
Alcoholic Beverages	1.4
Housing	1.8
Apparel	3.5
Public Transportation	7.4
Medical Care	2.8
Education	4.4
Tobacco and Smoking Products	2.4
Gasoline	33.3
New and Used Motor Vehicles	3.6

Source: Bureau of Labor Statistics

13. A study conducted by the Orentreich Foundation found that women who practiced transcendental meditation (T.M.) for 20 minutes a day had high levels of DHEA-S, a hormone that may help prevent breast cancer and osteoporosis. Suppose eight women are randomly selected to participate in a study. The DHEA-S levels of the participants are measured prior to practicing transcendental meditation and then measured one year after practicing transcendental meditation for 20 minutes a day. The following table is a summary of the results of the study.

Table for Exercise 13 – Study Results

Study Participant	DHEA-S Level Before T.M. (mg)	DHEA-S Level After T.M. (mg)
A	20	25
B	25	25
C	18	20
D	27	26
E	19	20
F	24	26
G	20	21
H	30	29

a. Using the sign test, do the data indicate that the DHEA-S level of women increases after practicing transcendental meditation for 20 minutes per day for one year at $\alpha = 0.05$?

b. What assumptions were necessary to perform the sign test?

c. Using the signed-rank test, do the data indicate that the DHEA-S level of women increases after practicing transcendental mediation for 20 minutes per day for one year at $\alpha = 0.05$?

d. What assumptions were necessary to perform the signed-rank test?

e. Which test do you think produces more accurate results? Why?

14. The management for a large grocery store chain would like to determine if a new cash register will enable cashiers to process a larger number of items on average than the cash register which they are currently using. Seven cashiers are randomly selected, and the number of grocery items which they can process in three minutes is measured for both the old cash register and the new cash register. The results of the test are as follows.

Table for Exercise 14 – Number of Grocery Items Processed in Three Minutes							
Cashier	1	2	3	4	5	6	7
Old Cash Register	60	70	55	75	62	52	58
New Cash Register	65	71	55	75	65	57	57

a. What assumption must be made in order to perform the test of hypothesis using the paired difference t-test?

b. Using the signed-rank test, do the data provide conclusive evidence that the new cash register enables cashiers to process a significantly larger number of items than the old cash register at $\alpha = 0.05$?

c. What assumptions were made in performing the signed-rank test?

d. How do the results of the signed-rank test compare with the paired difference t-test performed in Section 11.3, Exercise 9?

15. An auto dealer is marketing two different models of a high-end sedan. Since customers are particularly interested in the safety features of the sedans, the dealer would like to determine if there is a difference in the braking distance (the number of feet required to go from 60 mph to 0 mph) of the two sedans. Six drivers are randomly selected and asked to participate in a test to measure the braking distance for both models. Each driver is asked to drive both models and brake once they have reached exactly 60 mph. The distance required to come to a complete halt is then measured in feet. The results of the test are as follows.

Table for Exercise 15 – Braking Distance of High-End Sedans (in Feet)						
Driver	1	2	3	4	5	6
Model A	150	145	160	155	152	153
Model B	152	146	160	157	154	155

a. What assumption must be made in order to perform a test of hypothesis using the paired difference t-test?

b. Using the signed-rank test, do the data provide conclusive evidence that there is a significant difference in the median braking distance of the two sedans at $\alpha = 0.10$?

c. What assumptions were made in performing the signed-rank test?

d. How do the results of the sign test performed in Section 16.1, Exercise 21 and the signed-rank test performed in part **b.** compare with the paired difference t-test performed in Section 11.3, Exercise 10?

16.3 The Wilcoxon Rank-Sum Test

We discussed nonparametric procedures for testing claims about a paired difference experiment in the previous two sections. In this section we will discuss a nonparametric procedure for hypothesis tests in which an independent experimental design is used to compare two population medians.

In Chapter 11 we discussed the small sample *t*-test for comparing two population means when independent random samples are drawn from two separate populations. In order to perform that test we need to make the important assumption that the distributions of both populations of interest are approximately normal.

The **Wilcoxon rank-sum test** is a nonparametric technique which can be used to compare two populations when we are either unwilling or unable to make the assumption of normality. It may also be used in the situation where the level of measurement is only ordinal, meaning we can only rank the data. Although the Wilcoxon rank-sum test does not require the assumption of normality, it does require that the two samples are drawn in a random and independent manner and that the data are such that they can be ranked from smallest to largest.

The test procedure for the Wilcoxon rank-sum test will be illustrated in the following example.

Example 16.5

Audiophiles R Us (ARU) reviews high-end audio electronics as a service to customers and publishes their findings in *ARU Magazine*. One of the most popular issues is the one in which ARU reviews headphones. In its most recent magazine article about headphones, ARU reviewed two of the most popular high-end models of headphones on the market. Brand A headphones cost $449 and feature over the ear pads and crisp sound. Brand B headphones cost $300 and have on-ear pads and equally crisp sound. ARU has randomly selected 20 customers to evaluate the headphones (10 provided with Brand A headphones and 10 provided with Brand B headphones). The customers were told to provide a review of the headphones, giving a score between 0 and 100. The ratings are given in Table 16.6. Do the data provide sufficient evidence for ARU to conclude that there is a difference between Brand A and Brand B headphones at $\alpha = 0.05$?

Table 16.6 – Headphone Ratings			
Brand A	**Rank**	**Brand B**	**Rank**
93	15.5	99	19
100	20	84	8.5
75	3	76	4
92	14	93	15.5
55	1	87	12
98	18	95	17
85	10	77	5
84	8.5	79	6
82	7	61	2
90	13	86	11
Rank Sum	110		100

16

Solution

Histograms of the two ratings are provided in Figure 16.9 and Figure 16.10. The histograms indicate that the ratings for Brand A and Brand B headphones appear to follow a skewed distribution. Thus, the assumption of normality may not be reasonable for these data. If the assumption of normality is not met, the small sample t-test for comparing two population means will not be valid. However, independent random samples were drawn for each set of customers and the data can be ranked. Thus, the assumptions for the Wilcoxon rank-sum test are satisfied and we can proceed with the test procedure.

Figure 16.9   Figure 16.10

Step 1: State the hypotheses in plain English.

• **Null Hypothesis:** The ratings for Brand A and Brand B headphones are the same.

• **Alternative Hypothesis:** The ratings for Brand A and Brand B headphones are different.

Step 2: Select the appropriate statistical measure.

Because we are not making a hypothesis about a specific population parameter, we will delay the discussion of a statistical measure until **Step 6**.

Step 3: Determine whether the hypothesis should be one-sided or two-sided.

Since ARU wants to determine if there is a difference between the headphones, this is a two-tailed test, and the hypothesis will be two-sided.

Step 4: Specify the hypotheses using the appropriate statistical measure.

H_0: The distributions of the ratings for Brand A headphones and Brand B headphones are the same.

H_a: The distribution of the ratings for Brand A headphones is shifted to the right or to the left of the distribution of the ratings for Brand B headphones.

Step 5: Specify the level of the test.

The level of the test is specified in the problem to be $\alpha = 0.05$.

Step 6: Select the appropriate test statistic.

The Wilcoxon rank-sum test uses the idea that if the null hypothesis is true, both samples come from the same population and you should be able to combine them. After the samples are combined, rank the data and compute the sum of the ranks for each sample. In our example, the Wilcoxon rank-sum test compares the sum of the ranks, the rank sum, for Brand A headphone ratings to the rank sum for Brand B headphone ratings.

If the null hypothesis is true, we would expect the rank sums to have approximately the same value. If the rank sums are very different, it will cause us to doubt that the observations come from the same population. Thus, the rank sum will be the measure which will be used for our test statistic. How large a difference in the rank sums is necessary before they are considered significantly different? This question is answered by the critical values specified in Step 7.

If $n \leq 10$, the test statistic is given by the rank sum associated with the population from which the smallest sample is taken.

$$T = \text{rank sum of the population with the smallest sample size.}$$

If the sample sizes are equal and the hypothesis is one-sided, use the rank sum associated with the population in the alternative hypothesis which is specified to be shifted to the right or shifted to the left. If the sample sizes are equal and the hypothesis is two-sided, either rank sum can be used.

If $n > 10$, the test statistic is given in the procedure following this example. The distribution of the test statistic can be approximated by a standard normal distribution.

Step 7: Determine the critical value.

If $n \leq 10$, the critical values are determined from Appendix A, Table K for the Wilcoxon rank-sum test. The critical value is specified by the level of α, $n_1 =$ the size of the first sample, $n_2 =$ the size of the second sample, and whether the test is a one-tailed or two-tailed test.

- If the alternative hypothesis is one-sided and specifies that the distribution of Population X is shifted to the right of the distribution of Population Y, the null hypothesis is rejected if $T \geq T_U$ from Table K.

- If the alternative hypothesis is one-sided, and specifies that the distribution of Population X is shifted to the left of the distribution of Population Y, the null hypothesis is rejected if $T \leq T_L$ from Table K.

- If the alternative hypothesis is two-sided and specifies that the distribution of Population X is shifted to the left of the distribution of Population Y or the distribution of Population X is shifted to the right of the distribution of Population Y, the null hypothesis is rejected if $T \leq T_L$ or $T \geq T_U$ from Table K.

Note: The smaller sample size is associated with Population X.

For the headphones example, $n_1 = 10$, $n_2 = 10$, $\alpha = 0.05$, and the alternative hypothesis is that there is a difference between the headphone ratings. Thus, this is a two-tailed test where the distribution of the ratings for Brand A headphones is hypothesized to be shifted to the right or to the left of the distribution of the ratings for Brand B headphones. Using Table K, we find that the null hypothesis should be rejected if $T \leq 79$ or $T \geq 131$. The rejection region is displayed in Figure 16.11.

Figure 16.11

If $n > 10$, the critical values are determined in the usual manner for test statistics which have an approximate standard normal distribution under the null hypothesis. The box following this example describes the procedure for determining the rejection region in this case.

Step 8: Compute the test statistic.

For the headphones example, $n_1 = 10$ and $n_2 = 10$. Since the two sample sizes are the same, we will use the rank sum associated with the ratings for Brand A headphones. Since $n_1 = n_2$, we have the following.

$$T = \text{the rank sum for Brand A headphones} = 110$$

Note that since the sample sizes are the same and we have a two-sided alternative, either population could have been used to calculate the rank sum for the test statistic.

Step 9: Make the decision.

As shown in Figure 16.12, the value of the test statistic does not fall in the rejection region ($79 < T = 110 < 131$). Thus, we fail to reject the null hypothesis at $\alpha = 0.05$. The difference in the distributions of the populations is not likely to be attributed to anything other than ordinary sampling variation.

Figure 16.12

Step 10: State the conclusion in terms of the original question.

There is not enough evidence to conclude at the $\alpha = 0.05$ level that there is a significant difference in the ratings between Brand A and Brand B headphones.

Procedure

Wilcoxon Rank-Sum Test

Hypotheses:

H_0: The distributions of the two populations of interest are the same.

H_a: > One-Tailed: The distribution of Population X is shifted to the right of the distribution of Population Y.

    < One-Tailed: The distribution of Population X is shifted to the left of the distribution of Population Y.

    ≠ Two-Tailed: The distribution of Population X is shifted to the right or to the left of the distribution of Population Y.

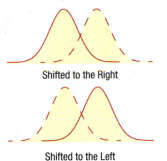

Shifted to the Right

Shifted to the Left

Test Statistic:

n_1 = the smaller of the sample sizes

n_2 = the larger of the sample sizes

If $n_1 \leq 10$, and the smaller sample size, n_1, is associated with Population X:

If H_a is > One-Tailed, then

$T = T_x$ = the rank sum of the sample with the fewest members.
    (If the sample sizes are the same, T_x = the rank sum of the population hypothesized to be shifted to the right.)

If H_a is < One-Tailed, then

$T = T_x$ = the rank sum of the sample with the fewest members.
    (If the sample sizes are the same, T_x = the rank sum of the population hypothesized to be shifted to the left.)

If H_a is ≠ Two-Tailed, then

$T = T_x$ = the rank sum of the sample with the fewest members.
    (If the sample sizes are the same, either rank sum can be used.)

If $n_1 > 10$, and the smaller sample size, n_1, is associated with Population X, then

$$z = \frac{T - \dfrac{n_1\left(n_1 + n_2 + 1\right)}{2}}{\sqrt{\dfrac{n_1 n_2\left(n_1 + n_2 + 1\right)}{12}}},$$ where T is defined just as when $n_1 \leq 10$.

Continued on the next page...

Procedure

Wilcoxon Rank-Sum Test (cont.)

Critical Value(s):

If $n_1 \leq 10$:

If H_a is > One-Tailed: then reject H_0 if $T \geq T_U$, the critical value in Table K.

If H_a is < One-Tailed: then reject H_0 if $T \leq T_L$, the critical value in Table K.

If H_a is $\neq$ Two-Tailed: then reject H_0 if $T \leq T_L$ or $T \geq T_U$, the critical values in Table K.

If $n_1 > 10$:

If H_a is > One-Tailed, then reject H_0 if $z \geq z_\alpha$.

If H_a is < One-Tailed, then reject H_0 if $z \leq -z_\alpha$.

If H_a is $\neq$ Two-Tailed, then reject H_0 if $z \leq -z_{\alpha/2}$ or if $z \geq z_{\alpha/2}$.

Assumptions:

The data are such that they can be ranked. The two samples are selected in an independent and random fashion.

16.3   Exercises

Basic Concepts

1. What type of data is the Wilcoxon rank-sum test used to analyze?

2. What is the parametric test used to analyze the type of data that can also be analyzed by the Wilcoxon rank-sum test? What assumptions are associated with this test, and why are they sometimes not reasonable?

3. What assumptions are required for the Wilcoxon rank-sum test?

4. What levels of measurement may data possess in order for the Wilcoxon rank-sum test to be performed?

5. Describe the procedure for ranking data in order to perform a Wilcoxon rank-sum test.

6. What are the null and alternative hypotheses associated with the Wilcoxon rank-sum test?

7. What is the test statistic associated with the Wilcoxon rank-sum test for small samples? What does the test statistic depend on and how small is a *small* sample?

8. What is the test statistic associated with the Wilcoxon rank-sum test for large samples?

9. Identify the critical values associated with the Wilcoxon rank-sum test for both large and small samples.

Exercises

10. A luxury car dealer is considering two possible locations for a new auto mall. The rent on the south side of town is cheaper. However, the dealer believes that the average household income is significantly higher on the north side of town. The dealer has decided that he will locate the new auto mall on the north side of town if the results of a study which he has commissioned show that the median household income is significantly higher on the north side of town. The results of the study are as follows.

Table for Exercise 10 – Household Incomes	
North Side ($)	South Side ($)
50,000	43,000
45,000	45,000
55,000	42,000
25,000	50,000
75,000	36,000
35,000	48,000
65,000	38,000
55,000	43,000
45,000	43,000

a. Use the Wilcoxon rank-sum test to determine if the auto dealer should locate the new auto mall on the north side of town. Use $\alpha = 0.05$.

b. What assumptions were made in performing the hypothesis test in part **a.**?

11. An internal auditor for Tiger Enterprises has been asked to determine if there is a difference in the amount charged for daily expenses by two top salesmen, Mr. Ellis and Mr. Ford. The auditor randomly selects seven days and determines the daily expenses for each of the salesmen.

Table for Exercise 11 – Daily Expenses	
Mr. Ellis ($)	Mr. Ford ($)
55	60
53	55
58	65
54	50
56	70
55	55
55	65

a. Using the Wilcoxon rank-sum test, can the auditor conclude that there is a difference in the median amount charged for daily expenses by the two top salesmen, Mr. Ellis and Mr. Ford? Use $\alpha = 0.05$.

b. What assumptions were made in performing the test in part **a.**?

12. The Armed Forces have two different programs for training aircraft personnel. A government regulatory agency has been commissioned to evaluate any differences which may exist between the two programs. The agency administers a standardized test to randomly selected groups of students from the two programs. The results of the test for the students in each of the programs are as follows.

Table for Exercise 12 – Standardized Test Scores	
Program A	Program B
85	87
95	96
75	78
100	100
70	74
90	92
80	82

    a. Using the Wilcoxon rank-sum test, can the agency conclude that there is a difference in the median test scores of students in the two programs? Use $\alpha = 0.10$.

    b. What assumptions were made in performing the test in part **a.**?

13. Tom Anderson, a supply clerk with the Navy, has been asked to determine if a new battery which has been offered to the Navy (at a reduced price) has a shorter life than the battery which they are currently using. He randomly selects batteries of each type and allows them to run continuously so that he can measure the time until failure for each battery. The results of the test are as follows.

Table for Exercise 13 – Time Until Failure for Batteries (Hours)	
New Battery	Old Battery
655	745
730	675
670	730
715	690
685	760
745	660

    a. Using the Wilcoxon rank-sum test, do the data suggest at $\alpha = 0.05$ that the median time until failure for the new battery is significantly less than the median time until failure for the old battery?

    b. What assumptions were made in performing the test in part **a.**?

14. A cereal manufacturer has advertised that its product, Fiber Oat Flakes, has a lower fat content than its competitor, Bran Flakes Plus. Because of the complaints from the manufacturer of Bran Flakes Plus, the FDA has decided to test the claim that Fiber Oat Flakes has a lower median fat content than Bran Flakes Plus. Several boxes of each cereal are selected and the fat content per serving is measured. The results of the study are as follows.

Table for Exercise 14 – Fat Content of Cereals (Grams)	
Fiber Oat Flakes	Bran Flakes Plus
5	6
6	8
4	4
7	9
3	3
5	7
5	5
6	8
4	4

a. Using the Wilcoxon rank-sum test, does the study performed by the FDA substantiate the claim made by the manufacturer of Fiber Oat Flakes at $\alpha = 0.05$?

b. What assumptions were made in performing the test in part **a.**?

15. A Hollywood studio believes that a movie which is considered a drama will draw a larger crowd on average than a movie which is a comedy. To test this theory, the studio randomly selects several movies which are classified as dramas and several movies which are classified as comedies and determines the box office revenue for each movie. The results of the survey are as follows.

Table for Exercise 15 – Box Office Revenues (Millions of Dollars)	
Drama	Comedy
180	150
240	190
120	110
220	170
140	130

a. Using the Wilcoxon rank-sum test, do the data substantiate the studio's belief that dramas will draw a larger crowd on average than comedies at $\alpha = 0.05$?

b. What assumptions were made in performing the test in part **a.**?

16

16. *Consumer Magazine* is reviewing the top selling amplifiers produced by two major stereo manufacturers. One of the most important qualities of the amplifiers is the maximum power output. Brand A has redone their internal design and claims to have a higher maximum power level than Brand B. To test this claim, *Consumer Magazine* randomly selects amplifiers from each brand and determines the maximum power output. The results of the test are as follows.

Table for Exercise 16 – Maximum Power Output (Watts)	
Brand A	**Brand B**
800	780
828	805
772	755
830	807
770	753
826	803
774	757

a. Using the Wilcoxon rank-sum test, do the data substantiate the claim that the Brand A amplifier has a higher median maximum power output than Brand B at $\alpha = 0.05$?

b. What assumptions were made in performing the test in part **a.**?

17. A state environmental board wants to compare pollution levels in two of its major cities. Sunshine City thrives on the tourist industry and Service City thrives on the service industry. The environmental board randomly selects several areas within the cities and measures the pollution levels in parts per million with the following results.

Table for Exercise 17 – Pollution Levels (ppm)	
Sunshine City	**Service City**
8.50	7.90
9.00	8.35
8.00	7.45
9.07	8.40
7.93	7.40
9.14	8.45
7.86	7.35
8.50	7.90

a. Using the Wilcoxon rank-sum test, can the state environmental board conclude at $\alpha = 0.05$ that Service City has a lower pollution level on average than Sunshine City?

b. What assumptions were made in performing the test in part **a.**?

16.4   The Rank Correlation Test

In Section 4.8 we studied the **correlation coefficient** as a measure of association between two random variables. The parametric correlation coefficient gives a direct correlation between the variables. In this section we will transform the data from two variables into ranks and develop a method for detecting an association between them.

Measure of Correlation

Let us denote the ranks assigned to data values x_i and y_i by $R(x_i)$ and $R(y_i)$, respectively. Then the **Spearman rank correlation coefficient**, also called **Spearman's rho**, is usually denoted by r_s, and is defined as follows.

Formula

Spearman Rank Correlation Coefficient

The Spearman rank correlation coefficient is given by

$$r_s = 1 - \frac{6\sum_{i=1}^{n} d_i^2}{n(n^2 - 1)},$$

where $d_i = R(x_i) - R(y_i)$.

This method is applicable when there are no ties in the rank data. The values of r_s are always between -1 and 1. The r_s statistic behaves like the parametric correlation coefficient (r). If there is a positive relationship, r_s will be positive but less than or equal to 1. If there is a negative relationship, r_s will be negative but greater than or equal to -1.

Example 16.6

Consider the following data set.

Table 16.7 – Original Data

x	y
7	4
5	7
8	9
9	8

Calculate the Spearman rank correlation coefficient, r_s, for these data.

Solution

Converting the data to ranks (from smallest to largest), we obtain

Table 16.8 – Ranked Data			
x	**R(x)**	**y**	**R(y)**
7	2	4	1
5	1	7	2
8	3	9	4
9	4	8	3

The value of r_s is given by

$$r_s = 1 - \frac{6\left((2-1)^2 + (1-2)^2 + (3-4)^2 + (4-3)^2\right)}{4(4^2-1)} = 1 - \frac{6(4)}{4(15)} = 0.60.$$

We want to test the null hypothesis that there is no correlation between the two variables against the alternative hypothesis that there is a correlation. More specifically, we test

H_0: $\rho_s = 0$ (There is no correlation between the two variables.)

H_a: $\rho_s \neq 0$ (There is a correlation between the two variables.)

where ρ is the Greek letter rho. Note that ρ_s represents Spearman's rho for the population and r_s is the corresponding sample statistic.

Note: The advantage of Spearman's rank correlation coefficient is that it can be used to test for a monotonic, nonlinear relationship (i.e. y increases as x increases or y decreases as x increases, but not necessarily linearly).

Procedure

Rank Correlation Procedure

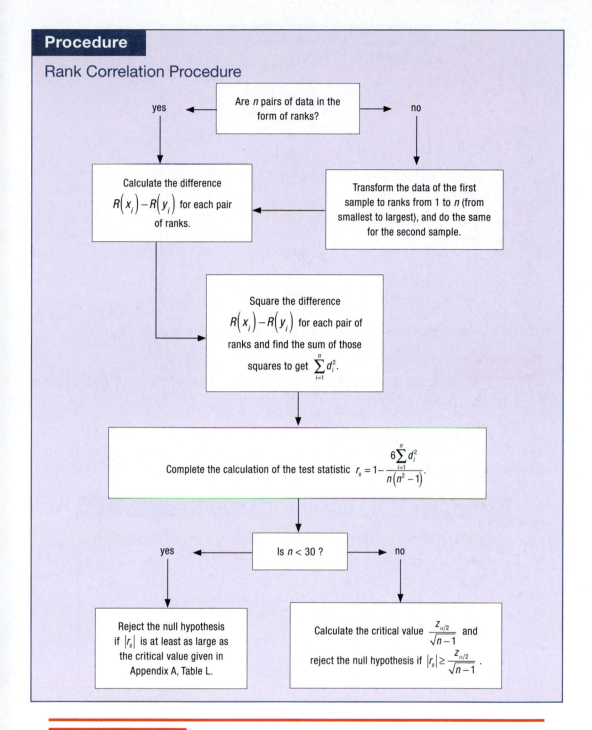

Example 16.7

The academic performances of 12 college graduates are observed to examine the relationship between their SAT scores and their GPAs. The SAT scores (combined mathematics and critical reading sections which could yield a maximum score of 1600) and the GPAs are given in the following table.

Table 16.9 – SAT Scores and GPAs		
Student	**SAT (x)**	**GPA (y)**
1	1210	4.00
2	1110	3.97
3	1140	3.92
4	1080	3.85
5	1020	3.72
6	1090	3.60
7	1120	3.51
8	1030	3.49
9	1050	3.45
10	1040	3.39
11	1070	3.28
12	1060	3.27

Test the hypothesis that there is a relationship between the variables at the $\alpha = 0.10$ level.

Solution

$$H_0: \rho_s = 0$$
$$H_a: \rho_s \neq 0$$

In order to test whether an association exists between these two variables at $\alpha = 0.10$ we prepare the following table.

Table 16.10 – SAT Scores and GPAs					
Student	**SAT (x)**	**GPA (y)**	**R (x)**	**R (y)**	$d_i^2 = \left[R(x_i) - R(y_i) \right]^2$
1	1210	4.00	12	12	0
2	1110	3.97	9	11	4
3	1140	3.92	11	10	1
4	1080	3.85	7	9	4
5	1020	3.72	1	8	49
6	1090	3.60	8	7	1
7	1120	3.51	10	6	16
8	1030	3.49	2	5	9
9	1050	3.45	4	4	0
10	1040	3.39	3	3	0
11	1070	3.28	6	2	16
12	1060	3.27	5	1	16
				Total	116

We then calculate the test statistic

$$r_s = 1 - \frac{6(116)}{12(12^2 - 1)} \approx 0.5944.$$

Since our sample size is 12 ($n < 30$), the value of the statistic is compared with the critical value obtained using Appendix A, Table L.

n	$\alpha = 0.10$	$\alpha = 0.05$	...	$\alpha = 0.01$
$\vdots$				
11	0.523	0.623		0.818
12	0.497	0.591		0.780
13	0.475	0.566		0.745
14	0.457	0.545		0.716
15	0.441	0.525		0.689
$\vdots$				

Note that Table L in Appendix A is constructed for a two-tailed test. So, we want to reject H_0 if $r_s \leq -0.497$ or $r_s \geq 0.497$. Since $0.5944 > 0.497$, we reject the null hypothesis at the 10% level of significance. Hence, there seems to be an association between SAT scores and GPAs.

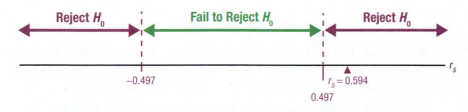

Figure 16.13

16.4   Exercises

Basic Concepts

1.   What is the correlation coefficient? How is this different from the Spearman rank correlation coefficient?

2.   What is the formula for calculating Spearman's rho?

3.   Can you calculate Spearman's rho if there are ties in the rank data?

4.   Identify the difference in notation between Spearman's rho for population and sample data.

5.   Explain the similarities in the behavior of the parametric correlation coefficient and Spearman's rho.

6.   Identify one main advantage of the Spearman's rank correlation coefficient versus the parametric correlation coefficient.

7.   Explain the procedure for ranking data when calculating Spearman's rho.

8.   What are the null and alternative hypotheses for the rank correlation test?

9.   Consider the value $r_s = 0.12$. Interpret this value in terms of the x and y variables used to calculate Spearman's rho.

Exercises

10. Chris is a new cashier assigned to a cash register in a supermarket. Each day a sample of purchases at that register is examined and a percent of pricing errors is recorded along with the total number of customers who used that register. Do the following data indicate an association between Chris' performance and how busy his register was? Use $\alpha = 0.05$.

Table for Exercise 10 – % Pricing Errors and Total Customers			
Number of Customers	**Errors (%)**	**Number of Customers**	**Errors (%)**
57	4.2	67	2.5
44	5.5	71	2.9
32	5.7	69	2.6
60	3.9	56	1.0
55	3.2	51	2.0
59	4.1	70	1.7
63	3.3		

11. Twelve new runners were randomly assigned to different training programs, where they were required to run a certain number of miles every week for a year prior to a major race. After the training, the participants ran the race and their finishing times were recorded.

Table for Exercise 11 – Miles of Training and Race Times			
Miles Logged	**Race Time (Minutes)**	**Miles Logged**	**Race Time (Minutes)**
35	198	30	189
25	165	29	240
45	155	42	224
60	148	24	201
70	135	19	246
21	243	55	166

a. With 95% confidence, is there evidence that the number of miles logged in a week during training affects the runner's race time?

b. Can the linear correlation coefficient, r, be calculated in order to fit a least squares regression line to the data in the table in an effort to predict the finish time of runners based on the number of miles logged during training? Why or why not?

12. The following data consist of college rankings of five universities by two different magazines. Is there a correlation between the rankings of the magazines? Use $\alpha = 0.10$.

Table for Exercise 12 – College Rankings by Magazines					
College	A	B	C	D	E
Magazine 1	1	4	2	3	5
Magazine 2	4	3	1	5	2

13. An anthropologist records the heights (in inches) of ten fathers and their sons. Do the following data support (at the 5% level) that taller fathers tend to have taller sons?

Table for Exercise 13 – Heights of Fathers and Sons (Inches)			
Son's Height	Father's Height	Son's Height	Father's Height
72	70	65	71
68	73	70	78
74	72	69	67
66	68	67	65
71	69	80	66

14. After a mother-daughter golf tournament, mothers and daughters were ranked among themselves. Do the following data show (at the 5% level) a correlation between the daughters' and mothers' golf skills?

Table for Exercise 14 – Golf Rankings			
Daughter's Ranking	Mother's Ranking	Daughter's Ranking	Mother's Ranking
1	5	5	3
9	4	3	6
10	8	7	7
2	2	6	10
4	1	8	9

16

16.5   The Runs Test for Randomness

Randomness is an important concept in probability and statistics. In this section we are going to discuss a method for determining whether a sequence of observations exhibits randomness. To illustrate this concept we will use the familiar coin-tossing experiment. One characteristic of the coin-tossing experiment is that in the long run there should be approximately equal numbers of heads and tails. In an ordered sequence, however, randomness implies more than compliance with this frequency criterion. For example, if the outcomes of 20 tosses of a coin were recorded as

$$H\,H\,H\,H\,H\,T\,T\,T\,T\,T\,T\,T\,T\,T\,T\,H\,H\,H\,H\,H,$$

we would suspect that the process was flawed. We would be equally surprised if the ordered outcomes were

$$H\,T\,H\,T\,H\,T\,H\,T\,H\,T\,H\,T\,H\,T\,H\,T\,H\,T\,H\,T,$$

but be reasonably happy with the sequence

$$H\,H\,T\,H\,T\,T\,T\,H\,T\,H\,H\,H\,T\,H\,T\,H\,H\,H\,T\,T\,H.$$

A characteristic that reflects our reservations about the first two sequences is the number of **runs**, where a run is a subsequence of one or more heads (or tails). In the first sequence, there are three runs: a run of 5 heads, then 10 tails, then 5 heads.

In the second sequence, there are 20 runs, each consisting of a single head or tail.

Intuitively, we feel that these two sequences have respectively too few and too many runs for a truly random sequence. In the last sequence there are 13 runs.

Given the number of times the coin was flipped, the number of runs does not seem excessively large or small. This intuitive notion of rejecting "randomness" in the sequence if there are too few or too many runs is the same notion we will use in the test of hypothesis.

In order to test the hypothesis

$$H_0\text{: The sequence is random.}$$

$$H_a\text{: The sequence is not random.}$$

we use a test based on the number of runs, R, in a sequence of N ordered observations.

Procedure

Runs Test for Randomness

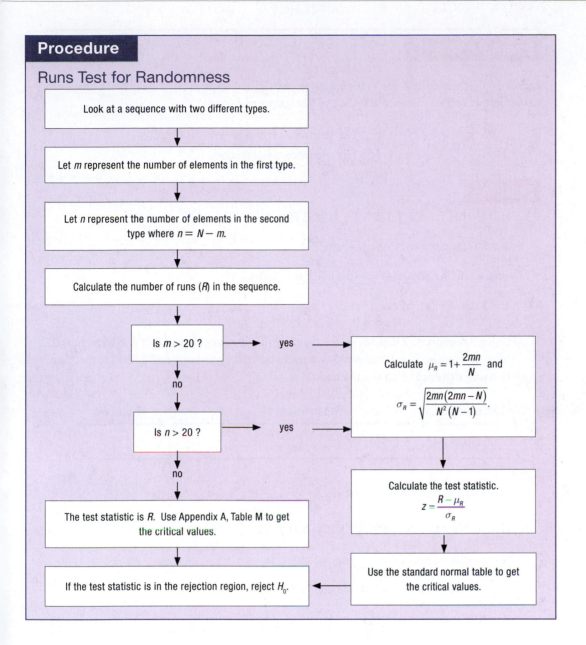

Look at a sequence with two different types.

↓

Let m represent the number of elements in the first type.

↓

Let n represent the number of elements in the second type where $n = N - m$.

↓

Calculate the number of runs (R) in the sequence.

↓

Is $m > 20$? — yes →

no
↓

Is $n > 20$? — yes →

no
↓

The test statistic is R. Use Appendix A, Table M to get the critical values.

Calculate $\mu_R = 1 + \dfrac{2mn}{N}$ and $\sigma_R = \sqrt{\dfrac{2mn(2mn - N)}{N^2(N-1)}}$.

↓

Calculate the test statistic.
$$z = \frac{R - \mu_R}{\sigma_R}$$

↓

Use the standard normal table to get the critical values.

↓

If the test statistic is in the rejection region, reject H_0.

931

16

Example 16.8

Let us apply the runs test for randomness to each of the sequences of heads and tails seen earlier in this section. In each test the hypotheses are as follows.

H_0: The sequence is random.

H_a: The sequence is not random.

Solution

a. H H H H H T T T T T T T T T T H H H H H

$N = 20$ (the number of observations)

m (the number of heads) $= n$ (the number of tails) $= 10$

$R = 3$ (the number of runs)

Using Appendix A, Table M, we see that the critical values are $R \leq 6$ or $R \geq 16$ at the 0.05 level of significance. Since $R = 3$, we reject the null hypothesis and conclude that there is strong evidence of non-randomness.

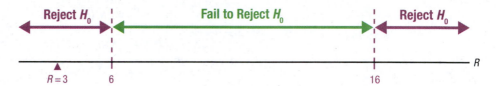

Figure 16.14

b. H T H T H T H T H T H T H T H T H T H T

$N = 20$ (the number of observations)

m (the number of heads) $= n$ (the number of tails) $= 10$

$R = 20$ (the number of runs)

Using Appendix A, Table M, we see that the critical values are $R \leq 6$ or $R \geq 16$ at the 0.05 level of significance. Since $R = 20$, we reject the null hypothesis and conclude that there is strong evidence of non-randomness.

Figure 16.15

c. H H T H T T T H T H H H T H T H H H T T H

$N = 20$ (the number of observations)

$m = 11$ (the number of heads)

$n = 9$ (the number of tails)

$R = 13$ (the number of runs)

Using Appendix A, Table M, we see that the critical values are $R \leq 6$ or $R \geq 16$ at the 0.05 level of significance. Since $R = 13$, we fail to reject the null hypothesis and conclude that there is not sufficient evidence of non-randomness.

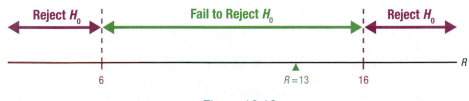

Figure 16.16

Example 16.9

The admissions office of a college records the gender of 28 applicants in the order the applications arrive. The data set is given as follows.

M F F M M M F F F F M M M M M M M M M M F M M M M M M M

Is this sequence random? Test at the 0.05 level.

Solution

To test the randomness of this sequence at the 5% level of significance, we obtain the following.

$N = 28$ (the number of observations)

$m = 21$ (the number of males)

$n = 7$ (the number of females)

$R = 7$ (the number of runs)

Since $m > 20$, we calculate μ_R and σ_R.

$$\mu_R = 1 + \frac{2mn}{N} \qquad\qquad \sigma_R = \sqrt{\frac{2mn(2mn - N)}{N^2(N-1)}}$$

$$= 1 + \frac{2(7)(21)}{28} \qquad\qquad = \sqrt{\frac{2(7)(21)(2(7)(21) - 28)}{28^2(28-1)}}$$

$$= 11.5 \qquad\qquad\qquad\qquad \approx 1.92$$

Thus, we have the following.

$$z = \frac{R - \mu_R}{\sigma_R}$$

$$= \frac{7 - 11.5}{1.92}$$

$$\approx -2.34$$

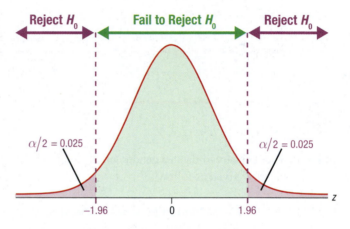

Figure 16.17

Since the test statistic is in the rejection region, we reject the null hypothesis that the gender of the arriving student applications follows a random sequence.

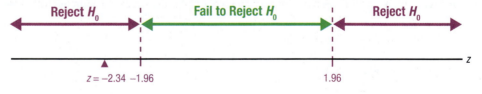

Figure 16.18

Example 16.10

Are the following data random? Test at $\alpha = 0.05$.

$$16, 25, 52, 11, 38, 47, 12, 98, 4$$

Solution

How do you test randomness with a numerical set? Create a new data set comparing each value to the median value. To do this, substitute each value in the original data set with an A if it is above the median value, a B if it is below the median value, and eliminate any values that equal the median.

H_0: The data are random.

H_a: The data are not random.

Median = 25

16	25	52	11	38	47	12	98	4
B	Ø	A	B	A	A	B	A	B

$m = 4$ (the number of A's)

$n = 4$ (the number of B's)

$R = 7$ (the number of runs)

Using Appendix A, Table M the rejection region is $R \le 1$ or $R \ge 9$ at the 0.05 level of significance.

Since $R = 7$, we fail to reject H_0 and conclude that there is not sufficient evidence of non-randomness.

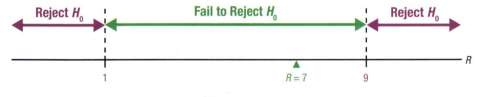

Figure 16.19

16

16.5   Exercises

Basic Concepts

1. Describe in your own words what is being tested with the runs test.

2. Consider the following sequence of 10 coin tosses.

   H, H, T, T, H, H, H, T, T, H

   Without performing any kind of test, do you believe this sequence is random? Explain why or why not.

3. What are the null and alternative hypotheses associated with the runs test?

4. What parameters need to be calculated in order to perform a runs test?

5. What is the rejection rule for a small sample runs test? How small is a *small* sample?

6. What is the rejection rule for a large sample runs test? How large is a *large* sample?

7. If a numerical set of data is under consideration, which parameter are the data points compared to in order to perform the runs test?

Exercises

8. Suppose that in your city the number of deaths due to traffic accidents involving drunk driving from 1999 to 2011 were 75, 91, 54, 85, 79, 63, 12, 55, 63, 49, 89, 98, and 71. Use the runs test to examine non-randomness at the 0.05 level.

9. A sociologist designs a study that involves a procedure of selecting families randomly from a phone book and then calling them to determine if they own or rent their residence. The results are recorded in the order of phone calls (O = Own, R = Rent).

   O O O R R O R O R R O R R R O R R R O O R R R O R

   Does the sociologist have a random sequence of residential data at the 0.05 level?

10. A car tire manufacturer keeps track of the tires produced by one of the production lines. They observe the following sequence (D for defective items and N for non-defective items).

    D D D N N D N D N D D D

    Test the quality control manager's claim that there is no pattern in producing defective tires at the 0.05 level.

11. A marathon runner tries to run every day except when it is raining during the month of July. He observes the rainy (R) days and sunny (S) days to be able to predict the weather as follows.

    S S S R R S S S R R R R S R S R R S S R S R S R R S R S R S S

    Are the rainy days randomly scattered in the month of July at the 0.05 level?

16.6   The Kruskal-Wallis Test

In this section we present a procedure where k random samples are obtained, one from each of the k possibly different populations, and we are interested in testing whether all of the populations have identical distributions. Suitable hypotheses for this test would be as follows.

H_0: The populations from which the samples are drawn have identical distributions.

H_a: Not all populations have the same distribution.

The **Kruskal-Wallis test** is a method that can be used instead of the ANOVA F-test. The Kruskal-Wallis test does not need the assumption of normality of the populations. It does require that independent, random samples be drawn.

The Kruskal-Wallis test is similar to the Wilcoxon rank-sum test in that the test statistic will be based on the sums of the ranks of the groups being compared.

The data consist of k random samples (not necessarily the same size) drawn from their respective populations. The data set may be arranged as follows.

Group 1	Group 2		Group k
$x_{1,1}$	$x_{2,1}$	$\cdots$	$x_{k,1}$
$x_{1,2}$	$x_{2,2}$	$\cdots$	$x_{k,2}$
$\vdots$	$\vdots$	$\cdots$	$\vdots$
x_{1,n_1}	x_{2,n_2}	$\cdots$	x_{k,n_k}

Let N be the total number of observations, that is, $N = \sum_{i=1}^{k} n_i$.

To compute the Kruskal-Wallis test statistic we first combine all N observations and order them from smallest to largest. Next, we assign rank 1 to the smallest of all observations, rank 2 to the second smallest, and so on, until the largest observation is assigned the rank N. Let us denote the rank of $x_{i,j}$ by r_{ij}, and define

$$R_i = \sum_{j=1}^{n_i} r_{ij} \qquad i = 1, 2, ..., k.$$

For example, R_1 is the sum of the ranks received by the Group 1 observations. The test statistic for the Kruskal-Wallis test is then given by the following formula.

$$H = \frac{12}{N(N+1)} \sum_{i=1}^{k} \frac{R_i^2}{n_i} - 3(N+1)$$

We reject the null hypothesis if $H \geq \chi_\alpha^2$, where χ_α^2 is obtained from the chi-square table (Appendix A, Table G) with $k - 1$ degrees of freedom and a predetermined value of α.

Example 16.11

Ris Bocaj, the owner of Master Deck Sealer (MDS), specializes in staining and sealing wooden decks of residential homes. The company has developed a new stain, called MDS1, that not only stains but seals the decks from inclement weather for a period of 10 years. Even though there are other products on the market, MDS believes that its product is significantly better than the competition. To test his claim, Ris conducted accelerated-climate tests in a laboratory, exposing the wood (treated with his MDS1 and two other popular brands) to inclement weather such as rain, snow, high temperatures, and other conditions that would eventually result in the wood needing to be re-treated. The number of months before the stain began to wear and needed to be re-treated is recorded in the following table.

Table 16.11 – Wood Stains: Months Before Re-Treatment		
MDS1	**Competition Brand 1**	**Competition Brand 2**
124	39	25
75	29	14
69	26	26
70	28	35
122	26	34
73		16
		12

Use the Kruskal-Wallis test to determine whether there is a difference between the deck stains using $\alpha = 0.05$.

Solution

The $N = 18$ data values are assigned overall ranks. In the following table, the ranks are shown in the column next to each observation.

Table 16.12 – Wood Stain Rankings					
MDS1	**Rank**	**Competition Brand 1**	**Rank**	**Competition Brand 2**	**Rank**
124	18	39	12	25	4
75	16	29	9	14	2
69	13	26	6	26	6
70	14	28	8	35	11
122	17	26	6	34	10
73	15			16	3
				12	1

The hypotheses of this test can be written as follows.

H_0: The three stains are equally durable.

H_a: At least one stain durability is different.

The sum of the ranks for each brand of stain are as follows.

MDS1: $18+16+13+14+17+15 = 93 = R_1$

Competition Brand 1: $12+9+6+8+6 = 41 = R_2$

Competition Brand 2: $4+2+6+11+10+3+1 = 37 = R_3$

Hence, we have the following test statistic.

$$H = \frac{12}{N(N+1)} \sum_{i=1}^{k} \frac{R_i^2}{n_i} - 3(N+1)$$

$$= \frac{12}{18(18+1)} \left(\frac{93^2}{6} + \frac{41^2}{5} + \frac{37^2}{7} \right) - 3(18+1)$$

$$\approx 12.238$$

There are $k - 1 = 3 - 1 = 2$ degrees of freedom. The critical value is given in Appendix A, Table G as $\chi_{0.05}^2 = 5.991$. Since $12.238 > 5.991$, we reject the null hypothesis and state that the durability of the stains does differ significantly.

Figure 16.20

Example 16.12

A traffic safety engineer records the braking distances of a test vehicle at a fixed speed using three different brake pads. Braking distances (in feet) and their respective ranks (in parentheses) for brake pads of types A, B, and C are given in the following table. Is there sufficient evidence at the 0.05 level to conclude that the braking distances for the three pads differ?

Table 16.13 – Braking Distances		
Brake Pad A	**Brake Pad B**	**Brake Pad C**
89 (1)	97 (5)	98 (6)
99 (7)	139 (15)	116 (11)
101 (8)	119 (12)	104 (10)
94 (2)	103 (9)	96 (4)
95 (3)	127 (14)	126 (13)
Rank Sum 21	55	44

16

Solution

The null and alternative hypotheses for this test can be written as follows.

H_0: The braking distances for the three pads are the same.

H_a: At least one of the braking distances is different.

In Table 16.13 we are given the ranks of the observations. We then need to determine R_1, R_2, and R_3, corresponding to the sums of the ranks assigned to the observations for brake pads A, B, and C. The sum of the ranks for each brake pad is as follows.

$$R_1 = 21$$
$$R_2 = 55$$
$$R_3 = 44$$

Having the ranks, we can now calculate the test statistic, H.

$$H = \frac{12}{N(N+1)} \sum_{i=1}^{k} \frac{R_i^2}{n_i} - 3(N+1)$$

$$= \frac{12}{15(15+1)} \left(\frac{21^2}{5} + \frac{55^2}{5} + \frac{44^2}{5} \right) - 3(15+1)$$

$$= 6.02$$

Referring to Appendix A, Table G, we see that $\chi^2_{0.05} = 5.991$. Thus, we want to reject the null hypothesis if the test statistic, H, is greater than or equal to 5.991. Since the test statistic exceeds the critical value (6.02 > 5.991), we reject the null hypothesis and conclude that the braking distances are sufficiently different.

Figure 16.21

16.6   Exercises

Basic Concepts

1. Which parametric test corresponds to the nonparametric Kruskal-Wallis test?

2. What are the null and alternative hypotheses associated with the Kruskal-Wallis test?

3. What are the assumptions associated with the Kruskal-Wallis test?

4. How is the Kruskal-Wallis test similar to the Wilcoxon rank-sum test?

5. What is the test statistic for the Kruskal-Wallis test? How is it calculated?

6. What is the rejection rule for the Kruskal-Wallis test?

7. How many populations can be compared using the Kruskal-Wallis test?

Exercises

8. An Internet service provider is considering four different servers for purchase. Potentially, the company would be purchasing hundreds of these servers, so it wants to make sure it is making the best decision. Initially, five of each type of server are borrowed, and each is randomly assigned to one of the 20 technicians (all technicians are similar in skill). Each server is then put through a series of tasks and rated using a standardized test. The higher the score on the test, the better the performance of the server. The data are as follows.

Table for Exercise 8 – Server Test Scores			
Server 1	Server 2	Server 3	Server 4
48.5	56.4	52.1	64.3
46.5	68.2	56.3	68.3
52.4	68.5	48.3	72.2
54.1	64.2	52.2	70.6
58.9	60.1	54.8	56.5

Perform a Kruskal-Wallis test on these data using $\alpha = 0.10$. Are there differences between the servers?

9. The following summary is obtained from an experiment where groups of cows were fed according to one of the four different feeding schedules, and their milk productions were recorded. The data given show the daily milk production in gallons for each cow. Test at $\alpha = 0.10$ to examine whether or not the milk production for all four schedules is the same.

Table for Exercise 9 – Milk Production by Schedule (Gallons)				
Schedule 1	11.5	12.7	12.9	10.1	10.5
Schedule 2	9.1	10.7	9.5	10.9	10.4
Schedule 3	12.4	11.9	10.0	11.4	12.1
Schedule 4	12.8	12.6	11.7	11.3	10.9

10. The following data set contains the reading speed (in words per minute) of second grade students.

Table for Exercise 10 – Reading Speeds (wpm)		
Public School	Private School	Home School
54	66	65
67	55	64
63	62	60
105	69	72
61	71	68

Is there sufficient evidence at the 0.01 level of significance to conclude that the reading speeds vary by school type?

T   Discovering Technology

Using Excel

Test Statistic for the Sign Test

When doing the sign test you are asked to find the test statistic.

If $n \leq 25$, $X =$ the number of times the less frequent sign occurs.

If $n > 25$, $z = \dfrac{X + 0.5 - \left(\dfrac{n}{2}\right)}{\dfrac{\sqrt{n}}{2}}$.

Compute the test statistic using the parameters in Example 16.2

1. In a new worksheet, enter the labels X, n, and z in cells A1, A2, and A3, respectively.

2. Put the values for X and n in cells B1 and B2. In this case, enter **15** in B1 and **50** in B2.

3. In cell B3, enter the formula for the z-test statistic.

$$\textbf{=(B1+0.5-(B2/2))/(sqrt(B2)/2)}$$

4. The result is -2.69 rounded to two decimal places. This is the value of the z-test statistic.

	A	B
1	X	15
2	n	50
3	z	-2.68701

Figure 16.22

If you need to work more of these problems you can just change the X and n values and z should dynamically change. Remember, though, this formula only applies when n is greater than 25.

Spearman Rank Correlation Coefficient

For this exercise, use the information from Example 16.7.

1. Enter the SAT scores and GPAs from Table 16.9 in columns A and B, respectively.

2. You can use the sort feature in Excel to help you determine the ranks for the two variables. Highlight the SAT scores and press **Ctrl + C** to copy them to the clipboard. Paste the scores in Column F. With the scores still highlighted, select **Sort Smallest to Largest** under the **Data** tab. The SAT scores are sorted from smallest to largest, and we can assign ranks.

3. Enter the ranks for the SAT scores in Column C based on the sorted data in Column F.

4. Copy the GPA data from Column B and paste the values in Column G. Again select Sort Smallest to Largest under the Data tab and enter the ranks for the GPAs in Column D.

5. In Column E, we will compute the values for d_i^2. In cell E2, enter the following formula.

$$=(C2-D2)\wedge 2$$

Copy the formula by clicking on the black box in the bottom right hand corner of the cell and dragging down to cell E13.

6. In cell E14, sum the values in Column E by entering the following formula.

$$=sum(E2{:}E13)$$

7. We are now ready to compute r_s using the formula $r_s = 1 - \dfrac{6\sum d_i^2}{n\left(n^2 - 1\right)}$. In cell E15, enter the following formula.

$$=1\text{-}6*E14/(12*(12\wedge2\text{-}1))$$

The result is Spearman's rho. Compare this value to that which we calculated in Example 16.7.

Using Minitab

Wilcoxon Signed-Rank Test

Use the information from Example 16.4 for this exercise.

1. Enter the 2006 data in Column C1 and the 2008 data in Column C2.

2. Choose **Edit**, and then **Command Line Editor**.

3. In the Command Line Editor dialog box, enter **LET C3=C1−C2**. Press **Submit Commands**. The differences are now calculated in Column C3.

4. Choose **Stat**, **Nonparametrics**, and **1-Sample Wilcoxon**.

5. Enter **C3** for Variables and select the radio button next to **Test median**. Choose **greater than** as the Alternative and press **OK**.

6. Observe the output screen for the Wilcoxon signed-rank test. The *P*-value is reported as 0.062 which is greater than the level of α. Thus, we fail to reject the null hypothesis.

```
Wilcoxon Signed Rank Test: C3

Test of median = 0.000000 versus median > 0.000000

          N for   Wilcoxon              Estimated
      N   Test   Statistic       P       Median
C3    9    8          29.5   0.062        2.000
```

Figure 16.23

Wilcoxon Rank-Sum Test

Use the information from Example 16.5 for this exercise.

1. Input the ratings for Brand A headphones in Column C1 and the ratings for Brand B headphones in Column C2.

2. Choose **Stat**, **Nonparametrics**, and **Mann-Whitney**.

3. Select **C1 (Brand A)** as the First Sample and **C2 (Brand B)** as the Second Sample. Ensure that the Confidence level is set to **95.0** and that the Alternative is **not equal**. Press **OK**.

4. Observe the results of the Wilcoxon rank-sum test (also known as the Mann-Whitney test). Compare the value of the test statistic, $W = 110.0$, to that which we calculated in Example 16.5. Minitab reports the P-value of this test to be 0.7337, causing us to fail to reject the null hypothesis.

```
Mann-Whitney Test and CI: Brand A, Brand B

            N   Median
Brand A    10    87.50
Brand B    10    85.00


Point estimate for ETA1-ETA2 is 3.00
95.5 Percent CI for ETA1-ETA2 is (-9.00,14.01)
W = 110.0
Test of ETA1 = ETA2 vs ETA1 not = ETA2 is significant at 0.7337
The test is significant at 0.7335 (adjusted for ties)
```

Figure 16.24

Runs Test for Randomness

Use the information from Example 16.10 for this exercise.

1. Enter the data into Column C1.

2. Choose **Stat**, **Nonparametrics**, and **Runs test**.

3. Select C1 as the Variables.

4. Notice that when conducting the runs test for randomness, Minitab counts observations above and below the mean rather than the median as we did in Example 16.10. You have the option to use some other value (such as the median) by choosing the radio button below Above and below the mean. Press **OK**.

5. Observe the outcome of the runs test. The P-value is indicated as 0.261, thus we will fail to reject the null hypothesis.

```
Runs Test: C1

Runs test for C1

Runs above and below K = 33.6667

The observed number of runs = 7
The expected number of runs = 5.44444
4 observations above K, 5 below
* N is small, so the following approximation may be invalid.
P-value = 0.261
```

Figure 16.25

The Kruskal-Wallis Test

Use the information from Example 16.11 for this exercise.

1. Enter the data for the time until re-treatment is needed for all three brands of stains in Column C1. Enter the brand corresponding to each data value in Column C2 as shown.

↓	C1	C2-T
	Time	Brands
1	124	MDS1
2	75	MDS1
3	69	MDS1
4	70	MDS1
5	122	MDS1
6	73	MDS1
7	39	BRAND1
8	29	BRAND1
9	26	BRAND1
10	28	BRAND1
11	26	BRAND1
12	25	BRAND2
13	14	BRAND2
14	26	BRAND2
15	35	BRAND2
16	34	BRAND2
17	16	BRAND2
18	12	BRAND2

Figure 16.26

2. Choose **Stat**, **Nonparametrics**, and **Kruskal-Wallis**. Enter **Time** for the Response and **Brands** for the Factor. Press **OK**.

3. Observe the results of the Kruskal-Wallis test. Compare the computed value of the test statistic to that which we found in Example 16.11. The *P*-value is given as 0.002, causing us to reject the null hypothesis at the 0.05 level of significance.

```
Kruskal-Wallis Test: Time versus Brands

Kruskal-Wallis Test on Time

Brands    N  Median  Ave Rank       Z
BRAND1    5   28.00       8.2   -0.64
BRAND2    7   25.00       5.3   -2.67
MDS1      6   74.00      15.5    3.37
Overall  18                9.5

H = 12.24  DF = 2  P = 0.002
H = 12.29  DF = 2  P = 0.002  (adjusted for ties)
```

Figure 16.27

R Chapter 16 Review

Key Terms and Ideas

- Parametric Statistics
- Nonparametric Statistical Methods
- Sign Test
- Hypothesis Test about a Population Median
- Wilcoxon Signed-Rank Test
- Rank Sum
- Wilcoxon Rank-Sum Test
- Rank Correlation Test

- Correlation Coefficient
- Spearman Rank Correlation Coefficient (Spearman's Rho)
- Runs Test for Randomness
- Runs
- Randomness
- Non-Randomness
- Kruskal-Wallis Test

Key Formulas

	Section
Test Statistic for the Sign Test, $n \leq 25$ X = the number of times the less frequent sign occurs	16.1
Test Statistic for the Sign Test, $n > 25$ $$z = \frac{X + 0.5 - \left(\dfrac{n}{2}\right)}{\dfrac{\sqrt{n}}{2}}$$	16.1
Test Statistic for the Wilcoxon Signed-Rank Test, $n \leq 25$ If H_a is > One-Tailed: $T = T_+$ = the sum of the ranks associated with the positive differences. If H_a is < One-Tailed: $T = T_-$ = the sum of the ranks associated with the negative differences. If H_a is ≠ Two-Tailed: $T = \text{Min}\left(T_+, T_-\right).$	16.2

Key Formulas (cont.)

Section

Test Statistic for the Wilcoxon Signed-Rank Test, $n > 25$

$$z = \frac{T - \dfrac{n(n+1)}{4}}{\sqrt{\dfrac{n(n+1)(2n+1)}{24}}}$$

16.2

where $T = \text{Min}(T_+,\, T_-)$.

Test Statistic for the Wilcoxon Rank-Sum Test, $n_1 \leq 10$

If H_a is $>$ One-Tailed:

$T = T_x =$ the rank sum of the sample with the fewest members. (If the sample sizes are the same, $T_x =$ the rank sum of the population hypothesized to be shifted to the right.)

If H_a is $<$ One-Tailed:

16.3

$T = T_x =$ the rank sum of the sample with the fewest members. (If the sample sizes are the same, $T_x =$ the rank sum of the population hypothesized to be shifted to the left.)

If H_a is $\neq$ Two-Tailed:

$T = T_x =$ the rank sum of the sample with the fewest members. (If the sample sizes are the same, either rank sum can be used.)

Test Statistic for the Wilcoxon Rank-Sum Test, $n_1 > 10$

$$z = \frac{T - \dfrac{n_1(n_1 + n_2 + 1)}{2}}{\sqrt{\dfrac{n_1 n_2 (n_1 + n_2 + 1)}{12}}}$$

16.3

Spearman Rank Correlation Coefficient (Spearman's Rho)

$$r_s = 1 - \frac{6\sum\limits_{i=1}^{n} d_i^2}{n(n^2 - 1)}$$

16.4

where $d_i = R(x_i) - R(y_i)$.

Key Formulas (cont.)

	Section

Test Statistic for the Runs Test for Randomness, $m > 20$ or $n > 20$

$$z = \frac{R - \mu_R}{\sigma_R}$$

16.5

where $\mu_R = 1 + \dfrac{2mn}{N}$ and $\sigma_R = \sqrt{\dfrac{2mn(2mn - N)}{N^2(N-1)}}$.

Test Statistic for the Kruskal-Wallis Test

$$H = \frac{12}{N(N+1)} \sum_{i=1}^{k} \frac{R_i^2}{n_i} - 3(N+1)$$

16.6

where $R_i = \sum_{j=1}^{n_i} r_{ij}$ and $N = \sum_{i=1}^{k} n_i$.

AE   Additional Exercises

1. A new method for temporarily relieving the lung congestion of cystic fibrosis patients has been introduced. The traditional method of relieving the congestion involves a series of manual techniques where the chest and back area are pounded and massaged. The new method is a mechanical vest which has been designed to perform the manual techniques. A study is conducted to measure the effectiveness of the new vest. Five cystic fibrosis patients are randomly selected and the diameter of the blood vessels in their lungs is measured after using the traditional treatment and after using the vest treatment. The larger the diameter of the blood vessels within the lungs, the better the treatment. If the study provides conclusive evidence that the vest is more effective than the manual method in increasing the diameter of the blood vessels, the hospital will recommend the vest to its patients because the vest allows the patients to be much more independent. The results of the study are as follows.

Table for Exercise 1 – Diameter of Lung Blood Vessels (in mm)					
Subject	1	2	3	4	5
After Traditional Treatment	0.5	0.4	0.7	0.6	0.2
After Vest Treatment	0.6	0.6	0.7	0.7	0.5

a. What assumption must be made in order to perform the test of hypothesis using the paired difference t-test?

b. Using the sign test, do the data provide conclusive evidence that the median diameter of blood vessels in the lungs is significantly larger after using the vest treatment than after using the traditional treatment at $\alpha = 0.01$?

c. What assumptions were made in performing the sign test?

d. Using the signed-rank test, do the data provide conclusive evidence that the median diameter of blood vessels in the lungs is significantly larger after using the vest treatment than after using the traditional treatment at $\alpha = 0.01$?

e. What assumptions were made in performing the signed-rank test?

f. Which test do you think produces more accurate results? Why?

g. Perform a paired difference t-test. How do the results of the sign test and the signed-rank test compare with the results of the t-test?

2. As private companies prepare to go public, many analysts attempt to predict whether the stock will have a positive or negative return in the first day of trading. One particular analyst believes the return after the first day of trading is positive, on average. Another analyst wishes to test this analyst's claim using 15 recently offered public stocks.

Table for Exercise 2 – Initial Public Offerings		
Company	Initial Offer Price ($)	Price After 1st Day of Trading ($)
Enduro Royalty Trust (NDRO)	22.00	21.26
ZELTIQ Aesthetics (ZLTQ)	13.00	15.50
Ubiquiti Networks (UBNT)	15.00	17.50
Tudou Holdings Limited (TUDO)	29.00	25.56
Carbonite (CARB)	10.00	12.35
SandRidge Permian Trust (PER)	18.00	18.00
American Capital Mortgage Investment (MTGE)	20.00	18.41
C&J Energy Services (CJES)	29.00	30.50
Chefs Warehouse Holdings (CHEF)	15.00	17.50
Spirit Airlines (SAVE)	12.00	11.55
Pandora Media (P)	16.00	17.42
Wesco Aircraft Holdings (WAIR)	15.00	14.92
American Midstream Partners (AMID)	21.00	20.95
Dunkin Brands Group (DNKN)	19.00	27.85
Skullcandy (SKUL)	20.00	20.00

Source: IPO Scoop.com

a. Use the sign test to test the hypothesis that the return after the first day of trading is positive. Test at the 0.05 level.

b. Suppose that an analyst claimed that the median return on a stock in the first day of trading is +$1.00. Perform a test of hypothesis to determine if the median return is different than what the analyst claims. Use $\alpha = 0.05$.

c. Use the Wilcoxon signed-rank test to determine if the price after the first day of trading is generally greater than the initial offer price. Test at $\alpha = 0.05$.

d. Analyze the results of the sign test performed in part **a.** and the signed-rank test performed in part **c.** in terms of the problem. Which test do you think yields more accurate results?

e. What concerns do you have with the tests performed in this problem?

3. A weight loss center is trying to determine which of its diets results in higher client satisfaction. The center polled 20 clients (10 were on Diet A and 10 on Diet B) and had them rate their satisfaction in the diets from 1 to 100.

Table for Exercise 3 – Diet Ratings	
Diet A	Diet B
84	94
77	81
89	95
98	93
97	97
100	99
75	82
85	92
96	95
78	89

a. Which nonparametric test do you think is most appropriate to test the claim that Diet B results in higher client satisfaction than Diet A? Explain why.

b. Write the null and alternative hypotheses for a rank-sum test to determine if Diet B results in higher client satisfaction than Diet A.

c. Using the rank-sum test, do the data provide sufficient evidence at $\alpha = 0.05$ that Diet B results in greater client satisfaction than Diet A?

d. What assumptions were made in the test performed in part **c.**?

4.   Are students with higher GPAs more likely to get a higher paying job upon graduation? Consider the following data regarding student GPA and starting salary.

GPA	Starting Salary ($)
Table for Exercise 4 – GPA and Starting Salary	
2.37	37,000
3.20	38,000
3.21	42,000
3.39	40,000
3.55	44,000
3.57	48,000
3.76	50,000
3.77	60,000
3.79	59,000
3.90	55,000

a.   Determine the ranks for the x-variable, GPA.

b.   Determine the ranks for the y-variable, Starting Salary.

c.   Calculate the Spearman rank correlation coefficient.

d.   Interpret the value of the coefficient. Is the relationship between these two variables positive or negative? Is this relationship what you expected? Explain.

e.   Comment on the strength of the relationship between these two variables.

f.   Is there evidence at $\alpha = 0.10$ that these two variables are related?

5. *Fortune* magazine releases a list of the world's most admired companies. In the survey they ask business people to vote for companies that they admire most from various industries. The table below lists the top 10 most admired companies of 2011 along with each company's Fortune 500 ranking. Note that the Fortune 500 ranks companies based on revenues.

Table for Exercise 5 – Top 10 Most Admired Companies and Profits		
Company	**Most Admired Ranking**	**Fortune 500 Ranking**
Apple	1	35
Google	2	92
Berkshire Hathaway	3	7
Southwest Airlines	4	205
Procter & Gamble	5	26
Coca-Cola	6	70
Amazon.com	7	78
FedEx	8	73
Microsoft	9	38
McDonald's	10	111

Source: CNN Money/Fortune Magazine

a. Compute the Spearman rank correlation coefficient. Interpret this value.

b. With 95% confidence, can we conclude that there is an association between company admiration and revenue?

6. The given table shows key dates of the Dow-Jones Industrial Average. Apply the runs test to check for randomness at the 0.05 level. (**Hint:** First find the median of the values, then label each value by A if it is above the median and B if it is below the median.)

Table for Exercise 6 – Significant Levels on the Dow (December 1974 to February 2009)		
Date	**Dow Jones Industrial Average**	**Significance**
December 6, 1974	577	The last Bear Market bottom
July 12, 1976	1011	Highest point between January, 1973 and October, 1982
August 12, 1982	776	The start of the "Reagan Bull"
August 25, 1987	2722	The 1987 high
October 19, 1987	1738	The (508 point) crash of 1987
February 2, 1994	3975	The top of the post 1987 crash recovery
November 23, 1994	3674	The start of the Clinton "super bull"
March 29, 1999	10,006	The first Dow close above 10,000
January 14, 2000	11,723	The "Clinton bull" high
March 17, 2000	10,630	The biggest one day gain (499 points)
March 20, 2001	9720	Dow closes below previous year low for the first time since 1982
September 11-14, 2001	9605	Terrorist attack closed the Dow for four days
September 17, 2001	8920	The biggest one day fall (685 points)
September 21, 2001	8235	The Dow's second worst week ever (−14.26%)
December 31, 2001	10,021	Dow up 21.7% from September 21 low but down 7.2% on the year
September 30, 2002	7591	New 2002 low – all treasury yields (except 30-year bond) at 2002 lows
October 9, 2002	7286	New 2002 low – Dow down 37.8% from the January, 2000 all-time high
October 31, 2002	8397	Dow up 806 points (10.6%) for October – first positive month since March
November 6, 2002	8771	Federal reserve cuts rates for the first time since December, 2001
December 31, 2002	8341	Dow down 16.8% for 2002 – first three consecutive year loss since 1939-41
May 23, 2003	8601	Senate passes bill raising the treasury debt limit
June 25, 2003	9011	Federal reserve cuts rates by 0.25%
December 31, 2003	10,453	Dow up 25.32% in 2003
October 3, 2006	11,727	Dow exceeds the previous all-time high in January, 2000
October 9, 2007	14,164	New all-time high on the Dow
July 2, 2008	11,215	Dow closes more than 20% below the October, 2007 high
February 27, 2009	7062	Dow closes more than 50% below the October, 2007 high

7. The irrational number π can be approximated by the rational number $\dfrac{22}{7}$. Test the randomness of odd and even digits in $\dfrac{22}{7}$ at the 0.05 level using the first nine digits.

8. A polling agency conducts exit interviews after an election. If R = Republican and D = Democrat, the first 20 voter responses in a random sample are as follows.

D D D R D R R D D D D R D D R R R D R D

Test for non-randomness using $\alpha = 0.05$.

9. Consider the following *U.S. News and World Report* college rankings for schools in the Big Ten, the Big 12, and the Atlantic Coast conferences.

Table for Exercise 9 – College Rankings by Conference		
Atlantic Coast	**Big Ten**	**Big 12**
29	55	101
10	62	45
101	42	58
71	28	143
68	45	90
31	71	94
55	45	101
25	12	75
38	68	101
101	71	132
36	75	160
25		97

Source: *U.S. News and World Report*, 2011

With $\alpha = 0.10$, use the Kruskal-Wallis test to determine if there is a significant difference in *U.S. News and World Report* rankings among the three athletic conferences.

10. Consider the following scores reported by *Condé Nast Traveler* in their annual list of the top cities to visit around the world. The results were determined from more than 8 million votes cast in the Readers' Choice Awards survey.

Table for Exercise 10 – Readers' Choice City Rankings, 2011					
Asia		**Europe**		**United States**	
Kyoto	82.3	Florence	85.0	Charleston, SC	84.7
Bangkok	81.6	Barcelona	82.8	San Francisco, CA	83.7
Hong Kong	81.1	Rome	82.4	Santa Fe, NM	83.0
Chiang Mai	80.8	Paris	81.9	Chicago, IL	82.2
Ubud	80.0	Bruges	81.7	Honolulu, HI	80.9
Singapore	78.4	Venice	81.7	New York, NY	80.8
Tokyo	76.8	Salzburg	81.4	Savannah, GA	79.1
Luang Prabang	76.4	Vienna	81.0	Carmel, CA	78.5
Thimphu	75.1	Prague	79.7	Seattle, WA	78.4
Shanghai	74.9	Siena	79.7	Boston, MA	78.0

Source: *Condé Nast Traveler*, 2011

a. Using the Kruskal-Wallis test and $\alpha = 0.05$, test to determine if there is a difference in rankings between Asia, Europe, and the United States.

b. What assumptions were made for the test performed in part **a.**?

Chapter Seventeen
Statistical Process Control

17

D   **Discovering the Real World**

Quality Control, Deming's 14 Points for Management, and Six Sigma Technology

Statistics and statistical methods have led to major discoveries and economic gains in agriculture, biology, business, engineering, and the sciences. Historically, the use of statistical techniques to improve quality began to evolve during World War II, due to the importance placed on war-time efficiency. Immediately after the war, manufacturers in the United States had a relatively easy time selling their goods throughout the world, regardless of quality, since there were very few international competitors. But by the 1960s, Japanese manufacturers began to challenge American companies. It was during this period that the Japanese gained a reputation for producing quality products. Also during this time, the modern field of **quality control**, known by names such as **total quality control**, **total quality management**, or simply **quality management**, started to emerge.

During the 1920s, Dr. Walter A. Shewhart of Bell Laboratories developed the concepts of **statistical quality control**. He introduced the concept of "controlling" the quality of a product as it was being manufactured rather than after it was manufactured. To control the quality, Shewhart developed charting techniques for controlling in-process manufacturing operations. Shewhart also introduced statistical sample inspection to estimate the quality of a product as it was being manufactured. This method replaced the older method of inspecting each item after it was manufactured.

World War II virtually destroyed Japanese production capabilities. After World War II, Dr. W. Edwards Deming and Joseph M. Juran convinced the management of several large Japanese companies that adoption of statistical methods would lead to increased quality, thereby building up manufacturing and improving the quality of living of the Japanese people. These manufacturers have done an excellent job of attaining high quality for their products, as evidenced by their cars, television sets, and many other electronic products. What made Japanese management different was its focus on process-oriented management. The strength of this system is that it acknowledges the ends (the results) but emphasizes the means (the process).

Deming's philosophy is an important framework for implementing quality and productivity improvement. His philosophy is summarized in his 14 Points for Management. **Deming's 14 Points** are as follows.

1. Create a constancy of purpose toward improvement of product and service, with the aim to become competitive and to stay in business, and to provide jobs.

2. Adopt the new philosophy. We are in a new economic age. Western management must awaken to the challenge, must learn their responsibilities, and take on leadership for change.

3. Cease dependence on inspection to achieve quality. Eliminate the need for inspection on a mass basis by building quality into the product in the first place.

4. End the practice of awarding business on the basis of price tag. Instead, minimize total cost. Move toward a single supplier for any one item, on a long-term relationship of loyalty and trust.

5. Improve constantly and forever the system of production and service, to improve quality and productivity, and thus constantly decrease costs.

6. Institute training on the job.

7. **Institute leadership.** The aim of supervision should be to help people and machines and gadgets to do a better job. Supervision of management is in need of overhaul, as well as supervision of production workers.

8. **Drive out fear,** so that everyone may work effectively for the company.

9. **Break down barriers between departments.** People in research, design, sales, and production must work as a team, to foresee problems of production and in use that may be encountered with the product or service.

10. Eliminate slogans, exhortations, and targets for work force asking for zero defects and new levels of productivity. Such exhortations only create adversarial relationships, as the bulk of the causes of low quality and low productivity belong to the system and thus lie beyond the power of the work force.

11. **a.** Eliminate work standards (quotas) on the factory floor. Substitute leadership.

  b. Eliminate management by objective. Eliminate management by numbers, numerical goals. Substitute leadership.

12. **a.** Remove barriers that rob the hourly worker of his right to pride of workmanship. The responsibility of supervisors must be changed from sheer numbers to quality.

  b. Remove barriers that rob people in management and in engineering of their right to pride of workmanship. This means, *inter alia*, abolishment of the annual or merit rating and of management by objective.

13. Institute a vigorous program of education and self-improvement.

14. Put everybody in the company to work to accomplish the transformation. The transformation is everybody's job.

Source: Deming, W. Edwards. *Out of the Crisis.* Cambridge, MA: MIT Press, 1982. 23-24. Print.

These 14 points apply anywhere, to small or large organizations, to the service or manufacturing industry.

The fathers of modern quality control are many and have come from various fields, but they have one common philosophy: confronting reality in manufacturing and service operations through the collection and analysis of data is the best way to improve quality.

In the 1980s, a program called **Six Sigma** was developed at Motorola after a Japanese firm took over the Motorola factory that manufactured television sets in the United States. The firm began to make drastic operational changes in response to Motorola having been consistently beaten in the television set market. With the new changes, the factory was soon producing TV sets with $1/20^{th}$ the number of defects they had produced under previous management. The improvement was a result of the implementation of Six Sigma methodology. Six Sigma methodology blends many of the key elements of past quality initiatives while adding its own special approach to business management. Six Sigma emphasizes the reduction in variation, a focus on doing the right things right, combining customer knowledge with the core process improvement efforts, and subsequent improvement in company sales and revenue growth. Basically, Six Sigma is driven by results and encourages companies to take a customer focus in order to improve their business processes.

Companies using the Six Sigma methodology have seen an enhanced ability to provide value to their customers. Internally, the companies have a better understanding of their key business process and these processes have undergone process flow improvements. Improved process flow means reduced cycle times, reduction to elimination of defects, and increased capacity and productivity rates.

17.1   Basic Charts and Diagrams Used in Quality Control

The quality movement is not simply about using statistical tools to monitor quality. The main thrust of the quality movement is the improvement of processes. These improvements are often facilitated by the use of statistical techniques. In this section, we will discuss some of the basic charts and diagrams that are used to monitor processes.

Flowcharts

A process cannot be improved unless everyone agrees on what the process is. One way of beginning to determine how a process should work is to use a flowchart to define how the process currently works.

One of the processes that some students are familiar with is the test preparation process.

A student wanting to improve test-taking skills would define their process and improve upon it over time. There are two types of charts that aid in process improvement: Pareto charts and run charts.

Pareto Charts

The **Pareto chart** (pronounced *pah-ray-toe*), named after Vilfredo Pareto, is a graphical tool for ranking the causes of problems from most significant to least significant. Pareto's research examined patterns of wealth and income in nineteenth-century England. He found that 20% of the population enjoyed 80% of the wealth, the top 10% of the population controlled 65% of the wealth, and the top 5% controlled 50% of the wealth. What really excited Pareto was that he saw this same pattern of imbalance repeated in different time periods and different countries.

What Pareto discovered with respect to wealth has also been seen in vastly different venues.

- 80% of the world's energy is consumed by 15% of the population.
- 20% of a country's population consumes 80% of the health resources.
- 20% of a company's customers account for 80% of a company's revenue.
- 80% of complaints come from 20% of customers.
- 80% of crimes are committed by 20% of criminals.

This 80/20 notion has been found to be very useful in analyzing process improvement. The notion has been deemed the **Pareto principle** (also known as the **80/20 rule** and the **law of the vital few**). Every process has problems to overcome. The causes and effects of these problems are not equally linked. If the Pareto principle holds, we would see that 20% of the causes produce 80% of the problems. Identifying the highly significant 20% of process problems should focus attention on these key areas and lead to substantial process improvement. While the split is not always 80/20, the Pareto chart is a visual method allowing us to identify which problems are most significant. The use of a Pareto chart also limits the tendency of people to focus on the most recent problems rather than on the most important problems.

Definition

A **Pareto chart** is a bar graph with the categories ranked from largest to smallest.

Procedure

Pareto Chart

A Pareto chart is constructed using the following steps.

1. Create a list of problems that will be the subject of the chart.

2. Determine what data need to be collected. These could be count data, percentages, or costs.

3. Determine the time frame to be used for collecting the data.

4. Use a check sheet to tally the data. Then add the numbers in each category.

5. List the items in decreasing order of the measure of comparison.

6. List the items (problems) being tracked on the horizontal axis. Label the vertical axis with frequencies, percentages, etc.

7. Analyze the chart(s). The largest bars represent the vital few problems. If there does not appear to be one or two major problems, re-check the categories to determine if another analysis is necessary.

Suppose you are a publishing company executive and you have just learned that your typesetting budget had been exceeded by 40%. By carefully studying typesetting budget overruns during the next six months you observe the following causes.

Note: The percentages given in Table 17.1 are approximate, as they are rounded to one decimal place.

Table 17.1 – Typesetting Budget Overrun Causes		
Causes of Problems	**Frequency of Occurrence**	**Percentage (%)**
Author late with corrections	48	41.0
Author late with original manuscript	35	29.9
Author made too many corrections	12	10.3
Book longer than planned	9	7.7
Proofreader late	4	3.4
Figures incorrectly done	3	2.6
Index compiler late	3	2.6
Permissions received late	1	0.9
Typesetter correction errors	1	0.9
Schedule changed by editor	1	0.9

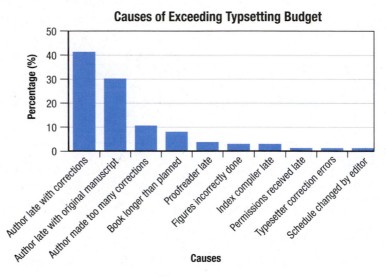

Figure 17.1

The Pareto chart in Figure 17.1 indicates that approximately 81.2% of the problems are associated with the author. The executives need to discuss issues with the author to determine if problems can be alleviated. The visual summary of these data is quite powerful and should provide a manager with an extremely sharp focus of what problems need to be attacked.

In addition to the main bar graph, the Pareto chart may also include an integrated line graph that shows the cumulative frequency (or cumulative relative frequency) of the observations at each bar on the bar graph. The line graph can be used to help determine which problems (the bars) belong to the **vital few** (the 20%) and the **trivial many** (the 80%).

Run Charts

In Section 2.7 we discussed time series plots. In these plots, the variable of interest is graphed against a time period plotted on the horizontal axis. A **run chart** is a time series plot. The purpose of these charts in a quality control context is to monitor quality characteristics. Suppose you were manufacturing ball bearings.

You would be concerned with quality characteristics such as the following.

- The average diameter of the ball bearings

- The number of surface defects

- The average compression force measured in Newtons (crushing strength)

Processes that are working acceptably produce measurable quality characteristics that are stable. When processes are not working properly, they are either becoming unstable or are unstable. To effectively manage a process (e.g. a production line manufacturing ball bearings), a manager would like to intervene when a process is becoming unstable. What would a process that is becoming unstable look like?

If you were manufacturing two-inch ball bearings, then a quality characteristic that you would be interested in is the diameter of the ball bearings being produced. Assume that each data point in the following diagrams represents the average diameter of a sample of ball bearings. Also, assume the samples of ball bearings are drawn every 15 minutes from the production line. The following run charts are graphs of the sample mean diameters for 40 samples of five ball bearings.

Figure 17.2

In Figure 17.2, the mean diameter of the ball bearings is trending downward. In Figure 17.3, the mean diameter of the ball bearings is trending upward. Both of these charts suggest that process output is unstable. These patterns indicate a gradual process shift that will eventually produce output that is unusable.

Figure 17.3

Figure 17.4

Notice how the data in the graph in Figure 17.4 move systematically up and down in a repeating pattern. This pattern (called a cycle) is also an indication of a process that is unstable.

17.1   Exercises

Basic Concepts

1. Describe the contributions made by Dr. Walter A. Shewhart in the field of quality control.

2. What contributions did W. Edwards Deming make in the field of quality control? Of Deming's 14 Points, give five that you believe are most important.

3. To which types of organizations do Deming's 14 Points apply?

4. What is the common philosophy of the fathers of modern quality control?

5. What is Six Sigma? Briefly describe the methodology behind Six Sigma.

6. What is a flowchart? Why are flowcharts important in quality control?

7. What is a Pareto chart? Explain how a Pareto chart can be used in the field of quality control.

8. What is a run chart? Is a run chart the same as a time series plot? Explain.

9. Explain how a run chart can be used in the field of quality control.

Exercises

10. Create a flowchart for each of the following processes.

    a. Getting ready for work in the morning

    b. Getting married

    c. Going on a week-long vacation in Bermuda

    d. Going to a job interview

11. Consider the following flowchart regarding incoming call routing for a software company.

a. Explain why it is important for a new receptionist working at the company to understand this flowchart.

b. In what ways do you think this flowchart could be improved?

12. Consider the following Pareto chart regarding reasons for a delay in processing credit card applications.

Reasons for a Delay in Processing Credit Card Applications

a. What percentage of delays is caused by a missing signature on the application?

b. What percentage of delays is caused by a missing signature or illegible writing on the application?

c. From the chart, what would you identify as the "vital few" problems?

d. How do you think the credit card company could attempt to correct these problems?

e. Does the 80/20 notion seem to apply here? Explain.

13. Consider the following data regarding customer complaints for a clothing store.

Table for Exercise 13 – Customer Complaints	
Complaint	**Frequency**
Not Enough Parking	80
Rude Personnel	50
Poor Lighting	42
Confusing Store Layout	28
Limited Sizes	15
Clothing Unattractive	10

a. Compute the relative frequencies for the complaints listed in the table.

b. Create a Pareto chart for the data.

c. Which problem(s) would you identify as the "vital few"?

d. Does the 80/20 notion seem to apply here? Explain.

e. How should the clothing store proceed in attempting to improve customer satisfaction? In your opinion, what should be done first? Explain why.

14. Consider the following data regarding the number of returned products for a large online retailer, by month, for the years 2003 and 2011.

Table for Exercise 14 – Number of Returned Products in 2003 and 2011		
Month	2003	2011
January	79	100
February	81	105
March	92	96
April	101	84
May	111	72
June	120	80
July	119	64
August	125	60
September	137	55
October	120	59
November	140	42
December	145	56

a. Create a run chart for the number of returned items, by month, in 2003.

b. Analyze the run chart for returned items in 2003. Is there a downward or upward trend or is the pattern cyclic?

c. Would you consider the process to be stable or unstable? Explain why.

d. Create a run chart for the number of returned items, by month, in 2011.

e. Analyze the run chart for returned items in 2011. Is there a downward or upward trend or is the pattern cyclic? Does the process appear to be stable or unstable? Explain.

f. Does it appear that the retailer has improved the process from 2003 to 2011? Explain.

15. Consider the following data regarding revenues, by quarter, for a popular local restaurant.

Year	Quarter	Revenue
Table for Exercise 15 – Quarterly Revenues (Thousands of Dollars)		
2008	1	270
2008	2	369
2008	3	468
2008	4	306
2009	1	285
2009	2	354
2009	3	525
2009	4	330
2010	1	261
2010	2	288
2010	3	375
2010	4	366
2011	1	303
2011	2	420
2011	3	471
2011	4	414

a. Create a run chart for the quarterly revenues.

b. Analyze the run chart. Is there a downward trend or an upward trend? Do revenues appear to be cyclical? Explain.

c. Do you think restaurant sales depend on the time of year?

d. Can you think of any reasons why there would be this type of trend for restaurant sales?

e. Does this process appear to be unstable? If so, suggest ways that quality control could help the restaurant manager control the process.

17.2   Basic Concepts

The scientific method for attaining quality relies on two basic concepts.

1. No matter what the specifications are for a product, the process that produces the product will create output that has **variation**. (For example, suppose a manufacturer desires to produce ball bearings with a diameter of two inches. If a process is set up to produce ball bearings with a diameter of two inches, then each item, when actually measured, will show deviation from the **ideal** of two inches.)

2. Improving a process requires removing variation from it. (Though the ideal would be to remove all variation, this cannot be achieved. The goal then is to move towards the ideal. This notion is known as **continuous improvement**.)

With these two powerful ideas in mind, let's look at some important definitions.

> ### Definitions
>
> A **control chart** for a process consists of values plotted over time. This chart has an upper bound and a lower bound called the **upper control limit (UCL)** and the **lower control limit (LCL)**, respectively. The process is **out of control** when a measurement falls either above the UCL or below the LCL. The control chart also contains a **centerline** that represents the average value of the quality characteristic corresponding to the in-control state.

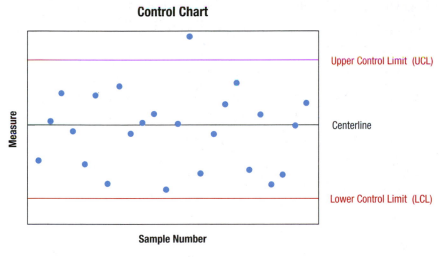

Control Chart

Figure 17.5

The control chart with its limits tells us when to stop the process (if it is out of control) and when not to interrupt the process. When data points fall within the UCL and LCL, we think the variation is due to **normal process variation** (also called **common cause variation** or **chance variation**). But when a point or points fall outside the control limits, the cause is said to be **assignable variation** (or **special cause variation**). This type of variation is not random and can be eliminated (or reduced) by investigating the problem and determining the root cause(s). Reducing system variation is the surest path toward continuous improvement. For assignable causes, the system should be stopped and the cause(s) should be found and removed before the process is resumed.

It is important to emphasize that a control chart focuses on the process, not on the product. It does not ensure good quality, but instead allows management to check a quality characteristic of the process at

The Highway Control Chart

When you are driving a car and you stay in your lane, you could say that you are operating the car "in control." The white lines that define your lane are similar to the UCL and LCL. The car would be expected to move around within the lane, which would be normal process variation. Veering outside your lane might have assignable causes such as cell phone usage, children fighting in the back seat, or any number of other distractions.

regular intervals in order to determine if the statistical distribution of the characteristic has changed. If it has, then modifications may be needed to correct the process.

17.2 Exercises

Basic Concepts

1. Identify and describe the two concepts on which the scientific method for attaining quality is based on.

2. What is a control chart? What is the basic purpose of control charts?

3. Identify and define the three basic components of a control chart.

4. What is normal process variation?

5. What is assignable variation?

6. Does a control chart give information about a process or a product? Explain.

7. What does it mean to say that a process is in control?

8. Does an in-control state guarantee quality output? Explain.

Exercises

9. Consider the following control chart.

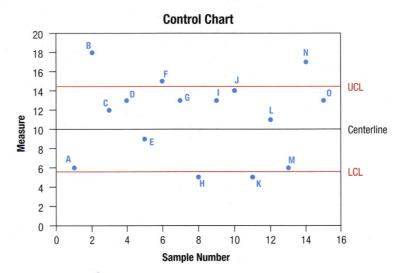

a. From the chart, estimate the values of the UCL, LCL, and centerline.

b. Interpret the estimated values of the UCL, LCL, and centerline.

c. Which points, if any, are out of control?

10. Consider the following values of the UCL, LCL, and centerline from a control chart.

13.56, 16.56, 10.56

    a. Identify which value is the LCL, which is the UCL, and which is the centerline.

    b. Plot the UCL, LCL, and centerline on a control chart.

    c. Identify three points that would be considered in control and three points that would be considered out of control.

    d. Plot these points on the chart that you made in part **b.**

11. Consider the following control chart.

    a. From the chart, estimate the values of the UCL, LCL, and centerline.

    b. Interpret the estimated values of the UCL, LCL, and centerline.

    c. Identify any points that can be attributed to assignable variation.

17.3   Monitoring with $\bar{x}$ and R Charts

Control charts provide a means of testing a hypothesis. The null hypothesis is that the process is in control (stable) and the alternative hypothesis is that it is out of control (unstable). It is customary to monitor both the mean and the variability of a process. The mean is important because it indicates whether or not the process is operating at its target level. The variability is important because it indicates if the process is operating consistently. An $\bar{x}$ **chart** is used to monitor the process mean and a **range chart (R chart)** can be used to monitor the variability of the process. So long as the quality characteristic being measured is within 3σ of the process mean, the process is thought to be **in control**. If the quality characteristic is not within 3σ of the process mean, the process is thought to be **out of control**.

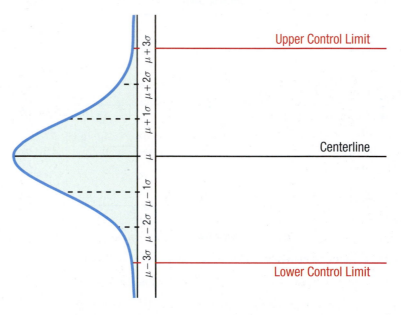

Figure 17.6

Assuming the sample statistics have an approximately normal distribution, we would expect 99.7% of the sample values to fall within 3σ of the mean. Only 0.3% of the values would fall outside the interval $\mu \pm 3\sigma$ due to random process variation. Thus, the upper and lower control limits represent the boundaries between variation which is considered random (normal process variation) and variation which is considered to be assignable (not caused by random variation). When sample statistics fall outside of $\mu \pm 3\sigma$, the process is said to be out of control.

The most common control chart, the $\bar{x}$ chart, plots the means of small samples taken from the process at regular intervals over time. Since each sample mean, $\bar{x}$, is an unbiased estimator of the process mean, μ, the $\bar{x}$ chart monitors the variation in the center of the process.

If the process mean and standard deviation are known, and we assume a normal distribution, then we can use μ as the centerline of the $\bar{x}$ chart and the upper and lower control limits are given by

$$\text{UCL} = \mu + 3\sigma_{\bar{x}} = \mu + \frac{3\sigma}{\sqrt{n}} \quad \text{and} \quad \text{LCL} = \mu - 3\sigma_{\bar{x}} = \mu - \frac{3\sigma}{\sqrt{n}},$$

respectively. Constructing the control limits in this way ensures that about 99.7% of all values of $\bar{x}$ will lie within the control limits when the process is working properly.

Formula

3σ Control Limits for $\bar{x}$ when Standard Values of the Process Mean and Standard Deviation are Known

$$\text{Upper Control Limit (UCL)} = \mu + 3\sigma_{\bar{x}} = \mu + \frac{3\sigma}{\sqrt{n}}$$

$$\text{Lower Control Limit (LCL)} = \mu - 3\sigma_{\bar{x}} = \mu - \frac{3\sigma}{\sqrt{n}}$$

$$\text{Centerline} = \mu$$

where μ is the process mean, σ is the process standard deviation, and n is the number of observations in each sample.

Of course, it is rarely the case that the process mean and standard deviation are known. When μ and σ are unknown, we estimate them based on samples taken when the process is believed to be in control. Taking at least 20 to 25 samples are recommended with a sample size of n. Once the data have been collected, we must calculate the two items that will be used to construct the control charts: $\bar{\bar{x}}$, the **grand mean** (think of this as the mean of the sample means), and the average range, $\bar{R}$ which is the average of the sample ranges (computed as the difference between the maximum and minimum data values in each sample). The grand mean can be used to estimate the population mean, and $\bar{R}$ can be used to estimate the population standard deviation when constructing an $\bar{x}$ chart when the process mean and standard deviation are unknown.

Formula

3σ Control Limits for $\bar{x}$ when the Standard Values of the Process Mean and Standard Deviation are Unknown

$$\text{UCL} = \bar{\bar{x}} + A\bar{R}$$

$$\text{LCL} = \bar{\bar{x}} - A\bar{R}$$

$$\text{Centerline} = \bar{\bar{x}}$$

where

$\bar{R}$ = the mean of the sample ranges,

$\bar{\bar{x}}$ = the mean of the sample means, and

A = a factor that relates 3σ to $\bar{R}$ and can be found using Table 17.2.

Table 17.2 – 3σ Factors for Computing Control Chart Limits

Sample Size	Mean Factor (A)	Lower Range (D_3)	Upper Range (D_4)
2	1.880	0.000	3.267
3	1.023	0.000	2.574
4	0.729	0.000	2.282
5	0.577	0.000	2.114
6	0.483	0.000	2.004
7	0.419	0.076	1.924
8	0.373	0.136	1.864
9	0.337	0.184	1.816
10	0.308	0.223	1.777
15	0.223	0.348	1.652
20	0.180	0.414	1.586
25	0.153	0.459	1.541
Over 25	$0.75\left(\dfrac{1}{\sqrt{n}}\right)$	$1.55 - 0.0015n$	$0.45 + 0.001n$

In addition to the process mean, managers often are concerned with the variability in the samples. Even if the $\bar{x}$ chart indicates that the process is in control, the variation within the samples could be too much. To check this, managers often employ a range chart (R chart). In a manner similar to the $\bar{x}$ chart, the upper and lower control limits for the R chart are calculated using

$$\text{Centerline} = \bar{R} = \frac{\sum_{i=1}^{m} R_i}{m}$$

$$\text{UCL}_R = \bar{R} + 3\sigma_R$$

$$\text{LCL}_R = \bar{R} - 3\sigma_R$$

where UCL_R and LCL_R are the upper and lower control limits of the R chart, respectively. R_i represents the range of the ith sample, σ_R is the population standard deviation of the sample ranges, and m is the number of samples. The population standard deviation of the sample ranges is often not available, so the upper and lower control limits for the R chart are constructed using factors D_3 and D_4, which are also found in Table 17.2.

Formula

3σ Control Chart for the Process Range

$$\text{Upper Control Limit (UCL)} = \bar{R}D_4$$

$$\text{Lower Control Limit (LCL)} = \bar{R}D_3$$

$$\text{Centerline} = \bar{R}$$

where $\bar{R}$ is the mean of the sample ranges and D_3 and D_4 are factors which can be found in Table 17.2.

Example 17.1

A manufacturing company uses a machine to punch out parts of a window hinge. Historically, the manufacturing process has produced hinges with a mean width of 0.45 inches with a process standard deviation of 0.11 inches. To monitor the production and to make sure the parts are acceptable for the next stage of home window assembly, a sample of three parts is taken each hour. The width of each part is then measured. As a result there are a total of eight samples with three observations in each sample. Table 17.3 contains the data for the width of the hinges. Determine if the process is in control using an $\bar{x}$ chart. Indicate which (if any) samples are out of control.

Table 17.3 – Sampling Data for Window Hinges (Inches)			
	Observations		
Sample	1	2	3
1	0.42	0.48	0.46
2	0.44	0.48	0.45
3	0.44	0.45	0.41
4	0.47	0.42	0.46
5	0.53	0.43	0.42
6	0.47	0.42	0.45
7	0.44	0.47	0.50
8	0.48	0.43	0.44

Solution

Control limits need to be established and plotted on the $\bar{x}$ chart. Typically, control limits for the $\bar{x}$ chart are three standard deviations above and below the process mean for the given quality characteristic.

In this problem, we have external standards to gauge our process. In particular, the process mean (0.45) and process standard deviation (0.11) are known. Thus, the control limits are determined as follows.

$$\text{UCL} = \mu + \frac{3\sigma}{\sqrt{n}} = 0.45 + 3\left(\frac{0.11}{\sqrt{3}}\right)$$
$$\approx 0.6405$$

$$\text{LCL} = \mu - \frac{3\sigma}{\sqrt{n}} = 0.45 - 3\left(\frac{0.11}{\sqrt{3}}\right)$$
$$\approx 0.2595$$

$$\text{Centerline} = \mu = 0.45$$

Table 17.4 – Sampling Data for Window Hinges (Inches)				
	Observations			
Sample	**1**	**2**	**3**	**Sample Mean**
1	0.42	0.48	0.46	0.453
2	0.44	0.48	0.45	0.457
3	0.44	0.45	0.41	0.433
4	0.47	0.42	0.46	0.450
5	0.53	0.43	0.42	0.460
6	0.47	0.42	0.45	0.447
7	0.44	0.47	0.50	0.470
8	0.48	0.43	0.44	0.450

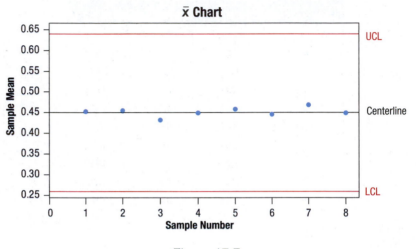

Figure 17.7

The mean for each sample is calculated and displayed in Table 17.4. The sample means are plotted in Figure 17.7 along with the UCL, LCL, and centerline. Note that according to the $\bar{x}$ chart, since all of the sample means fall within the control limits, the process is in control.

In the previous example, it was assumed that the process mean and standard deviation were known. However, in most instances, this is not the case. In the next example, we will illustrate how to construct control charts when the process mean and variation are not known.

Example 17.2

Using the data in Example 17.1 and assuming that we do not know the mean or standard deviation of the process, construct the $\bar{x}$ chart and R chart for this process.

Solution

In this example, we do not know the process standard deviation. However, we are going to use another measure of variation, the sample range, to compute the UCL and LCL. Since we do not know the true process mean, we are going to estimate that mean with $\bar{\bar{x}}$, the mean of the sample means.

Before calculating the control limits for the $\bar{x}$ chart and R chart, we need to perform some preliminary calculations. We need to calculate the mean and range for each sample. That is, the ith sample mean and range are calculated as follows.

$$\bar{x}_i = \frac{\sum\limits_{i=1}^{n} x_i}{n}$$

R_i = Maximum Observation in i^{th} Sample – Minimum Observation in i^{th} Sample

Note that n is the number of observations in each sample.

The means and ranges for each sample are presented in Table 17.5.

Table 17.5 – Sampling Data for Window Hinges (Inches)					
	Observations				
Sample	1	2	3	Mean	Range
1	0.42	0.48	0.46	0.453	0.06
2	0.44	0.48	0.45	0.457	0.04
3	0.44	0.45	0.41	0.433	0.04
4	0.47	0.42	0.46	0.450	0.05
5	0.53	0.43	0.42	0.460	0.11
6	0.47	0.42	0.45	0.447	0.05
7	0.44	0.47	0.50	0.470	0.06
8	0.48	0.43	0.44	0.450	0.05

Using the mean and range for each sample, we can calculate the grand mean and average range of the process. They are given by

$$\bar{\bar{x}} = \frac{\sum\limits_{i=1}^{m} \bar{x}_i}{m}$$

and

$$\bar{R} = \frac{\sum\limits_{i=1}^{m} R_i}{m}$$

where m is the number of samples.

Using the data in Table 17.5 we find the grand mean and the average range to be the following.

$$\bar{\bar{x}} = 0.4525 \quad \text{(mean of the sample means)}$$
$$\bar{R} = 0.0575 \quad \text{(mean of the sample ranges)}$$

Table 17.6 provides 3σ factors for computing control chart limits.

Table 17.6 – 3σ Factors for Computing Control Chart Limits

Sample Size	Mean Factor (A)	Lower Range (D_3)	Upper Range (D_4)
2	1.880	0.000	3.267
3	1.023	0.000	2.574
4	0.729	0.000	2.282
5	0.577	0.000	2.114
6	0.483	0.000	2.004
7	0.419	0.076	1.924
8	0.373	0.136	1.864
9	0.337	0.184	1.816
10	0.308	0.223	1.777
15	0.223	0.348	1.652
20	0.180	0.414	1.586
25	0.153	0.459	1.541
Over 25	$0.75\left(\dfrac{1}{\sqrt{n}}\right)$	$1.55 - 0.0015n$	$0.45 + 0.001n$

For a sample size of 3, the mean factor, $A = 1.023$. Using a mean value of 0.4525 and an average range of 0.0575, the control limits for the $\bar{x}$ chart are determined as follows.

$$\text{UCL} = \bar{\bar{x}} + A\bar{R}$$
$$= 0.4525 + 1.023(0.0575)$$
$$\approx 0.5113$$
$$\text{LCL} = \bar{\bar{x}} - A\bar{R}$$
$$= 0.4525 - 1.023(0.0575)$$
$$\approx 0.3937$$
$$\text{Centerline} = \bar{\bar{x}} = 0.4525$$

$\bar{x}$ Chart

Figure 17.8

Computing the control limits for the R chart, we have

$$\text{UCL} = \bar{R}D_4$$
$$= 0.0575(2.574)$$
$$\approx 0.1480$$
$$\text{LCL} = \bar{R}D_3$$
$$= 0.0575(0)$$
$$= 0$$
$$\text{Centerline} = \bar{R} = 0.0575.$$

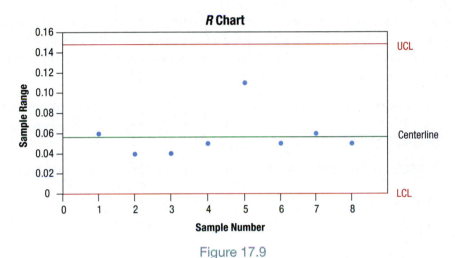

Figure 17.9

Examining the $\bar{x}$ chart and R chart, one can see that both the process mean and variability are within the control limits. That is, each of the sample means and sample ranges falls within the control limits of the $\bar{x}$ chart and R chart, respectively, indicating that the process is in control.

17.3 Exercises

Basic Concepts

1. How are control charts associated with hypothesis testing? Identify the null and alternative hypotheses that can be tested using control charts.

2. What is an $\bar{x}$ chart?

3. What is an R chart?

4. What does the Central Limit Theorem have to do with statistical quality control?

5. What is a common interval used to measure in-control and out-of-control processes? What is the probability that random variation would cause a sample statistic to fall outside this interval?

6. How are the upper and lower control limits for an $\bar{x}$ chart calculated if the process mean and standard deviation are known?

7. How do the calculations for the control limits change for an $\bar{x}$ chart when the process mean and standard deviation are unknown?

8. When studying a control chart, how do you determine if the process is out of control?

9. Consider the process in Example 17.1. Give an example of something that might cause this process to be out of control.

10. Why do managers study $\bar{x}$ charts and R charts together?

Exercises

11. An automobile manufacturer requires the fuel injections to be adjusted so that its cars get an average of 30 miles per gallon with a standard deviation of 3 miles per gallon over a long period of time. Calculate the upper and lower control limits for an $\bar{x}$ chart if the quality control department starts sampling 64 cars each day during the month.

12. A pharmaceutical manufacturer requires the average active ingredient of allergy pills to be 0.03 grams, with a standard deviation of 0.002 grams. An FDA inspector inspects 10 batches of 100 pills and finds the following sample means.

| 0.032 | 0.028 | 0.031 | 0.032 | 0.026 | 0.027 | 0.030 | 0.033 | 0.034 | 0.026 |

    a. Determine which batches, if any, are out of control using an $\bar{x}$ chart.

    b. Does the process appear to be in statistical control? Explain.

13. A manufacturer of auto windows employs a constant quality control technique where the thickness of glass is checked every hour. A perfect piece of glass will have a thickness of 4 mm. From past experience, it is known that the standard deviation of thickness is 0.25 mm. The result of one shift's production is given in the following table.

Table for Exercise 13 – Glass Thickness (mm)											
Sample	Observations					Sample	Observations				
1	4	3	4	2	4	9	5	4	3	2	5
2	5	3	5	4	2	10	4	4	4	4	4
3	3	3	3	3	4	11	1	6	4	4	2
4	4	5	5	4	5	12	3	3	4	5	4
5	4	2	2	3	2	13	4	4	5	6	5
6	4	5	2	5	4	14	3	2	5	4	2
7	3	5	4	4	5	15	3	3	3	2	1
8	2	4	4	4	3	16	4	4	4	5	3

    a. Construct an $\bar{x}$ chart for these data.

    b. Construct an R chart for these data.

    c. According to the $\bar{x}$ chart constructed in part a., which samples, if any, are out of control?

    d. According to the R chart constructed in part b., which samples, if any, are out of control?

14. Princeton Manufacturing produces air conditioning units designed to maintain 45 degrees. Samples of 10 units are taken to monitor the process, and it is found that the units are maintaining 45 degrees as designed. The mean of the sample ranges is found to be 2 degrees.

   a. Find the UCL, LCL, and centerline for the $\bar{x}$ chart.

   b. Find the UCL, LCL, and centerline for the R chart.

15. A trucking company tries to deliver its freight in 24 hours. Ten samples of 20 customers are taken with the following sample means.

| 22.6 | 24.5 | 24.1 | 23.8 | 25.3 | 25.0 | 23.8 | 23.6 | 23.0 | 25.2 |

   The average range for these deliveries is 5.8 hours.

   a. Compute the 3σ control limits for the mean delivery time. Which samples, if any, are out of control?

   b. Compute the 3σ control limits for the process range.

16. Natural Life produces a variety of natural food products. The quality control department samples one cereal to ensure proper net weight. In the past when taking samples of 15 boxes, the average range was 0.45 ounces. Find the upper and lower control limits for an R chart.

17. A paper products manufacturer makes 60-inch cores that are later cut into smaller lengths in the production of bathroom tissue. To monitor the production and to make sure the cores are acceptable for the cutting state, a sample of 25 cores is taken each hour of the day. Along with the core length, the range of the core length is recorded for each sample (as shown in the following table). Determine the upper and lower control limits for an R chart and indicate which samples, if any, are out of control.

Table for Exercise 17 – Core Lengths			
Sample Number	Sample Range	Sample Number	Sample Range
1	0.10	13	0.19
2	0.20	14	0.08
3	0.22	15	0.10
4	0.08	16	0.07
5	0.06	17	0.16
6	0.23	18	0.19
7	0.20	19	0.21
8	0.09	20	0.14
9	0.25	21	0.16
10	0.17	22	0.19
11	0.14	23	0.12
12	0.18	24	0.13

18. A manufacturer of small electric motors has a model that draws 1300 watts when working properly. To ensure conformity with this standard, samples of 20 motors are taken each hour during the day shift. Along with the average wattage, the range of wattage is recorded for each sample (as shown in the following table). Determine the upper and lower control limits for an R chart and indicate which samples, if any, are out of control.

Table for Exercise 18 – Electric Motor Wattage			
Sample Number	Sample Range	Sample Number	Sample Range
1	8.8	5	4.1
2	12.2	6	9.6
3	11.6	7	8.8
4	8.0	8	5.3

17.4   Monitoring with a *p* Chart

The $\bar{x}$ chart and the R chart, which are control charts for variables, cannot be used when we are sampling attributes. Variables are some measurable characteristic of a product or service such as the length of a rod, diameter of a ball bearing, or ounces in a bottle of soda. Attributes, however, are characteristics associated with a product or service such as the number of leaking containers, number of scratches on a car's paint job, or number of errors on an invoice. **Attribute sampling** involves counting the defective units in a sample. With a *p* **chart**, the total number of defective units is divided by the total number of items sampled and multiplied by 100 to find the percent defective.

When there is a "standard" percent defective, or the population proportion is known, the control limits for a 3σ *p* chart are calculated as follows.

Formula

3σ Control Chart for *p* when the Process Proportion is Known

$$\text{Upper Control Limit (UCL)} = p + 3\sigma_p = p + 3\sqrt{\frac{p(1-p)}{n}}$$

$$\text{Lower Control Limit (LCL)} = p - 3\sigma_p = p - 3\sqrt{\frac{p(1-p)}{n}}$$

$$\text{Centerline} = \mu = p$$

where p = standard process percent defective and n = the sample size.

When the process proportion is unknown, the proportion is estimated using the sample data.

Formula

3σ Control Chart for *p* when the Process Proportion is Unknown

$$\text{Upper Control Limit (UCL)} = \bar{p} + 3\sqrt{\frac{\bar{p}(1-\bar{p})}{n}}$$

$$\text{Lower Control Limit (LCL)} = \bar{p} - 3\sqrt{\frac{\bar{p}(1-\bar{p})}{n}}$$

$$\text{Centerline} = \bar{p}$$

where $\bar{p} = \dfrac{\text{total number of defects found in all samples}}{\text{total number of items sampled}}$ and n = the sample size.

Example 17.3

Silvia Garcia is monitoring the operation of a machine that makes radar components. Historically, she expects about 2% defectives and some chance variation. After studying the process, Silvia decides to construct a 3σ p chart. She takes 10 daily samples of 200 components. Her results are shown in Table 17.7. What does her p chart look like, and which samples (if any) are out of control?

Table 17.7 – Sampling Data for Components			
Sample Number	Number Defective	Sample Number	Number Defective
1	9	6	15
2	7	7	11
3	7	8	6
4	8	9	9
5	9	10	8

Solution

Since a "standard" percent defective has been specified in the problem, we will not have to estimate the process percent defective with the sample data. Silvia calculates the percent defective for each sample. For example, in Sample 1, the percent defective is $\frac{9}{200}(100) = 4.5\%$. The results are shown in Table 17.8. Using the data in Table 17.8 with the UCL and LCL, we construct the p chart.

Table 17.8 – Sampling Percent Defectives			
Sample Number	Percent Defective	Sample Number	Percent Defective
1	4.5	6	7.5
2	3.5	7	5.5
3	3.5	8	3.0
4	4.0	9	4.5
5	4.5	10	4.0

The 3σ control limits are set as follows.

$$\text{UCL} = p + 3\sqrt{\frac{p(1-p)}{n}}$$

$$= 0.02 + 3\sqrt{\frac{0.02(0.98)}{200}}$$

$$\approx 0.0497 = 4.97\%$$

$$\text{LCL} = p - 3\sqrt{\frac{p(1-p)}{n}}$$

$$= 0.02 - 3\sqrt{\frac{0.02(0.98)}{200}}$$

$$\approx -0.0097 \approx 0\%$$

Note: Since the LCL was negative, we set LCL = 0, since proportions are never negative.

Figure 17.10

From the *p* chart, we can see that samples 6 and 7 are out of control.

Example 17.4

In Example 17.3, a standard process value was known. Use the sample data in Example 17.3 to create a 3σ control chart for *p* (*p* chart) assuming there is no standard process value.

Solution

In many instances, the process proportion of interest is not known and has to be estimated from the data. In this example, there are 10 samples of size 200 and we wish to know the proportion of defective output the process is producing. To estimate the process proportion, we need to pool the data from all 10 samples.

$$\bar{p} = \frac{\text{total number of defects found in all samples}}{\text{total number of items sampled}}$$

$$= \frac{89}{2000}$$

$$= 0.0445$$

Using the estimate for *p* from the sample data, the control limits are calculated as follows.

17

$$UCL = \overline{p} + 3\sqrt{\frac{\overline{p}(1-\overline{p})}{n}}$$

$$= 0.0445 + 3\sqrt{\frac{0.0445(1-0.0445)}{200}}$$

$$\approx 0.0882 = 8.82\%$$

$$LCL = \overline{p} - 3\sqrt{\frac{\overline{p}(1-\overline{p})}{n}}$$

$$= 0.0445 - 3\sqrt{\frac{0.0445(1-0.0445)}{200}}$$

$$\approx 0.0008 = 0.08\%$$

Using the data in Table 17.8 and the UCL and LCL allows Silvia to draw the *p* chart and to plot each sample, as shown in Figure 17.11.

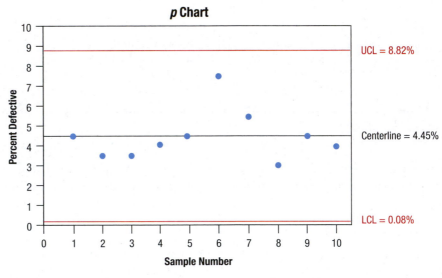

Figure 17.11

From the *p* chart, we can see that when the standard percent defective is unknown, all of the samples fall within the control limits.

17.4   Exercises

Basic Concepts

1. Explain the difference between control charts for attributes and control charts for variables.

2. What is a *p* chart?

3. How are the upper and lower control limits calculated for a *p* chart when the process proportion is known?

4. Is the allowable variation for a process involving a *p* chart larger, smaller, or the same as the allowable variation for a process involving a mean chart or range chart? Explain.

5. Suppose the LCL for a process is computed to be −0.07. What value should be used for the LCL in the p chart? Explain.

6. How does the procedure for constructing a *p* chart change if the process proportion is not known?

7. What is $\bar{p}$? What other hypothesis testing procedure uses the concept of $\bar{p}$? Are these measures the same? Explain.

8. When examining a *p* chart, how do you determine if samples are out of control?

Exercises

9. In a paper products plant, 100 product samples are taken each hour and tested for being either acceptable or defective. In the past, 1% defective was considered normal. Find the upper and lower control limits for a 3σ *p* chart.

10. To monitor the production of sheet metal screws by a particular machine in a large manufacturing company, a sample of 100 screws is examined each hour for three shifts of eight hours each. Each screw is inspected and designated as conforming or nonconforming according to specifications. Historically, the proportion of nonconforming screws has been 5%. Use the following results of one day's sampling to construct a 3σ *p* chart. Which samples, if any, are out of control?

Table for Exercise 10 – Nonconforming Screws					
Sample Number	Number Defective	Sample Number	Number Defective	Sample Number	Number Defective
1	4	9	10	17	9
2	7	10	5	18	11
3	9	11	5	19	14
4	10	12	4	20	5
5	8	13	12	21	6
6	6	14	6	22	12
7	5	15	7	23	15
8	1	16	13	24	5

17

11. A production process involves the manufacture of rubber gaskets for windows. When these gaskets are inspected, they are classified as conforming or nonconforming based on a number of different characteristics, such as thickness, consistency, overall size, and so on. To monitor the percentage of nonconforming gaskets being produced, a sample of 25 gaskets is inspected each hour. Management predetermines the acceptable fraction of nonconforming gaskets as 10%.

   a. Determine the UCL, LCL, and centerline.

   b. Use the following table to plot the samples on your control chart.

Table for Exercise 11 – Nonconforming Gaskets			
Sample Number	Percent Defective	Sample Number	Percent Defective
1	16	13	12
2	16	14	8
3	16	15	8
4	12	16	12
5	8	17	8
6	8	18	12
7	4	19	12
8	0	20	4
9	8	21	8
10	4	22	12
11	4	23	16
12	4	24	4

   c. Are any samples out of control? If so, identify which ones.

12. The academic dean decides to sample 200 students each semester to study the drop rate at his institution. The numbers of drops for the last eight semesters are shown in the following table. Find the upper and lower control limits and construct a p chart. Indicate which semesters, if any, are out of control.

Table for Exercise 12 – Number of Drops			
Semester Number	Number of Drops	Semester Number	Number of Drops
1	10	5	8
2	12	6	6
3	14	7	13
4	9	8	15

13. The Thompson Company makes voltage protectors at its Midland, Georgia plant. During each shift, 10 protectors are tested until failure, with some rated defective and others rated non-defective. The numbers of defective protectors for the last 20 shifts are given in the following table. Find the upper and lower control limits and construct a p chart. Indicate which shifts, if any, are out of control.

Table for Exercise 13 – Defective Voltage Protectors			
Sample Number	Number Defective	Sample Number	Number Defective
1	1	11	1
2	1	12	1
3	0	13	1
4	2	14	2
5	3	15	0
6	1	16	0
7	0	17	2
8	2	18	1
9	3	19	1
10	0	20	1

| T | **Discovering Technology** |

Using Excel

Pareto Chart

This exercise will illustrate how to create a Pareto chart in Microsoft Excel with the integrated line graph of cumulative percentages. Note that the line graph portion of this chart is optional. Use the information from Table 17.1 for this exercise.

1. Enter the data from the first two columns of Table 17.1 into an Excel worksheet. "Causes of Problems" should be in Column A and "Frequency of Occurrence" should be in Column B.

2. In Column C, we need to compute the percentages. In cell C1 enter the title Percentage (%). In cell C2 compute the percentage for the author being late with corrections by entering the following formula.

$$=\textbf{B2/117*100}$$

Copy this formula in cells C3 through C11 to calculate the remaining percentages.

3. In order to create a Pareto chart, we need to calculate the cumulative frequencies. We will do this in Column D. In cell D1 enter the title **Cumulative %**. In cell D2, enter the formula =**sum(C2)**. In cell D3 enter the formula =**sum(C2:C3)**. The cumulative frequency in cell D4 is given by =**sum(C2:C4)**. Continue to enter the formulas for the cumulative frequencies in Column D until you have the sum of 100, given by =**sum(C2:C11)** in cell D11.

4. Highlight the data in columns A, C, and D. You can do this by using the Ctrl key, since we do not need the "Frequency of Occurrence" column.

5. Under the Insert tab, choose **Column**, and **Clustered Column**. A side-by-side bar graph is created.

6. Right click on one of the bars that represents the cumulative % data, and select **Change Series Chart Type**. When the dialog box appears, select **Line with Markers**.

7. A Pareto chart is created. You can customize the look of the Pareto chart by adding a title or axis labels, changing the size of the markers on the line, or removing the legend.

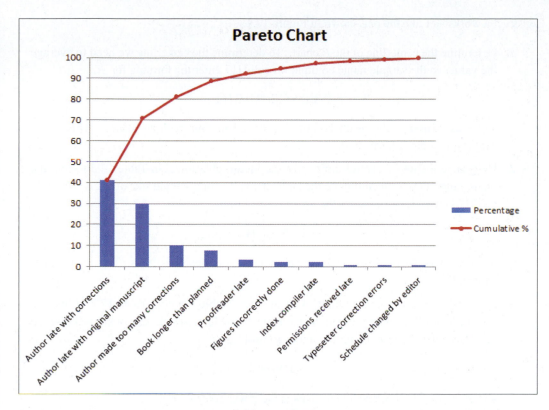

Figure 17.12

R Chart

A steel manufacturer has decided to use an *R* chart to monitor the variability of their 52 pound steel rods. The shift manager randomly samples 6 rods at 14 successive time periods and checks the weight. The data below contain the results for each sample.

	A	B	C	D	E	F	G	H
1				Observation				
2	Period	1	2	3	4	5	6	Sample Range
3	1	51.99	51.99	51.97	52.01	52.03	52.05	0.08
4	2	52.04	52.00	51.98	52.04	51.99	52.05	0.07
5	3	52.02	51.96	52.01	52.00	52.01	52.01	0.06
6	4	52.01	52.03	51.95	52.04	51.99	52.03	0.09
7	5	51.98	52.02	51.95	51.97	52.04	51.97	0.09
8	6	52.03	52.00	52.01	52.05	52.04	52.04	0.05
9	7	51.99	52.02	51.95	51.96	52.04	52.00	0.09
10	8	51.95	52.05	51.98	52.04	52.01	51.98	0.10
11	9	51.96	52.03	51.97	52.03	52.04	52.01	0.08
12	10	51.98	52.00	51.97	51.99	52.02	51.98	0.05
13	11	51.99	52.00	51.99	51.97	52.01	52.05	0.08
14	12	51.95	52.05	51.99	51.95	52.01	52.04	0.10
15	13	52.02	51.97	51.96	51.95	52.00	52.03	0.08
16	14	51.96	52.02	51.96	52.03	52.00	52.05	0.09

Figure 17.13

Construct a worksheet like the one shown in Figure 17.13.

1. Determine the centerline of the R chart. To determine the centerline we need to average the values in the sample range column. In cell H17 enter the formula for $\bar{R}$.

$$=AVERAGE(H3:H16)$$

and press **Enter**. The result is 0.0793 rounded to four decimal places. This is the centerline of the control chart.

2. Determine the upper control limit. Use the factor table found in Table 17.2 to find the upper range (D_4). According to the table, the value for a sample size of six is 2.004. In cell H18 enter the formula for the upper control limit:

$$=H17*2.004$$

and press **Enter**. The result is 0.1589 rounded to four decimal places. This is the upper control limit.

3. Determine the lower control limit. Use the factor table found in Table 17.2 to find the lower range (D_3). According to the table, the value for a sample size of six is 0.000. In cell H19 enter the formula for the lower control limit:

$$=H17*0$$

and press **Enter**. The result is 0. This is the lower control limit.

	A	B	C	D	E	F	G	H
1				Observation				
2	Period	1	2	3	4	5	6	Sample Range
3	1	51.99	51.99	51.97	52.01	52.03	52.05	0.08
4	2	52.04	52.00	51.98	52.04	51.99	52.05	0.07
5	3	52.02	51.96	52.01	52.00	52.01	52.01	0.06
6	4	52.01	52.03	51.95	52.04	51.99	52.03	0.09
7	5	51.98	52.02	51.95	51.97	52.04	51.97	0.09
8	6	52.03	52.00	52.01	52.05	52.04	52.04	0.05
9	7	51.99	52.02	51.95	51.96	52.04	52.00	0.09
10	8	51.95	52.05	51.98	52.04	52.01	51.98	0.10
11	9	51.96	52.03	51.97	52.03	52.04	52.01	0.08
12	10	51.98	52.00	51.97	51.99	52.02	51.98	0.05
13	11	51.99	52.00	51.99	51.97	52.01	52.05	0.08
14	12	51.95	52.05	51.99	51.95	52.01	52.04	0.10
15	13	52.02	51.97	51.96	51.95	52.00	52.03	0.08
16	14	51.96	52.02	51.96	52.03	52.00	52.05	0.09
17							R-bar	0.079285714
18							UCL	0.158888571
19							LCL	0

Figure 17.14

4. Create a scatterplot of the data. Highlight the data points in Column H (H3:H16). Under the **Insert** tab, choose **Scatter**, and **Scatter with only Markers**. A scatterplot is generated. Under the **Layout** tab, choose **Chart Title**, and **Above Chart**. Make the title of the chart **Control Chart**. Next, choose **Legend**, and **None** to turn off the legend.

5. Select the *y*-axis of the control chart, right click, and select **Format Axis**. Using this dialog box, we will change the scale of the *y*-axis so that the upper and lower control limits can be displayed on the chart. In the "Axis Options" window, select the radio buttons next to "Fixed" for maximum value and enter **0.18**. Press **Close**.

6. Next we need to draw the centerline and upper and lower control limits. Under the **Insert** tab, choose **Shapes**, and a line with no arrow. Draw a line in the chart area which intersects the *y*-axis at 0.1589 and extends to the right border of the graph. (Note: holding the shift key down as you draw the line ensures that the line is horizontal.) Once the line is drawn, under the **Format** tab, choose **Shape Outline**, **Weight**, and **1½ pt**. This is the upper control limit. With this line selected, press **Ctrl+C** to copy the line and then **Ctrl+V** to paste it. Now you have a copy of the line.

7. Drag this copy of the line so that it intersects with 0.0793 on the *y*-axis. This is the centerline. Make another copy of the line and drag it so that it intersects the *y*-axis at 0. This is the lower control limit.

8. Now we need to make labels for the lines. Under the **Insert** tab, choose **Shapes**, and **Text Box**. Draw a text box on the right-hand side of the graphing area just above the upper control limit. In the text box, label the line **UCL**. Make a copy of this text box by using **Ctrl+C** and paste it using **Ctrl+V**. Write **Centerline** in this text box and place it above the centerline of the control chart. Paste another copy of the text box using **Ctrl+V** and label the lower control limit **LCL**. For each of the text boxes, be sure that under the **Format** tab you choose **No Fill** for **Shape Fill** and **No Outline** for **Shape Outline**. You now have an *R* chart. Notice that the process appears to be in control since all observations fall between the UCL and the LCL.

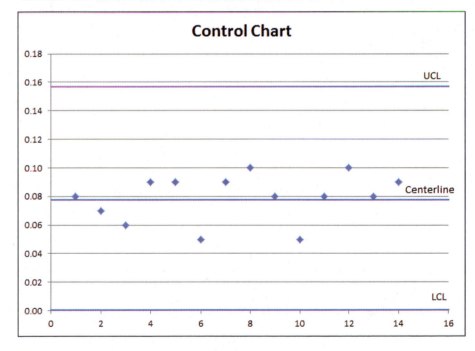

Figure 17.15

Using Minitab

$\bar{x}$ Chart

Use the information from Example 17.1 for this exercise.

1. Enter the data from Table 17.3 into columns C1 through C4. Sample should be in Column C1, Observation 1 should be in Column C2, Observation 2 should be in Column C3, and Observation 3 should be in Column C4.

2. Choose **Stat**, **Control Charts**, **Variables Charts for Subgroups**, and **Xbar**.

3. In the "Xbar Chart" dialog box, be sure that the option **Observations for a subgroup are in one row of columns** is selected from the drop down menu at the top. Note that you could also make the chart if all of the observations were in a single column. Select **C2-C4** as the columns to be charted.

4. Choose **Xbar Options**. Enter 0.45 for the "Mean" and 0.11 for the "Standard deviation." Press **OK**. Press **OK** again.

5. Observe the $\bar{x}$ chart. Compare the values of the UCL, $\bar{\bar{x}}$, and the LCL to those which we found in Example 17.1. Notice that the process appears to be in control, since there are no observations above the UCL or below the LCL.

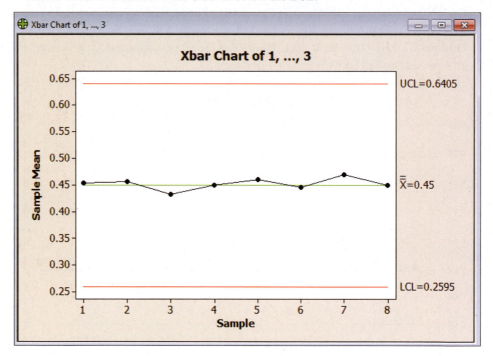

Figure 17.16

p Chart

Use the information from Example 17.3 for this exercise.

1. Enter the data from Table 17.7 into columns C1 and C2. The Sample Number should be in Column C1 and the Number Defective should be in Column C2.

2. Choose **Stat**, **Control Charts**, **Attributes Charts**, and **P**.

3. In the "P Chart" dialog box, select **C2** for "Variables" and enter **200** as the "Subgroup size."

4. Click on the **P Chart Options** button. Enter **0.02** for the proportion since we are assuming the population proportion is known. Press **OK**. Press **OK** again.

5. Observe the *p* chart. Compare the values for the UCL and the LCL to those which we found in Example 17.3. Notice that there are two points marked in red. These two samples are out of control.

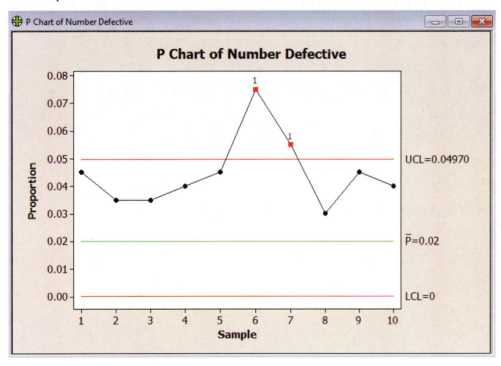

Figure 17.17

R   Chapter 17 Review

Key Terms and Ideas

- Quality Control
- Total Quality Control
- Total Quality Management
- Quality Management
- Statistical Quality Control
- Deming's 14 Points of Quality Control
- Six Sigma
- Flowchart
- Pareto Chart
- Pareto Principle
- 80/20 Rule
- Vital Few
- Trivial Many
- Run Chart
- Variation
- Continuous Improvement
- Control Chart

- Upper Control Limit (UCL)
- Lower Control Limit (LCL)
- Centerline
- Normal Process Variation
- Common Cause Variation
- Chance Variation
- Assignable Variation
- Special Cause Variation
- Quality Theory
- $\bar{x}$ Chart
- Range Chart (R Chart)
- In Control State
- Out of Control State
- Grand Mean $\left(\bar{\bar{x}} \right)$
- Attribute Sampling
- p Chart

Key Formulas

	Section

Control Limits for $\bar{x}$ when the Process Mean and Standard Deviation are Known

$$UCL = \mu + 3\sigma_{\bar{x}} = \mu + \frac{3\sigma}{\sqrt{n}}$$

$$LCL = \mu - 3\sigma_{\bar{x}} = \mu - \frac{3\sigma}{\sqrt{n}}$$

$$\text{Centerline} = \mu$$

17.3

where μ is the process mean, σ is the process standard deviation, and n is the number of observations in each sample.

Control Limits for $\bar{x}$ when the Process Mean and Standard Deviation are Unknown

$$UCL = \bar{\bar{x}} + A\bar{R}$$

$$LCL = \bar{\bar{x}} - A\bar{R}$$

$$\text{Centerline} = \bar{\bar{x}}$$

17.3

where

$$\bar{R} = \text{the mean of the sample ranges,}$$

$$\bar{\bar{x}} = \text{the mean of the sample means, and}$$

$$A = \text{a factor that relates } 3\sigma \text{ to } \bar{R}.$$

Control Limits for the Process Range

$$\text{Upper Control Limit (UCL)} = \bar{R}D_4$$

$$\text{Lower Control Limit (LCL)} = \bar{R}D_3$$

$$\text{Centerline} = \bar{R}$$

17.3

where $\bar{R}$ is the mean of the sample ranges and D_3 and D_4 are relational factors.

Control Limits for p when the Process Proportion is Known

$$UCL = p + 3\sigma_p = p + 3\sqrt{\frac{p(1-p)}{n}}$$

$$LCL = p - 3\sigma_p = p - 3\sqrt{\frac{p(1-p)}{n}}$$

$$\text{Centerline} = \mu = p$$

17.4

where p is the standard process percent defective and n is the sample size.

AE   Additional Exercises

1. Edwards Electrical Company produces voltage regulators designed to maintain 220 volts. Samples of 10 units are taken from production to monitor the process and the voltage regulators are found to maintain 220 volts as designed. The mean of the sample ranges is found to be 4 volts.

   a. Determine the UCL, LCL, and centerline for an $\bar{x}$ chart.

   b. Determine the UCL, LCL, and centerline for an R chart.

2. A steel products manufacturer makes 20-foot lengths of pipe that are later cut into smaller lengths in the production process. To monitor the production and to make sure the pipe is acceptable for the cutting state, a sample of 20 pipes is taken each hour of the day. Along with the pipe length, the range of pipe length is recorded for each sample. Determine the upper and lower control limits for an R chart and indicate which samples, if any, are out of control.

Table for Exercise 2 – Pipe Length Ranges					
Sample Number	Sample Range	Sample Number	Sample Range	Sample Number	Sample Range
1	0.03	9	0.08	17	0.05
2	0.07	10	0.06	18	0.06
3	0.07	11	0.05	19	0.07
4	0.03	12	0.06	20	0.05
5	0.02	13	0.06	21	0.05
6	0.08	14	0.03	22	0.06
7	0.07	15	0.03	23	0.04
8	0.03	16	0.02	24	0.04

3. The league director decides to sample 200 matches each season to study forfeits in league games. The numbers of forfeits for the last eight seasons are shown in the following table. Find the upper and lower control limits and construct a p chart. Indicate which seasons, if any, are out of control.

Table for Exercise 3 – Game Forfeits			
Season Number	Number Forfeited	Season Number	Number Forfeited
1	8	5	7
2	10	6	5
3	9	7	12
4	8	8	16

4. A tire manufacturer randomly samples 20 tires at the end of each shift to determine if the tires are defective. The numbers of defectives in 12 shifts are given in the following table. Construct an appropriate control chart to determine if the tire manufacturing process is in control. Identify any shifts that are out of control.

Table for Exercise 4 – Defective Tires	
Shift	**Number of Defectives**
1	4
2	2
3	0
4	5
5	2
6	3
7	14
8	2
9	3
10	4
11	12
12	3

5. The vice president of audit at a Fortune 500 firm customarily checks the financial statements for errors in 15 departments. The following table contains information about mistakes made on financial statements. Construct an appropriate control chart to determine if the VP should be concerned about the number of mistakes being made on the financial statements.

Table for Exercise 5 – Mistakes on Financial Statements			
Sample Size	**Number of Mistakes**	**Sample Size**	**Number of Mistakes**
15	0	15	0
15	0	15	3
15	3	15	8
15	2	15	7
15	6	15	0
15	3	15	1
15	4	15	5
15	8	15	4
15	2		

6. 15 samples of five items each were taken to monitor the amount of fill for a 12-ounce bottle of soda. As each bottle is selected from the line, it is measured. The data from the process is given in the following table. Create an $\bar{x}$ chart and an R chart to determine if the process is in control.

Table for Exercise 6 – Fill Amounts for 12-Ounce Bottles					
	Observations				
Sample	1	2	3	4	5
1	13.11	11.17	11.35	13.71	13.00
2	13.19	13.50	13.63	12.36	11.98
3	12.83	12.12	12.65	12.22	11.43
4	11.62	13.72	13.73	11.36	13.93
5	13.65	12.44	13.15	12.42	11.21
6	12.60	11.07	11.17	12.88	12.33
7	11.07	13.22	11.41	13.21	13.14
8	11.56	12.16	12.47	13.41	13.13
9	13.23	12.05	11.25	12.55	13.00
10	12.30	12.18	12.00	13.59	12.18
11	14.00	12.33	11.52	13.29	13.51
12	12.05	11.52	12.80	12.21	12.67
13	13.29	11.91	11.11	11.32	13.85
14	11.21	11.89	13.07	11.22	13.09
15	13.37	13.25	11.48	13.46	11.26

7. A company packages salt pellets for in-ground well water softeners in bags with a 40-pound label weight. During a typical day's operation of the filling process, 10 samples of five bag fills are selected and measured. Using the data in the following table, create an $\bar{x}$ and an R chart to determine if the process is in control.

Table for Exercise 7 – Bag Weights (Pounds)					
	Observations				
Sample	1	2	3	4	5
1	41.41	40.63	38.83	40.57	39.94
2	40.31	40.20	41.28	40.13	42.53
3	41.64	41.02	39.49	41.64	39.14
4	40.34	39.23	41.96	42.26	40.97
5	40.77	42.73	41.83	42.59	40.43
6	42.54	42.87	40.04	40.00	40.84
7	39.84	42.65	42.86	41.75	39.68
8	42.25	40.96	39.11	41.66	39.69
9	42.94	38.52	41.50	39.13	40.57
10	39.42	39.98	38.90	41.17	41.03

Appendix A
Statistical Tables

A   Standard Normal Distribution

Numerical entries represent the probability that a standard normal random variable is between $-\infty$ and z where $z = \dfrac{x - \mu}{\sigma}$.

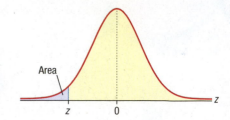

Area

z	0.09	0.08	0.07	0.06	0.05	0.04	0.03	0.02	0.01	0.00
−3.4	0.0002	0.0003	0.0003	0.0003	0.0003	0.0003	0.0003	0.0003	0.0003	0.0003
−3.3	0.0003	0.0004	0.0004	0.0004	0.0004	0.0004	0.0004	0.0005	0.0005	0.0005
−3.2	0.0005	0.0005	0.0005	0.0006	0.0006	0.0006	0.0006	0.0006	0.0007	0.0007
−3.1	0.0007	0.0007	0.0008	0.0008	0.0008	0.0008	0.0009	0.0009	0.0009	0.0010
−3.0	0.0010	0.0010	0.0011	0.0011	0.0011	0.0012	0.0012	0.0013	0.0013	0.0013
−2.9	0.0014	0.0014	0.0015	0.0015	0.0016	0.0016	0.0017	0.0018	0.0018	0.0019
−2.8	0.0019	0.0020	0.0021	0.0021	0.0022	0.0023	0.0023	0.0024	0.0025	0.0026
−2.7	0.0026	0.0027	0.0028	0.0029	0.0030	0.0031	0.0032	0.0033	0.0034	0.0035
−2.6	0.0036	0.0037	0.0038	0.0039	0.0040	0.0041	0.0043	0.0044	0.0045	0.0047
−2.5	0.0048	0.0049	0.0051	0.0052	0.0054	0.0055	0.0057	0.0059	0.0060	0.0062
−2.4	0.0064	0.0066	0.0068	0.0069	0.0071	0.0073	0.0075	0.0078	0.0080	0.0082
−2.3	0.0084	0.0087	0.0089	0.0091	0.0094	0.0096	0.0099	0.0102	0.0104	0.0107
−2.2	0.0110	0.0113	0.0116	0.0119	0.0122	0.0125	0.0129	0.0132	0.0136	0.0139
−2.1	0.0143	0.0146	0.0150	0.0154	0.0158	0.0162	0.0166	0.0170	0.0174	0.0179
−2.0	0.0183	0.0188	0.0192	0.0197	0.0202	0.0207	0.0212	0.0217	0.0222	0.0228
−1.9	0.0233	0.0239	0.0244	0.0250	0.0256	0.0262	0.0268	0.0274	0.0281	0.0287
−1.8	0.0294	0.0301	0.0307	0.0314	0.0322	0.0329	0.0336	0.0344	0.0351	0.0359
−1.7	0.0367	0.0375	0.0384	0.0392	0.0401	0.0409	0.0418	0.0427	0.0436	0.0446
−1.6	0.0455	0.0465	0.0475	0.0485	0.0495	0.0505	0.0516	0.0526	0.0537	0.0548
−1.5	0.0559	0.0571	0.0582	0.0594	0.0606	0.0618	0.0630	0.0643	0.0655	0.0668
−1.4	0.0681	0.0694	0.0708	0.0721	0.0735	0.0749	0.0764	0.0778	0.0793	0.0808
−1.3	0.0823	0.0838	0.0853	0.0869	0.0885	0.0901	0.0918	0.0934	0.0951	0.0968
−1.2	0.0985	0.1003	0.1020	0.1038	0.1056	0.1075	0.1093	0.1112	0.1131	0.1151
−1.1	0.1170	0.1190	0.1210	0.1230	0.1251	0.1271	0.1292	0.1314	0.1335	0.1357
−1.0	0.1379	0.1401	0.1423	0.1446	0.1469	0.1492	0.1515	0.1539	0.1562	0.1587
−0.9	0.1611	0.1635	0.1660	0.1685	0.1711	0.1736	0.1762	0.1788	0.1814	0.1841
−0.8	0.1867	0.1894	0.1922	0.1949	0.1977	0.2005	0.2033	0.2061	0.2090	0.2119
−0.7	0.2148	0.2177	0.2206	0.2236	0.2266	0.2296	0.2327	0.2358	0.2389	0.2420
−0.6	0.2451	0.2483	0.2514	0.2546	0.2578	0.2611	0.2643	0.2676	0.2709	0.2743
−0.5	0.2776	0.2810	0.2843	0.2877	0.2912	0.2946	0.2981	0.3015	0.3050	0.3085
−0.4	0.3121	0.3156	0.3192	0.3228	0.3264	0.3300	0.3336	0.3372	0.3409	0.3446
−0.3	0.3483	0.3520	0.3557	0.3594	0.3632	0.3669	0.3707	0.3745	0.3783	0.3821
−0.2	0.3859	0.3897	0.3936	0.3974	0.4013	0.4052	0.4090	0.4129	0.4168	0.4207
−0.1	0.4247	0.4286	0.4325	0.4364	0.4404	0.4443	0.4483	0.4522	0.4562	0.4602
−0.0	0.4641	0.4681	0.4721	0.4761	0.4801	0.4840	0.4880	0.4920	0.4960	0.5000

B   Standard Normal Distribution

Numerical entries represent the probability that a standard normal random variable is between $-\infty$ and z where $z = \dfrac{x - \mu}{\sigma}$.

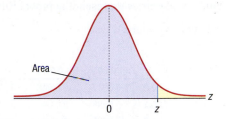

Area

0   z   z

z	0.00	0.01	0.02	0.03	0.04	0.05	0.06	0.07	0.08	0.09
0.0	0.5000	0.5040	0.5080	0.5120	0.5160	0.5199	0.5239	0.5279	0.5319	0.5359
0.1	0.5398	0.5438	0.5478	0.5517	0.5557	0.5596	0.5636	0.5675	0.5714	0.5753
0.2	0.5793	0.5832	0.5871	0.5910	0.5948	0.5987	0.6026	0.6064	0.6103	0.6141
0.3	0.6179	0.6217	0.6255	0.6293	0.6331	0.6368	0.6406	0.6443	0.6480	0.6517
0.4	0.6554	0.6591	0.6628	0.6664	0.6700	0.6736	0.6772	0.6808	0.6844	0.6879
0.5	0.6915	0.6950	0.6985	0.7019	0.7054	0.7088	0.7123	0.7157	0.7190	0.7224
0.6	0.7257	0.7291	0.7324	0.7357	0.7389	0.7422	0.7454	0.7486	0.7517	0.7549
0.7	0.7580	0.7611	0.7642	0.7673	0.7704	0.7734	0.7764	0.7794	0.7823	0.7852
0.8	0.7881	0.7910	0.7939	0.7967	0.7995	0.8023	0.8051	0.8078	0.8106	0.8133
0.9	0.8159	0.8186	0.8212	0.8238	0.8264	0.8289	0.8315	0.8340	0.8365	0.8389
1.0	0.8413	0.8438	0.8461	0.8485	0.8508	0.8531	0.8554	0.8577	0.8599	0.8621
1.1	0.8643	0.8665	0.8686	0.8708	0.8729	0.8749	0.8770	0.8790	0.8810	0.8830
1.2	0.8849	0.8869	0.8888	0.8907	0.8925	0.8944	0.8962	0.8980	0.8997	0.9015
1.3	0.9032	0.9049	0.9066	0.9082	0.9099	0.9115	0.9131	0.9147	0.9162	0.9177
1.4	0.9192	0.9207	0.9222	0.9236	0.9251	0.9265	0.9279	0.9292	0.9306	0.9319
1.5	0.9332	0.9345	0.9357	0.9370	0.9382	0.9394	0.9406	0.9418	0.9429	0.9441
1.6	0.9452	0.9463	0.9474	0.9484	0.9495	0.9505	0.9515	0.9525	0.9535	0.9545
1.7	0.9554	0.9564	0.9573	0.9582	0.9591	0.9599	0.9608	0.9616	0.9625	0.9633
1.8	0.9641	0.9649	0.9656	0.9664	0.9671	0.9678	0.9686	0.9693	0.9699	0.9706
1.9	0.9713	0.9719	0.9726	0.9732	0.9738	0.9744	0.9750	0.9756	0.9761	0.9767
2.0	0.9772	0.9778	0.9783	0.9788	0.9793	0.9798	0.9803	0.9808	0.9812	0.9817
2.1	0.9821	0.9826	0.9830	0.9834	0.9838	0.9842	0.9846	0.9850	0.9854	0.9857
2.2	0.9861	0.9864	0.9868	0.9871	0.9875	0.9878	0.9881	0.9884	0.9887	0.9890
2.3	0.9893	0.9896	0.9898	0.9901	0.9904	0.9906	0.9909	0.9911	0.9913	0.9916
2.4	0.9918	0.9920	0.9922	0.9925	0.9927	0.9929	0.9931	0.9932	0.9934	0.9936
2.5	0.9938	0.9940	0.9941	0.9943	0.9945	0.9946	0.9948	0.9949	0.9951	0.9952
2.6	0.9953	0.9955	0.9956	0.9957	0.9959	0.9960	0.9961	0.9962	0.9963	0.9964
2.7	0.9965	0.9966	0.9967	0.9968	0.9969	0.9970	0.9971	0.9972	0.9973	0.9974
2.8	0.9974	0.9975	0.9976	0.9977	0.9977	0.9978	0.9979	0.9979	0.9980	0.9981
2.9	0.9981	0.9982	0.9982	0.9983	0.9984	0.9984	0.9985	0.9985	0.9986	0.9986
3.0	0.9987	0.9987	0.9987	0.9988	0.9988	0.9989	0.9989	0.9989	0.9990	0.9990
3.1	0.9990	0.9991	0.9991	0.9991	0.9992	0.9992	0.9992	0.9992	0.9993	0.9993
3.2	0.9993	0.9993	0.9994	0.9994	0.9994	0.9994	0.9994	0.9995	0.9995	0.9995
3.3	0.9995	0.9995	0.9995	0.9996	0.9996	0.9996	0.9996	0.9996	0.9996	0.9997
3.4	0.9997	0.9997	0.9997	0.9997	0.9997	0.9997	0.9997	0.9997	0.9997	0.9998

C   Standard Normal Distribution

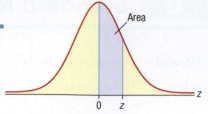

Area

Numerical entries represent the probability that a standard normal random variable is between 0 and z where $z = \dfrac{x - \mu}{\sigma}$.

z	0.00	0.01	0.02	0.03	0.04	0.05	0.06	0.07	0.08	0.09
0.0	0.0000	0.0040	0.0080	0.0120	0.0160	0.0199	0.0239	0.0279	0.0319	0.0359
0.1	0.0398	0.0438	0.0478	0.0517	0.0557	0.0596	0.0636	0.0675	0.0714	0.0753
0.2	0.0793	0.0832	0.0871	0.0910	0.0948	0.0987	0.1026	0.1064	0.1103	0.1141
0.3	0.1179	0.1217	0.1255	0.1293	0.1331	0.1368	0.1406	0.1443	0.1480	0.1517
0.4	0.1554	0.1591	0.1628	0.1664	0.1700	0.1736	0.1772	0.1808	0.1844	0.1879
0.5	0.1915	0.1950	0.1985	0.2019	0.2054	0.2088	0.2123	0.2157	0.2190	0.2224
0.6	0.2257	0.2291	0.2324	0.2357	0.2389	0.2422	0.2454	0.2486	0.2517	0.2549
0.7	0.2580	0.2611	0.2642	0.2673	0.2704	0.2734	0.2764	0.2794	0.2823	0.2852
0.8	0.2881	0.2910	0.2939	0.2967	0.2995	0.3023	0.3051	0.3078	0.3106	0.3133
0.9	0.3159	0.3186	0.3212	0.3238	0.3264	0.3289	0.3315	0.3340	0.3365	0.3389
1.0	0.3413	0.3438	0.3461	0.3485	0.3508	0.3531	0.3554	0.3577	0.3599	0.3621
1.1	0.3643	0.3665	0.3686	0.3708	0.3729	0.3749	0.3770	0.3790	0.3810	0.3830
1.2	0.3849	0.3869	0.3888	0.3907	0.3925	0.3944	0.3962	0.3980	0.3997	0.4015
1.3	0.4032	0.4049	0.4066	0.4082	0.4099	0.4115	0.4131	0.4147	0.4162	0.4177
1.4	0.4192	0.4207	0.4222	0.4236	0.4251	0.4265	0.4279	0.4292	0.4306	0.4319
1.5	0.4332	0.4345	0.4357	0.4370	0.4382	0.4394	0.4406	0.4418	0.4429	0.4441
1.6	0.4452	0.4463	0.4474	0.4484	0.4495	0.4505	0.4515	0.4525	0.4535	0.4545
1.7	0.4554	0.4564	0.4573	0.4582	0.4591	0.4599	0.4608	0.4616	0.4625	0.4633
1.8	0.4641	0.4649	0.4656	0.4664	0.4671	0.4678	0.4686	0.4693	0.4699	0.4706
1.9	0.4713	0.4719	0.4726	0.4732	0.4738	0.4744	0.4750	0.4756	0.4761	0.4767
2.0	0.4772	0.4778	0.4783	0.4788	0.4793	0.4798	0.4803	0.4808	0.4812	0.4817
2.1	0.4821	0.4826	0.4830	0.4834	0.4838	0.4842	0.4846	0.4850	0.4854	0.4857
2.2	0.4861	0.4864	0.4868	0.4871	0.4875	0.4878	0.4881	0.4884	0.4887	0.4890
2.3	0.4893	0.4896	0.4898	0.4901	0.4904	0.4906	0.4909	0.4911	0.4913	0.4916
2.4	0.4918	0.4920	0.4922	0.4925	0.4927	0.4929	0.4931	0.4932	0.4934	0.4936
2.5	0.4938	0.4940	0.4941	0.4943	0.4945	0.4946	0.4948	0.4949	0.4951	0.4952
2.6	0.4953	0.4955	0.4956	0.4957	0.4959	0.4960	0.4961	0.4962	0.4963	0.4964
2.7	0.4965	0.4966	0.4967	0.4968	0.4969	0.4970	0.4971	0.4972	0.4973	0.4974
2.8	0.4974	0.4975	0.4976	0.4977	0.4977	0.4978	0.4979	0.4979	0.4980	0.4981
2.9	0.4981	0.4982	0.4982	0.4983	0.4984	0.4984	0.4985	0.4985	0.4986	0.4986
3.0	0.4987	0.4987	0.4987	0.4988	0.4988	0.4989	0.4989	0.4989	0.4990	0.4990
3.1	0.4990	0.4991	0.4991	0.4991	0.4992	0.4992	0.4992	0.4992	0.4993	0.4993
3.2	0.4993	0.4993	0.4994	0.4994	0.4994	0.4994	0.4994	0.4995	0.4995	0.4995
3.3	0.4995	0.4995	0.4995	0.4996	0.4996	0.4996	0.4996	0.4996	0.4996	0.4997
3.4	0.4997	0.4997	0.4997	0.4997	0.4997	0.4997	0.4997	0.4997	0.4997	0.4998

Critical Values of z

Level of Confidence	$z_{\alpha/2}$
0.80	1.28
0.90	1.645
0.95	1.96
0.99	2.575

D   Critical Values of *t*

Numerical entries represent the value of *t* such that the area to the right of the *t* is equal to α.

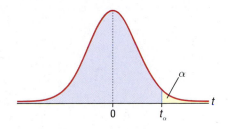

Degrees of Freedom	Area to the Right of the Critical Value					
	$t_{0.200}$	$t_{0.100}$	$t_{0.050}$	$t_{0.025}$	$t_{0.010}$	$t_{0.005}$
1	1.376	3.078	6.314	12.706	31.821	63.657
2	1.061	1.886	2.920	4.303	6.965	9.925
3	0.978	1.638	2.353	3.182	4.541	5.841
4	0.941	1.533	2.132	2.776	3.747	4.604
5	0.920	1.476	2.015	2.571	3.365	4.032
6	0.906	1.440	1.943	2.447	3.143	3.707
7	0.896	1.415	1.895	2.365	2.998	3.499
8	0.889	1.397	1.860	2.306	2.896	3.355
9	0.883	1.383	1.833	2.262	2.821	3.250
10	0.879	1.372	1.812	2.228	2.764	3.169
11	0.876	1.363	1.796	2.201	2.718	3.106
12	0.873	1.356	1.782	2.179	2.681	3.055
13	0.870	1.350	1.771	2.160	2.650	3.012
14	0.868	1.345	1.761	2.145	2.624	2.977
15	0.866	1.341	1.753	2.131	2.602	2.947
16	0.865	1.337	1.746	2.120	2.583	2.921
17	0.863	1.333	1.740	2.110	2.567	2.898
18	0.862	1.330	1.734	2.101	2.552	2.878
19	0.861	1.328	1.729	2.093	2.539	2.861
20	0.860	1.325	1.725	2.086	2.528	2.845
21	0.859	1.323	1.721	2.080	2.518	2.831
22	0.858	1.321	1.717	2.074	2.508	2.819
23	0.858	1.319	1.714	2.069	2.500	2.807
24	0.857	1.318	1.711	2.064	2.492	2.797
25	0.856	1.316	1.708	2.060	2.485	2.787
26	0.856	1.315	1.706	2.056	2.479	2.779
27	0.855	1.314	1.703	2.052	2.473	2.771
28	0.855	1.313	1.701	2.048	2.467	2.763
29	0.854	1.311	1.699	2.045	2.462	2.756
30	0.854	1.310	1.697	2.042	2.457	2.750
40	0.851	1.303	1.684	2.021	2.423	2.704
50	0.849	1.299	1.676	2.009	2.403	2.678
60	0.848	1.296	1.671	2.000	2.390	2.660
70	0.847	1.294	1.667	1.994	2.381	2.648
80	0.846	1.292	1.664	1.990	2.374	2.639
90	0.846	1.291	1.662	1.987	2.368	2.632
100	0.845	1.290	1.660	1.984	2.364	2.626
120	0.845	1.289	1.658	1.980	2.358	2.617
∞	0.842	1.282	1.645	1.96	2.326	2.576

E Cumulative Binomial Probabilities

Numerical entries represent $P(X \le x)$.

n	x	0.1	0.2	0.3	0.4	0.5	0.6	0.7	0.8	0.9
1	0	0.9000	0.8000	0.7000	0.6000	0.5000	0.4000	0.3000	0.2000	0.1000
	1	1.0000	1.0000	1.0000	1.0000	1.0000	1.0000	1.0000	1.0000	1.0000
2	0	0.8100	0.6400	0.4900	0.3600	0.2500	0.1600	0.0900	0.0400	0.0100
	1	0.9900	0.9600	0.9100	0.8400	0.7500	0.6400	0.5100	0.3600	0.1900
	2	1.0000	1.0000	1.0000	1.0000	1.0000	1.0000	1.0000	1.0000	1.0000
3	0	0.7290	0.5120	0.3430	0.2160	0.1250	0.0640	0.0270	0.0080	0.0010
	1	0.9720	0.8960	0.7840	0.6480	0.5000	0.3520	0.2160	0.1040	0.0280
	2	0.9990	0.9920	0.9730	0.9360	0.8750	0.7840	0.6570	0.4880	0.2710
	3	1.0000	1.0000	1.0000	1.0000	1.0000	1.0000	1.0000	1.0000	1.0000
4	0	0.6561	0.4096	0.2401	0.1296	0.0625	0.0256	0.0081	0.0016	0.0001
	1	0.9477	0.8192	0.6517	0.4752	0.3125	0.1792	0.0837	0.0272	0.0037
	2	0.9963	0.9728	0.9163	0.8208	0.6875	0.5248	0.3483	0.1808	0.0523
	3	0.9999	0.9984	0.9919	0.9744	0.9375	0.8704	0.7599	0.5904	0.3439
	4	1.0000	1.0000	1.0000	1.0000	1.0000	1.0000	1.0000	1.0000	1.0000
5	0	0.5905	0.3277	0.1681	0.0778	0.0313	0.0102	0.0024	0.0003	0.0000
	1	0.9185	0.7373	0.5282	0.3370	0.1875	0.0870	0.0308	0.0067	0.0005
	2	0.9914	0.9421	0.8369	0.6826	0.5000	0.3174	0.1631	0.0579	0.0086
	3	0.9995	0.9933	0.9692	0.9130	0.8125	0.6630	0.4718	0.2627	0.0815
	4	1.0000	0.9997	0.9976	0.9898	0.9688	0.9222	0.8319	0.6723	0.4095
	5	1.0000	1.0000	1.0000	1.0000	1.0000	1.0000	1.0000	1.0000	1.0000
6	0	0.5314	0.2621	0.1176	0.0467	0.0156	0.0041	0.0007	0.0001	0.0000
	1	0.8857	0.6554	0.4202	0.2333	0.1094	0.0410	0.0109	0.0016	0.0001
	2	0.9842	0.9011	0.7443	0.5443	0.3438	0.1792	0.0705	0.0170	0.0013
	3	0.9987	0.9830	0.9295	0.8208	0.6563	0.4557	0.2557	0.0989	0.0159
	4	0.9999	0.9984	0.9891	0.9590	0.8906	0.7667	0.5798	0.3446	0.1143
	5	1.0000	0.9999	0.9993	0.9959	0.9844	0.9533	0.8824	0.7379	0.4686
	6	1.0000	1.0000	1.0000	1.0000	1.0000	1.0000	1.0000	1.0000	1.0000
7	0	0.4783	0.2097	0.0824	0.0280	0.0078	0.0016	0.0002	0.0000	0.0000
	1	0.8503	0.5767	0.3294	0.1586	0.0625	0.0188	0.0038	0.0004	0.0000
	2	0.9743	0.8520	0.6471	0.4199	0.2266	0.0963	0.0288	0.0047	0.0002
	3	0.9973	0.9667	0.8740	0.7102	0.5000	0.2898	0.1260	0.0333	0.0027
	4	0.9998	0.9953	0.9712	0.9037	0.7734	0.5801	0.3529	0.1480	0.0257
	5	1.0000	0.9996	0.9962	0.9812	0.9375	0.8414	0.6706	0.4233	0.1497
	6	1.0000	1.0000	0.9998	0.9984	0.9922	0.9720	0.9176	0.7903	0.5217
	7	1.0000	1.0000	1.0000	1.0000	1.0000	1.0000	1.0000	1.0000	1.0000

Cumulative Binomial Probabilities (cont.)

						p				
n	*x*	**0.1**	**0.2**	**0.3**	**0.4**	**0.5**	**0.6**	**0.7**	**0.8**	**0.9**
8	0	0.4305	0.1678	0.0576	0.0168	0.0039	0.0007	0.0001	0.0000	0.0000
	1	0.8131	0.5033	0.2553	0.1064	0.0352	0.0085	0.0013	0.0001	0.0000
	2	0.9619	0.7969	0.5518	0.3154	0.1445	0.0498	0.0113	0.0012	0.0000
	3	0.9950	0.9437	0.8059	0.5941	0.3633	0.1737	0.0580	0.0104	0.0004
	4	0.9996	0.9896	0.9420	0.8263	0.6367	0.4059	0.1941	0.0563	0.0050
	5	1.0000	0.9988	0.9887	0.9502	0.8555	0.6846	0.4482	0.2031	0.0381
	6	1.0000	0.9999	0.9987	0.9915	0.9648	0.8936	0.7447	0.4967	0.1869
	7	1.0000	1.0000	0.9999	0.9993	0.9961	0.9832	0.9424	0.8322	0.5695
	8	1.0000	1.0000	1.0000	1.0000	1.0000	1.0000	1.0000	1.0000	1.0000
9	0	0.3874	0.1342	0.0404	0.0101	0.0020	0.0003	0.0000	0.0000	0.0000
	1	0.7748	0.4362	0.1960	0.0705	0.0195	0.0038	0.0004	0.0000	0.0000
	2	0.9470	0.7382	0.4628	0.2318	0.0898	0.0250	0.0043	0.0003	0.0000
	3	0.9917	0.9144	0.7297	0.4826	0.2539	0.0994	0.0253	0.0031	0.0001
	4	0.9991	0.9804	0.9012	0.7334	0.5000	0.2666	0.0988	0.0196	0.0009
	5	0.9999	0.9969	0.9747	0.9006	0.7461	0.5174	0.2703	0.0856	0.0083
	6	1.0000	0.9997	0.9957	0.9750	0.9102	0.7682	0.5372	0.2618	0.0530
	7	1.0000	1.0000	0.9996	0.9962	0.9805	0.9295	0.8040	0.5638	0.2252
	8	1.0000	1.0000	1.0000	0.9997	0.9980	0.9899	0.9596	0.8658	0.6126
	9	1.0000	1.0000	1.0000	1.0000	1.0000	1.0000	1.0000	1.0000	1.0000
10	0	0.3487	0.1074	0.0282	0.0060	0.0010	0.0001	0.0000	0.0000	0.0000
	1	0.7361	0.3758	0.1493	0.0464	0.0107	0.0017	0.0001	0.0000	0.0000
	2	0.9298	0.6778	0.3828	0.1673	0.0547	0.0123	0.0016	0.0001	0.0000
	3	0.9872	0.8791	0.6496	0.3823	0.1719	0.0548	0.0106	0.0009	0.0000
	4	0.9984	0.9672	0.8497	0.6331	0.3770	0.1662	0.0473	0.0064	0.0001
	5	0.9999	0.9936	0.9527	0.8338	0.6230	0.3669	0.1503	0.0328	0.0016
	6	1.0000	0.9991	0.9894	0.9452	0.8281	0.6177	0.3504	0.1209	0.0128
	7	1.0000	0.9999	0.9984	0.9877	0.9453	0.8327	0.6172	0.3222	0.0702
	8	1.0000	1.0000	0.9999	0.9983	0.9893	0.9536	0.8507	0.6242	0.2639
	9	1.0000	1.0000	1.0000	0.9999	0.9990	0.9940	0.9718	0.8926	0.6513
	10	1.0000	1.0000	1.0000	1.0000	1.0000	1.0000	1.0000	1.0000	1.0000
11	0	0.3138	0.0859	0.0198	0.0036	0.0005	0.0000	0.0000	0.0000	0.0000
	1	0.6974	0.3221	0.1130	0.0302	0.0059	0.0007	0.0000	0.0000	0.0000
	2	0.9104	0.6174	0.3127	0.1189	0.0327	0.0059	0.0006	0.0000	0.0000
	3	0.9815	0.8389	0.5696	0.2963	0.1133	0.0293	0.0043	0.0002	0.0000
	4	0.9972	0.9496	0.7897	0.5328	0.2744	0.0994	0.0216	0.0020	0.0000
	5	0.9997	0.9883	0.9218	0.7535	0.5000	0.2465	0.0782	0.0117	0.0003
	6	1.0000	0.9980	0.9784	0.9006	0.7256	0.4672	0.2103	0.0504	0.0028
	7	1.0000	0.9998	0.9957	0.9707	0.8867	0.7037	0.4304	0.1611	0.0185
	8	1.0000	1.0000	0.9994	0.9941	0.9673	0.8811	0.6873	0.3826	0.0896
	9	1.0000	1.0000	1.0000	0.9993	0.9941	0.9698	0.8870	0.6779	0.3026
	10	1.0000	1.0000	1.0000	1.0000	0.9995	0.9964	0.9802	0.9141	0.6862
	11	1.0000	1.0000	1.0000	1.0000	1.0000	1.0000	1.0000	1.0000	1.0000

Cumulative Binomial Probabilities (cont.)

						p				
n	x	0.1	0.2	0.3	0.4	0.5	0.6	0.7	0.8	0.9
12	0	0.2824	0.0687	0.0138	0.0022	0.0002	0.0000	0.0000	0.0000	0.0000
	1	0.6590	0.2749	0.0850	0.0196	0.0032	0.0003	0.0000	0.0000	0.0000
	2	0.8891	0.5583	0.2528	0.0834	0.0193	0.0028	0.0002	0.0000	0.0000
	3	0.9744	0.7946	0.4925	0.2253	0.0730	0.0153	0.0017	0.0001	0.0000
	4	0.9957	0.9274	0.7237	0.4382	0.1938	0.0573	0.0095	0.0006	0.0000
	5	0.9995	0.9806	0.8822	0.6652	0.3872	0.1582	0.0386	0.0039	0.0001
	6	0.9999	0.9961	0.9614	0.8418	0.6128	0.3348	0.1178	0.0194	0.0005
	7	1.0000	0.9994	0.9905	0.9427	0.8062	0.5618	0.2763	0.0726	0.0043
	8	1.0000	0.9999	0.9983	0.9847	0.9270	0.7747	0.5075	0.2054	0.0256
	9	1.0000	1.0000	0.9998	0.9972	0.9807	0.9166	0.7472	0.4417	0.1109
	10	1.0000	1.0000	1.0000	0.9997	0.9968	0.9804	0.9150	0.7251	0.3410
	11	1.0000	1.0000	1.0000	1.0000	0.9998	0.9978	0.9862	0.9313	0.7176
	12	1.0000	1.0000	1.0000	1.0000	1.0000	1.0000	1.0000	1.0000	1.0000
13	0	0.2542	0.0550	0.0097	0.0013	0.0001	0.0000	0.0000	0.0000	0.0000
	1	0.6213	0.2336	0.0637	0.0126	0.0017	0.0001	0.0000	0.0000	0.0000
	2	0.8661	0.5017	0.2025	0.0579	0.0112	0.0013	0.0001	0.0000	0.0000
	3	0.9658	0.7473	0.4206	0.1686	0.0461	0.0078	0.0007	0.0000	0.0000
	4	0.9935	0.9009	0.6543	0.3530	0.1334	0.0321	0.0040	0.0002	0.0000
	5	0.9991	0.9700	0.8346	0.5744	0.2905	0.0977	0.0182	0.0012	0.0000
	6	0.9999	0.9930	0.9376	0.7712	0.5000	0.2288	0.0624	0.0070	0.0001
	7	1.0000	0.9988	0.9818	0.9023	0.7095	0.4256	0.1654	0.0300	0.0009
	8	1.0000	0.9998	0.9960	0.9679	0.8666	0.6470	0.3457	0.0991	0.0065
	9	1.0000	1.0000	0.9993	0.9922	0.9539	0.8314	0.5794	0.2527	0.0342
	10	1.0000	1.0000	0.9999	0.9987	0.9888	0.9421	0.7975	0.4983	0.1339
	11	1.0000	1.0000	1.0000	0.9999	0.9983	0.9874	0.9363	0.7664	0.3787
	12	1.0000	1.0000	1.0000	1.0000	0.9999	0.9987	0.9903	0.9450	0.7458
	13	1.0000	1.0000	1.0000	1.0000	1.0000	1.0000	1.0000	1.0000	1.0000
14	0	0.2288	0.0440	0.0068	0.0008	0.0001	0.0000	0.0000	0.0000	0.0000
	1	0.5846	0.1979	0.0475	0.0081	0.0009	0.0001	0.0000	0.0000	0.0000
	2	0.8416	0.4481	0.1608	0.0398	0.0065	0.0006	0.0000	0.0000	0.0000
	3	0.9559	0.6982	0.3552	0.1243	0.0287	0.0039	0.0002	0.0000	0.0000
	4	0.9908	0.8702	0.5842	0.2793	0.0898	0.0175	0.0017	0.0000	0.0000
	5	0.9985	0.9561	0.7805	0.4859	0.2120	0.0583	0.0083	0.0004	0.0000
	6	0.9998	0.9884	0.9067	0.6925	0.3953	0.1501	0.0315	0.0024	0.0000
	7	1.0000	0.9976	0.9685	0.8499	0.6047	0.3075	0.0933	0.0116	0.0002
	8	1.0000	0.9996	0.9917	0.9417	0.7880	0.5141	0.2195	0.0439	0.0015
	9	1.0000	1.0000	0.9983	0.9825	0.9102	0.7207	0.4158	0.1298	0.0092
	10	1.0000	1.0000	0.9998	0.9961	0.9713	0.8757	0.6448	0.3018	0.0441
	11	1.0000	1.0000	1.0000	0.9994	0.9935	0.9602	0.8392	0.5519	0.1584
	12	1.0000	1.0000	1.0000	0.9999	0.9991	0.9919	0.9525	0.8021	0.4154
	13	1.0000	1.0000	1.0000	1.0000	0.9999	0.9992	0.9932	0.9560	0.7712
	14	1.0000	1.0000	1.0000	1.0000	1.0000	1.0000	1.0000	1.0000	1.0000

E   Cumulative Binomial Probabilities (cont.)

						p				
n	x	0.1	0.2	0.3	0.4	0.5	0.6	0.7	0.8	0.9
15	0	0.2059	0.0352	0.0047	0.0005	0.0000	0.0000	0.0000	0.0000	0.0000
	1	0.5490	0.1671	0.0353	0.0052	0.0005	0.0000	0.0000	0.0000	0.0000
	2	0.8159	0.3980	0.1268	0.0271	0.0037	0.0003	0.0000	0.0000	0.0000
	3	0.9444	0.6482	0.2969	0.0905	0.0176	0.0019	0.0001	0.0000	0.0000
	4	0.9873	0.8358	0.5155	0.2173	0.0592	0.0093	0.0007	0.0000	0.0000
	5	0.9978	0.9389	0.7216	0.4032	0.1509	0.0338	0.0037	0.0001	0.0000
	6	0.9997	0.9819	0.8689	0.6098	0.3036	0.0950	0.0152	0.0008	0.0000
	7	1.0000	0.9958	0.9500	0.7869	0.5000	0.2131	0.0500	0.0042	0.0000
	8	1.0000	0.9992	0.9848	0.9050	0.6964	0.3902	0.1311	0.0181	0.0003
	9	1.0000	0.9999	0.9963	0.9662	0.8491	0.5968	0.2784	0.0611	0.0022
	10	1.0000	1.0000	0.9993	0.9907	0.9408	0.7827	0.4845	0.1642	0.0127
	11	1.0000	1.0000	0.9999	0.9981	0.9824	0.9095	0.7031	0.3518	0.0556
	12	1.0000	1.0000	1.0000	0.9997	0.9963	0.9729	0.8732	0.6020	0.1841
	13	1.0000	1.0000	1.0000	1.0000	0.9995	0.9948	0.9647	0.8329	0.4510
	14	1.0000	1.0000	1.0000	1.0000	1.0000	0.9995	0.9953	0.9648	0.7941
	15	1.0000	1.0000	1.0000	1.0000	1.0000	1.0000	1.0000	1.0000	1.0000
16	0	0.1853	0.0281	0.0033	0.0003	0.0000	0.0000	0.0000	0.0000	0.0000
	1	0.5147	0.1407	0.0261	0.0033	0.0003	0.0000	0.0000	0.0000	0.0000
	2	0.7892	0.3518	0.0994	0.0183	0.0021	0.0001	0.0000	0.0000	0.0000
	3	0.9316	0.5981	0.2459	0.0651	0.0106	0.0009	0.0000	0.0000	0.0000
	4	0.9830	0.7982	0.4499	0.1666	0.0384	0.0049	0.0003	0.0000	0.0000
	5	0.9967	0.9183	0.6598	0.3288	0.1051	0.0191	0.0016	0.0000	0.0000
	6	0.9995	0.9733	0.8247	0.5272	0.2272	0.0583	0.0071	0.0002	0.0000
	7	0.9999	0.9930	0.9256	0.7161	0.4018	0.1423	0.0257	0.0015	0.0000
	8	1.0000	0.9985	0.9743	0.8577	0.5982	0.2839	0.0744	0.0070	0.0001
	9	1.0000	0.9998	0.9929	0.9417	0.7728	0.4728	0.1753	0.0267	0.0005
	10	1.0000	1.0000	0.9984	0.9809	0.8949	0.6712	0.3402	0.0817	0.0033
	11	1.0000	1.0000	0.9997	0.9951	0.9616	0.8334	0.5501	0.2018	0.0170
	12	1.0000	1.0000	1.0000	0.9991	0.9894	0.9349	0.7541	0.4019	0.0684
	13	1.0000	1.0000	1.0000	0.9999	0.9979	0.9817	0.9006	0.6482	0.2108
	14	1.0000	1.0000	1.0000	1.0000	0.9997	0.9967	0.9739	0.8593	0.4853
	15	1.0000	1.0000	1.0000	1.0000	1.0000	0.9997	0.9967	0.9719	0.8147
	16	1.0000	1.0000	1.0000	1.0000	1.0000	1.0000	1.0000	1.0000	1.0000

E ## Cumulative Binomial Probabilities (cont.)

						p				
n	x	0.1	0.2	0.3	0.4	0.5	0.6	0.7	0.8	0.9
17	0	0.1668	0.0225	0.0023	0.0002	0.0000	0.0000	0.0000	0.0000	0.0000
	1	0.4818	0.1182	0.0193	0.0021	0.0001	0.0000	0.0000	0.0000	0.0000
	2	0.7618	0.3096	0.0774	0.0123	0.0012	0.0001	0.0000	0.0000	0.0000
	3	0.9174	0.5489	0.2019	0.0464	0.0064	0.0005	0.0000	0.0000	0.0000
	4	0.9779	0.7582	0.3887	0.1260	0.0245	0.0025	0.0001	0.0000	0.0000
	5	0.9953	0.8943	0.5968	0.2639	0.0717	0.0106	0.0007	0.0000	0.0000
	6	0.9992	0.9623	0.7752	0.4478	0.1662	0.0348	0.0032	0.0001	0.0000
	7	0.9999	0.9891	0.8954	0.6405	0.3145	0.0919	0.0127	0.0005	0.0000
	8	1.0000	0.9974	0.9597	0.8011	0.5000	0.1989	0.0403	0.0026	0.0000
	9	1.0000	0.9995	0.9873	0.9081	0.6855	0.3595	0.1046	0.0109	0.0001
	10	1.0000	0.9999	0.9968	0.9652	0.8338	0.5522	0.2248	0.0377	0.0008
	11	1.0000	1.0000	0.9993	0.9894	0.9283	0.7361	0.4032	0.1057	0.0047
	12	1.0000	1.0000	0.9999	0.9975	0.9755	0.8740	0.6113	0.2418	0.0221
	13	1.0000	1.0000	1.0000	0.9995	0.9936	0.9536	0.7981	0.4511	0.0826
	14	1.0000	1.0000	1.0000	0.9999	0.9988	0.9877	0.9226	0.6904	0.2382
	15	1.0000	1.0000	1.0000	1.0000	0.9999	0.9979	0.9807	0.8818	0.5182
	16	1.0000	1.0000	1.0000	1.0000	1.0000	0.9998	0.9977	0.9775	0.8332
	17	1.0000	1.0000	1.0000	1.0000	1.0000	1.0000	1.0000	1.0000	1.0000
18	0	0.1501	0.0180	0.0016	0.0001	0.0000	0.0000	0.0000	0.0000	0.0000
	1	0.4503	0.0991	0.0142	0.0013	0.0001	0.0000	0.0000	0.0000	0.0000
	2	0.7338	0.2713	0.0600	0.0082	0.0007	0.0000	0.0000	0.0000	0.0000
	3	0.9018	0.5010	0.1646	0.0328	0.0038	0.0002	0.0000	0.0000	0.0000
	4	0.9718	0.7164	0.3327	0.0942	0.0154	0.0013	0.0000	0.0000	0.0000
	5	0.9936	0.8671	0.5344	0.2088	0.0481	0.0058	0.0003	0.0000	0.0000
	6	0.9988	0.9487	0.7217	0.3743	0.1189	0.0203	0.0014	0.0000	0.0000
	7	0.9998	0.9837	0.8593	0.5634	0.2403	0.0576	0.0061	0.0002	0.0000
	8	1.0000	0.9957	0.9404	0.7368	0.4073	0.1347	0.0210	0.0009	0.0000
	9	1.0000	0.9991	0.9790	0.8653	0.5927	0.2632	0.0596	0.0043	0.0000
	10	1.0000	0.9998	0.9939	0.9424	0.7597	0.4366	0.1407	0.0163	0.0002
	11	1.0000	1.0000	0.9986	0.9797	0.8811	0.6257	0.2783	0.0513	0.0012
	12	1.0000	1.0000	0.9997	0.9942	0.9519	0.7912	0.4656	0.1329	0.0064
	13	1.0000	1.0000	1.0000	0.9987	0.9846	0.9058	0.6673	0.2836	0.0282
	14	1.0000	1.0000	1.0000	0.9998	0.9962	0.9672	0.8354	0.4990	0.0982
	15	1.0000	1.0000	1.0000	1.0000	0.9993	0.9918	0.9400	0.7287	0.2662
	16	1.0000	1.0000	1.0000	1.0000	0.9999	0.9987	0.9858	0.9009	0.5497
	17	1.0000	1.0000	1.0000	1.0000	1.0000	0.9999	0.9984	0.9820	0.8499
	18	1.0000	1.0000	1.0000	1.0000	1.0000	1.0000	1.0000	1.0000	1.0000

Cumulative Binomial Probabilities (cont.)

						p				
n	*x*	0.1	0.2	0.3	0.4	0.5	0.6	0.7	0.8	0.9
19	0	0.1351	0.0144	0.0011	0.0001	0.0000	0.0000	0.0000	0.0000	0.0000
	1	0.4203	0.0829	0.0104	0.0008	0.0000	0.0000	0.0000	0.0000	0.0000
	2	0.7054	0.2369	0.0462	0.0055	0.0004	0.0000	0.0000	0.0000	0.0000
	3	0.8850	0.4551	0.1332	0.0230	0.0022	0.0001	0.0000	0.0000	0.0000
	4	0.9648	0.6733	0.2822	0.0696	0.0096	0.0006	0.0000	0.0000	0.0000
	5	0.9914	0.8369	0.4739	0.1629	0.0318	0.0031	0.0001	0.0000	0.0000
	6	0.9983	0.9324	0.6655	0.3081	0.0835	0.0116	0.0006	0.0000	0.0000
	7	0.9997	0.9767	0.8180	0.4878	0.1796	0.0352	0.0028	0.0000	0.0000
	8	1.0000	0.9933	0.9161	0.6675	0.3238	0.0885	0.0105	0.0003	0.0000
	9	1.0000	0.9984	0.9674	0.8139	0.5000	0.1861	0.0326	0.0016	0.0000
	10	1.0000	0.9997	0.9895	0.9115	0.6762	0.3325	0.0839	0.0067	0.0000
	11	1.0000	1.0000	0.9972	0.9648	0.8204	0.5122	0.1820	0.0233	0.0003
	12	1.0000	1.0000	0.9994	0.9884	0.9165	0.6919	0.3345	0.0676	0.0017
	13	1.0000	1.0000	0.9999	0.9969	0.9682	0.8371	0.5261	0.1631	0.0086
	14	1.0000	1.0000	1.0000	0.9994	0.9904	0.9304	0.7178	0.3267	0.0352
	15	1.0000	1.0000	1.0000	0.9999	0.9978	0.9770	0.8668	0.5449	0.1150
	16	1.0000	1.0000	1.0000	1.0000	0.9996	0.9945	0.9538	0.7631	0.2946
	17	1.0000	1.0000	1.0000	1.0000	1.0000	0.9992	0.9896	0.9171	0.5797
	18	1.0000	1.0000	1.0000	1.0000	1.0000	0.9999	0.9989	0.9856	0.8649
	19	1.0000	1.0000	1.0000	1.0000	1.0000	1.0000	1.0000	1.0000	1.0000
20	0	0.1216	0.0115	0.0008	0.0000	0.0000	0.0000	0.0000	0.0000	0.0000
	1	0.3917	0.0692	0.0076	0.0005	0.0000	0.0000	0.0000	0.0000	0.0000
	2	0.6769	0.2061	0.0355	0.0036	0.0002	0.0000	0.0000	0.0000	0.0000
	3	0.8670	0.4114	0.1071	0.0160	0.0013	0.0000	0.0000	0.0000	0.0000
	4	0.9568	0.6296	0.2375	0.0510	0.0059	0.0003	0.0000	0.0000	0.0000
	5	0.9887	0.8042	0.4164	0.1256	0.0207	0.0016	0.0000	0.0000	0.0000
	6	0.9976	0.9133	0.6080	0.2500	0.0577	0.0065	0.0003	0.0000	0.0000
	7	0.9996	0.9679	0.7723	0.4159	0.1316	0.0210	0.0013	0.0000	0.0000
	8	0.9999	0.9900	0.8867	0.5956	0.2517	0.0565	0.0051	0.0001	0.0000
	9	1.0000	0.9974	0.9520	0.7553	0.4119	0.1275	0.0171	0.0006	0.0000
	10	1.0000	0.9994	0.9829	0.8725	0.5881	0.2447	0.0480	0.0026	0.0000
	11	1.0000	0.9999	0.9949	0.9435	0.7483	0.4044	0.1133	0.0100	0.0001
	12	1.0000	1.0000	0.9987	0.9790	0.8684	0.5841	0.2277	0.0321	0.0004
	13	1.0000	1.0000	0.9997	0.9935	0.9423	0.7500	0.3920	0.0867	0.0024
	14	1.0000	1.0000	1.0000	0.9984	0.9793	0.8744	0.5836	0.1958	0.0113
	15	1.0000	1.0000	1.0000	0.9997	0.9941	0.9490	0.7625	0.3704	0.0432
	16	1.0000	1.0000	1.0000	1.0000	0.9987	0.9840	0.8929	0.5886	0.1330
	17	1.0000	1.0000	1.0000	1.0000	0.9998	0.9964	0.9645	0.7939	0.3231
	18	1.0000	1.0000	1.0000	1.0000	1.0000	0.9995	0.9924	0.9308	0.6083
	19	1.0000	1.0000	1.0000	1.0000	1.0000	1.0000	0.9992	0.9885	0.8784
	20	1.0000	1.0000	1.0000	1.0000	1.0000	1.0000	1.0000	1.0000	1.0000

F   Poisson Probabilities

Numerical entries represent $P(X = x)$.

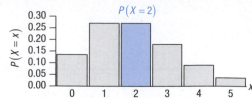

x	0.02	0.03	0.04	0.05	0.06	λ 0.07	0.08	0.09	0.10	0.20	0.30
0	0.9802	0.9704	0.9608	0.9512	0.9418	0.9324	0.9231	0.9139	0.9048	0.8187	0.7408
1	0.0196	0.0291	0.0384	0.0476	0.0565	0.0653	0.0738	0.0823	0.0905	0.1637	0.2222
2	0.0002	0.0004	0.0008	0.0012	0.0017	0.0023	0.0030	0.0037	0.0045	0.0164	0.0333
3	0.0000	0.0000	0.0000	0.0000	0.0000	0.0001	0.0001	0.0001	0.0002	0.0011	0.0033
4	0.0000	0.0000	0.0000	0.0000	0.0000	0.0000	0.0000	0.0000	0.0000	0.0001	0.0003

x	0.40	0.50	0.60	0.70	0.80	0.90	1.00	1.10	1.20	1.30	1.40
0	0.6703	0.6065	0.5488	0.4966	0.4493	0.4066	0.3679	0.3329	0.3012	0.2725	0.2466
1	0.2681	0.3033	0.3293	0.3476	0.3595	0.3659	0.3679	0.3662	0.3614	0.3543	0.3452
2	0.0536	0.0758	0.0988	0.1217	0.1438	0.1647	0.1839	0.2014	0.2169	0.2303	0.2417
3	0.0072	0.0126	0.0198	0.0284	0.0383	0.0494	0.0613	0.0738	0.0867	0.0998	0.1128
4	0.0007	0.0016	0.0030	0.0050	0.0077	0.0111	0.0153	0.0203	0.0260	0.0324	0.0395
5	0.0001	0.0002	0.0004	0.0007	0.0012	0.0020	0.0031	0.0045	0.0062	0.0084	0.0111
6	0.0000	0.0000	0.0000	0.0001	0.0002	0.0003	0.0005	0.0008	0.0012	0.0018	0.0026
7	0.0000	0.0000	0.0000	0.0000	0.0000	0.0000	0.0001	0.0001	0.0002	0.0003	0.0005
8	0.0000	0.0000	0.0000	0.0000	0.0000	0.0000	0.0000	0.0000	0.0000	0.0001	0.0001

x	1.50	1.60	1.70	1.80	1.90	2.00	2.10	2.20	2.30	2.40	2.50
0	0.2231	0.2019	0.1827	0.1653	0.1496	0.1353	0.1225	0.1108	0.1003	0.0907	0.0821
1	0.3347	0.3230	0.3106	0.2975	0.2842	0.2707	0.2572	0.2438	0.2306	0.2177	0.2052
2	0.2510	0.2584	0.2640	0.2678	0.2700	0.2707	0.2700	0.2681	0.2652	0.2613	0.2565
3	0.1255	0.1378	0.1496	0.1607	0.1710	0.1804	0.1890	0.1966	0.2033	0.2090	0.2138
4	0.0471	0.0551	0.0636	0.0723	0.0812	0.0902	0.0992	0.1082	0.1169	0.1254	0.1336
5	0.0141	0.0176	0.0216	0.0260	0.0309	0.0361	0.0417	0.0476	0.0538	0.0602	0.0668
6	0.0035	0.0047	0.0061	0.0078	0.0098	0.0120	0.0146	0.0174	0.0206	0.0241	0.0278
7	0.0008	0.0011	0.0015	0.0020	0.0027	0.0034	0.0044	0.0055	0.0068	0.0083	0.0099
8	0.0001	0.0002	0.0003	0.0005	0.0006	0.0009	0.0011	0.0015	0.0019	0.0025	0.0031
9	0.0000	0.0000	0.0001	0.0001	0.0001	0.0002	0.0003	0.0004	0.0005	0.0007	0.0009
10	0.0000	0.0000	0.0000	0.0000	0.0000	0.0000	0.0001	0.0001	0.0001	0.0002	0.0002

x	2.60	2.70	2.80	2.90	3.00	3.10	3.20	3.30	3.40	3.50	3.60
0	0.0743	0.0672	0.0608	0.0550	0.0498	0.0450	0.0408	0.0369	0.0334	0.0302	0.0273
1	0.1931	0.1815	0.1703	0.1596	0.1494	0.1397	0.1304	0.1217	0.1135	0.1057	0.0984
2	0.2510	0.2450	0.2384	0.2314	0.2240	0.2165	0.2087	0.2008	0.1929	0.1850	0.1771
3	0.2176	0.2205	0.2225	0.2237	0.2240	0.2237	0.2226	0.2209	0.2186	0.2158	0.2125
4	0.1414	0.1488	0.1557	0.1622	0.1680	0.1733	0.1781	0.1823	0.1858	0.1888	0.1912
5	0.0735	0.0804	0.0872	0.0940	0.1008	0.1075	0.1140	0.1203	0.1264	0.1322	0.1377
6	0.0319	0.0362	0.0407	0.0455	0.0504	0.0555	0.0608	0.0662	0.0716	0.0771	0.0826
7	0.0118	0.0139	0.0163	0.0188	0.0216	0.0246	0.0278	0.0312	0.0348	0.0385	0.0425
8	0.0038	0.0047	0.0057	0.0068	0.0081	0.0095	0.0111	0.0129	0.0148	0.0169	0.0191
9	0.0011	0.0014	0.0018	0.0022	0.0027	0.0033	0.0040	0.0047	0.0056	0.0066	0.0076
10	0.0003	0.0004	0.0005	0.0006	0.0008	0.0010	0.0013	0.0016	0.0019	0.0023	0.0028
11	0.0001	0.0001	0.0001	0.0002	0.0002	0.0003	0.0004	0.0005	0.0006	0.0007	0.0009
12	0.0000	0.0000	0.0000	0.0000	0.0001	0.0001	0.0001	0.0001	0.0002	0.0002	0.0003
13	0.0000	0.0000	0.0000	0.0000	0.0000	0.0000	0.0000	0.0000	0.0000	0.0001	0.0001

F **Poisson Probabilities** (cont.)

x	3.70	3.80	3.90	4.00	4.10	4.20	4.30	4.40	4.50	4.60	4.70
0	0.0247	0.0224	0.0202	0.0183	0.0166	0.0150	0.0136	0.0123	0.0111	0.0101	0.0091
1	0.0915	0.0850	0.0789	0.0733	0.0679	0.0630	0.0583	0.0540	0.0500	0.0462	0.0427
2	0.1692	0.1615	0.1539	0.1465	0.1393	0.1323	0.1254	0.1188	0.1125	0.1063	0.1005
3	0.2087	0.2046	0.2001	0.1954	0.1904	0.1852	0.1798	0.1743	0.1687	0.1631	0.1574
4	0.1931	0.1944	0.1951	0.1954	0.1951	0.1944	0.1933	0.1917	0.1898	0.1875	0.1849
5	0.1429	0.1477	0.1522	0.1563	0.1600	0.1633	0.1662	0.1687	0.1708	0.1725	0.1738
6	0.0881	0.0936	0.0989	0.1042	0.1093	0.1143	0.1191	0.1237	0.1281	0.1323	0.1362
7	0.0466	0.0508	0.0551	0.0595	0.0640	0.0686	0.0732	0.0778	0.0824	0.0869	0.0914
8	0.0215	0.0241	0.0269	0.0298	0.0328	0.0360	0.0393	0.0428	0.0463	0.0500	0.0537
9	0.0089	0.0102	0.0116	0.0132	0.0150	0.0168	0.0188	0.0209	0.0232	0.0255	0.0281
10	0.0033	0.0039	0.0045	0.0053	0.0061	0.0071	0.0081	0.0092	0.0104	0.0118	0.0132
11	0.0011	0.0013	0.0016	0.0019	0.0023	0.0027	0.0032	0.0037	0.0043	0.0049	0.0056
12	0.0003	0.0004	0.0005	0.0006	0.0008	0.0009	0.0011	0.0013	0.0016	0.0019	0.0022
13	0.0001	0.0001	0.0002	0.0002	0.0002	0.0003	0.0004	0.0005	0.0006	0.0007	0.0008
14	0.0000	0.0000	0.0000	0.0001	0.0001	0.0001	0.0001	0.0001	0.0002	0.0002	0.0003
15	0.0000	0.0000	0.0000	0.0000	0.0000	0.0000	0.0000	0.0000	0.0001	0.0001	0.0001

x	4.80	4.90	5.00	5.10	5.20	5.30	5.40	5.50	5.60	5.70	5.80
0	0.0082	0.0074	0.0067	0.0061	0.0055	0.0050	0.0045	0.0041	0.0037	0.0033	0.0030
1	0.0395	0.0365	0.0337	0.0311	0.0287	0.0265	0.0244	0.0225	0.0207	0.0191	0.0176
2	0.0948	0.0894	0.0842	0.0793	0.0746	0.0701	0.0659	0.0618	0.0580	0.0544	0.0509
3	0.1517	0.1460	0.1404	0.1348	0.1293	0.1239	0.1185	0.1133	0.1082	0.1033	0.0985
4	0.1820	0.1789	0.1755	0.1719	0.1681	0.1641	0.1600	0.1558	0.1515	0.1472	0.1428
5	0.1747	0.1753	0.1755	0.1753	0.1748	0.1740	0.1728	0.1714	0.1697	0.1678	0.1656
6	0.1398	0.1432	0.1462	0.1490	0.1515	0.1537	0.1555	0.1571	0.1584	0.1594	0.1601
7	0.0959	0.1002	0.1044	0.1086	0.1125	0.1163	0.1200	0.1234	0.1267	0.1298	0.1326
8	0.0575	0.0614	0.0653	0.0692	0.0731	0.0771	0.0810	0.0849	0.0887	0.0925	0.0962
9	0.0307	0.0334	0.0363	0.0392	0.0423	0.0454	0.0486	0.0519	0.0552	0.0586	0.0620
10	0.0147	0.0164	0.0181	0.0200	0.0220	0.0241	0.0262	0.0285	0.0309	0.0334	0.0359
11	0.0064	0.0073	0.0082	0.0093	0.0104	0.0116	0.0129	0.0143	0.0157	0.0173	0.0190
12	0.0026	0.0030	0.0034	0.0039	0.0045	0.0051	0.0058	0.0065	0.0073	0.0082	0.0092
13	0.0009	0.0011	0.0013	0.0015	0.0018	0.0021	0.0024	0.0028	0.0032	0.0036	0.0041
14	0.0003	0.0004	0.0005	0.0006	0.0007	0.0008	0.0009	0.0011	0.0013	0.0015	0.0017
15	0.0001	0.0001	0.0002	0.0002	0.0002	0.0003	0.0003	0.0004	0.0005	0.0006	0.0007
16	0.0000	0.0000	0.0000	0.0001	0.0001	0.0001	0.0001	0.0001	0.0002	0.0002	0.0002
17	0.0000	0.0000	0.0000	0.0000	0.0000	0.0000	0.0000	0.0000	0.0001	0.0001	0.0001

x	5.90	6.00	6.10	6.20	6.30	6.40	6.50	6.60	6.70	6.80	6.90
0	0.0027	0.0025	0.0022	0.0020	0.0018	0.0017	0.0015	0.0014	0.0012	0.0011	0.0010
1	0.0162	0.0149	0.0137	0.0126	0.0116	0.0106	0.0098	0.0090	0.0082	0.0076	0.0070
2	0.0477	0.0446	0.0417	0.0390	0.0364	0.0340	0.0318	0.0296	0.0276	0.0258	0.0240
3	0.0938	0.0892	0.0848	0.0806	0.0765	0.0726	0.0688	0.0652	0.0617	0.0584	0.0552
4	0.1383	0.1339	0.1294	0.1249	0.1205	0.1162	0.1118	0.1076	0.1034	0.0992	0.0952
5	0.1632	0.1606	0.1579	0.1549	0.1519	0.1487	0.1454	0.1420	0.1385	0.1349	0.1314

Poisson Probabilities (cont.)

x	5.90	6.00	6.10	6.20	6.30	λ 6.40	6.50	6.60	6.70	6.80	6.90
6	0.1605	0.1606	0.1605	0.1601	0.1595	0.1586	0.1575	0.1562	0.1546	0.1529	0.1511
7	0.1353	0.1377	0.1399	0.1418	0.1435	0.1450	0.1462	0.1472	0.1480	0.1486	0.1489
8	0.0998	0.1033	0.1066	0.1099	0.1130	0.1160	0.1188	0.1215	0.1240	0.1263	0.1284
9	0.0654	0.0688	0.0723	0.0757	0.0791	0.0825	0.0858	0.0891	0.0923	0.0954	0.0985
10	0.0386	0.0413	0.0441	0.0469	0.0498	0.0528	0.0558	0.0588	0.0618	0.0649	0.0679
11	0.0207	0.0225	0.0244	0.0265	0.0285	0.0307	0.0330	0.0353	0.0377	0.0401	0.0426
12	0.0102	0.0113	0.0124	0.0137	0.0150	0.0164	0.0179	0.0194	0.0210	0.0227	0.0245
13	0.0046	0.0052	0.0058	0.0065	0.0073	0.0081	0.0089	0.0099	0.0108	0.0119	0.0130
14	0.0019	0.0022	0.0025	0.0029	0.0033	0.0037	0.0041	0.0046	0.0052	0.0058	0.0064
15	0.0008	0.0009	0.0010	0.0012	0.0014	0.0016	0.0018	0.0020	0.0023	0.0026	0.0029
16	0.0003	0.0003	0.0004	0.0005	0.0005	0.0006	0.0007	0.0008	0.0010	0.0011	0.0013
17	0.0001	0.0001	0.0001	0.0002	0.0002	0.0002	0.0003	0.0003	0.0004	0.0004	0.0005
18	0.0000	0.0000	0.0000	0.0001	0.0001	0.0001	0.0001	0.0001	0.0001	0.0002	0.0002
19	0.0000	0.0000	0.0000	0.0000	0.0000	0.0000	0.0000	0.0000	0.0001	0.0001	0.0001

x	7.00	7.10	7.20	7.30	7.40	7.50	7.60	7.70	7.80	7.90	8.00
0	0.0009	0.0008	0.0007	0.0007	0.0006	0.0006	0.0005	0.0005	0.0004	0.0004	0.0003
1	0.0064	0.0059	0.0054	0.0049	0.0045	0.0041	0.0038	0.0035	0.0032	0.0029	0.0027
2	0.0223	0.0208	0.0194	0.0180	0.0167	0.0156	0.0145	0.0134	0.0125	0.0116	0.0107
3	0.0521	0.0492	0.0464	0.0438	0.0413	0.0389	0.0366	0.0345	0.0324	0.0305	0.0286
4	0.0912	0.0874	0.0836	0.0799	0.0764	0.0729	0.0696	0.0663	0.0632	0.0602	0.0573
5	0.1277	0.1241	0.1204	0.1167	0.1130	0.1094	0.1057	0.1021	0.0986	0.0951	0.0916
6	0.1490	0.1468	0.1445	0.1420	0.1394	0.1367	0.1339	0.1311	0.1282	0.1252	0.1221
7	0.1490	0.1489	0.1486	0.1481	0.1474	0.1465	0.1454	0.1442	0.1428	0.1413	0.1396
8	0.1304	0.1321	0.1337	0.1351	0.1363	0.1373	0.1381	0.1388	0.1392	0.1395	0.1396
9	0.1014	0.1042	0.1070	0.1096	0.1121	0.1144	0.1167	0.1187	0.1207	0.1224	0.1241
10	0.0710	0.0740	0.0770	0.0800	0.0829	0.0858	0.0887	0.0914	0.0941	0.0967	0.0993
11	0.0452	0.0478	0.0504	0.0531	0.0558	0.0585	0.0613	0.0640	0.0667	0.0695	0.0722
12	0.0263	0.0283	0.0303	0.0323	0.0344	0.0366	0.0388	0.0411	0.0434	0.0457	0.0481
13	0.0142	0.0154	0.0168	0.0181	0.0196	0.0211	0.0227	0.0243	0.0260	0.0278	0.0296
14	0.0071	0.0078	0.0086	0.0095	0.0104	0.0113	0.0123	0.0134	0.0145	0.0157	0.0169
15	0.0033	0.0037	0.0041	0.0046	0.0051	0.0057	0.0062	0.0069	0.0075	0.0083	0.0090
16	0.0014	0.0016	0.0019	0.0021	0.0024	0.0026	0.0030	0.0033	0.0037	0.0041	0.0045
17	0.0006	0.0007	0.0008	0.0009	0.0010	0.0012	0.0013	0.0015	0.0017	0.0019	0.0021
18	0.0002	0.0003	0.0003	0.0004	0.0004	0.0005	0.0006	0.0006	0.0007	0.0008	0.0009
19	0.0001	0.0001	0.0001	0.0001	0.0002	0.0002	0.0002	0.0003	0.0003	0.0003	0.0004
20	0.0000	0.0000	0.0000	0.0001	0.0001	0.0001	0.0001	0.0001	0.0001	0.0001	0.0002
21	0.0000	0.0000	0.0000	0.0000	0.0000	0.0000	0.0000	0.0000	0.0000	0.0001	0.0001

x	8.10	8.20	8.30	8.40	8.50	8.60	8.70	8.80	8.90	9.00	9.10
0	0.0003	0.0003	0.0002	0.0002	0.0002	0.0002	0.0002	0.0002	0.0001	0.0001	0.0001
1	0.0025	0.0023	0.0021	0.0019	0.0017	0.0016	0.0014	0.0013	0.0012	0.0011	0.0010
2	0.0100	0.0092	0.0086	0.0079	0.0074	0.0068	0.0063	0.0058	0.0054	0.0050	0.0046
3	0.0269	0.0252	0.0237	0.0222	0.0208	0.0195	0.0183	0.0171	0.0160	0.0150	0.0140
4	0.0544	0.0517	0.0491	0.0466	0.0443	0.0420	0.0398	0.0377	0.0357	0.0337	0.0319

Poisson Probabilities (cont.)

						λ					
x	8.10	8.20	8.30	8.40	8.50	8.60	8.70	8.80	8.90	9.00	9.10
5	0.0882	0.0849	0.0816	0.0784	0.0752	0.0722	0.0692	0.0663	0.0635	0.0607	0.0581
6	0.1191	0.1160	0.1128	0.1097	0.1066	0.1034	0.1003	0.0972	0.0941	0.0911	0.0881
7	0.1378	0.1358	0.1338	0.1317	0.1294	0.1271	0.1247	0.1222	0.1197	0.1171	0.1145
8	0.1395	0.1392	0.1388	0.1382	0.1375	0.1366	0.1356	0.1344	0.1332	0.1318	0.1302
9	0.1256	0.1269	0.1280	0.1290	0.1299	0.1306	0.1311	0.1315	0.1317	0.1318	0.1317
10	0.1017	0.1040	0.1063	0.1084	0.1104	0.1123	0.1140	0.1157	0.1172	0.1186	0.1198
11	0.0749	0.0776	0.0802	0.0828	0.0853	0.0878	0.0902	0.0925	0.0948	0.0970	0.0991
12	0.0505	0.0530	0.0555	0.0579	0.0604	0.0629	0.0654	0.0679	0.0703	0.0728	0.0752
13	0.0315	0.0334	0.0354	0.0374	0.0395	0.0416	0.0438	0.0459	0.0481	0.0504	0.0526
14	0.0182	0.0196	0.0210	0.0225	0.0240	0.0256	0.0272	0.0289	0.0306	0.0324	0.0342
15	0.0098	0.0107	0.0116	0.0126	0.0136	0.0147	0.0158	0.0169	0.0182	0.0194	0.0208
16	0.0050	0.0055	0.0060	0.0066	0.0072	0.0079	0.0086	0.0093	0.0101	0.0109	0.0118
17	0.0024	0.0026	0.0029	0.0033	0.0036	0.0040	0.0044	0.0048	0.0053	0.0058	0.0063
18	0.0011	0.0012	0.0014	0.0015	0.0017	0.0019	0.0021	0.0024	0.0026	0.0029	0.0032
19	0.0005	0.0005	0.0006	0.0007	0.0008	0.0009	0.0010	0.0011	0.0012	0.0014	0.0015
20	0.0002	0.0002	0.0002	0.0003	0.0003	0.0004	0.0004	0.0005	0.0005	0.0006	0.0007
21	0.0001	0.0001	0.0001	0.0001	0.0001	0.0002	0.0002	0.0002	0.0002	0.0003	0.0003
22	0.0000	0.0000	0.0000	0.0000	0.0001	0.0001	0.0001	0.0001	0.0001	0.0001	0.0001
x	9.20	9.30	9.40	9.50	9.60	9.70	9.80	9.90	10.00	11.00	12.00
0	0.0001	0.0001	0.0001	0.0001	0.0001	0.0001	0.0001	0.0001	0.0000	0.0000	0.0000
1	0.0009	0.0009	0.0008	0.0007	0.0007	0.0006	0.0005	0.0005	0.0005	0.0002	0.0001
2	0.0043	0.0040	0.0037	0.0034	0.0031	0.0029	0.0027	0.0025	0.0023	0.0010	0.0004
3	0.0131	0.0123	0.0115	0.0107	0.0100	0.0093	0.0087	0.0081	0.0076	0.0037	0.0018
4	0.0302	0.0285	0.0269	0.0254	0.0240	0.0226	0.0213	0.0201	0.0189	0.0102	0.0053
5	0.0555	0.0530	0.0506	0.0483	0.0460	0.0439	0.0418	0.0398	0.0378	0.0224	0.0127
6	0.0851	0.0822	0.0793	0.0764	0.0736	0.0709	0.0682	0.0656	0.0631	0.0411	0.0255
7	0.1118	0.1091	0.1064	0.1037	0.1010	0.0982	0.0955	0.0928	0.0901	0.0646	0.0437
8	0.1286	0.1269	0.1251	0.1232	0.1212	0.1191	0.1170	0.1148	0.1126	0.0888	0.0655
9	0.1315	0.1311	0.1306	0.1300	0.1293	0.1284	0.1274	0.1263	0.1251	0.1085	0.0874
10	0.1210	0.1219	0.1228	0.1235	0.1241	0.1245	0.1249	0.1250	0.1251	0.1194	0.1048
11	0.1012	0.1031	0.1049	0.1067	0.1083	0.1098	0.1112	0.1125	0.1137	0.1194	0.1144
12	0.0776	0.0799	0.0822	0.0844	0.0866	0.0888	0.0908	0.0928	0.0948	0.1094	0.1144
13	0.0549	0.0572	0.0594	0.0617	0.0640	0.0662	0.0685	0.0707	0.0729	0.0926	0.1056
14	0.0361	0.0380	0.0399	0.0419	0.0439	0.0459	0.0479	0.0500	0.0521	0.0728	0.0905
15	0.0221	0.0235	0.0250	0.0265	0.0281	0.0297	0.0313	0.0330	0.0347	0.0534	0.0724
16	0.0127	0.0137	0.0147	0.0157	0.0168	0.0180	0.0192	0.0204	0.0217	0.0367	0.0543
17	0.0069	0.0075	0.0081	0.0088	0.0095	0.0103	0.0111	0.0119	0.0128	0.0237	0.0383
18	0.0035	0.0039	0.0042	0.0046	0.0051	0.0055	0.0060	0.0065	0.0071	0.0145	0.0255
19	0.0017	0.0019	0.0021	0.0023	0.0026	0.0028	0.0031	0.0034	0.0037	0.0084	0.0161
20	0.0008	0.0009	0.0010	0.0011	0.0012	0.0014	0.0015	0.0017	0.0019	0.0046	0.0097
21	0.0003	0.0004	0.0004	0.0005	0.0006	0.0006	0.0007	0.0008	0.0009	0.0024	0.0055
22	0.0001	0.0002	0.0002	0.0002	0.0002	0.0003	0.0003	0.0004	0.0004	0.0012	0.0030
23	0.0001	0.0001	0.0001	0.0001	0.0001	0.0001	0.0001	0.0002	0.0002	0.0006	0.0016
24	0.0000	0.0000	0.0000	0.0000	0.0000	0.0000	0.0001	0.0001	0.0001	0.0003	0.0008

Poisson Probabilities (cont.)

						λ					
x	13.00	14.00	15.00	16.00	17.00	18.00	19.00	20.00	21.00	22.00	23.00
0	0.0000	0.0000	0.0000	0.0000	0.0000	0.0000	0.0000	0.0000	0.0000	0.0000	0.0000
1	0.0000	0.0000	0.0000	0.0000	0.0000	0.0000	0.0000	0.0000	0.0000	0.0000	0.0000
2	0.0002	0.0001	0.0000	0.0000	0.0000	0.0000	0.0000	0.0000	0.0000	0.0000	0.0000
3	0.0008	0.0004	0.0002	0.0001	0.0000	0.0000	0.0000	0.0000	0.0000	0.0000	0.0000
4	0.0027	0.0013	0.0006	0.0003	0.0001	0.0001	0.0000	0.0000	0.0000	0.0000	0.0000
5	0.0070	0.0037	0.0019	0.0010	0.0005	0.0002	0.0001	0.0001	0.0000	0.0000	0.0000
6	0.0152	0.0087	0.0048	0.0026	0.0014	0.0007	0.0004	0.0002	0.0001	0.0000	0.0000
7	0.0281	0.0174	0.0104	0.0060	0.0034	0.0019	0.0010	0.0005	0.0003	0.0001	0.0001
8	0.0457	0.0304	0.0194	0.0120	0.0072	0.0042	0.0024	0.0013	0.0007	0.0004	0.0002
9	0.0661	0.0473	0.0324	0.0213	0.0135	0.0083	0.0050	0.0029	0.0017	0.0009	0.0005
10	0.0859	0.0663	0.0486	0.0341	0.0230	0.0150	0.0095	0.0058	0.0035	0.0020	0.0012
11	0.1015	0.0844	0.0663	0.0496	0.0355	0.0245	0.0164	0.0106	0.0067	0.0041	0.0024
12	0.1099	0.0984	0.0829	0.0661	0.0504	0.0368	0.0259	0.0176	0.0116	0.0075	0.0047
13	0.1099	0.1060	0.0956	0.0814	0.0658	0.0509	0.0378	0.0271	0.0188	0.0127	0.0083
14	0.1021	0.1060	0.1024	0.0930	0.0800	0.0655	0.0514	0.0387	0.0282	0.0199	0.0136
15	0.0885	0.0989	0.1024	0.0992	0.0906	0.0786	0.0650	0.0516	0.0395	0.0292	0.0209
16	0.0719	0.0866	0.0960	0.0992	0.0963	0.0884	0.0772	0.0646	0.0518	0.0401	0.0301
17	0.0550	0.0713	0.0847	0.0934	0.0963	0.0936	0.0863	0.0760	0.0640	0.0520	0.0407
18	0.0397	0.0554	0.0706	0.0830	0.0909	0.0936	0.0911	0.0844	0.0747	0.0635	0.0520
19	0.0272	0.0409	0.0557	0.0699	0.0814	0.0887	0.0911	0.0888	0.0826	0.0735	0.0629
20	0.0177	0.0286	0.0418	0.0559	0.0692	0.0798	0.0866	0.0888	0.0867	0.0809	0.0724
21	0.0109	0.0191	0.0299	0.0426	0.0560	0.0684	0.0783	0.0846	0.0867	0.0847	0.0793
22	0.0065	0.0121	0.0204	0.0310	0.0433	0.0560	0.0676	0.0769	0.0828	0.0847	0.0829
23	0.0037	0.0074	0.0133	0.0216	0.0320	0.0438	0.0559	0.0669	0.0756	0.0810	0.0829
24	0.0020	0.0043	0.0083	0.0144	0.0226	0.0328	0.0442	0.0557	0.0661	0.0743	0.0794
25	0.0010	0.0024	0.0050	0.0092	0.0154	0.0237	0.0336	0.0446	0.0555	0.0654	0.0731
26	0.0005	0.0013	0.0029	0.0057	0.0101	0.0164	0.0246	0.0343	0.0449	0.0553	0.0646
27	0.0002	0.0007	0.0016	0.0034	0.0063	0.0109	0.0173	0.0254	0.0349	0.0451	0.0551
28	0.0001	0.0003	0.0009	0.0019	0.0038	0.0070	0.0117	0.0181	0.0262	0.0354	0.0452
29	0.0001	0.0002	0.0004	0.0011	0.0023	0.0044	0.0077	0.0125	0.0190	0.0269	0.0359
30	0.0000	0.0001	0.0002	0.0006	0.0013	0.0026	0.0049	0.0083	0.0133	0.0197	0.0275
31	0.0000	0.0000	0.0001	0.0003	0.0007	0.0015	0.0030	0.0054	0.0090	0.0140	0.0204
32	0.0000	0.0000	0.0001	0.0001	0.0004	0.0009	0.0018	0.0034	0.0059	0.0096	0.0147
33	0.0000	0.0000	0.0000	0.0001	0.0002	0.0005	0.0010	0.0020	0.0038	0.0064	0.0102
34	0.0000	0.0000	0.0000	0.0000	0.0001	0.0002	0.0006	0.0012	0.0023	0.0041	0.0069
35	0.0000	0.0000	0.0000	0.0000	0.0000	0.0001	0.0003	0.0007	0.0014	0.0026	0.0045
36	0.0000	0.0000	0.0000	0.0000	0.0000	0.0001	0.0002	0.0004	0.0008	0.0016	0.0029
37	0.0000	0.0000	0.0000	0.0000	0.0000	0.0000	0.0001	0.0002	0.0005	0.0009	0.0018
38	0.0000	0.0000	0.0000	0.0000	0.0000	0.0000	0.0000	0.0001	0.0003	0.0005	0.0011
39	0.0000	0.0000	0.0000	0.0000	0.0000	0.0000	0.0000	0.0001	0.0001	0.0003	0.0006
40	0.0000	0.0000	0.0000	0.0000	0.0000	0.0000	0.0000	0.0000	0.0001	0.0002	0.0004
41	0.0000	0.0000	0.0000	0.0000	0.0000	0.0000	0.0000	0.0000	0.0000	0.0001	0.0002
42	0.0000	0.0000	0.0000	0.0000	0.0000	0.0000	0.0000	0.0000	0.0000	0.0000	0.0001
43	0.0000	0.0000	0.0000	0.0000	0.0000	0.0000	0.0000	0.0000	0.0000	0.0000	0.0001

G   Critical Values of χ^2

Numerical entries represent the value of χ_α^2.

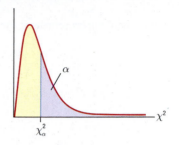

Area to the Right of the Critical Value

df	$\chi_{0.995}^2$	$\chi_{0.990}^2$	$\chi_{0.975}^2$	$\chi_{0.950}^2$	$\chi_{0.900}^2$	$\chi_{0.100}^2$	$\chi_{0.050}^2$	$\chi_{0.025}^2$	$\chi_{0.010}^2$	$\chi_{0.005}^2$
1	0.000	0.000	0.001	0.004	0.016	2.706	3.841	5.024	6.635	7.879
2	0.010	0.020	0.051	0.103	0.211	4.605	5.991	7.378	9.210	10.597
3	0.072	0.115	0.216	0.352	0.584	6.251	7.815	9.348	11.345	12.838
4	0.207	0.297	0.484	0.711	1.064	7.779	9.488	11.143	13.277	14.860
5	0.412	0.554	0.831	1.145	1.610	9.236	11.070	12.833	15.086	16.750
6	0.676	0.872	1.237	1.635	2.204	10.645	12.592	14.449	16.812	18.548
7	0.989	1.239	1.690	2.167	2.833	12.017	14.067	16.013	18.475	20.278
8	1.344	1.646	2.180	2.733	3.490	13.362	15.507	17.535	20.090	21.955
9	1.735	2.088	2.700	3.325	4.168	14.684	16.919	19.023	21.666	23.589
10	2.156	2.558	3.247	3.940	4.865	15.987	18.307	20.483	23.209	25.188
11	2.603	3.053	3.816	4.575	5.578	17.275	19.675	21.920	24.725	26.757
12	3.074	3.571	4.404	5.226	6.304	18.549	21.026	23.337	26.217	28.300
13	3.565	4.107	5.009	5.892	7.042	19.812	22.362	24.736	27.688	29.819
14	4.075	4.660	5.629	6.571	7.790	21.064	23.685	26.119	29.141	31.319
15	4.601	5.229	6.262	7.261	8.547	22.307	24.996	27.488	30.578	32.801
16	5.142	5.812	6.908	7.962	9.312	23.542	26.296	28.845	32.000	34.267
17	5.697	6.408	7.564	8.672	10.085	24.769	27.587	30.191	33.409	35.718
18	6.265	7.015	8.231	9.390	10.865	25.989	28.869	31.526	34.805	37.156
19	6.844	7.633	8.907	10.117	11.651	27.204	30.144	32.852	36.191	38.582
20	7.434	8.260	9.591	10.851	12.443	28.412	31.410	34.170	37.566	39.997
21	8.034	8.897	10.283	11.591	13.240	29.615	32.671	35.479	38.932	41.401
22	8.643	9.542	10.982	12.338	14.041	30.813	33.924	36.781	40.289	42.796
23	9.260	10.196	11.689	13.091	14.848	32.007	35.172	38.076	41.638	44.181
24	9.886	10.856	12.401	13.848	15.659	33.196	36.415	39.364	42.980	45.559
25	10.520	11.524	13.120	14.611	16.473	34.382	37.652	40.646	44.314	46.928
26	11.160	12.198	13.844	15.379	17.292	35.563	38.885	41.923	45.642	48.290
27	11.808	12.879	14.573	16.151	18.114	36.741	40.113	43.195	46.963	49.645
28	12.461	13.565	15.308	16.928	18.939	37.916	41.337	44.461	48.278	50.993
29	13.121	14.256	16.047	17.708	19.768	39.087	42.557	45.722	49.588	52.336
30	13.787	14.953	16.791	18.493	20.599	40.256	43.773	46.979	50.892	53.672
40	20.707	22.164	24.433	26.509	29.051	51.805	55.758	59.342	63.691	66.766
50	27.991	29.707	32.357	34.764	37.689	63.167	67.505	71.420	76.154	79.490
60	35.534	37.485	40.482	43.188	46.459	74.397	79.082	83.298	88.379	91.952
70	43.275	45.442	48.758	51.739	55.329	85.527	90.531	95.023	100.425	104.215
80	51.172	53.540	57.153	60.391	64.278	96.578	101.879	106.629	112.329	116.321
90	59.196	61.754	65.647	69.126	73.291	107.565	113.145	118.136	124.116	128.299
100	67.328	70.065	74.222	77.929	82.358	118.498	124.342	129.561	135.807	140.169

H   Critical Values of the *F*-Distribution ($\alpha = 0.100$)

Numerical entries represent the value of F_α.

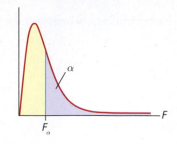

Numerator Degrees of Freedom

	1	2	3	4	5	6	7	8	9
1	39.8635	49.5000	53.5932	55.8330	57.2401	58.2044	58.9060	59.4390	59.8576
2	8.5263	9.0000	9.1618	9.2434	9.2926	9.3255	9.3491	9.3668	9.3805
3	5.5383	5.4624	5.3908	5.3426	5.3092	5.2847	5.2662	5.2517	5.2400
4	4.5448	4.3246	4.1909	4.1072	4.0506	4.0097	3.9790	3.9549	3.9357
5	4.0604	3.7797	3.6195	3.5202	3.4530	3.4045	3.3679	3.3393	3.3163
6	3.7759	3.4633	3.2888	3.1808	3.1075	3.0546	3.0145	2.9830	2.9577
7	3.5894	3.2574	3.0741	2.9605	2.8833	2.8274	2.7849	2.7516	2.7247
8	3.4579	3.1131	2.9238	2.8064	2.7264	2.6683	2.6241	2.5893	2.5612
9	3.3603	3.0065	2.8129	2.6927	2.6106	2.5509	2.5053	2.4694	2.4403
10	3.2850	2.9245	2.7277	2.6053	2.5216	2.4606	2.4140	2.3772	2.3473
11	3.2252	2.8595	2.6602	2.5362	2.4512	2.3891	2.3416	2.3040	2.2735
12	3.1765	2.8068	2.6055	2.4801	2.3940	2.3310	2.2828	2.2446	2.2135
13	3.1362	2.7632	2.5603	2.4337	2.3467	2.2830	2.2341	2.1953	2.1638
14	3.1022	2.7265	2.5222	2.3947	2.3069	2.2426	2.1931	2.1539	2.1220
15	3.0732	2.6952	2.4898	2.3614	2.2730	2.2081	2.1582	2.1185	2.0862
16	3.0481	2.6682	2.4618	2.3327	2.2438	2.1783	2.1280	2.0880	2.0553
17	3.0262	2.6446	2.4374	2.3077	2.2183	2.1524	2.1017	2.0613	2.0284
18	3.0070	2.6239	2.4160	2.2858	2.1958	2.1296	2.0785	2.0379	2.0047
19	2.9899	2.6056	2.3970	2.2663	2.1760	2.1094	2.0580	2.0171	1.9836
20	2.9747	2.5893	2.3801	2.2489	2.1582	2.0913	2.0397	1.9985	1.9649
21	2.9610	2.5746	2.3649	2.2333	2.1423	2.0751	2.0233	1.9819	1.9480
22	2.9486	2.5613	2.3512	2.2193	2.1279	2.0605	2.0084	1.9668	1.9327
23	2.9374	2.5493	2.3387	2.2065	2.1149	2.0472	1.9949	1.9531	1.9189
24	2.9271	2.5383	2.3274	2.1949	2.1030	2.0351	1.9826	1.9407	1.9063
25	2.9177	2.5283	2.3170	2.1842	2.0922	2.0241	1.9714	1.9292	1.8947
26	2.9091	2.5191	2.3075	2.1745	2.0822	2.0139	1.9610	1.9188	1.8841
27	2.9012	2.5106	2.2987	2.1655	2.0730	2.0045	1.9515	1.9091	1.8743
28	2.8938	2.5028	2.2906	2.1571	2.0645	1.9959	1.9427	1.9001	1.8652
29	2.8870	2.4955	2.2831	2.1494	2.0566	1.9878	1.9345	1.8918	1.8568
30	2.8807	2.4887	2.2761	2.1422	2.0492	1.9803	1.9269	1.8841	1.8490
40	2.8354	2.4404	2.2261	2.0909	1.9968	1.9269	1.8725	1.8289	1.7929
60	2.7911	2.3933	2.1774	2.0410	1.9457	1.8747	1.8194	1.7748	1.7380
120	2.7478	2.3473	2.1300	1.9923	1.8959	1.8238	1.7675	1.7220	1.6842
∞	2.7055	2.3026	2.0838	1.9449	1.8473	1.7741	1.7167	1.6702	1.6315

Denominator Degrees of Freedom

H   Critical Values of the *F*-Distribution ($\alpha = 0.100$) (cont.)

Numerator Degrees of Freedom

	10	11	12	13	14	15	16	17	18
1	60.1950	60.4727	60.7052	60.9028	61.0727	61.2203	61.3499	61.4644	61.5664
2	9.3916	9.4006	9.4081	9.4145	9.4200	9.4247	9.4289	9.4325	9.4358
3	5.2304	5.2224	5.2156	5.2098	5.2047	5.2003	5.1964	5.1929	5.1898
4	3.9199	3.9067	3.8955	3.8859	3.8776	3.8704	3.8639	3.8582	3.8531
5	3.2974	3.2816	3.2682	3.2567	3.2468	3.2380	3.2303	3.2234	3.2172
6	2.9369	2.9195	2.9047	2.8920	2.8809	2.8712	2.8626	2.8550	2.8481
7	2.7025	2.6839	2.6681	2.6545	2.6426	2.6322	2.6230	2.6148	2.6074
8	2.5380	2.5186	2.5020	2.4876	2.4752	2.4642	2.4545	2.4458	2.4380
9	2.4163	2.3961	2.3789	2.3640	2.3510	2.3396	2.3295	2.3205	2.3123
10	2.3226	2.3018	2.2841	2.2687	2.2553	2.2435	2.2330	2.2237	2.2153
11	2.2482	2.2269	2.2087	2.1930	2.1792	2.1671	2.1563	2.1467	2.1380
12	2.1878	2.1660	2.1474	2.1313	2.1173	2.1049	2.0938	2.0839	2.0750
13	2.1376	2.1155	2.0966	2.0802	2.0658	2.0532	2.0419	2.0318	2.0227
14	2.0954	2.0729	2.0537	2.0370	2.0224	2.0095	1.9981	1.9878	1.9785
15	2.0593	2.0366	2.0171	2.0001	1.9853	1.9722	1.9605	1.9501	1.9407
16	2.0281	2.0051	1.9854	1.9682	1.9532	1.9399	1.9281	1.9175	1.9079
17	2.0009	1.9777	1.9577	1.9404	1.9252	1.9117	1.8997	1.8889	1.8792
18	1.9770	1.9535	1.9333	1.9158	1.9004	1.8868	1.8747	1.8638	1.8539
19	1.9557	1.9321	1.9117	1.8940	1.8785	1.8647	1.8524	1.8414	1.8314
20	1.9367	1.9129	1.8924	1.8745	1.8588	1.8449	1.8325	1.8214	1.8113
21	1.9197	1.8956	1.8750	1.8570	1.8412	1.8271	1.8146	1.8034	1.7932
22	1.9043	1.8801	1.8593	1.8411	1.8252	1.8111	1.7984	1.7871	1.7768
23	1.8903	1.8659	1.8450	1.8267	1.8107	1.7964	1.7837	1.7723	1.7619
24	1.8775	1.8530	1.8319	1.8136	1.7974	1.7831	1.7703	1.7587	1.7483
25	1.8658	1.8412	1.8200	1.8015	1.7853	1.7708	1.7579	1.7463	1.7358
26	1.8550	1.8303	1.8090	1.7904	1.7741	1.7596	1.7466	1.7349	1.7243
27	1.8451	1.8203	1.7989	1.7802	1.7638	1.7492	1.7361	1.7243	1.7137
28	1.8359	1.8110	1.7895	1.7708	1.7542	1.7395	1.7264	1.7146	1.7039
29	1.8274	1.8024	1.7808	1.7620	1.7454	1.7306	1.7174	1.7055	1.6947
30	1.8195	1.7944	1.7727	1.7538	1.7371	1.7223	1.7090	1.6970	1.6862
40	1.7627	1.7369	1.7146	1.6950	1.6778	1.6624	1.6486	1.6362	1.6249
60	1.7070	1.6805	1.6574	1.6372	1.6193	1.6034	1.5890	1.5760	1.5642
120	1.6524	1.6250	1.6012	1.5803	1.5617	1.5450	1.5300	1.5164	1.5039
∞	1.5987	1.5705	1.5458	1.5240	1.5046	1.4871	1.4714	1.4570	1.4439

Denominator Degrees of Freedom

H   Critical Values of the *F*-Distribution (α = 0.100) (cont.)

Numerator Degrees of Freedom

	19	20	24	30	40	60	120
1	61.6579	61.7403	62.0020	62.2650	62.5291	62.7943	63.0606
2	9.4387	9.4413	9.4496	9.4579	9.4662	9.4746	9.4829
3	5.1870	5.1845	5.1764	5.1681	5.1597	5.1512	5.1425
4	3.8485	3.8443	3.8310	3.8174	3.8036	3.7896	3.7753
5	3.2117	3.2067	3.1905	3.1741	3.1573	3.1402	3.1228
6	2.8419	2.8363	2.8183	2.8000	2.7812	2.7620	2.7423
7	2.6008	2.5947	2.5753	2.5555	2.5351	2.5142	2.4928
8	2.4310	2.4246	2.4041	2.3830	2.3614	2.3391	2.3162
9	2.3050	2.2983	2.2768	2.2547	2.2320	2.2085	2.1843
10	2.2077	2.2007	2.1784	2.1554	2.1317	2.1072	2.0818
11	2.1302	2.1230	2.1000	2.0762	2.0516	2.0261	1.9997
12	2.0670	2.0597	2.0360	2.0115	1.9861	1.9597	1.9323
13	2.0145	2.0070	1.9827	1.9576	1.9315	1.9043	1.8759
14	1.9701	1.9625	1.9377	1.9119	1.8852	1.8572	1.8280
15	1.9321	1.9243	1.8990	1.8728	1.8454	1.8168	1.7867
16	1.8992	1.8913	1.8656	1.8388	1.8108	1.7816	1.7507
17	1.8704	1.8624	1.8362	1.8090	1.7805	1.7506	1.7191
18	1.8450	1.8368	1.8103	1.7827	1.7537	1.7232	1.6910
19	1.8224	1.8142	1.7873	1.7592	1.7298	1.6988	1.6659
20	1.8022	1.7938	1.7667	1.7382	1.7083	1.6768	1.6433
21	1.7840	1.7756	1.7481	1.7193	1.6890	1.6569	1.6228
22	1.7675	1.7590	1.7312	1.7021	1.6714	1.6389	1.6041
23	1.7525	1.7439	1.7159	1.6864	1.6554	1.6224	1.5871
24	1.7388	1.7302	1.7019	1.6721	1.6407	1.6073	1.5715
25	1.7263	1.7175	1.6890	1.6589	1.6272	1.5934	1.5570
26	1.7147	1.7059	1.6771	1.6468	1.6147	1.5805	1.5437
27	1.7040	1.6951	1.6662	1.6356	1.6032	1.5686	1.5313
28	1.6941	1.6852	1.6560	1.6252	1.5925	1.5575	1.5198
29	1.6849	1.6759	1.6465	1.6155	1.5825	1.5472	1.5090
30	1.6763	1.6673	1.6377	1.6065	1.5732	1.5376	1.4989
40	1.6146	1.6052	1.5741	1.5411	1.5056	1.4672	1.4248
60	1.5534	1.5435	1.5107	1.4755	1.4373	1.3952	1.3476
120	1.4926	1.4821	1.4472	1.4094	1.3676	1.3203	1.2646
∞	1.4318	1.4206	1.3832	1.3419	1.2951	1.2400	1.1686

Denominator Degrees of Freedom

Critical Values of the *F*-Distribution (α = 0.050)

Numerator Degrees of Freedom

	1	2	3	4	5	6	7	8	9
1	161.4476	199.5000	215.7073	224.5832	230.1619	233.9860	236.7684	238.8827	240.5433
2	18.5128	19.0000	19.1643	19.2468	19.2964	19.3295	19.3532	19.3710	19.3848
3	10.1280	9.5521	9.2766	9.1172	9.0135	8.9406	8.8867	8.8452	8.8123
4	7.7086	6.9443	6.5914	6.3882	6.2561	6.1631	6.0942	6.0410	5.9988
5	6.6079	5.7861	5.4095	5.1922	5.0503	4.9503	4.8759	4.8183	4.7725
6	5.9874	5.1433	4.7571	4.5337	4.3874	4.2839	4.2067	4.1468	4.0990
7	5.5914	4.7374	4.3468	4.1203	3.9715	3.8660	3.7870	3.7257	3.6767
8	5.3177	4.4590	4.0662	3.8379	3.6875	3.5806	3.5005	3.4381	3.3881
9	5.1174	4.2565	3.8625	3.6331	3.4817	3.3738	3.2927	3.2296	3.1789
10	4.9646	4.1028	3.7083	3.4780	3.3258	3.2172	3.1355	3.0717	3.0204
11	4.8443	3.9823	3.5874	3.3567	3.2039	3.0946	3.0123	2.9480	2.8962
12	4.7472	3.8853	3.4903	3.2592	3.1059	2.9961	2.9134	2.8486	2.7964
13	4.6672	3.8056	3.4105	3.1791	3.0254	2.9153	2.8321	2.7669	2.7144
14	4.6001	3.7389	3.3439	3.1122	2.9582	2.8477	2.7642	2.6987	2.6458
15	4.5431	3.6823	3.2874	3.0556	2.9013	2.7905	2.7066	2.6408	2.5876
16	4.4940	3.6337	3.2389	3.0069	2.8524	2.7413	2.6572	2.5911	2.5377
17	4.4513	3.5915	3.1968	2.9647	2.8100	2.6987	2.6143	2.5480	2.4943
18	4.4139	3.5546	3.1599	2.9277	2.7729	2.6613	2.5767	2.5102	2.4563
19	4.3807	3.5219	3.1274	2.8951	2.7401	2.6283	2.5435	2.4768	2.4227
20	4.3512	3.4928	3.0984	2.8661	2.7109	2.5990	2.5140	2.4471	2.3928
21	4.3248	3.4668	3.0725	2.8401	2.6848	2.5727	2.4876	2.4205	2.3660
22	4.3009	3.4434	3.0491	2.8167	2.6613	2.5491	2.4638	2.3965	2.3419
23	4.2793	3.4221	3.0280	2.7955	2.6400	2.5277	2.4422	2.3748	2.3201
24	4.2597	3.4028	3.0088	2.7763	2.6207	2.5082	2.4226	2.3551	2.3002
25	4.2417	3.3852	2.9912	2.7587	2.6030	2.4904	2.4047	2.3371	2.2821
26	4.2252	3.3690	2.9752	2.7426	2.5868	2.4741	2.3883	2.3205	2.2655
27	4.2100	3.3541	2.9604	2.7278	2.5719	2.4591	2.3732	2.3053	2.2501
28	4.1960	3.3404	2.9467	2.7141	2.5581	2.4453	2.3593	2.2913	2.2360
29	4.1830	3.3277	2.9340	2.7014	2.5454	2.4324	2.3463	2.2783	2.2229
30	4.1709	3.3158	2.9223	2.6896	2.5336	2.4205	2.3343	2.2662	2.2107
40	4.0847	3.2317	2.8387	2.6060	2.4495	2.3359	2.2490	2.1802	2.1240
60	4.0012	3.1504	2.7581	2.5252	2.3683	2.2541	2.1665	2.0970	2.0401
120	3.9201	3.0718	2.6802	2.4472	2.2899	2.1750	2.0868	2.0164	1.9588
∞	3.8415	2.9957	2.6049	2.3719	2.2141	2.0986	2.0096	1.9384	1.8799

Denominator Degrees of Freedom

H   Critical Values of the *F*-Distribution ($\alpha = 0.050$) (cont.)

Numerator Degrees of Freedom

	10	11	12	13	14	15	16	17	18
1	241.8817	242.9835	243.9060	244.6898	245.3640	245.9499	246.4639	246.9184	247.3232
2	19.3959	19.4050	19.4125	19.4189	19.4244	19.4291	19.4333	19.4370	19.4402
3	8.7855	8.7633	8.7446	8.7287	8.7149	8.7029	8.6923	8.6829	8.6745
4	5.9644	5.9358	5.9117	5.8911	5.8733	5.8578	5.8441	5.8320	5.8211
5	4.7351	4.7040	4.6777	4.6552	4.6358	4.6188	4.6038	4.5904	4.5785
6	4.0600	4.0274	3.9999	3.9764	3.9559	3.9381	3.9223	3.9083	3.8957
7	3.6365	3.6030	3.5747	3.5503	3.5292	3.5107	3.4944	3.4799	3.4669
8	3.3472	3.3130	3.2839	3.2590	3.2374	3.2184	3.2016	3.1867	3.1733
9	3.1373	3.1025	3.0729	3.0475	3.0255	3.0061	2.9890	2.9737	2.9600
10	2.9782	2.9430	2.9130	2.8872	2.8647	2.8450	2.8276	2.8120	2.7980
11	2.8536	2.8179	2.7876	2.7614	2.7386	2.7186	2.7009	2.6851	2.6709
12	2.7534	2.7173	2.6866	2.6602	2.6371	2.6169	2.5989	2.5828	2.5684
13	2.6710	2.6347	2.6037	2.5769	2.5536	2.5331	2.5149	2.4987	2.4841
14	2.6022	2.5655	2.5342	2.5073	2.4837	2.4630	2.4446	2.4282	2.4134
15	2.5437	2.5068	2.4753	2.4481	2.4244	2.4034	2.3849	2.3683	2.3533
16	2.4935	2.4564	2.4247	2.3973	2.3733	2.3522	2.3335	2.3167	2.3016
17	2.4499	2.4126	2.3807	2.3531	2.3290	2.3077	2.2888	2.2719	2.2567
18	2.4117	2.3742	2.3421	2.3143	2.2900	2.2686	2.2496	2.2325	2.2172
19	2.3779	2.3402	2.3080	2.2800	2.2556	2.2341	2.2149	2.1977	2.1823
20	2.3479	2.3100	2.2776	2.2495	2.2250	2.2033	2.1840	2.1667	2.1511
21	2.3210	2.2829	2.2504	2.2222	2.1975	2.1757	2.1563	2.1389	2.1232
22	2.2967	2.2585	2.2258	2.1975	2.1727	2.1508	2.1313	2.1138	2.0980
23	2.2747	2.2364	2.2036	2.1752	2.1502	2.1282	2.1086	2.0910	2.0751
24	2.2547	2.2163	2.1834	2.1548	2.1298	2.1077	2.0880	2.0703	2.0543
25	2.2365	2.1979	2.1649	2.1362	2.1111	2.0889	2.0691	2.0513	2.0353
26	2.2197	2.1811	2.1479	2.1192	2.0939	2.0716	2.0518	2.0339	2.0178
27	2.2043	2.1655	2.1323	2.1035	2.0781	2.0558	2.0358	2.0179	2.0017
28	2.1900	2.1512	2.1179	2.0889	2.0635	2.0411	2.0210	2.0030	1.9868
29	2.1768	2.1379	2.1045	2.0755	2.0500	2.0275	2.0073	1.9893	1.9730
30	2.1646	2.1256	2.0921	2.0630	2.0374	2.0148	1.9946	1.9765	1.9601
40	2.0772	2.0376	2.0035	1.9738	1.9476	1.9245	1.9037	1.8851	1.8682
60	1.9926	1.9522	1.9174	1.8870	1.8602	1.8364	1.8151	1.7959	1.7784
120	1.9105	1.8693	1.8337	1.8026	1.7750	1.7505	1.7285	1.7085	1.6904
∞	1.8307	1.7887	1.7522	1.7202	1.6918	1.6664	1.6435	1.6228	1.6039

Denominator Degrees of Freedom

H   Critical Values of the *F*-Distribution (α = 0.050) (cont.)

Numerator Degrees of Freedom

	19	20	24	30	40	60	120
1	247.6861	248.0131	249.0518	250.0951	251.1432	252.1957	253.2529
2	19.4431	19.4458	19.4541	19.4624	19.4707	19.4791	19.4874
3	8.6670	8.6602	8.6385	8.6166	8.5944	8.5720	8.5494
4	5.8114	5.8025	5.7744	5.7459	5.7170	5.6877	5.6581
5	4.5678	4.5581	4.5272	4.4957	4.4638	4.4314	4.3985
6	3.8844	3.8742	3.8415	3.8082	3.7743	3.7398	3.7047
7	3.4551	3.4445	3.4105	3.3758	3.3404	3.3043	3.2674
8	3.1613	3.1503	3.1152	3.0794	3.0428	3.0053	2.9669
9	2.9477	2.9365	2.9005	2.8637	2.8259	2.7872	2.7475
10	2.7854	2.7740	2.7372	2.6996	2.6609	2.6211	2.5801
11	2.6581	2.6464	2.6090	2.5705	2.5309	2.4901	2.4480
12	2.5554	2.5436	2.5055	2.4663	2.4259	2.3842	2.3410
13	2.4709	2.4589	2.4202	2.3803	2.3392	2.2966	2.2524
14	2.4000	2.3879	2.3487	2.3082	2.2664	2.2229	2.1778
15	2.3398	2.3275	2.2878	2.2468	2.2043	2.1601	2.1141
16	2.2880	2.2756	2.2354	2.1938	2.1507	2.1058	2.0589
17	2.2429	2.2304	2.1898	2.1477	2.1040	2.0584	2.0107
18	2.2033	2.1906	2.1497	2.1071	2.0629	2.0166	1.9681
19	2.1683	2.1555	2.1141	2.0712	2.0264	1.9795	1.9302
20	2.1370	2.1242	2.0825	2.0391	1.9938	1.9464	1.8963
21	2.1090	2.0960	2.0540	2.0102	1.9645	1.9165	1.8657
22	2.0837	2.0707	2.0283	1.9842	1.9380	1.8894	1.8380
23	2.0608	2.0476	2.0050	1.9605	1.9139	1.8648	1.8128
24	2.0399	2.0267	1.9838	1.9390	1.8920	1.8424	1.7896
25	2.0207	2.0075	1.9643	1.9192	1.8718	1.8217	1.7684
26	2.0032	1.9898	1.9464	1.9010	1.8533	1.8027	1.7488
27	1.9870	1.9736	1.9299	1.8842	1.8361	1.7851	1.7306
28	1.9720	1.9586	1.9147	1.8687	1.8203	1.7689	1.7138
29	1.9581	1.9446	1.9005	1.8543	1.8055	1.7537	1.6981
30	1.9452	1.9317	1.8874	1.8409	1.7918	1.7396	1.6835
40	1.8529	1.8389	1.7929	1.7444	1.6928	1.6373	1.5766
60	1.7625	1.7480	1.7001	1.6491	1.5943	1.5343	1.4673
120	1.6739	1.6587	1.6084	1.5543	1.4952	1.4290	1.3519
∞	1.5865	1.5705	1.5173	1.4591	1.3940	1.3180	1.2214

Denominator Degrees of Freedom

H | Critical Values of the *F*-Distribution ($\alpha = 0.025$)

Numerator Degrees of Freedom

	1	2	3	4	5	6	7	8	9
1	647.7890	799.5000	864.1630	899.5833	921.8479	937.1111	948.2169	956.6562	963.2846
2	38.5063	39.0000	39.1655	39.2484	39.2982	39.3315	39.3552	39.3730	39.3869
3	17.4434	16.0441	15.4392	15.1010	14.8848	14.7347	14.6244	14.5399	14.4731
4	12.2179	10.6491	9.9792	9.6045	9.3645	9.1973	9.0741	8.9796	8.9047
5	10.0070	8.4336	7.7636	7.3879	7.1464	6.9777	6.8531	6.7572	6.6811
6	8.8131	7.2599	6.5988	6.2272	5.9876	5.8198	5.6955	5.5996	5.5234
7	8.0727	6.5415	5.8898	5.5226	5.2852	5.1186	4.9949	4.8993	4.8232
8	7.5709	6.0595	5.4160	5.0526	4.8173	4.6517	4.5286	4.4333	4.3572
9	7.2093	5.7147	5.0781	4.7181	4.4844	4.3197	4.1970	4.1020	4.0260
10	6.9367	5.4564	4.8256	4.4683	4.2361	4.0721	3.9498	3.8549	3.7790
11	6.7241	5.2559	4.6300	4.2751	4.0440	3.8807	3.7586	3.6638	3.5879
12	6.5538	5.0959	4.4742	4.1212	3.8911	3.7283	3.6065	3.5118	3.4358
13	6.4143	4.9653	4.3472	3.9959	3.7667	3.6043	3.4827	3.3880	3.3120
14	6.2979	4.8567	4.2417	3.8919	3.6634	3.5014	3.3799	3.2853	3.2093
15	6.1995	4.7650	4.1528	3.8043	3.5764	3.4147	3.2934	3.1987	3.1227
16	6.1151	4.6867	4.0768	3.7294	3.5021	3.3406	3.2194	3.1248	3.0488
17	6.0420	4.6189	4.0112	3.6648	3.4379	3.2767	3.1556	3.0610	2.9849
18	5.9781	4.5597	3.9539	3.6083	3.3820	3.2209	3.0999	3.0053	2.9291
19	5.9216	4.5075	3.9034	3.5587	3.3327	3.1718	3.0509	2.9563	2.8801
20	5.8715	4.4613	3.8587	3.5147	3.2891	3.1283	3.0074	2.9128	2.8365
21	5.8266	4.4199	3.8188	3.4754	3.2501	3.0895	2.9686	2.8740	2.7977
22	5.7863	4.3828	3.7829	3.4401	3.2151	3.0546	2.9338	2.8392	2.7628
23	5.7498	4.3492	3.7505	3.4083	3.1835	3.0232	2.9023	2.8077	2.7313
24	5.7166	4.3187	3.7211	3.3794	3.1548	2.9946	2.8738	2.7791	2.7027
25	5.6864	4.2909	3.6943	3.3530	3.1287	2.9685	2.8478	2.7531	2.6766
26	5.6586	4.2655	3.6697	3.3289	3.1048	2.9447	2.8240	2.7293	2.6528
27	5.6331	4.2421	3.6472	3.3067	3.0828	2.9228	2.8021	2.7074	2.6309
28	5.6096	4.2205	3.6264	3.2863	3.0626	2.9027	2.7820	2.6872	2.6106
29	5.5878	4.2006	3.6072	3.2674	3.0438	2.8840	2.7633	2.6686	2.5919
30	5.5675	4.1821	3.5894	3.2499	3.0265	2.8667	2.7460	2.6513	2.5746
40	5.4239	4.0510	3.4633	3.1261	2.9037	2.7444	2.6238	2.5289	2.4519
60	5.2856	3.9253	3.3425	3.0077	2.7863	2.6274	2.5068	2.4117	2.3344
120	5.1523	3.8046	3.2269	2.8943	2.6740	2.5154	2.3948	2.2994	2.2217
∞	5.0239	3.6889	3.1161	2.7858	2.5665	2.4082	2.2876	2.1918	2.1137

Denominator Degrees of Freedom

H   Critical Values of the *F*-Distribution (α = 0.025) (cont.)

Numerator Degrees of Freedom

	10	11	12	13	14	15	16	17	18
1	968.6274	973.0252	976.7079	979.8368	982.5278	984.8668	986.9187	988.7331	990.3490
2	39.3980	39.4071	39.4146	39.4210	39.4265	39.4313	39.4354	39.4391	39.4424
3	14.4189	14.3742	14.3366	14.3045	14.2768	14.2527	14.2315	14.2127	14.1960
4	8.8439	8.7935	8.7512	8.7150	8.6838	8.6565	8.6326	8.6113	8.5924
5	6.6192	6.5678	6.5245	6.4876	6.4556	6.4277	6.4032	6.3814	6.3619
6	5.4613	5.4098	5.3662	5.3290	5.2968	5.2687	5.2439	5.2218	5.2021
7	4.7611	4.7095	4.6658	4.6285	4.5961	4.5678	4.5428	4.5206	4.5008
8	4.2951	4.2434	4.1997	4.1622	4.1297	4.1012	4.0761	4.0538	4.0338
9	3.9639	3.9121	3.8682	3.8306	3.7980	3.7694	3.7441	3.7216	3.7015
10	3.7168	3.6649	3.6209	3.5832	3.5504	3.5217	3.4963	3.4737	3.4534
11	3.5257	3.4737	3.4296	3.3917	3.3588	3.3299	3.3044	3.2816	3.2612
12	3.3736	3.3215	3.2773	3.2393	3.2062	3.1772	3.1515	3.1286	3.1081
13	3.2497	3.1975	3.1532	3.1150	3.0819	3.0527	3.0269	3.0039	2.9832
14	3.1469	3.0946	3.0502	3.0119	2.9786	2.9493	2.9234	2.9003	2.8795
15	3.0602	3.0078	2.9633	2.9249	2.8915	2.8621	2.8360	2.8128	2.7919
16	2.9862	2.9337	2.8890	2.8506	2.8170	2.7875	2.7614	2.7380	2.7170
17	2.9222	2.8696	2.8249	2.7863	2.7526	2.7230	2.6968	2.6733	2.6522
18	2.8664	2.8137	2.7689	2.7302	2.6964	2.6667	2.6404	2.6168	2.5956
19	2.8172	2.7645	2.7196	2.6808	2.6469	2.6171	2.5907	2.5670	2.5457
20	2.7737	2.7209	2.6758	2.6369	2.6030	2.5731	2.5465	2.5228	2.5014
21	2.7348	2.6819	2.6368	2.5978	2.5638	2.5338	2.5071	2.4833	2.4618
22	2.6998	2.6469	2.6017	2.5626	2.5285	2.4984	2.4717	2.4478	2.4262
23	2.6682	2.6152	2.5699	2.5308	2.4966	2.4665	2.4396	2.4157	2.3940
24	2.6396	2.5865	2.5411	2.5019	2.4677	2.4374	2.4105	2.3865	2.3648
25	2.6135	2.5603	2.5149	2.4756	2.4413	2.4110	2.3840	2.3599	2.3381
26	2.5896	2.5363	2.4908	2.4515	2.4171	2.3867	2.3597	2.3355	2.3137
27	2.5676	2.5143	2.4688	2.4293	2.3949	2.3644	2.3373	2.3131	2.2912
28	2.5473	2.4940	2.4484	2.4089	2.3743	2.3438	2.3167	2.2924	2.2704
29	2.5286	2.4752	2.4295	2.3900	2.3554	2.3248	2.2976	2.2732	2.2512
30	2.5112	2.4577	2.4120	2.3724	2.3378	2.3072	2.2799	2.2554	2.2334
40	2.3882	2.3343	2.2882	2.2481	2.2130	2.1819	2.1542	2.1293	2.1068
60	2.2702	2.2159	2.1692	2.1286	2.0929	2.0613	2.0330	2.0076	1.9846
120	2.1570	2.1021	2.0548	2.0136	1.9773	1.9450	1.9161	1.8900	1.8663
∞	2.0483	1.9927	1.9447	1.9028	1.8657	1.8326	1.8028	1.7760	1.7515

Denominator Degrees of Freedom

Critical Values of the *F*-Distribution (α = 0.025) (cont.)

Numerator Degrees of Freedom

	19	20	24	30	40	60	120
1	991.7973	993.1028	997.2492	1001.4144	1005.5981	1009.8001	1014.0202
2	39.4453	39.4479	39.4562	39.4646	39.4729	39.4812	39.4896
3	14.1810	14.1674	14.1241	14.0805	14.0365	13.9921	13.9473
4	8.5753	8.5599	8.5109	8.4613	8.4111	8.3604	8.3092
5	6.3444	6.3286	6.2780	6.2269	6.1750	6.1225	6.0693
6	5.1844	5.1684	5.1172	5.0652	5.0125	4.9589	4.9044
7	4.4829	4.4667	4.4150	4.3624	4.3089	4.2544	4.1989
8	4.0158	3.9995	3.9472	3.8940	3.8398	3.7844	3.7279
9	3.6833	3.6669	3.6142	3.5604	3.5055	3.4493	3.3918
10	3.4351	3.4185	3.3654	3.3110	3.2554	3.1984	3.1399
11	3.2428	3.2261	3.1725	3.1176	3.0613	3.0035	2.9441
12	3.0896	3.0728	3.0187	2.9633	2.9063	2.8478	2.7874
13	2.9646	2.9477	2.8932	2.8372	2.7797	2.7204	2.6590
14	2.8607	2.8437	2.7888	2.7324	2.6742	2.6142	2.5519
15	2.7730	2.7559	2.7006	2.6437	2.5850	2.5242	2.4611
16	2.6980	2.6808	2.6252	2.5678	2.5085	2.4471	2.3831
17	2.6331	2.6158	2.5598	2.5020	2.4422	2.3801	2.3153
18	2.5764	2.5590	2.5027	2.4445	2.3842	2.3214	2.2558
19	2.5265	2.5089	2.4523	2.3937	2.3329	2.2696	2.2032
20	2.4821	2.4645	2.4076	2.3486	2.2873	2.2234	2.1562
21	2.4424	2.4247	2.3675	2.3082	2.2465	2.1819	2.1141
22	2.4067	2.3890	2.3315	2.2718	2.2097	2.1446	2.0760
23	2.3745	2.3567	2.2989	2.2389	2.1763	2.1107	2.0415
24	2.3452	2.3273	2.2693	2.2090	2.1460	2.0799	2.0099
25	2.3184	2.3005	2.2422	2.1816	2.1183	2.0516	1.9811
26	2.2939	2.2759	2.2174	2.1565	2.0928	2.0257	1.9545
27	2.2713	2.2533	2.1946	2.1334	2.0693	2.0018	1.9299
28	2.2505	2.2324	2.1735	2.1121	2.0477	1.9797	1.9072
29	2.2313	2.2131	2.1540	2.0923	2.0276	1.9591	1.8861
30	2.2134	2.1952	2.1359	2.0739	2.0089	1.9400	1.8664
40	2.0864	2.0677	2.0069	1.9429	1.8752	1.8028	1.7242
60	1.9636	1.9445	1.8817	1.8152	1.7440	1.6668	1.5810
120	1.8447	1.8249	1.7597	1.6899	1.6141	1.5299	1.4327
∞	1.7291	1.7085	1.6402	1.5660	1.4836	1.3883	1.2685

Denominator Degrees of Freedom

H Critical Values of the *F*-Distribution ($\alpha = 0.010$)

Numerator Degrees of Freedom

	1	2	3	4	5	6	7	8	9
1	4052.1807	4999.5000	5403.3520	5624.5833	5763.6496	5858.9861	5928.3557	5981.0703	6022.4732
2	98.5025	99.0000	99.1662	99.2494	99.2993	99.3326	99.3564	99.3742	99.3881
3	34.1162	30.8165	29.4567	28.7099	28.2371	27.9107	27.6717	27.4892	27.3452
4	21.1977	18.0000	16.6944	15.9770	15.5219	15.2069	14.9758	14.7989	14.6591
5	16.2582	13.2739	12.0600	11.3919	10.9670	10.6723	10.4555	10.2893	10.1578
6	13.7450	10.9248	9.7795	9.1483	8.7459	8.4661	8.2600	8.1017	7.9761
7	12.2464	9.5466	8.4513	7.8466	7.4604	7.1914	6.9928	6.8400	6.7188
8	11.2586	8.6491	7.5910	7.0061	6.6318	6.3707	6.1776	6.0289	5.9106
9	10.5614	8.0215	6.9919	6.4221	6.0569	5.8018	5.6129	5.4671	5.3511
10	10.0443	7.5594	6.5523	5.9943	5.6363	5.3858	5.2001	5.0567	4.9424
11	9.6460	7.2057	6.2167	5.6683	5.3160	5.0692	4.8861	4.7445	4.6315
12	9.3302	6.9266	5.9525	5.4120	5.0643	4.8206	4.6395	4.4994	4.3875
13	9.0738	6.7010	5.7394	5.2053	4.8616	4.6204	4.4410	4.3021	4.1911
14	8.8616	6.5149	5.5639	5.0354	4.6950	4.4558	4.2779	4.1399	4.0297
15	8.6831	6.3589	5.4170	4.8932	4.5556	4.3183	4.1415	4.0045	3.8948
16	8.5310	6.2262	5.2922	4.7726	4.4374	4.2016	4.0259	3.8896	3.7804
17	8.3997	6.1121	5.1850	4.6690	4.3359	4.1015	3.9267	3.7910	3.6822
18	8.2854	6.0129	5.0919	4.5790	4.2479	4.0146	3.8406	3.7054	3.5971
19	8.1849	5.9259	5.0103	4.5003	4.1708	3.9386	3.7653	3.6305	3.5225
20	8.0960	5.8489	4.9382	4.4307	4.1027	3.8714	3.6987	3.5644	3.4567
21	8.0166	5.7804	4.8740	4.3688	4.0421	3.8117	3.6396	3.5056	3.3981
22	7.9454	5.7190	4.8166	4.3134	3.9880	3.7583	3.5867	3.4530	3.3458
23	7.8811	5.6637	4.7649	4.2636	3.9392	3.7102	3.5390	3.4057	3.2986
24	7.8229	5.6136	4.7181	4.2184	3.8951	3.6667	3.4959	3.3629	3.2560
25	7.7698	5.5680	4.6755	4.1774	3.8550	3.6272	3.4568	3.3239	3.2172
26	7.7213	5.5263	4.6366	4.1400	3.8183	3.5911	3.4210	3.2884	3.1818
27	7.6767	5.4881	4.6009	4.1056	3.7848	3.5580	3.3882	3.2558	3.1494
28	7.6356	5.4529	4.5681	4.0740	3.7539	3.5276	3.3581	3.2259	3.1195
29	7.5977	5.4204	4.5378	4.0449	3.7254	3.4995	3.3303	3.1982	3.0920
30	7.5625	5.3903	4.5097	4.0179	3.6990	3.4735	3.3045	3.1726	3.0665
40	7.3141	5.1785	4.3126	3.8283	3.5138	3.2910	3.1238	2.9930	2.8876
60	7.0771	4.9774	4.1259	3.6490	3.3389	3.1187	2.9530	2.8233	2.7185
120	6.8509	4.7865	3.9491	3.4795	3.1735	2.9559	2.7918	2.6629	2.5586
∞	6.6349	4.6052	3.7816	3.3192	3.0173	2.8020	2.6393	2.5113	2.4074

Denominator Degrees of Freedom

Critical Values of the *F*-Distribution (α = 0.010) (cont.)

Numerator Degrees of Freedom

	10	11	12	13	14	15	16	17	18
1	6055.8467	6083.3168	6106.3207	6125.8647	6142.6740	6157.2846	6170.1012	6181.4348	6191.5287
2	99.3992	99.4083	99.4159	99.4223	99.4278	99.4325	99.4367	99.4404	99.4436
3	27.2287	27.1326	27.0518	26.9831	26.9238	26.8722	26.8269	26.7867	26.7509
4	14.5459	14.4523	14.3736	14.3065	14.2486	14.1982	14.1539	14.1146	14.0795
5	10.0510	9.9626	9.8883	9.8248	9.7700	9.7222	9.6802	9.6429	9.6096
6	7.8741	7.7896	7.7183	7.6575	7.6049	7.5590	7.5186	7.4827	7.4507
7	6.6201	6.5382	6.4691	6.4100	6.3590	6.3143	6.2750	6.2401	6.2089
8	5.8143	5.7343	5.6667	5.6089	5.5589	5.5151	5.4766	5.4423	5.4116
9	5.2565	5.1779	5.1114	5.0545	5.0052	4.9621	4.9240	4.8902	4.8599
10	4.8491	4.7715	4.7059	4.6496	4.6008	4.5581	4.5204	4.4869	4.4569
11	4.5393	4.4624	4.3974	4.3416	4.2932	4.2509	4.2134	4.1801	4.1503
12	4.2961	4.2198	4.1553	4.0999	4.0518	4.0096	3.9724	3.9392	3.9095
13	4.1003	4.0245	3.9603	3.9052	3.8573	3.8154	3.7783	3.7452	3.7156
14	3.9394	3.8640	3.8001	3.7452	3.6975	3.6557	3.6187	3.5857	3.5561
15	3.8049	3.7299	3.6662	3.6115	3.5639	3.5222	3.4852	3.4523	3.4228
16	3.6909	3.6162	3.5527	3.4981	3.4506	3.4089	3.3720	3.3391	3.3096
17	3.5931	3.5185	3.4552	3.4007	3.3533	3.3117	3.2748	3.2419	3.2124
18	3.5082	3.4338	3.3706	3.3162	3.2689	3.2273	3.1904	3.1575	3.1280
19	3.4338	3.3596	3.2965	3.2422	3.1949	3.1533	3.1165	3.0836	3.0541
20	3.3682	3.2941	3.2311	3.1769	3.1296	3.0880	3.0512	3.0183	2.9887
21	3.3098	3.2359	3.1730	3.1187	3.0715	3.0300	2.9931	2.9602	2.9306
22	3.2576	3.1837	3.1209	3.0667	3.0195	2.9779	2.9411	2.9082	2.8786
23	3.2106	3.1368	3.0740	3.0199	2.9727	2.9311	2.8943	2.8613	2.8317
24	3.1681	3.0944	3.0316	2.9775	2.9303	2.8887	2.8519	2.8189	2.7892
25	3.1294	3.0558	2.9931	2.9389	2.8917	2.8502	2.8133	2.7803	2.7506
26	3.0941	3.0205	2.9578	2.9038	2.8566	2.8150	2.7781	2.7451	2.7153
27	3.0618	2.9882	2.9256	2.8715	2.8243	2.7827	2.7458	2.7127	2.6830
28	3.0320	2.9585	2.8959	2.8418	2.7946	2.7530	2.7160	2.6830	2.6532
29	3.0045	2.9311	2.8685	2.8144	2.7672	2.7256	2.6886	2.6555	2.6257
30	2.9791	2.9057	2.8431	2.7890	2.7418	2.7002	2.6632	2.6301	2.6003
40	2.8005	2.7274	2.6648	2.6107	2.5634	2.5216	2.4844	2.4511	2.4210
60	2.6318	2.5587	2.4961	2.4419	2.3943	2.3523	2.3148	2.2811	2.2507
120	2.4721	2.3990	2.3363	2.2818	2.2339	2.1915	2.1536	2.1194	2.0885
∞	2.3209	2.2477	2.1848	2.1299	2.0815	2.0385	2.0000	1.9652	1.9336

Denominator Degrees of Freedom

H   Critical Values of the *F*-Distribution ($\alpha = 0.010$) (cont.)

Numerator Degrees of Freedom

	19	20	24	30	40	60	120
1	6200.5756	6208.7302	6234.6309	6260.6486	6286.7821	6313.0301	6339.3913
2	99.4465	99.4492	99.4575	99.4658	99.4742	99.4825	99.4908
3	26.7188	26.6898	26.5975	26.5045	26.4108	26.3164	26.2211
4	14.0480	14.0196	13.9291	13.8377	13.7454	13.6522	13.5581
5	9.5797	9.5526	9.4665	9.3793	9.2912	9.2020	9.1118
6	7.4219	7.3958	7.3127	7.2285	7.1432	7.0567	6.9690
7	6.1808	6.1554	6.0743	5.9920	5.9084	5.8236	5.7373
8	5.3840	5.3591	5.2793	5.1981	5.1156	5.0316	4.9461
9	4.8327	4.8080	4.7290	4.6486	4.5666	4.4831	4.3978
10	4.4299	4.4054	4.3269	4.2469	4.1653	4.0819	3.9965
11	4.1234	4.0990	4.0209	3.9411	3.8596	3.7761	3.6904
12	3.8827	3.8584	3.7805	3.7008	3.6192	3.5355	3.4494
13	3.6888	3.6646	3.5868	3.5070	3.4253	3.3413	3.2548
14	3.5294	3.5052	3.4274	3.3476	3.2656	3.1813	3.0942
15	3.3961	3.3719	3.2940	3.2141	3.1319	3.0471	2.9595
16	3.2829	3.2587	3.1808	3.1007	3.0182	2.9330	2.8447
17	3.1857	3.1615	3.0835	3.0032	2.9205	2.8348	2.7459
18	3.1013	3.0771	2.9990	2.9185	2.8354	2.7493	2.6597
19	3.0274	3.0031	2.9249	2.8442	2.7608	2.6742	2.5839
20	2.9620	2.9377	2.8594	2.7785	2.6947	2.6077	2.5168
21	2.9039	2.8796	2.8010	2.7200	2.6359	2.5484	2.4568
22	2.8518	2.8274	2.7488	2.6675	2.5831	2.4951	2.4029
23	2.8049	2.7805	2.7017	2.6202	2.5355	2.4471	2.3542
24	2.7624	2.7380	2.6591	2.5773	2.4923	2.4035	2.3100
25	2.7238	2.6993	2.6203	2.5383	2.4530	2.3637	2.2696
26	2.6885	2.6640	2.5848	2.5026	2.4170	2.3273	2.2325
27	2.6561	2.6316	2.5522	2.4699	2.3840	2.2938	2.1985
28	2.6263	2.6017	2.5223	2.4397	2.3535	2.2629	2.1670
29	2.5987	2.5742	2.4946	2.4118	2.3253	2.2344	2.1379
30	2.5732	2.5487	2.4689	2.3860	2.2992	2.2079	2.1108
40	2.3937	2.3689	2.2880	2.2034	2.1142	2.0194	1.9172
60	2.2230	2.1978	2.1154	2.0285	1.9360	1.8363	1.7263
120	2.0604	2.0346	1.9500	1.8600	1.7628	1.6557	1.5330
∞	1.9048	1.8783	1.7908	1.6964	1.5923	1.4730	1.3246

Denominator Degrees of Freedom

H Critical Values of the *F*-Distribution ($\alpha = 0.005$)

Numerator Degrees of Freedom

	1	2	3	4	5	6	7	8	9
1	16210.7227	19999.5000	21614.7414	22499.5833	23055.7982	23437.1111	23714.5658	23925.4062	24091.0041
2	198.5013	199.0000	199.1664	199.2497	199.2996	199.3330	199.3568	199.3746	199.3885
3	55.5520	49.7993	47.4672	46.1946	45.3916	44.8385	44.4341	44.1256	43.8824
4	31.3328	26.2843	24.2591	23.1545	22.4564	21.9746	21.6217	21.3520	21.1391
5	22.7848	18.3138	16.5298	15.5561	14.9396	14.5133	14.2004	13.9610	13.7716
6	18.6350	14.5441	12.9166	12.0275	11.4637	11.0730	10.7859	10.5658	10.3915
7	16.2356	12.4040	10.8824	10.0505	9.5221	9.1553	8.8854	8.6781	8.5138
8	14.6882	11.0424	9.5965	8.8051	8.3018	7.9520	7.6941	7.4959	7.3386
9	13.6136	10.1067	8.7171	7.9559	7.4712	7.1339	6.8849	6.6933	6.5411
10	12.8265	9.4270	8.0807	7.3428	6.8724	6.5446	6.3025	6.1159	5.9676
11	12.2263	8.9122	7.6004	6.8809	6.4217	6.1016	5.8648	5.6821	5.5368
12	11.7542	8.5096	7.2258	6.5211	6.0711	5.7570	5.5245	5.3451	5.2021
13	11.3735	8.1865	6.9258	6.2335	5.7910	5.4819	5.2529	5.0761	4.9351
14	11.0603	7.9216	6.6804	5.9984	5.5623	5.2574	5.0313	4.8566	4.7173
15	10.7980	7.7008	6.4760	5.8029	5.3721	5.0708	4.8473	4.6744	4.5364
16	10.5755	7.5138	6.3034	5.6378	5.2117	4.9134	4.6920	4.5207	4.3838
17	10.3842	7.3536	6.1556	5.4967	5.0746	4.7789	4.5594	4.3894	4.2535
18	10.2181	7.2148	6.0278	5.3746	4.9560	4.6627	4.4448	4.2759	4.1410
19	10.0725	7.0935	5.9161	5.2681	4.8526	4.5614	4.3448	4.1770	4.0428
20	9.9439	6.9865	5.8177	5.1743	4.7616	4.4721	4.2569	4.0900	3.9564
21	9.8295	6.8914	5.7304	5.0911	4.6809	4.3931	4.1789	4.0128	3.8799
22	9.7271	6.8064	5.6524	5.0168	4.6088	4.3225	4.1094	3.9440	3.8116
23	9.6348	6.7300	5.5823	4.9500	4.5441	4.2591	4.0469	3.8822	3.7502
24	9.5513	6.6609	5.5190	4.8898	4.4857	4.2019	3.9905	3.8264	3.6949
25	9.4753	6.5982	5.4615	4.8351	4.4327	4.1500	3.9394	3.7758	3.6447
26	9.4059	6.5409	5.4091	4.7852	4.3844	4.1027	3.8928	3.7297	3.5989
27	9.3423	6.4885	5.3611	4.7396	4.3402	4.0594	3.8501	3.6875	3.5571
28	9.2838	6.4403	5.3170	4.6977	4.2996	4.0197	3.8110	3.6487	3.5186
29	9.2297	6.3958	5.2764	4.6591	4.2622	3.9831	3.7749	3.6131	3.4832
30	9.1797	6.3547	5.2388	4.6234	4.2276	3.9492	3.7416	3.5801	3.4505
40	8.8279	6.0664	4.9758	4.3738	3.9860	3.7129	3.5088	3.3498	3.2220
60	8.4946	5.7950	4.7290	4.1399	3.7599	3.4918	3.2911	3.1344	3.0083
120	8.1788	5.5393	4.4972	3.9207	3.5482	3.2849	3.0874	2.9330	2.8083
∞	7.8795	5.2983	4.2794	3.7151	3.3499	3.0913	2.8968	2.7444	2.6211

Denominator Degrees of Freedom

H   Critical Values of the *F*-Distribution (α = 0.005) (cont.)

Numerator Degrees of Freedom

	10	11	12	13	14	15	16	17	18
1	24224.4868	24334.3581	24426.3662	24504.5356	24571.7673	24630.2051	24681.4673	24726.7982	24767.1704
2	199.3996	199.4087	199.4163	199.4227	199.4282	199.4329	199.4371	199.4408	199.4440
3	43.6858	43.5236	43.3874	43.2715	43.1716	43.0847	43.0083	42.9407	42.8804
4	20.9667	20.8243	20.7047	20.6027	20.5148	20.4383	20.3710	20.3113	20.2581
5	13.6182	13.4912	13.3845	13.2934	13.2148	13.1463	13.0861	13.0327	12.9850
6	10.2500	10.1329	10.0343	9.9501	9.8774	9.8140	9.7582	9.7086	9.6644
7	8.3803	8.2697	8.1764	8.0967	8.0279	7.9678	7.9148	7.8678	7.8258
8	7.2106	7.1045	7.0149	6.9384	6.8721	6.8143	6.7633	6.7180	6.6775
9	6.4172	6.3142	6.2274	6.1530	6.0887	6.0325	5.9829	5.9388	5.8994
10	5.8467	5.7462	5.6613	5.5887	5.5257	5.4707	5.4221	5.3789	5.3403
11	5.4183	5.3197	5.2363	5.1649	5.1031	5.0489	5.0011	4.9586	4.9205
12	5.0855	4.9884	4.9062	4.8358	4.7748	4.7213	4.6741	4.6321	4.5945
13	4.8199	4.7240	4.6429	4.5733	4.5129	4.4600	4.4132	4.3716	4.3344
14	4.6034	4.5085	4.4281	4.3591	4.2993	4.2468	4.2005	4.1592	4.1221
15	4.4235	4.3295	4.2497	4.1813	4.1219	4.0698	4.0237	3.9827	3.9459
16	4.2719	4.1785	4.0994	4.0314	3.9723	3.9205	3.8747	3.8338	3.7972
17	4.1424	4.0496	3.9709	3.9033	3.8445	3.7929	3.7473	3.7066	3.6701
18	4.0305	3.9382	3.8599	3.7926	3.7341	3.6827	3.6373	3.5967	3.5603
19	3.9329	3.8410	3.7631	3.6961	3.6378	3.5866	3.5412	3.5008	3.4645
20	3.8470	3.7555	3.6779	3.6111	3.5530	3.5020	3.4568	3.4164	3.3802
21	3.7709	3.6798	3.6024	3.5358	3.4779	3.4270	3.3818	3.3416	3.3054
22	3.7030	3.6122	3.5350	3.4686	3.4108	3.3600	3.3150	3.2748	3.2387
23	3.6420	3.5515	3.4745	3.4083	3.3506	3.2999	3.2549	3.2148	3.1787
24	3.5870	3.4967	3.4199	3.3538	3.2962	3.2456	3.2007	3.1606	3.1246
25	3.5370	3.4470	3.3704	3.3044	3.2469	3.1963	3.1515	3.1114	3.0754
26	3.4916	3.4017	3.3252	3.2594	3.2020	3.1515	3.1067	3.0666	3.0306
27	3.4499	3.3602	3.2839	3.2182	3.1608	3.1104	3.0656	3.0256	2.9896
28	3.4117	3.3222	3.2460	3.1803	3.1231	3.0727	3.0279	2.9879	2.9520
29	3.3765	3.2871	3.2110	3.1454	3.0882	3.0379	2.9932	2.9532	2.9173
30	3.3440	3.2547	3.1787	3.1132	3.0560	3.0057	2.9611	2.9211	2.8852
40	3.1167	3.0284	2.9531	2.8880	2.8312	2.7811	2.7365	2.6966	2.6607
60	2.9042	2.8166	2.7419	2.6771	2.6205	2.5705	2.5259	2.4859	2.4498
120	2.7052	2.6183	2.5439	2.4794	2.4228	2.3727	2.3280	2.2878	2.2514
∞	2.5188	2.4325	2.3583	2.2938	2.2371	2.1868	2.1417	2.1011	2.0643

Denominator Degrees of Freedom

Critical Values of the *F*-Distribution (α = 0.005) (cont.)

Numerator Degrees of Freedom

	19	20	24	30	40	60	120
1	24803.3549	24835.9709	24939.5653	25043.6277	25148.1532	25253.1369	25358.5735
2	199.4470	199.4496	199.4579	199.4663	199.4746	199.4829	199.4912
3	42.8263	42.7775	42.6222	42.4658	42.3082	42.1494	41.9895
4	20.2104	20.1673	20.0300	19.8915	19.7518	19.6107	19.4684
5	12.9422	12.9035	12.7802	12.6556	12.5297	12.4024	12.2737
6	9.6247	9.5888	9.4742	9.3582	9.2408	9.1219	9.0015
7	7.7881	7.7540	7.6450	7.5345	7.4224	7.3088	7.1933
8	6.6411	6.6082	6.5029	6.3961	6.2875	6.1772	6.0649
9	5.8639	5.8318	5.7292	5.6248	5.5186	5.4104	5.3001
10	5.3055	5.2740	5.1732	5.0706	4.9659	4.8592	4.7501
11	4.8863	4.8552	4.7557	4.6543	4.5508	4.4450	4.3367
12	4.5606	4.5299	4.4314	4.3309	4.2282	4.1229	4.0149
13	4.3008	4.2703	4.1726	4.0727	3.9704	3.8655	3.7577
14	4.0888	4.0585	3.9614	3.8619	3.7600	3.6552	3.5473
15	3.9127	3.8826	3.7859	3.6867	3.5850	3.4803	3.3722
16	3.7641	3.7342	3.6378	3.5389	3.4372	3.3324	3.2240
17	3.6372	3.6073	3.5112	3.4124	3.3108	3.2058	3.0971
18	3.5275	3.4977	3.4017	3.3030	3.2014	3.0962	2.9871
19	3.4318	3.4020	3.3062	3.2075	3.1058	3.0004	2.8908
20	3.3475	3.3178	3.2220	3.1234	3.0215	2.9159	2.8058
21	3.2728	3.2431	3.1474	3.0488	2.9467	2.8408	2.7302
22	3.2060	3.1764	3.0807	2.9821	2.8799	2.7736	2.6625
23	3.1461	3.1165	3.0208	2.9221	2.8197	2.7132	2.6015
24	3.0920	3.0624	2.9667	2.8679	2.7654	2.6585	2.5463
25	3.0429	3.0133	2.9176	2.8187	2.7160	2.6088	2.4961
26	2.9981	2.9685	2.8728	2.7738	2.6709	2.5633	2.4501
27	2.9571	2.9275	2.8318	2.7327	2.6296	2.5217	2.4079
28	2.9194	2.8899	2.7941	2.6949	2.5916	2.4834	2.3690
29	2.8847	2.8551	2.7594	2.6600	2.5565	2.4479	2.3331
30	2.8526	2.8230	2.7272	2.6278	2.5241	2.4151	2.2998
40	2.6281	2.5984	2.5020	2.4015	2.2958	2.1838	2.0636
60	2.4171	2.3872	2.2898	2.1874	2.0789	1.9622	1.8341
120	2.2183	2.1881	2.0890	1.9840	1.8709	1.7469	1.6055
∞	2.0307	1.9999	1.8983	1.7891	1.6692	1.5326	1.3638

Denominator Degrees of Freedom

Critical Values for the Sign Test

Note: * denotes that it is not possible to have values in the critical region.

α for a one-tailed test	0.005	0.01	0.025	0.05
α for a two-tailed test	0.01	0.02	0.05	0.10
n				
5	*	*	*	0
6	*	*	0	0
7	*	0	0	0
8	0	0	0	1
9	0	0	1	1
10	0	0	1	1
11	0	1	1	2
12	1	1	2	2
13	1	1	2	3
14	1	2	2	3
15	2	2	3	3
16	2	2	3	4
17	2	3	4	4
18	3	3	4	5
19	3	4	4	5
20	3	4	5	5
21	4	4	5	6
22	4	5	5	6
23	4	5	6	7
24	5	5	6	7
25	5	6	7	7

$$\text{For } n > 25, \ z = \frac{X + 0.5 - \dfrac{n}{2}}{\dfrac{\sqrt{n}}{2}}.$$

J Critical Values for the Wilcoxon Signed-Rank Test

Note: * denotes that it is not possible to have values in the critical region.

α for a one-tailed test	0.005	0.01	0.025	0.05
α for a two-tailed test	0.01	0.02	0.05	0.10
n				
5	*	*	*	1
6	*	*	1	2
7	*	0	2	4
8	0	2	4	6
9	2	3	6	8
10	3	5	8	11
11	5	7	11	14
12	7	10	14	17
13	10	13	17	21
14	13	16	21	26
15	16	20	25	30
16	19	24	30	36
17	23	28	35	41
18	28	33	40	47
19	32	38	46	54
20	37	43	52	60
21	43	49	59	68
22	49	56	66	75
23	55	62	73	83
24	61	69	81	92
25	68	77	90	101
26	76	85	98	110
27	84	93	107	120
28	92	102	117	130
29	100	111	127	141
30	109	120	137	152

Critical Values for the Wilcoxon Rank-Sum Test

For one-tailed tests with $\alpha = 0.025$ or two-tailed tests with $\alpha = 0.05$

n_1	3		4		5		6		7		8		9		10	
n_2	T_L	T_U	T_L	T_U	T_L	T_U	T_L	T_U	T_L	T_U	T_L	T_U	T_L	T_U	T_L	T_U
3	5	16	6	18	6	21	7	23	7	26	8	28	8	31	9	33
4	6	18	11	25	12	28	12	32	13	35	14	38	15	41	16	44
5	6	21	12	28	18	37	19	41	20	45	21	49	22	53	24	56
6	7	23	12	32	19	41	26	52	28	56	29	61	31	65	32	70
7	7	26	13	35	20	45	28	56	37	68	39	73	41	78	43	83
8	8	28	14	38	21	49	29	61	39	73	49	87	51	93	54	98
9	8	31	15	41	22	53	31	65	41	78	51	93	63	108	66	114
10	9	33	16	44	24	56	32	70	43	83	54	98	66	114	79	131

For one-tailed tests with $\alpha = 0.05$ or two-tailed tests with $\alpha = 0.10$

n_1	3		4		5		6		7		8		9		10	
n_2	T_L	T_U	T_L	T_U	T_L	T_U	T_L	T_U	T_L	T_U	T_L	T_U	T_L	T_U	T_L	T_U
3	6	15	7	17	7	20	8	22	9	24	9	27	10	29	11	31
4	7	17	12	24	13	27	14	30	15	33	16	36	17	39	18	42
5	7	20	13	27	19	36	20	40	22	43	24	46	25	50	26	54
6	8	22	14	30	20	40	28	50	30	54	32	58	33	63	35	67
7	9	24	15	33	22	43	30	54	39	66	41	71	43	76	46	80
8	9	27	16	36	24	46	32	58	41	71	52	84	54	90	57	95
9	10	29	17	39	25	50	33	63	43	76	54	90	66	105	69	111
10	11	31	18	42	26	54	35	67	46	80	57	95	69	111	83	127

L   Critical Values of Spearman's Rank Correlation Coefficient, r_s

Note: * denotes that it is not possible to have values in the critical region.

n	$\alpha = 0.10$	$\alpha = 0.05$	$\alpha = 0.02$	$\alpha = 0.01$
5	0.900	*	*	*
6	0.829	0.886	0.943	*
7	0.714	0.786	0.893	*
8	0.643	0.738	0.833	0.881
9	0.600	0.683	0.783	0.833
10	0.564	0.648	0.745	0.794
11	0.523	0.623	0.736	0.818
12	0.497	0.591	0.703	0.780
13	0.475	0.566	0.673	0.745
14	0.457	0.545	0.646	0.716
15	0.441	0.525	0.623	0.689
16	0.425	0.507	0.601	0.666
17	0.412	0.490	0.582	0.645
18	0.399	0.476	0.564	0.625
19	0.388	0.462	0.549	0.608
20	0.377	0.450	0.534	0.591
21	0.368	0.438	0.521	0.576
22	0.359	0.428	0.508	0.562
23	0.351	0.418	0.496	0.549
24	0.343	0.409	0.485	0.537
25	0.336	0.400	0.475	0.526
26	0.329	0.392	0.465	0.515
27	0.323	0.385	0.456	0.505
28	0.317	0.377	0.448	0.496
29	0.311	0.370	0.440	0.487
30	0.305	0.364	0.432	0.478

M   Critical Values for the Number of Runs ($\alpha = 0.05$)

Value of n

Value of m	2	3	4	5	6	7	8	9	10	11	12	13	14	15	16	17	18	19	20
2	1 6	1 6	1 6	1 6	1 6	1 6	1 6	1 6	1 6	1 6	2 6	2 6	2 6	2 6	2 6	2 6	2 6	2 6	2 6
3	1 6	1 8	1 8	1 8	2 8	2 8	2 8	2 8	2 8	2 8	2 8	2 8	2 8	3 8	3 8	3 8	3 8	3 8	3 8
4	1 6	1 8	1 9	2 9	2 9	2 10	3 10	3 10	3 10	3 10	3 10	3 10	3 10	3 10	4 10	4 10	4 10	4 10	4 10
5	1 6	1 8	2 9	2 10	3 10	3 11	3 11	3 12	3 12	4 12	4 12	4 12	4 12	4 12	4 12	4 12	5 12	5 12	5 12
6	1 6	2 8	2 9	3 10	3 11	3 12	3 12	4 13	4 13	4 13	4 13	5 14	5 14	5 14	5 14	5 14	5 14	6 14	6 14
7	1 6	2 8	2 10	3 11	3 12	3 13	4 13	4 14	5 14	5 14	5 14	5 15	5 15	6 15	6 16	6 16	6 16	6 16	6 16
8	1 6	2 8	3 10	3 11	3 12	4 13	4 14	5 14	5 15	5 15	6 16	6 16	6 16	6 16	6 17	7 17	7 17	7 17	7 17
9	1 6	2 8	3 10	3 12	4 13	4 14	5 14	5 15	5 16	6 16	6 16	6 17	7 17	7 18	7 18	7 18	8 18	8 18	8 18
10	1 6	2 8	3 10	3 12	4 13	5 14	5 15	5 16	6 16	6 17	7 17	7 18	7 18	7 18	8 19	8 19	8 19	8 20	9 20
11	1 6	2 8	3 10	4 12	4 13	5 14	5 15	6 16	6 17	7 17	7 18	7 19	8 19	8 19	8 20	9 20	9 20	9 21	9 21
12	2 6	2 8	3 10	4 12	4 13	5 14	6 16	6 16	7 17	7 18	7 19	8 19	8 20	8 20	9 21	9 21	9 21	10 22	10 22
13	2 6	2 8	3 10	4 12	5 14	5 15	6 16	6 17	7 18	7 19	8 19	8 20	9 20	9 21	9 21	10 22	10 22	10 23	10 23
14	2 6	2 8	3 10	4 12	5 14	5 15	6 16	7 17	7 18	8 19	8 20	9 20	9 21	9 22	10 22	10 23	10 23	11 23	11 24
15	2 6	3 8	3 10	4 12	5 14	6 15	6 16	7 18	7 18	8 19	8 20	9 21	9 22	10 22	10 23	11 23	11 24	11 24	12 25
16	2 6	3 8	4 10	4 12	5 14	6 16	6 17	7 18	8 19	8 20	9 21	9 21	10 22	10 23	11 23	11 24	11 25	12 25	12 25
17	2 6	3 8	4 10	4 12	5 14	6 16	7 17	7 18	8 19	9 20	9 21	10 22	10 23	11 23	11 24	11 25	12 25	12 26	13 26
18	2 6	3 8	4 10	5 12	5 14	6 16	7 17	8 18	8 19	9 20	9 21	10 22	10 23	11 24	11 25	12 25	12 26	13 26	13 27
19	2 6	3 8	4 10	5 12	6 14	6 16	7 17	8 18	8 20	9 21	10 22	10 23	11 23	11 24	12 25	12 26	13 26	13 27	13 27
20	2 6	3 8	4 10	5 12	6 14	6 16	7 17	8 18	9 20	9 21	10 22	10 23	11 24	12 25	12 25	13 26	13 27	13 27	14 28

Appendix B

Getting Started with Microsoft Excel

The Basics of Excel 2010

Microsoft Excel is a spreadsheet program which allows users to track and analyze data. Spreadsheets such as those created with Microsoft Excel are widely used in the business world to perform various tasks such as accounting, budgeting, billing, reporting, planning, and tracking. When you open Excel, you will see the rows run top to bottom and are labeled with numbers. The columns run left to right and are labeled with letters. There are three tabs at the bottom of a blank worksheet when it is opened. Each of these is a blank spreadsheet. These tabs can be renamed so that the data you require can be easily found.

Figure B.1

Cells

Cells in Microsoft Excel may contain data and/or text. The cells' locations are described by their positions in terms of rows and columns. Rows are listed along the left side of the worksheet and are labeled with numbers. Rows run top to bottom in the spreadsheet. Columns are listed across the top of the worksheet and are labeled with letters. Columns run left to right in the spreadsheet.

A thick black border will outline the active cell. To help you identify the cell address, the column and row headers are also highlighted. Another way to identify the active cell's address is to look at the window just above the header for Column A. A column letter followed by a row number describes the active cell position. For example, B2 would be referring to the cell found in the second row of Column B.

Figure B.2

To the right of the cell address window is the formula window. This displays the contents of the active cell. In the formula window type an = sign. This is the command to start a formula. Cells or data can be added, subtracted, multiplied, divided, and manipulated in any combination through formulas.

To change the active cell, move the mouse to the desired cell and click. The border will now be surrounding the new cell and the address will have changed in the active cell address window. The arrow keys can also control navigation of the active cell.

As you enter data in the active cell, it is displayed in the cell and in the formula window at the top of the worksheet.

Figure B.3

When you finish entering data in a cell, press Enter. The active cell will now move one row down. The answer to the formula, rather than the formula itself, will be displayed in the cell.

Excel will try to anticipate a reoccurring word or phrase within a column. When you type a word in a cell, you can then move to a lower cell in that column. If you type the first letter of the word again, Excel will fill in the rest of the word or phrase. To accept the automatic completion of the word, just hit the Enter key or click in a new cell. If you do not want to use the automatic completion of the word, simply keep typing in the cell and the new word or phrase will appear.

Figure B.4

Filling Cells

Excel has a helpful tool when entering a series of data. We will refer to this as filling cells. Try entering 1 into cell A1 and 2 into cell A2. Then highlight both of the cells. You will see a small black square at the bottom right of the highlighted section.

Figure B.5

If you move your cursor over this black square it should change into a narrow plus sign. Now click the left mouse button and drag the mouse down to include cells A3 through A10. When you release the mouse button, the cells should now be filled with values from 1 to 10. Also notice that at the bottom of the screen Excel gives some summary statistics about the highlighted cells. The average, number of values, and sum are displayed for any highlighted cells.

Figure B.6

The fill cells tool will work for other values as well. Excel finds the relationship between two data points and replicates this in the fill cells. Now try this with multiples of 5. Enter 5 in cell A1 and 10 in cell A2. Highlight cells A1 and A2 again and put your cursor over the square in the bottom right of the highlighted section. Click the left mouse button and drag down to cell A10. Release the button. Now the values should have changed to 5 through 50.

Figure B.7

You can also use this feature with sequential labels such as days of the week or months of the year.

Figure B.8

Now that you have the basic skills to work with Excel, start a new worksheet and we will try some examples.

Enter the labels Checking Balance and Savings Balance in cells A1 and A2, respectively. You will notice that the label you typed in cell A1 is cut off. We need to resize the columns to allow for the entire label. Move your cursor to the line that separates the column labels for A and B. Your cursor will change to an arrow pointing off to the left and right. You can click and drag the cursor to the right to increase the column width, or simply double click to have the width auto size to fit the label. Repeat this for Column B.

Next select the column labels A and B and you should see both columns highlighted. Under the Home tab there is an area labeled "Number". The buttons in this area format any numbers you enter in the selected cells. With columns A and B highlighted, press the $ button. Now all of the values put in these columns will be formatted as dollars.

Next we will use the fill tool to get some values to work with. Start by entering 100 in cell A2 and 200 in cell A3. Now move to cell B2, enter 1000, and enter 2000 in cell B3. Now highlight cells A2 through B3 by clicking in A2 and dragging down and to the right to include B3. With those cells highlighted, click on the Fill tool at the bottom right of the highlighted section. Click and drag down to row 11. When you release the mouse button, the cells should fill with data and end with $10,000.00 in cell B11. You should have a screen that looks like the following.

Figure B.9

Formulas can be applied to manipulate data between cells. We will see, however, that special attention needs to be paid to the copying of formulas. We will continue working with the above worksheet with the checking and savings balances while we try some formulas.

Formulas and Relative Addressing

We want to see what your total bank balance would be if you had the amount listed in a particular row in your checking and savings accounts. In cell C2, type = to indicate you are entering a formula, and then click in cell A2, type +, and click in cell B2. Press Enter. You have just added cells A2 and B2 together, and the sum is displayed in cell C2. Under the Home tab, the "Clipboard" contains buttons allowing you to cut, copy, and paste cells. With cell C2 selected, either press the copy 📋 Copy ▾ button or use the keyboard shortcut Ctrl+C to copy the formula. Highlight cells C3 through C11. Under the Home tab, press the Paste button or use the keyboard shortcut Ctrl+V to paste the formula into these cells. Notice that the value $1,100.00 that was displayed in C2 was not copied, but the formula that made up that value. The formula changed for each of the rows and substituted the new row value in the formula. This is called relative addressing. The formula changes relative to its address. The last cell should show the value $11,000.00.

Suppose we wanted to see what the annual interest would be on our savings account at the levels listed in Column B. Create a new label in cell A13 called Interest Rate. Now in cell B13 input the interest rate 0.045. Format the cell for percentages rather than currency. To do this, highlight cell B13, and under the Home tab, select the % button next to the $ button that we used to originally format the cells. The number will most likely be displayed as 5%. We can adjust the number of decimals displayed by clicking on the 🔲 button, also located under the Home tab. The value should now display as 4.5%.

Now we want to use this value to compute interest for each level of savings. Label cell D1 Savings Interest. Resize the cell to fit the label. In cell D2 we will put our formula. Type = to start the formula, and then select cell B13 to get the percentage rate. Type * to indicate multiplication, and select cell B2. The resulting value will be the annual interest gained on $1,000.00 in savings at 4.5% APR.

Since we have the formula, we can copy it to the rest of the table and get the interest income for each of the savings levels. With cell D2 selected, click in the lower right hand corner of the cell and drag down to D3. Is this the result you expected? Probably not. Remember the rules of relative addressing. As you copy the formula down one row, the formula values change by one row. So in cell D3 the formula is pulling from cell B14 for the interest rate and cell D2 for the principal amount. You will notice that cell B14 is empty. Thus, Excel computes this formula as being equal to zero.

Absolute Addressing

We want to use cell B13 for each of these formulas. So how do we lock in the address of cell B13? We use absolute addressing. The $ symbol is the key to locking the position in the formula. In cell D2 we need to change the formula to read =B$13*B2. We do not need to put a $ in front of the first "B" because it refers to the column, and we are not interested in values in any other columns. We are only copying the formula between rows.

Go back and copy the new formula into cells D3 through D11. The results should now look like the following.

Figure B.10

Charts

Suppose you have the following data about ticket sales from your traveling circus. Adult tickets are $20 and child tickets are $12. The sales for the last week are listed in the following spreadsheet.

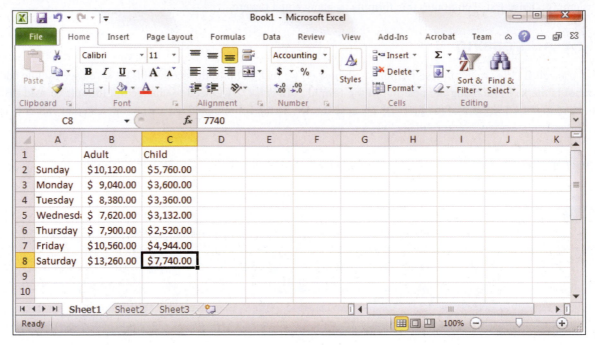

Figure B.11

From these data you might see how the ticket sales change daily and wish to compare child and adult ticket sales. This information might be easier to understand if it were graphically displayed on a chart. Enter the data as seen above in a new worksheet and use a bar chart for this particular example.

To create a chart we need to select the data to be graphed. Of course we want to chart the ticket sales for both the adults and children. In a bar chart the dollar amount will determine the height of the bar. What about the days of the week? These should also be included since they are the labels for the bars.

Highlight the data in columns A, B, and C from rows 1 to 8. (Excel can interpret the first row or column in a set of data as labels and not part of the data.) Under the Insert tab, there is a "Charts" group of icons.

Figure B.12

Choose Column, and the top left graph (the first one listed under the 2-D column heading). This is a clustered column graph. Excel creates a side-by-side bar graph based on the highlighted data. After the chart is created a set of tabs labeled "Chart Tools" appears. These tabs can be used to edit the chart that has been made. For example, under the Layout tab, select Chart Title, and Above Chart. Give the graph a title of Ticket Sales. Under the Layout tab you can edit the chart appearance by adding axis titles, data labels, or gridlines. Microsoft Excel makes it easy to create the chart you want.

Figure B.13

Installing Data Analysis Tools

To get the most out of Excel as a statistical tool, you will want to use the Analysis ToolPak. When Microsoft Excel is first installed, this feature is not included. The ToolPak is an application add-in. To install the Analysis ToolPak, under the File tab, select Options, and then Add-Ins on the left-hand side. When Add-Ins is selected, at the bottom of the dialog box there is a drop down menu that has "Excel Add-ins" selected. Press the Go button next to this drop down menu. A list of available add-ins will appear. Check the boxes next to Analysis ToolPak and Analysis ToolPak – VBA. Press OK. The data analysis tools will be installed.

Figure B.14

With these tools installed you can perform many useful analyses in statistics. The tool is found under the Data tab once it is installed. The button is located in the "Analysis" group of icons, and is labeled Data Analysis.

Figure B.15

You can utilize this tool to perform many important statistical analyses including regression analysis, constructing histograms, computing descriptive statistics, and performing hypothesis tests. Many of the Discovering Technology sections throughout the textbook use the functions contained in the data analysis add-in.

Appendix C
Getting Started with Minitab

The Basics of Minitab 16

Minitab is a more advanced statistical software program which many businesses use for statistical analysis and quality control. In addition to these types of analytics, Minitab performs many of the basic functions of a spreadsheet program such as Microsoft Excel.

Opening a new Minitab file provides a blank worksheet. You can enter data into a cell by typing into the dark-bordered cell. To move to another cell, you can use the up/down/left/right arrow keys on the keyboard or simply click with your mouse. The first row of the worksheet is reserved for column titles. Any numerical data values should go in the white cells rather than the grey. Cut, copy, and paste functions operate in Minitab in the same way as other spreadsheet programs. Data can be copied by highlighting the appropriate cell and selecting Edit, Copy Cells, or by using the keyboard shortcut Ctrl+C. The data can be pasted using Paste Cells in the Edit menu or by using the keyboard shortcut Ctrl+V. Cells can also be copied by clicking in the lower right corner of the selected cell and dragging to cover all cells for which you wish to have that value.

When you open Minitab you will notice a menu bar at the top of the screen. On that menu bar you will see menus such as Edit, Data, Calc, Stat, and Graph. The Stat menu contains many of the tools we utilize in this text. If you click on the Stat menu, you will see a list of concepts, some of which should be familiar such as Basic Statistics, Regression, and ANOVA. Clicking on any of these reveals other possibilities. Clicking on Basic Statistics leads to Display Descriptive Statistics, clicking on Regression allows you to make a Fitted Line Plot, and clicking on ANOVA allows you to perform a one or two-way analysis of variance.

If you choose Calc on the menu bar, you will see a list including options such as Random Data and Probability Distributions. Choosing Random Data will allow you to generate data according to some specified distribution. Clicking on Probability Distributions allows you to calculate probabilities for the binomial or Poisson distributions, among others.

Both Stat and Calc require inputs into the cells before clicking on that menu item. The Discovering Technology sections at the end of the chapters throughout the text explain how to provide these inputs for the particular application and give further explanation as to how to use the dialog boxes required along the way.

Above the worksheet is the Session window. The results of each session are displayed there. To print what is in the window, you can go to File, Print Session Window, and then press OK.

Cells

To create a new worksheet, you can select File, New, and select Minitab Worksheet. This opens a new worksheet to use in the current project. Columns run left to right and are labeled with the letter C and a number. Rows run top to bottom and are labeled with numbers. To refer to a specific cell, we will use the format "Column name, Row name". For example, if we were referring to the third row in the second column, we would use "C2, 3" to reference the cell. Minitab provides a row that is not labeled for you to insert your column headings.

The cells in Minitab may contain data or text. If you are inputting text in the cells other than the cell for the column heading, the column label will add "-T" to the column name. For example, if we typed the days of the week in the first eight rows of Column C1, the column name would change to "C1-T".

	C1-T	C2	C3
	Days of the Week		
1	Monday		
2	Tuesday		
3	Wednesday		
4	Thursday		
5	Friday		
6	Saturday		
7	Sunday		
8	Monday		
9			
10			

Worksheet 1 ***

Figure C.1

Filling Cells and Formulas

Now, let's input some data values into a new project to show some of the calculation features of Minitab. Choose File, New, and Minitab Project. Click OK.

Minitab allows you to fill cells when entering a series of data. Enter 100 into cell "C1, 1" and 200 into cell "C1, 2". Highlight both of the cells and locate the small black square in the bottom right corner of the highlighted section. When you move your cursor over the square, a plus sign should appear. Click the left mouse button and drag the mouse down to include "C1, 3" through "C1, 10". The cells should now be filled with values from 100 to 1000.

We can perform some basic functions on these values. From the menu bar, choose Calc and Column Statistics. This allows you to find the sum, the mean, the standard deviation, and many other summary measures corresponding to a particular column. Choose the mean and click inside the box next to "Input variable". While your cursor is in this box, all the columns that contain data will be listed in the box to the left. You can click on the column in this box and press Select to specify C1 as the "Input variable". Click OK and the mean of C1 will appear in the session window.

Figure C.2

To perform calculations on the data in the cells, select Calc and Calculator from the menu bar. In the "Store result in variable" box, enter C2. You will see a drop down menu of functions, as well as a list of functions on the right-hand side of the dialog box. In the list of functions, scroll down and select the function Round. Press Select and the text ROUND(number,decimals) will appear in the "Expression" box. For the parameter "number" enter C1*7/3, and for the parameter "decimals" enter 1. This will multiply the data in Column C1 by $\frac{7}{3}$, round the result to one decimal place, and display the final value in Column C2.

Figure C.3

Press OK, and you will see the values of this expression displayed in Column C2, rounded to one decimal place.

Worksheet 1 ***			
↓	C1	C2	C3
1	100	233.3	
2	200	466.7	
3	300	700.0	
4	400	933.3	
5	500	1166.7	
6	600	1400.0	
7	700	1633.3	
8	800	1866.7	
9	900	2100.0	
10	1000	2333.3	
11			
12			

Figure C.4

Graphs

Suppose you had the following data about a firm's sales per quarter for a particular year.

Worksheet 1 ***			
↓	C1-T	C2	C3
	Quarter	Sales	
1	First	356210	
2	Second	349800	
3	Third	370355	
4	Fourth	402775	
5			
6			
7			
8			
9			
10			

Figure C.5

To graphically display this information, select Graph and Time Series Plot. Select the Simple time series plot and press OK. Put the cursor in the "Series" dialog box, and the available columns with data will appear in the box on the left. Since we only have one column with quantitative data in this case, only C2 appears. Select C2 in the box and press Select. Press OK. A time series plot is created.

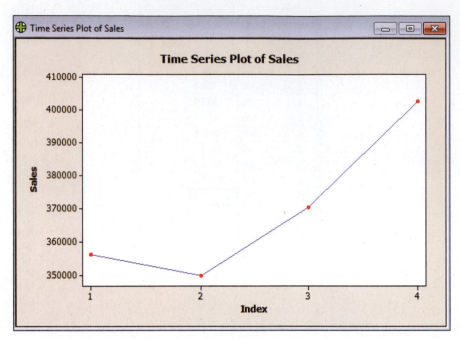

Figure C.6

Data Analysis

One of the most important uses of Minitab is statistical analysis. Start a new Minitab project and enter the following data for the number of phone calls a company receives each hour of the work day.

Worksheet 1 ***	C1	C2	C3
↓	Phone Calls		
1	10		
2	27		
3	21		
4	24		
5	30		
6	23		
7	36		
8	30		
9	15		
10			

Figure C.7

Select Stat, Basic Statistics, and Display Descriptive Statistics. In the "Variables" dialog box, enter C1 and press OK. Basic descriptive statistics are displayed about the data in C1. This includes the number of data values, the mean, the standard deviation, the minimum value, the first quartile, the median, the third quartile, and the maximum value.

```
Descriptive Statistics: Phone Calls

Variable      N   N*    Mean   SE Mean   StDev   Minimum      Q1   Median     Q3
Phone Calls   9   0    24.00      2.67    8.00     10.00   18.00    24.00  30.00

Variable      Maximum
Phone Calls     36.00
```

Figure C.8

There are many other useful statistical features of Minitab that you will encounter throughout the text. Refer to the Discovering Technology sections at the end of the chapters throughout the text for detailed instructions on using Minitab for statistical analysis.

Answer Key

1   Decision Making Using Statistics

Additional Exercises

EXERCISES

19. a. Internet users   **b.** The amount of time users view photos of products on websites   **c.** Inferential   **21. a.** Elderly citizens   **b.** Self-esteem after reading a news story   **c.** 276 elderly citizens   **d.** Answers will vary.   **e.** Inferential
23. a. Mobile phone users   **b.** The percentage of people who use their mobile phones to access the Internet for different reasons   **c.** 500 American adults 18 years of age and older   **d.** Answers will vary.   **e.** Inferential   **25. a.** Married couples   **b.** The percentage of married couples that met online   **c.** 7000 adults married in the past 5 years   **d.** Answers will vary.   **27. a.** U.S. states with coastlines   **b.** Coastline length

2   Data, Reality, and Problem Solving

2.1   Quality of Measurements

EXERCISES

5. a. Well-defined   **b.** Well-defined   **c.** Not well-defined   **d.** Well-defined   **e.** Not well-defined   **7.** There is no well-defined scale to measure cleanliness or aesthetics. Answers will vary.

2.2   Science and Data

EXERCISES

5. Answers will vary.   **7.** Inferential. Data were collected from a sample and the information was used to make a general claim about the population of the company's bulbs.

2.4   Collecting Data

EXERCISES

15. a. By randomly assigning women to two groups and using one of the groups as a "control" group, the experiment should produce data that will reveal the impact of the different diets.   **b.** Difference in diet (The first group received a 1200 calorie per day diet for the entire period whereas the second group received a 420 calorie per day diet for 16 weeks and then were shifted to a 1200 calorie per day diet for the rest of the experimental period.)   **c.** Weight loss.   **d.** Yes, the women receiving the 1200 calorie per day diet represent a control group.   **e.** Observational studies are subject to self-selection bias. We would not necessarily know the cause of the weight reduction.   **17. a.** Phase 1: Gather information about the phenomenon being studied.   **b.** Controlled experiment.   **c.** Number of major attacks of Multiple Sclerosis.   **d.** Bovine myelin.   **e.** Fifteen individuals in the early stages of MS fed bovine myelin.   **f.** Fifteen individuals in the early stages of MS given a placebo.   **19.** Jacob's knee could feel better simply because he took a week break from playing basketball. Answers will vary.   **21.** There are many factors that affect whether or not someone is happy. Additionally, since both questions require a yes or no reply there is no way to quantify happiness or going to church. Answers will vary.   **23.** Generally people that have more money will seek the help of a financial advisor since advisors are paid for their services. Answers will vary.

2.5 Levels of Measurement

EXERCISES

9. a. Ratio   **b.** Nominal   **c.** Ordinal   **d.** Ratio

2.6 Data Classifications

EXCERSISES

7. a. Discrete   **b.** Continuous   **c.** Discrete   **d.** Continuous   **e.** Discrete   **9. a.** Gender, Political Affiliation, Opinion on how President Clinton was handling the situation in Bosnia, News media performance.   **b.** Gender, Political Affiliation, Opinion, News media performance are qualitative variables. There are no quantitative variables.   **c.** Gender – Nominal; Political Affiliation – Nominal; Opinion – Nominal; News media performance – Ordinal   **d.** There may be an inordinate number of one party affiliation (Republicans or Democrats) selected for the survey. In general there are no "controls" on the survey.

2.7 Time Series Data

EXERCISES

7. a. Yes   **b.** Nonstationary

2.8 Cross-Sectional Data

EXERCISES

3. The table contains time series data. Answers will vary.

Additional Exercises

EXERCISES

1. a. Standardized aptitude test, GPA, etc.   **b.** Number of student suspensions, number of student detentions, etc.
c. Survey students to rate teacher preparedness: Always prepared, Sometimes prepared, Never prepared; or measure the number of times teachers are unprepared on a spot check by the principal.   **d.** Number of days missed, number of hours missed (for both teachers and students).   **e.** Survey students, survey teachers: e.g. Rate cafeteria food as Excellent, Very Good, Good, Poor, Very Poor.   **3. a.** Types of investments available, rate of return, amount, etc.   **b.** Rates of return on various types of investments: Returns for diversified portfolio of stocks: e.g. S&P 500, Dow Jones Index Returns for diversified portfolio of bonds.   **5.** One example: Make a list of standard grocery items for several randomly selected families. Have families buy groceries from one store and then buy exactly the same groceries from another store and then compare prices.   **7.** Survey of customer satisfaction: How would you rate the service you received when purchasing the car: Poor, Average, Good? How would you rate the service after you purchased the car: Poor, Average, Good? How would you rate the car's performance: Poor, Average, Good? How satisfied are you with your purchase: Not Satisfied, Satisfied, Very Satisfied?   **9. a.** Interval **b.** Interval   **c.** Ordinal   **d.** Nominal   **e.** Ratio   **11. a.** The level of measurement for millions of barrels of oil per day is ratio.   **b.** Time series, Nonstationary

13. Answers will vary.
15. a. Nominal   **b.** Time Series   **c.** If respondents were owners or renters, answers will vary.

3   Organizing, Displaying, and Interpreting Data

3.1   Frequency Distributions

EXERCISES

7. a. Nominal   **b.** Qualitative

c.

Air Conditioner	16
Lawn Mower	10
Fan	7
Washing Machine	6
Miscellaneous	9

9. a. Answers may vary.

Type of Complaint	March	July
Comfort	17	28
Price	11	15
Service	18	14
Schedule	29	33

b. Answers may vary.

Type of Complaint	March	July
Plane	22	34
Personnel	8	3
Building/ Equipment	17	16
Other	28	37

c. No. Another person would not necessarily have assigned the various complaints to the same categories. Results may vary depending on who prepared the data.   **d.** Yes. Given that you assign a complaint to only one category, the categories are mutually exclusive. Given that you assign each complaint to a category, the categories are exhaustive.

3.3   Displaying Qualitative Data Graphically

EXERCISES

9. a. Bars are not proportional. The bar for the Dodge Intrepid (27) should be shorter than the bar for the Ford Taurus (28). The graph also lacks labels on the horizontal axis. Answers may vary.   **b.** It is hard to tell if bars are proportional because there are no horizontal axis labels. The names of the vehicles would be easier to read if they were listed outside the graph area. Horizontal and vertical axis labels are needed. Answers will vary.

11. a.

b.

c.

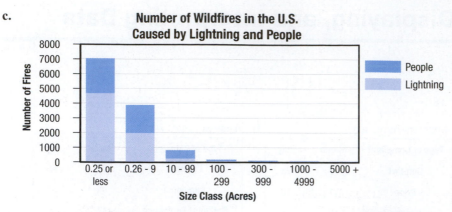

d. **Number of Wildfires in the U.S. Caused by Lightning**

e. **Number of Wildfires in the U.S. Caused by People**

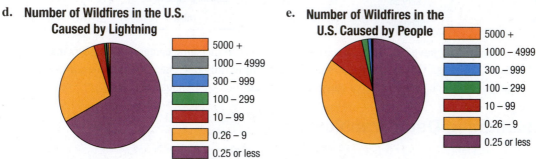

f. A majority of fires both lightning-caused and people-caused occur in the 0.25 or less acre size class. For the 0.25 or less acre size class, about twice as many fires are caused by lightning than by people. However, in the 0.26-9 acre size class, the number of wildfires caused by lightning and by people are almost the same. The number of wildfires caused by both lightning and people is much smaller for the size classes above 0.26-9 acres.

13. a.

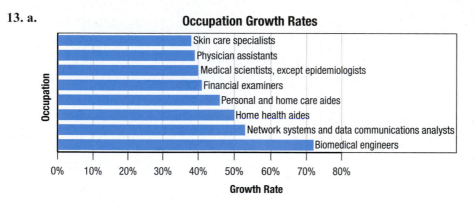

b. Biomedical engineers have the highest projected growth rate. The rest of the growth rates appear to be between 35% and 55% whereas biomedical engineers is above 70%.

3.4 Constructing Frequency Distributions for Quantitative Data

EXERCISES

7.

Days Traveling	Frequency	Relative Frequency	Cumulative Frequency
0 – 6	15	0.20	15
7 – 13	21	0.28	36
14 – 20	27	0.26	63
21 – 27	9	0.12	72
28 – 34	2	0.03	74
35 and above	1	0.01	75

9.	Average Temp (°F)	Frequency	Relative Frequency	Cumulative Frequency
	40 – 49	3	0.20	3
	50 – 59	7	0.467	10
	60 – 69	4	0.267	14
	70 – 79	1	0.067	15

3.5 Histograms

EXERCISES

5. a.

Closing Price	Frequency
$0 - $44.00	10
$45.00 - $89.00	3
$90.00 - $134.00	0
$135.00 - $179.00	0
$180.00 - $224.00	1

b.

7. a., b.

% of Calories From Fat	Frequency	Relative Frequency
15% – 19%	1	0.04
20% – 24%	2	0.08
25% – 29%	5	0.20
30% – 34%	8	0.32
35% – 39%	3	0.12
40% – 44%	3	0.12
45% – 49%	3	0.12

c.

d. A majority of the subjects in the sample consumed between 30% and 34.9% of calories from fat per day. The sample percentages appear to have a bell-shaped distribution. Relatively few of the subjects consumed less than 20% of calories per day from fat. Answers may vary.

3.6 The Stem-and-Leaf Display

EXERCISES

5. a. Ratio **b.**

Stem	Leaf
12	3 5 5
13	5 5 5
14	7 7 8 8 8 9
15	6 6 6 6 8 8
16	9
17	8 8 9 9
18	9 9 9
19	8 9 9
20	
21	4 5 5
22	
23	5 9
24	8 8 9
25	6 7 8 8 8 9 9 9
26	5 8 9 9
27	
28	8

c. There appear to be two different clusters of daily rates for semi-private rooms. One of the clusters seems to center around $150 per day. The other cluster seems to center around $250 per day. One explanation of this could be the location of the surveyed hospitals, it could be that some of the hospitals surveyed were in large metropolitan areas and some were in smaller cities or suburban areas. Answers will vary.

7. a.

Stem	Leaf
2	3 6 7 8 8 8 9 9
3	1 3 4 6 6 7 7 8
4	0 2 6

b. The majority of the miles per gallon are between 25 and 38. Miles per gallon ratings above 40 are uncommon. Answers will vary.

3.7 The Ordered Array

EXERCISES

3. a. −4, 0, 10, 10, 16, 19, 19, 20, 20, 22, 23, 24, 24, 27, 33, 37 **b.** Only 1 quarter showed a loss, and most quarters showed growth between 20% and 30%. Answers will vary.

3.8 Dot Plots

EXERCISES

5.

3.9 Plotting Time Series Data

EXERCISES

5. a. All of the rates seem to change in the same way. The longer the period of time, the higher the interest rate charged. Answers will vary. **b.** Stationary. Answers will vary. **7. a.** Year – Ordinal; Number – Ratio; Rate – Ratio

b.

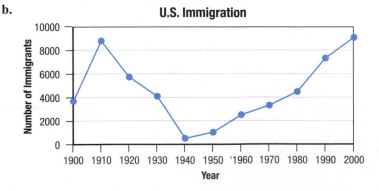

U.S. Immigration

c. 146.61% **d.** −39.62%. Though the number of immigrants has increased drastically, the general population of the U.S. has increased also, so the rate of annual immigration (per 1000 people) has actually decreased from 1900 to 1990. Answers will vary.

Additional Exercises

EXERCISES

1. a. Time sequence plot or line graph of both the median family income and percent change in median family income.

b.

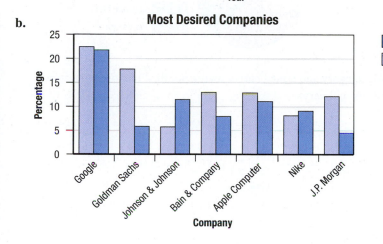

Median Family Income

c. Median household income has an upward trend, but this could likely be due to inflation. 2009 is the only year in which median household income decreased from the previous year. Answers will vary. **3. a.** A side-by-side bar chart showing the percentage of both men and women who ranked each company in their top 5. Answers will vary.

b.

Most Desired Companies

Percent of Women
Percent of Men

c. Google appears to be the most desirable company for both men and women. Goldman Sachs appears to be desirable for men, but not so much for women. Johnson & Johnson appears to be popular among women, but not so much among men. Answers will vary. **5. a.** Bar charts and pie charts would both be appropriate for displaying the data. Answers will vary. **b.** Answers will vary.

c. Answers will vary.

Radio Formats 6 AM - 10 AM and 10 AM - 3 PM

Mon-Fri 10 AM-3 PM
Mon-Fri 6 AM-10 AM

7. a. Spending as a % of sales is a more useful measure of research and development expenditures for comparative purposes because it standardizes research and development expenditures. **b.** A bar chart for R&D expenditures, pie charts representing industries and headquarters locations of the top 20 R&D spenders, a histogram to represent spending as a percent of sales, answers will vary. **c.**

d.

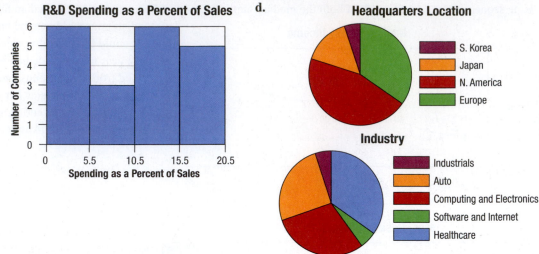

9. a. A line graph displaying the percentage of voters in each category and a pie chart showing political identification in a particular year might be helpful in visualizing the data. Answers will vary.

b.

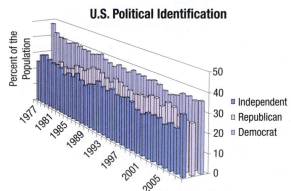

c. The time series appears to be stationary. Recently it appears that more people identify themselves as democrats than republicans or independents. Answers will vary.

4   Numerical Descriptive Statistics

4.1   Measures of Location

EXERCISES

13. mean = 15, median = 15, mode = 11, 20% trimmed mean = 15   **15.** mean = 12, median = 12, mode = 12, 10% trimmed mean = 12   **17. a.** mean or median   **b.** mode   **c.** median or mode   **d.** median (some very high home prices may skew data)   **19.** The tour guide was "accurate" in saying that the average daily temperature is 60 degrees. However, this obviously does not completely describe the temperatures since they have a range of 120 degrees per day. To more accurately represent the temperatures the tour guide should have also specified how much the daily temperatures vary.   **21. a.** Ratio   **b.** mean = 29.22, 10% trimmed mean = 20.755, 20% trimmed mean = 20.97   **c.** The 10% trimmed mean because there is one outlier. Answers will vary.   **23. a.** mean = 38.12   **b.** median = 21.4   **c.** bimodal, mode 1 = 14.50, mode 2 = 8.50   **d.** 25.7563   **e.** The skewness of the distribution will be towards the left.

4.2   Measures of Dispersion

EXERCISES

9. 24   **11. a.** 5.8222   **b.** 2.4129 years   **c.** 8   **d.** How often the car was driven, which area of the country the car was driven (e.g. a car driven in the North may have lower time until failure than the car driven in the South because of the winters, etc.) Answers will vary.   **13. a.** 0.2917   **b.** 0.5401 or 54.01 percentage points   **c.** 1.5%   **d.** Credit score, financial stability of the bank, etc. Answers will vary.   **15. a.** Average market value, standard deviation of market value, maximum value, minimum value, range, etc. Answers will vary.   **b.** Portfolio A: Mean market value = $150,000, Standard deviation = $7905.6942, Maximum value = $160,000, Minimum value = $140,000, Range = $20,000, Portfolio B: Mean market value = $150,000, Standard deviation = $37,249.1611, Maximum value = $195,000, Minimum value = $100,000, Range = $95,000   **c.** Portfolio A is less risky because it has the smaller standard deviation.   **17. a.** mean of original data = 749.1, var of original data = 536.9889, mean of adjusted data = 729.1, var of adjusted data = 536.9889   **b.** mean adjusted data = mean of original data − 20, the variances are the same.   **c.** These results hold in general. If you subtract a constant from each data point, the mean of the adjusted data is equal to the mean of the original data minus the constant. The variance of the adjusted data is the same as the variance of the original data (i.e. the way in which the data vary does not change if you simply subtract a constant from each data point.)   **19.** 718 to 1226   **21. a.** $3750 to $5250   **b.** $3000 to $6000   **c.** $2250 to $6750   **d.** In order to use the empirical rule, we must assume that the distribution of the daily sales is approximately bell-shaped.

4.3   Measures of Relative Position

EXERCISES

11. a. 1   **b.** 6   **c.** 25th percentile: approximately 25% of the salespeople sold one or fewer copiers in a day. 90th percentile: approximately 90% of salespeople sold 6 or fewer copiers in a day.   **d.** 83rd percentile   **e.** 33rd percentile   **13. a.** .185   **b.** .255   **c.** .285   **d.** 1st quartile: approximately 25% of the batting averages are at or below .185, 2nd quartile: approximately 50% of the batting averages are at or below .255, 3rd quartile: approximately 75% of the batting averages are at or below .285.   **e.** 0.1   **f.**

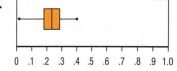

0.02 is an outlier. This likely represents the pitcher's batting average. **g.** −2.61   **h.** 1.14   **i.** The player with the .020 batting average has a batting average 2.61 standard deviations below the mean. The player with a .330 batting average has a batting average 1.14 standard deviations above the mean.   **j.** 70th percentile   **k.** 15th percentile   **15.** First exam: $z = 0.10$, Second exam: $z = -0.43$. On the first exam, the student's score was 0.1 standard deviations above the mean; but on the second exam, the student's score was 0.43 standard deviations below the mean. Thus, although the student achieved a higher absolute score on the second exam, the student performed relatively better on the first exam than on the second.

4.5 The Coefficient of Variation

EXERCISES

5. a. 12 **b.** 10 **c.** Since the coefficient of variation is smaller for Machine Y than Machine X, the standard deviation in the diameter of the bolts produced is smaller relative to the average diameter for Machine Y than for Machine X. Thus, Machine Y more consistently produces bolts of the correct diameter.

4.6 Analyzing Grouped Data

EXERCISES

3. Mean = 17.5 days, Variance = 55.3043 days^2

4.7 Proportions

EXERCISES

5. a. 0.5 **b.** 0.3415 **c.** 0.4661 **d.** 0.6341 **e.** It appears that the supplier executives are optimistic, while the original equipment managers appear to be skeptical about the economic recovery. Answers will vary. **7. a.** 28% **b.** 22%
c. Americans are not in the habit of saving money in case of a financial emergency. Around a quarter would not be able to cope at all. Answers will vary. **9. a.** 835 **b.** 2884

4.8 Measures of Association between Two Variables

EXERCISES

11. a. Yes, the pattern roughly follows a straight line. The pattern is upward sloping; as x increases, y generally increases. The data values are widely dispersed. The fanning of the data could be considered a significant deviation from the pattern.
b. Yes, the pattern roughly follows a straight line. The pattern is downward sloping; as x increases, y decreases. The data values are tightly clustered; in fact, they exactly fall in a straight line. There are no significant deviations from the pattern.
13. a. Yes, the data collected seem to be appropriate to study the relationship between training and countertop defects. An example of a bias could be that some employees could have more experience than others aside from the training. The data are collected using an observational study. Answers will vary.
b.

c. Yes, the pattern roughly follows a straight line. The pattern is downward sloping; as the number of hours of training increases the number of defects tends to decrease. The data values are tightly clustered. There is one significant deviation from the pattern: the employee with only one hour of training who had only one defect.

15.

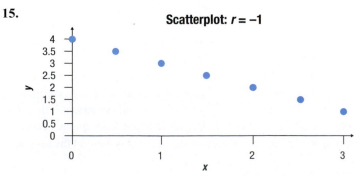

Answers may vary, but points should be plotted in a downward sloping, perfectly straight line. **17. a.** Tightly clustered in a positive linear fashion. **b.** Loosely clustered in a positive linear fashion. **c.** Tightly clustered in a negative linear fashion. **d.** Loosely clustered in a negative linear fashion. **e.** Loosely clustered in a positive linear fashion.

19. a.

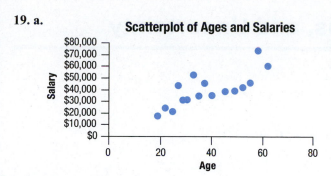

Scatterplot of Ages and Salaries

b. $r = 0.7707$ **c.** The correlation coefficient indicates a moderate positive linear relationship. This seems consistent with the scatterplot.   **21. a.** Summer   **b.** Valentine's Day, Christmas   **c.** Winter

Additional Exercises

EXERCISES

1. No. Because you did not want any variation in the length of the boards. Answers will vary.   **3. a.** 0.3438
b. Mean = 2.4531, Standard Deviation = 2.5754   **c.** 0 to 5.0285   **d.** 87.5%   **e.** The empirical rule predicts that 68.26% of the data falls within one standard deviation of the mean. The percent of the data in this problem falling within one standard deviation is 87.5%, which is not very close to the empirical rule. Answers may vary.   **5. a.** Machine A = 3.1429, Machine B = 3.1429
b. Machine A = 5.8095, Machine B = 0.8095   **c.** Machine A = 2.4103, Machine B = 0.8997   **d.** Machine B is probably a better machine because the average number of defects produced by the 2 machines is the same but Machine B is much more consistent in the number of defective circuit boards it produces. Answers may vary.   **7. a.** 92.0929   **b.** 9.1383
c. 9   **d.** 12   **e.** The literacy rates are normally distributed.   **9. a.** Yes, the variables measured seem appropriate to study the relationship between reaction time and the amount of drug in the bloodstream. Biases could include differences in reaction time without the drug. The data are ratio data. Answers will vary.

b.

Scatterplot of Drug Amounts and Reaction Times

c. Yes, the pattern roughly follows a straight line. The pattern is upward sloping; as the amount of drug increases, the reaction time tends to increase. The data values are tightly clustered. There is one significant deviation from the pattern: the person who was administered 8 mg of the drug had a reaction time of only 0.6 sec. Answers will vary.   **11. a.** Strong positive linear relationship.   **b.** Weak positive linear relationship.   **c.** Strong negative linear relationship.   **d.** Weak negative linear relationship.   **e.** Weak positive linear relationship.

13. a.

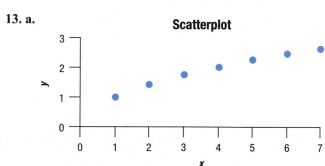

Scatterplot

b. $r = 0.9916$ **c.** There is a strong positive relationship between x and y. However, the scatterplot does not appear to be linear. It appears that y is the square root of x. Answers will vary.

5 Probability, Randomness, and Uncertainty

5.1 Probability Concepts

EXERCISES

7. a. S = {Brand A, Brand B, Brand C, Brand D}   **b.** M = {Brand B, Brand C, Brand D}

5.2 Interpreting Probability

EXERCISES

9. a. $P(1) = 0.4242$, $P(2) = 0.2121$, $P(3) = 0.1515$, $P(4) = 0.0606$, $P(5) = 0.0909$, $P(6) = 0.0606$ **b.** 1

11. 0.88   **13.** 0.9   **15. a.** subjective   **b.** relative frequency   **c.** subjective   **d.** relative frequency   **e.** classical

17. a. S = {MM, MF, FM, FF}   **b.** $\frac{1}{4} = 0.25$ **c.** $\frac{3}{4} = 0.75$ **19. a.** $\frac{1}{3} \approx 0.3333$ **b.** $\frac{2}{3} \approx 0.6667$

5.4 Laws of Probability

EXERCISES

7. a. Yes   **b.** No, probabilities cannot be greater than 1.   **c.** Yes   **d.** No, probabilities cannot be less than 0.   **e.** Yes
9. a. The event cannot occur.   **b.** The event is certain to occur.   **c.** Relative Frequency interpretation: If an experiment is performed 100 times, the event will occur, on average, 45 times.   **d.** Relative Frequency interpretation: If an experiment is performed 100 times, the event will occur, on average, 65 times.   **e.** Not a valid probability because it is negative.
11. a. 0.9696   **b.** 0.0523   **13. a.** 0.27   **b.** 0.08   **c.** 0.165   **d.** 0.4   **e.** 0.3   **f.** 0.51   **g.** Relative frequency   **h.** No, the wife could have more than $150,000 in insurance and the husband could have between $50,000 and $100,000 of insurance. Answers may vary.

5.5 Conditional Probability

EXERCISES

3. a. 0.3194   **b.** 0.3472   **c.** 0.6327   **d.** 0.6032   **5. a.** 0.8571   **b.** 0.1429   **7. a.** 0.005   **b.** 0.0417

5.6 Independence

EXERCISES

7. No, the events are dependent. If A = husband has more than $150,000 insurance and B = wife has more than $50,000 insurance, $P(A|B) \neq P(A)$.   **9.** 0.0001   **11. a.** 0.9980   **b.** 0.0020   **c.** 0.000001   **13. a.** 0.0004   **b.** 0.0004

5.7 Bayes' Theorem

EXERCISES

5. P(Dem|Favor) = 0.7404   **7.** P(Def|Insp) = 0.2069   **9.** P(Woman's name|Man chosen) = 0.7347

5.8 Counting

EXERCISES

5. 120   **7. a.** 1   **b.** 6   **c.** 120   **d.** 5040   **9.** 665,280   **11.** 50,400   **13.** 55

Additional Exercises

EXERCISES

1. a. $S = \{$MMM, MMF, MFM, MFF, FMM, FMF, FFM, FFF$\}$ **b.** $\frac{1}{8} = 0.125$ **c.** $\frac{7}{8} = 0.875$ **3. a.** $\frac{18}{38} = \frac{9}{19} \approx 0.4737$

b. $\frac{12}{38} = \frac{6}{19} \approx 0.3158$ **c.** $\frac{2}{38} = \frac{1}{19} \approx 0.0526$ **d.** $\frac{1}{38} \approx 0.0263$ **e.** $\frac{35}{38} \approx 0.9211$ **5.** 0.9989 **7. a.** 1 to 5 **b.** 1 to 1

c. 7 to 1 **d.** $\frac{8}{11} \approx 0.7273$ **9.** $\frac{11}{14} \approx 0.7857$ **11. a.** 8% **b.** 16% **c.** 14% **d.** 84%

6 Discrete Probability Distributions: Information about the Future

6.1 Types of Random Variables

EXERCISES

5. a. Discrete **b.** Continuous **c.** Discrete **d.** Continuous **e.** Discrete **7. a.** Discrete **b.** Continuous **c.** Continuous
d. Discrete **e.** Continuous

6.2 Discrete Probability Distributions

EXERCISES

7. No. The sum of the probabilities is greater than 1. **9.** No. Probabilities cannot be negative. **11.** Yes **13.** No. The sum
of the probabilities is $\frac{15}{16}$ which is less than 1.

x	P(X = x)
1	$\frac{1}{16}$
2	$\frac{2}{16}$
3	$\frac{3}{16}$
4	$\frac{4}{16}$
5	$\frac{5}{16}$

6.3 Expected Value and Variance

EXERCISES

7.

x	p(x)	xp(x)	$\frac{(x-\mu)^2}{p(x)}$
400	0.0	0	0.0
420	0.1	42	291.6
440	0.1	44	115.6
460	0.2	92	39.2
480	0.2	96	7.2
500	0.4	200	270.4
Total	1.0	474	724.0

$E(X) = 474$

$\sigma^2 = 724$

$\sigma = 26.9072$

9.

x	p(x)	xp(x)	$\frac{(x-\mu)^2}{p(x)}$
1	0.1	0.1	0.484
2	0.2	0.4	0.288
3	0.3	0.9	0.012
4	0.2	0.8	0.128
5	0.2	1.0	0.648
Total	1.0	3.2	1.560

a. $E(X) = 3.2$
b. $\sigma^2 = 1.56$
c. $\sigma = 1.2490$
d. $P(X = 5) = 0.2$
e. $P(X \geq 2) = 0.9$
f. $P(X \leq 3) = 0.6$
g. $P(X < 2) = 0.1$

11. a.

x	p(x)	xp(x)	$\frac{(x-\mu)^2}{p(x)}$
$50,000	0.4	20,000	518,400,000
-$10,000	0.6	-6,000	345,600,000
Total	1.0	14,000	864,000,000

b. $14,000   **c.** $29,393.88   **13. a.** Cereal A = $200,000; Cereal B = $276,000.   **b.** Cereal A = $188,414.40; Cereal B = $309,619.10   **c.** Cereal B has a greater value for expected sales, but also a much greater standard deviation. The difference in the expected sales is much smaller than the difference in the standard deviation, so Cereal A is probably the best choice. Answers will vary.

6.4 The Discrete Uniform Distribution

EXERCISES

5. a. HH1, HH2, HH3, HH4, HH5, HH6, HT1, HT2, HT3, HT4, HT5, HT6, TH1, TH2, TH3, TH4, TH5, TH6, TT1, TT2, TT3, TT4, TT5, TT6   **b.** X = Sum of the number of heads on the two coins and number of dots on the die. $X = \{1, 2, 3, 4, 5, 6, 7, 8\}$
c.

x	$P(X = x)$
1	$\frac{1}{24}$
2	$\frac{3}{24}$
3	$\frac{4}{24}$
4	$\frac{4}{24}$
5	$\frac{4}{24}$
6	$\frac{4}{24}$
7	$\frac{3}{24}$
8	$\frac{1}{24}$

d. $E(X) = 4.5$

6.5 The Binomial Distribution

EXERCISES

7. a. 5   **b.** 45   **c.** 15   **d.** 1   **9. a.** $E(X) = 0.9$ **b.** $\sigma = 0.9$ **c.** $P(X = 2) = 0.1722$ **d.** $P(X \leq 3) = 0.9917$
e. $P(X \geq 2) = 0.2252$ **f.** $P(X < 5) = 0.9991$ **11. a.** Binomial distribution with $n = 10$ and $p = 0.10$ **b.** $E(X) = 1$
c. $\sigma = 0.9487$ **d.** $P(X = 1) = 0.3874$ **e.** $P(X = 5) = 0.0015$ **f.** $P(X \geq 3) = 0.0702$ **13. a.** Binomial distribution with

$n = 7$ and $p = 0.1$ **b.** $P(X = 0) = 0.4783$. There is a 47.83% chance that none of the plants will strike. $P(X = 4) = 0.0026$. There is a 0.26% chance that exactly 4 of the plants will strike. $P(X = 7) = 0$. There is a negligible chance that all 7 plants will strike. **c.** $E(X) = 0.7$ **d.** $\sigma = 0.794$. The standard deviation is larger than the expected value. The standard deviation is expressed as the number of plants that strike. Answers may vary. **15. a.** $P(X = 2) = 0.375$ **b.** $P(X = 4) = 0.0625$

17. a. $P(X \le 1) = 0.8290$ **b.** $E(X) = 0.75$ **19. a.** $\dfrac{2}{9}$ or 0.2222 **b.** $P(X = 5) = 0.0389$ **c.** $P(X = 0) = 0.0810$
d. $E(X) = 2.2222, \sigma^2 = 1.7284$

6.6 The Poisson Distribution

EXERCISES

7. $P(X = 2) = 0.0842$ **9. a.** $P(X = 0) = 0.1353$ **b.** $P(X = 0) = 0.0003$ **c.** $\mu = \lambda = 2$ **d.** $\mu = \lambda = 8$ **e.** $\sigma = 2.8284$
f. $P(X \ge 4) = 0.9576$ **11. a.** $\lambda = 20$ (If 5 people arrive on average in 15 minutes, then 20 will arrive on average in 60 minutes.) **b.** $P(X = 0) = 0$ **c.** $P(X > 6) = 0.2378$ **13.** $P(X = 6) = 0.0771$

6.7 The Hypergeometric Distribution

EXERCISES

5. a. X has a hypergeometric distribution with $N = 50, A = 3$ and $n = 10$. **b.** $E(X) = 0.6$ **c.** $\sigma = 0.6785$
d. $P(X \ge 1) = 0.4959$ **e.** $P(X \le 2) = 0.9939$ **f.** $P(X > 3) = 0$ **7. a.** $E(X) = 5$ **b.** $\sigma = 1.5076$ **c.** $P(X = 10) = 0.0006$
d. $P(X = 0) = 0.0006$

Additional Exercises

EXERCISES

1. a. $E(X) = 2.3$ **b.** $\sigma^2 = 1.41$ **c.** $\sigma = 1.1874$ **d.** $P(X = 4.0) = 0.15$ **e.** $P(X \ge 2.0) = 0.75$ **f.** $P(X \le 1.0) = 0.25$
g. $P(X > 3.0) = 0.15$ **3. a.** $P(4 \le X \le 6) = 0.6563$ **b.** $P(X \ge 8) = 0.0547$ **c.** $P(X = 1) = 0.0098$ **5. a.** $E(X) = 1.6667,$
$\sigma^2 = 1.3889$ **b.** $E(X) = 3.25, \sigma^2 = 2.4375$ **c.** $E(X) = 8.8, \sigma^2 = 1.056$ **d.** $E(X) = 0.8333, \sigma^2 = 0.4419$ **e.** $E(X) = 3.5,$
$\sigma^2 = 2.9167$ **7.** $P(X \ge 1) = 0.9615$ **9.** \$0.25 **11.** $P(X \ge 1) = 0.9933$ **13.** $P(X \ge 3) = 0.2962$ **15. a.** Binomial distribution with $n =$ the number of buildings inspected and $p = 0.5$. The binomial is used rather than the hypergeometric because it is not known how many buildings in the population of new buildings have violations. Answers may vary. **b.** $P(X = 3) = 0.25$

7 Continuous Random Variables

7.1 The Uniform Distribution

EXERCISES

7. a. $\mu = 60$ **b.** $\sigma = 1.7321$ **c.** 0.3333 **d.** 0.5 **e.** 0.1667 **f.** 0 **9. a.** $\mu = 8{:}15$ AM **b.** $\sigma = 0.1443$ **c.** 0.3333
d. 0.1667 **e.** 0.5 **f.** 0

7.2 The Normal Distribution

EXERCISES

9.   **11.**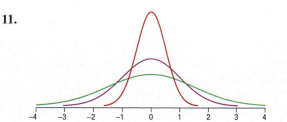

7.3 The Standard Normal Distribution

EXERCISES

5. a. 0.2486   **b.** 0.4500   **c.** 0.4750   **d.** 0.4950   **7. a.** 0.6680   **b.** 0.6710   **c.** 0.9631   **d.** 0.9422

9. a. $P(z \leq -0.44) = 0.3300$ 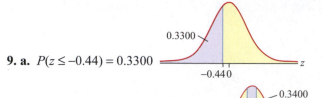   **b.** $P(z \geq 0.44) = 0.3300$

c. $P(-0.44 \leq z \leq 0.44) = 0.3400$ **d.** $P(z \leq -0.67) = 0.2514$

e. $P(z \geq 0.67) = 0.2514$ **f.** $P(-0.67 \leq z \leq 0.67) = 0.4972$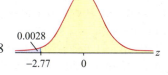

11. a. $P(0 \leq z \leq 0.79) = 0.2852$ **b.** $P(-1.57 \leq z \leq 2.33) = 0.9319$

c. $P(z \geq 1.89) = 0.0294$ **d.** $P(z \leq -2.77) = 0.0028$

13. 1.645   **15.** 1.28   **17.** −2.33   **19.** 1.14   **21.** 1.645   **23. a.** 0.7333   **b.** 0.0548   **c.** 0.0228   **25.** 0.4101
27. a. $631   **b.** 0.2206   **c.** 0.1190   **d.** 0.4235   **e.** Answers will vary.   **29. a.** 0.0668   **b.** 0.0668   **c.** 0.6826
31. a. 92.24   **b.** No. The score must be at least a 92.24 to be in the top 10% of scores.   **c.** 75.28   **d.** The student who
scored a 65 would receive an F because the score is less than 71.76 and thus is in the lowest 10% of the scores.
33. a. At least 131   **b.** 0.13%   **c.** 6.5%   **d.** At least 135

7.4 Approximations to Other Distributions

EXERCISES

5. a. 45   **b.** 2.1213   **c.** 0   **d.** 0.9830   **e.** 0.0011   **7. a.** 120   **b.** 6.9282   **c.** 0   **d.** 0.5264   **e.** 0.0853   **9. a.** 2.2361
b. 0.0318   **c.** 0.0222   **d.** 0.0318 is the more exact probability. When using the normal approximation, the probability is
underestimated. Answers will vary.   **11. a.** 30   **b.** 5.4772   **c.** 0.9977   **d.** 0.4641   **e.** 0.8413

Additional Exercises

EXERCISES

1. a. 20   **b.** 8.6603   **c.** 0.1667   **d.** 57.73%   **e.** According to the empirical rule 68% of the results will fall within one
standard deviation of the mean. The discrepancy is due to the fact that the empirical rule applies to distributions that are
bell-shaped. The uniform distribution is not bell-shaped.

3. a. $P(0 \le z \le 1.00) = 0.3413$ **b.** $P(-2.50 \le z \le 3.01) = 0.9925$

c. $P(z \ge 3.25) = 0.0006$ **d.** $P(z \le -2.50) = 0.0062$

5. 1.28   **7. a.** 5   **b.** 2.2349   **c.** Using the normal approximation: 0.9778; Using tables: 0.9933   **d.** Using the normal approximation: 0.9418; Using tables: 0.9596   **e.** Yes, $np = 5$, which is almost too small for the approximation to the binomial to produce accurate results. Answers may vary.   **9.** $\mu \approx 507$, $\sigma \approx 114$ **11.** 535 days   **13.** $\mu \approx 5.71$ ml   **15.** The bulbs should be replaced after approximately 397 hours.   **17.** $\mu = \$61,000$, $\sigma = \$12,000$

8   Samples and Sampling Distributions

8.1   Random Samples

EXERCISES

9. a. Yes, readers voluntarily sent in their responses.   **b.** People with strong opinions would be more likely to reply and "Anger" is not well-defined and reader response may be influenced by whether or not they are angry at the moment, for example. Answers will vary.   **c.** No. All Americans do not read the magazine, and even if they did, it is likely that only those readers with strong opinions would have responded to the survey. Answers will vary.   **11.** No. This is a voluntary survey so only those with strong opinions may reply. People without phones or who are unable or unwilling to pay for the call are excluded. Answers will vary.

8.3   The Distribution of the Sample Mean

EXERCISES

11. a. $\mu_{\bar{x}} = 50$, $\sigma_{\bar{x}} = 1.5811$ **b.** $\mu_{\bar{x}} = 50$, $\sigma_{\bar{x}} = 1.3484$ **c.** $\mu_{\bar{x}} = 50$, $\sigma_{\bar{x}} = 1$ **d.** The standard error decreases, reflecting the additional information provided by a larger sample size.   **13.** 0.0174   **15. a.** 1   **b.** 1   **c.** 1   **d.** 0.0793   **17. a.** 0.2912   **b.** 0.0006   **19. a.** 0.0016   **b.** 0.0016   **c.** 0.9232   **21. a.** 0.2981   **b.** 0.0170   **c.** 0.6261

8.4   The Distribution of the Sample Proportion

EXERCISES

7. a. 0.3   **b.** 0.7   **9. a.** $\mu_{\hat{p}} = 0.35$, $\sigma_{\hat{p}} = 0.0774$ **b.** $\mu_{\hat{p}} = 0.35$, $\sigma_{\hat{p}} = 0.0661$ **c.** $\mu_{\hat{p}} = 0.35$, $\sigma_{\hat{p}} = 0.0551$ **d.** The standard deviation decreases, reflecting the additional information provided by a larger sample size.   **11.** 0.1151   **13. a.** 0.8461   **b.** 0.0011   **c.** 0.4793   **d.** 0.4989   **15. a.** 0.0023   **b.** 0.0384   **17. a.** 0.9767   **b.** 0.6536

8.5   Other Forms of Sampling

EXERCISES

9. A systematic sample is not a random sample because it has potential for bias if there is some inherent data pattern.
11. a. Families in the state of Florida   **b.** The number of children per family   **c.** Ratio   **d.** Simple random sampling: develop sampling frame, choose participants randomly from sampling frame; Cluster sampling: create clusters, randomly select clusters, survey all members of the chosen clusters; Stratified sampling: create strata, randomly select participants from each

stratum such that the population characteristics are adequately represented. Answers will vary.   **e.** Cluster sampling, because travel costs would likely be minimized. Answers will vary.   **13. a.** Convenience sampling   **b.** All of the people surveyed may not be residents of Orlando, Florida because there are a large number of tourists visiting the area.   **c.** No, the sample was not representative of the population of interest. Answers will vary.

Additional Exercises

EXERCISES

1. a. Voluntary sampling   **b.** All Americans do not watch the news program. It is likely that only those with strong opinions responded.   **c.** No. Answers will vary.   **3.** 0   **5. a.** Convenience sample   **b.** Pre-med majors may be over-represented since there are more pre-med majors in a biology class than in most other classes. Answers will vary.   **c.** No, because the sample is biased.   **7. a.** $\mu_{\bar{x}} = 15\%$, $\sigma_{\bar{x}} = 6.957\%$ **b.** 8.043% to 21.957%   **9. a.** 0.0031   **b.** Not necessarily. The samples may not have been representative of the population and employees may have not been honest about the amount of time they spend texting at work. Answers will vary.   **c.** Survey respondents might not be completely truthful when answering the survey since it is a representation of job performance. Answers will vary.   **11. a.** 0.0132   **b.** 0   **c.** No. Noise in excess of 103 decibels only occurs 1.32% of the time.   **13. a.** $\mu_{\hat{p}} = 0.90$, $\sigma_{\hat{p}} = 0.0134$ **b.** $\mu_{\hat{p}} = 0.85$, $\sigma_{\hat{p}} = 0.0179$ **c.** 0.8638   **d.** 0.9750   **15. a.** 0.9951   **b.** 0.8788   **c.** 0.1212   **17. a.** $P(\bar{x} > 35) \approx 1$, indicating that the researchers are likely correct in claiming that the U.S. average is greater than 35 hours (i.e. 50.4 hours).   **b.** $\sigma \approx 16.5$

Estimating Means and Proportions: Single Samples

9.3   Interval Estimation of the Population Mean

EXERCISES

11. a. 1.96   **b.** 2.575   **c.** 1.645   **13. a.** 2.33   **b.** 1.88   **c.** 1.75   **15.** (237, 263)   **17.** (4.7, 5.3)   **19.** (4964, 5036)   **21.** (29,232, 30,768)   **23.** (8.5, 10.7)

9.4   Interval Estimation of the Population Mean: Small Samples, σ Unknown

EXERCISES

7. 2.518   **9. a.** 2.201   **b.** 2.898   **c.** 1.721   **11.** (77.3579, 83.3887)   **13. a.** (5.6, 7.2)   **b.** We are 95% confident that the true average length of stay for the hospital's abdominal surgery patients is between 5.6 days and 7.2 days. We are assuming that the lengths of stay are approximately normally distributed.   **15. a.** (101.36, 128.64)   **b.** We are 90% confident that the true average price of a regular room with a king size bed in the resort community is between $101.36 and $128.64. We are assuming the prices are normally distributed.

9.5   Precision and Sample Size: Means

EXERCISES

9. $n = 98$ **11.** $n = 31$

9.6   Estimating the Population Proportion

EXERCISES

5. a. 65%   **b.** (62.52, 67.48)   **7. a.** (0.2176, 0.3324)   **b.** No, 0.40 falls outside the confidence interval.   **9. a.** (0.3856, 0.5644)   **b.** Yes, 0.665 falls above the interval.   **11. a.** (0.0625, 0.1475)   **b.** Yes, 0.05 falls below the interval.

9.7 Precision and Sample Size: Proportions

EXERCISES

5. $n = 944$ **7.** $n = 41$

Additional Exercises

EXERCISES

1. $n = 27$ **3. a.** $n = 68$ **b.** $(0.2598, 0.3402)$ We are 95% confident that the true proportion of Fontana residents who think safety is a significant factor in their decision about whether or not to ride a bus is between 0.2598 and 0.3402.   **5.** $(9.57, 9.93)$
7. $n = 1038$ **9.** $(29{,}630, 32{,}570)$ **11.** $(621, 679)$ **13.** $(0.3965, 0.4835)$

10 Hypothesis Testing: Single Samples

10.2 Reaching a Conclusion

EXERCISES

7. a. $H_0: \mu \le 30{,}000$; $H_a: \mu > 30{,}000$ Type I error: They will risk needing to replace tires. Type II error: The company will research ways to make their tires last longer, even though that may be unnecessary.   **b.** $H_0: \mu \le 240$; $H_a: \mu > 240$ Type I error: Mrs. Russell doesn't research ways to improve the light bulbs even though she may need to. Type II error: Mrs. Russell will do research to improve the light bulbs, even though that may be unnecessary.   **c.** $H_0: \mu \ge 7$; $H_a: \mu < 7$ Type I error: The grocery store may not look for ways to improve the checkout process, even though it takes longer than 7 minutes. Type II error: The grocery store will look for ways to improve the checkout process, even though it takes less than 7 minutes.

10.4 Testing a Hypothesis about a Population Mean

EXERCISES

11. a. $z = -1.645$ **b.** $z = 2.05$ **c.** $z = 1.75$ and -1.75 **13.** $H_0: \mu \le 4100$; $H_a: \mu > 4100$, Critical value $= 2.33$, $z = 0.80$, Fail to reject H_0.   **15. a.** The homes that were covered by insurance in the area damaged by hurricane Andrew.   **b.** $\mu =$ the average size of the claim.   **c.** $H_0: \mu \le 24{,}000$; $H_a: \mu > 24{,}000$ **d.** Critical value $= 2.33$, $z = 13.37$, Reject H_0.   **e.** Yes, there is overwhelming evidence that the home damage caused by hurricane Andrew is greater than the historical average.   **17. a.** Yes. $H_0: \mu = 45$; $H_a: \mu \ne 45$, $z = 2.24$, P-value $= 0.025$, Reject H_0.   **b.** 43.25 minutes   **19.** Yes. $H_0: \mu \ge 1$; $H_a: \mu < 1$, Critical value $= -1.645$, $z = -2.24$, Reject H_0.   **21.** Yes. $H_0: \mu \ge 5$; $H_a: \mu < 5$ Critical value $= -2.33$, $z = -8.66$, Reject H_0.   **23.** Yes. $H_0: \mu \ge 10$; $H_a: \mu < 10$, Critical value $= -1.645$, $z = -44.27$, Reject H_0.   **25. a.** $t = -3.106$ **b.** $t = 2.776$ **c.** $t = 1.711$ and -1.711 **27.** $H_0: \mu \le 100$; $H_a: \mu > 100$, Critical value $= 1.895$, $t = 0.736$, P-value > 0.10 (tables), P-value $= 0.2429$ (exact), Fail to reject H_0.   **29. a.** NarStor hard drives   **b.** Time until failure   **c.** Ratio   **d.** No. $H_0: \mu \ge 14{,}400$; $H_a: \mu < 14{,}400$ Critical value $= -2.602$, $t = -1.727$, Fail to reject H_0.   **e.** We must assume that the time until failure of the NarStor hard drives has an approximately normal distribution.   **31. a.** Yes. $H_0: \mu \le 1000$; $H_a: \mu > 1000$, Critical value $= 1.711$, $t = 5$, Reject H_0.   **b.** The life of the light bulbs has an approximately normal distribution.   **c.** $P < 0.01$ (tables), $P = 0.00002$ (exact)   **33. a.** Yes. $H_0: \mu \le 100$; $H_a: \mu > 100$, Critical value $= 1.318$, $t = 1.522$, Reject H_0.   **b.** The number of shoppers per day at the neighboring shopping center has an approximately normal distribution.   **35. a.** Time taken for the moves of top international chess players   **b.** Yes. $H_0: \mu \ge 5$; $H_a: \mu < 5$, Critical value $= -1.645$, $z = -7.07$, Reject H_0.   **37. a.** Reject H_0 **b.** Reject H_0 **c.** Fail to reject H_0 **d.** Reject H_0 **39. a.** P-value $= 0.0571$, Fail to reject H_0.   **b.** P-value $= 0.0122$, Reject H_0.   **c.** P-value $= 0.0574$, Fail to reject H_0.   **41. a.** P-value $= 0.0086$, Reject H_0.   **b.** P-value $= 0.0233$, Fail to reject H_0.   **c.** P-value $= 0.0652$, Fail to reject H_0.   **43. a.** 0.0314   **b.** Yes, there is sufficient evidence to support the claim that the boots can remain immersed for more than 12 hours without leaking.

10.5 Practical Significance versus Statistical Significance

EXERCISES

3. The test is statistically significant since the null hypothesis that the job should take 45 minutes to complete was rejected. However, the average of the sample was only 2 minutes greater than what the supervisor believed, which might not make a huge difference in the context of the wage negotiations. Answers will vary. **5.** The test is statistically significant since the null hypothesis that the average waiting time is less than or equal to 45 minutes was rejected. The test is also practically significant since we are dealing with customer complaints, and, based on the sample, customers are having to wait 10 minutes longer than they should to talk to an IRS representative. Answers will vary.

10.6 The Relationship between Confidence Interval Estimation and Hypothesis Testing

EXERCISES

3. a. (40.09, 44.91) Reject H_0 **b.** Heights of children are approximately normally distributed. **5. a.** (1121.28, 1213.40) **b.** No

10.7 Testing a Hypothesis about a Population Proportion

EXERCISES

5. a. $z = -1.645$ **b.** $z = 2.33$ **c.** $z = 1.645$ and -1.645 **7.** Yes. $H_0: p \leq 0.38$; $H_a: p > 0.38$, Critical value $= 2.33$, $z = 4.04$, Reject H_0. **9. a.** Yes. $H_0: p = 0.013$; $H_a: p \neq 0.013$, Critical values $= -1.96$, 1.96, $z = 2.75$, Reject H_0. **b.** Teenagers may not be honest when answering a survey like this. Answers will vary. **11. a.** No. $H_0: p \geq 0.20$; $H_a: p < 0.20$, Critical value $= -2.33$, $z = -1.77$, Fail to reject H_0. **b.** No, $np_0 \geq 5$ and $n(1 - p_0) \geq 5$. **13. a.** No. $H_0: p \leq 0.05$; $H_a: p > 0.05$, Critical value $= 1.28$, $z = -0.46$, Fail to reject H_0. **b.** Yes, $np_0 \geq 5$ and $n(1 - p_0) \geq 5$, but $np_0 = 5$, which could be a concern. Answers may vary. **15.** No. $H_0: p \leq 0.15$; $H_a: p > 0.15$ Critical value $= 2.33$, $z = 0.31$, Fail to reject H_0. **17. a.** No. $H_0: p \leq 0.002$; $H_a: p > 0.002$, Critical value $= 1.645$, $z = 0.12$, Fail to reject H_0. **b.** 0.4522 **c.** No **19. a.** $H_0: p \geq 0.40$; $H_a: p < 0.40$, $z = -0.73$, P-value $= 0.2327$ **b.** No **21.** Yes. $H_0: p \geq 0.32$; $H_a: p < 0.32$, Critical value $= -1.645$, $z = -2.12$, Reject H_0. **23. a.** Yes. $H_0: p \leq 0.49$; $H_a: p > 0.49$, Critical value $= 1.645$, $z = 3.82$, Reject H_0. **b.** 110 people

10.8 Testing a Hypothesis about a Population Variance

EXERCISES

7. a. $df = 19$, $\chi^2 = 36.191$ **b.** $df = 23$, $\chi^2 = 35.172$ **c.** $df = 4$, $\chi^2 = 14.860$ **9. a.** (0.0557, 0.0941) We are 95% confident that the standard deviation of the bolt diameters is between 0.0557 inch and 0.0941 inch. **b.** Yes. $H_0: \sigma^2 \leq 0.0025$; $H_a: \sigma^2 > 0.0025$, Critical value $= 42.557$, $\chi^2 = 56.84$, Reject H_0. **c.** The diameters of the bolts have an approximately normal distribution. **11. a.** Yes. $H_0: \sigma^2 \leq 0.0625$; $H_a: \sigma^2 > 0.0625$, Critical value $= 42.980$, $\chi^2 = 47.04$, Reject H_0. **b.** (0.2841, 0.4608) We are 90% confident that the standard deviation of the share price of the bond fund is between $0.28 and $0.46. **c.** The share prices of the bond fund have an approximately normal distribution.

Additional Exercises

EXERCISES

1. a. $H_0: \mu = 18$; $H_a: \mu \neq 18$ **b.** The company believes that the average time to replace a set of 4 tires has changed when in fact the average time is unchanged. **c.** The company believes that the average time to replace a set of 4 tires remains unchanged when in fact the average time has changed. **3.** Yes. $H_0: \mu \geq 3.5$; $H_a: \mu < 3.5$, Critical value $= -1.28$, $z = -4.34$, Reject H_0. **5.** Yes. $H_0: \mu \geq 13.20$; $H_a: \mu < 13.20$, $z = -2.53$, P-value $= 0.0057$, Reject H_0. **7.** No. $H_0: \mu \leq 895$; $H_a: \mu > 895$, Critical value $= 1.28$, $z = 1.19$, Fail to reject H_0. **9. a.** No. $H_0: p \geq 0.003$; $H_a: p < 0.003$, Critical value $= -1.645$, $z = -1.42$, Fail to reject H_0. **b.** 0.0778 **c.** Yes, H_0 would be rejected. **11.** No. $H_0: p \leq 0.2632$; $H_a: p > 0.2632$, Critical value $= 1.645$, $z = 0.90$, Fail to reject H_0. **13. a.** P-value $= 0.0062$, Reject H_0. **b.** P-value $= 0.0256$, Fail to reject H_0.

c. P-value $= 0.0002$, Reject H_0.   **15. a.** P-value $= 0.0464$, Reject H_0.   **b.** P-value $= 0.0050$, Reject H_0.   **c.** P-value $= 0.0510$, Fail to reject H_0.   **17.** Yes. $H_0: \sigma^2 \leq 0.00156$; $H_a: \sigma^2 > 0.00156$, Critical value $= 118.498$, $\chi^2 = 134.615$, Reject H_0.
19. a. No. $H_0: p \leq 0.90$; $H_a: p > 0.90$, Critical value $= 1.28$, $z = 1.25$, Fail to reject H_0.   **b.** 0.1056 **21. a.** No.
$H_0: \sigma^2 \geq 0.01$; $H_a: \sigma^2 < 0.01$, Critical value $= 3.325$, $\chi^2 = 1.44$, Fail to reject H_0.   **b.** $(0.0008, 0.0053)$

11   Inferences about Means and Proportions: Two Samples

11.1   Comparing Two Population Means

EXERCISES

7. a. $z = -1.75$ **b.** $z = 1.41$ **c.** $z = 2.33$ and -2.33 **9. a.** $(-7.50, -2.50)$ We are 95% confident that Mr. Ellis' expenses are between \$2.50 and \$7.50 less than Mr. Ford's.   **b.** Yes. $H_0: \mu_1 - \mu_2 = 0$; $H_a: \mu_1 - \mu_2 \neq 0$, Critical values $= -1.96, 1.96, z = -3.93$, Reject H_0.   **c.** The confidence interval only contains negative values indicating that with 95% confidence the expenses for Mr. Ellis will always be less than those of Mr. Ford.   **11. a.** Yes. $H_0: \mu_1 - \mu_2 \geq 0$; $H_a: \mu_1 - \mu_2 < 0$, Critical value $= -1.28$, $z = -1.283$, Reject H_0.   **b.** P-value $= 0.0997$ **c.** Yes, we would fail to reject H_0 at $\alpha = 0.05$.

11.2   Comparing Two Population Means: Small Samples, σ_1 and σ_2 Unknown

EXERCISES

7. a. $df = 36$, $t = -2.028$ **b.** $df = 23$, $t = 2.807$ **c.** $df = 28$, $t = 1.701$ and -1.701 **9. a.** $(-1.96, -0.04)$ We are 90% confident that the fat content in Fiber Oat Flakes is between 1.96 and 0.04 grams less than the fat content in Bran Flakes Plus.   **b.** Yes. $H_0: \mu_1 - \mu_2 \geq 0$; $H_a: \mu_1 - \mu_2 < 0$, Critical value $= -1.311$, $t = -1.778$, Reject H_0.   **c.** An independent experimental design is used, both populations are approximately normal, population variances are equal.   **11. a.** $(-2.70, 62.70)$ We are 95% confident that dramas generate between \$2.7 million less and \$62.7 million more in box office revenues than comedies.   **b.** No. $H_0: \mu_1 - \mu_2 \leq 0$; $H_a: \mu_1 - \mu_2 > 0$, Critical value $= 2.479$, $t = 1.886$, Fail to reject H_0.   **c.** P-value $= 0.0353$ **d.** An independent experimental design is used, both populations are approximately normal, population variances are equal.   **13. a.** An independent experimental design is used, both populations are approximately normal, population variances are equal.   **b.** Yes. $H_0: \mu_1 - \mu_2 \leq 0$; $H_a: \mu_1 - \mu_2 > 0$, Critical value $= 2.500$, $t = 2.703$, Reject H_0.   **c.** $H_0: \mu_1 - \mu_2 \leq 0$; $H_a: \mu_1 - \mu_2 > 0$, Critical value $= 2.518$, $t = 2.778$, Reject H_0.   **d.** H_0 is rejected regardless of the variance assumption. Answers will vary.

11.3   Paired Difference

EXERCISES

7. a. $df = 11$, $t = -3.106$ **b.** $df = 4$, $t = 2.776$ **c.** $df = 24$, $t = 1.711$ and -1.711 **9. a.** Yes. The same cashier is using the old register and the new register so the samples can be paired. Answers will vary.   **b.** The differences have an approximately normal distribution.   **c.** Answers will vary.   **d.** $(-4.15, 0.43)$ With 95% confidence, cashiers using the old cash register process between 4.15 fewer and 0.43 more items than using the new cash register.   **e.** Yes. $H_0: \mu_d \geq 0$; $H_a: \mu_d < 0$, Critical value $= -1.943$, $t = -1.983$, Reject H_0.

11.4   Compairing Two Population Proportions

EXERCISES

7. a. $z = -2.33$ **b.** $z = 1.645$ **c.** $z = 1.645$ and -1.645 **9. a.** Yes, the sample sizes are sufficiently large. $H_0: p_1 - p_2 \geq 0$; $H_a: p_1 - p_2 < 0$, Critical value $= -1.28$, $z = -0.91$, Fail to reject H_0. There is not sufficient evidence to support the fund-raiser's theory.   **b.** P-value $= 0.1814$. This is the probability of making a Type I error. Answers will vary.   **c.** $(-0.1098, 0.0403)$ We are 95% confident that the proportion of men who answered "Yes" when asked to donate to a worthy cause is between 0.1098 less than and 0.0403 greater than the proportion of women who answered "Yes."   **11. a.** Yes, the sample sizes are sufficiently large. (−0.261, −0.149) We are 99% confident that the proportion of people age 30 or younger that believe alcoholic beverage commercials are targeted at teenagers is between 0.149 and 0.261 lower than the proportion of people in the older than 30 age group. **b.** Yes. $H_0: p_1 - p_2 \geq 0; H_a: p_1 - p_2 < 0$, Critical value = −2.33, $z = -9.22$, Reject H_0.

Additional Exercises

EXERCISES

1. Yes. $H_0: \mu_1 - \mu_2 \geq 0; H_a: \mu_1 - \mu_2 < 0$, Critical value = −1.645, $z = -2.59$, Reject H_0. **3. a.** Yes. The cholesterol levels are measured for the same person before and after the diet so the samples can be paired. Answers will vary. **b.** The differences have an approximately normal distribution. **c.** Answers will vary. **d.** No. $H_0: \mu_d \leq 0; H_a: \mu_d > 0$, Critical value = 3.143, $t = 1.769$, Fail to reject H_0. **5.** $H_0: \mu_1 - \mu_2 = 0; H_a: \mu_1 - \mu_2 \neq 0$, Critical values = −2.878, 2.878, $t = -1.388$, Fail to reject H_0. **7. a.** No. $H_0: \mu_1 - \mu_2 \leq 0; H_a: \mu_1 - \mu_2 > 0$, Critical value = 1.645, $z = 2.00$, Reject H_0. **b.** (0.08, 0.52) **9.** $H_0: \mu_1 - \mu_2 \leq 10$; $H_a: \mu_1 - \mu_2 > 10$, Critical value = 2.014, $t = 0.380$, Fail to reject H_0. There is not sufficient evidence that Arrangement A has a lower mean net annual income by at least \$10. **11.** $H_0: \mu_1 - \mu_2 = 0; H_a: \mu_1 - \mu_2 \neq 0$, Critical values = −2.101, 2.101, $t = 8.771$, Reject H_0. There is sufficient evidence of a difference in the average weights. **13. a.** $H_0: p \geq 0.10; H_a: p < 0.10$, Critical value = −1.645, 2009: $z = -16.29$, Reject H_0, 2010: $z = -5.84$, Reject H_0, 2011: $z = 0.53$, Fail to reject H_0. **b.** (−0.0316, −0.0133) We are 95% confident that the proportion of travelers using tablets or e-readers in 2011 is between 1.33% and 3.16% greater than the proportion of travelers using tablets or e-readers in 2010. **c.** Yes. $H_0: p_1 - p_2 \geq 0; H_a: p_1 - p_2 < 0$, Critical value = −1.645, $z = -4.77$, Reject H_0. **15. a.** Yes. $H_0: p_1 - p_2 = 0; H_a: p_1 - p_2 \neq 0$, Critical values = −1.96, 1.96, $z = 4.78$, Reject H_0. **b.** (0.0450, 0.0750) We are 95% confident that the percentage of people with maxed out credit cards with IQs of 90 is between 4.55% and 7.5% greater than the percentage of people with maxed out credit cards with IQs greater than 125. **c.** An independent experimental design was used.

12   Analysis of Variance (ANOVA)

12.1   Are Means from Samples Significantly Different?

EXERCISES

9. a. Probably not, the boxes overlap quite a bit. Variation appears to be large in comparison to the difference in the middle values. Answers will vary. **b.** Probably, the variation in scores for juniors is larger than for freshmen but the middle value is greater for juniors than for freshmen. Answers will vary. **c.** Probably not, the boxes overlap quite a bit. Variation appears to be large in comparison to the difference in the middle values. Answers will vary. **11. a.** The experimental units are the items produced and the treatment is the shift. **b.** First shift: mean = 158, median = 161; Second shift: mean = 191.8571, median = 182; Third shift: mean = 111.5714, median = 111 **c.** First shift: Min = 127, Max = 181, $Q_1 = 140$, $Q_3 = 173$; Second shift: Min = 162, Max = 224, $Q_1 = 168$, $Q_3 = 219$; Third shift: Min = 77, Max = 147, $Q_1 = 77$, $Q_3 = 145$

d.

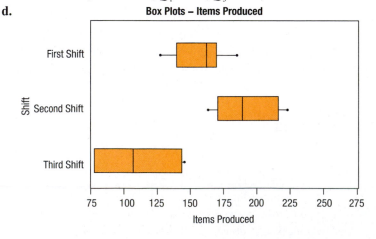

Box Plots – Items Produced

e. Probably, answers will vary. **f.** Yes, answers will vary. **g.** Probably, answers will vary. **h.** Second shift, answers will vary.

12.2 Analysis of Variance (ANOVA)

EXERCISES

7. a. 35   **b.** 2   **c.** 9.0278   **d.** 13.2879   **9. a.** 153.8095   **b.** 7   **c.** 3   **d.** 21   **e.** Total $df = 20$, SST $df = 2$, SSE $df = 18$, $20 = 2 + 18$, so the relationship holds.

12.3 Assumptions of the Test

EXERCISES

9. Sample 1: No, the histogram represents data drawn from a population that has a negative exponential distribution. Middle histogram Sample 2: No, the histogram represents data drawn from a population that has an unknown distribution. Sample 3: Yes, the histogram represents data drawn from a population that has a normal distribution.   **11.** No, because the boxes are not basically the same width. The box plot at the bottom appears to be less than half the size of the box plot in the middle, indicating that the variation for this population is considerably smaller than that of the other two populations. Answers may vary.

12.4 The *F*-Test

EXERCISES

11. a. Yes. H_0: $\mu_1 = \mu_2 = \mu_3 = \mu_4$; H_a: at least one μ_i is different, $F = 16.8582$, $F_\alpha = 3.2389$, Reject H_0.   **b.** Hourly wages for employees are approximately normally distributed with equal variances. Observations were collected in an independent and random fashion. Answers will vary.   **13. a.** No. H_0: $\mu_1 = \mu_2 = \mu_3 = \mu_4$; H_a: at least one μ_i is different, $F = 3.4667$, $F_\alpha = 4.0662$, Fail to reject H_0.   **b.** Maximum heart rates for each workout are approximately normally distributed with equal variances. Observations were selected in an independent and random fashion. Answers will vary.   **15. a.** Yes. H_0: $\mu_1 = \mu_2 = \mu_3$; H_a: at least one μ_i is different, $F = 5.1012$, $F_\alpha = 2.5893$, Reject H_0.   **b.** Dividends per share for each industry are approximately normally distributed with equal variances. Observations were selected in an independent and random fashion. Answers will vary.   **c.** It appears that the transportation industry pays lower dividends per share than the banking and energy industries, but the ANOVA test does not tell us which population mean(s) differ significantly. Answers will vary.

12.5 Two-Way ANOVA: Randomized Block Design

EXERCISES

13. a. Yes, because the dealer believes that the average gas mileage of a particular car will vary depending on the person who is driving the car due to different driving styles. Blocking will reduce the variation in gas mileage which is not due to the type of car.   **b.** Yes. H_0: $\mu_1 = \mu_2 = \mu_3 = \mu_4$; H_a: at least one μ_i is different, $F = 696.8608$, $F_\alpha = 3.2874$, Reject H_0.   **c.** Yes. H_0: $\mu_1 = \mu_2 = \mu_3 = \mu_4 = \mu_5 = \mu_6$, H_a: at least one μ_i is different, $F = 101.7798$, $F_\alpha = 2.9013$, Reject H_0.   **15. a.** Yes, because the FAA believes that the number of on-time arrivals varies by airport. Blocking will reduce the variation in on-time arrivals which is not due to airline.   **b.** Yes. H_0: $\mu_1 = \mu_2 = \mu_3 = \mu_4$; H_a: at least one μ_i is different, $F = 58.8261$, $F_\alpha = 6.9919$, Reject H_0.   **c.** Yes. H_0: $\mu_1 = \mu_2 = \mu_3 = \mu_4$; H_a: at least one μ_i is different, $F = 15.2609$, $F_\alpha = 6.9919$, Reject H_0.   **17. a.** So that any variation not due to the type of device used to measure systolic blood pressure can be reduced.   **b.** Yes. H_0: $\mu_1 = \mu_2 = \mu_3 = \mu_4$; H_a: at least one μ_i is different, $F = 9.9883$, $F_\alpha = 3.2874$, Reject H_0.   **c.** Yes. H_0: $\mu_1 = \mu_2 = \mu_3 = \mu_4 = \mu_5 = \mu_6$; H_a: at least one μ_i is different, $F = 135.8303$, $F_\alpha = 2.9013$, Reject H_0.

12.6 Two-Way ANOVA: The Factorial Design

EXERCISES

13. a. Yes, there appears to be interaction between airport location and major rental car company for all three cities.   **b.** Yes. H_0: There is no interaction; H_a: There is interaction, $F = 13.8127$, $F_\alpha = 2.9277$, Reject H_0.   **c.** We cannot test for effect of company on average daily rental rates because there is interaction.   **15. a.** There appears to be slight interaction between operator and machine. If there was no interaction, the lines would be parallel. Answers may vary.   **b.** No, there is not significant interaction. This agrees with part **a.**, we only thought the interaction was slight. H_0: There is no interaction;

H_a: There is interaction, $F = 0.0789$, $F_\alpha = 2.2858$, Fail to reject H_0. **c.** Yes. H_0: $\mu_1 = \mu_2 = \mu_3$; H_a: at least one μ_i is different, $F = 4.5$, $F_\alpha = 2.6240$, Reject H_0. **d.** Yes. H_0: $\mu_1 = \mu_2 = \mu_3$; H_a: at least one μ_i is different, $F = 14.2895$, $F_\alpha = 2.6240$, Reject H_0. **17. a.**

Source	SS	df	MS	F
Power	1.270	1	1.270	14.7246
Knowledge	0.250	1	0.250	2.8986
Interaction	0.010	1	0.010	0.1159
Error	4.140	48	0.0863	
Total	5.670	51		

b. No. H_0: There is no interaction; H_a: There is interaction, $F = 0.1159$, $F_\alpha = 2.8131$, Fail to reject H_0. **c.** Yes. H_0: $\mu_1 = \mu_2$; H_a: at least one μ_i is different, $F = 14.7246$, $F_\alpha = 4.0427$, Reject H_0. **d.** No. H_0: $\mu_1 = \mu_2$; H_a: at least one μ_i is different, $F = 2.8986$, $F_\alpha = 4.0427$, Fail to reject H_0.

Additional Exercises

EXERCISES

1. a. Dividing the students of each class into blocks categorized as Below Average, Average, and Above Average. Answers will vary. **b.** Dividing participants into blocks, categorized as <100 lb overweight, 50-100 lb overweight, and 0-50 lb overweight. Answers will vary. **c.** Dividing the persons into blocks categorized as Low IQ, Average IQ, and High IQ. Answers will vary. **3. a.** Yes. H_0: $\mu_1 = \mu_2 = \mu_3$; H_a: at least one μ_i is different, $F = 15.6429$, $F_\alpha = 6.3589$, Reject H_0. **b.** Fat contents for each brand of margarine are approximately normally distributed with equal variances. Observations were collected in an independent and random fashion. Answers will vary. **c.** The data are obtained from servings. There are no variables associated with the servings that we can block with. Answers will vary. **5. a.** $\bar{x} = 3.3064$, SST $= 20.1246$ **b.** $F = 35.3327$ **c.** Yes. H_0: $\mu_1 = \mu_2 = \mu_3$; H_a: at least one μ_i is different, $F = 35.3327$, $F_\alpha = 4.7623$, Reject H_0. **d.** The distributions of all of the populations of interest are approximately normal with equal variances, each of the k samples must be selected independently from each other and in a random fashion. They cannot be checked in this instance because the raw data are not available. **7. a.** H_0: The average median starting salary for all four types of majors is the same; H_a: At least one of the median starting salaries is different. **b.** Yes, there is sufficient evidence that at least one of the starting salaries is different. H_0: $\mu_1 = \mu_2 = \mu_3 = \mu_4$; H_a: at least one μ_i is different, $F = 17.1205$, $F_\alpha = 4.4156$, Reject H_0. **c.** H_0: The average median mid-career salary for all four types of majors is the same; H_a: At least one of the median mid-career salaries is different. **d.** Yes, there is sufficient evidence that at least one of the mid-career salaries is different. H_0: $\mu_1 = \mu_2 = \mu_3 = \mu_4$; H_a: at least one μ_i is different, $F = 10.9840$, $F_\alpha = 4.4156$, Reject H_0.

9. a.

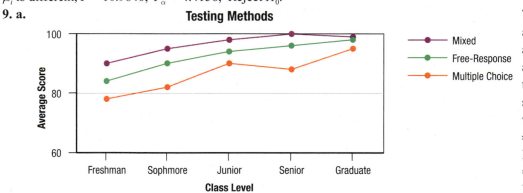

The mixed question tests appear to result in the highest average test scores. It also appears that graduate students typically have higher test scores, on average. Answers will vary. **b.** Yes. H_0: $\mu_1 = \mu_2 = \mu_3 = \mu_4 = \mu_5$; H_a: at least one μ_i is different, $F = 22.0786$, $F_\alpha = 3.8379$, Reject H_0. **c.** Yes, because the blocking effects were significant. Answers will vary.

11. a.

Source	SS	df	MS	F
Block	2154.1333	9	239.3481	96.8876
Treatment	30.2	2	15.1	6.1124
Error	44.4667	18	2.4704	
Total	2228.8	29		

b. Yes, H_0: $\mu_1 = \mu_2 = \mu_3 = \mu_4$; H_a: at least one μ_i is different, $F = 6.1124$, $F_\alpha = 3.5546$, Reject H_0. **c.** $P = 0.0094$. The probability of a Type I error for this hypothesis test is 0.0094.

13. a. H_0: $\mu_1 = \mu_2 = \mu_3$; H_a: at least one μ_i is different.

b.

Source	SS	df	MS	F
Block	1103	3	367.6667	31.2908
Treatment	158.1667	2	79.0833	6.7305
Error	70.5	6	11.75	
Total	1331.6667	11		

c. Yes, at least one of the revenues is significantly different. $H_0: \mu_1 = \mu_2 = \mu_3$; H_a: at least one μ_i is different, $F = 6.7305$, $F_\alpha = 5.1433$, Reject H_0. **d.** Yes, $H_0: \mu_1 = \mu_2 = \mu_3 = \mu_4$; H_a: at least one μ_i is different, $F = 31.2908$, $F_\alpha = 4.7571$, Reject H_0.

13   Regression, Inference, and Model Building

13.1   Building a Simple Linear Regression Model

EXERCISES

11. a. β_0 **b.** β_1 **c.** Yes, $\hat{y}_i$ is the estimated value of y for some given value of x. Answers will vary.
13. a. Salary **b.** Years of experience **c.** 25,689.10 **d.** 2148.35 **e.** $57,914.35

13.2   Defining a Linear Relationship

EXERCISES

5. a. $y = 2 + 3x$
b. $y = 4 + 8x$
c. $y = 9 - 2x$
d. $y = x$

7. a.

x	ŷ
2	15
5	45
7	65
9	85
10	95

b. Positive **c.** We would expect r to be positive, since the variables have a positive relationship. **9. a.** Estimated Selling Price = $b_0 + b_1$(Square Footage)
b.

Housing Prices and Square Footage

c.

Predicted Selling Price (Thousands of Dollars)	Error	Squared Error
201.45	−1.55	2.4025
227.91	0.09	0.0081
234.35	0.65	0.4225
273.13	11.87	140.8969
276.35	−37.35	1395.0225
297.35	−4.35	18.9225
304.35	−19.35	374.4225
314.15	50.85	2585.7225
323.25	−28.25	798.0625
325.07	−35.07	1229.9049
367.91	17.09	292.0681
416.35	88.65	7858.8225
444.35	−19.35	374.4225
472.35	−57.35	3289.0225

d. $\text{SSE} = 18{,}360.123$

13.3 Finding the Least Squares Line

EXERCISES

7. a.

b. $\hat{y} = 0.3 + 1.1x$

c.

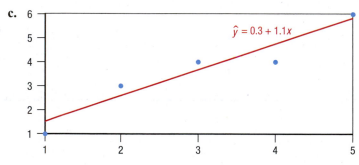

$\hat{y} = 0.3 + 1.1x$

d. −0.4, 0.5, 0.4, −0.7, 0.2   **9.** The least squares line in Exercise 7 fits the data better because it has a smaller SSE (The SSE for Exercise 7 is 1.1 and the SSE for Exercise 8 is 6.9176.).   **11. a.** Points scored   **b.** Minutes played   **c.** 225.3 points   **d.** 64.5 points   **e.** 0.645   **f.** −97.2   **g.** Possibly because you have to be in the game to score points. However, we cannot conclude that there is a causality from the estimated regression equation. Answers will vary.

13.4 The Importance of Errors

EXERCISES

11. a.

Predicted y	Error	Squared Error
124.8372	−14.8372	220.1425
131.1462	3.8538	14.8518
145.8672	4.1328	17.0800
143.7642	5.2358	27.4136
147.9702	10.0298	100.5969
177.4122	−8.4122	70.7651

b. 450.8499   **c.** 112.7125   **d.** 10.6166   **e.** The standard error of the model is relatively small in comparison with the predicted y-values, so yes. Answers will vary.

13.5   Evaluating the Fit of a Model

EXERCISES

13. a. Ratio   **b.** Savings   **c.** Income

d.

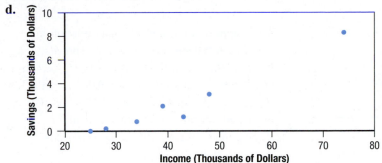

Yes, the data points in the scatterplot appear approximately linear with a positive slope trend.
e. $\hat{y} = -4.8658 + 0.171x$ **f.** \$3684.20   **g.** Savings is estimated to increase by approximately \$171 for each additional \$1000 earned in annual income.
h. $R^2 = 0.9495$ Approximately 94.95% of the variation in savings is explained by the variation annual income.

15. a. Controlled experiment

b.

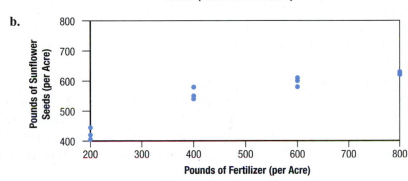

c. Pounds of sunflower seeds because the researchers are interested in the effect of fertilizer on the yield of sunflower seeds.
d. $\hat{y} = 389 + 0.323x$ **e.** As the amount of fertilizer per acre increases by one pound, the yield of sunflower seeds is predicted to increase by 0.323 of a pound.   **f.** $R^2 = 0.8456$ Approximately 84.56% of the variation in the sunflower yield can be explained by the variation in the amount of fertilizer used.   **g.** 550.5 seeds

13.6   Fitting a Linear Time Trend

EXERCISES

7. a. Independent variable: Month; Dependent variable: Sales   **b.** $\hat{y} = 403.7576 + 40.9091x$ **c.** MSE = 2622.248, $s_e = 51.2079$ **d.** \$935,575.90   **e.** Approximately 90.12% of the variation in sales is explained by the time trend model, therefore the model seems to fit the data accurately.

13.8   Inference: How Good is the Estimate of β_1?

EXERCISES

7. a. −1391.087   **b.** 124.6005   **c.** (−1809.1701, −973.0039)   **d.** We are 99% confident that the true decrease in asking price for a car for each additional year is between \$973.00 and \$1809.17.

13.9   Testing a Hypothesis Concerning β_1

EXERCISES

7. a.

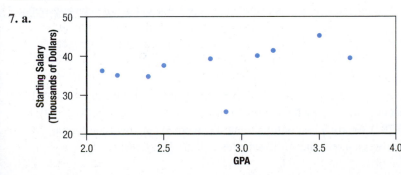

The two variables appear to have a positive correlation.   **b.** $\hat{y} = 24.9542 + 4.4175x$ **c.** No, $H_0: \beta_1 = 0$; $H_a: \beta_1 \neq 0$, $t = 1.471$, P-value = 0.1794, Fail to reject H_0.   **d.** \$35,997.95   **e.** For each additional one point increase in GPA, starting salary is expected to increase by approximately \$4417.50.   **f.** Approximately 21.3%   **g.** The error term is a normally distributed random variable, the expected value of the error term is zero, the variance of the error term is constant, and the errors are independent.

9. a. $\hat{y} = 42.5154 + 0.3914x$ **b.** Approximately 20.76% **c.** Yes, $H_0: \beta_1 = 0$; $H_a: \beta_1 \neq 0$, $t = 2.231$, P-value = 0.0379, Reject H_0. **d.** 72

13.10 Inference Concering the Model's Prediction

EXERCISES

5. a.

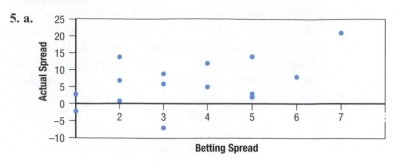

There appears to be a weak linear relationship between actual spreads and betting spreads. **b.** $\hat{y} = -0.3834 + 1.9198x$ **c.** No, $H_0: \beta_1 = 0$; $H_a: \beta_1 \neq 0$, $t = 2.043$, P-value = 0.0619, Fail to reject H_0. **d.** Approximately 24.3% **e.** The estimated increase in the actual spread for a one point increase in the betting spread is approximately 1.9198 points.

f. (−0.1104, 3.9500) We are 95% confident that the change in actual spread for a one point increase in betting spread is between −0.1104 and 3.9500 points. **g.** 9.2156 **h.** (−5.27, 23.70) We are 95% confident that the actual spread when the betting spread is 5 is between −5.27 and 23.70. **i.** (4.59, 13.85)

7. a.

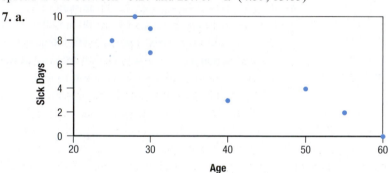

There appears to be a moderate negative linear relationship between age and sick days. **b.** $s_e = 1.4765$. This is the estimated standard deviation of the errors associated with the model. **c.** With each additional year of age, the number of sick says is expected to decrease by approximately 0.247. **d.** Approximately 85.8% **e.** Yes, $H_0: \beta_1 = 0$; $H_a: \beta_1 \neq 0$, $t = -6.01$, P-value = 0.001, Reject H_0.

f. (−0.3473, −0.1464) We are 95% confident that the true decrease in sick days for a one year increase in age is between 0.3473 and 0.1464 days. **g.** Approximately 6.5 days **h.** (5.184, 7.911) We are 95% confident that the average number of sick days for a 35-year-old employee is between 5.184 and 7.911. **i.** (2.686, 10.409) We are 95% confident that a new 35-year-old employee will take between 2.686 and 10.409 sick days. **j.** 4 days

Additional Exercises

EXERCISES

1. a.

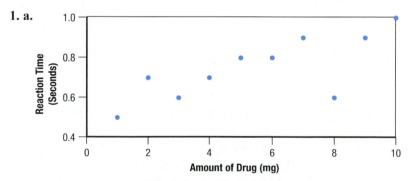

There appears to be a moderate positive linear relationship between amount of drug and reaction time. **b.** $s_e = 0.1101$. This is the estimated standard deviation of the errors associated with the model. **c.** Reaction time increases by approximately 0.0394 seconds for each additional milligram of the drug. **d.** Approximately 56.9%. Other factors could include age, weight, etc. Answers will vary.

e. $H_0: \beta_1 = 0$; $H_a: \beta_1 \neq 0$, $t = 3.25$, P-value = 0.012, Reject H_0 at the 0.05 level, Fail to reject H_0 at the 0.01 level. **f.** (0.011, 0.067) We are 95% confident that the true change in reaction time for each additional milligram of the drug is between 0.011 and 0.067 seconds. **g.** 0.6909 seconds **h.** (0.600, 0.782) We are 95% confident that the average reaction time for an individual with 4 mg of the drug in the bloodstream is between 0.600 and 0.782 seconds. **i.** (0.421, 0.961)

3. a.

There appears to be a strong negative relationship between FICO score and interest rate. **b.** $\hat{y} = 16.2146 - 0.0129x$ **c.** 0.3202 **d.** Yes. $H_0: \beta_1 = 0$; $H_a: \beta_1 \neq 0$, $t = -10.281$, P-value = 0.0000, Reject H_0. **e.** Interest rate decreases by approximately 0.0129 percent for each additional one point increase in FICO score.

f. (−0.0155, −0.0102) We are 95% confident that the true change in interest rate for each additional one point increase in FICO score is between −0.0155 and −0.0102 percentage points. **g.** $R^2 = 0.8545$. Approximately 85.45% of the variation in interest rate is explained by the variation in FICO scores. **h.** $r = -0.9244$. There is a strong negative linear relationship between FICO score and interest rate. **i.** 6.8621% **j.** (6.613, 7.148) We are 95% confident that the average interest rate for a FICO score of 725 is between 6.613 and 7.148 percent. **k.** (5.864, 7.898) We are 95% confident that a person with a FICO score of 725 will receive an interest rate between 5.864 and 7.898 percent.

5. a.

There appears to be a strong positive linear relationship between home size and annual energy usage. **b.** $\hat{y} = 6774.4571 + 2.8406x$ **c.** $R^2 = 0.9982$ Approximately 99.82% of the variation in annual energy usage is explained by the variation in home size. **d.** Yes. $H_0: \beta_1 = 0$; $H_a: \beta_1 \neq 0$, $t = 112.114$, P-value = 0.0000, Reject H_0. **e.** For each additional square foot of home size, annual energy usage is expected to increase by approximately 2.8406 kWh.

f. (2.7695, 2.9117) We are 99% confident that the true increase in annual energy usage for each additional square foot of home size is between 2.7695 and 2.9117 kWh. **g.** $r = 0.9991$. There is a strong positive linear relationship between home size and annual energy usage. **h.** 15,864.3771 kWh **i.** (15,572.2, 15,976.5) We are 95% confident that the James family will use on average between 15,572.2 and 15,976.5 kWh in their first year in the home.

14 Multiple Regression

14.1 Multiple Regression

EXERCISES

7. a. $b_0 = 11214.1992$, $b_1 = 2854.8913$, $b_2 = 839.6360$ **b.** $\hat{y} = 11214.1992 + 2854.8913x_1 + 839.6360x_2$ **c.** Age, median income for your location, industry, etc. Answers will vary.
9. a. Weight $= \beta_0 + \beta_1(\text{Height}) + \beta_2(\text{Age}) + \beta_3(\text{Calories}) + \beta_4(\text{Exercise}) + \varepsilon_i$ **b.** Height, Age, Calories, Exercise **c.** Positive: Height, Age, Calories; Negative: Exercise **d.** Medical history (diabetes, thyroid issues, etc.), Smoker/non-smoker, etc. Answers will vary.

14.2 The Coefficient of Determination and Adjusted R^2

EXERCISES

11. a. $R_a^2 = 0.6607$ **b.** The adjusted R^2 for this model is larger than the adjusted R^2 from Exercise 10. The adjusted R^2 value indicates that approximately 66.07% of the variation in price is explained by the variation in the independent variables, compared with only 29.19% in the previous model. Answers will vary. **c.** Yes, because the adjusted R^2 value is significantly larger for this model. Answers will vary. **13. a.** Revenue $= \beta_0 + \beta_1(\text{Television}) + \beta_2(\text{Newspaper}) + \beta_3(\text{Mail}) + \varepsilon_i$

b. $\hat{y} = 73.9320 + 2.3830x_1 + 1.4544x_2 + 1.8160x_3$ **c.** For each additional \$1000 spent on TV advertising, revenue is expected to increase by approximately \$2383, assuming newspaper and mail advertising expenditures remain constant. **d.** $R_a^2 = 0.8865$. Approximately 88.65% of the variation in revenue is explained by the variation in the three independent variables.   **e.** The adjusted R^2 value in this model is larger than the R^2 value in the previous model, so this model appears to be more useful. Answers may vary.   **f.** $R^2 = 0.9352$. The adjusted R^2 value should be used to compare this model to the simple model because additional independent variables have been added.

14.3   Interpreting the Coefficients of the Multiple Regression Model

EXERCISES

5. a. Yes, the coefficient is positive, meaning that as years of education increases, so does estimated salary. The magnitude of \$2854.89 for each additional year of education seems reasonable.   **b.** Yes, the coefficient is positive, meaning that as years of experience increases, so does estimated salary. The magnitude of \$839.64 for each additional year of experience seems reasonable.   **c.** For each additional year of experience, annual salary is expected to increase by \$839.64, assuming years of education remains constant.   **d.** \$38,251.52   **e.** His annual salary would be expected to increase by \$839.64, assuming years of education remains constant.   **f.** The employee with the master's degree is expected to earn approximately \$5709.78 more than the employee with the bachelor's degree.

14.4   The *F*-Distribution

EXERCISES

9. a. 2.3830   **b.** 3.0794   **c.** 3.8940   **d.** 5.1981   **11. a.** 3.1533   **b.** 3.0045   **c.** 2.4035   **d.** 3.1730

14.5   Inference Concerning the Multiple Regression Model and its Coefficients

EXERCISES

17. a. $H_0: \beta_1 = \beta_2 = 0$; H_a: At least one $\beta_i \neq 0$.   **b.** $F = 10.8947$ **c.** Yes. P-value $= 0.0001$, Reject H_0.   **d.** (1466.6644, 4243.1181) We are 95% confident that the true increase in annual salary for each additional year of education is between \$1466.66 and \$4243.12.   **e.** $H_0: \beta_1 = 0$; $H_a: \beta_1 \neq 0$ **f.** Yes. $t = 4.140$, P-value $= 0.0001$, Reject H_0.
19. a. Rent $= \beta_0 + \beta_1(\text{Population}) + \beta_2(\text{Income}) + \varepsilon_i$ **b.** We would expect both coefficients to be positive since larger cities tend to have more expensive rental rates and more expensive rental rates would be expected in areas with greater incomes. Answers will vary.   **c.** $b_0 = 138.5023$; $b_1 = 0.1199$; $b_2 = 16.8207$; $\hat{y} = 138.5023 + 0.1199x_1 + 16.8207x_2$. A city with 0 population and 0 income would have an expected monthly rent of about \$138.50. For each additional 1000 people in the city, monthly rent is expected to increase by about \$0.12. For each additional \$1000 in average median income, monthly rent is expected to increase by about \$16.82.   **d.** Yes; $H_0: \beta_1 = \beta_2 = 0$; H_a: At least one $\beta_i \neq 0$; $F = 30.6224$; P-value $= 0.0000$; Reject H_0.   **e.** (11.4624, 22.1789); We are 95% confident that the true increase in monthly rent for each additional \$1000 increase in median income is between \$11.46 and \$22.18.   **f.** Population: No; $H_0: \beta_1 = 0$; $H_a: \beta_1 \neq 0$; $t = 1.423$; P-value $= 0.1803$; Fail to reject H_0. Income: Yes; $\beta_2 = 0$; $H_a: \beta_2 \neq 0$; $t = 6.840$; P-value $= 0.0000$; Reject H_0.   **g.** Yes, the population variable should be removed, as it is not a significant predictor of monthly rent. Answers may vary.

14.6   Inference Concerning the Model's Prediction

EXERCISES

7. a.

There appears to be a positive relationship between each parent's height and the child height. The mother-daughter plot appears more linear than the father-daughter plot. **b.** $\hat{y} = -4.6456 + 0.5939x_1 + 0.4523x_2$ **c.** Yes; H_0: $\beta_1 = \beta_2 = 0$; H_a: At least one $\beta_i \neq 0$; $F = 11.2521$; P-value $= 0.0028$; Reject H_0.   **d.** No; H_0: $\beta_2 = 0$; H_a: $\beta_2 \neq 0$; $t = 2.176$; P-value $= 0.0546$; Fail to reject H_0.   **e.** No; H_0: $\beta_1 = 0$; H_a: $\beta_1 \neq 0$; $t = 2.628$; P-value $= 0.0253$; Fail to reject H_0.   **f.** For each additional inch in the mother's height, the daughter's height is expected to increase by approximately 0.5939 inch. For each additional inch in the father's height, the daughter's height is expected to increase by approximately 0.4523 inch.   **g.** Mother: (0.0903, 1.0974); We are 95% confident that for each additional inch in the mother's height, the daughter will be between 0.0903 and 1.0974 inches taller. Father: (−0.0108, 0.9153); We are 95% confident that for each additional inch in the father's height, the daughter will be between 0.0108 in. shorter and 0.9153 in. taller.

h. 66.8297 inches, or 5 feet 6.8297 inches   **i.** (61.533, 72.126); We are 95% confident that a particular daughter whose mother is 5 foot 4 and father is 6 foot 2 will be between 61.533 and 72.126 inches tall.   **j.** (64.785, 68.875)

9. a. $\hat{y} = -15.2395 + 0.1319x_1 + 0.0869x_2$ **b.** Yes; H_0: $\beta_1 = \beta_2 = 0$; H_a: At least one $\beta_i \neq 0$; $F = 21.6521$; P-value $= 0.0000$; Reject H_0.   **c.** Approximately 59.89%. This is slightly lower than the percentage for the model with the 3 independent variables (59.99%). However, the difference is likely due to the additional independent variable in the previous model.
d. The model without the first downs variable is likely the better model. Though the R^2 value is lower, the adjusted R^2 value is larger, indicating that the difference is likely due to the addition of the first downs variable, which is not useful in predicting points scored. Answers will vary.   **e.** Approximately 4 (3.6890)   **f.** (−2.96, 11.16); We are 95% confident that the average points scored is between 0 and 11 points (since points scored cannot be negative in football) when the offense has 102 rushing yards and 63 passing yards.   **g.** (−12.70, 20.91); We are 95% confident that in this particular game against Miami, Buffalo will score between 0 and 21 points (since points scored cannot be negative in football).

14.7   Models with Qualitative Independent Variables

EXERCISES

9. a. Yes, the overall model is significant. H_0: $\beta_1 = \beta_2 = \beta_3 = 0$; H_a: At least one $\beta_i \neq 0$, $F = 235.1310$, P-value $= 0.0000$, Reject H_0.   **b.** Approximately 51.06%   **c.** Yes. H_0: $\beta_3 = 0$; H_a: $\beta_3 \neq 0$, $t = -3.051$, P-value $= 0.0024$, Reject H_0.   **d.** Yes, all of the independent variables appear to be significant in predicting GPA. However, the R^2 value is only about 51%. This indicates that there may be other factors influencing GPA that have not been accounted for. Answers will vary.   **e.** Answers will vary. Quantitative: hours of study time, number of credit hours, SAT score, etc. Qualitative: major, gender, extracurricular activities, etc.

11. a.

School	Private
1	1
2	0
3	0
4	1
5	1
6	0
7	1
8	1
9	0
10	0
11	1
12	0
13	1
14	0
15	1
16	0
17	0
18	1
19	0
20	1

b. $\hat{y} = 105.9195 + 2.5203x_1 - 0.0033x_2 - 65.3808x_3$ **c.** Yes. $H_0: \beta_1 = \beta_2 = \beta_3 = 0$; H_a: At least one $\beta_i \neq 0$, $F = 35.6335$, P-value $= 0.0000$, Reject H_0. **d.** The coefficient for police is positive, indicating that as the number of police increases, so does the number of crimes. This is not what would be expected. The coefficient for enrollment is negative, indicating that as enrollment increases, the number of crimes decreases. This is also surprising, as one would think that a larger student body would result in increased crimes. The coefficient for private is negative, indicating that private schools tend to have less crimes that public schools. This is somewhat expected since private schools are more expensive and generally smaller in size. Answers will vary. **e.** $H_0: \beta_3 = 0$; $H_a: \beta_3 \neq 0$, $t = -2.624$, P-value $= 0.0184$, Reject H_0 at the 0.05 level, so it supports the officials' belief, Fail to reject H_0 at the 0.01 level, so yes, the decision would change.

14.8 Pitfalls and Concerns in Regression Modeling

EXERCISES

11. a. Age and experience are likely correlated since the older you are, the more experience you tend to have. Answers will vary. **b.** By using statistical software to find the correlation coefficient for each pair of variables. The closer the correlation coefficients are to 1 or −1, the more likely it is that collinearity exists. Answers will vary. **c.** $58,921.91 **d.** Yes, because the overall model is not significant. None of the independent variables appear to be useful in predicting salary. Also, the age and experience values for which we are predicting salary are outside the range of the observed data, so extrapolation could be an issue. Answers will vary. **13.** There appears to be a strong positive correlation between x_1 and x_3. There also appears to be a moderate positive correlation between x_1 and x_4 and x_3 and x_4. Answers will vary.

Additional Exercises

EXERCISES

1. a.

Observation	Degree
1	0
2	1
3	0
4	0
5	1
6	1
7	1
8	1
9	0
10	1
11	0
12	1
13	0
14	0
15	1
16	0
17	1
18	1
19	0
20	0

b. $\hat{y} = 10419.8636 + 1501.2807x_1 + 13047.4495x_2$ **c.** \$1501.28 **d.** Yes. H_0: $\beta_2 = 0$; H_a: $\beta_2 \neq 0$, $t = 3.845$, P-value = 0.0000, Reject H_0. This indicates that a master's degree is significant in predicting salary, and the estimated increase in annual salary for people with master's degrees is \$13,047.45, which is greater than \$10,000. **e.** There are many other factors that influence annual salary. Answers will vary. **3. a.**

The relationship appears to be positive, but it does not look linear. Answers will vary.

b. $\hat{y} = 5.9860 + 6.3361x_1$, $R^2 = 0.7971$, $s_e^2 = 163.5177$ **c.** Linear: 107.3638, Polynomial: 91.09. The observed value when the additive was 16.4 was a yield of 94, which is far closer to the polynomial result. Answers will vary. **d.** The polynomial model has a higher R^2 value and a lower s_e^2 value, indicating that the polynomial model fits the data better than the linear model. Answers may vary. **e.** The polynomial model because it has a higher R^2 value and a lower s_e^2 value. Answers will vary. **5. a.** Answers may vary. GPA = $\beta_0 + \beta_1$(Xoom) + β_2(Galaxy) + β_3(iPad) + ε **b.** Answers may vary. $\hat{y} = 3.4867 - 0.2761$(Xoom) $- 0.2503$(Galaxy) $- 0.4182$(iPad) **c.** No. H_0: $\beta_1 = \beta_2 = \beta_3 = 0$; H_a: At least one $\beta_i \neq 0$, $F = 0.4740$, P-value = 0.7031, Fail to reject H_0. Answers may vary. **d.** Answers may vary. $\hat{y} = 3.5473 + 0.1115(<\$30,000) - 0.6059(\$30,000 - \$49,999) - 0.3448(\$50,000 - \$74,999)$, This model is also not significant at the 0.05 level. There is not sufficient evidence that type of tablet or income is a significant predictor of GPA. **e.** No, none of the tablet types are significant in predicting GPA. Answers will vary. **7. a.** Benefit = $\beta_0 + \beta_1$(Family Size) + β_2(Income) + ε_i **b.** Yes. $\hat{y} = 40.7903 + 3.6594x_1 + 0.1461x_2$, H_0: $\beta_1 = \beta_2 = 0$; H_a: At least one $\beta_i \neq 0$, $F = 397.9462$, P-value = 0.0000, Reject H_0. **c.** Family size is significant at the 0.10 level ($t = 1.702$, P-value = 0.0972), and monthly income is significant at the 0.01 level ($t = 27.955$, P-value = 0.0000). Monthly income appears to be a much better predictor for benefits than family size. Answers may vary. **d.** (410.70, 430.77), We are 95% confident that the average monthly benefit for a 4-person family with a monthly income of \$2500 is between \$410.70 and \$420.77. **e.** (338.93, 502.53), We are 99% confident that the monthly benefit for a particular 4-person family with a monthly income of \$2500 is between \$338.93 and \$502.53. **f.** The prediction interval is wider than the confidence interval by 143.53. This is due to the increased confidence level (99% vs. 95%) and the prediction interval accounts for individual variation. Answers may vary.

15   Looking for Relationships in Qualitative Data

15.1   The Chi-Square Distribution

EXERCISES

9. a. 41.401   **b.** 28.845   **c.** 5.024   **d.** 107.565   **e.** 24.769   **11. a.** 66.766   **b.** 27.488   **c.** 7.378   **d.** 33.196
e. 63.167   **13. a.** 0.042982   **b.** 30.0874   **c.** 16.013

15.2   The Chi-Square Test for Goodness of Fit

EXERCISES

9. a. No. $H_0: p_1 = p_2 = \ldots = p_5 = \dfrac{1}{5}$; H_a: Any possible difference. $\chi_\alpha^2 = 9.488$, $\chi^2 = 7.647$, Fail to reject H_0.   **b.** The distribution
of the number of service calls has a multinomial probability distribution.   **11. a.** Yes. $H_0: p_1 = p_2 = \ldots = p_{12} = \dfrac{1}{12}$; H_a:
Any possible difference. $\chi_\alpha^2 = 24.735$, $\chi^2 = 26.377$, Reject H_0.   **b.** The distribution of fatal accidents has a multinomial

probability distribution.   **13. a.** Yes. $H_0: p_1 = 0.15, p_2 = p_3 = 0.30, p_4 = 0.25$, H_a: Any possible difference. $\chi_\alpha^2 = 11.345$,
$\chi^2 = 28.000$. Reject H_0.   **b.** The distribution of the survey responses has a multinomial probability distribution.

15.3   The Chi-Square Test for Association between Two Qualitative Variables

EXERCISES

13. a. Yes. H_0: Age and political affiliation are independent; H_a: Age and political affiliation are dependent. $\chi_\alpha^2 = 9.488$,
$\chi^2 = 31.881$. Reject H_0.   **b.** Each variable satisfies the properties of a multinomial distribution.   **b.** The distribution of
the survey responses has a multinomial probability distribution.   **15. a.** No. H_0: Education and crime are independent;
H_a: Education and crime are dependent. $\chi_\alpha^2 = 6.251$, $\chi^2 = 2.026$. Fail to reject H_0.   **b.** Each variable satisfies the properties
of a multinomial distribution.   **17. a.** No. H_0: Causes of fires and region are independent. H_a: Causes of fires and region are
dependent. $\chi_\alpha^2 = 30.578$, $\chi^2 = 1.963$. Fail to reject H_0.   **b.** Each variable satisfies the properties of a multinomial distribution.

Additional Exercises

EXERCISES

1. a. Yes. $H_0: p_1 = 0.50, p_2 = p_3 = 0.25$; H_a: Any possible difference. $\chi_\alpha^2 = 5.991$, $\chi^2 = 68.400$. Reject H_0.   **b.** The underlying
distribution is a multinomial probability distribution.   **3.** Yes. $H_0: p_1 = 0.51, p_2 = 0.33, p_3 = 0.16$; H_a: Any possible difference.
$\chi_\alpha^2 = 5.991$, $\chi^2 = 10.861$. Reject H_0.   **5.** No. $H_0: p_1 = 0.226, p_2 = 0.611, p_3 = 0.075, p_4 = 0.088$; H_a: Any possible difference.
$\chi_\alpha^2 = 11.345$, $\chi^2 = 8.012$. Fail to reject H_0.   **7.** Yes. H_0: Opinion and region are independent; H_a: Opinion and region are
dependent. $\chi_\alpha^2 = 16.812$, $\chi^2 = 96.416$. Reject H_0.

16   Nonparametrics

16.1   The Sign Test

EXERCISES

15. H_0: Median $\leq \$25{,}000$; H_a: Median $> \$25{,}000$. Critical value $= -1.28$. $z = -3.24$. Reject H_0.   **17. a.** H_0: Median $\geq 14{,}000$
hrs; H_a: Median $< 14{,}000$ hrs. Critical value $= 2$. $X = 6$. Fail to reject H_0.   **b.** The data are randomly selected.   **19.** H_0: # of
Positive Signs $\leq$ # of Negative Signs; H_a: # of Positive Signs $>$ # of Negative Signs. Critical value $= 6$. $X = 9$. Fail to reject
H_0.   **21.** H_0: # of Negative Signs $=$ # of Positive Signs; H_a: # of Negative Signs $\neq$ # of Positive Signs. Critical value $= 0$.

$X = 0$. Reject H_0.

16.2   The Wilcoxon Signed-Rank Test

EXERCISES

11.

Mutual Fund	Price	Rank
American Funds	24.4	7.5
Columbia Management	9.41	2.5
Morgan Stanley	23.74	6
Fidelity Investments	24.40	7.5
John Hancock	9.41	2.5
DWS Investments	15.57	5
UBS	12.15	4
Prudential Investments	9.23	1
Value Line Funds	32.82	9
The Vanguard Group	34.72	10

13. a. H_0: DHEA-S is not increased; H_a: DHEA-S is increased. Critical value = 0. $X = 2$. Fail to reject H_0.   **b.** The data are randomly selected.   **c.** H_0: DHEA-S is not increased; H_a: DHEA-S is increased. Critical value = 4. $T_x = 5$. Fail to reject H_0.   **d.** Pairs of data have been randomly selected and are such that the absolute values of their differences can be ranked.   **e.** The signed-rank test because the magnitudes of the differences are not ignored. Answers will vary.   **15. a.** The paired differences have an approximately normal distribution.   **b.** H_0: Model A = Model B; H_a: Model A ≠ Model B. Critical value = 1. $T_x = 0$. Reject H_0.   **c.** Pairs of data have been randomly selected and are such that the absolute values of their differences can be ranked.   **d.** In all three tests the null hypothesis is rejected in favor of the alternative.

16.3   The Wilcoxon Rank-Sum Test

EXERCISES

11. a. H_0: Mr. Ellis = Mr. Ford, H_a: Mr. Ellis ≠ Mr. Ford. Critical value = 37, 68. $T_x = 42$, 63. Fail to reject H_0.   **b.** The data are such that they can be ranked. The two samples are selected in an independent and random fashion.   **13. a.** H_0: New Battery ≥ Old Battery; H_a: New Battery < Old Battery. Critical value = 28. $T_x = 35$. Fail to reject H_0.   **b.** The data are such that they can be ranked. The two samples are selected in an independent and random fashion.   **15. a.** H_0: Dramas ≤ Comedies; H_a: Dramas > Comedies. Critical value = 36. $T_x = 32$. Fail to reject H_0.   **b.** The data are such that they can be ranked. The two samples are selected in an independent and random fashion.   **17. a.** H_0: Service City ≥ Sunshine City; H_a: Service City < Sunshine. Critical value = 52. $T_x = 47$. Fail to reject H_0.   **b.** The data are such that they can be ranked. The two samples are selected in an independent and random fashion.

16.4   The Rank Correlation Test

EXERCISES

11. a. H_0: $\rho_s = 0$, H_a: $\rho_s \neq 0$. $r_s = -0.8042$. Critical values = -0.591, 0.591. Reject H_0.   **b.** Yes, but the assumption must be made that the relationship between the variables is linear. Answers will vary.   **13.** H_0: $\rho_s = 0$, H_a: $\rho_s \neq 0$. $r_s = 0.0182$. Critical values = -0.648, 0.648. Fail to reject H_0.

16.5   The Runs Test for Randomness

EXERCISES

9. H_0: The sequence is random; H_a: The sequence is not random. $N = 26$, $m = 10$, $n = 16$, $R = 14$. Critical values = 8, 19. Fail to reject H_0.   **11.** H_0: The sequence is random; H_a: The sequence is not random. $N = 31$, $m = 15$, $n = 16$, $R = 19$. Critical values = 10, 23. Fail to reject H_0.

16.6   The Kruskal-Wallis Test

EXERCISES

9. H_0: The milk production for all schedules is the same; H_a: The milk production for at least one of the schedules is different. $H = 7.0234$. Critical value = 6.2514. Reject H_0.

Additional Exercises

EXERCISES

1. a. The paired differences have an approximately normal distribution. **b.** H_0: Vest Treatment $\leq$ Traditional Treatment; H_a: Vest Treatment > Traditional Treatment. $X = 0$. Fail to reject H_0. **c.** The data are randomly selected. **d.** H_0: Vest Treatment $\leq$ Traditional Treatment; H_a: Vest Treatment > Traditional Treatment. $T_+ = 0$. Fail to reject H_0. **e.** Pairs of data have been randomly selected and are such that the absolute values of their differences can be ranked. **f.** The signed-rank test because the magnitudes of the differences are not ignored. Answers will vary. **g.** There is not sufficient evidence that the diameter of blood vessels in the lungs is significantly larger after using the vest treatment when performing a paired difference test. There is sufficient evidence using the nonparametric methods. Answers will vary. H_0: Vest Treatment $\leq$ Traditional Treatment; H_a: Vest Treatment > Traditional Treatment. $t = -2.75$. Critical value $= -3.747$. Fail to reject H_0.
3. a. The Wilcoxon rank-sum test because we are not dealing with paired data. **b.** H_0: Diet B $\leq$ Diet A; H_a: Diet B > Diet A **c.** H_0: Diet B $\leq$ Diet A; H_a: Diet B > Diet A. Critical value $= 131$. $T_x = 113$. Fail to reject H_0. **d.** The data are such that they can be ranked. The two samples are selected in an independent and random fashion. **5. a.** $r_s = 0.2727$ these two variables have a weak positive relationship. **b.** H_0: $\rho_s = 0$; H_a: $\rho_s \neq 0$. $r_s = 0.2727$. Critical values $= -0.648, 0.648$. Fail to reject H_0.
7. H_0: The sequence is random; H_a: The sequence is not random. $N = 9$, $m = 5$, $n = 4$, $R = 4$. Critical values $= 2, 9$. Fail to reject H_0. **9.** H_0: The rankings are the same for the three conferences; H_a: The ranking of at least one conference is different. $H = 13.5755$. Critical value $= 4.6052$. Reject H_0.

---

17 Statistical Process Control

17.1 Basic Charts and Diagrams Used in Quality Control

EXERCISES

11. a. In order to assist customers in the best possible manner, the receptionist should understand how to handle various customer calls. Answers will vary. **b.** The flowchart appears to be incomplete, since it does not tell you what to do in the case of something other than shipping questions, billing questions, problems with software function, and product information. The arrow following the "other" box goes nowhere. Answers will vary.

13. a.

Complaint	Relative Frequency (%)
Not Enough Parking	35.6
Rude Personnel	22.2
Poor Lighting	18.7
Confusing Store Layout	12.4
Limited Sizes	6.7
Clothing Unattractive	4.4

b.

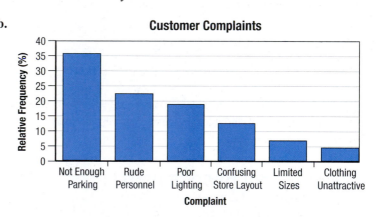

c. Not enough parking, rude personnel, poor lighting **d.** No, 80% of the problems are explained by 50% of the causes. Answers will vary. **e.** The store should implement more parking, better training for its employees, and better lighting. The parking issue should be addressed first because it is generating the largest percentage of complaints. Answers will vary.

15. a.

Quarterly Revenues

b. The revenues appear to be cyclical. Answers will vary. **c.** Based on the run chart, yes. Answers will vary. **d.** Families probably tend to eat out more when the weather is nicer. Answers will vary. **e.** There is a cyclical pattern, so the process could be considered unstable. The restaurant could offer coupons or incentives in months where revenues tend to be lower. Answers will vary.

17.2 Basic Concepts

EXERCISES

9. a. UCL = 14.2, LCL = 5.8, Centerline = 10.0 **b.** The average value of the measurements is 10. Measurements between 5.8 and 14.2 are considered in control, and can be attributed to normal process variation. **c.** B, F, H, K, N **11. a.** UCL = 36, LCL = 16, Centerline = 26 **b.** The average value of the measurements is 26. Measurements between 16 and 36 are considered in control, and can be attributed to normal process variation. **c.** G, K, M

17.3 Monitoring with $\bar{x}$ and *R* Charts

EXERCISES

11. UCL = 31.125, LCL = 28.875 **13. a.** UCL = 4.3354, LCL = 3.6646, Centerline = 4 **b.** UCL = 4.6244, LCL = 0, Centerline = 2.1875 **c.** 1, 3, 4, 5, 8, 11, 13, 14, and 15 **d.** 13 **15. a.** UCL = 25.134, LCL = 23.046, samples 1, 4, 9, and 10 are out of control **b.** UCL = 9.1988, LCL = 2.4012 **17.** UCL = 0.235, LCL = 0.070, samples 5 and 9 are out of control.

17.4 Monitoring with a *p* Chart

EXERCISES

9. UCL = 3.98%, LCL = 0% **11. a.** UCL = 28%, LCL = 0%, Centerline = 10%

b.

p Chart

UCL = 28%
Centerline = 10%
LCL = 0%

c. No, no samples are out of control.
13. UCL = 41.77%, LCL = 0%, Centerline 11.5%, no samples are out of control.

Additional Exercises

EXERCISES

1. a. UCL = 221.232, LCL = 218.768, Centerline = 220 **b.** UCL = 7.108, LCL = 0.892, Centerline = 4 **3.** UCL = 9.17%, LCL = 0.21%, there are no seasons out of control. **5.** UCL = 54.03%, LCL = 0%, Centerline = 21.96%, there are no samples out of control, so the VP should not be concerned about the number of mistakes being made on financial statements.
7. $\bar{x}$ chart: UCL = 42.4908, LCL = 39.1800, Centerline = 40.8354, no samples are out of control. *R* chart: UCL = 6.0651, LCL = 0, Centerline = 2.869, no samples are out of control.

Index

X

$\bar{x}$ chart 974
  mean and standard deviation known
    centerline 975
    lower control limit (LCL) 975
    upper control limit (UCL) 975
  mean and standard deviation unknown
    centerline 975
    lower control limit (LCL) 975
    upper control limit (UCL) 975

Y

y-intercept 684
  least squares line 696

Z

z
  distribution 351
  score 171, 357
  test statistic 499
  transformation 500

Numerical entries represent the value of *t* such that the area to the right of the *t* is equal to α.

Degrees of Freedom	Area to the Right of the Critical Value					
	$t_{0.200}$	$t_{0.100}$	$t_{0.050}$	$t_{0.025}$	$t_{0.010}$	$t_{0.005}$
1	1.376	3.078	6.314	12.706	31.821	63.657
2	1.061	1.886	2.920	4.303	6.965	9.925
3	0.978	1.638	2.353	3.182	4.541	5.841
4	0.941	1.533	2.132	2.776	3.747	4.604
5	0.920	1.476	2.015	2.571	3.365	4.032
6	0.906	1.440	1.943	2.447	3.143	3.707
7	0.896	1.415	1.895	2.365	2.998	3.499
8	0.889	1.397	1.860	2.306	2.896	3.355
9	0.883	1.383	1.833	2.262	2.821	3.250
10	0.879	1.372	1.812	2.228	2.764	3.169
11	0.876	1.363	1.796	2.201	2.718	3.106
12	0.873	1.356	1.782	2.179	2.681	3.055
13	0.870	1.350	1.771	2.160	2.650	3.012
14	0.868	1.345	1.761	2.145	2.624	2.977
15	0.866	1.341	1.753	2.131	2.602	2.947
16	0.865	1.337	1.746	2.120	2.583	2.921
17	0.863	1.333	1.740	2.110	2.567	2.898
18	0.862	1.330	1.734	2.101	2.552	2.878
19	0.861	1.328	1.729	2.093	2.539	2.861
20	0.860	1.325	1.725	2.086	2.528	2.845
21	0.859	1.323	1.721	2.080	2.518	2.831
22	0.858	1.321	1.717	2.074	2.508	2.819
23	0.858	1.319	1.714	2.069	2.500	2.807
24	0.857	1.318	1.711	2.064	2.492	2.797
25	0.856	1.316	1.708	2.060	2.485	2.787
26	0.856	1.315	1.706	2.056	2.479	2.779
27	0.855	1.314	1.703	2.052	2.473	2.771
28	0.855	1.313	1.701	2.048	2.467	2.763
29	0.854	1.311	1.699	2.045	2.462	2.756
30	0.854	1.310	1.697	2.042	2.457	2.750
40	0.851	1.303	1.684	2.021	2.423	2.704
50	0.849	1.299	1.676	2.009	2.403	2.678
60	0.848	1.296	1.671	2.000	2.390	2.660
70	0.847	1.294	1.667	1.994	2.381	2.648
80	0.846	1.292	1.664	1.990	2.374	2.639
90	0.846	1.291	1.662	1.987	2.368	2.632
100	0.845	1.290	1.660	1.984	2.364	2.626
120	0.845	1.289	1.658	1.980	2.358	2.617
∞	0.842	1.282	1.645	1.96	2.326	2.576

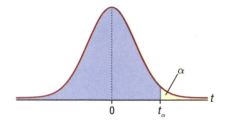

Critical Values of *z*

Level of Confidence	$z_{\alpha/2}$
0.80	1.28
0.90	1.645
0.95	1.96
0.99	2.575